Particle Dynamics with Aggregation and Fragmentation

Particle Dynamics with Aggregation and Fragmentation

An Introduction for Scientists and Engineers

Culbert B. Laney
Affiliate Faculty, Department of Mechanical Engineering, George Mason University and Senior Project Engineer, The Aerospace Corporation

OXFORD
UNIVERSITY PRESS

OXFORD
UNIVERSITY PRESS

Great Clarendon Street, Oxford, OX2 6DP,
United Kingdom

Oxford University Press is a department of the University of Oxford.
It furthers the University's objective of excellence in research, scholarship,
and education by publishing worldwide. Oxford is a registered trade mark of
Oxford University Press in the UK and in certain other countries

Published in the United States of America by Oxford University Press
198 Madison Avenue, New York, NY 10016, United States of America

British Library Cataloguing in Publication Data
Data available

Library of Congress Control Number: 2024915860

ISBN 9780198920663

DOI: 10.1093/9780198920687.001.0001

Printed and bound by
CPI Group (UK) Ltd, Croydon, CR0 4YY

Preface

This book concerns particle systems with random or seemingly random properties such as mass, momentum, and energy. This book builds on a number of research efforts done over the course of my career. From 1991 to 1996, I was a professor of Aerospace Engineering Sciences at the University of Colorado, Boulder, where I specialized in computational fluid dynamics. In 1996, I transitioned from academia to industry. From 1997 to 2003, I led a project which required modeling aggregation and fragmentation of various liquids, solids, and powders. In this project, the materials initially assumed simple geometric shapes such as spheres, cylinders, or cylindrical shells. The resulting airborne particles were used to initialize a computational fluid dynamics code, where turbulent entrainment and mixing led to additional rounds of aggregation and fragmentation. From 2004 to 2007, I led a second project which required modeling aggregation and fragmentation of various solids and powders. As before, the materials initially assumed simple geometric shapes such as cylindrical shells. Also as before, the resulting airborne fragments were used to initialize a computational fluid dynamics code, where turbulent entrainment and mixing led to combustion.

Starting in 2008, I was assigned to help oversee a large research and development portfolio. During this assignment, I obtained a bird's eye view of many different efforts in many different fields. Because of my background, I took a particular interest in projects that involved aggregation, fragmentation, and turbulence. These glimpses reinforced an earlier impression that aggregation, fragmentation, and turbulence often have a surprising amount in common, regardless of the exact circumstances. Kolmogorov's 5/3 rule is a standard result for turbulent kinetic energy. As I broadened my perspective, it became increasingly clear that aggregation and fragmentation produced analogous results for particle sizes, shapes, velocities, and kinetic energies. In some cases, these discoveries were entirely empirical, i.e., they were based on curve fits to observations or experimental results. However, in other cases, these results were proven theoretically, although the practical implications of the underlying assumptions were not always completely clear.

From 2014 to 2021, I was given the opportunity to dedicate a substantial fraction of my time to overseeing one particular fragmentation program, which used shock tubes to load samples of common building materials such as glass, cinder block, and concrete. In some cases, the conditions were identical, except for differences in the sample material. These results again indicated that different materials behaved surprisingly similarly, after accounting for a few simple factors such as geometry, density, strength, and strain rate. In addition to small-scale work, this program used a large shock tube to load realistic full-scale building elements such as plate glass windows, clay tile roofs, and concrete spandrel panels. The results were captured with high-speed video. Results from this experimental program are used occasionally throughout the book to illustrate some of the basic principles.

To complement this experimental program, in my personal time, I began an effort to better understand why different types of aggregation, fragmentation, and turbulence often behaved similarly. While my original goal was to produce a single journal article, the effort continuously expanded as a literature search uncovered more and more relevant material in more and more different fields. After several years, I found I had explored dozens of different topic areas spread across aerospace engineering, chemical engineering, mechanical engineering, mining engineering, earth sciences, astrophysics, and many other fields. While I felt I had still just scratched the surface, I realized that organizing what I had already learned would require an enormous amount of effort. After many more years of work including collaborative feedback from fellow researchers, as well as independent peer reviews of various drafts, the book finally assumed its present shape.

My first book, *Computational Gas dynamics*, was published in 1998. That book concerned a relatively narrow body of research done in aerodynamics and applied mathematics completed over a relatively short period of time. In fact, the vast majority of papers surveyed in that book were written within a 20 year period. Because the field was relatively small at the time, it was possible to consult almost every significant relevant journal article, technical report, and conference paper. That book was published right as the field matured, i.e., just as basic research results were being actively incorporated into large-scale commercial codes for the first time. By contrast, the present book concerns a wide-ranging body of research done in many different fields over many decades. Because of this, it was impossible to consult all of the significant relevant journal articles, technical reports, and conference papers. Despite its size and long history, the majority of the relevant work on particle aggregation and fragmentation is still basic research. In other words, the majority of the material covered in this book has never fully transitioned into commercial applications. One goal of this book is to help break this longstanding log jam or, at least, to shed new light on what the main holdups are.

As another major difference, my first book concerned deterministic problems exclusively, while this second book concerns random and seemingly random problems exclusively. My first exposure to random and seemingly random problems occurred in graduate school, where I was taught turbulence by two of the leading experts of the day, John Lumley and Stephen Pope. However, their instruction had little immediate impact on my research. Random and seemingly random behaviors occurred only in limited regions in the problems I was studying at the time, such as thin, turbulent boundary layers on the surfaces of airfoils. Thus, my early research focused on deterministic phenomenology, with the small exceptional regions relegated to short subroutines written by others. This changed once I transitioned from academia to industry, where I was suddenly faced with problems dominated by random and seemingly random behaviors—including aggregation, fragmentation, and turbulence—which could occur anywhere and at any scale. While this book focuses on traditional particle aggregation and fragmentation, as the title indicates, it also considers turbulence. In fact, this book was directly influenced by and includes a number of quotes from turbulence books such as those written by Stephen Pope, W. David McComb, and Peter Davidson.

Acknowledgements

I would like to thank those people who took the time and trouble to review the book at various stages. The book benefitted enormously from their encouragement and suggestions. This list includes Dennis Grady, Falk Wittel, Ferenc Kun, Judith Curry, Gary Dilts, Eli Ben-Naim, Charles Needham, and several anonymous reviewers. I would also like to thank those people who granted me permission to use the graphics, videos, and other materials which appear throughout the book. This list includes Matt Barsotti, Eddie O'Hare, and Mohsen Sanai. In addition, I would like to thank Mark Sward, whose far-sighted approach to research and development (R&D) management helped make this book possible.

I would also like to thank those people who have supported me throughout my career including Melvin Laney, Carolyn Laney, Lisa Lane Laney, John Toll, Deborah Toll, Carolyn Sahni, Reed Whittemore, Donald Kleine, Charles Fefferman, John Howarth, Faith Gabelnick, Daniel Sweet, Richard Sebastian, John Hanna, Tom Coleman, Lars Wahlbin, Robert Kulp, Richard Seebass, Florence Padgett, Phil Hookham, Ken Kreyenhagen, Gene Sevin, Ed Conrad, and Peter Nanos. I would especially like to thank my Ph.D. thesis advisor, David Caughey, a gifted researcher and role model.

While most of the influences on this book should be obvious from the main text, I would like to acknowledge three researchers whose contributions might not be so obvious. First, I would like to note the influence of Paul Woodward of the University of Minnesota, who gave an eye-opening seminar at the University of Colorado in the 1990s. This was my first exposure to the seeming conundrum that computational methods for inviscid flows can capture many of the characteristic features of turbulence flows, as observed in direct numerical simulation, despite the lack of physical viscous terms or fully resolved small scales. In my subsequent career, I successfully used this approach, sometimes called Monotonically Integrated Large Eddy Simulation (MILES), and related approaches such as the Discrete Element Method (DEM), in a number of different applications. These approaches, which temper the appealing rigor of determinism with the more forgiving nature of chaos, form a substantial part of this book.

Second, I would like to note the influence of Gerard Faeth of the University of Michigan, who produced an important series of articles on liquid atomization starting in 1992, which continued to appear belatedly even after his death in 2003. While this book could only include a few highlights, these papers collectively offer state-of-the-art engineering solutions suitable for a surprising variety of liquid fragmentation problems. While the general consensus is that most researchers do their best work while young, Gerard Faeth is an example of someone who produced work of the first rank later in life, well after attaining emeritus status. Now that I myself am older, this inspires me to believe that research productivity is more a matter of attitude than age.

Third and finally, I would like to note the influence of a talk given by Kary Mullis at Fort Belvoir, Virginia, on October 30, 2009. In this talk, Dr Mullis described the circumstances that led him to win the Nobel Prize for Chemistry in 1993 as follows.

> I didn't have a grant. I didn't have time limits. I had no preliminary sense that my idea would work. I wasn't writing monthly reports. I had at least three serious factual mistakes in my head that would have been flagged in a grant application ... The real trick was that my company had a policy that said scientists could spend 10% of their time doing whatever they liked. I had a day job that justified my salary and existence. But it was the side project—done completely on my own initiative with no oversight or support—that provided the innovation.

With this as inspiration, I found I was able to write this book working after hours and on weekends over a period of ten years, while holding a series of demanding full-time industrial R&D positions.

Contents

About the Author xiv
About the Book ("Frequently Asked Questions") xv

1. Introduction **1**
- 1.1 Basic Definitions 2
- 1.2 Particle Types 6
- 1.3 Particle Shapes 7
- 1.4 Particle Sizes 10
- 1.5 Universal Size Distributions 12
- 1.6 Conclusions 15
- References 17

2. Basic Properties **22**
- 2.1 Chaos 22
- 2.2 Self-Organization 29
- 2.3 Self-Organized Criticality 31
- 2.4 Homogeneity and Isotropy for Scalar Properties 34
- 2.5 Homogeneity and Isotropy for Vector Properties 35
- 2.6 Conclusions 37
- References 38

3. Distributions and Averages **43**
- 3.1 Number Distributions 43
- 3.2 Concentration Distributions 48
- 3.3 Probability Density Functions 48
- 3.4 Cumulative Distribution Functions 50
- 3.5 Averages 53
- 3.6 Self-Similar Size Distributions 57
- 3.7 Common Size Distributions 57
- 3.8 Conclusions 63
- References 66

4. Computational Methods **68**
- 4.1 High-Fidelity Computational Methods for Aggregation 69
- 4.2 Examples of High-Fidelity Computational Results for Aggregation 70
- 4.3 Simplified Computational Methods for Aggregation 71
- 4.4 Examples of Simplified Computational Results for Aggregation 73
- 4.5 High-Fidelity Computational Methods for Fragmentation 74
- 4.6 Examples of High-Fidelity Computational Results for Fragmentation 76
- 4.7 Simplified Computational Methods for Fragmentation 80
- 4.8 Examples of Simplified Computational Results for Fragmentation 82
- 4.9 Conclusions 93
- References 95

5. Analytic, Semi-Empirical, and Empirical Methods 104
5.1 Mean-Field Theory 104
5.2 Stochastic Geometry 107
5.3 Maximum Entropy Theory 111
5.4 Dimensional Analysis 114
5.5 Semi-Empirical Models 115
5.6 Empirical Models 116
5.7 Examples of Analytic and Empirical Model Results for Aggregation 124
5.8 Examples of Analytic and Empirical Model Results for Fragmentation 127
5.9 Examples of Analytic Model Results for Aggregation and Fragmentation 130
5.10 Experimental Calibration 132
5.11 Conclusions 134
References 135

6. Particle Shape Distributions 147
6.1 Different Particles Have the Same Shape 148
6.2 Different Particles Have the Same Shape on Average 149
6.3 Different Particles Have Different Shapes 158
6.4 Conclusions 165
References 166

7. Particle Velocity Distributions 170
7.1 Joint Probability Density Functions 170
7.2 Velocity Distributions for Aggregation 172
7.3 Velocity Distributions for Fragmentation 174
7.4 Conclusions 177
References 179

8. Particle Energy Distributions 181
8.1 Variances and Covariances 181
8.2 Structure Factors 185
8.3 Structure Functions 186
8.4 Granular Gases 187
8.5 Fluid Turbulence 193
8.6 Relationships Between Kinetic Energy and Size Distributions 200
8.7 Conclusions 202
References 203

9. Aggregation Cascades Part 1: Mean-Field Governing Equations 209
9.1 Discrete Governing Equation 209
9.2 Continuous Governing Equation 210
9.3 Kernels 212
9.4 Initial Conditions 217
9.5 Boundary Conditions 218
9.6 Conservation of Mass 221
9.7 Uniqueness 221
9.8 Conclusions 222
References 222

10. Aggregation Cascades Part 2: Full Mean-Field Solutions **227**
10.1 Moments 228
10.2 Common Size Distributions 229
10.3 Gelation—Part 1 230
10.4 Non-Stationary Closed-Form Solutions for Constant Kernels 231
10.5 Non-Stationary Closed-Form Solutions for Sum Kernels 237
10.6 Non-Stationary Closed-Form Solutions for Product Kernels 239
10.7 Computational Validation 242
10.8 Experimental Validation 243
10.9 Gelation—Part 2 250
10.10 Conclusions 251
References 253

11. Aggregation Cascades Part 3: Late-Time Mean-Field Solutions **259**
11.1 Boundary Conditions 261
11.2 Common Size Distributions 261
11.3 Size Distributions for Non-Local Aggregation 262
11.4 Size Distributions for Local Aggregation 269
11.5 Average Particle Sizes 275
11.6 Computational Validation 278
11.7 Experimental Validation 278
11.8 Aggregation Combined with Fragmentation 284
11.9 Conclusions 284
References 288

12. Fragmentation Cascades Part 1: Mean-Field Theory for Binary Fragmentation **294**
12.1 Governing Equations 295
12.2 Kernels 299
12.3 Shattering 299
12.4 Size Distributions and Moments 300
12.5 Full Mean-Field Solutions 301
12.6 Late-Time Mean-Field Solutions 310
12.7 Experimental Validation 311
12.8 Conclusions 316
References 317

13. Fragmentation Cascades Part 2: Mean-Field Theory for Random Fragmentation **321**
13.1 Governing Equations 321
13.2 Breakage Functions 327
13.3 Kernels 331
13.4 Shattering 334
13.5 Distributions and Moments 335
13.6 Full Mean-Field Solutions 336
13.7 Late-Time Mean-Field Solutions 347
13.8 Experimental Validation 349
13.9 Energy-Size Reduction Relationships 354
13.10 Conclusions 357
References 358

14. Fragmentation Cascades Part 3: Mean-Field Theory for Collisional Fragmentation **365**
14.1 Governing Equations 366
14.2 Shattering 370
14.3 Full Mean-Field Solutions 370
14.4 Late-Time Mean-Field Solutions 372
14.5 Experimental Validation 380
14.6 Conclusions 382
References 383

15. Unitary Fragmentation Part 1: Basic Principles **387**
15.1 Challenges 387
15.2 Strain Rates 396
15.3 Conclusions 401
References 402

16. Unitary Fragmentation Part 2: Average Sizes **406**
16.1 Mean-Field Theory Models 406
16.2 Non-Mean-Field-Theory Models 408
16.3 Experimental Validation Part I: Primary Fragmentation in Cylindrical Geometries 417
16.4 Experimental Validation Part II: Primary Fragmentation in Spherical Geometries 422
16.5 Experimental Validation Part III: Primary Fragmentation in Cartesian Geometries 429
16.6 Experimental Validation Part IV: Secondary Fragmentation 435
16.7 Conclusions 445
References 446

17. Unitary Fragmentation Part 3: Size Distributions **453**
17.1 Common Size Distributions 453
17.2 Mean-Field Theory 454
17.3 Stochastic Geometry 455
17.4 Maximum Entropy Theory 458
17.5 Experimental Validation of Primary Fragmentation Models 459
17.6 Experimental Validation of Secondary Fragmentation Models 464
17.7 Conclusions 473
References 473

18. Maximum Entropy Theory for Discrete Particle Systems **480**
18.1 Combinatorics and Constrained Optimization 482
18.2 Conclusions 487
References 488

19. Maximum Entropy Theory for Continuous Particle Systems **490**
19.1 Families of Size Distributions 491
19.2 Specific Size Distributions 497
19.3 Velocity Distributions 498

19.4 Validation 499
19.5 Conclusions 500
References 501

20. Conclusions **505**
References 514

Appendix A: Different Forms of Common Size Distributions *515*
A.1 Sizes Distributions that Look the Same but are Actually Different *515*
A.2 Size Distributions that Look the Same and Actually are the Same *516*
A.3 Size Distributions that Look Different but are Actually the Same *516*
A.4 Simple Ways of Deciding if Size Distributions are the Same or Different *517*
A.5 Rosin–Rammler Size Distributions *517*
A.6 Weibull Size Distributions ($l = n$) *519*
A.7 Gamma Size Distributions ($l = 1$ *or* m) *521*
A.8 Power Law Size Distributions ($l = 0$) *522*
A.9 Conversion Techniques *525*
References *526*

Index 529

About the Author

The author is a Senior Project Engineer with The Aerospace Corporation, where he supports the United States' Department of Energy. In addition, he is an Affiliate Faculty member with the Department of Mechanical Engineering at George Mason University. Prior to this, the author served in a series of other research positions in industry and academia. In the 1990s, he was an Assistant Professor of Aerospace Engineering Sciences at the University of Colorado, Boulder, where he specialized in high-speed gas flows with shocks, and wrote the textbook *Computational Gasdynamics*. In the 2000s, he was a Chief Scientist at the Northrop-Grumman Corporation, where he helped the United States Department of Defense oversee a nearly $1B research and development portfolio dedicated to countering weapons of mass destruction. In the 2010s, he was a Chief Nuclear Engineer at Science Applications International Corporation (SAIC), where participated in a major multi-year effort to revitalize a legacy research program rooted in the Manhattan Project. The author obtained a B.S. in Mathematics from the University of Maryland in 1984, an M.S. in Computer Science from Cornell University in 1989, and a Ph.D. in Applied Mathematics from Cornell University in 1991. The author has over 25 years of experience modeling aggregation and fragmentation in a variety of different applications.

About the Book ("Frequently Asked Questions")

1. What fields and applications does this book cover?
This book includes material drawn from many different fields and applications including the following.

Aerospace Engineering. Specific applications include aircraft vulnerability to hail and projectiles; satellite vulnerability to space debris; and turbulent gas dynamics.
Astrophysics. Specific applications include: the formation and evolution of asteroids, planets, stars, dust clouds, etc.; and cratering and ejecta due to high-speed impacts.
Biochemistry/Chemistry/Chemical Engineering. Specific applications include polymers, colloids, granular gases, and dense granular systems such as sand piles.
Biosciences. Specific applications include biologic clustering such as schools of fish and flocks of birds.
Civil Engineering. Specific applications include structural vulnerability to terrestrial debris such as that produced by hurricanes, volcanoes, and stand-off explosions.
Earth Sciences/Atmospheric Sciences. Specific applications include aerosol dynamics; cloud microphysics; and the formation and evolution of raindrops and snowflakes.
Mechanical Engineering. Specific applications include fragmentation of rapidly expanding circular rings and cylindrical shells; liquid atomization and sprays; turbulence.
Mining Engineering. Specific applications include drilling, blasting, crushing, grinding, and milling of rocks and minerals.
Oceanography. Specific applications include hydrosol dynamics, i.e., the evolution of small particles suspended in water such as algae, plankton, microplastics, minerals, etc.
Process Engineering/Product Engineering. Specific applications include the production of energetic materials, pharmaceuticals, etc. including inhalers, conveyer belts, etc.

Some of the applications listed previously are studied in several different fields. In particular, many of the applications listed are often considered part of physics or applied mathematics. Thus, relevant papers often appear in journals published by the American Institute of Physics, the Institute of Physics, and the Society for Industrial and Applied Mathematics.

This book does not include material specifically relevant to nuclear engineering, nuclear physics, or particle physics. In these fields, the term "particle" generally refers to *subatomic* particles such as photons, neutrons, fermions, bosons, etc. By contrast, in this book, the term "particle" generally refers to *superatomic* particles, i.e., particles composed of one or more complete atoms.

2. Why does this book cover so many different fields and applications?
Despite its long history, particle dynamics with aggregation and fragmentation has not evolved into a fully independent field of study. Rather it is an adjunct to many different fields, such as those listed previously. In some cases, different fields have obtained the same results independently, either nearly simultaneously or decades apart. In other cases, major results in one field are completely unknown in other fields. As one of its goals, this book tries to bring together results from as many different fields and applications as possible, to provide as complete a picture as possible. This book includes an unusually large number of direct quotes from the research literature. Because this book surveys so many different fields, no single author could possibly master all of them. Rather than paraphrase and risk distorting the original sense, this book tries to let the experts in each field express themselves directly.

3. What if I'm only interested in one or two of the fields and applications listed previously?
This book was written to appeal to a wide range of scientists and engineers. The main body of the text is mostly generic. It requires knowing only basic calculus, physics, and probability theory. Occasionally, out of necessity, the main text gives examples involving specific applications to illustrate general principles. However, in most cases, examples involving specific applications are separated from the main text and placed into grey boxes. This allows readers to skim or skip these examples, according to their level of interest. In a few cases, the examples given in grey boxes concern important general-interest topics. These examples are separated out simply to avoid interrupting the narrative flow of the main text. These examples, which should be read carefully rather than skimmed or skipped, will hopefully be obvious from the context.

4. What if I'm not interested in any of the fields and applications listed previously?
The list given previously is by no means comprehensive. Even if a specific field or application does not appear on this list, this book may still be relevant if the following eight criteria are met. First and most obviously, the problem must involve a physical system of particles. In this book, the term "particle" can refer to a single object such as a molecule or a rock. Alternatively, it can refer to a cluster of objects such as a chemically bonded chain of molecules as found in polymer physics or a gravitationally coalesced rubble pile as found in astrophysics. While the term "particle" usually refers to an inanimate object, it may also refer to a living organism such as a bird or fish. Alternatively, it may refer to a group of living organisms such as a flock of birds or a school of fish. The material presented in this book has direct applications to problems involving graphs and trees. However, because those applications do not involve anything like a physical system of particles, they are not covered in this book.

Second, the problem must involve particles whose mass, momentum, energy, etc. vary over time. This variation may be due to instantaneous events, such as random binary collisions. Alternatively, this variation may be due to processes that occur over an extended period of time, such as stochastic condensation. The term *aggregation* refers to events or processes which causes particle sizes to increase, e.g., low-speed "sticky" collisions between two particles. Conversely, the term *fragmentation* refers to events or processes which cause particle sizes to decrease, e.g., high-speed "shattering" collisions between two particles. To put particle aggregation and fragmentation into context, this book will compare and contrast with cases where the particle mass is fixed, especially ideal gases composed of identical atoms or molecules.

Third, the problem must involve a large number of particles whose masses, momenta, and energies vary randomly or seemingly randomly. In other words, this book only concerns cases that can be well-represented using standard probability theory, such as probability density functions and cumulative distribution functions. In this book, the term *seemingly random* generally refers to chaos. Chaotic systems also have other properties, related to their sensitivity to initial conditions, but their seeming randomness is their most obvious feature. While chaotic systems are not truly random, this book assumes they can be described using standard probability theory. This book generally assumes large numbers of particles. However, as exceptions, fragmentating particles systems may start with as a few as one particle and aggregating particle systems may end with as few as one particle.

Fourth, the problem must involve particles that all have the same composition. The particles can have almost any composition. They can be solids, liquids, or gases. They can even be living organisms. However, this book does not allow mixtures of different types of particles. If different types of particles are present in the same region, they will be assumed to evolve independently of one another. Thus, as noted earlier, this book does not apply to typical problems in nuclear engineering, nuclear physics, and particle physics, which generally involve combinations of subatomic particles. Instead, this book concerns single materials composed of one or more entire atoms, i.e., the pieces and the whole are made of the same substance.

Fifth, the problem must involve particles that reside in a vacuum or in a relatively quiescent medium composed of a single substance. This book allows particles to reside within a wide variety of different media. For example, solid particles may reside within almost any gas or liquid medium. But, just like the particles must be composed of a single substance, the medium must also be composed of a single substance. If the medium is a fluid, i.e., a liquid or a gas, the fluid dynamics of the medium may play an important role in the particle dynamics. However, in this book, the dynamics of the particles always dominates the dynamics of the fluid medium. In other words, in this book, the particle properties are primarily determined by the mass, momentum, and energy of the particles themselves rather than by the mass, momentum, and energy of any surrounding fluid. The opposite case—where the fluid dynamics is on the same order as or dominates the particle dynamics—is called *multiphase flow*. This book does not consider multiphase flows. Thus, with a few exceptions, this book will *not* consider gas bubbles in liquid media. Because gas bubbles weigh so little, it is rare for the mass, momentum, and energy of gas bubbles to dominate the mass, momentum, and energy of the liquid medium.

Sixth, as noted earlier, the problem must involve particles that are composed of at least one whole atom. Other than that, this book allows particles to have any mass. They may be as small as a single atom or as large as an interstellar dust cloud. In fact, in general, the particle systems in this book have an extremely wide range of different sizes, spanning several orders-of-magnitude or more, especially at late times. Along with particle composition, particle size plays a critical role in determining the influence of interparticle forces. For example, particles the size of dust may experience van der Waals forces while particles the size of planets may experience gravity. Because this book allows a wide variety of different particle sizes, it allows a commensurately wide variety of different particle-particle and particle-medium forces.

Seventh, the problem must feature particle momenta and energies in a range that allows aggregation, fragmentation, or both to occur. For example, in those cases dominated by binary collisions, particles must travel slowly enough to experience aggregation or rapidly enough to experience fragmentation. For another example, in those cases dominated by thermal degradation, the particles must have enough vibrational energy to break apart. Other than that, this book allows particles to have almost any momentum or energy.

Eighth and finally, the problem must involve particles that interact as if they have simple compact shapes. Here, the term *compact* means that each particle's dimensions are roughly equal to one another, while the term *simple* means that each particle's shape is roughly geometrical. In some cases, particles literally have simple compact shapes, e.g., liquid droplets are often nearly spherical. However, particles may still interact with each other as if they have simple compact shapes, even when this is not literally true. For example, imagine coins being flipped end-over-end. Even though the coins are not spheres, they may interact with each other as if they were. By a similar token, rotating particles with complex fractal shapes, e.g., colloids and crosslinked polymers, may interact with each other as if they have simple compact shapes. Finally, particles with high aspect ratios such as extended vortex filaments in turbulent flows or long molecular chains in polymer physics may interact as if they have simple compact shapes. Thus, this book allows particles to have almost any shape *provided* the non-compact aspects of particle shape do not substantially affect the particle interactions. For example, long filaments and chains should not routinely become entangled. While this book does not claim to cover every possible case, it is surprising how often statistical results derived assuming simple compact particle shapes apply more generally.

5. How can this book cover so many different fields and applications at once? How can particles as small as atoms have anything in common with particles as large as planets?

One of the themes of this book is that results found in one field may apply to many other fields. This is true despite profound differences in the particle compositions, shapes, sizes, velocities, etc. Why is this? In theory, understanding a given aggregation and fragmentation problem requires conducting multiple random trials, while tracking the state of every particle at every time in every trial including their position, mass, momentum, energy, rotation, and shape. Of course, it is nearly impossible to capture this much detail, except in very simple cases. Thus, in practice, most studies—including those based on analytical models, computational models, experiments, and observations—focus on only one or two aspects. For example, many studies focus just on particle size distributions.

But what can really be learned from studying only one or two parts of a complex problem? Is it even possible to understand just one aspect of a complex of problem without understanding the entire problem? Based on over 100 years of research into aggregation and fragmentation, the answer is a qualified yes. As one general strategy, an approximate solution for a single unknown of interest—such as the particle size distribution—can be obtained by making reasonable assumptions about the other parts of the problem. This book argues that, in many cases, these assumptions are often best expressed in terms of chaos, isotropy, homogeneity, and degrees of freedom. In this schematic, the assumptions are completely generic – they apply regardless of the size, shape, composition, etc. of the particles.

Many studies of aggregation and fragmentation find specific distributions for size, shape, and/or velocity. In fact, certain specific distributions appear again and again under widely varying circumstances. As a result, they are sometimes called *universals*. These are the result of the underlying conditions, real or assumed. For analytic models, the underlying assumptions are typically stated in somewhat abstract terms—e.g., an entropy constraint or a volumetric collision rate—making it difficult to determine which universal, if any, applies in a given set of circumstances. One of the goals of this book is to, as much as possible, clarify and generalize the underlying assumptions. By considering the full range of reasonable possibilities, this book finds that aggregation and fragmentation may admit dozens of different universal distributions. This is true even though only a fraction of these may be commonly observed in a given application.

While it would obviously be better to obtain a unique universal for a given application, multiple solutions are the inevitable result of focusing on just one aspect of a complex multifaceted problem. Aggregation and fragmentation affect almost all aspects of a particle system—including sizes, velocities, shapes, and rotations—all at once. Unless the entire problem is solved, it may be impossible to say for certain which if any universal sizes, velocities, shapes, etc. occur. If a problem is decomposed into parts, and each part is solved separately, the resulting solutions may be ambiguous. This ambiguity can be resolved only by integrating the parts back into a coherent whole.

6. a. I'm only interested in aggregation. Can I skip the parts about fragmentation?
b. I'm only interested in fragmentation. Can I skip the parts about aggregation?

The answer to the first question is a qualified yes. The answer to the second question is a qualified no. At a minimum, everyone should read the first five chapters of the book. These chapters provide a broad overview of computational, experimental, and analytical results for particle aggregation and fragmentation, with a focus on results that apply regardless of the specific application.

Other than this, readers who are only interested in aggregation can safely read the rest of the book selectively, skimming or skipping the parts on fragmentation. As a caveat, particle systems dominated by aggregation may still experience lesser amounts of fragmentation. Just because the particle sizes in a system increase over time does not automatically mean that only aggregation and no fragmentation occurs. Understanding the dual action of aggregation and fragmentation requires reading the book as a whole.

On the other hand, readers who are only interested in fragmentation are strongly encouraged to read the book in its entirety, including the parts on aggregation. Aggregation is a useful and arguably essential introduction to fragmentation. The simplest form of fragmentation is sometimes called *reverse aggregation*. As the name implies, understating regular forward aggregation is helpful for understanding reverse aggregation. By the same token, understanding reverse aggregation is helpful for understanding more complex types of fragmentation. As an additional point, particle systems dominated by fragmentation may still experience lesser amounts of aggregation. Just because the particle sizes in a system decrease over time does not automatically mean that only fragmentation and no aggregation occurs. Understanding the dual action of aggregation and fragmentation requires reading the whole book.

7. I'm only interested in computational models. Can I skip the parts about analytic models?

Everyone should at least read the introduction to analytic, empirical and semi-empirical methods given in Chapter 5. While the rest is optional, you may be surprised at how much you'll learn from the more in-depth discussion of analytic models given in Chapters 9 through 20. There are two main reasons why analytic models are so important.

- *Practicality.* Computational models have long since supplanted analytic models for most types of problems. However, this is not the case for aggregation, fragmentation, and turbulence, where spatial scales typically span multiple orders of magnitude. For these types of problems, first-principles computational methods are typically too expensive. As a practical matter, even the most efficient computational methods running on the fastest supercomputers generally must rely on analytic, semi-empirical, or empirical models, at least to some extent.
- *Pedagogy.* Analytic methods are a time-tested way of building physical intuition. Thus, analytic methods still commonly appear in introductory texts, even in applications where computational methods long ago came to dominate. Furthermore, analytic methods tend to provide general-purpose results of interest to a wide variety of readers, while computational methods tend to provide specific solutions to specific problems of interest to only a few readers. Finally, this book focuses on analytic methods to avoid having to delve into the endless complex details of typical computational methods.

This book compares and contrasts several different analytic modeling approaches including dimensional analysis, maximum entropy theory, stochastic geometry, and mean-field theory. While most analytic approaches are limited to late-time solutions, mean-field theory can be used to obtain full time-dependent solutions. However, obtaining closed-form solutions with mean-field theory typically requires using non-trivial mathematical techniques such as generating functions. In this regard, while many existing references provide the relevant derivations, few explain what the results mean in clear, concrete language. This book provides mainly abbreviated proofs that focus, as much as possible, on building physical intuition. For readers interested in seeing full rigorous proofs, the book provides extensive references. To make analytic models more palatable to computationally oriented readers, this book reviews their storied histories. In many cases, analytic models for aggregation and fragmentation were originally developed by world-famous scientists and mathematicians. In other cases, the foundational work was done by researchers who are well-known in some fields but not others, including a number of Nobel prize winners. By reviewing the relevant history, this book hopes the put readers from different fields on an even footing.

8. a. I'm only interested in solids. Can I skip the parts about fluids?
b. I'm only interested in fluids. Can I skip the parts about solids?

The answer to both questions is a qualified yes. With the partial exception of particle shapes, the main text is more or less agnostic on the topic of fluids vs. solids. This distinction only becomes critical for specific applications. As noted earlier, specific applications appear mainly in examples, which are separated from the main text and placed into grey boxes. Thus, readers interested in solids can easily skip the examples involving fluids and vice versa.

That having been said, readers interested only in solids should be forewarned that, as one of its themes, this book argues that particle dynamics often resembles fluid dynamics. In particular, at low particle densities—where particles interact primarily through body forces or sporadic binary collisions—particle dynamics often resembles gas dynamics. At higher particle densities—where particles remain in extended contact with their neighbors—particle dynamics often resembles fluid mechanics. In some cases, fluid-like behavior occurs because the particles are fluids, or the medium containing the particles is a fluid. However, fluid-like behaviors occur even when the particles are solids resident in a vacuum. For example, particle systems may develop size discontinuities, reminiscent of shock waves in high-speed gases. In fact, some of the simple model problems studied in aggregation and fragmentation are similar or identical to those studied in gas dynamics. As another example, in some cases, particles in aggregating and fragmenting particle systems interact entirely through binary collisions, just like particles in ideal gases. However, while ideal gases only allow perfectly elastic collisions, aggregating particle systems allow perfectly inelastic collisions, in which two particles merge to form a single larger particle. As a final example, particles in aggregating and fragmenting particle systems may interact through partially elastic collisions, causing them to self-organize into clusters that resemble the eddies seen in turbulent flows.

Because of these fluid-like behaviors, this book is able to draw on a number of results from fluid dynamics. As one example, this book argues that Maxwell's concept of degrees of freedom, a key element of ideal gas theory, also applies to aggregation and fragmentation. This concept is covered in most basic physics textbooks and should be familiar to most readers. As another example, this book argues that spatial homogeneity and isotropy, key statistical concepts used in fluid turbulence theory, also apply to aggregating and fragmenting particle systems. Roughly speaking, a particle system is spatially homogenous and isotropic when it is well-mixed and uniform. As a third example, this book argues that chaos, a fundamental property of turbulent flows, also applies to aggregation and fragmentation. While chaos theory has its roots in the 19th century, it was not well understood until digital computers became widely available in in the 1960s and 1970s. However, much of the basic theory behind fluid turbulence—as well as particle aggregation and fragmentation—was developed well before the computer age. Even after all of this time, many treatments of turbulence, aggregation, and fragmentation mention chaos only in passing. By contrast, this book puts chaos front and center. In fluid dynamics, researchers have used these four concepts—chaos, isotropy, homogeneity, degrees of freedom—either implicitly or explicitly to derive universal turbulent kinetic energy spectra. The most famous example is Kolmogorov's 5/3 law. Similarly, in aggregation and fragmentation, researchers have used these four concepts—chaos, isotropy, homogeneity, degrees of freedom—either implicitly or explicitly to derive universal size distributions.

9. Why the focus on particle size distributions? What about other particle properties?
Chapters 6, 7, and 8 provide substantial treatments of particle shape, velocity, and energy distributions, respectively. However, much of the rest of the book concerns particle size distributions. In fact, historically, the vast majority of research on aggregation and fragmentation concerns particle size distributions. This research dates back over 100 years. As one landmark, Arthur O. Gates, an industrial mining engineer with ties to Purdue University, published the first well-known journal article on fragment size distributions in 1915. As another landmark, Marian von Smoluchowski, a physics professor at Jagiellonian University, published the first well-known journal article on aggregate size distributions in 1916. The former article was deeply empirical while the latter was deeply theoretical. This diversity

of fields and approaches, as well as the focus on particle size distributions, have been hallmarks of aggregation and fragmentation research ever since. This book, to some extent, represents a belated celebration of the 100th anniversary of these landmark papers. For a timelier celebration see, for example, the "Special Issue on Marian Smoluchowski's 1916 Paper—'A Century of Inspiration'" published in 2017 in *Journal of Physics A: Mathematical and Theoretical.*

10. Why does the book return to certain subjects repeatedly? Why not provide a single complete discussion of each major subject?
This book introduces certain subjects in stages, returning or "circling back" several times to provide additional information. This approach was inspired by Peter Davidson's book *Turbulence: An Introduction for Scientists and Engineers, 2nd edition*, Oxford University Press, 2015. As in that earlier work, the "circling back" tactic is intended to make this book more approachable to a broad range of readers. Advanced readers may prefer to have the full depth and rigor on each topic collected in one place. However, there are already specialized books and survey articles available to such readers. This book was meant to fill a gap, namely, to provide an introductory general-purpose treatment.

1

Introduction

This book concerns particle dynamics with aggregation and fragmentation. More specifically, this book concerns particle systems with random or seemingly random properties such as velocity, shape, and size. Particle dynamics with aggregation and fragmentation occurs in many if not most scientific and engineering disciplines, ranging from colloids and polymers with scales as small as 10^{-10} meters, to star-forming molecular clouds with scales as large as 10^{18} meters. The purpose of this book is to explain the key analytical, computational, and experimental approaches used to characterize the random or seemingly random outcomes, especially those that apply widely or universally regardless of the specific application.

Most previous books on aggregation and fragmentation are meant for specialists in a single field. For example, King et al. (2012) describes fragmentation modeling for the mineral processing industry. For another example, Litster (2016) describes aggregation and fragmentation modeling for the formulated products industry including agricultural products, energetic materials, and pharmaceuticals. As a final example, Bertoin (2006) and Banasiak et al. (2019a, b) describe aggregation and fragmentation modeling from the mathematician's point of view, i.e., they emphasize formal proofs and lemmas. While these books overlap each other, they are intended for fundamentally different audiences. Thus, these books are little known outside of their respective fields, despite the fact that much of the material is of general interest.

By contrast with most previous books, this book is meant for scientists and engineers in a large range of different fields. By using a broad interdisciplinary approach, this book hopes to introduce practitioners to the widest possible range of key results, including those from outside of their specialty areas. While occasional mathematical proofs are given, they are intuitive rather than rigorous. Combining results from different fields helps to provide the most complete possible treatment. In some cases, results that have long been considered strictly empirical in one field turn out to have been given a solid theoretical foundation in some other field. By focusing on general results rather than specific applications, this book hopes to offer practitioners a deeper appreciation for the fundamental physical principles involved. This book will discuss many types of models including empirical, semi-empirical, analytic, and computational models. However, semi-empirical and computational models are, by their very nature, specific to certain problems. Thus, much of the book focuses on analytic models, which offer the shortest and clearest path to the key results of broad general interest.

This book contains numerous practical examples, including comparisons between theory, computation, and experiment. Such comparisons are sometimes missing from theoretical and computational treatments, which invites skepticism about their practical value and validity. Of course, real-life examples must be drawn from some specific topic area. To explain such examples requires delving, at least to some extent, into the unique technical aspects of each problem. For readers with limited time and interest, this book separates most such examples from the main text using shaded grey boxes. This allows readers to quickly

Particle Dynamics with Aggregation and Fragmentation. Culbert B. Laney, Oxford University Press. © Culbert B. Laney (2025).
DOI: 10.1093/9780198920687.003.0001

skim or entirely skip examples in favor of the general material in the main text. On the other hand, some examples are integrated into the main text, especially in this first chapter. These examples illustrate principles of general interest. Thus, these examples merit attention regardless of the reader's specialty area.

1.1 Basic Definitions

As defined in this book, aggregation and fragmentation both involve *particles*, a.k.a., *objects*, *components*, *constituents*, or *bodies*. More specifically, *aggregation* refers to events, including but not limited to binary collisions, that cause two particles to form a single particle. Alternative terms include *accretion*, *agglomeration*, *assemblage*, *clustering*, *coagulation*, *coalescence*, *collection*, and *merger*. *Fragmentation* refers to events, including but not limited to binary collisions, that cause one particle to separate into two or more particles. Alternative terms include *atomization*, *breakage*, *breakup*, *disintegration*, *rupturing*, *shattering*, *shredding*, and *spallation*. The particles that participate in fragmentation are often called *parents* or *mothers* while the particles produced by fragmentation are often called *children*, *daughters*, or *progeny*.

For the most part, this book concerns systems with large numbers of particles. However, this book also includes cases that either start with or end with only one particle. In particular, while aggregation may start with a large number of small particles, it can end with as few as one large particle. This special case is known as *unitary aggregation*. In *complete* unitary cases, the single large particle incorporates *all* of the original small particles. For example, this occurs in explosive consolidation of powders; see, e.g., Meyers et al. (1999). In *incomplete* unitary cases, the single large particle incorporates some, but not all, of the original small particles. For example, this occurs in *polymer gelation*; see, e.g., Hiemenz & Lodge (2007). In such cases, the large object is sometimes called the *gel* and the smaller particles are sometimes called the *sol*; see, e.g., Lushnikov (2006a, b, 2013). Figure 1.1 compares

(a)

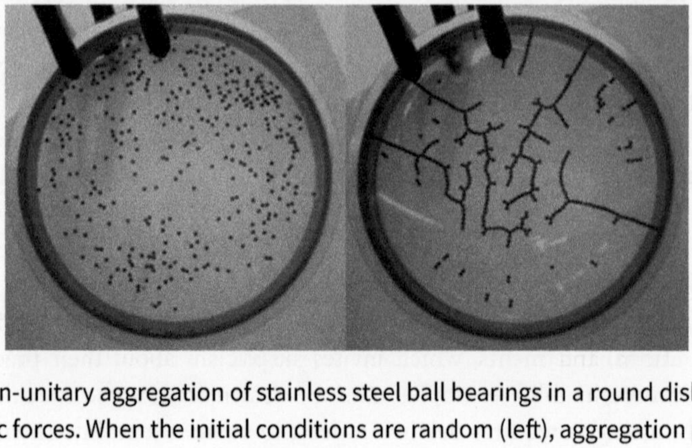

Figure 1.1a Non-unitary aggregation of stainless steel ball bearings in a round dish driven by electromagnetic forces. When the initial conditions are random (left), aggregation produces a wide range of seemingly random particle shapes and sizes including individual ball bearings, pairs of ball bearings, long linear chains, and large fractal shapes (right).
Figure reprinted with permission from Hübler & Crutchfield (2010). Copyright 2010 Wiley Periodicals Inc.

(b)

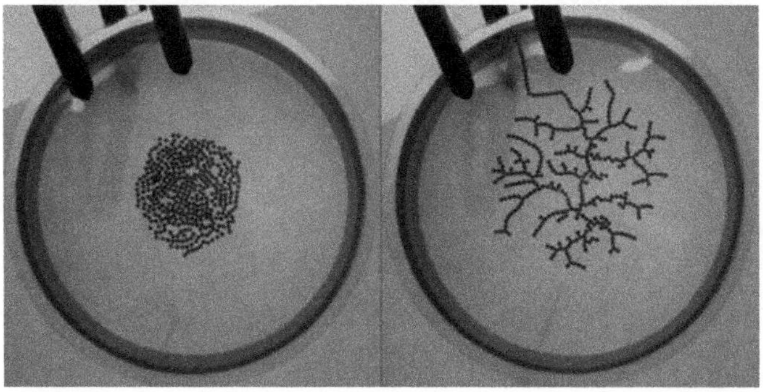

Figure 1.1b Unitary aggregation of stainless steel ball bearings in a round dish driven by electromagnetic forces. When the initial conditions are mostly deterministic (left), aggregation produces a single large object that incorporates all of the ball bearings into a single seemingly random fractal shape (right). In general, this book concentrates on particle system that "forget" their initial conditions. However, as this example shows, "forgetting" may depend at least partly on randomness in the initial conditions.

non-unitary to unitary aggregation; the former case leads to seemingly random sizes; both cases lead to seemingly random shapes.

By a similar token, while fragmentation may end with a large number of small particles, it can start with as few as one large particle. This special case is known as *unitary fragmentation.* In *complete* unitary cases, the original large object is fully converted to small particles. For example, Figure 1.2 shows complete unitary fragmentation of a clay tile roof due to the passage of a long-duration airblast. This leads to seemingly random fragment shapes, sizes, velocities, rotation rates, and so forth. In *incomplete* unitary cases, the original object survives in a reduced form due to, e.g., cratering, chipping, etc. Suppose the products of unitary fragmentation, themselves, experience fragmentation. In other words, suppose that unitary fragmentation involves two stages: *primary* and *secondary fragmentation.* Secondary fragmentation in solids is especially likely when primary fragmentation causes cracking, crumbling, plastic distortion, and other macro- and microscopic damage. For example, Figure 1.3 shows secondary fragmentation driven by aerodynamic forces. In this example, damaged concrete fragments progressively shed material as they travel through the air. While such *aerodynamic stripping* is rare for solids, it is common for liquids; see, e.g., Ashgriz (2011). For another example, Figure 1.4 shows secondary fragmentation driven by high-speed collisions between primary fragments.

Particle systems are either *open* or *closed* depending on whether they *do* or *do not* exchange mass, momentum, and energy with their surroundings, respectively. In unitary aggregation, particle systems are often open, e.g., a single large object (the gel) may act like a sink of mass, momentum, and energy. This sink generally forms spontaneously after some finite amount of time. Similarly, in unitary fragmentation, particle systems are often open, i.e., a single large object acts like a source of mass, momentum, and energy for a particle system.

(a) Single clay tile.

(b) Clay tile roof mounted in a shock tube

(c) Clay tile roof fragmenting due to blast.

(d) Side view of a typical fragment.

(e) Tile fragments sorted into size bins.

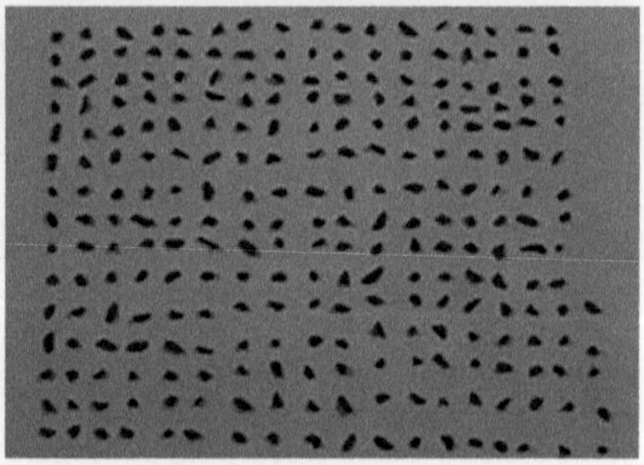

(f) Array of tile fragments from a size bin.

Figure 1.2 Complete unitary fragmentation of a clay tile roof due to the passage of a long-duration airblast. The collected tile fragments were roughly sorted by size by passing them through a series of increasingly finer sieves. The contents of each sieve were emptied into clear plastic storage bins as seen in part (e). The contents of each bin were arrayed on a flat surface and photographed as seen in part (f). The exact size of the arrayed fragments was estimated using image analysis software.

Figure reprinted with permission from Bewick et al. (2019).

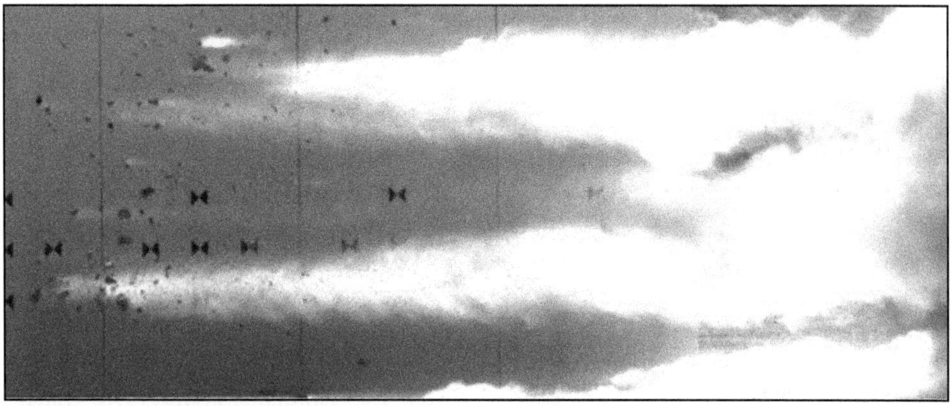

Figure 1.3 Secondary fragmentation due to aerodynamic forces. Solid concrete fragments, weakened and damaged during formation, progressively shed material as they travel from right to left, leaving dusty wakes.
Figure courtesy of Applied Research Associates (ARA).

Figure 1.4 Secondary fragmentation due to collisions. Fragments from a plate glass window experience mid-air collisions, which causes the fragments to break up into even smaller pieces.
Figure reprinted with permission from Bewick et al. (2017).

In *aggregation cascades*, two particles merge to form a single larger particle, and that larger particle merges with another particle to form an even larger particle, and so forth, over and over again. Left to their own devices, aggregation cascades are naturally unsteady,[1] with particle growth limited only by the availability of mass, momentum, and energy. However, steady states may occur if sinks preferentially remove existing large particles and sources preferentially introduce new small particles. For example, for clouds in the atmosphere, sources of small particles typically include condensation while sinks of large particles typically include gravitational settling; see, e.g., Khvorostyanov & Curry (2014). The end products of aggregation cascades are often weak, porous, and unstable. Thus, aggregation cascades may be followed by another stage: either *secondary aggregation* or *secondary fragmentation*. In secondary

[1] In this chapter, the terms "steady" and "unsteady" are meant in an average statistical sense. The next chapter will introduce the more accurate terms "stationary" and "non-stationary."

aggregation, a.k.a., *compaction* or *restructuring*, clusters reconfigure themselves into stronger, denser forms. In *secondary fragmentation*, clusters break into smaller, stronger pieces.

In *fragmentation cascades*, a particle breaks to form smaller particles; and some or all of those particles break to form even smaller particles, and so forth, over and over again. Left to their own devices, fragmentation cascades are naturally unsteady, with particle shrinkage limited only by the availability of mass, momentum, and energy. However, steady states may occur if sources preferentially introduce new large particles and sinks preferentially remove existing small particles. For example, in the asteroid belt, sources may include collisions between asteroids and minor planets such as Ceres, which chip away material from the surfaces of the minor planets, while sinks may include solar radiation that preferentially drives small fragments out of the asteroid belt; see, e.g., Dohnanyi (1969, 1970), Tanaka et al. (1996), O'Brien & Greenberg (2003), Birnstiel et al. (2011) and Pan & Schlichting (2012).

Mixed cascades involve both aggregation (e.g., due to collisions between small, slow particles) and fragmentation (e.g., due to collisions between large, fast particles). While aggregation and fragmentation cascades are naturally unsteady, mixed cascades can be naturally steady. If fragmentation is the exact inverse of aggregation—e.g., if every fragmentation event creates two particles out of one and every aggregation event creates one particle out of two—then steady states occur when the aggregation and fragmentation rates are equal. However, if fragmentation is not the exact inverse of aggregation—e.g., if fragmentation creates many particles out of one—then steady states occur only when the aggregation rate is proportionally larger than the fragmentation rate. Instead of painstakingly adjusting aggregation and fragmentation rates, it is common to use aggregation cascades with large-size sinks and small-size sources to effectively mimic steady mixed cascades; see, e.g., Connaughton et al. (2018).

To summarize, this section has introduced a number of key terms including aggregation, unitary aggregation, primary aggregation, secondary aggregation, aggregation cascades, fragmentation, unitary fragmentation, primary fragmentation, secondary fragmentation, and fragmentation cascades. In the remainder of this book, the term "aggregation" may refer to any form of aggregation but, in most cases, it refers to aggregation cascades. The exact meaning should be clear from the context. Similarly, the term "fragmentation" may refer to any type of fragmentation but, in most cases, it refers to either fragmentation cascades or unitary fragmentation. The exact meaning should be clear from the context.

1.2 Particle Types

In *discrete systems*, particles can only obtain a finite number of different sizes. For example, in the simplest case, all particles are composed of identical monomers; see, e.g., Figure 1.1. In such cases, particles composed of one or more monomers are often referred to as *clusters*. In *continuous systems*, particles can obtain an infinite continuum of different sizes; see, e.g., Figures 1.2, 1.3, and 1.4. Most of this book concerns ordinary *bonded particles*, composed of either continuous material or constituents held together by body or surface forces, e.g., gravity, electrostatic forces, van der Waals forces, etc. However, for discrete systems, this book will also consider *unbonded particles*, e.g., loose collections of monomers not held together by body or surface forces. Ordinary bonded particles tend to be long-lived and to have clear, distinct edges. By contrast, unbonded particles tend to be short-lived and to

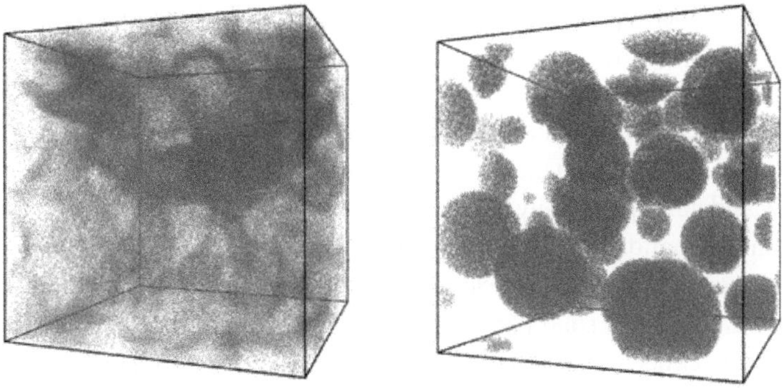

(a) Unbonded particle system (b) Bonded particle system.

Figure 1.5 Example of unbonded and bonded particles in a discrete particle system. The unbonded particle system seen in (a) is called a granular gas.

have nebulous, indistinct edges; see Figure 1.5. The lifetimes of unbonded particles tend to decrease as their translation or rotation rates increase. This vague, ephemeral quality makes it difficult to determine the exact size and shape of unbonded particles.

1.3 Particle Shapes

In continuous systems, the particle shapes are always *regular*, e.g., spheres, ellipsoids, cubes, jagged triangles, etc. However, in discrete systems, the particles shapes may be either regular or *fractal*. Figure 1.6 illustrates the difference between regular and fractal shapes.

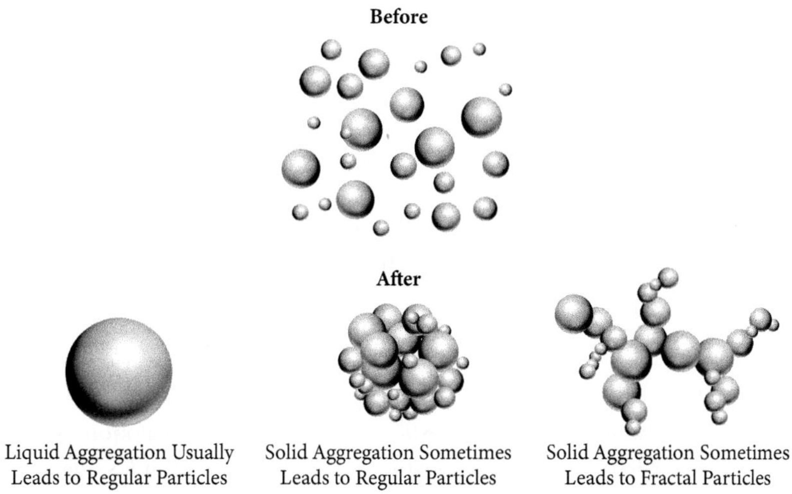

Figure 1.6 Liquid aggregation typically leads to regular particles, e.g., spheres. However, solid aggregation may lead to either regular or fractal particles.

Sornette (2006) says that fractal particles have "a fine structure, too much irregularity to be described in traditional geometric language, both locally and globally, some form of self-similarity, perhaps approximate or statistical, a fractal dimension (somehow defined) which is greater than its topological dimension, and a simple definition, usually recursive." In this context, *self-similarity* means that fractals can be "subdivided into parts, each of which is (at least approximately) a reduced-size copy of the whole." Mandelbrot (1975, 1982) provided the first general description of fractals, including those produced by solid aggregation. This discovery led to a surge of interest in fractal aggregates and, to a lesser extent, aggregates in general. Mandelbrot personally participated in much of this pioneering aggregation work; see, e.g., Family & Landau (1984), Pietronero & Tosatti (1986), Fleischmann et al. (1990).

As seen in Figure 1.6, liquid particles are typically continuous. Thus, liquid particles tend to be regular. By contrast, solid particles may be either continuous or discrete. Thus, solid particles may be either regular or fractal; see Figures 1.1 to 1.6. For solid fragmentation, children tend to inherit regular vs. fractal from their parents. More specifically, if the parent particle is regular, then usually so are the children. Similarly, if the parent particle is fractal, then usually so are the children, except when a parent decomposes into its constituent monomers.

In general, different particles have different shapes. For example, in Figure 1.7a, the particles are rectangular boxes with randomly varying length-to-width and width-to-height ratios. However, in the simplest case, all particles have roughly the same shape, i.e., the bigger particles are just magnified versions of the smaller particles, at least on average. For example, in Figure 1.7b, the particles are cubes with fixed length-to-width and width-to-height ratios of one.

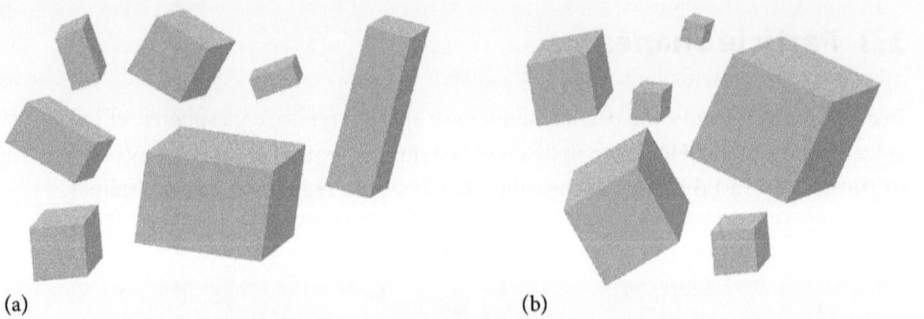

Figure 1.7 (a) In general, different particles have different shapes. (b) In the simplest cases, different particles have roughly the same shape, at least on average.

If all of the particles in a given system have the same shape on average, then the particle size can be characterized by a length scale called the *diameter* D such that:

$$M = c_1 \rho D^m \tag{1.1}$$

where M is the particle mass, c_1 is a constant factor, ρ is the particle density, and m is an exponent. In the future, Equation (1.1) will sometimes be written as $M = \text{const.}\ \rho D^m$. In general, this book will use the notation "const." to avoid introducing a specific notation such as c_1 for each and every constant. In places where many constants appear, it can be tedious and confusing to introduce a specific notation for each one.

Let $d = 1, 2,$ or 3 be the *spatial dimension*. For regular particles, this book assumes that m is an integer between 1 and d. More specifically, if $d = 3$, then $m = 1, 2,$ or 3. Similarly, if $d = 2$, then $m = 1$ or 2. Figure 1.8 shows two examples of m and d for regular particles.

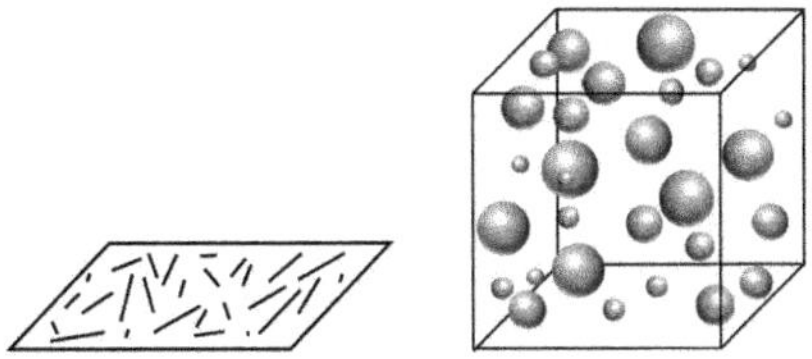

Figure 1.8 Two examples of the particle dimension m and spatial dimension d.

For fractal particles, this book assumes that m is a real number such that $1 < m < d$. In this case, m is usually called the *fractal dimension*; see, e.g., Bushell et al. (2002).

When aggregation creates fractal shapes, those shapes tend to be sensitive to the exact conditions. For example, Figure 1.9 shows that the fractal dimension depends on the aggregation rate. Figure 1.9 concerns *colloids*, i.e., small clusters of atoms or molecules in a fluid medium. Colloids form due to collisions caused by Brownian motion, turbulent flow, etc. and attractive forces such as van der Waals forces, electrostatic forces, etc. By some definitions, colloidal particles are less than 1µm in diameter, meaning they stay in suspension indefinitely with minimal gravitational settling. Examples of colloids include paint, glue, milk, detergents, cosmetics, and other common household items.

In Figure 1.9, the aggregation rate is varied in two different ways. First, if colloidal clusters can only bond with individual atoms, the aggregation rate is slower. However, if colloidal clusters can bond with each other, the aggregation rate is faster. Second, if the atomic bonding time approaches or exceeds the time between collisions, the aggregation rate is slower; this is called *reaction-limited*. However, if the atomic bonding time is less than the time between collisions, the aggregation rate is faster; this is called *diffusion-limited*. Reaction-limited aggregation typically occurs when short-range body forces repel particles as they approach each other; this dramatically slows or, in many cases, entirely prevents bonding. Diffusion-limited aggregation typically occurs in cases dominated by Brownian motion. Bushell et al. (2002) report that Diffusion-Limited Cluster-Cluster Aggregation (DLCA) tends to obtain $m \sim 1.8$ because "a collision between clusters always results in formation of a bond. This ... produces quite tenuous, wispy structures." This is consistent with the result given in the lower right corner of Figure 1.9. By contrast, Reaction-Limited Cluster-Cluster (RLCA) aggregation tends to obtain $m \sim 2.1$ because "collisions ... almost never form bonds, so that all physically possible conformations between clusters have an equal chance of forming a bond and thus a new aggregate. This ... produces structures that are still quite tenuous ... but noticeably more compact and stronger looking than DLCA aggregates." This is consistent with the result given in the lower left corner of Figure 1.9.

So far, this section has assumed that all the particles in a given system have the same shape, at least on average. However, in general, different particles have completely different shapes. For regular particles with different shapes, Equation (1.1) typically still applies with a single integer value of m. For example, consider unitary fragments from a spherical or cylindrical shell. If most of the fragments have two smooth surfaces, inherited from the inner and outer surface of the shell, then most of the fragments formed in two dimensions. In this case, it is reasonable to assume $m = 2$. If most of the fragments are rough on all sides, with no smooth surfaces, then most of the fragments formed in three dimensions. In this case, it is

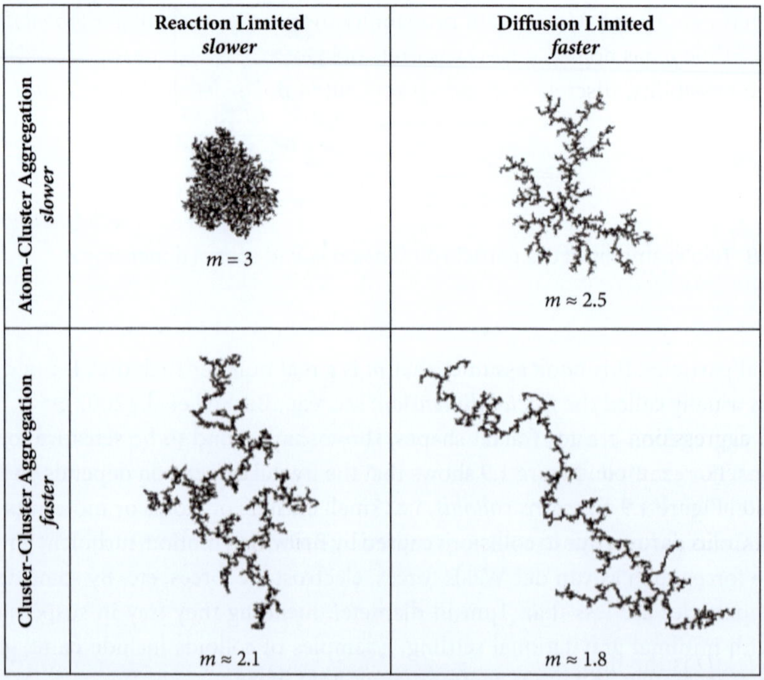

Figure 1.9 Aggregations of atoms in a colloid are highly sensitive to aggregation rates. Schaefer (1988) says that the "fractal dimensions listed are for 3D clusters even though their 2D analogies are displayed."
Figure reproduced from Schaefer (1988) with permission from Springer Nature.

reasonable to assume $m = 3$. For fractal particles with different fractal shapes, Equation (1.1) typically still applies but with a range of different values of *m*. For example, for interstellar dust aggregation due to Brownian motion, Kempf et al. (1999) found that "the distribution of the fractal dimension for fixed times is remarkably broad and about 10% of the aggregates have a fractal dimension larger than two." In such cases, this book will try to give a sense of the range of values for *m*. This simple approach will tide us over until the more detailed treatment of particle shapes given in Chapter 6.

1.4 Particle Sizes

As mentioned earlier, aggregation and fragmentation often exhibit random or seemingly random behaviors. In such cases, the results are usually represented probabilistically. This section will introduce probabilistic descriptions of particle masses and diameters. Later chapters will provide additional details including theoretical derivations of particle size distributions. For discrete particle systems, the particle mass distribution may be expressed as f_i, where f_i is the number fraction of particles of size *i*. In the simplest case, these are *i* identical monomers of mass M_1. Similarly, for continuous particle systems, the particle mass distribution may be expressed as $f(M)$, where $f(M)dM$ is the number fraction of particles with masses between M and $M + dM$. In probability theory, $f(M)$ is called a *probability density function*. While f_i has no units, $f(M)$ has units of inverse mass.

As one possibility, discrete particle systems often obtain *power law size distributions* as follows:

$$f_i = AM_i^{r-1} \tag{1.2}$$

where M_i is a set of masses and i is an integer index. In the simplest case, $M_i = iM_1$. Similarly, as one possibility, continuous particle systems often obtain power law size distributions where:

$$f(M) = AM^{r-1}. \tag{1.3}$$

In these expressions, f and A may be functions of time, space, or other factors; however, the exponent r is a constant. Instead of mass, power law size distributions may also be expressed in terms of diameter, surface area, or any other particle size metric. For example, for continuous particle systems, power law size distributions can be expressed as follows:

$$f(D) = BD^{n-1} \tag{1.4}$$

where $f(D)dD$ is the number fraction of particles with diameters between D and $D + dD$. Notice that Equations (1.3) and (1.4) are related as follows:

$$A = (1/m)(c_1\rho)^{-n/m} B \tag{1.5}$$

and:

$$r = n/m. \tag{1.6}$$

This is true because:

$$f(D)dD = f(M)dM. \tag{1.7}$$

In addition, this is true because according to Equation (1.1):

$$dD = (1/m)(c_1\rho)^{-1/m} M^{1/m-1} dM. \tag{1.8}$$

To make it easy to convert between mass and diameter, this book will generally write power law size distributions in the following form:

$$f_i = AM_i^{n/m-1} \tag{1.9}$$

for discrete particle systems and

$$f(M) = AM^{n/m-1} \tag{1.10}$$

for continuous particle systems. In other words, this book will generally use the notation n/m rather than r for the power law exponent. Equation (1.10) implies both Equation (1.3) for mass and Equation (1.4) for diameter. In that sense, Equation (1.10) is a two-in-one expression.

1.5 Universal Size Distributions

If the power law exponent n/m is a fixed rational number such as ±1/3, ±1/2, or ±1, which applies regardless of the exact conditions, then the particle size distributions may be called *ideal* or *universal*. For example, Figure 1.10 shows some experimental results for small high-speed projectiles impacting large stationary targets. If the projectile has a low enough kinetic energy, the target experiences only localized cratering, chipping, or scoring. However, if the projectile has a high enough kinetic energy, the target is catastrophically destroyed. In either case, the midsized target fragments obtain a universal power law size distribution with $n/m \approx -5/6$, regardless of the exact circumstances. This size distribution is sometimes called the *Seebaugh–Schoutens size distribution*; see Seebaugh (1976, 1977) and Schoutens (1979).

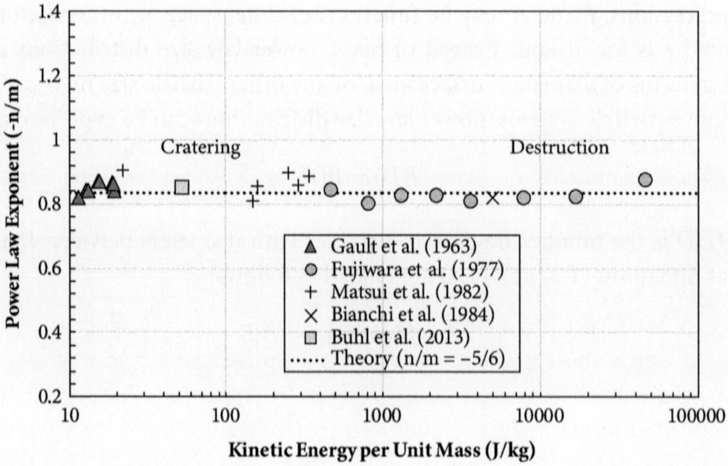

Figure 1.10 In some cases, for a target impacted by a compact high-speed projectile, the mid-sized fragments obtain a universal power law size distribution. These experimental results involve both primary and secondary fragmentation. For Matsui et al. (1982), this figure only includes projectile kinetic energies per unit target mass below 350 J/kg.

If the power law exponent n/m is a real number which varies continuously depending on the exact conditions, then the particle size distributions may be called *non-ideal* or *non-universal*. For example, Figure 1.11 shows some experimental results for small high-speed projectiles impacting large stationary targets. Unlike before, the midsized target fragments obtain a range of different power law exponents, depending on the projectile's kinetic energy, shape, impact angle, and other properties. If the projectile imparts all of its kinetic energy to the target, the results will tend to fall on or near the dashed line shown in Figure 1.11. However, if the projectile only imparts some of its kinetic energy to the target, the results will tend to fall below the dashed line shown in Figure 1.11. More specifically, in some of the tests done by Takagi et al. (1984), the projectile rebounded or the target split, which prevented the full conversion of the projectile's kinetic energy into target fragmentation. In addition, Takagi et al. (1984) used cylindrical projectiles, meaning the energy transfer depended partly on the projectiles' orientation at the moment of impact. By contrast, Durda et al. (2015) and Michikami et al. (2016) used spherical projectiles, meaning the energy transfer did not depend on the projectiles' orientation.

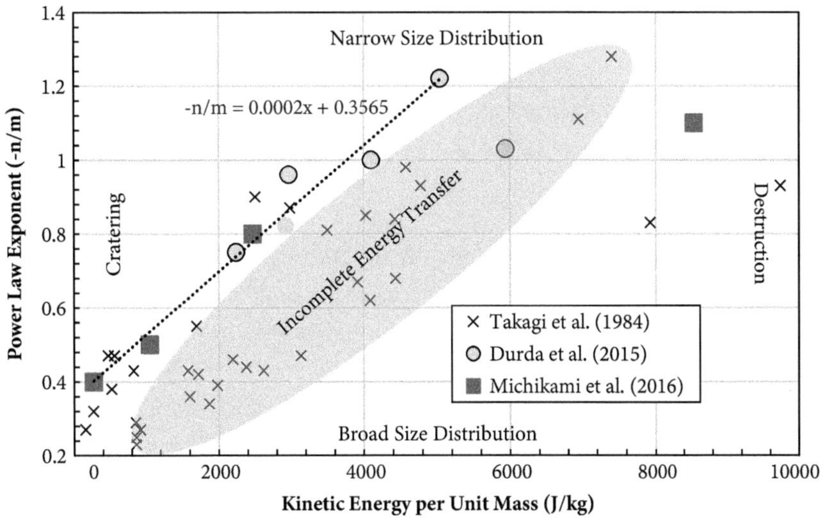

Figure 1.11 In some cases, for a target impacted by a compact high-speed projectile, the mid-sized fragments obtain a non-universal power law size distribution. These experimental results involve only primary fragmentation

In Figure 1.10, the target fragments were collected after they impacted adjacent hard surfaces. In other words, these fragments were typically produced in two stages—primary and secondary—as discussed in Section 1.1. By contrast, in Figure 1.11, the target fragments were collected soon after formation using a soft-capture technique. For example, Michikami et al. (2016) used a small "cardboard box whose interior was covered with 1cm thick urethane sheets." As a result, these fragments were typically produced in a single stage. To summarize, all of the experiments in Figures 1.10 and 1.11 experienced non-ideal primary breakup. However, in Figure 1.10, the results of the first non-ideal stage were overwritten by a second ideal stage. Example 1.1 gives another example of this behavior.

Example 1.1 The effects of secondary fragmentation

Bergstrom (1963) conducted experiments in which glass spheres with nominal diameters between 1/8″ and 1″ were slowly "squeezed between a parallel pair of extremely hard tungsten carbide platens." At first, the sphere and platen deformed, leading them to accumulate elastic potential energy. However, eventually, the sphere broke "into several large lune-shaped fragments located near the exterior of the sphere plus a large number of smaller fragments arising from the highly stressed inner diametrical core of the sphere." The large "lune-shaped fragments were propelled radially outwards in the equatorial plane at essentially equal velocities of about 40 ft/sec." These large "flying fragments possessed up to 90% of the mass of the sphere" and their kinetic energy "was about 45% of the total [stored] strain energy required to cause the fracture of the sphere." In fact, the kinetic energy was easily "large enough to cause secondary fracture if the fragments impacted against a hard, stationary surface."

Most experiments characterize fragments in only one or two different ways. However, in his experiments, Bergstrom (1963) characterized fragments in three different ways as follows:

- *High-speed photography*. Among other things, this showed that the primary fragmentation was "extremely sudden," i.e., it occurred in less than 4×10^{-4} s.
- *"Soft" physical collection*. The fragments were captured by a thick gelatin matrix, meaning that only primary fragmentation occurred.
- *"Hard" physical collection*. The fragments were captured by a hard steel retaining ring, meaning that both primary and secondary fragmentation occurred.

When only primary fragmentation occurred, the fragment size distribution was substantially non-ideal, i.e., the probability density function had several large humps. However, when both primary and secondary fragmentation occurred, the fragment size distribution was nearly ideal. More specifically, $f(D)$ was approximately constant, with only relatively minor oscillations for the largest fragments. This is equivalent to a power law size distribution $f(D) = BD^{n-1}$ where B is a constant and $n = 1$. Alternatively, if $m = 3$, this is equivalent to a power law size distribution $f(M) = AM^{n/m-1}$ where A is a constant and $n/m = 1/3$. In short, when the fragments were captured in a steel retaining ring, the results of the first non-ideal primary stage were overwritten by an ideal secondary stage. For additional information on these tests see, e.g., Fuerstenau & Abouzeid (2002) and Section 2.4 of Grady (2017).
[The author wishes to thank Dennis Grady for suggesting this example.]

So far, this section has focused on fragmentation. However, the same principles apply for aggregation. For example, Lin et al. (1989a, 1989b, 1990) found that the RLCA of colloidal gold, silica, and polystyrene obtained a universal power law size distribution with $n/m \approx -1/2$, regardless of the exact experimental conditions. On the other hand, Olivier & Sorensen (1990) found that DLCA of colloidal gold obtained a wide variety of different non-ideal size distributions, depending on the exact conditions; see also Wang & Sorensen (2001). As one explanation, in the slower reaction-limited experiments, the results were produced in a single stage that resulted in relatively strong dense clusters with $m \approx 2.1$. However, in the faster diffusion-limited experiments, the results were produced in two stages: a primary aggregation stage that resulted in relatively weak porous clusters with $m \approx 1.2$–1.6; and a secondary aggregation stage that resulted in stronger, denser clusters with $m \approx 1.6$–1.9; see also, e.g., Gardner et al. (1998), Jungblut et al. (2019a, b). In other words, in Olivier & Sorensen (1990), the results of the first ideal stage were overwritten by a second non-ideal stage.

According Babick (2016), secondary aggregation in colloids involves processes such as "the breakage of particle bonds, the restructuring within the aggregation (consolidation), the alteration of contact zones due to sintering, or the slow formation of covalent bonds between particle surfaces in contact." Babick (2016) warns that secondary aggregation, when it occurs, does "not possess the same universal character as [primary] aggregation." Kim et al. (2006)

showed that primary and secondary aggregation can alternate, i.e., particles can grow due to diffusion-limited aggregation, then shrink due to "percolation" or "restructuring," then grow further due to diffusion-limited aggregation, then shrink again. This, again, is non-ideal behavior.

The next chapter will explain in detail the conditions under which universal size distributions do and do not occur. For the time being, as a rough intuitive explanation, universal size distributions tend to occur if all of the particles in a given size range aggregate or fragment in a random or seemingly random fashion, under similar conditions, and according to similar rules. Conversely, non-universal size distributions tend to occur if different particles in a given size range aggregate or fragment in a deterministic (non-random) fashion, under different conditions, or according to different rules.

1.6 Conclusions

This book focuses on general results for aggregation and fragmentation that apply to many different fields of application. More specifically, this book focuses on three main problems:

- *Aggregation cascades.* These are processes which cause a particle system to form larger particles, usually due to repeated random or seemingly random events such as low-speed binary collisions,[2] which take place over an extended period of time.
- *Fragmentation cascades.* These are processes which cause a particle system to form smaller particles, due to repeated random or seemingly random events such as high-speed binary collisions, which take place over an extended period of time.
- *Unitary fragmentation.* These are processes or events which cause a single object to form a system of particles, due to random or seemingly random phenomena such as crack propagation, which take place over an extended period of time or nearly instantaneously.

Fragmentation cascades always involve unitary fragmentation, e.g., one or both particles participating in a high-speed binary collision experience unitary fragmentation. Vice versa, unitary fragmentation often involves or is followed by fragmentation cascades. For simplicity, this book will often refer to aggregation cascades as "aggregation." Similarly, this book will often refer to fragmentation cascades and unitary fragmentation as "fragmentation."

This book assumes that aggregation and fragmentation produce random or seemingly random outcomes. In some cases, this leads to universal probabilistic solutions for size, shape, speed, and other particle properties. Even problems that lack general universal probabilistic solutions may still be *locally universal* in time, space, or both. In other words, in many cases, non-universal solutions may be constructed by integrating or otherwise blending different universal solutions. In many cases, such blends lead to a simple linear dependence on key parameters, similar to that seen in Figure 1.11. While computational models

[2] Binary collisions involve exactly two particles. In theory, aggregation can occur due to collisions between three or more particles. However, in practice, such simultaneity is impossible—two particles will always meet at least slightly before any others. Thus, aggregation can often be treated as if it occurs due to a series of binary collisions. However, this argument may not apply to other types of interactions, e.g., when multiple particles simultaneously experience body forces.

tend to obtain specific results that apply only to a single problem, analytic models may obtain general results. Thus, this book will often focus on general analytic results. However, even when focused on analytic models, this book frequently introduces experimental and computational results for comparison.

Aggregating and fragmenting particle systems may be completely random, completely non-random (deterministic), or somewhere in between. For example, suppose a pane of glass is pre-scored into rectangular sections, e.g., by cutting straight grooves into one surface. In such cases, a mild insult such as bending may cause the glass to break in a completely non-random (deterministic) fashion. However, a somewhat more vigorous insult, such as dropping the pane of glass onto the floor, may lead to an intermediate result, with some breaks occurring along score lines and other breaks occurring seemingly at random. In any modeling effort, care must be taken to ensure that the model accurately reflects both the non-random and random elements present in the problem of interest. Computational methods for aggregation and fragmentation usually have some degree of numerical dissipation, which can artificially damp out random or seeming-random behaviors. By contrast, analytic methods for aggregation and fragmentation generally assume random or seemingly random behaviors. However, because analytic methods generally represent random systems by deterministic equations, the solutions can easily end up becoming associated with deterministic rather than random systems, regardless of the modeler's original intent.

What follows is an outline of the rest of the book. Chapter 2 discusses why aggregation and fragmentation tend to produce random or seemingly random outcomes. In addition, Chapter 2 discusses the conditions under which universal probabilistic solutions for size, shape, speed, and other particle properties tend to occur. Chapter 3 discusses probabilistic distributions and averages for particle sizes, velocities, shapes, and so forth. Chapters 4 and 5 introduce computational and analytic methods, respectively, for aggregation and fragmentation problems. Chapters 4 and 5 emphasize the special measures that must be taken to properly account for the random or seemingly random behaviors that typify aggregating and fragmenting particle systems. To keep these introductory discussions within some bounds, Chapters 4 and 5 focus on particle size distributions. On the other hand, Chapters 6, 7, and 8 discuss particle shapes, velocities, and kinetic energies, respectively. Chapters 6 and 7 focus on regular bonded particle systems while Chapter 8 focuses on self-organized unbonded particle systems. Following that, Chapters 9, 10, and 11 discuss analytic models for aggregation cascades with a focus on mean-field theory; Chapters 12, 13, and 14 discuss analytic models for fragmentation cascades with a focus on mean-field theory; and Chapters 15, 16, and 17 discuss analytic models for unitary fragmentation. To complete the discussion of analytic methods, Chapters 18 and 19 discuss maximum entropy theory for discrete and continuous particle systems, respectively. Maximum entropy theory is a simple analytic modeling approach which applies equally to aggregation cascades, fragmentation cascades, and unitary fragmentation. Finally, Chapter 20 summarizes the book's main themes and conclusions.

Even though aggregation and fragmentation have been studied for over 100 years, there is still no fully standard treatment. This book takes one possible approach. Readers who want to see other possible approaches should consult books such as Ramkrishna (2000), Bertoin (2006), Krapivsky et al. (2010), Yeoh et al. (2014), Litster (2016), Matsoukas (2018), Banasiak et al. (2019a, b), Chakraborty (2019), and Podgórska (2020). In addition to these books, which consider both aggregation and fragmentation, there are a number books dedicated mainly to fragmentation; see, e.g., Antoun et al. (2003), Lin (2003), Grady (2006), Ashgriz (2011), King et al. (2012), Yarin et al. (2017), Grady (2017), Lefebvre & McDonell (2017), Basu et al. (2018), and Lambert et al. (2019). While they are mostly older, there are also a number of books dedicated mainly to aggregation: see, e.g., Jullien & Botet (1987), Sonntag & Strenge (1987), Vicsek (1992), Elimelech et al. (1995), Privman (1997), and Meakin (1998). In addition to the books just cited, aggregation and fragmentation commonly appear in application-oriented textbooks in fields such as astrophysics, cloud physics, colloid science, polymer science, chemical engineering, mining engineering, and so forth. Readers interested in such books can look ahead to Tables 5.2 and 5.4, which collectively reference over 50 examples.

References

T. Antoun, D. R. Curran, S. V. Razorenov, L. Seaman, G. I. Kanel, & A. V. Utkin, *Spall Fracture*, Springer, 2003, https://doi.org/10.1007/b97226

N. Ashgriz, editor, *Handbook of Atomization and Sprays: Theory and Applications*, Springer, 2011, https://doi.org/10.1007/978-1-4419-7264-4

F. Babick, *Suspensions of Colloidal Particles and Aggregates*, Springer, 2016, https://doi.org/10.1007/978-3-319-30663-6

J. Banasiak, W. Lamb, & P. Laurencot, *Analytic Methods for Coagulation-Fragmentation Models, Volume I*, CRC, 2019, https://doi.org/10.1201/9781315154428

J. Banasiak, W. Lamb, & P. Laurencot, *Analytic Methods for Coagulation-Fragmentation Models, Volume II*, CRC, 2019, https://doi.org/10.1201/9780429280320

Basu, A. K. Agarwal, & C. Patel, editors, *Droplets and Sprays: Applications for Combustion Propulsion*, Springer, 2018, https://doi.org/10.1007/978-981-10-7449-3_1

B. H. Bergstrom, Energy and Size Distribution Aspects of Single Particle Crushing, in C. Fairhurst, editor, *Rock Mechanics: Proceedings of the Fifth Symposium on Rock Mechanics held at the University of Minnesota May 1962*, Pergamon Press, 1963

J. Bertoin, *Random Fragmentation and Coagulation Processes*, Cambridge University Press, 2006, https://doi.org/10.1017/CBO9780511617768

B. Bewick, G. Rolater, K. Marchand, A. Ziemba, M. Sanai, & C. Laney, *Fragmentation of Solid Materials Using Shock Tubes. Part 2: First Test Series in a Large Diameter Shock Tube*, Defense Threat Reduction Agency Technical Report DTRA-TR-17-50, December 2017, https://apps.dtic.mil/sti/citations/AD1043784

B. Bewick, G. Rolater, K. Marchand, A. Ziemba, M. Sanai, & C. Laney, *Fragmentation of Solid Materials Using Shock Tubes. Part 4: Second Test Series in a Large Diameter Shock Tube*, Defense Threat Reduction Agency Technical Report DTRA-TR-19-5, January 2019, https://apps.dtic.mil/sti/citations/AD1065932

R. Bianchi, F. Capaccioni, P. Cerroni, M. Coradini, E. Flamini, P. Hurren, et al., Experimental Simulation of Asteroidal Fragmentation by Macroscopic Hypervelocity Impacts Against Free Falling Bodies, *Astronomy and Astrophysics*, 139(1), 1–6, 1984, http://adsabsharvardedu/abs/1984A&A ... 139 ... 1B

T. Birnstiel, C. W. Ormel, & C. P. Dullemond, Dust Size Distributions in Coagulation/Fragmentation Equilibrium: Numerical Solutions and Analytical Fits, *Astronomy & Astrophysics*, 525, A11, 2011, https://doi.org/10.1051/0004-6361/201015228

E. Buhl, A. Kowitz, D. Elbeshausen, F. Sommer, G. Dresen, M. H. Poelchau, et al., Particle Size Distribution and Strain Rate Attenuation in Hypervelocity Impact and Shock Recovery Experiments, *Journal of Structural Geology*, 56, 20–33, 2013, https://doi.org/10.1016/j.jsg.2013.08.007

G. C. Bushell, Y. D. Yan, D. Woodfield, J. Raper, & R. Amal, On Techniques for the Measurement of the Mass Fractal Dimension of Aggregates, *Advances in Colloid and Interface Science*, 95(1), 1–50, 2002, https://doi.org/10.1016/S0001-8686(00)00078-6

J. Chakraborty, *Engineering of Submicron Particles: Fundamental Concepts and Models*, Wiley, 2019, https://doi.org/10.1002/9781119296447

C. Connaughton, A. Dutta, R. Rajesh, N. Siddharth, & O. Zaboronski, Stationary Mass Distribution and Nonlocality in Models of Coalescence and Shattering, *Physical Review E*, 97(2), 022137, 2018, https://doi.org/10.1103/PhysRevE.97.022137

J. S. Dohnanyi, Collisional Model of Asteroids and their Debris, *Journal of Geophysical Research*, 74(10), 2531–2554, 1969, https://doi.org/10.1029/JB074i010p02531

J. S. Dohnanyi, On the Origin and Distribution of Meteoroids, *Journal of Geophysical Research*, 75(17), 3468–3493, 1970, https://doi.org/10.1029/JB075i017p03468

D. D. Durda, A. Campo-Bagatin, R. A. Alemañ, G. J. Flynn, M. M. Strait, A. N. Clayton, et al., The Shapes of Fragments from Catastrophic Disruption Events: Effects of Target Shape and Impact Speed, *Planetary and Space Science*, 107, 77–83, 2015, https://doi.org/10.1016/j.pss.2014.10.006

M. Elimelech, J. Gregory, X. Jia, & R. Williams, *Particle Deposition and Aggregation: Measurement, Modeling, and Simulation*, Butterworth-Heineman, 1995, https://doi.org/10.1016/C2013-0-04548-3

F. Family & D. P. Landau, *Kinetics of Aggregation and Gelation*, Elsevier, 1984, https://doi.org/10.1016/B978-0-444-86912-8.50007-9

M. Fleischmann, D. J. Tildesley, & R. C. Ball, *Fractals in the Natural Sciences*, Princeton University Press, 1990, https://doi.org/10.1515/9781400861040

D. W. Fuerstenau & A.-Z. M. Abouzeid, The Energy Efficiency of Ball Milling in Comminution, *International Journal of Mineral Processing*, 67(1–4), 161–185, 2002, https://doi.org/10.1016/S0301-7516(02)00039-X

A. Fujiwara, G. Kamimoto, & A. Tsukamoto, Destruction of Basaltic Bodies by High-Velocity Impact, *Icarus*, 31(2), 277–288, 1977, https://doi.org/10.1016/0019-1035(77)90038-0

K. H. Gardner, T. L. Theis, & T. C. Young, Colloid Aggregation: Numerical Solution and Measurements, *Colloids and Surfaces A: Physicochemical and Engineering Aspects*, 141(2), 237–252, 1998, https://doi.org/10.1016/S0927-7757(98)00383-5

D. E. Gault, E. M. Shoemaker, & H. J. Moore, *Spray Ejected from the Lunar Surface by Meteroid Impact*, NASA Technical Note D-1767, April 1963, https://ntrs.nasa.gov/search.jsp?R=19630004711

D. E. Grady, *Fragmentation of Rings and Shells: The Legacy of N.F. Mott*, Springer, 2006, https://doi.org/10.1007/b138675

D. E. Grady, *Physics of Shock and Impact Volume 1: Fundamentals and Dynamic Failure*, IOP Publishing, 2017, https://doi.org/10.1088/978-0-7503-1254-7ch2

P. C. Hiemenz & P. Lodge, *Polymer Chemistry*, 2nd edition, CRC Press, 2007, https://doi.org/10.1201/9781420018271

A. Hübler & J. P. Crutchfield, Order and Disorder in Open Systems, *Complexity*, 16(1), 6–9, 2010, https://doi.org/10.1002/cplx.20344

R. Jullien & R. Botet, *Aggregation and Fractal Aggregates*, World Scientific, 1987

S. Jungblut, J.-O. Joswig, & A. Eychmüller, Diffusion-Limited Cluster Aggregation: Impact of Rotational Diffusion, *The Journal of Physical Chemistry C*, 123(1), 950–954, 2019, https://doi.org/10.1021/acs.jpcc.8b10805

S. Jungblut, J.-O. Joswig, & A. Eychmüller, Diffusion- and Reaction-Limited Cluster Aggregation Revisited, *Physical Chemistry Chemical Physics*, 21, 5723–5729, 2019, https://doi.org/10.1039/c9cp00549h

S. Kempf, S. Pfalzner, & T. K. Henning, N-Particle-Simulations of Dust Growth I. Growth Driven by Brownian Motion, *Icarus*, 141(2), 388–398, 1999, https://doi.org/10.1006/icar.1999.6171

V. I. Khvorostyanov & J. A. Curry, *Thermodynamics, Kinetics, and Microphysics of Clouds*, Cambridge University Press, 2014, https://doi.org/10.1017/CBO9781139060004

W. Kim, C. M. Sorensen, D. Fry, & A. Chakrabarti, Soot Aggregates, Superaggregates and Gel-Like Networks in Laminar Diffusion Flames, *Journal of Aerosol Science*, 37(3), 386–401, 2006, https://doi.org/10.1016/j.jaerosci.2005.05.022

R. P. King, C. L. Schneider, & E. A. King, *Modeling and Simulation of Mineral Processing Systems*, 2nd edition, Society for Mining, Metallurgy, and Exploration (SME), 2012

P. L. Krapivsky, S. Redner, & E. Ben-Naim, *A Kinetic View of Statistical Physics*, Cambridge University Press, 2010, https://doi.org/10.1017/CBO9780511780516

F. Kun, F. K. Wittel, H. J. Herrmann, B. H. Kröplin, & K. J. Måløy, Scaling Behavior of Fragment Shapes, *Physical Review Letters*, 96(2), 025504, 2006, https://doi.org/10.1103/PhysRevLett.96.025504

D. E. Lambert, C. L. Pasiliao, B. Erzar, B. Revil-Baudard, & O. Cazacu, editors, *Dynamic Damage and Fragmentation*, Wiley, 2019, https://doi.org/10.1002/9781119579311

A. H. Lefebvre & V. G. McDonell, *Atomization and Sprays*, 2nd edition, CRC Press, 2017, https://doi.org/10.1201/9781315120911

M. Y. Lin, H. M. Lindsay, D. A. Weitz, R. C. Ball, R. Klein, & P. Meakin, Universality in Colloid Aggregation, *Nature*, 339, 360–362, 1989a, https://doi.org/10.1038/339360a0

M. Y. Lin, H. M. Lindsay, D. A. Weitz, R. C. Ball, R, Klein, & P. Meakin, Universality of Fractal Aggregates as Probed by Light Scattering, *Proceedings of the Royal Society A: Mathematical and Physical Sciences*, 423(1864), 71–87, 1989b, https://doi.org/10.1098/rspa.1989.0042

M. Y. Lin, M. H. Lindsay, D. A. Weitz, R. C. Ball, R. Klein, & P. Meakin, Universal Reaction-Limited Colloid Aggregation, *Physical Review A*, 41(4), 2005–2020, 1990, https://doi.org/10.1103/PhysRevA.41.2005

S. P. Lin, *Breakup of Liquid Sheets and Jets*, Cambridge University Press, 2003, https://doi.org/10.1017/CBO9780511547096

J. Litster, *Design and Processing of Particulate Products*, Cambridge University Press, 2016, https://doi.org/10.1017/CBO9781139017558

A. A. Lushnikov, Gelation in Coagulating Systems, *Physica D: Nonlinear Phenomena*, 222 (1–2), 37–53, 2006a, https://doi.org/10.1016/j.physd.2006.08.002

A. A. Lushnikov, Exact Kinetics of Sol-Gel Transition in a Coagulating Mixture, *Physical Review E*, 73(3), 036111, 2006b, https://doi.org/10.1103/PhysRevE.73.036111

A. A. Lushnikov, Postcritical Behavior of a Gelling System, *Physical Review E*, 88(5), 052120, 2013, https://doi.org/10.1103/PhysRevE.88.052120

B. B. Mandelbrot, *Les Objets Fractals: Forme, Hasard et Dimension [Fractals: Form, Chance and Dimension]*, Flammarion, 1975

B. B. Mandelbrot, *The Fractal Geometry of Nature*, W. H. Freeman, 1982

T. Matsoukas, *Generalized Statistical Thermodynamics: Thermodynamics of Probability Distributions and Stochastic Processes*, Springer, 2018, https://doi.org/10.1007/978-3-030-04149-6

T. Matsui, T. Waza, K. Kani, & S. Suzuki, Laboratory Simulation of Planetesimal Collision, *Journal of Geophysical Research*, 87(B13), 10,968–10,982, 1982, https://doi.org/10.1029/JB087iB13p10968

P. Meakin, *Fractals, Scaling and Growth Far from Equilibrium*, Cambridge University Press, 1998

M. A. Meyers, D. J. Benson, & E. A. Olevsky, Shock Consolidation: Microstructurally-Based Analysis and Computational Modeling, *Acta Materialia*, 47(7), 2089–2108, 1999, https://doi.org/10.1016/S1359-6454(99)00083-X

T. Michikami, A. Hagermann, T. Kadokawa, A. Yoshida, A. Shimada, S. Hasegawa, et al., Fragment Shapes in Impact Experiments Ranging from Cratering to Catastrophic Disruption, *Icarus*, 264, 316–330, 2016, https://doi.org/10.1016/j.icarus.2015.09.038

D. P. O'Brien & R. Greenberg, Steady-State Size Distributions for Collisional Populations: Analytical Solution with Size-Dependent Strength, *Icarus*, 164(2), 334–345, 2003, https://doi.org/10.1016/S0019-1035(03)00145-3

B. J. Olivier & C. M. Sorensen, Variable Aggregation Rates in Collodial Gold: Kernel Homogeneity Dependence on Aggregant Concentration, *Physical Review A*, 41(4), 2093–2100, 1990, https://doi.org/10.1103/PhysRevA.41.2093

M. Pan & H. E. Schlichting, Self-Consistent Size and Velocity Distributions of Collisional Cascades, *The Astrophysical Journal*, 747(2), 113, 2012, https://doi.org/10.1088/0004-637X/747/2/113

S. N. Pathak, Z. Jabeen, D. Das, & R. Rajesh, Energy Decay in Three-Dimensional Freely Cooling Granular Gas, *Physical Review Letters*, 112, 038001, 2014, https://doi.org/10.1103/PhysRevLett.112.038001

L. Pietronero & E. Tosatti, editors, *Fractals in Physics*, Elsevier, 1986, https://doi.org/10.1016/C2009-0-14331-3

W. Podgórska, *Multiphase Particulate Systems in Turbulent Flows: Fluid-Liquid and Solid-Liquid Dispersions*, CRC Press, 2020, https://doi.org/10.1201/9781315118383

V. Privman, editor, *Nonequilibrium Statistical Mechanics in One Dimension*, Cambridge University Press, 1997, https://doi.org/10.1017/CBO9780511564284

D. Ramkrishna, *Population Balances: Theory and Applications to Particulate Systems in Engineering*, Elsevier, 2000, https://doi.org/10.1016/B978-0-12-576970-9.X5000-0

D. W. Schaefer, Fractal Models and the Structure of Materials, *MRS Bulletin*, 13(2), 22–27, 1988, https://doi.org/10.1557/S088376940006632X

J. E. Schoutens, editor, Empirical Analysis of Nuclear and High-Explosive Cratering and Ejecta, In *Nuclear Geoplosics Sourcebook, Volume IV, Part II, Empirical Analysis of Nuclear and High-Explosive Cratering and Ejecta*, Defense Nuclear Agency 6501H(4–2), March 1979, https://apps.dtic.mil/sti/citations/ADA095097

W. R. Seebaugh, *A Dynamic Crater Ejecta Model*, Defense Nuclear Agency 4144F, October 1976, https://apps.dtic.mil/sti/citations/ADA042546

W. R. Seebaugh, A Dynamic Crater Ejecta Model, in D. J. Roddy, R. O. Pepin, & R. B. Merrill, editors, *Impact and Explosion Cratering*, Pergammon Press, 1977, http://adsabs.harvard.edu/full/1977iecp.symp.1043S

H. Sonntag & K. Strenge, *Coagulation Kinetics and Structure Formation*, Springer, 1987, https://doi.org/10.1007/978-1-4757-0617-8

D. Sornette, *Critical Phenomena in Natural Sciences: Chaos, Fractals, Self-Organziation and Disorder*, 2nd edition, Springer, 2006, https://doi.org/10.1007/3-540-33182-4

Y. Takagi, H. Mizutani, & S.-I. Kawakami, Impact Fragmentation Experiments of Basalts and Pyrophyllites, *Icarus*, 59(3), 462–477, 1984, https://doi.org/10.1016/0019-1035(84)90114-3

H. Tanaka, S. Inaba, & K. Nakazawa, Steady-State Size Distribution for the Self-Similar Collision Cascade, *Icarus*, 123(2), 450–455, 1996, https://doi.org/10.1006/icar.1996.0170

T. Vicsek, *Fractal Growth Phenomena*, 2nd edition, World Scientific, 1992, https://doi.org/10.1142/1407

G. M. Wang & C. M. Sorensen, Aggregation Kernel Homogeneity for Fractal Aggregate Aerosols in the Slip Regime, *Aerosol Science and Technology*, 34(3), 297–306, 2001, https://doi.org/10.1080/02786820118313

A. L. Yarin, I. V. Roisman, & C. Tropea, *Collision Phenomena in Liquids and Solids*, Cambridge University Press, 2017, https://doi.org/10.1017/9781316556580

G. H. Yeoh, C. P. Cheung, & J. Tu, *Multiphase Flow Analysis Using Population Balance Modeling: Bubbles, Drops and Particles*, Butterworth-Heienemann, 2014, https://doi.org/10.1016/C2011-0-05568-0

2
Basic Properties

This chapter discusses why aggregation and fragmentation tend to produce random or seemingly random outcomes. In addition, this chapter discusses the conditions under which universal probabilistic solutions for size, shape, speed, and other particle properties may occur. In pursuit of these goals, this chapter introduces four basic properties—chaos, self-organization, homogeneity, and isotropy—which typify many aggregating and fragmenting particle systems.

2.1 Chaos

As defined in this book, chaotic particle systems have the properties listed in Table 2.1. Many treatments assume that property (1.) implies properties (2.) and (3.). However, this book will consider these properties separately. Property (4.), known as *molecular chaos*, is optional and represents an especially strong form of chaos. Boltzmann (1872) first suggested that the velocities of two colliding particles were *statistically independent*. Readers who are unfamiliar with this term can look ahead to the definition given in Section 7.1. Tame (2019) says "of course this is not entirely true, since two molecules which have just collided will have related positions and momenta ... As the collision between any two chosen molecules becomes rarer, however, Boltzmann's assumptions become more valid." While Boltzmann (1872) called this property *Stoßzahlansatz*, Jeans (1906) and most subsequent English-speaking researchers called it molecular chaos.

In this book, chaos arises from repeated collisions, attractive long-range body forces, crack propagation, and so forth; see Examples 2.1 through 2.5. These examples use terms like *correlated*, *uncorrelated*, and *autocorrelated*. Readers who are unfamiliar with these terms can look ahead to the definitions given in Section 8.1.

Table 2.1 Properties of chaotic particle systems.

1. *Locally Sensitive to Initial Conditions*	Interactions between particles are highly sensitive to their initial conditions, i.e., for each interaction, slightly different initial conditions diverge exponentially in phase space.
2. *Seemingly Random*	Particle states fluctuate in an apparently random way, even when all of the particle interactions are deterministic, and even when the overall system averages do not fluctuate.
3. *Globally Insensitive to Initial Conditions: Forgets Initial Conditions*	As times passes, particle properties become increasingly uncorrelated in time. Eventually, the state of each particle becomes statistically independent of its initial state.
4. *Molecular Chaos (Optional)*	The velocities of adjacent particles are statistically independent. More generally, at any given time, all components of velocity are statistically independent, both for one particle and for pairs of adjacent particles.

Particle Dynamics with Aggregation and Fragmentation. Culbert B. Laney, Oxford University Press. © Culbert B. Laney (2025).
DOI: 10.1093/9780198920687.003.0002

Example 2.1 Identical particles that interact through perfectly elastic collisions are naturally chaotic

Consider a *molecular gas* composed of identical hard spheres that interact through instantaneous, binary, perfectly elastic collisions. In this context, "perfectly elastic" means that each collision obeys conservation of mass, momentum, and energy. The following shows that molecular gases obtain all of the properties of chaos.

1. *Locally sensitive to Initial Conditions*	For equal-sized particles, each collision is sensitive to its initial conditions; see Figure 2.1. This sensitivity is multiplied exponentially by subsequent collisions. For example, Poincaré (1907) says "if the first collision multiplied the deviation by a very large number, A, after n collisions it will be multiplied by A^n." For a more rigorous and modern explanation see, e.g., Dorfman (1999).
2. *Seemingly Random*	The positions and velocities of each particle fluctuate in an apparently random way due to repeated collisions. For example, Soto (2016) says "a molecule in a gas performs a random walk as a consequence of collisions with other molecules."
3. *Globally Insensitive to Initial Conditions: Forgets Initial Conditions*	In some cases, the velocity autocorrelation function decays exponentially with time; see, e.g., Eder (1977) and Bouchaud & Doussal (1985). However, in other cases, instead of "an exponential decay of the correlation function in time ... molecules keep a memory of their previous states of motion for much longer;" see Battimelli & Ciccotti (2018).
4. *Molecular Chaos (Optional)*	According to Tame (2019), "in general, even very minor changes in the initial positions and velocities of only one or two molecules in a simulation of thousands will give essentially uncorrelated trajectories within a few nanoseconds. The *Stoßzahlansatz* [molecular chaos] turns out to be an excellent assumption."

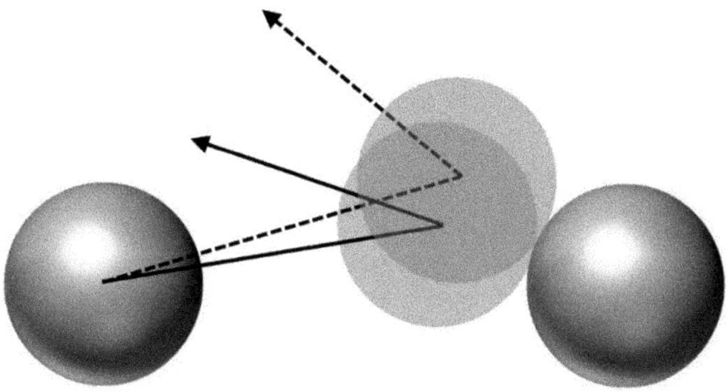

Figure 2.1 Collisions between identical spheres magnify differences in initial conditions.

Example 2.2 Identical particles that interact through partially elastic collisions are naturally chaotic, provided the collision rate is high enough

Consider a *granular gas* composed of identical hard spheres interacting through instantaneous, binary, partially elastic collisions. In this context, "partially elastic" means that each collision obeys conservation of mass but not conservation of momentum or energy due, e.g., to surface friction. While perfectly elastic collisions are typical of microscopic particles, partially elastic collisions are typical of macroscopic particles. As one of their most interesting features, granular gases spontaneously form unbonded particles; see Figures 1.5 and 2.2. The following shows that granular gases obtain all of the properties of chaos, provided the particle concentrations and collision rates are high enough.

1. *Locally Sensitive to Initial Conditions*	The collision rate tends to increase with particle concentration. In other words, the more particles in a given volume, the more often they collide, the more sensitive they are to their initial conditions. At high enough particle concentrations, the sensitivity may be exponential; see, e.g., McNamara & Mareschal (2001), Eckmann et al. (2005).
2. *Seemingly Random*	The positions and velocities of each particle fluctuate in an apparently random way due to repeated collisions; see, e.g., Brudeau & Viot (2009).
3. *Globally Insensitive to Initial Conditions: Forgets Initial Conditions*	At high enough particle concentrations, the velocity autocorrelation tends to decay at a less-than-exponential rate in time; see, e.g., Ahmad & Puri (2007).
4. *Molecular Chaos (Optional)*	Baxter & Olafsen (2007) showed that molecular chaos may occur due to external "energy injection ... randomized in both space and time;" see also, e.g., Brudeau & Viot (2009) and Windows-Yule & Parker (2013).

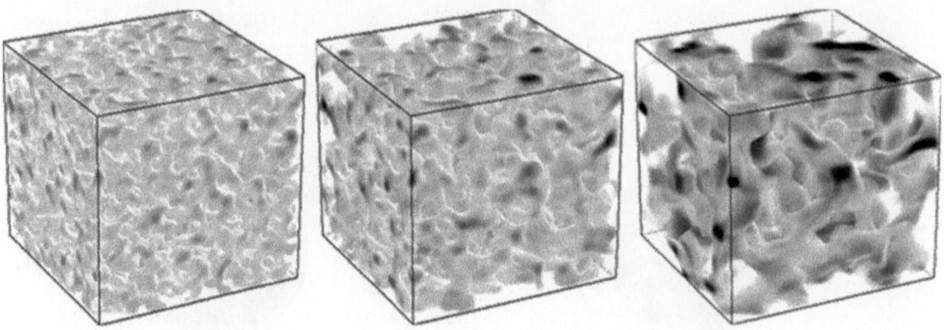

Figure 2.2 Three-dimensional map of monomer densities in a granular gas. Time increases from left to right. "Darker regions represent larger local average density."
Figure reprinted with permission from Hummel et al. (2016).

Example 2.3 Non-identical particles that interact through perfectly elastic collisions are naturally chaotic, provided the collision rate is high enough

Consider a particle system composed of two types of identical hard spheres: a large number of small, fast spheres; and a small number of large, slow spheres. In such cases, the large particles experience *Brownian motion* due to collisions with the small particles. Friedlander (2000) describes Brownian motion as "a haphazard dancing motion resulting from the fluctuating forces exerted on ... [the larger molecules] by the surrounding [smaller] molecules." While collisions between equal-sized particles are sensitive to the initial conditions, collisions between unequal-sized particles are less sensitive. In the most extreme case, a small sphere may see the surface of a large sphere as almost flat; then the collisions are completely insensitive; see Figure 2.3.

Cecconi et al. (2007) compared Brownian motion for circles vs. squares in two dimensions, i.e., for collisions involving round vs. flat particle surfaces. Under the same circumstances, circles obtained exponential sensitivity to initial conditions while squares did not. However, if completely flat surfaces do not occur, the following shows that Brownian motion obtains all of the properties of chaos, provided the concentrations of small particles are high enough, and the collision rates between large and small particles are high enough.

1. *Locally Sensitive to Initial Conditions*	The large particles are sensitive to their initial conditions provided the concentration of small particles is large enough; see, e.g., Gaspard & Van Beijeren (2002). The more small particles that exist in a given volume, the more often they collide with the large particles, the more sensitive the large particles are to their initial conditions.
2. *Seemingly Random*	According to Mo & Raizen (2019), "Brownian motion is the random movement of particles agitated by the thermal motion of the molecules in a fluid ... Instead of studying Brownian motion at the single collision level, most successful theories begin from a statistical point of view by averaging over many collisions in a clever way."
3. *Globally Insensitive to Initial Conditions: Forgets Initial Conditions*	After enough time has passed, the large particles obtain a uniformly random spatial distribution, regardless of initial conditions. This process is called *diffusion*. In addition, the velocity autocorrelation function of the large particles decays gradually with time; see, e.g., Bian et al. (2016) and Mo & Raizen (2019).
4. *Molecular Chaos (Optional)*	Theers et al. (2016) found that large "rigid spherical colloidal particles dissolved in a fluid" obtained molecular chaos during an initial period with "uncorrelated fluid-particle motion and local interactions ... dominated by ballistic fluid-colloid collisions." However, after this initial period, absent external energy input, the large particles did not experience molecular chaos.

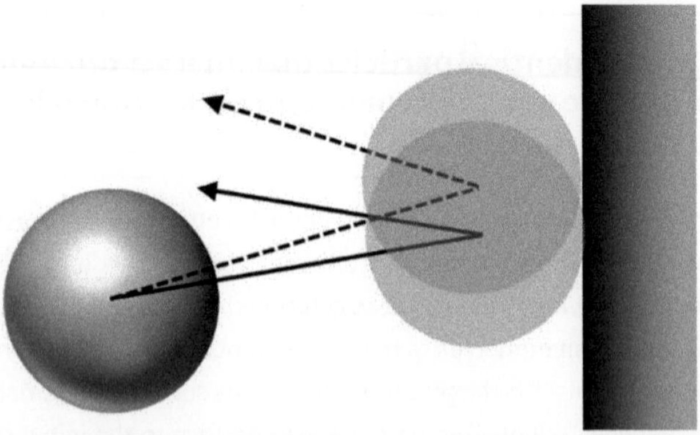

Figure 2.3 Collisions between a sphere and a flat surface do not magnify differences in initial conditions.

Example 2.4 Particles that interact through long-range body forces are naturally chaotic, at least intermittently

Consider a collisionless system of three particles interacting through gravity. Based on numerical simulations, Chernin & Valtonen (1998) showed that such three-body problems obtain all of the properties of chaos.

1. *Locally Sensitive to Initial Conditions*	"Dynamical instability … due to nonlinear interactions of the gravitating bodies … manifests itself in high sensitivity of the system behavior to the fine details of the initial conditions … [such that] two initially almost identical trajectories diverge in the phase space and also in the momentum (velocities) subspace."
2. *Seemingly Random*	"Dynamical chaos appears as apparently random motions in a deterministic dynamical system, i.e., in a system with no external random forcing … The evolution with random alternations of almost deterministic states and the states of developed chaos is similar to the phenomenon of intermittency discovered in … turbulence."
3. *Globally Insensitive to Initial Conditions: Forgets Initial Conditions*	The temporal "autocorrelation functions rapidly … decreased" as the bodies crossed each other, leading the system to "forget its initial conditions."
4. *Molecular Chaos (Optional)*	Depending on circumstances, if the initial conditions obtain molecular chaos, the system will obtain molecular chaos indefinitely; see, e.g., Boers & Pickl (2016).

Chernin & Valtonen (1998) argue that such three-body problems "reveal all the complexity of chaotic behavior" typical of N-body systems. This is true even though three-body problems have between 2 and 18 degrees of freedom while N-body problems have up to 6N degrees of freedom.

Example 2.5 Fragmentation of brittle solids via crack propagation is naturally chaotic, provided the concentration of cracks is sufficiently large, i.e., provided the number of fragments created per unit volume is sufficiently large

Levy & Molinari (2010) say that "microcracks originate at seemingly random sites, open, propagate at large speeds, and eventually coalesce with one another to generate fragments ... Since each damaged site releases its own stress waves, a chaotic network of interactions emerges." More specifically:

1. *Locally Sensitive to Initial Conditions*	Deshpande et al. (2001) say "for the case where the [crack] nucleation sites were randomly perturbed ... after an initial delay ... the deviation of the trajectories grows exponentially." Miklashevich (2005) argues that chaos occurs only at the "crack tip ... associated with a quasi-particle," i.e., instability and chaos are confined to "the first stage of crack development."
2. *Seemingly Random*	Zhou et al. (1996) say "it is impossible for a crack to branch completely symmetrically—the choice of path is made randomly." Ponson (2016) argues that cracks often "follow trajectories close to a directed random walk: At any time during failure, the crack has the same probability to propagate upward as downward, irrespective of the prior propagation direction."
3. *Globally Insensitive to Initial Conditions: Forgets Initial Conditions*	Ray (2018) says "the development of fracture in heterogeneous systems is a phenomenon correlated over large length and time scales ... On the other hand, when the concentration of the defects [i.e., crack initiation points] is large, the correlation in the sequence of breaking events goes away."

The preceding examples can be generalized to include most kinds of particle aggregation and fragmentation. More specifically, consider a system of particles interacting through binary collisions. Further, suppose a fragmenting collision converts two parents with statistically independent properties into a large number of children with random, statistically independent properties. Then each fragmentation event tends to decrease correlations between particles and, thus, increase chaos. On the other hand, suppose an aggregating collision converts two particles with random, statistically independent properties into a single particle with a deterministic average velocity and momentum. In other words, suppose the two particles stick together. These are sometimes called *perfectly inelastic* collisions. Each such aggregation event tends to increase correlations between formerly separate particles and, thus, decrease chaos. Having said this, aggregating particle systems may still be chaotic indefinitely if: (a.) the particles experience frequent elastic or semi-elastic collisions in between inelastic collisions as discussed in Examples 2.1, 2.2, and 2.3; (b.) the particles interact through long-range body forces in between inelastic collisions as discussed in Example 2.4; or (c.) there is an external source of randomized mass, momentum, and/or energy such as vibration or shaking as discussed in Example 2.2. If none of these conditions is met, then an aggregating particle system will become less and less chaotic as time increases. In this case, highly random initial conditions may still lead to chaos for an extended period of time; see Example 2.6.

Example 2.6 Identical ball bearings interacting via electromagnetic forces

Jun & Hübler (2005) and Hübler & Crutchfield (2010) studied particle systems composed of "400–1,200 stainless-steel ball bearings ... in a glass or acrylic dish that contains dielectric liquid (castor oil)." In these experiments, a circular electrode lined the perimeter of the round dish. In addition, a second needle-shaped electrode was "placed above the center of the dish" which "sprayed electrons quasi-homogeneously over the surface of the oil." A battery connected to the electrodes induced electromagnetic forces between the ball bearings. At first, the ball bearings tended to gather in dense clusters. However, over a period of hours or days, the ball bearings gradually rearranged themselves into one or more extended linear or fractal shapes. If the ball bearings are initially placed in widely spaced random positions, the system seems to obtain all of the properties of chaos; see Figure 1.1a. However, if the ball bearings are initially placed in tightly spaced nearly regular positions, the final results seem to depend on the initial conditions, which violates Property (3.) of chaotic systems; see Figure 1.1b. Still, the single resulting particle cluster obtains a seemingly random fractal shape. This example shows that, while chaotic systems "forget" their initial conditions, the initial conditions may help determine whether or not the particle system is chaotic in the first place.

Lorenz (1963) discovered chaos while studying a simple system of three differential equations, meant to model atmospheric convection. To his surprise, he found that "slightly differing initial states can evolve into considerably different states." This became known as the *butterfly effect* and later, following Li & Yorke (1975), *chaos*. Following Lorenz (1963), much of the work done under the heading of chaos theory concerns simple periodic or oscillatory systems such as driven damped pendulums, spring-block systems, and recirculating flows. In general, such systems tend to converge on lower-dimensional regions of phase space called *attractors*. Without chaos, these attractors tend to obtain regular geometric shapes, e.g., oval, torus. However, with chaos, these attractors tend to obtain irregular, fractal shapes; see, e.g., the beautiful color-coded phase diagrams in Schuster & Just (2005). The particle systems studied in this book generally do *not* experience periodic or oscillatory behaviors; thus, the concept of attractors does not apply.

While most observers credit Lorenz (1963) with discovering chaos, Kautz (2011) notes that Maxwell (1873), Poincaré (1907), and others clearly described chaos well before Lorenz, albeit in the specific context of the kinetic theory of molecular gases. Regarding such applications, Hilborn (2000) says the following.

> Chaotic behavior ... looks like ... a system with many, many degrees of freedom, each "doing its own thing." ... Most real systems in biology, chemistry, physics, and engineering

> are made of billions and billions of atoms and molecules ... To be somewhat more technical, we could say that these complex systems have many degrees of freedom ... [According to one argument] it is the activity of these many degrees of freedom that leads to the apparently random behavior ... The crucial importance of chaos is that it provides an alternative explanation for this apparent randomness ... that [does not depend on] complexity. Chaotic behavior shows up in systems that are ... relatively simple—only a few degrees of freedom are active.

The concept of degrees of freedom, as mentioned here, helps to explain why many chaotic particle systems obtain universal behaviors, such as the ideal gas law for molecular gases.

As this discussion shows, some sources use the term "chaos" to refer to simple systems with only a few elements, such as driven damped pendulums. However, other sources use the term "chaos" to refer to more complicated systems with large numbers of elements. To avoid confusion, some sources refer to the latter as *complex systems*. For example, Kaneko & Tsuda (2001) say "nowadays many scientists profess the study of complex systems ... We believe that it is impossible to discuss complex systems without mentioning the concept of chaos. The emergence of the concept of chaos is one of the biggest scientific achievements of this century, and has inspired interest in complex systems by radically changing many fundamental beliefs in science." In the remainder of this book, the term "chaos" will be used exclusively to refer to dynamic particle systems, which can be seen as a specific type of complex system.

2.2 Self-Organization

As defined in this book, self-organized systems have the properties listed in Table 2.2. Many treatments assume that properties (2) and (3) imply property (1). However, this book will consider these properties separately. Regarding property (1), Zhang (2016) says that such ordering involves a "dimensional reduction of state space or the reduction of degrees of freedom." Regarding property (2), self-organized structures are sometimes called *dissipative structures*; see, e.g., Prigogine & Nicolis (1971) and Prigogine (1975). Regarding property (3), Zhang (2016) says that self-organization "is often triggered by random fluctuations that are amplified by positive feedback." If these seemingly random fluctuations are due to chaos, self-organization draws *order out of chaos*; see, e.g., Prigogine & Stengers (1984).

There are two main types of self-organization seen in the research literature: deliberate and spontaneous. Pask (1964) says that deliberately self-organized systems have "a controller

Table 2.2 Properties of spontaneously self-organized systems. For purposes of this book, the term "constituent" usually refers to particles.

1. *Ordered*	Order or structure arises spontaneously via continual interactions between constituents. Then the system is *far from equilibrium.*
2. *Dissipation*	Constituent interactions involve friction, viscosity, inelasticity, plasticity, or other forms of dissipation.
3. *Chaos or Randomness*	Constituent states fluctuate in random or apparently random ways because of chaos or external sources of mass, momentum, or energy.

[with] ... a hierarchical structure wherein there are levels of control that aim to achieve different levels of goal." By contrast, Zhang (2016) says that spontaneously self-organized systems are "not directed or controlled by any agent or subsystem inside or outside the system." Example 2.7 gives a historical overview of these two types of self-organization. From now on, in this book, the term "self-organization" will always refer to "spontaneous self-organization."

As Example 2.7 indicates, many self-organized particle systems involve living organisms or, at least, life-like entities. For example, Aschwanden et al. (2018) list almost 70 different self-organized systems such as "feeding aggregations of bark beetles" and "swarm raids of army ants." For another example, Ginot et al. (2018) found that "a remarkable feature of active matter is the propensity to self-organize" including systems as diverse as "proteins and colloids," "living polymers and wormlike micelles," "microtubule bundles," and "swarming bacteria, birds or fishes." In fact, for all such "self-propelled entities," Ginot et al. (2018) argue that a combination of aggregation and fragmentation leads to universal power law size distributions with $n/m \approx -5/6$. However, self-organization is not limited to living organisms and life-like entities. For example, Example 2.8 shows that granular gases have all of the properties of spontaneously self-organized systems. As other examples, Aschwanden et al. (2018) list "Saturn rings and moons," "star formation," and "a diffuse aggregation process of magnetic flux tubes in the upper levels of the solar convection zone."

Example 2.7 Historical overview of deliberate vs. spontaneous self-organization

The concept of *deliberate* self-organization was first developed by a community of interdisciplinary researchers that included psychiatrists such as Ross Ashby (1947, 1962), psychologists such as Gordon Pask (1961, 1964), biologists such as Ludwig von Bertalanffy (1964, 1968), electrical engineers such as Heinz von Foerster (1960), and mathematicians such as Norbert Wiener (1961, 1962). This community, sometimes referred to by terms such as *cybernetics* or *general systems theory*, focuses on the relationships between living and non-living systems. For example, Pask (1964) used self-organization to describe "automata and computing mechanisms which may be either fabricated artifacts or living organisms" including "learning behavior ... generated by populations of evolving automata."

The concept of *spontaneous* self-organization was developed by a different community of interdisciplinary researchers that included chemists and physicists such as: (a.) Manfred Eigen; see, e.g., Eigen (1971a, b), Eigen & Schuster (1979), Epstein & Eigen (1979); (b.) Ilya Prigogine; see, e.g., Prigogine & Nicolis (1971), Nicolis & Prigogine (1971, 1977), Glansdorff & Prigogine (1971), Prigogine & Lefever (1975); (c.) Hermann Haken; see, e.g., Haken & Graham (1971), Haken (1977); and (d.) Werner Ebeling; see, e.g., Ebeling & Feistel (1974) and Ebeling & Mahnke (1979). The earliest research done in this community used the generic term *organization*; see, e.g., Prigogine (1955, 1961, 1968). However, starting with Eigen (1971a, b), the term *self-organization* gradually became standard. Like the early work on deliberate self-organization, the early work on spontaneous self-organization often focused on the relationships between living and non-living systems. For example, Ebeling & Mahnke (1979) explored how deoxyribonucleic acid (DNA) might form through natural selection among "identical copies (replication) and error copies (mutation)."

Among the earliest researchers involved in studying spontaneous self-organization, Ilya Prigogine is perhaps the best known. Feistel & Ebeling (2011) note that "when in 1970, Ilya Prigogine ... presented his great lecture on self-organization ..., this was appreciated by the audience as a kind of theoretical revolution ... The lecture room ... was overcrowded, the excitement was enormous." This was followed by "the 'magic' year of 1971 when ... seminal publications" by Eigen (1971a, b), Glansdorff & Prigogine (1971) and Haken & Graham (1971) "appeared almost simultaneously." In 1977, Ilya Prigogine received the Nobel Prize in Chemistry for "for his contributions to non-equilibrium thermodynamics, including the theory of dissipative structures," and for demonstrating that "it is possible to create order from disorder."

Example 2.8 Identical particles that interact through partially elastic collisions are spontaneously self-organized, provided the collision rate is high enough

As discussed earlier in Example 2.2, consider a granular gas composed of identical hard spheres interacting through instantaneous, binary, partially elastic collisions. In this context, "partially elastic" means that each collision obeys conservation of mass but not conservation of momentum or energy due, e.g., to surface friction. The following shows that granular gases obtain all of the properties of self-organization provided the spherical particle concentrations and collision rates are high enough.

1. *Ordered*	After enough time has passed, the spheres typically self-organize into unbonded clusters, independent of their initial conditions, provided the sphere concentrations and collision rates are high enough; see, e.g., Figures 1.5 and 2.2.
2. *Dissipation*	According to Goldhirsch (1999), "the dissipative nature of the particle collisions ... is the source of clustering." Similarly, Brilliantov & Pöschel (2004) say that "slightly altering the collision law of the particles, that is, making them (even slightly!) dissipative, causes drastic changes ... [leading to] self-organized spatio-temporal structures."
3. *Chaos or Randomness*	Repeated collisions may cause the positions and velocities of each sphere to fluctuate in an apparently random way. External sources of mass, momentum, and energy may accentuate such fluctuations.

2.3 Self-Organized Criticality

As defined in this book, systems with *self-organized criticality* have the properties listed in Table 2.3. Notice that properties (1.), (2.), and (3.) are also properties of self-organization. Thus, according to the definitions used in this book, *self-organized criticality is a special case*

Table 2.3 Properties of self-organized criticality. For purposes of this book, the terms "constituent" and "structure" typically refer to monomers and clusters, respectively.

1. *Ordered*	Order or structure arises spontaneously via continual interactions between constituents.
2. *Dissipation*	Constituent interactions involve friction, viscosity, inelasticity, or other forms of dissipation.
3. *Chaos or Randomness*	Constituent states fluctuate in sudden or apparently random ways because of chaos or random external sources of mass, momentum, or energy.
4. *"Avalanches" or "Catastrophes"*	The system routinely stores and suddenly releases energy in a way that substantially alters the system state, either locally or globally.
5. *Power Laws (Optional)*	The structures in the system exhibit power law size distributions, velocity distributions, etc. with either real or rational exponents.
6. *Fractal Shapes (Optional)*	The structures in the system exhibit fractal shapes such as those seen in Figures 1.1, 1.5, 1.6, and 1.9.

of self-organization. As an additional element, according to property (4.), self-organized criticality only occurs in systems that experience events known as *catastrophes* or *avalanches*, which involve the sudden release of stored potential energy. As one optional additional element, according to property (5.), self-organized criticality leads to power law size distributions. As a second optional additional element, according to property (6.), self-organized criticality leads to fractal shapes. Pruessner (2012) warns that "a precise definition of the term 'self-organized criticality' is hampered by its indiscriminate use, which has blurred its meaning." In particular, some researchers consider power-law size distributions to be mandatory rather than optional. For example, Aschwanden (2011) says that "the power-law shape of size distributions became the hallmark and principal diagnostic of self-organized criticality phenomena." In addition, some researchers consider fractal shapes to be mandatory rather than optional. For example, Pruessner (2012) says that, when self-organized criticality was first introduced, "it was suggested to be *the* explanation of fractal structures surrounding us everywhere in space and time." Finally, some researchers consider catastrophes to be optional rather than mandatory; see, e.g., Saunders & Ho (1994), Brown (1995), Fong & Key (1999).

Like fractals and chaos, self-organized criticality only became widely known in the computer age. For example, Pruessner (2012) says "the primacy of the computer model is manifest in the first publication and throughout the history of self-organized criticality, which has evolved with and revolved around such computer models. This has led to a plethora of computer 'models,' many of which are not intended to model much except themselves."

The concept of self-organized criticality was first developed by Bak et al. (1987, 1988); see also Bak (1996). To illustrate the concept, they used a simple example of a dense granular flow, in which a slow steady driver occasionally causes a sudden response in the form of small, medium, or large avalanches. More specifically:

> Consider a simple pile of sand. Suppose we start from scratch and build the pile by randomly adding sand, a grain at a time. The pile will grow, and the slope will increase.

> Eventually, the slope will reach a critical value ... If more sand is added it will slide off ... One would expect a scaling relation for the size of [sliding] bulk clusters to obey a [power law] ... where the critical exponent can be thought of as the fractal dimension of the clusters.

In essence, the surface of the sandpile sporadically shifts between a granular solid which stores mass and energy from an external source and a dense granular flow which transports and dissipates the stored mass and energy. Building on the sandpile example, sporadic sudden responses of all types are sometimes called *avalanches*. Alternatively, one-time or sporadic sudden responses are sometimes called *catastrophes*; see, e.g., Thom (1972), Zeeman (1976), Poston & Stewart (1978), Saunders (1980).

Like self-organization, self-organized criticality often leads to universal size distributions. In fact, in one of the original papers on self-organized criticality, Bak et al. (1988) said that power law size distributions could be "grouped into universality classes within which the exponents are the same, depending on symmetry, dimension, and so on." Further research has confirmed this. In particular, Pruessner (2012) characterizes universals in self-organized criticality as follows: (a) "universal quantities are first and foremost critical exponents;" (b) "boundary conditions can alter the universal behavior;" and (c) "a universality class is .... normally specific to a specific spatial dimension."

In this book, self-organized criticality mainly occurs in two problems, namely, dense granular flows and unitary fragmentation. First, consider dense granular flows, where particles interact through extended contacts with multiple neighbors. As noted by Oyama et al. (2019), dense granular flows experience "jamming" and "intermittent drops in the time evolution of the potential energy ... which are one typical feature of avalanche dynamics." Informally speaking, dense granular flows routinely trap and release mass, momentum, and energy. For comparison, consider granular gases where particles interact primarily through instantaneous binary collisions; see Examples 2.2 and 2.8. Unlike dense granular flows, granular gases cannot trap or release potential energy. This is entirely a matter of particle concentration. In other words, the only difference between dense granular flows and granular gases is the number and mass of particles per unit volume. While this book mainly focuses on low-density particle systems such as granular gases, it will occasionally discuss high-density particle systems such as dense granular flows; see, e.g., Section 8.4 and Chapter 13.

Second, consider unitary fragmentation, as discussed in Chapter 1. In this case, from a rather simplistic point of view, a virgin object absorbs a large amount of energy from, e.g., an adjacent explosion or high-velocity impact. Following this, the object releases the absorbed energy by partially or fully disintegrating into fragments. As an example, Oddershede et al. (1993) conducted a series of simple unitary fragmentation tests designed to study self-organized criticality. In some cases, these tests obtained universal power law size distributions with $n/m \approx -1/6$. As another example, Aschwanden et al. (2016) surveyed a wide variety of different astrophysical systems, including those featuring unitary fragmentation, chosen to illustrate self-organized criticality. Based on this survey, as well as dimensional analysis, they suggested a universal power law size distribution with $n/m \approx -2/3$. As noted earlier, according to the definitions used in this book, self-organized criticality is a special case of self-organization. For simplicity, this book will hereafter use the more general term self-organization rather than the less general term self-organized criticality.

2.4 Homogeneity and Isotropy for Scalar Properties

The solutions to a deterministic problem, which do not involve randomness or chaos, may be steady and uniform. In other words, the solutions may be entirely independent of time and space. However, the solutions to random or chaotic problems are never truly steady or uniform. Instead, the solution is always fluctuating, at least on the smallest time and length scales. Thus, instead of being steady and uniform, the solutions to random and chaotic problems are *homogeneous* and *isotropic*. The terms "homogeneous" and "isotropic" can be defined based on either scalar or vector properties. This section gives the scalar definitions. The next section gives the vector definitions. Let X be a non-negative random scalar property such as the particle mass or diameter. Let $\mathbf{r}$ refer to space and t refer to time. If X depends on $\mathbf{r}$ or t or both, then X may be called a *random scalar process*. As one possible way to describe a random scalar process, a *one-point, one-time probability density function* $f(x; \mathbf{r}, t)$ may be defined such that for any $x_{\min}$ and $x_{\max}$:

$$\int_{x_{\min}}^{x_{\max}} f(x; \mathbf{r}, t) dx \tag{2.1}$$

is the probability that $X(\mathbf{r}, t)$ is in the range $[x_{\min}, x_{\max})$. Notice that:

$$\int_{0}^{\infty} f(x; \mathbf{r}, t) dx = 1. \tag{2.2}$$

One-point, one-time probability density functions have no way of knowing how different points in space and time depend on each other. As an alternative, a *one-point, two-time joint probability density function* $f(x_1, x_2; \mathbf{r}, t_1, t_2)$ may be defined such that for any $x_{1\min}, x_{1\max}, x_{2\min}, x_{2\max}$:

$$\int_{x_{2\min}}^{x_{2\max}} \int_{x_{1\min}}^{x_{1\max}} f(x_1, x_2; \mathbf{r}, t_1, t_2)\, dx_1 dx_2 \tag{2.3}$$

is the probability that $X(\mathbf{r}, t_1)$ is in the range $[x_{1\min}, x_{1\max})$ at time t_1 at a given point $\mathbf{r}$ and $X(\mathbf{r}, t_2)$ is in the range $[x_{2\min}, x_{2\max})$ at time t_2 at a given point $\mathbf{r}$. Notice that:

$$\int_{0}^{\infty} \int_{0}^{\infty} f(x_1, x_2; \mathbf{r}, t_1, t_2)\, dx_1 dx_2 = 1. \tag{2.4}$$

In addition to two-time joint probability density functions, there are also three-time, four-time, five-time, etc. joint probability density functions.

A system is called *temporally homogeneous* or *stationary* if all possible joint probability density functions are the same regardless of offsets in time. For example, for any Δt:

$$f(x_1, x_2; \boldsymbol{r}, t_1 + \Delta t, t_2 + \Delta t) = f(x_1, x_2; \boldsymbol{r}, t_1, t_2). \tag{2.5}$$

Similarly, *two-point, one-time joint probability density functions* $f(x_1, x_2; \boldsymbol{r}_1, \boldsymbol{r}_2, t)$ may be defined such that for any $x_{1\,\min}$, $x_{1\,\max}$, $x_{2\,\min}$, $x_{2\,\max}$:

$$\int_{x_{2\,\min}}^{x_{2\,\max}} \int_{x_{1\,\min}}^{x_{1\,\max}} f(x_1, x_2; \boldsymbol{r}_1, \boldsymbol{r}_2, t)\, dx_1 dx_2 \tag{2.6}$$

is the probability that $X(\boldsymbol{r}_1, t)$ is in the range $[x_{1\,\min}, x_{1\,\max})$ at point $\boldsymbol{r}_1$ at a given time t and $X(\boldsymbol{r}_2, t)$ is in the range $[x_{2\,\min}, x_{2\,\max})$ at point $\boldsymbol{r}_2$ at a given time t. Notice that:

$$\int_0^{\infty} \int_0^{\infty} f(x_1, x_2; \boldsymbol{r}_1, \boldsymbol{r}_2, t)\, dx_1 dx_2 = 1. \tag{2.7}$$

In addition to two-point joint probability density functions, there are also three-point, four-point, five-point, etc. joint probability density functions.

A system is called *spatially homogeneous* if all possible joint probability density functions are the same regardless of offsets in space. For example, for any $\Delta\boldsymbol{r}$:

$$f(x_1, x_2; \boldsymbol{r}_1 + \Delta\boldsymbol{r}, \boldsymbol{r}_2 + \Delta\boldsymbol{r}, t) = f(x_1, x_2; \boldsymbol{r}_1, \boldsymbol{r}_2, t). \tag{2.8}$$

In addition, a system is called *isotropic* if all possible joint probability density functions are the same regardless of rotations or reflections of the coordinate system. Put simply, a system is spatially homogeneous if the statistics are invariant to shifts in position and isotropic if the statistics are invariant to rotations and reflections.

2.5 Homogeneity and Isotropy for Vector Properties

Let $\boldsymbol{V}$ be a random or seemingly random vector property, such as the particle velocity, with negative or non-negative components V_i, $i = 1, \ldots, d$. If $\boldsymbol{V}$ depends on $\boldsymbol{r}$ or t or both, then $\boldsymbol{V}$ may be called a *random vector process*. A *one-point, one-time probability density function* $f(v_i; \boldsymbol{r}, t)$ may be defined such that for any $v_{i\,\min}$ and $v_{i\,\max}$:

$$\int_{v_{i\,\min}}^{v_{i\,\max}} f(v_i; \boldsymbol{r}, t)\, dv_i \tag{2.9}$$

is the probability that $V_i(\boldsymbol{r}, t)$ is in the range $[v_{i\,\min}, v_{i\,\max})$. Notice that:

$$\int_{-\infty}^{\infty} f(v_i; \boldsymbol{r}, t)\, dv_i = 1. \tag{2.10}$$

As another alternative, a *one-point, one-time joint probability density function* $f(v_i, v_j; \boldsymbol{r}, t)$ may be defined such that for any $v_{i\ \min}$, $v_{i\ \max}$, $v_{j\ \min}$, and $v_{j\ \max}$:

$$\int_{v_{j\ \min}}^{v_{j\ \max}} \int_{v_{i\ \min}}^{v_{i\ \max}} f(v_i, v_j; \boldsymbol{r}, t)\, dv_i dv_j \tag{2.11}$$

is the probability that $V_i(\boldsymbol{r}, t)$ is in the range $[v_{i\ \min}, v_{i\ \max})$ and $V_j\ (\boldsymbol{r}, t)$ is in the range $[v_{j\ \min}, v_{j\ \max})$. Notice that:

$$\int_{-\infty}^{\infty} \int_{-\infty}^{\infty} f(v_i, v_j; \boldsymbol{r}, t)\, dv_i dv_j = 1. \tag{2.12}$$

The previous probability density functions have no way of knowing how different points in space and time depend on each other. As an alternative, a *one-point, two-time joint probability density function* $f(v_i, v_j; \boldsymbol{r}, t_1, t_2)$ may be defined such that for any $v_{i\ \min}$, $v_{i\ \max}$, $v_{j\ \min}$, and $v_{j\ \max}$:

$$\int_{v_{j\ \min}}^{v_{j\ \max}} \int_{v_{i\ \min}}^{v_{i\ \max}} f(v_i, v_j; \boldsymbol{r}, t_1, t_2)\, dv_i dv_j \tag{2.13}$$

is the probability that $V_i(\boldsymbol{r}, t_1)$ is in the range $[v_{i\ \min}, v_{i\ \max})$ and $V_j(\boldsymbol{r}, t_2)$ is in the range $[v_{j\ \min}, v_{j\ \max})$. In addition to two-time joint probability density functions, there are also three-time, four-time, five-time, etc. joint probability density functions.

A system may be called *temporally homogeneous* or *stationary* if all possible probability density functions are the same regardless of offsets in time. For example, for any Δt:

$$f(v_i, v_j; \boldsymbol{r}, t_1 + \Delta t, t_2 + \Delta t) = f(v_i, v_j; \boldsymbol{r}, t_1, t_2) \tag{2.14}$$

Pope (2000) says "a process is statistically stationary if all multi-time statistics are invariant under a shift in time ... Even though the... variables vary with time, the statistics are independent of time." Similarly, a *two-point, one-time joint probability density function* $f(v_i, v_j; \boldsymbol{r}_1, \boldsymbol{r}_2, t)$ may be defined such that for any $v_{i\ \min}$, $v_{i\ \max}$, $v_{j\ \min}$, and $v_{j\ \max}$:

$$\int_{v_{j\ \min}}^{v_{j\ \max}} \int_{v_{i\ \min}}^{v_{i\ \max}} f(v_i, v_j; \boldsymbol{r}_1, \boldsymbol{r}_2, t)\, dv_i dv_j \tag{2.15}$$

is the probability that $V_i(\boldsymbol{r}_1, t)$ is in the range $[v_{i\ \min}, v_{i\ \max})$ and $V_j(\boldsymbol{r}_2, t)$ is in the range $[v_{j\ \min}, v_{j\ \max})$. In addition to two-point joint probability density functions, there are also three-point, four-point, five-point, etc. joint probability density functions. A system may be called *spatially homogeneous* if all possible joint probability density functions are the same regardless of offsets in space. For example, for any $\Delta \boldsymbol{r}$:

$$f(v_i, v_j; \boldsymbol{r}_1 + \Delta\boldsymbol{r}, \boldsymbol{r}_2 + \Delta\boldsymbol{r}, t) = f(v_i, v_j; \boldsymbol{r}_1, \boldsymbol{r}_2, t). \tag{2.16}$$

A system may be called *isotropic* if all possible joint probability density functions are the same regardless of rotations or reflections of the coordinate system.

In the remainder of the book, the exact meaning of "homogeneous" and "isotropic" will generally depend on the context. For example:

- When the topic is particle size, the terms "homogeneous" and "isotropic" will refer to scalar metrics for particle size such as diameter, area, and mass.
- When the topic is particle shape, the terms "homogeneous" and "isotropic" will refer scalar metrics for particle shape such as aspect ratios.
- When the topic is particle velocity, the terms "homogeneous" and "isotropic" will refer to the velocity vector.

In unitary fragmentation, before multiple particles form, the terms "homogeneous" and "isotropic" refer to material properties such as strength, or loading properties such as stress, which ultimately determine the particle properties. Where convenient, this book will freely assume that *all* of the relevant properties in a given system are homogeneous and isotropic.

The remainder of this book will usually describe temporally homogeneous properties by one-point probability density functions, i.e., $f(x; \boldsymbol{r})$ for scalars or $f(v_i; \boldsymbol{r})$ for vectors. Similarly, the remainder of this book will usually describe spatially homogeneous and isotropic properties by one-time probability density functions, i.e., $f(x; t)$ for scalars or $f(v_i; t)$ for vectors. Finally, the remainder of this book will usually describe temporally homogeneous, spatially homogeneous, and isotropic properties by plain probability density functions, i.e., $f(x)$ for scalars or $f(v_i)$ for vectors as discussed earlier in Section 1.4.

2.6 Conclusions

This chapter introduced four common properties of aggregating and fragmenting particle systems, namely, chaos, self-organization, homogeneity, and isotropy. While the other properties are optional, this book generally assumes that particle systems are chaotic, unless specifically stated otherwise. As discussed earlier, while fragmentation naturally tends to increase chaos, aggregation naturally tends to decrease chaos. However, aggregating particle systems still commonly obtain chaos depending on factors such as the following.

- *Initial conditions.* These include the degree of randomness in the initial particle positions, masses, momenta, energies, etc.
- *External conditions.* These include the degree of randomness introduced by external drivers such as vibration, rotation, shaking, stirring, swirling, heating, etc.
- *Volumetric fraction of particles.* This is preferably large enough to allow frequent collisions but small enough to ensure that particles do not remain in extended contact.

This book does not necessarily assume homogeneity and isotropy. However, as it turns out, the same things listed previously that promote chaos—initial conditions, external conditions, and frequent elastic or partially elastic collisions—also tend to promote homogeneity and isotropy. From one point of view, this is because they promote mixing. In fact, homogeneous and isotropic systems are sometimes called *well-mixed*; see, e.g., Krapivksy (2015) and Matsoukas (2016). While aggregation naturally tends to increase homogeneity and isotropy, fragmentation naturally tends to decrease homogeneity and isotropy. This is because each fragmentation event creates a set of particles in a local region whose properties differ substantially from the other particles in the system. When they occur, homogeneity and isotropy tend to lead to universal distributions or spectra for size, shape, velocity, kinetic energy, etc. As a shorthand, this book sometimes says that such cases are *ideal*, while all other cases are *non-ideal*.

References

S. R. Ahmad & S. Puri, Velocity Distributions and Aging in a Cooling Granular Gas, *Physical Review E*, 75, 031302, 2007, https://doi.org/10.1103/PhysRevE.75.031302

M. Aschwanden, *Self-Organized Criticality in Astrophysics: The Statistics of Nonlinear Processes in the Universe*, Springer, 2011, https://doi.org/10.1007/978-3-642-15001-2

M. J. Aschwanden, N. B. Crosby, M. Dimitropoulou, M. K. Georgoulis, S. Hergarten, J. McAteer, et al., 25 Years of Self-Organized Criticality: Solar and Astrophysics, *Space Science Reviews*, 198(1), 47–166, 2016, https://doi.org/10.1007/s11214-014-0054-6

M. J. Aschwanden, F. Scholkmann, W. Béthune, W. Schmutz, V. Abramenko, M. C. M. Cheung, et al., Order out of Randomness: Self-Organization Processes in Astrophysics, *Space Science Reviews*, 214, 55, 2018, https://doi.org/10.1007/s11214-018-0489-2

W. R. Ashby, Principles of the Self-Organizing Dynamic System, *Journal of General Psychology*, 37(2), 125–128, 1947, https://doi.org/10.1080/00221309.1947.9918144

W. R. Ashby, Principles of the Self-Organizing System, in H. Von Foerster & G. W. Zopf, Jr., editors, *Principles of Self-Organization*, Pergamon Press, 1962

P. Bak, C. Tang, & K. Wiesenfeld, Self-Organized Criticality: An Explanation of $1/f$ Noise, *Physical Review Letters*, 59(4), 381–384, 1987, https://doi.org/10.1103/PhysRevLett.59.381

P. Bak, C. Tang, & K. Wiesenfeld, Self-Organized Criticality, *Physical Review A*, 38(1), 364–374, 1988, https://doi.org/10.1103/PhysRevA.38.364

P. Bak, *How Nature Works: The Science of Self-Organized Criticality*, Springer, 1996, https://doi.org/10.1007/978-1-4757-5426-1

G. Battimelli & G. Ciccotti, Berni Alder and the Pioneering Times of Molecular Simulation, *The European Physical Journal H*, 43(3), 303–335, 2018, https://doi.org/10.1140/epjh/e2018-90027-5

G. W. Baxter & J. S. Olafsen, Experimental Evidence for Molecular Chaos in Granular Gases, *Physical Review Letters*, 99, 028001, 2007, https://doi.org/10.1103/PhysRevLett.99.028001

X. Bian, C. Kim, & G. E. Karniadakis, 111 Years of Brownian Motion, *Soft Matter*, 12(30), 6331–6346, 2016, https://doi.org/10.1039/c6sm01153e

N. Boers & P. Pickl, On Mean Field Limits for Dynamical Systems, *Journal of Statistical Physics*, 164(1), 1–16, 2016, https://doi.org/10.1007/s10955-015-1351-5

L. Boltzmann, Weitere Studien über das Wärmegleichgewicht unter Gasmolekülen, *Sitzungsberichte Akademie der Wissenschaften*, 66, 275–370, 1872

J.-P. Bouchaud & P. Le Doussal, Numerical Study of a D-Dimensional Periodic Lorenz Gas with Universal Properties, *Journal of Statistical Physics*, 41(1–2), 225–248, 1985, https://doi.org/10.1007/BF01020610

N. V. Brilliantov & T. Pöschel, *Kinetic Theory of Granular Gases*, Oxford University Press, 2004, https://doi.org/10.1093/acprof:oso/9780198530381.001.0001

C. Brown, *Chaos and Catastrophe Theories*, Sage Publications, 1995, https://doi.org/10.4135/9781412983624

A. Burdeau & P. Viot, Quasi-Gaussian Velocity Distribution of a Vibrated Granular Bilayer System, *Physical Review E*, 79, 061306, 2009, https://doi.org/10.1103/PhysRevE.79.061306

F. Cecconi, M. Cencini, & A. Vulpiani, Transport Properties of Chaotic and Non-Chaotic Many Particle Systems, *Journal of Statistical Mechanics: Theory and Experiment*, P12001, 2007, https://doi.org/10.1088/1742-5468/2007/12/P12001

A. D. Chernin & M. J. Valtonen, Intermittent Chaos in Three-Body Dynamics, *New Astronomy Reviews*, 42(1), 41–66, 1998, https://doi.org/10.1016/S1387-6473(98)00021-9

V. S. Deshpande, A. Needleman, & E. Van der Giessen, Dislocation Dynamics is Chaotic, *Scripta Materialia*, 45(9), 1047–1053, 2001, https://doi.org/10.1016/S1359-6462(01)01135-6

J. R. Dorfman, *An Introduction to Chaos in Nonequilbrium Statistical Mechanics*, Cambridge University Press, 1999, https://doi.org/10.1017/CBO9780511628870.001

W. Ebeling & R. Feistel, Zur Kinetik molekularer Replikationsprozesse mit Selektionscharakter [On the Kinetics of Molecular Replication Processes with Selection], *Studia Biophysica*, 46(3), 183–195, 1974

W. Ebeling & R. Mahnke, Kinetics of Molecular Replication and Selection, *Zagadnienia Biofizyki Współczesnej [Problems in Contemporary Biophysics]*, 4, 119–128, 1979

J.-P. Eckmann, C. Forster, H. A. Posch, & E. Zabey, Lyapunov Modes in Hard-Disk Systems, *Journal of Statistical Physics*, 118(5–6), 813–847, 2005, https://doi.org/10.1007/s10955-004-2687-4

O. J. Eder, The Velocity Autocorrelation Function and the Diffusion Coefficient for a Dilute Hard Sphere Gas, *The Journal of Chemical Physics*, 66(9), 3866–3869, 1977, https://doi.org/10.1063/1.434461

M. Eigen, Self-Organization of Matter and the Evolution of Biological Macromolecules, *Die Naturwissenschaften*, 58(10), 465–523, 1971a, https://doi.org/10.1007/BF00623322

M. Eigen, Molecular Self-Organization and the Early Stages of Evolution, *Quarterly Review of Biophysics*, 4(2–3), 149–212, 1971b, https://doi.org/10.1017/S0033583500000627

M. Eigen & P. Schuster, *The Hypercycle: A Principle of Natural Self-Organization*, Springer, 1979

I. R. Epstein & M. Eigen, Selection and Self-Organization of Self-Reproducing Macromolecules under the Constraint of Constant Flux, *Biophysical Chemistry*, 10(2), 153–160, 1979, https://doi.org/10.1016/0301-4622(79)85035-8

R. Feistel & W. Ebeling, *Physics of Self-Organization and Evolution*, Wiley, 2011, https://doi.org/10.1002/9783527636792

C. F. C. M. Fong & D. D. Kee, *Perturbation Methods, Instability, Catastrophe and Chaos*, World Scientific, 1999, https://doi.org/10.1142/3989

S. K. Friedlander, *Smoke, Dust and Haze*, 2nd edition, Oxford University Press, 2000

P. Gaspard & H. van Beijeren, When do Tracer Particles Dominate the Lyapunov Spectrum?, *Journal of Statistical Physics*, 109(3–4), 671–704, 2002, https://doi.org/10.1023/A:1020466732292

F. Ginot, I. Theurkauff, F. Detcheverry, C. Ybert, & C. Cottin-Bizonne, Aggregation-Fragmentation and Individual Dynamics of Active Clusters, *Nature Communications*, 9, 696, 2018, https://doi.org/10.1038/s41467-017-02625-7

P. Glansdorff & I. Prigogine, *Thermodynamics of Structure, Stability and Fluctuations*, Wiley, 1971

I. Goldhirsch, Scales and Kinetics of Granular Flows, *Chaos*, 9(3), 659–672, 1999, https://doi.org/10.1063/1.166440

H. Haken & R. Graham, Synergetik: Die Lehre vom Zusammenwirken in Wissenschaft und Technik, *Umschau*, 6, 191–195, 1971

H. Haken, *Synergetics: An Introduction: Nonequilibrium Phase Transitions and Self-Organization in Physics, Chemistry, and Biology*, Springer, 1977, https://doi.org/10.1007/978-3-642-96469-5

R. C. Hilborn, *Chaos and Nonlinear Dynamics: An Introduction for Scientists and Engineers*, 2nd edition, Oxford University Press, 2000, https://10.1093/acprof:oso/9780198507239.001.0001

A. Hübler & J. P. Crutchfield, Order and Disorder in Open Systems, *Complexity*, 16(1), 6–9, 2010, https://doi.org/10.1002/cplx.20344

M. Hummel, J. P. D. Clewett, & M. G. Mazza, A Universal Scaling Law for the Evolution of Granular Gases, *Europhysics Letters*, 114(1), 10002, 2016, https://doi.org/10.1209/0295-5075/114/10002

J. H. Jeans, *The Dynamical Theory of Gas*, Cambridge University Press, 1906, https://doi.org/10.1017/CBO9780511694370

J. K. Jun & A. H. Hübler, Formation and Structure of Ramified Charge Transportation Networks in an Electromechanical System, *Proceedings of the National Academy of Sciences*, 102(3), 536–540, 2005, https://doi.org/10.1073/pnas.0406025102

K. Kaneko & I. Tsuda, *Complex Systems: Chaos and Beyond: A Constructive Approach with Applications in Life Sciences*, Springer, 2001, https://doi.org/10.1007/978-3-642-56861-9

R. Kautz, *Chaos: The Science of Predictable Random Motion*, Oxford University Press, 2011

P. L. Krapivsky, Aggregation Driven by a Localized Source, *Journal of Physics A: Mathematical and Theoretical*, 48, 245002, 2015, https://doi.org/10.1088/1751-8113/48/24/245002

S. Levy & J. F. Molinari, Dynamic Fragmentation of Ceramics, Signature of Defects and Scaling of Fragment Sizes, *Journal of the Mechanics and Physics of Solids*, 58(1), 12–26, 2010, https://doi.org/10.1016/j.jmps.2009.09.002

T.-Y. Li & J. A. Yorke, Period Three Implies Chaos, *The American Mathematical Monthly*, 82(1), 985–992, 1975, https://doi.org/10.1080/00029890.1975.11994008

E. N. Lorenz, Deterministic Nonperiodic Flow, *Journal of Atmospheric Sciences*, 20(2), 130–141, 1963, https://doi.org/10.1175/1520-0469(1963)020<0130:DNF>2.0.CO;2

T. Matsoukas, Thermodynamics of Irreversible Aggregation, in M. Papadrakakis, V. Papadopoulos, G. Stefanou, & V. Plevris, editors, *Proceedings of the VII European Congress on Computational Methods in Applied Sciences and Engineering*, ECCOMAS, 2016, https://doi.org/10.7712/100016.1797.10608

J. C. Maxwell, *Essay for the Eranus Club on Science and Free Will*, 1873; see L. Campbell & W. Garnett, *The Life of James Clerk Maxwell*, Cambridge University Press, 1882, https://doi.org/10.1017/CBO9780511709050

S. McNamara & M. Mareschal, Lyapunov Spectrum of Granular Gases, *Physical Review E*, 63, 061306, 2001, https://doi.org/10.1103/PhysRevE.63.061306

I. Miklashevich, Crack Trajectory Instability: Propagation in Inhomogeneous Medium, *Theoretical and Applied Fracture Mechanics*, 43(3), 360–368, 2005, https://doi.org/10.1016/j.tafmec.2005.03.006

J. Mo & M. G. Raizen, Highly Resolved Brownian Motion in Space and in Time, *Annual Review of Fluid Mechanics*, 51, 403–428, 2019, https://doi.org/10.1146/annurev-fluid-010518-040527

G. Nicolis & I. Prigogine, Fluctuations in Nonequilibrium Systems, *Proceedings of the National Academy of Sciences*, 68(9), 2102–2017, 1971, https://doi.org/10.1073/pnas.68.9.2102

G. Nicolis & I. Prigogine, *Self-Organization in Nonequilibrium Systems: From Dissipative Structures to Order through Fluctuations*, Wiley, 1977

L. Oddershede, P. Dimon, & J. Bohr, Self-Organized Criticality in Fragmenting, *Physical Review Letters*, 71(19), 3107–3110, 1993, https://doi.org/10.1103/PhysRevLett.71.3107

N. Oyama, H. Mizuno, & K. Saitoh, Avalanche Interpretation of the Power-Law Energy Spectrum in Three-Dimensional Dense Granular Flow, *Physical Review Letters*, 122, 188004, 2019, https://doi.org/10.1103/PhysRevLett.122.188004

G. Pask, *An Approach to Cybernetics*, Hutchinson, 1961

G. Pask, A Discussion of Artificial Intelligence and Self-Organization, *Advances in Computers*, 5, 109–226, 1964, https://doi.org/10.1016/S0065-2458(08)60353-8

H. Poincaré, Le Hasard (Chance), *Le Revue du Mois*, (3), 257–276, 1907; see also J. Newman, editor, *The World of Mathematics, Volume 2*, Dover Publications, 1956

L. Ponson, Statistical Aspects in Crack Growth Phenomena: How the Fluctuations Reveal the Failure Mechanisms, *International Journal of Fracture*, 201(1), 11–27, 2016, https://doi.org/10.1007/s10704-016-0117-7

S. B. Pope, *Turbulent Flows*, Cambridge University Press, 2000, https://doi.org/10.1017/CBO9780511840531.001

T. Poston & I. Stewart, *Catastrophe Theory and its Applications*, Pittman, 1978

I. Prigogine, *Introduction to Thermodynamics of Irreversible Processes*, 1st edition, Charles C. Thompson, 1955; 2nd edition, Wiley, 1961; 3rd edition, Wiley 1968

I. Prigogine, Dissipative Structures, Dynamics, and Entropy, *International Journal of Quantum Chemistry*, 9(S9), 443–456, 1975, https://doi.org/10.1002/qua.560090854

I. Prigogine, Order Through Fluctuation: Self-organization and Social System, in E. Jantsch, editor, *Evolution and Consciousness: Human Systems in Transition*, Addison-Wesley, 1976

I. Prigogine & G. Nicolis, Biological Order, Structure, and Instabilities, *Quarterly Reviews of Biophysics*, 4(2–3), 107–148, 1971, https://doi.org/10.1017/S0033583500000615

I. Prigogine & R. Lefever, Stability and Self-Organization in Open Systems, in G. Nicolis & R. Lefever, editors, *Membranes, Dissipative Structures, and Evolution*, Wiley, 1975, https://doi.org/10.1002/9780470143810.ch1

I. Prigogine & I. Stengers, *Order out of Chaos: Man's New Dialog with Nature*, Bantam Books, 1984

G. Pruessner, *Self-Organized Criticality: Theory, Models, and Characterisation*, Cambridge University Press, 2012, https://doi.org/10.1017/CBO9780511977671

P. Ray, Statistical Physics Perspective of Fracture in Brittle and Quasi-Brittle Materials, *Philosophical Transactions of the Royal Society A: Mathematical, Physical and Engineering Sciences*, 377(2136), 20170396, 2018, https://doi.org/10.1098/rsta.2017.0396

P. T. Saunders, *An Introduction to Catastrophe Theory*, Cambridge University Press, 1980, https://doi.org/10.1017/CBO9781139171533

P. T. Saunders & M. W. Ho, Self-Organization, Catastrophe Theory and the Problem of Segmentation, in R. K. Mishra, D. Maaß, & E. Zwierlein, editors, *On Self-Organization: An Interdisciplinary Search for a Unifying Principle*, Springer-Verlag, 1994, https://doi.org/10.1007/978-3-642-45726-5_10

H. G. Schuster & W. Just, *Deterministic Chaos: An Introduction*, 4th edition, Wiley, 2005, https://doi.org/10.1002/3527604804

R. Soto, *Kinetic Theory and Transport Phenomena*, Oxford University Press, 2016, https://doi.org/10.1093/acprof:oso/9780198716051.001.0001

J. Tame, *Approaches to Entropy*, Springer, 2019, https://doi.org/10.1007/978-981-13-2315-7

M. Theers, E. Westphal, G. Gompper, & R. G. Winkler, From Local to Hydrodynamic Friction in Brownian Motion: A Multiparticle Collision Dynamics Simulation Study, *Physical Review E*, 93, 032604, 2016, https://doi.org/10.1103/PhysRevE.93.032604

R. Thom, *Stabilité Structurelle et Morphogénèse; Essai d'une Théorie Générale des Modèles*, W. A. Benjamin, 1972. See also R. Thom, *Structural Stability and Morphogenesis: An Outline of a General Theory of Models*, CRC Press, 1989, https://doi.org/10.1201/9780429493027

L. von Bertalanffy, Basic Concepts in Quantitative Biology of Metabolism, *Helgoländer Wissenschaftliche Meeresuntersuchungen*, 9(1–4), 5–37, 1964, https://doi.org/10.1007/BF01610024

L. von Bertalanffy, *General System Theory: Foundations, Development, Applications*, George Braziller, 1968

H. von Foerster, On Self-Organizing Systems and Their Environments, in M. C. Yovits & S. Cameron, editors, *Self-Organizing Systems*, Pergamon Press, 1960

N. Wiener, *Cybernetics: Or Control and Communication in the Animal and the Machine*, 2nd edition, MIT Press, 1961

N. Wiener, *The Mathematics of Self-Organizing Systems: Recent Developments in Information and Decision Processes*, Macmillan, 1962

C. R. K. Windows-Yule & D. J. Parker, Boltzmann Statistics in a Three-Dimensional Vibrofluidized Granular Bed: Idealizing the Experimental System, *Physical Review E*, 87, 022211, 2013, https://doi.org/10.1103/PhysRevE.87.022211

E. C. Zeeman, Catastrophe Theory, *Scientific American*, 234(4), 65–83, 1976, https://doi.org/10.1038/scientificamerican0476-65

W. J. Zhang, *Selforganizology: The Science of Self-Organization*, World Scientific, 2016, https://doi.org/10.1142/9685

S. J. Zhou, P. S. Lomdahl, R. Thomson, & B. L. Holian, Dynamic Crack Processes via Molecular Dynamics, *Physical Review Letters*, 76(13), 2318–2321, 1996, https://doi.org/10.1103/PhysRevLett.76.2318

3
Distributions and Averages

This chapter concerns probabilistic distributions and averages. These may be used to describe particle sizes, speeds, shapes, and other properties of aggregating and fragmenting particle systems. Most of the specific examples given in this chapter involve size distributions. This is because aggregation and fragmentation research traditionally focuses on particle size distributions. In addition, this is because particle size distributions pose special challenges not faced by other distributions. In particular, it is often necessary to transform between different metrics of particle size such as mass and diameter. By contrast, there is generally no need to transform between different metrics of particle speed, shape, or other properties.

3.1 Number Distributions

Let x refer to any non-negative scalar particle property such as mass, diameter, speed, and so forth. Assume x is statistically independent of other properties. Suppose that N random samples of x are taken from a particle system and sorted into non-overlapping bins $[x_{i\,\mathrm{min}}, x_{i\,\mathrm{max}})$, where:

$$0 \le x_{1\,\mathrm{min}} < x_{1\,\mathrm{max}} = x_{2\,\mathrm{min}} < x_{2\,\mathrm{max}} = x_{3\,\mathrm{min}} < \cdots < x_{N_{\mathrm{bins}}\,\mathrm{max}} < \infty.$$

For the time being, for simplicity, assume the particle system is temporally homogeneous, spatially homogenous, and isotropic so that there is no need to account for when or where the samples were taken. Define the *number distribution* $n(x)$ such that:

$$\int_{x_{i\,\mathrm{min}}}^{x_{i\,\mathrm{max}}} n(x)dx \tag{3.1}$$

is the number of samples with $x_{i\,\mathrm{min}} \le x < x_{i\,\mathrm{max}}$. Then:

$$N = \int_{0}^{\infty} n(x)dx \tag{3.2}$$

is the total number of samples. If the samples are chosen in an unbiased fashion and the number distribution is reasonably smooth, Equations (3.1) and (3.2) uniquely define $n(x)$ in the limit as $N \to \infty$, $N_{\mathrm{bins}} \to \infty$, and $x_{i\,\mathrm{max}} - x_{i\,\mathrm{min}} \to 0$. As seen in Examples 3.1 and 3.2, counting-based measurement techniques naturally yield number distributions. If a particle

Particle Dynamics with Aggregation and Fragmentation. Culbert B. Laney, Oxford University Press. © Culbert B. Laney (2025).
DOI: 10.1093/9780198920687.003.0003

system is spatially homogeneous and isotropic, but not temporally homogeneous, then N samples taken at time t lead to $n(x, t)$. Similarly, if a particle system is temporally homogeneous, but not spatially homogeneous and isotropic, then N samples taken at position $\boldsymbol{r}$ lead to $n(x, \boldsymbol{r})$. In the most general case, N sample particles taken at time t and position $\boldsymbol{r}$ lead to $n(x, \boldsymbol{r}, t)$. A single experiment can obtain, at most, one particle at a given location $\boldsymbol{r}$ and time t. Thus, to obtain N samples, it is necessary to repeat the same experiment at least N times. Such a collection of identical experiments with random or seemingly random outcomes is called an *ensemble*.

Example 3.1 Physical collection

As one common way of measuring size distributions, *physical collection* refers to any measurement technique that involves physically removing samples from a particle system. Possible collection techniques include: drawing particles into a tube or pipette; allowing slowly moving particles to settle onto a horizontal witness plate under the influence of gravity; allowing rapidly moving particles to impact a vertical witness plate; and capturing the particles with panels or boxes, including those internally lined with a soft fabric or foam to prevent secondary breakup. In the latter case, as a high-accuracy and high-cost option, each collected particle can be measured individually. As a low-accuracy and low-cost option, the collected particles can be bulk sieved; see Example 3.2.

As one advantage, physical collection is flexible. For intact solid particles, sample measurements can be done at any time—even weeks, months, or years after collection. This allows for iterative improvements to, and easy comparisons between, different measurement techniques. As one disadvantage, physical collection does not work well for fragile particles, such as heavily fractured solids, which may not survive even the gentlest collection and measurement techniques. As another disadvantage, physical collection only allows for measuring particle shapes and sizes. In most cases, information about the particle positions, velocities, and rotations is lost.

Example 3.2 Sieving techniques for estimating number-weighted size distributions

Following physical collection, solid particle samples may be passed through a series of sieves. In this approach, the coarsest sieve captures the largest particles, the next coarsest sieve captures the next largest particles, and so forth. The known sizes of the meshes bound the possible sizes of the particles trapped in each sieve. Sieving can be used to measure particles diameters as small as 1 μm; see, e.g., Kawara et al. (2016). For diameters less than 100 μm, the solid particles are generally mixed with a liquid, whose flow forces the particles through the sieves. The smaller the particles, the more rapid the flow required. However, for diameters greater than 100 μm, solid particles will generally pass through sieves due to shaking, tapping, vibration, rotation, etc. without the assistance of a liquid flow; see Figure 3.1a. The contents of each sieve are often stored in size bins; see Figure 3.1b.

As one possibility, suppose the number of particles in each size bin is counted. If there are a small number of particles, they can be counted by hand. However, if there are a large number of particles, counting is generally done using automated procedures and/or specialized equipment; see, e.g., Figure 3.1c. The average mass of the particles in each size bin can be determined by dividing the total mass of the particles in each size bin by the number of particles in each size bin. For example, Figure 3.2 shows a mass histogram, i.e., the number of particles in each size bin vs. the average weight of the particles in each size bin. Curve fitting this size histogram yields an estimate for $n(M)$.

While sieving is meant to sort particles by size, it also partially sorts particles by shape. More specifically, the probability that a given particle passes through a given mesh depends partly on its aspect ratio, i.e., the largest linear dimension divided by the smallest linear dimension. In this regard, Matthews (1991) says the following:

> **Particles whose . . . [aspect] ratios approach 1.0 are . . . more likely to pass a given sieve than those with . . .[higher aspect] ratios. The probability of a particle being retained on a sieve whose size is larger than the true sieve size of the particle is, therefore, dependent on the particle shape and the ability of the particle to achieve the proper orientation for passage during the times allotted for sieving . . . Sieve size is therefore a combined measure of the least cross-sectional area and maximum axial length, further complicated by orientation effect . . . The combination of an area measurement with a probability-related length measure leaves us with a 2.5-dimensional measure, not quite as good as a three-dimensional measure, but better than a simple one- or two-dimensional measure . . .**

To overcome this, the same procedures and equipment used to count the number of particles may be used to characterize the sizes and shapes of each particle; see Figure 3.1c. As a limitation, just as particles must survive physical collection, particles must also survive measurement. In the case of unitary fragmentation, any event violent enough to break a solid object into thousands of small pieces may also be violent enough to severely weaken those small pieces. For example, Sommer et al. (2013) report that physically collected sandstone fragments suffered from cracks, splits, and other damage. As a result, "despite careful handling during measurement of the . . . size distribution, the . . . material tends to [further] fragment, which causes an increase in the fraction of the finer particles."

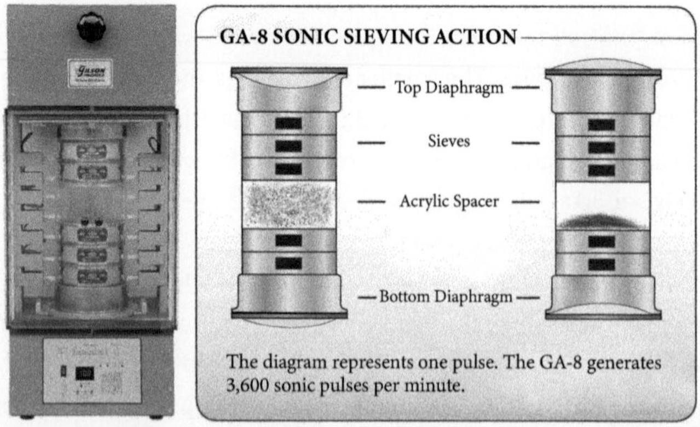

Figure 3.1a In a sifter, particles pass through a stack of progressively finer sieves. On the left is a photo of a Gilson GA-8 Gilsonic UltraSiever® Sonic Sifter. On the right is an illustration showing how this sifter operates. Sifting reduces the errors that result from estimating three-dimensional particle sizes from two-dimensional cross sections, as seen in Figure 3.1c.
Figure courtesy of Gilson Company Inc.

Figure 3.1b Particles trapped in each sieve are stored in plastic bins. The size of the particles held in each bin increases from front to back.
Figure courtesy of Protection Engineering Consultants (PEC).

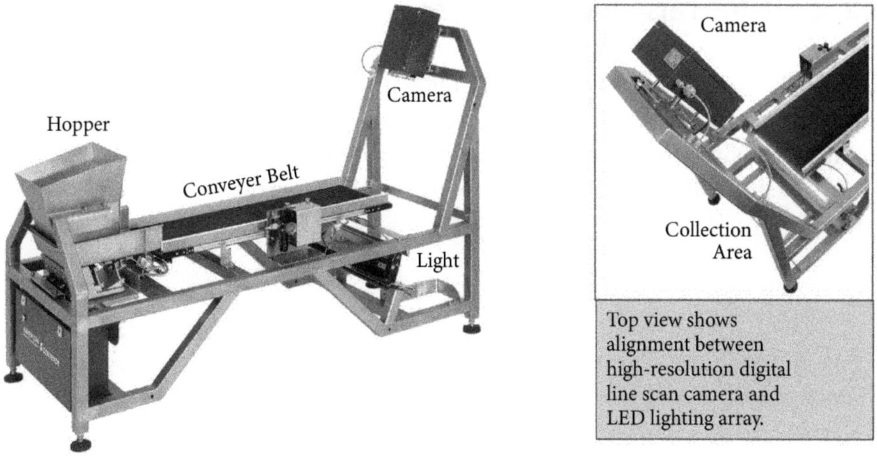

Figure 3.1c W. S. Tyler Computerized Particle Analyzer (CPA-4). To count the number of particles, the contents of each bin are placed in a hopper and transported a few at a time down a conveyer belt. A high-resolution digital camera located at the end of the conveyer belt records each particle. In addition to counting particles, this system can also characterize particle sizes and shapes based on a cross-sectional view.
Figure courtesy of Nexopart GmbH & Co.

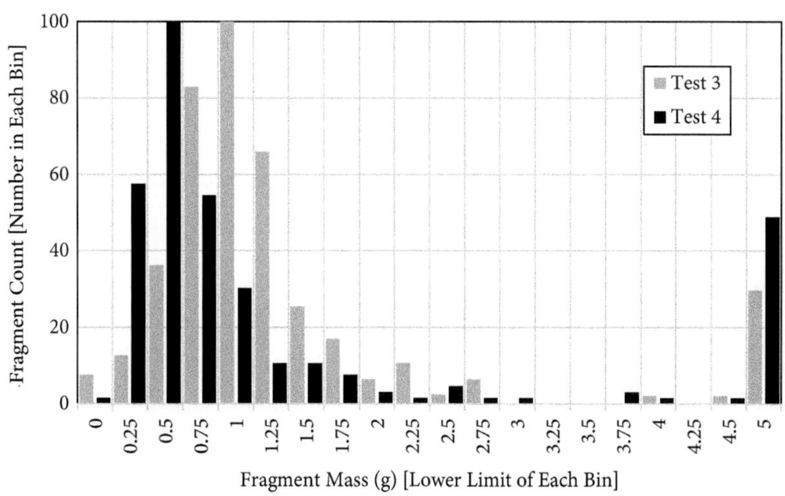

Figure 3.2 Number of particles in each mass bin. In this example, the particles are fragments from small masonry plates shattered by a shock tube. Test 3, shown in grey, involved lower incident overpressures and thus produced larger fragments on average than Test 4, shown in black.
Figure reprinted with permission from Bewick et al. (2017).

3.2 Concentration Distributions

Suppose that N samples of x are taken from a particle system. Furthermore, suppose the samples consist of *all* of the particles in a given spatial region of volume $\mathcal{V}$. If the particle system is temporally homogeneous, spatially homogenous, and isotropic, then the *concentration* (a.k.a., *number density*) *distribution* of the particles in the volume $\mathcal{V}$ may be defined as follows:

$$c(x) = \frac{1}{\mathcal{V}} n(x) \tag{3.3}$$

where $n(x)$ is the number distribution as defined in the last section. Notice that:

$$C = \frac{N}{\mathcal{V}} = \int_0^\infty c(x)dx \tag{3.4}$$

is the *total concentration* of particles in the volume $\mathcal{V}$. Similar definitions apply to $c(x, t)$, $c(x, \mathbf{r})$, $c(x, \mathbf{r}, t)$; see the discussion in the previous section. Because concentration helps determine collision rates, concentration is a natural choice for describing particle systems that evolve through collisions. At first glance, based on Equations (3.3) and (3.4), the concentration distribution appears to be just a trivial rescaling of the number distribution. However, in reality, n and c are substantially different. As one difference, n is an *extensive quantity*, e.g., it always depends on the sample size N. By contrast, c is an *intensive quantity*, e.g., it does not depend on N provided that N is large enough. As another difference, while n may be based on random partial samples, c must be based on the complete set of particles in a given volume $\mathcal{V}$. As a result, closely spaced particles obtain higher concentrations than widely spaced particles, even when the total number of particles is the same.

3.3 Probability Density Functions

Suppose that N samples of x are selected randomly from a given particle system. If the particle system is temporally homogeneous, spatially homogenous, and isotropic, the *probability density function* $f(x)$ may be defined such that:

$$\int_{x_{i\,\min}}^{x_{i\,\max}} f(x)dx \tag{3.5}$$

is the number fraction of samples with $x_{i\,\min} \leq x < x_{i\,\max}$. If the samples are chosen in an unbiased fashion and the probability density function is reasonably smooth, Equation (3.5) uniquely defines $f(x)$ in the limit as $N \to \infty$, $N_{\text{bins}} \to \infty$, and $x_{i\,\max} - x_{i\,\min} \to 0$. Alternatively, if Y is a particle property that depends on x, the Y-weighted probability density function $f_Y(x)$ may be defined such that:

$$\int_{x_{i\,\min}}^{x_{i\,\max}} f_Y(x)dx \tag{3.6}$$

is the Y fraction of samples with $x_{i\,\min} \leq x < x_{i\,\max}$. For example, suppose $x = D$ and $Y = M$, i.e., suppose $f_Y(x) = f_M(D)$. Then Equation (3.6) obtains the mass fraction of particles with diameters between $D_{i\,\min}$ and $D_{i\,\max}$. Example 3.3 shows that traditional sieving techniques lead naturally to estimates of $f_M(D)$. While preferable in theory, estimating $f(M)$ requires a substantial amount of extra effort to count, instead of just weighing, the particles in each size bin. In standard probability theory, $\int_{-\infty}^{\infty} f(x)dx = 1$. As a modified definition, this book requires:

$$\int_0^{\infty} f(x)dx = \int_0^{\infty} f_Y(x)dx = 1 \tag{3.7}$$

which simply says that all particles have sizes, speeds, etc. between 0 and ∞. Notice that f is related f_Y to as follows:

$$f_Y(x) = \frac{Y(x)f(x)}{\int_0^{\infty} Y(x')f(x')\,dx'}. \tag{3.8}$$

In addition, notice that f is related to n and c as follows:

$$f(x) = \frac{n(x)}{N} = \frac{c(x)}{C}. \tag{3.9}$$

Based on this expression, probability density functions may appear to be just trivial rescalings of number and concentration distributions. However, in reality, n, c, and f are substantially different from each other. For example, unlike n but like c, f is an intensive quantity, i.e., it does not depend on N for large enough N. For another example, unlike c but like n, f can be based on random samples, i.e., it does not require collecting a complete set of particles in a given volume. In that sense, f combines the best properties of n and c. As perhaps its greatest benefit, f is a part of standard probability theory; see, e.g., Ross (2018) and Rossi (2018). More specifically:

$$\int_{x_{i\,\min}}^{x_{i\,\max}} f(x)dx = P\left[x_{i\,\min} \leq X < x_{i\,\max}\right] \tag{3.10}$$

where $P\left[x_{i\,\min} \leq X < x_{i\,\max}\right]$ is the probability that a given random variable X is in the range $[x_{i\,\min}, x_{i\,\max})$. Thus, far from being a trivial renormalization of number or concentration distributions, probability density functions derive from the branch of mathematics used to describe and analyze random and seemingly random phenomena of all types.

The previous definitions assume that the particle system is temporally homogeneous, spatially homogenous, and isotropic. If a particle system is spatially homogeneous and isotropic, but not temporally homogeneous, then N samples taken at time t lead to $f(x, t)$ or $f_Y(x, t)$. Similarly, if a particle system is temporally homogeneous but not spatially homogeneous and isotropic, then N samples taken at position $\boldsymbol{r}$ lead to $f(x, \boldsymbol{r})$ or $f_Y(x, \boldsymbol{r})$. In the most general case, N samples taken at time t and position $\boldsymbol{r}$ lead to $f(x, \boldsymbol{r}, t)$ and $f_Y(x, \boldsymbol{r}, t)$. A single experiment can obtain, at most, one particle at a given location $\boldsymbol{r}$ and time t. Thus, to obtain N samples, it is necessary to repeat the same experiment at least N times. As noted earlier, such a collection of identical random or seemingly random experiments is called an *ensemble*. In this context, the term "experiment" may refer to an actual physical experiment. Alternatively, it may refer to a computation or any other way of realizing a specific random or seemingly random outcome.

3.4 Cumulative Distribution Functions

The definitions given so far require introducing bins. In some cases, the bin limits $x_{i\ min}$ and $x_{i\ max}$ are dictated by the measurement technique employed; see Example 3.2. However, in other cases, when information about each individual particle is available, the bin limits may be chosen arbitrarily. In theory, distributions are uniquely defined in the limit of infinitely many samples sorted into infinitely many bins with infinitesimal widths. However, for a finite number of samples sorted into a finite number of bins with finite widths, any specific arbitrary choice of bins may influence the results. As an alternative which avoids binning, the *cumulative distribution function* $F(x)$, is the number fraction of samples with values less than or equal to x. Like f, F is a standard part of probability and statistics. More specifically:

$$F(x) = P[X \leq x] \tag{3.11}$$

where $P[X \leq x]$ is the probability that a random variable X is less than or equal to x. As an alternative to $F(x)$, the Y-weighted cumulative distribution function $F_Y(x)$ is the Y fraction of samples with values less than or equal to x. Notice that:

$$F(0) = F_Y(0) = 0 \tag{3.12}$$

$$F(\infty) = F_Y(\infty) = 1. \tag{3.13}$$

These expressions simply say that all particles have values of x between 0 and ∞. In addition, notice that F and F_Y are related to f and f_Y as follows:

$$F(x) = \int_0^x f(x')\,dx' \qquad F_Y(x) = \int_0^x f_Y(x')\,dx' \tag{3.14}$$

and conversely:

$$f(x) = \frac{dF(x)}{dx} \qquad f_Y(x) = \frac{dF_Y(x)}{dx}. \tag{3.15}$$

The previous definitions assume that the particle system is temporally homogeneous, spatially homogenous, and isotropic. If this is not true, then it is possible to define $F(x, t)$, $F_Y(x, t)$, $F(x, \mathbf{r})$, $F_Y(x, \mathbf{r})$, $F(x, \mathbf{r}, t)$, and $F_Y(x, \mathbf{r}, t)$ as described in the last section. Example 3.3 shows that traditional sieving techniques lead naturally to estimates of $F_M(D)$. Example 3.4 argues that, in practice, most size distributions are only valid between some finite minimum and maximum.

The definitions of n, c, f, f_Y, F, F_Y given previously assume that x can assume any real value between 0 and ∞. However, in discrete systems, particle mass can only assume certain discrete values. As seen in Example 3.5, this sometimes allows for substantial simplifications in the descriptions of size distributions. Such simplifications do not apply to speed or any other particle property, which may assume any real value between 0 and ∞, regardless of whether the particle system is discrete or continuous.

Example 3.3 Sieving techniques for estimating mass-weighted size distributions

As Example 3.2 shows, sieving techniques can be used to estimate $f(M)$ or $F(M)$. However, such estimates require time-consuming manual procedures, or expensive hardware and software, to count the particles in each sieve. On the other hand, sieving techniques can be used to estimate $f_M(D)$ or $F_M(D)$ almost effortlessly. Instead of counting all of the particles in each sieve, it is only necessary to weigh all of the particles in each sieve. In this case, D corresponds to the known mesh size for each sieve. To explain further, let $\mathcal{M}(D)$ be the total mass of particles with diameters less than or equal to D. Then:

$$F_M(D) = \mathcal{M}(D)/M_{\text{total}} \tag{3.16}$$

where M_{total} is the total mass of all particles. Estimates for $F_M(D)$ can be converted to estimates for $F(M)$. However, this requires making potentially questionable assumptions about the shapes of the particles. Thus, estimates for $F(M)$ obtained by bulk sieving and weighing the particles, as described in this example, may not be as accurate as estimates obtained by counting and sizing the individual particles, as described in the last example.

Example 3.4 Issues with characterizing large and small particles

First, consider experiments with small numbers of particles, e.g., hundreds or thousands. In such cases, number-weighted distributions are clearly preferable to mass-weighted distributions. In general, the largest particles are severely under-sampled, i.e., they vary dramatically from experiment to experiment. However, because large particles constitute only a small fraction of the overall number, they do not substantially affect number-weighted curve fits. On the other hand, because the large particles constitute a large fraction of the overall mass, they substantially affect mass-weighted curve fits. Now, consider experiments with large numbers of particles, e.g., millions or billions. In general, the smallest particles are severely over-sampled, i.e., they constitute the vast majority of the particles. Because small particles only constitute a small fraction of the overall mass, they do not substantially affect mass-weighted curve fits. On the other hand, because the small particles constitute a large fraction of the overall number, they substantially affect number-weighted curve fits. The only solution is to count every particle, no matter how small, which may not always be possible. Thus, mass-weighted distributions are clearly preferable to number-weighted distributions. Regardless of the system size, mass-weighted curve fits can still suffer if the largest particles deviate from the smaller particles. As a simple fix, consider a curve fit $\mathcal{M}(D)$ based on the smaller particles. Assuming that the total mass M_{total} is known, extrapolating to the larger particles leads to the *size modulus* or the *100% passing diameter*:

$$\mathcal{M}(D_{100}) = M_{\text{total}}. \tag{3.17}$$

By this definition, D_{100} is what the largest particle diameter would be if all the particles had the same size distribution. This fix works for mass-weighted curve fits because it is generally possible to know the total mass of particles. However, this fix does not work for number-weighted curve fits because it is not generally possible to know the total number of particles.

Example 3.5 Size distributions for discrete particle systems

In the simplest case, the particle mass in a discrete particle system must be $M_1, 2M_1, 3M_1$, etc. where M_1 is the monomer mass. This allows for a simplified description of size distributions. In particular, suppose that N samples are taken from a discrete particle system that is temporally homogeneous, spatially homogenous, and isotropic.

1. *Number distributions*. Let n_i be the number of samples with mass iM_1. Then the total number of samples is:

$$N = \sum_{i=1}^{\infty} n_i. \tag{3.18}$$

2. *Concentration distributions*. If the samples include all of the particles in a given spatial region of volume $\mathcal{V}$, let $c_i = n_i/\mathcal{V}$. Then the total concentration is:

$$C = \frac{N}{\mathcal{V}} = \sum_{i=1}^{\infty} c_i. \tag{3.19}$$

3. *Fractions*. Let f_i be the number fraction of samples with mass iM_1, i.e., let $f_i = n_i/N$. Alternatively, let $f_{M,i}$ be the mass fraction of samples with mass iM_1, i.e., let:

$$f_{M,i} = \frac{in_i}{\sum_{j=1}^{\infty} jn_j}. \tag{3.20}$$

Then $\sum_{i=1}^{\infty} f_i = \sum_{i=1}^{\infty} f_{M,i} = 1$.

4. *Cumulative fractions*. Let F_i be the number fraction of samples with masses less than or equal to iM_1. Alternatively, let $F_{M,i}$ be the mass fraction of samples with masses less than or equal to iM_1. Then $F_0 = F_{M,0} = 0$ and $F_\infty = F_{M,\infty} = 1$. In addition:

$$F_i = \sum_{j=1}^{i} f_j \quad F_{M,i} = \sum_{j=1}^{i} f_{M,j} \tag{3.21}$$

$$f_i = F_i - F_{i-1} \quad f_{M,i} = F_{M,i} - F_{M,i-1}. \tag{3.22}$$

The previous expressions assume that the particle system is temporally homogeneous, spatially homogenous, and isotropic. If the particle system is spatially homogeneous and isotopic, but not temporally homogeneous, then N samples taken at time t lead to $n_i(t)$, $c_i(t)$, $f_i(t)$, and so forth. Similarly, if a particle system is temporally homogeneous but not spatially homogeneous and isotropic, then N samples taken at position $\boldsymbol{r}$ lead to $n_i(\boldsymbol{r})$, $c_i(\boldsymbol{r})$, $f_i(\boldsymbol{r})$, and so forth. In the most general case, N samples taken at time t and position $\boldsymbol{r}$ lead to $n_i(\boldsymbol{r}, t)$, $c_i(\boldsymbol{r}, t)$, $f_i(\boldsymbol{r}, t)$, and so forth.

3.5 Averages

Suppose that N samples of $X(t)$ are taken at a fixed time t over some volume $\mathcal{V}$. As one possible average:

$$X_{\text{avg}}(t) \cong \frac{1}{N}\sum_{k=1}^{N} x_k(t) \tag{3.23}$$

where $x_k(t)$ is the k-th sample of $X(t)$ taken at time t. This expression assumes that N is large enough that adding or subtracting a few particles does not significantly affect the average. However, if a given experiment does not have enough particles in a given volume $\mathcal{V}$, it may be necessary to repeat the same experiment over and over again until enough samples are obtained. In the limit of infinitely many samples, the *spatial average* $X_{\text{avg}}(t)$ is defined as follows:

$$X_{\text{avg}}(t) = \frac{1}{\mathcal{V}}\int_{\mathcal{V}}\int_{0}^{\infty} xf(x; \boldsymbol{r}, t)dx\, d\boldsymbol{r}. \tag{3.24}$$

Alternatively, suppose that N samples of $X(\boldsymbol{r})$ are taken at a given spatial location $\boldsymbol{r}$ over a time interval $[t_{\min}, t_{\max})$. As one possible average:

$$\bar{X}(\boldsymbol{r}) \cong \frac{1}{N}\sum_{k=1}^{N} x_k(\boldsymbol{r}) \tag{3.25}$$

where $x_k(\boldsymbol{r})$ is the k-th sample of $X(\boldsymbol{r})$ taken at location $\boldsymbol{r}$. This expression assumes that N is large enough that adding or subtracting a few particles does not significantly affect the average. However, if a given experiment does not have enough particles in a given time interval, it may be necessary to repeat the same experiment over and over again until enough samples are obtained. In the limit of infinitely many samples, the *temporal average* $\bar{X}(\mathbf{r})$ is defined as follows:

$$\bar{X}(\boldsymbol{r}) = \frac{1}{t_{\max} - t_{\min}} \int_{t_{\min}}^{t_{\max}} \int_{0}^{\infty} x f(x; \boldsymbol{r}, t) dx\, dt. \tag{3.26}$$

Finally, suppose that N samples of $X(\boldsymbol{r}, t)$ are taken at a given spatial location $\boldsymbol{r}$ and at a given time t. Then one possible average is defined as follows:

$$\langle X\rangle(\boldsymbol{r}, t) \cong \frac{1}{N}\sum_{k=1}^{N} x_k(\boldsymbol{r}, t) \tag{3.27}$$

where $x_k(\boldsymbol{r}, t)$ is the k-th sample of $X(\boldsymbol{r}, t)$ taken at location $\boldsymbol{r}$ and time t. A single experiment can obtain, at most, one particle at a given location $\boldsymbol{r}$ and time t. Thus, it is always necessary to repeat the same experiment over and over again until enough samples are obtained. As noted before, a collection of such experiments is called an *ensemble*. In the limit of infinitely many samples, the *ensemble average* $\langle X\rangle(\boldsymbol{r}, t)$ is defined as follows:

$$\langle X\rangle(\boldsymbol{r}, t) = \int_{0}^{\infty} x f(x; \boldsymbol{r}, t) dx. \tag{3.28}$$

If a particle system is temporally homogeneous, spatially homogeneous, and isotropic, then X_{avg}, $\bar{X}$, and $\langle X\rangle$ do not depend on $\boldsymbol{r}$ or t. In such cases:

$$\bar{X} = \langle X\rangle = X_{\text{avg}}. \tag{3.29}$$

In such cases, this book will treat all averages as spatial averages, even if the samples were taken at different times or in different experiments. Examples 3.6 and 3.7 compare number-weighted spatial averages X_{avg} to other possible spatial averages such as $X_{M,\text{avg}}$. Similar alternatives exist for number-weighted temporal averages $\bar{X}$ and number-weighted ensemble averages $\langle X\rangle$.

Example 3.6 Average diameters

Suppose that N samples with diameters D_k are taken from a particle system. For simplicity, assume the particle system is temporally homogeneous, spatially homogeneous, and isotropic so that there is no need to account for when or where the samples were taken. Some possible average diameters are as follows:

$$D_{\text{avg}} \cong \frac{1}{N}\sum_{k=1}^{N} D_k \tag{3.34}$$

$$\frac{1}{D'_{\text{avg}}} \cong \frac{1}{N}\sum_{k=1}^{N} \frac{1}{D_k} \tag{3.35}$$

$$D_{M,\text{avg}} \cong \frac{\sum_{k=1}^{N} M_k D_k}{\sum_{k=1}^{N} M_k} = \frac{\sum_{k=1}^{N} D_k^{m+1}}{\sum_{k=1}^{N} D_k^{m}} \tag{3.36}$$

$$\frac{1}{D'_{M,\text{avg}}} \cong \frac{\sum_{k=1}^{N} M_k / D_k}{\sum_{k=1}^{N} M_k} = \frac{\sum_{k=1}^{N} D_k^{m-1}}{\sum_{k=1}^{N} D_k^{m}} \tag{3.37}$$

where the last two equations assume $M = c_1 \rho D^m$. As seen in Table 3.1, in the limit of infinitely many samples, these equations can be rewritten in terms of $n(D)$, $c(D)$, $f(D)$, and $f_M(D)$. In practical applications, the equations given in Table 3.1 are not an especially efficient way to estimate averages. However, these equations are useful for other purposes and will be used throughout the book. For future reference, D_{avg} is commonly known as the *count mean diameter*, $D_{M,\text{avg}}$ is commonly known as the *mass mean diameter* and $D'_{M,\text{avg}}$ is commonly known as the *Sauter mean diameter*. The count mean diameter should not be confused with the count median diameter, defined as the diameter at which $F(D) = 0.5$. Similarly, the mass mean diameter should not be confused with the mass median diameter, defined as the diameter at which $F_M(D) = 0.5$. In most cases, this book uses mean rather than median diameters.

Table 3.1 Integrals of particle diameter distributions. All integrals are from 0 to ∞.

$\int c(D)dD = C$	$\int Dc(D)dD = CD_{\text{avg}}$	$\int \frac{c(D)}{D}dD = \frac{C}{D'_{\text{avg}}}$
$\int n(D)dD = N$	$\int Dn(D)dD = ND_{\text{avg}}$	$\int \frac{n(D)}{D}dD = \frac{N}{D'_{\text{avg}}}$
$\int f(D)dD = 1$	$\int Df(D)dD = D_{\text{avg}}$	$\int \frac{f(D)}{D}dD = \frac{1}{D'_{\text{avg}}}$
$\int f_M(D)dD = 1$	$\int Df_M(D)dD = D_{M,\text{avg}}$	$\int \frac{f_M(D)}{D}dD = \frac{1}{D'_{M,\text{avg}}}$

Example 3.7 Average masses

Suppose that N samples with masses M_k are taken from a particle system. For simplicity, assume the particle system is temporally homogeneous, spatially homogeneous, and isotropic so that there is no need to account for when or where the samples were taken. Some possible average masses are as follows:

$$M_{\text{avg}} = M'_{M,\,\text{avg}} \cong \frac{1}{N}\sum_{k=1}^{N} M_k \tag{3.38}$$

$$\frac{1}{M'_{\text{avg}}} \cong \frac{1}{N}\sum_{k=1}^{N}\frac{1}{M_k} \tag{3.39}$$

$$M_{M,\,\text{avg}} \cong \frac{\sum_{k=1}^{N} M_k^2}{\sum_{k=1}^{N} M_k} \tag{3.40}$$

As seen in Table 3.2, in the limit of infinitely many samples, these expressions can be rewritten in terms of size distributions such as $n(M)$, $c(M)$, $f(M)$, and $f_M(M)$. In practical applications, the equations given in Table 3.2 are not an especially efficient way to estimate averages. However, these equations are useful for other purposes and will be used throughout the book.

Table 3.2 Integrals of particle mass distributions. All integrals are from 0 to ∞.

$\int c(M)dM = C$	$\int Mc(M)dM = CM_{\text{avg}}$ $= M_{\text{total}}/\mathcal{V}^{o} = \rho_{\text{total}}$	$\int \frac{c(M)}{M}dM = \frac{C}{M'_{\text{avg}}}$
$\int n(M)dM = N$	$\int Mn(M)dX = NM_{\text{avg}}$ $= M_{\text{total}}$	$\int \frac{n(M)}{M}dM = \frac{N}{M'_{\text{avg}}}$
$\int f(M)dM = 1$	$\int Mf(M)dM = M_{\text{avg}}$ $= M_{\text{total}}/N$	$\int \frac{f(M)}{M}dM = \frac{1}{M'_{\text{avg}}}$
$\int f_M(M)dM = 1$	$\int Mf_M(M)dM = M_{M,\,\text{avg}}$	$\int \frac{f_M(M)}{M}dM = \frac{1}{M'_{M,\,\text{avg}}}$

3.6 Self-Similar Size Distributions

If a particle system is spatially homogeneous and isotropic, but not temporally homogeneous, size distributions can often be written in *scaled* or *self-similar forms* such as the following:

$$n(x,t) = \frac{N(t)}{X_{\text{char}}(t)}\tilde{n}\left[\frac{x}{X_{\text{char}}(t)}\right] \tag{3.30}$$

$$c(x,t) = \frac{C(t)}{X_{\text{char}}(t)}\tilde{c}\left[\frac{x}{X_{\text{char}}(t)}\right] \tag{3.31}$$

$$f(x,t) = \frac{1}{X_{\text{char}}(t)}\tilde{f}\left[\frac{x}{X_{\text{char}}(t)}\right]; \quad f_Y(x,t) = \frac{1}{X_{\text{char}}(t)}\tilde{f}_Y\left[\frac{x}{X_{\text{char}}(t)}\right] \tag{3.32}$$

$$F(x,t) = \tilde{F}\left[\frac{x}{X_{\text{char}}(t)}\right]; \quad F_Y(x,t) = \tilde{F}_Y\left[\frac{x}{X_{\text{char}}(t)}\right] \tag{3.33}$$

where $X_{\text{char}}(t)$ is a characteristic value such as a maximum, minimum, initial, or average value. If a particle system is temporally homogeneous, in addition to being spatially homogeneous and isotropic, the same equations apply except that the time dependencies drop out. The concept of self-similarity is closely related to the concept of universals. More specifically, if self-similarity applies, then $\tilde{n}$, $\tilde{c}$, $\tilde{f}$, $\tilde{f}_Y$, $\tilde{F}$, $\tilde{F}_Y$ may be universal functions. In such cases, $X_{\text{char}}(t)$, $N(t)$, and $C(t)$ account for all of the non-universal, problem-specific aspects of the solution, such as the effects of arbitrary initial conditions. Bennett (1936) first suggested self-similarity in the context of fragmentation. However, self-similarity occurs in a wide variety of other applications, not limited to particle aggregation and fragmentation; see, e.g., Embrechts & Maejima (2002), Barenblatt (1996, 2003) and Zohuri (2015).

3.7 Common Size Distributions

This section will introduce seven common families of size distributions, namely, Rosin–Rammler, Weibull, gamma, power law, exponential, log normal, and root normal size distributions. These same distributions may also apply to other particle properties such as speed and shape. For simplicity, this section will assume the following:

- The particle system is spatially homogeneous but *not* temporally homogeneous.
- The independent variable used to express the size distribution is either mass or diameter.
- The dependent variable used to express the size distribution is f or, in one case, f_M.

First, consider *Rosin–Rammler size distributions*. In self-similar form, these may be written as follows:

$$f(D,t) = \frac{A_D}{D_{\text{char}}(t)}\left[\frac{D}{D_{\text{char}}(t)}\right]^{n-1} \exp\left\{-b_D\left[\frac{D}{D_{\text{char}}(t)}\right]^{l}\right\} \tag{3.41a}$$

$$f(M,t) = \frac{A_M}{M_{\text{char}}(t)}\left[\frac{M}{M_{\text{char}}(t)}\right]^{n/m-1} \exp\left\{-b_M\left[\frac{M}{M_{\text{char}}(t)}\right]^{l/m}\right\}. \tag{3.41b}$$

In these expressions, A_D, A_M, b_D, and b_M are unitless constants. In addition, n, l, n/m, and l/m are unitless exponents. Finally, m is the particle mass dimension as defined in Equation (1.1). Rosin & Rammler (1927a, b) first suggested size distributions of this general form. As shown in Example 3.8, A_D and b_D generally depend on n and l. Similarly, A_M and b_M generally depend on n/m and l/m. Thus, as shown in Example 3.9, Rosin–Rammler size distributions generally have three free parameters, namely, the exponents n, l, and a characteristic size. Assume that $n \neq 1$ and $n/m \neq 1$. If $l = 0$, Rosin–Rammler size distributions become *power law size distributions*; see Section 1.4. On the other hand, if $l > 0$, Rosin–Rammler size distributions are power law size distributions with tapering applied to the largest particles. Similarly, if $l < 0$, Rosin–Rammler size distributions are power law size distributions with tapering applied to the smallest particles.

As a simpler alterative, if $l = n$, Rosin–Rammler size distributions become a type of *Weibull size distribution*. In self-similar form, these may be written as follows:

$$f(D,t) = \frac{A_D}{D_{\text{char}}(t)}\left[\frac{D}{D_{\text{char}}(t)}\right]^{n-1} \exp\left\{-b_D\left[\frac{D}{D_{\text{char}}(t)}\right]^{n}\right\} \tag{3.42a}$$

$$f(M,t) = \frac{A_M}{M_{\text{char}}(t)}\left[\frac{M}{M_{\text{char}}(t)}\right]^{n/m-1} \exp\left\{-b_M\left[\frac{M}{M_{\text{char}}(t)}\right]^{n/m}\right\}. \tag{3.42b}$$

Rosin, Rammler, & Sperling (1933) and Bennett (1936) first suggested size distributions of this general form; see also Weibull (1939a, b, 1951) and Mott & Linfoot (1943). As shown in Example 3.9, Weibull size distributions generally have two free parameters, namely, n and a characteristic size. Like Rosin–Rammler size distributions, Weibull size distributions are power law size distributions with tapering applied to the largest or smallest particles; see Figure 3.3.

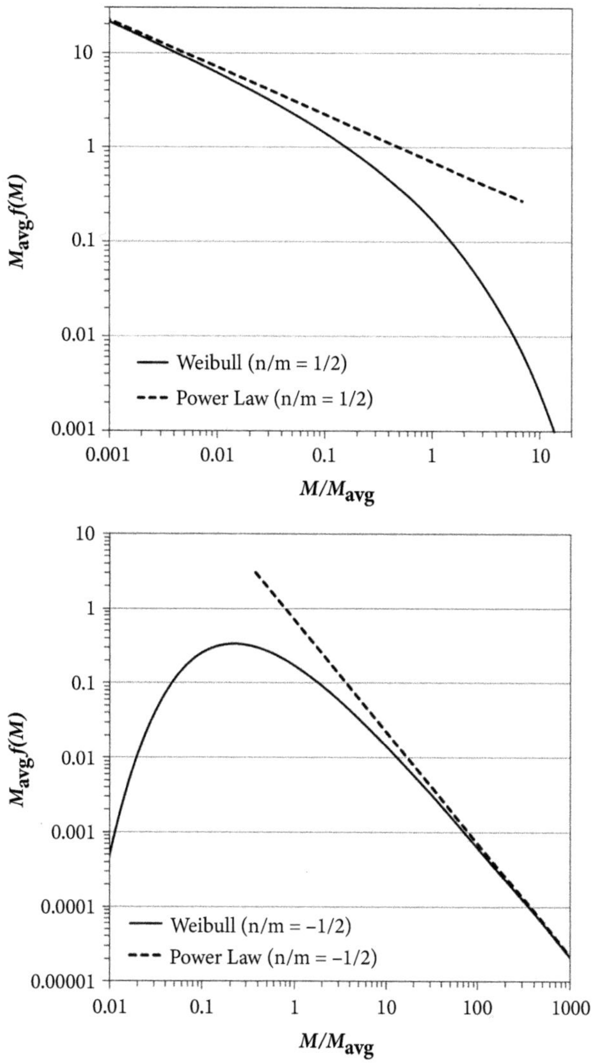

Figure 3.3 Weibull vs. power law size distributions for positive n (top) and negative n (bottom).

Example 3.8 Relationships between the parameters in Rosin–Rammler size distributions

First, consider Rosin–Rammler diameter distributions as given in Equation (3.41a). If $D_{char} = D_{avg}$ then by definition:

$$\int_0^\infty f(D,t)dD = 1$$

$$\int_0^\infty Df(D,t)dD = D_{avg}(t).$$

If $n > 0, l > 0$ or $n < 0, l < 0$, these imply the following:

$$A_D = \frac{|l| b_D^{n/l}}{\Gamma(n/l)} \tag{3.43}$$

$$b_D = \left\{ \frac{\Gamma[(n+1)/l]}{\Gamma(n/l)} \right\}^l \tag{3.44}$$

where $\Gamma(x)$ is the gamma function defined as follows:

$$\Gamma(x) = \int_0^\infty y^{x-1} \exp(-y)\, dy. \tag{3.45}$$

Similarly, consider Rosin–Rammler mass distributions as given in Equation (3.41b). If $M_{\text{char}} = M_{\text{avg}}$ then by definition:

$$\int_0^\infty f(M,t)dM = 1$$

$$\int_0^\infty Mf(M,t)dM = M_{\text{avg}}(t).$$

If $n > 0, l > 0$ or $n < 0, l < 0$, these imply the following:

$$A_M = \frac{|l| b_M^{n/l}}{m\Gamma(n/l)} \tag{3.46}$$

$$b_M = \left\{ \frac{\Gamma[(n+m)/l]}{\Gamma(n/l)} \right\}^{l/m}. \tag{3.47}$$

If Equations (3.43) to (3.47) apply, then Equation (3.41a) is exactly the same as Equation (3.41b). Appendix A provides many similar expressions for Rosin–Rammler size distributions including for cases where $D_{\text{char}} = D'_{\text{avg}}$ and $M_{\text{char}} = M'_{\text{avg}}$.

Example 3.9 Number of free parameters in Rosin–Rammler size distributions

So far, this chapter has assumed that there is an infinite range of particle sizes. Alternatively, suppose that a Rosin–Rammler size distribution applies only over a finite range such as:

$$M_{\min} \leq M \leq M_{\max}.$$

If $n > 0, l > 0$ or $n < 0, l < 0$, then $M_{\min}$ may be set equal to zero and $M_{\max}$ may be set equal to infinity, as seen in the last example. However, if $n < 0$ and $l \geq 0$, then $M_{\min}$ must be strictly greater than zero. Similarly, if $n > 0$ and $l \leq 0$, then M_{max} must be strictly less than infinity. At first glance, Equation (3.41b) involves up to seven free parameters, namely, A_M, b_M, l, n, M_{char}, M_{min}, and M_{max}. However, by definition:

$$\int_{M_{min}}^{M_{max}} f(M, t)dM = 1. \tag{3.48}$$

This means that A_M is not a free parameter. Instead, A_M is a function of the other parameters; see, e.g., Equation (3.46). In addition, suppose that $M_{char}(t) = M_{avg}(t)$. Then by definition:

$$\int_{M_{min}}^{M_{max}} Mf(M, t)dM = M_{avg}(t). \tag{3.49}$$

This means that b_M is not a free parameter. Instead, b_M is a function of the other parameters; see, e.g., Equation (3.47). Ideally, M_{min} and M_{max} should be chosen on a physical basis, i.e., they should reflect objective properties of the particle system. For example, M_{min} could be the monomer mass or the smallest mass emitted by a source in an aggregating particle system. If M_{min} and M_{max} reflect objective properties of the particle system, then they are not truly free parameters. This means there are only three free parameters, namely l, n, and M_{char}.

In addition to self-similar forms such as Equation (3.41), Rosin–Rammler size distributions can also be written in regular forms such as the following:

$$f(M, t) = \tilde{A}_M(t)M^{n/m-1}\exp\left[\tilde{b}_M(t)M^{l/m}\right]. \tag{3.50}$$

At first glance, Equation (3.50) involves six free parameters, namely, $\tilde{A}_M$, $\tilde{b}_M$, l, n, M_{min}, M_{max}. However, as before, Equation (3.48) implies that $\tilde{A}_M$ is not a free parameter. In addition, as before, if M_{min} and M_{max} reflect objective properties of the particle system then they are not truly free parameters. This means there are only three free parameters, namely, l, n, and $\tilde{b}_M$. Thus, while regular and self-similar forms involve somewhat different parameters, they both have exactly three free parameters. As described in the main text, Weibull size distributions are Rosin–Rammler size distributions with $l = n$, which reduces the number of free parameters from three to two.

If $l = m$, Rosin–Rammler size distributions become a type of *gamma size distribution*. In self-similar form, these may be written as follows:

$$f(D, t) = \frac{A_D}{D_{char}(t)}\left[\frac{D}{D_{char}(t)}\right]^{n-1}\exp\left\{-b_D\left[\frac{D}{D_{char}(t)}\right]^m\right\} \tag{3.51a}$$

$$f(M, t) = \frac{A_M}{M_{char}(t)}\left[\frac{M}{M_{char}(t)}\right]^{n/m-1}\exp\left\{-b_M\frac{M}{M_{char}(t)}\right\}. \tag{3.51b}$$

Similarly, if $l = 1$, Rosin–Rammler distributions become another type of gamma size distribution. In self-similar form, these may be written as follows:

$$f(D,t) = \frac{A_D}{D_{\text{char}}(t)} \left[\frac{D}{D_{\text{char}}(t)} \right]^{n-1} \exp\left\{ -b_D \frac{D}{D_{\text{char}}(t)} \right\} \tag{3.52a}$$

$$f(M,t) = \frac{A_M}{M_{\text{char}}(t)} \left[\frac{M}{M_{\text{char}}(t)} \right]^{n/m-1} \exp\left\{ -b_M \left[\frac{M}{M_{\text{char}}(t)} \right]^{1/m} \right\}. \tag{3.52b}$$

Gamma size distributions generally have two free parameters, namely, n and a characteristic size. Gamma distributions are a standard part of probability and statistics; see, e.g., Ross (2018) and Rossi (2018). In this context, Equation (3.51b) is a gamma distribution for mass and Equation (3.52a) is a gamma distribution for diameter. Unless stated otherwise, this book will use the term "gamma size distribution" to refer to Equation (3.51).

If $n = l = m$, then Rosin–Rammler size distributions become a type of *exponential size distribution*. In self-similar form with $D_{\text{char}} = D_{\text{avg}}$ and $M_{\text{char}} = M_{\text{avg}}$, these may be written as follows:

$$f(D,t) = \Gamma\left(1 + \frac{1}{m}\right)^m \frac{m}{D_{\text{avg}}(t)} \left[\frac{D}{D_{\text{avg}}(t)} \right]^{m-1} \exp\left\{ -\Gamma\left(1 + \frac{1}{m}\right)^m \left[\frac{D}{D_{\text{avg}}(t)} \right]^m \right\} \tag{3.53a}$$

$$f(M,t) = \frac{1}{M_{\text{avg}}(t)} \exp\left[\frac{M}{M_{\text{avg}}(t)} \right]. \tag{3.53b}$$

Similarly, if $n = l = 1$, then Rosin–Rammler size distributions become another type of exponential size distribution. In self-similar form with $D_{\text{char}} = D_{\text{avg}}$ and $M_{\text{char}} = M_{\text{avg}}$, these may be written as follows:

$$f(D,t) = \frac{1}{D_{\text{avg}}(t)} \exp\left[\frac{D}{D_{\text{avg}}(t)} \right] \tag{3.54a}$$

$$f(M,t) = \frac{\Gamma(1+m)^{1/m}}{m M_{\text{avg}}(t)} \left[\frac{M}{M_{\text{avg}}(t)} \right]^{n/m-1} \exp\left\{ -\Gamma(1+m)^{1/m} \left[\frac{M}{M_{\text{avg}}(t)} \right]^{1/m} \right\}. \tag{3.54b}$$

Exponential size distributions generally have one free parameter, namely a characteristic size. Exponential distributions are a standard part of probability and statistics; see, e.g., Ross (2018) and Rossi (2018). In this context, Equation (3.53b) is an exponential distribution for mass and Equation (3.54a) is an exponential distribution for diameter. As it turns out, if $n > 0$, $l > 0$ or $n < 0, l < 0$, Rosin–Rammler size distributions can always be rewritten as exponential size distributions for some particle size metric; see Chapter 19. Thus, Rosin–Rammler size distributions with $n > 0, l > 0$ or $n < 0, l < 0$ are equivalent to the exponential distributions studied in traditional probability and statistics. However, unless stated otherwise, this book will reserve the term "exponential size distribution" to refer to Equation (3.53).

While less common than Rosin–Rammler size distributions, some researchers have suggested that aggregation and fragmentation lead to log normal size distributions. Log normal distributions are a standard part of probability and statistics; see, e.g., Ross (2018) and Rossi (2018). In self-similar form, one type of log normal distribution may be written as follows:

$$f(M,t) = \frac{1}{\sqrt{2\pi\sigma_M^2}} \frac{1}{M} \exp\left\{ -\frac{1}{2\sigma_M^2} \left[\ln\left(\frac{M}{M_{\text{char}}(t)} \right) - a_M \right]^2 \right\} \tag{3.55}$$

where a_M and σ_M are unitless constants.

Alternatively, in self-similar form, another type of log normal size distribution may be written as follows:

$$f(D,t) = \frac{1}{\sqrt{2\pi\sigma_D^2}}\frac{1}{D}\exp\left\{-\frac{1}{2\sigma_D^2}\left[\ln\left(\frac{D}{D_{\text{char}}(t)}\right) - a_D\right]^2\right\} \tag{3.56}$$

where a_D and σ_D are unitless constants. Although they may look similar, Equations (3.55) and (3.56) are actually different size distributions. Unless stated otherwise, this book will reserve the term "log normal size distribution" to refer to Equation (3.55).

As a final possibility, some researchers have suggested that aggregation and fragmentation lead to root normal size distributions. These are generally written in the following self-similar form:

$$f_M(D,t) = \frac{A_{M,D}}{D_{M,\text{avg}}(t)}\frac{1}{\sqrt{2\pi\sigma_{M,D}^2}}\sqrt{\frac{D_{M,\text{avg}}(t)}{D}}\exp\left[-\frac{1}{2\sigma_{M,D}^2}\left(\sqrt{\frac{D}{D_{M,\text{avg}}(t)}} - a_{M,D}\right)^2\right] \tag{3.57}$$

where $a_{M,D}$ and $\sigma_{M,D}$ are unitless constants. Laney (2015a) showed that:

$$A_{M,D} = \text{erfc}\left(\frac{a_{M,D}}{\sqrt{2}\sigma_{M,D}}\right) \tag{3.58}$$

$$a_{M,D}^2 + \sigma_{M,D}^2 + \frac{2}{\sqrt{2\pi}}A_{M,D}a_{M,D}\sigma_{M,D}\exp\left(-\frac{a_{M,D}^2}{2\sigma_{M,D}^2}\right) = 1 \tag{3.59}$$

where erfc(x) is the complementary error function. Root normal size distributions generally have two free parameters, namely, $\sigma_{M,D}$ and $D_{M,\text{avg}}$. Unlike the other size distributions discussed previously, root normal size distributions are not a part of traditional probability and statistics. Instead, root normal size distributions were first suggested by Tate & Marshall (1953a,b) specifically for liquid sprays. As it turns out, in most cases, root normal size distributions are approximately equal to Weibull or gamma size distributions. More specifically, $\sigma_{M,D}$ can be chosen so that root normal size distributions are approximately equal to Weibull size distributions for small enough n/m and large enough particles; see Figure 3.4. Similarly, $\sigma_{M,D}$ can be chosen so that root normal size distributions are approximately equal to gamma size distributions for large enough n/m and large enough particles. For more information about root normal size distributions, see, e.g., Laney (2015a, b; 2016).

3.8 Conclusions

This chapter described probabilistic distributions and averages of the sort commonly used to describe aggregation and fragmentation, with an emphasis on size distributions and average sizes. Sections 3.1 to 3.4 discussed general forms including number distributions, concentration distributions, probability density functions, and cumulative distribution functions. While this discussion focused on distributions for particle size, the same general forms may apply to other particle properties such as speed. Table 3.3 summarizes the commonly observed forms for size distributions in aggregating and fragmenting particle systems.

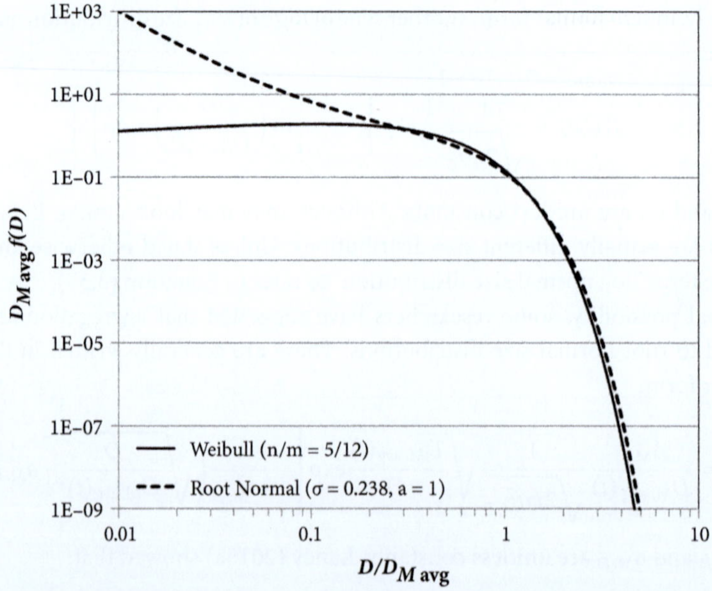

Figure 3.4 Root normal size distributions are approximately equal to Weibull size distributions for large enough particle sizes. In this figure, the root normal size distribution has been renormalized to compensate for the effects of the singularity at zero diameter.

In most cases, this book will use $c(M)$ or $f(M)$ to describe particle size distributions, as indicated by the shaded rows in Table 3.3. However, for dense particle systems, this book will sometimes use $f_M(D)$ or $F_M(D)$. Regarding the independent variable, this book prefers to use mass M because the diameter D does not always have a clear physical meaning. For example, as noted by Matthews (1991), the diameter D obtained through sieving is a "combined measure of the least cross-sectional area and maximum axial length, further complicated by orientation effect." Regarding the dependent variable, this book prefers to use number-weighted rather than mass-weighted distributions for three main reasons:

- Number-weighted distributions such as c, f, and F are a natural choice for many of the problems discussed in this book, including aggregation and collisional fragmentation.
- Number-weighted distributions such as f and F are a standard part of probability theory, while mass-weighted distributions such as f_M and F_M are not.
- Number-weighted distributions such as c, f, and F often provide easily recognizable universals, while mass-weighted distribution such as f_M and F_M generally do not.

Having said this, $f_M(D)$ or $F_M(D)$ may still be the best option for characterizing dense particle systems, which tend to contain extremely large numbers of particles, as described earlier in Examples 3.3 and 3.4.

Table 3.4 summarizes seven families of size distributions commonly seen in aggregation and fragmentation, as discussed earlier in Section 3.7. Of course, this table only hints at the range of possibilities. Appendix A discusses some additional possibilities. Because the same

Table 3.3a Common forms for discrete size distributions.

Notation	Name	Units
n_i	number (of particles of mass iM_1)	(unitless)
c_i	concentration (of particles of mass iM_1)	inverse volume, e.g., cm^{-3}
f_i	number fraction (of particles of mass iM_1)	(unitless)
F_i	cumulative number fraction	(unitless)

Table 3.3b Common forms for continuous size distributions.

Notation	Name	Units
$n(M)$	number distribution	inverse mass, e.g., g^{-1}
$n(D)$	number distribution	inverse length, e.g., cm^{-1}
$c(M)$	concentration distribution	inverse mass × volume, e.g., $g^{-1}cm^{-3}$
$c(D)$	concentration distribution	inverse length × volume, e.g., cm^{-4}
$f(M)$	(number weighted) pdf	inverse mass, e.g., g^{-1}
$f(D)$	(number weighted) pdf	inverse length, e.g., cm^{-1}
$f_M(D)$	(mass weighted) pdf	inverse length, e.g., cm^{-1}
$F(M)$	(number weighted) cdf	(unitless)
$F(D)$	(number weighted) cdf	(unitless)
$F_M(D)$	(mass weighted) cdf	(unitless)
$\mathcal{M}(D)$	cumulative mass distribution	mass, e.g., g

(pdf = probability density function)
(cdf = cumulative distribution function)

Table 3.4 Common families of size distributions.

Eqn.	Name	No. Free Params.	Comments
(3.41)	Rosin–Rammler	3	generalization of Weibull, gamma, power law distributions
(3.42)	Weibull	2	commonly observed in aggregation and fragmentation
(3.51)	Gamma	2	commonly observed in aggregation
(1.10)	Power Law	1	commonly observed in aggregation and fragmentation
(3.53)	Exponential	1	intersection of Weibull and gamma distributions
(3.55)	Log Normal	2	average of Weibull and gamma distributions
(3.57)	Root Normal	2	approximates Weibull and gamma distributions

family can be written in so many different ways, size distributions that look similar may actually be different, and size distributions that look different may actually be the same. This chapter gave some examples, e.g., Equation (3.51b) looks similar to Equation (3.52a) but is actually different. Appendix A gives some additional examples.

For Weibull, gamma, and power law size distributions, Appendix A shows that the size spread decreases as the exponent $|n|/m$ increases. Thus, aggregating and fragmenting systems with significant amounts of chaos or randomness will tend to produce wide particle size spreads with small values of $|n|/m$. Conversely, aggregating and fragmentating systems without significant amounts of chaos or randomness will tend to produce narrow particle size spreads with large values of $|n|/m$. The last chapter gave a rigorous definition of chaos. However, as a simpler and more practical definition, this book will generally assume that an aggregating or fragmentating particle system is chaotic if it can be described by a Weibull, gamma, or power law size distribution such that:

$$-2 \leq n/m \leq 2. \tag{3.60}$$

These limits cover the vast majority of the size distributions observed in the relevant research literature, while excluding cases that are deterministic, or only marginally chaotic or random.

References

G. I. Barenblatt, *Scaling, Self-Similarity, and Intermediate Asymptotics*, Cambridge University Press, 1996, https://doi.org/10.1017/CBO9781107050242

G. I. Barenblatt, *Scaling*, Cambridge University Press, 2003, https://doi.org/10.1017/CBO9780511814921

J. G. Bennett, Broken Coal, *Journal of the Institute of Fuel*, 10, 22–39, 1936

B. Bewick, G. Rolater, J. Bui, M. Barsotti, A. Ziemba, M. Sanai, et al., *Fragmentation of Solid Materials Using Shock Tubes. Part 1: First Test Series in a Small Diameter Shock Tube*, Defense Threat Reduction Agency Technical Report DTRA-TR-17-21, January 2017, http://apps.dtic.mil/sti/citations/AD1025099

P. Embrechts & M. Maejima, *Selfsimilar Processes*, Princeton University Press, 2002, https://www.jstor.org/stable/j.ctt7t1hk

N. Kawara, M. Kumita, H. Kurachi, T. Seto, S. Kamba, T. Kondo, & Y. Otani, Sieving of Aerosol Particles with Metal Screens, *Aerosol Science and Technology*, 50(6), 535–541, 2016, https://doi.org/10.1080/02786826.2016.1167275

C. B. Laney, *Improved Root Normal Size Distributions for Liquid Atomization*, Defense Threat Reduction Agency Technical Report DTRA-TR-16-03, November 2015a, https://apps.dtic.mil/sti/citations/ADA624919

C. B. Laney, *Transformation and Self-Similarity Properties of Gamma and Weibull Fragment Size Distributions*, Defense Threat Reduction Agency Technical Report DTRA-TR-16-06, December 2015b, https://apps.dtic.mil/sti/citations/ADA624878

C. B. Laney, *On the Relationships Between Liquid and Solid Fragmentation*, Defense Threat Reduction Agency Technical Report DTRA-TR-16-19, January 2016, https://apps.dtic.mil/sti/citations/AD1004358

M. D. Matthews, The Effect of Grain Shape and Density on Size Measurement, in J. P. M. Syvitski, editor, *Principles, Methods, and Applications of Particle Size Analysis*, Cambridge University Press, 1991, https://doi.org/10.1017/CBO9780511626142.004

N. F. Mott & E. H. Linfoot, *A Theory of Fragmentation*, U. K. Ministry of Supply, AC3348, January 1943; see also section II.1 of Grady (2006), https://doi.org/10.1007/b138675

S. Ross, *A First Course in Probability*, 10th edition, Pearson, 2018

R. J. Rossi, *Mathematical Statistics: An Introduction to Likelihood Based Inference*, Wiley, 2018, https://doi.org/10.1002/9781118771075

P. Rosin & E. Rammler, Feinheit und Struktur des Kohlenstaubs [Fineness and Structure of Coal Dust], *Zeitschrift des Vereins Deutscher Ingenieure*, 71, 1–7, 1927a

P. Rosin & E. Rammler, Feinheit und Struktur des Kohlenstaubes unter dem Einfluß von Mühlen-und Kohlenart [Fineness and Structure of Coal Dust under the Influence of Mill and Coal Type], *Zement*, 16(35), 820–823, 16(36), 840–844, 16(38), 897–901, 1927b

P. Rosin, E. Rammler, & K. Sperling, Korngrößenprobleme des Kohlenstaubes und ihre Bedeutung für die Vermahlung [Particle Size Problems of Coal Dust and their Importance for Grinding], *Bericht C 52 des Reichskohlenrates* [Report C52 of the National Coal Council], VDI–Verlag, 1933

F. Sommer, F. Reiser, A. Dufresne, M. H. Poelchau, T. Hoerth, A. Deutsch, et al., Ejection Behavior Characteristics in Experimental Cratering in Sandstone Targets, *Meteoritics & Planetary Science*, 48(1), 33–49, 2013, https://doi.org/10.1111/maps.12017

R. W. Tate & W. R. Marshall, Atomization by Centrifugal Pressure Nozzles – Part I, *Chemical Engineering Progress*, 49(4), 169–174, 1953a, https://archive.org/details/sim_chemicalengineering-progress_1953-04_49_4/page/169

R. W. Tate & W. R. Marshall Jr., Atomization by Centrifugal Pressure Nozzles – Part II, *Chemical Engineering Progress*, 49(5), 226–234, 1953b, https://archive.org/details/sim_chemical-engineering-progress_1953-05_49_5/page/226

W. Weibull, A Statistical Theory of the Strength of Materials, *Ingenïors Vetenskaps Akademiens Handligar*, 151, 1–45, 1939a, https://www.scribd.com/document/353394483

W. Weibull, The Phenomenon of Rupture in Solids, *Ingenïors Vetenskaps Akademiens Handligar*, 153, 1–55, 1939b, https://www.scribd.com/document/348631487

W. Weibull, A Statistical Distribution Function of Wide Applicability, *Journal of Applied Mechanics*, 18(3), 293–297, 1951, https://doi.org/10.1115/1.4010337

B. Zohuri, *Dimensional Analysis and Self-Similarity Methods for Engineers and Scientists*, Springer, 2015, https://doi.org/10.1007/978-3-319-13476-5

4
Computational Methods

This chapter surveys computational methods for aggregation and fragmentation, including molecular dynamics methods, finite element methods, and finite volume methods. A description of each type of method could fill an entire book and, in fact, there are many such books available. However, most of those books focus on other problems rather than on aggregation and fragmentation per se. This chapter attempts to fill that gap, albeit with a focus on results rather than on the complex details of the computational methods themselves. In general, computational methods represent the spatial domain by a set of points called *nodes*. Alternatively or in addition, they represent the spatial domain by a set of cells called *finite elements* or *finite volumes*.

Computational methods for aggregation and fragmentation may be classified in the following four ways: (1) Eulerian vs. Lagrangian; (2) mesh-based vs. mesh-free; (3) deterministic vs. non-deterministic; and (4) high-fidelity vs. simplified. The remainder of this introduction will briefly discuss each of these four categories. First, in *Eulerian* methods, the nodes and cells are fixed. In *Lagrangian* methods, the nodes and cells move along with the material in the system. All of the computational methods considered in this chapter are either fully Lagrangian or some blend of Eulerian and Lagrangian. Second, in *mesh-based methods*, the nodes and cells maintain recognizable geometric shapes and patterns despite any distortion that may occur, e.g., an element or volume cannot fold back on itself due to material motion. By contrast, in *mesh-free methods*, there are no cells and the nodes can move in any arbitrary fashion. In this case, the nodes are often called *particles.* These numerical particles may or may not correspond to actual physical particles. Mesh-free methods are an obvious choice for particle systems and an increasingly common choice for other types of systems; see, e.g., Chen et al. (2017). This chapter will discuss two mesh-based and seven meshless methods.

Third, *deterministic* methods *do not* involve randomness or seeming-randomness while *non-deterministic* methods *do* involve randomness or seeming-randomness. In deterministic computational methods, a given set of initial conditions always evolves into the same final conditions. This unique solution is typically highly sensitive to the specific details of the problem including the initial conditions. By contrast, in non-deterministic methods, the same set of initial conditions may evolve into different final conditions. The suite of possible solutions, expressed probabilistically, often has universal features. These universal features are insensitive to the specific details of the problem including the initial conditions. As discussed in Chapter 2, aggregation and fragmentation as treated in this book are non-deterministic. Thus, this chapter will discuss only non-deterministic methods. This includes methods that were originally deterministic but can be modified to be non-deterministic.

Fourth, computational methods may either be *high-fidelity* or *simplified.* Deterministic problems are usually solved using high-fidelity computational methods with local stability,[1]

[1] Insensitive to small perturbations and lacking spurious high-frequency oscillations.

Particle Dynamics with Aggregation and Fragmentation. Culbert B. Laney, Oxford University Press. © Culbert B. Laney (2025).
DOI: 10.1093/9780198920687.003.0004

high formal accuracy,[2] rapid convergence,[3] and realistic first-principles physics; see, e.g., Laney (1998) and Ramani & Shkoller (2020). By contrast, non-deterministic problems are often solved using simplified computational methods with local instability, low formal accuracy, slow convergence, and notional physics. Most applications-oriented computational methods include a mixture of high-fidelity and simplified elements. This chapter will distinguish between high-fidelity and simplified methods based entirely on how they treat aggregation and fragmentation. By this criterion, this chapter will discuss four high-fidelity and five simplified methods.

4.1 High-Fidelity Computational Methods for Aggregation

This section briefly describes two high-fidelity computational methods for particle aggregation. *Molecular Dynamics (MD)* methods solve the governing equations—including conservation of mass, momentum, and energy—for systems of particles interacting through collisions and *short-range body forces*, such as van der Waals forces. More specifically, MD methods solve Newton's second law of motion:

$$\frac{d}{dt}(M\boldsymbol{V}) = \boldsymbol{F} \tag{4.1}$$

where M is particle mass, $\boldsymbol{V}$ is the particle velocity, $M\boldsymbol{V}$ is the particle momentum, and $\boldsymbol{F}$ is the total force on the particle. The governing equations are integrated in time using the Runge–Kutta method or some other numerical method; see, e.g., Tadmor & Miller (2011). Gibson et al. (1960) and Rahman (1964) developed MD methods to simulate molecular interactions in liquids. However, they have since been extended to model various forms of particle aggregation. As a signature feature, MD methods typically include *thermostat algorithms*, which use boundary conditions to ensure a uniform average kinetic energy. As described by Hünenberger (2005), this "avoid[s] steady energy drifts caused by the accumulation of numerical errors during MD simulations." As shown in Examples 2.1, 2.2, and 2.3, particles that interact through collisions are naturally chaotic. Thus, when used to model collisional particle systems, MD methods are also naturally chaotic. However, numerical dissipation may cause them to underestimate the true degree of chaos. As a partial remedy, researchers often deliberately introduce randomness. For example, Bastea (2006) describes MD simulations "initiated from a configuration of nonoverlapping ... but otherwise randomly distributed ... particles."

As an alternative to MD methods, *N-body methods* solve the equations of motion—including conservation of mass, momentum, and energy—for systems of particles interacting through collisions and *long-range body forces*, such as gravity. As with MD methods, the governing equations are integrated in time using the Runge–Kutta method or some other numerical method; see, e.g., Aarseth (2003). Von Hoerner (1960, 1963) and Aarseth (1963) developed N-body methods to model collisionless clusters of stars and galaxies. However, N-body methods have since been extended to include small collisional bodies, e.g., dust

[2] Error decreases as function of the square, cube, etc. of the time step, cell size, etc. where the solution is smooth.

[3] Error decreases rapidly as a function of the time step, cell size, etc.; e.g., small coefficient on Δt^2 error term.

aggregation during planetesimal formation. As shown in Example 2.4, particles that interact through long-range body forces are naturally chaotic. Thus N-body methods are also naturally chaotic. However, numerical dissipation may cause them to underestimate the true degree of chaos. As a partial remedy, researchers often deliberately introduce randomness. For example, Singh & Mazza (2018) initialized their calculations "by randomly placing the particles in a 3D domain with initial velocities distributed according to a Gaussian with zero mean."

4.2 Examples of High-Fidelity Computational Results for Aggregation

This section gives examples of results from MD and N-body methods for aggregation. These methods simulate all of the particle properties including size, shape, velocity, and energy. However, for simplicity, this section will only give results for particle size. Chapters 6, 7, and 8 will give results for particle shape, velocity, and energy, respectively. If the particle system is spatially homogeneous and isotropic, MD and N-body methods generally obtain seemingly ideal size distributions. In other words, as shown in Table 4.1, n/m appears to be a universal rational number such as $\pm 1/3$, $\pm 1/2$, or ± 1 regardless of the exact details. In this regard, for colloids where atoms and molecules interact through the Lennard–Jones potential, Markutsya et al. (2008) say "the cluster size distributions ... appear to fall on universal curves." Of course, it is hard to be completely certain, since numerical calculations produce real numbers, generally within some error bounds, rather than rational numbers for n/m. While the calculations listed in Table 4.1 focus on aggregation, they generally also include some fragmentation, which occurs due to highly energetic collisions. Fragmentation generates chaos and reduces the spread between the largest and smallest particles, both of which have numerical benefits.

The calculations listed in Table 4.1 obtain different results mainly because they assume different interparticle forces. For example, in Chakrabarti et al. (2004), the particles interact via "a screened electrostatic repulsion and a van der Waals attraction," a.k.a. "the [Derjaguin–Landau–Verwey–Overbeek] DLVO potential," For another example, in Singh & Mazza (2018), the particles interact via "long-range electrostatic forces [where] conservation of charge results in [a] ... statistically equal number of pairwise attractions and repulsions." As a third example, Bushuev et al. (2017) used three different "atomic charge sets ... to calculate the energy of the Coulombic interactions" including the "[Complete Neglect of Differential Overlap] (CNDO) charge set" and the "modified [Becke, 3-parameter, Lee–Yang–Parr] B3LYP charge set," which resulted in three different particle size distributions.

In Table 4.1, the MD results were generated with standard MD codes for discrete particle systems such as DL_Poly Classic or LAMMPS. Because these codes were not originally designed for modeling aggregation, they cannot account for variable particle shapes and sizes. Instead, they assume that all of the particles are identical spherical monomers. As a result, cluster shapes and sizes are often judged based on monomer proximity. For example, Bastea (2006) used an "aggregation criterion that places two [monomers] in the same cluster

Table 4.1 Size distributions for late-time aggregation obtained with high-fidelity computational methods. While dominated by aggregation, these results include fragmentation.

a. Exponential size distributions. Gamma and Weibull size distributions are both equal to exponential size distributions when $n/m = 1$.

$r = n/m$	m	d	Field	References
$1^{1,*}$	$\approx 2.0^{a}$	3	Colloid Science	Bastea (2006)
$1^{1,*}$	$\approx 2.6^{a}$	3	Colloid Science	Markutsya et al. (2008)

b. Power law size distributions.

$r = n/m$	m	d	Field	References
-0.34 ± 0.038^{2}	$<3^{b}$	3	Granular Gases	Singh & Mazza (2018)
$-0.5^{1,*}$	$1.6\text{–}2.0^{a}$	2	Crystallization	Chakrabarti et al. (2004)
-1.01 ± 0.09^{1}	$\approx 2.0^{a}$	3	Crystallization	Bushuev et al. (2017)
$-1.5^{2,*}$	$<3^{b}$	3	Astrophysics	Wallace & Quinn (2019)
-1.5 ± 0.21^{1}	$\approx 2.0^{a}$	3	Crystallization	Bushuev et al. (2017)

(1) MD method.
(2) N-Body method.
(*) The original paper lacks error bounds.
(a) Different particles have the same fractal shape on average where m is a fixed real number.
(b) Different particles have different fractal shapes where m is a variable number less than d.

when their relative distance is smaller than a certain value." Similarly, Singh & Mazza (2018) "divide the simulation domain into equal-size boxes ... The boxes which have [a monomer] density higher than a threshold value are then labeled as an 'occupied site' ... A cluster is then defined as a region of such connected occupied sites."

4.3 Simplified Computational Methods for Aggregation

High-fidelity computational methods for particle aggregation tend to be expensive. For example, Wallace & Quinn (2019) say their MD simulations "required approximately 130,000 [central processing unit] CPU hours and over 50 days of wall clock time." However, in many cases, simplified computational methods can obtain similar results at a dramatically lower cost. As one possible simplification, instead of realistically curved trajectories, suppose that each particle travels along a series of straight lines. In this case, the governing equation can be written as follows:

$$\frac{d}{dt}(MV) = \boldsymbol{F}_{\text{decl}} + \boldsymbol{F}_{\text{defl}} \tag{4.2}$$

where $\boldsymbol{F}_{\text{decl}}$ is a deterministic *deceleration force* that continuously affects the particle speed but not direction and $\boldsymbol{F}_{\text{defl}}$ is a random or seemingly random *deflection force* that sporadically and instantaneously affects both the particle speed and direction.

In *Brownian Dynamics (BD) methods*, the deceleration force is non-zero and the deflection force is random. For example, the deflection force may cause normally distributed displacements in the *x*-, *y*, and *z*-directions at each time step. In such cases, Equation (4.2) is called the *Langevin equation* and the resulting piecewise-linear particle paths are called *random walks*. Langevin (1908) originally developed BD methods to model Brownian motion; see also Example 2.3. BD methods have since been modified to model aggregating particle systems. For example, consider colloid formation, as discussed earlier in Sections 1.3 and 1.5. According to Satoh (2011), "if a colloidal dispersion is dilute enough ... the Brownian motion of particles is generally described by the ....Langevin equation [which includes a] ... random force inducing the Brownian motion of the particles ... The force inducing the random motion of the particle should be independent of the particle, i.e., the particle position and velocity ... The random displacements can be ... generated by sampling according to the normal distributions." For another example, consider soot aggregation in a flame. According to Wang et al. (2019), "the primary particles are allowed to move in random directions, collide and form aggregates ... The Langevin equations used to describe the translational and rotational motion of Brownian particles and the momentum exchange during collision are not solved simultaneously."

As a simpler approach, in *Monte-Carlo (MC) methods*, the deceleration force is zero and the deflection force is random. MC methods generally involve more deflections, with less time and distance between deflections, than BD methods. Because the deceleration force is zero, all of the particles travel at constant speeds along straight lines between deflections, i.e., they obtain *ballistic trajectories*. Originally developed for the Manhattan Project, Monte-Carlo methods were quickly applied to particle systems involving perfectly elastic collisions; see, e.g., King (1951) and Metropolis et al. (1953). MC methods have since been modified to model particle systems involving perfectly inelastic collisions, i.e., aggregation. For example, to model colloid formation, Pierce et al. (2004) suggested a MC method in which randomly selected particles are either left alone or "moved one small monomer diameter" in a random direction, with a probability inversely proportional to the particle diameter. Aggregation occurs whenever such actions result in a collision.

As a third option, in *Event-Driven Molecular Dynamics (EDMD) methods*, the deceleration force is zero and the deflection force is seemingly random consistent with instantaneous binary collisions. In comparing EDMD to MD methods, Brilliantov & Pöschel (2004) say that if "the typical duration of a collision is much shorter than the time between successive collisions ... [then] pairwise collisions of particles may be considered as instantaneous events and each of these events may be considered separately ... Since time-consuming numerical integration is avoided [there is] an enormous increase of simulation speed." Alder & Wainwright (1957, 1959) first developed EDMD methods for astrophysical applications. In its original form, the particle collisions were assumed to be perfectly elastic. However, EDMD methods have since been modified to model partially elastic collisions such as those seen in granular gases. In addition, EDMD methods have since been modified to model perfectly inelastic collisions between ballistic particles, a.k.a., *ballistic aggregation*. As shown earlier in Examples 2.1, 2.2, and 2.3, particles that interact through instantaneous binary collisions are naturally chaotic. Thus, EDMD methods are also naturally chaotic, although numerical dissipation may cause them to underestimate the true degree of chaos. As a partial remedy, researchers often deliberately introduce randomness. For example, Trizac & Hansen

(1995, 1996) say "particles are given initial velocities sampled from some distribution" such as an exponential distribution.

In addition to simplifying the particle trajectories, many models also simplify the particle shapes. For example, Trizac & Hansen (1995, 1996) used an EDMD method to study polymer formation. To reduce the computation time, they replaced the true fractal particle shapes, such as "growing polymer coils," by round two-dimensional disks. During each aggregation event, two disks merge to become a single larger disk.

4.4 Examples of Simplified Computational Results for Aggregation

This section gives examples of results from BD, MC, and EDMD methods for particle aggregation. When the particle system is spatially homogeneous and isotropic, simplified computational methods generally obtain seemingly ideal size distributions, just like high-fidelity computational methods. In other words, as shown in Table 4.2, n/m appears to be a universal rational number such as $\pm 1/2$ or ± 1 regardless of the exact details. In this regard, using an EDMD method, Trizac & Hansen (1996) found that "the mass distribution is ... universal to a high degree of accuracy." As before, it is hard to be completely certain, since numerical calculations produce real numbers, generally within some error bounds, rather than rational numbers for n/m.

Table 4.2 Size distributions for late-time aggregation of colloids obtained with simplified computational methods.

a. Gamma size distributions.

$r = n/m$	m	d	Field	References
$1^{1,*}$	2	2	Polymer Science	Trizac & Hansen (1996)
$1^{2,*}$	$\approx 2.6^{a}$	3	Colloid Science	Markutsya et al. (2008)
$1.35\ (\cong 4/3)^{3,*}$	$1.4–1.54^{a}$	2	Colloid Science	Pierce et al. (2004)

b. Power law size distributions.

$r = n/m$	m	d	Field	References
$-1.20\ (\cong -7/6)^{3,*}$	$1.8–2.1^{a}$	3	Colloid Science	Rottereau et al. (2004)
$0.22 \pm 0.08\ (\cong 1/6)^{2}$	$<3^{b}$	3	Astrophysics	Kempf et al. (1999)
$0.55 \pm 0.09\ (\cong 1/2)^{2}$	$<3^{b}$	3	Astrophysics	Kempf et al. (1999)
$0.8\ (\cong 5/6)^{1,*}$	$<3^{b}$	3	Granular Gases	Pathak et al. (2014)

(1) EDMD method.
(2) BD method.
(3) MC method.
(*) The original paper lacks error bounds.
(a) Different particles have the same fractal shape on average where m is a fixed real number.
(b) Different particles have different fractal shapes where m is a variable number less than d. Kempf et al. (1999) says that "the distribution of the fractal dimension for fixed times is remarkably broad and about 10% of the aggregates have a fractal dimension larger than two."

In a direct comparison between MD and BD methods, Markutsya et al. (2008) found that if M_0 is a reference mass, such size distributions may be written in a self-similar form as follows:

$$c_i(t) \approx \frac{C_0 M_0}{M_{M,\,\mathrm{avg}}^2(t)} \exp\left[-\frac{M_0 i}{M_{M,\,\mathrm{avg}}(t)}\right].$$

While MD and BD methods obtained the same result for the universal aspects of the problem, they obtained different results for the non-universal aspects. In particular, Markutsya et al. (2008) found that "the agreement between the MD and BD simulations is very poor for the monomer frequency" $c_1(t)$ and for "the mass-average cluster size" $M_{M,\,\mathrm{avg}}(t)$.

4.5 High-Fidelity Computational Methods for Fragmentation

This section briefly describes two high-fidelity computational methods for unitary fragmentation. *Finite Element Methods (FEMs)* solve the governing equations—including conservation of mass, momentum, and energy—for unitary materials using meshes composed of hexahedral elements, tetrahedral elements, and so forth. According to Tanner (2011), "in the FEM, the conservation equations are reformulated as a weighted residual problem ... The spatial discretization is obtained by approximating the solution on each element by a piecewise smooth polynomial." According to Reddy (2015), "over each finite element, the unknown variables are approximated by a linear combination of algebraic polynomials and undetermined parameters, and algebraic relations among the parameters are obtained by satisfying the governing equations, in a weighted-integral sense, over each element. The undetermined parameters represent the values of the unknown variables at a finite number of preselected points, called *nodes*, in the element."

Turner et al. (1956) and Clough (1960) first developed the Finite Element Method (FEM) for aeronautical applications such as wing flutter; see also, e.g., Clough (1990, 2004) and Felippa (2001, 2004). While originally designed for modeling intact solids, finite element methods have since been modified to model solid and liquid breakup. As noted by Terranova et al. (2018), Lagrangian finite element methods may experience "the case where an element is highly distorted or folds backs on itself ... leading to a negative volume error and termination of the finite element program." To prevent this, finite element programs commonly remove elements that exceed a user-specified strain, e.g., 100–150%, where strain is a metric of distortion. This technique is known as *erosion*. If the rate-of-strain is sufficiently high, erosion will gradually convert the original unitary body into multiple smaller pieces. From one point of view, the surviving elements represent fragments large enough to be captured on the mesh, while the deleted elements represent fragments *not* large enough to be captured on the mesh. In essence, subgrid-scale fragments are removed by a sink, as discussed earlier in Section 1.1.

As one possible alternative to erosion, some finite element codes change tactics and use meshfree methods to simulate fragmentation, such as the Smoothed Particle Hydrodynamics methods discussed next; see, e.g., Moxnes et al. (2014), Terranova et al. (2018). As another alternative, some finite element codes use *node splitting*, which allows adjacent elements to separate; see, e.g., Holmen et al. (2016), Rasico et al. (2018). As a third alternative, applicable only to brittle solids which fragment through crack propagation, some finite element codes use the Extended Finite Element Method (X-FEM). According to Khoei (2015), in X-FEM, "a discontinuous function such as the Heaviside step function and the two-dimensional elastic asymptotic crack tip displacement fields ... enables the domain to be modeled by finite elements without explicitly meshing ... crack[s] ... [or other] strong (displacement) discontinuities."

For brittle solids, some researchers such as Daphalapurkar et al. (2011) suggest imposing random distributions of microscopic flaws in the initial conditions, based on the actual material properties as determined using electron microscopy, which serve as crack initiation points. However, as noted by Ray (2018), "recent experiments have shown that fracture in brittle materials is preceded by numerous microfractures inside the materials. These tiny ruptures are commonly imperceptible except that the acoustic emissions from these ruptures can be magnified and recorded with the help of ultrasonic detectors." In other words, at large enough rates-of-strain, materials dynamically form new microflaws, regardless of their initial states. As a result, to represent rapidly evolving flaw patterns, and to produce realistic levels of chaos, researchers tend to impose notional rather that realistic random variations in the initial material strength; see, e.g., Levy & Molinari (2010), Levy et al. (2012), Cereceda et al. (2017, 2018) and Ma et al. (2018a, b).

Turning from mesh-based to mesh-free methods, *Smoothed Particle Hydrodynamics (SPH) methods* solve the governing equations—including conservation of mass, momentum, and energy—of unitary materials using moving points. According to Randles & Libersky (1996), "the foundation of SPH is interpolation theory ... Computationally, information is known only at discrete points, so that the integral [forms of conservation of mass, momentum, and energy] are evaluated as sums over neighboring particles. The reason that an underlying grid is not needed is that functions are evaluated using their values at the discrete neighboring points (particles) and an interpolation kernel." According to Shadloo et al. (2016), "in order to narrow down the contributions coming from other particles [and] ... reduc[e] the computation time, SPH only includes the effects of nearby particles within a smoothing radius ... also called a *support domain* ... [which] is a localized domain over which the [interpolation] kernel is nonzero."

Lucy (1977) and Gingold & Monaghan (1977) first developed SPH methods "for the numerical solution of two- and three-dimensional gas dynamics problems of astronomical interest." While originally designed for gas dynamics, SPH methods have since been modified to model unitary fragmentation. For example, for brittle solids, Randles & Libersky (1996) note that "since SPH is an interpolating meshless particle code involving only a region of neighboring particles, fracture can easily be facilitated by simply diminishing the communication between damaged particles." More specifically, Ma et al. (2011) say that communication is reduced based on "a damage model that describes the evolution of the material from an intact state to a fractured state [based on an] ... elasto-plastic damage model ... able to account for both the shear-induced damage and the tension-induced fractures." Like most particle-based methods, SPH methods are well-adapted for chaotic systems. However, numerical dissipation may cause them to underestimate the true degree of chaos. As a partial remedy, Monaghan (2005) suggests assigning an initial "set of flaws (cracks) ... to the SPH particles at random, according to the Weibull distribution. Depending on the flaw, tension may or may not cause it to grow."

In addition to unitary fragmentation of solids, SPH can be used to model unitary fragmentation of liquids, a.k.a., *atomization*. In this case, there are multiple possible fragmentation mechanisms including cavitation, turbulence, and surface instabilities such as Rayleigh–Taylor and Kelvin–Helmholtz instabilities; see, e.g., Shinjo (2018), Mithun et al. (2018), Magnotti & Genzale (2019), Sun et al. (2019). Many of these behaviors are naturally chaotic; see, e.g., Manneville (2010) and Lefebvre & McDonell (2017). Regarding one particular form of chaos, known as turbulence, Koch et al. (2017) say the following:

> ... It has to be emphasized that no turbulence models [can be] used ... [Instead a] fine spatial resolution of approximately dx = 5 μm [must be used so that] most of the turbulent length scales are captured and resolved directly, i.e. virtually a direct numerical simulation (DNS) is performed. At the present stage of development of the SPH no methodology for generating artificial, synthetic turbulence is available.

However, once the driving mechanisms such as turbulence are sufficiently resolved, determining fragmentation is relatively straightforward. For example, Braun et al. (2019) describe an approach in which SPH "particles are marked as either belonging to the gaseous or to the liquid phase. The assignment to a certain phase does not change during the simulation." To identify fragments—including compact liquid droplets and extended liquid ligaments—they post-process the results using a cluster finding algorithm, based on simple rules such as "particles are ... connected if their distance is less than 1.5 times the mean particle distance." Because liquid fragments are so fragile, secondary breakup may begin even before primary breakup has finished. In practice, determining the secondary breakup of each liquid fragment, which can easily number in the millions, is just as computationally demanding as determining the primary breakup of the original liquid body. Thus, for the time being, SPH is limited to primary breakup.

4.6 Examples of High-Fidelity Computational Results for Fragmentation

This section gives examples of results from FEM and SPH methods for unitary fragmentation. Computational methods generally produce results for all of the fragment properties including sizes, shapes, velocities, and energies. However, for simplicity, this section will only give results for particle sizes. Chapters 6, 7, and 8 will give results for particle shapes, velocities, and energies, respectively. When fragmenting systems are spatially homogeneous and isotropic, high-fidelity computational methods generally obtain seemingly ideal size distributions. In other words, as shown in Table 4.3, n/m appears to be a universal rational number such as $\pm 1/3$, $\pm 1/2$, or ± 1 regardless of the exact details. Of course, it is hard to be completely certain, since numerical calculations produce real rather than rational numbers for n/m.

Example 4.1 gives additional details for one of the calculations listed in Table 4.3, including a discussion of finite element material models suitable for concrete. As this example shows, finite element calculations are often deterministic. However, using various stratagems such as imposing random variations in the initial material strength, they may become at least partly non-deterministic, resulting in improved agreement with experimental results.

Table 4.3 Size distributions for late-time unitary fragmentation obtained using high-fidelity computational methods.

a. Weibull size distributions. Large fragments only.

$r = n/m$	m	d	Field	References
0.32 ± 0.03[1, a]	2	3	Small-Scale Explosions	Kong et al. (2013)
0.5[1,*, b]	2	3	Small-Scale Explosions	Randles & Libersky (1996)
2[1,*, c]	1	2	Small-Scale Explosions	Randles & Liberskey (1996)

b. Power law size distributions. Medium and large fragments.

$r = n/m$	m	d	Field	References
-0.32 ± 0.02[1]	2, 3	3	Small-Scale Explosions	Myagkov (2017)
-1 ± 0.05[2]	3	3	Small-Scale Explosions	O'Hare et al. (2019)

(1) SPH
(2) FEM
(*) The original paper lacks error bounds.
(a) Kong et al. (2013) showed that their SPH calculations agreed with experimental results for explosively driven thin (6 mm) steel cylinders reported by Arnold & Rottenkolber (2008). Arnold & Rottenkolber showed that their experimental results can be fit by a Weibull size distribution.
(b) Randles & Libersky (1996) showed that their SPH calculations agreed with experimental results for explosively driven thick (1.9cm) iron cylinders reported by Mock & Holt (1983). Laney (2016) showed that these experimental results can be fit by a Weibull size distribution.
(c) Randles & Libersky (1996) showed that their SPH calculations agreed with experimental results for thin rings reported by Grady & Benson (1983). Zhou et al. (2006) showed that these experimental results can be fit by a Weibull size distribution.

Example 4.1 Thick panel fragmentation

O'Hare et al. (2019) used LS-DYNA, a well-known commercial finite element code, to model fragmentation of an 8-ft × 8-ft × 5-in reinforced concrete structural panel impacted by planar airblast; see Figure 4.1. The airblast had a peak overpressure of 40 psi and a positive-phase pulse duration of 100 ms.

As part of the standard software distribution, LS-DYNA includes dozens of different material models; however, only two are believed to be well-suited for treating concrete subjected to airblast, namely, the Karagozian & Case model (a.k.a., Material Model 72R3) and the Continuous Surface Cap Model (a.k.a., Material Model 159). These material models are both extremely complex, have many semi-empirical features, and have been iteratively refined over a period of decades. Like many material models, Material Model 72R3 may lead to significant mesh size dependencies. In particular, in this case, O'Hare et al. (2019) found that "as mesh size changed, fracture energy was adversely altered." By contrast, according to Murray (2007), Material Model 159 regulates mesh sensitivity by "maintain[ing] constant fracture energy regardless of element size." This is sometimes called the *smeared crack approach*; see, e.g., Kong et al. (2018). As a result, O'Hare et al. (2019) chose to use Material Model 159.

O'Hare et al. invoked the following three options in Material Model 159:

- According to Murray (2007), Material Model 159 requires "37 input parameters with a minimum of 19 parameters that must be fit to data." As an option, the user can specify just the unconfined compressive strength and concrete aggregate size. Based on this, the material model generates default values for all other input parameters. In using this option, O'Hare specified an unconfined compressive strength of 4,000 psi.
- According to Kong et al. (2018) "the dynamic behavior of concrete may be far different even though the unconfined compression strength is identical." As a second option, Material Model 159 can vary the concrete strength—especially those aspects directly related to fracture and fragmentation such as tensile yield strength and fracture energy—as a function of strain rate. According to Murray (2007) this option can account for the fact that "the strength of concrete increases with increasing strain ... [where] rate effects are more pronounced in tension than in compression."
- According to Murray (2007), Material Model 159 can use element erosion to account for "damage-based softening [in which] an element loses all strength and stiffness." As Kong et al. (2018) notes, "an erosion algorithm is essential ... to capture the physical fracture and failure of .... concrete slab[s];" see also the earlier discussion of element erosion in Section 4.5.

With these three options in place, O'Hare et al. (2019) found that tetrahedral elements incurred excessive erosion and computational expense; see Figure 4.2. Thus, they chose hexahedral elements. Beyond this, O'Hare et al. (2019) used trial-and-error parameter sweeps over loading, element formulations, and so forth to optimize agreement between the computational and experimental results for the fragment size distribution. Because of rate effects and element erosion models, these parameter sweeps indirectly varied the material strength and failure thresholds.

Because LS-DYNA does not have a native capability for characterizing fragments, the results were "post-processed through [a custom] ... numerical fragment-tracking algorithm ... [which] identified the number and mass of fragments." As shown in Figure 4.3, LS-DYNA obtained good agreement with the experimental results obtained using physical collection, except for the largest fragments. In terms of the largest fragments, the full panel simulations obtained the best result, the half-panel simulations obtained the next best result, and quarter-panel simulations obtained the worst results.

The LS-DYNA results tend to change continuously as the material properties and initial conditions change. This is partly because these calculations were semi-deterministic, with only incidental randomness and chaos due, e.g., to roundoff error. However, three concrete panel fragmentation experiments done under varying conditions—including different mountings, materials, and loadings—obtained about the same fragment size distribution; see Bewick et al. (2017, 2019). In other words, once modified to be fully non-deterministic, the LS-DYNA results should change *discontinuously* rather than *continuously* as the material properties and initial conditions change. Introducing chaos or randomness in the initial conditions or elsewhere, as discussed in the main text, can help high-fidelity FEM calculations better capture such behaviors. However, many FEM studies including O'Hare et al. (2019) emphasize material modeling rather than chaos or randomness.

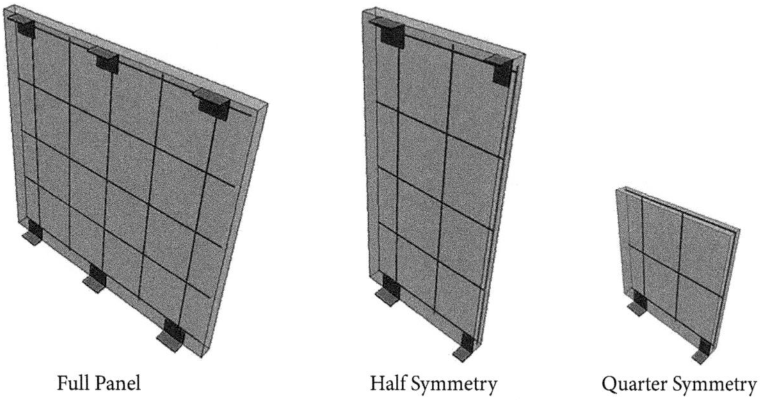

Figure 4.1 Images of a reinforced concrete structural panel as modeled in the LS-DYNA finite element and multiphysics code. Modeling the problem assuming half or quarter symmetry reduces the computational expense by a factor of roughly two or four, respectively, relative to the full panel. However, assuming half or quarter symmetry also reduces the accuracy of the results relative to the full panel.
Reprinted with permission from O'Hare et al. (2019).

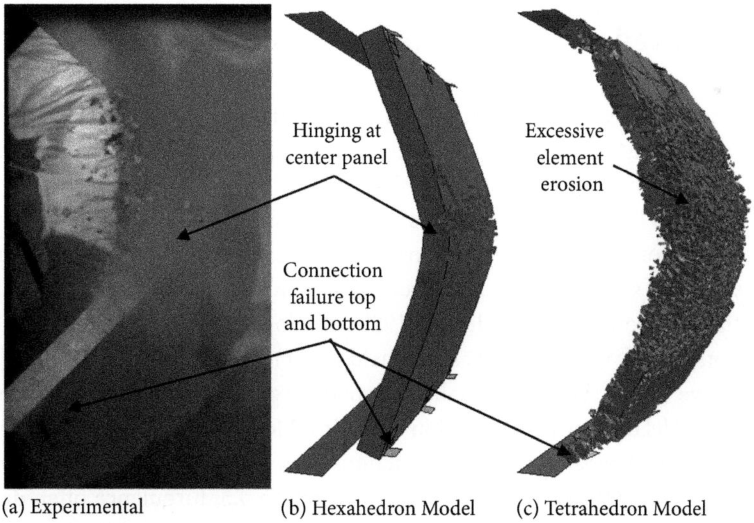

Figure 4.2 Computational (FEM) vs. experimental results for the response of a concrete panel impacted by planar airblast. These results show that the shape of the finite elements affects the degree of erosion.
Reprinted with permission from O'Hare et al. (2019).

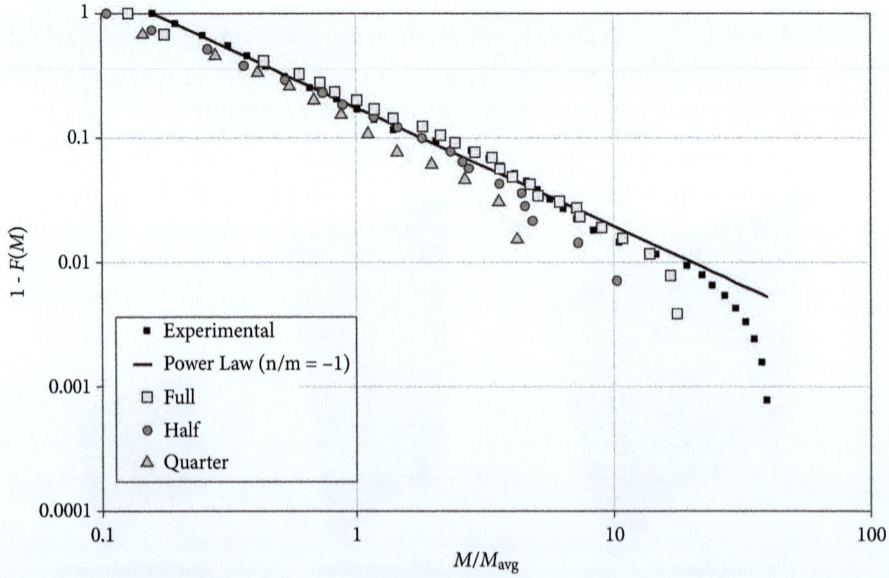

Figure 4.3 Computational (FEM) vs. experimental results for the fragment size distribution from a concrete panel impacted by planar airblast. Assuming half-symmetry or quarter-symmetry reduces the agreement between the computational and experimental results.

Computational results taken from O'Hare et al. (2019). Experimental results taken from Bewick et al. (2019). Figure reprinted with permission from O'Hare et al. (2019).

4.7 Simplified Computational Methods for Fragmentation

This section briefly describes two simplified computational methods for unitary fragmentation. *Finite Volume Methods (FVMs)* solve the governing equations—including conservation of mass, momentum, and energy—for unitary liquids using meshes. According to Tanner (2011), "in the FVM, a control volume is associated with each node, usually in such a way that the node becomes the center of the control volume. The integral [form of the governing] equations are then solved over each control volume [in a way] that guarantees that the physical quantities are conserved." Godunov (1959) developed the first FVM for high-speed gas flows. For other early work see, e.g., McDonald (1971), MacCormack & Paullay (1972) and Rizzi & Inouye (1973). While originally designed for gas dynamics, FVMs have since been modified to model unitary fragmentation of liquids. More specifically, liquid fragments may be described using probability density functions such as $f(D; \boldsymbol{r}, t)$, using an approach pioneered by Williams (1958). As noted earlier in Section 4.5, turbulence often drives liquid fragmentation. In typical applications, FVMs estimate turbulence and fragmentation at subgrid scales using grid-scale information. As discussed in the next chapter, these subgrid-scale turbulence and fragmentation models are generally analytic or semi-empirical. Because of the subgrid scale models, FVMs can use relatively coarse grids, which dramatically lower the computational expense relative to other alternatives such as SPH; see the earlier discussion in Section 4.5. While FVMs are generally considered to be high-fidelity, they are classified here as simplified because of the way they typically treat fragmentation. While FVMs are generally considered to be deterministic, they can be modified to be at least partly non-deterministic; see, e.g., the discussion of Monotonically Integrated Large Eddy Simulation (MILES) in Grinstein (2016).

Turning from mesh-based to mesh-free methods, *Discrete Element Methods (DEMs)* solve the equations of motion—including conservation of mass, momentum, and energy—for systems of particles interacting through *notional short-range body forces*. In its original form, as described by Matuttis & Chen (2014), DEMs treat systems of identical spheres interacting through a repulsive "spring-dashpot element in the normal direction and a slider-spring ... in the tangential direction." More specifically, the repulsive spring-dashpot elements involve velocity-dependent viscous damping similar to a "mass dangling from a spring in water" while the slider-spring elements involve Coulomb friction similar to a "block sliding on a surface under the influence of a spring." Cundall (1971) and Cundall & Strack (1979) first developed DEMs as a variant of MD methods suitable for dense particle flows. By using notional rather than realistic interparticle forces, DEMs offer substantially greater computational efficiency than MD methods in cases with moderate to large particle densities; see, e.g., Spellings et al. (2017).

While originally designed for dense particle flows, DEMs have since been modified to model unitary fragmentation of brittle solids. Instead of springs with dangling ends, which allow particles to interact with any other particle, suppose all of the springs are attached at both ends, which forces particles to interact only with specific neighbors. In other words, instead of a set of *free* particles, consider a set of *bound* particles. If the springs can break under sufficient tension or torque, DEMs can be used to model unitary fragmentation. More generally, suppose unitary objects are represented by an array of small disks or spheres connected by springs, beams, or other elements that can sustain tension, torque, and so forth up to some breaking point. For example, Carmona et al. (2008) modeled high-speed impact fragmentation of solids using "an agglomeration of [small] spheres of two different sizes ... connected by beam-truss elements that can elongate, shear, bend and torque," including beam breakage due to bending and stretching, and "contact forces from sphere-sphere contacts."

DEMs were first used to model unitary fragmentation in the 1980s; see, e.g., Zubelewicz & Mróz (1983), Bažant (1986), Zubelewicz & Bažant (1987), Trent et al. (1987), Butkovich et al. (1988), and Lorig & Cundall (1989). By the 1990s, DEMs were routinely used to predict fragment size distributions for two-dimensional unitary solids; see, e.g., Potapov et al. (1995a, b), Potapov & Campbell (1997), Kun & Herrmann (1996a, b; 1999), Kun et al. (1999). By the 2000s, with increasing computational resources, DEMs were routinely used to predict fragment size distributions for three-dimensional unitary solids. When DEMs are applied to unitary fragmentation, chaos naturally arises through collisions, body forces, and something akin to crack propagation; see Examples 2.1 to 2.5. Although DEMs are naturally chaotic, numerical dissipation often causes them to underestimate the true degree of chaos. As a partial remedy, researchers often randomly vary the initial particle masses, locations, and velocities; see, e.g., Wittel et al. (2004, 2005, 2006).

4.8 Examples of Simplified Computational Results for Fragmentation

This section will give examples of results from DEMs. This section will not give examples of results from FVMs. This is because FVMs typically rely on analytic and semi-empirical fragmentation models, which will be discussed in the next chapter. When fragmenting systems are spatially homogeneous and isotropic, DEMs generally obtain seemingly ideal size distributions, just like high-fidelity computational methods such as FEMs and SPHs. In other words, as shown in Table 4.4, n/m appears to be a universal rational number such as $\pm 1/3$, $\pm 1/2$, or ± 1 regardless of the exact details. Of course, it is hard to be completely certain, since numerical calculations produce real rather than rational numbers for n/m. However, Wittel et al. (2005) say that "universality prevails for large enough input energies when the system falls apart into small enough pieces ... The exponent of the power law is universal in the sense that it does not depend on specific material properties." Similarly, Carmona et al. (2014) say that "the most remarkable feature of fragmentation phenomena [as modeled with DEM] is that the value of the power law exponent of the size distribution of pieces shows an astonishing robustness, being independent of the way of loading, of material properties, and relevant length scales. During the past decades the understanding of the observed universality has been the main driving force of fragmentation research." Examples 4.2 and 4.3 give additional details for two of the calculations listed in Table 4.4, including discussions of how the fragment size distributions can differ for small, medium, and large particles.

Table 4.4 Size distributions for late-time unitary fragmentation obtained using Discrete Element Methods (DEMs). All results are power law size distributions for medium fragments.

$r = n/m$	m	d	Field	References
-0.35 ± 0.03	2, 3	3	Aerospace/Space Debris	Wittel et al. (2004, 2005, 2006)
-0.5	2	2	Mining Engineering	Åström et al. (2004)
-0.67	3	3	Mining Engineering	Wittel (2010)
-0.835 ± 0.08	2	2	Formulated Products & Processes	Behera et al. (2005)
-1 ± 0.05	2	2	Mining Engineering, Astrophysics	Kun et al. (1999)

Example 4.2 Thin shell fragmentation

Wittel et al. (2004, 2005, 2006) used DEM to study fragmentation of oval and spherical shells, modeled as a network of small spherical particles bound together by "springs having linear elastic behavior [until their] ... deformation exceeds a fixed breaking threshold ... As a result of successive spring breakings, cracks nucleate, grow, and merge on the spherical surface, giving rise to the complete breakup of the shell into pieces ...." The springs were given notional properties. In particular, the spring constant in Hooke's law, a.k.a. Young's modulus, was set equal to 1 GPa.

To give the calculations a stochastic character, Wittel et al. (2004) "constructed a three-dimensional discrete element model such that the surface of the unit sphere is discretized into randomly shaped triangles ... by throwing points randomly and independently on the surface ... Disorder is introduced in the model solely by the randomness of the tessellation so that the masses of the nodes and the lengths and the cross sections of the springs are determined by the tessellation." As noted by Wittel et al. (2005), two types of loadings were considered:

(a) *Impact loading* which "realizes the limiting case of instantaneous [total] energy input by giving a fixed initial radially oriented velocity to all material elements."
(b) *Pressure pulse* which imposes a fixed internal pressure. This "gives rise to an expansion of the system ... with a continuous increase of the [total] imparted energy ... and force acting on the shell."

In both cases, the shell expands and fragments homogenously across its entire surface.

To validate the computational results, Wittel et al. (2004, 2005, 2006) conducted a series of experiments involving hollow eggshells. More specifically, to validate the impact loading, they "catapult[ed] the object with a set of rubber bands, against a hard wall on the bottom of a soft plastic bag, prohibiting secondary fragmentation and assuring that no fragments were lost for further evaluation." In these tests the eggshells tended to buckle progressively, starting from the impact point. As seen in Figure 4.4, the intact portion of the shell maintained a near-constant speed during impact, which implies that the majority of the shell tended to experience the same lateral (bending) rates-of-strain during fragmentation, possibly excluding the leading and trailing edges.

To validate the pressure pulse loading, Wittel et al. (2004, 2005, 2006) exploded eggshells after filling them with a hydrogen-oxygen gas mixture. They note that "the experiments are carried out inside a soft plastic bag so that no secondary fragmentations due to fragment-wall collisions occur; furthermore, no pieces were lost after explosion." In these tests, the eggshells tended to fracture progressively, starting at one end. As seen in Figure 4.5, the fractures initiated and propagated at a roughly constant rate, which implies that the majority of shell fragments formed at about the same tensile rates-of-strain.

For impact loading, the computational and experimental results both obtained seemingly ideal power law size distributions with $n/m \approx -1/3$, except for the very largest fragments; see Figure 4.6. For the pressure pulse loading, the computational and experimental results both obtained seemingly ideal power law size distributions with $n/m \approx -1/3$, $1/3$, and -3 for small, medium, and large fragments, respectively; however, the computational and experimental results did not agree on the dividing lines between small, medium, and large fragments; see Figure 4.7.

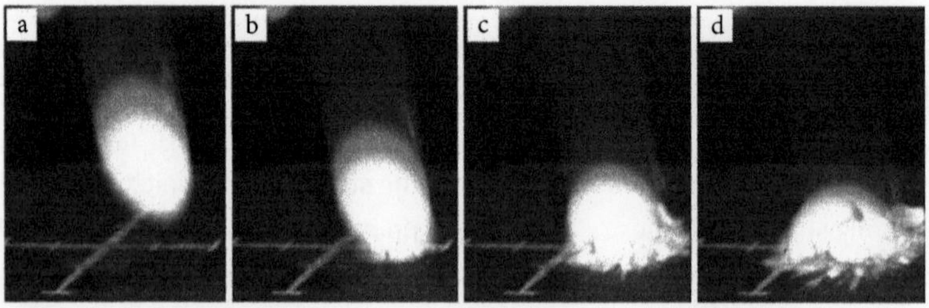

Figure 4.4 Experimental photographs showing the impact of an eggshell with a hard surface. Photographs taken at fixed time intervals.

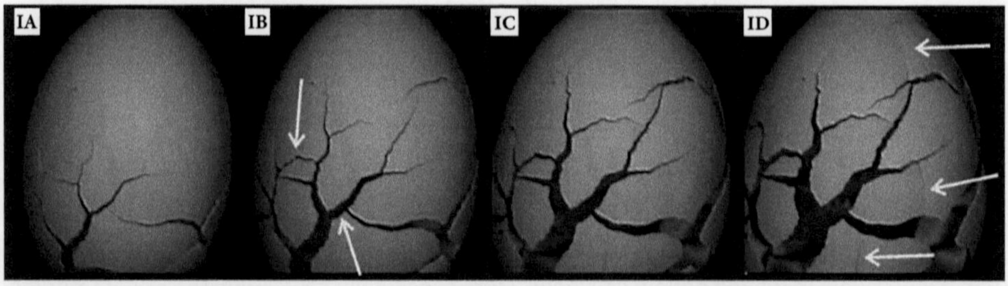

Figure 4.5 Experimental photographs showing an exploding eggshell. Photographs taken at fixed time intervals.

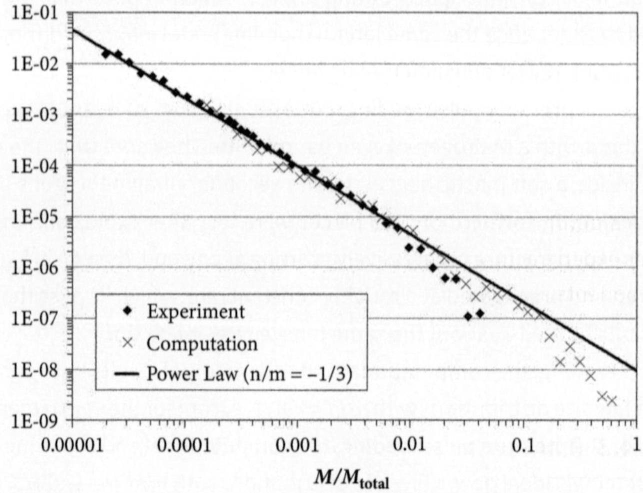

Figure 4.6 Computational (DEM) vs. experimental results for the fragment size distributions produced by eggshells impacting solid surfaces.

Computational and experimental results taken from Wittel et al. (2004, 2005, 2006).

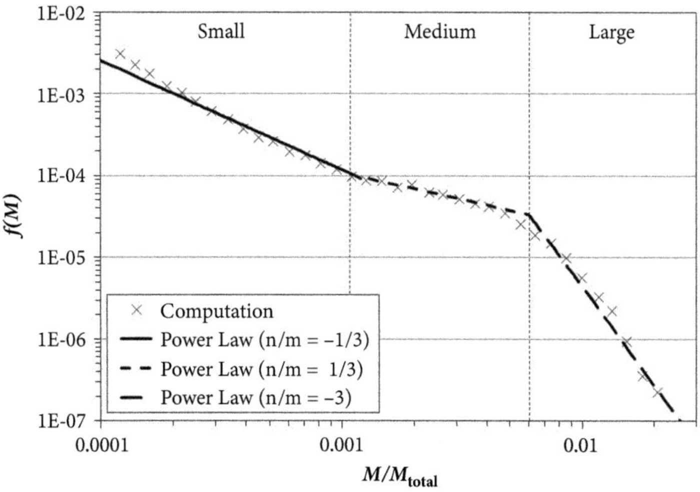

(a) Computational (DEM) results.

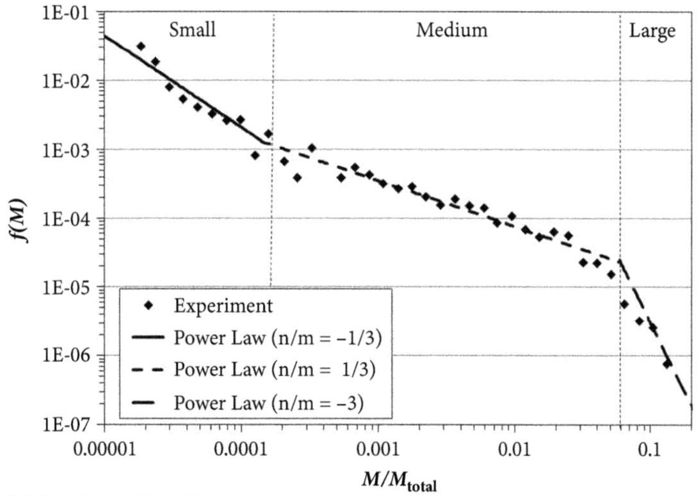

(b) Experimental results.

Figure 4.7 Computational (DEM) vs. experimental results for the fragment size distributions from eggshells experiencing internal gas explosions. In the notation used in this book, the power law exponents are $n/m - 1$. Thus $f(M)$ decreases with M for all $n/m < 1$.
Computational and experimental results taken from Wittel et al. (2004, 2005, 2006).

Example 4.3 Thin plate fragmentation

Linna et al. (2004) and Åström et al. (2004) used DEM to study fragmentation of thin rectangular plates, modeled as a "simple discretized 2D lattice whose sites are connected by ... linear elastic beams which may be stretched up to a predetermined tensile strength, a[fter] which they break instantly and irreversibly ... Loading is performed by expanding the periodic box [lattice] up to a maximum strain."

To give the calculations a stochastic character, Åström et al. (2004) used the following techniques:

(a) "introduce[d] random fluctuations in the system by geometrically distorting the lattice sites;"
(b) "used uniformly distributed uncorrelated variations in the Young's modulus of the beams" and
(c) "introduce[d] a small fraction of [randomly] prebroken bonds to model microcracks."

They note that "other types of disorder were also tested without any significant changes in the results. We have not attempted a systematic investigation of different types of disorder, however."

To a reasonable approximation, the calculations obtained ideal power law size distributions with $n/m = -0.5$; see Figure 4.8a. As noted by Åström et al. (2004), this result agrees with empirical curve fits and dimensional analysis of the sort long used within the mining engineering community.

To validate the computational result, Åström et al. (2004) conducted a series of drop impact experiments involving gypsum disks with diameters of 10.65 cm and thicknesses of 2.28 cm. A "tiny parachute" was used to ensure that the flat face of the disk landed exactly parallel to the floor. If the disks were dropped from a height of 7 m, the experiments obtained power law size distributions with $n/m \approx -0.33$, regardless of fragment size; see Figure 4.8b. However, if the disks were dropped from a height of 0.75 m, the experiments obtained different power law size distributions depending on the size range. In particular, in the experiments, the best-fit power law obtained an exponent of with $n/m \approx -0.33$ for small fragments, $n/m \approx -0.17$ for medium fragments, and a large value of n/m for large fragments, where the exact value is difficult to estimate given that there are really only two data points; see Figure 4.8.b.

Despite attempts to "tune the disorder," the computational results are fundamentally different than the experimental results. This is partly because, to reduce the computational expense, the DEM calculations assumed that the fragments formed in two dimensions. By contrast, in the experimental results, the smallest fragments formed in three dimensions, i.e., they were rough on all sides. At a drop height of 0.75 m, the largest fragments apparently formed in two dimensions and the medium fragments apparently formed in something between two and three dimensions. These results are consistent with a large body of research. For example, in one early paper, Mott & Linfoot (1943) suggested that explosively driven steel shells obtain Weibull size distributions with $n/m = 1/2$ for fragments that form in two dimensions, and $n/m = 1/3$ for fragments that form in three dimensions.

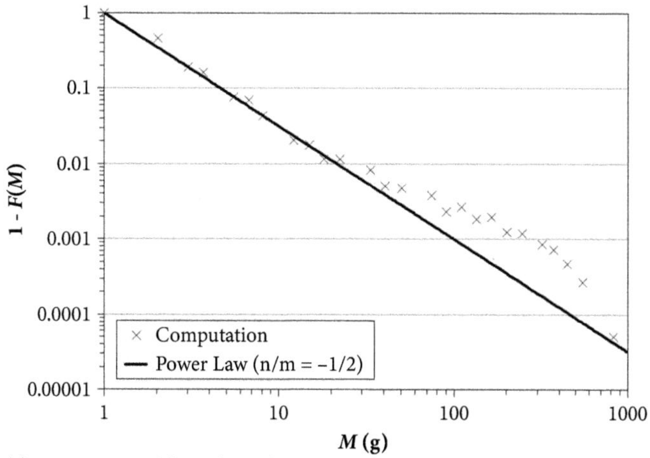

(a) Computational (DEM) results.

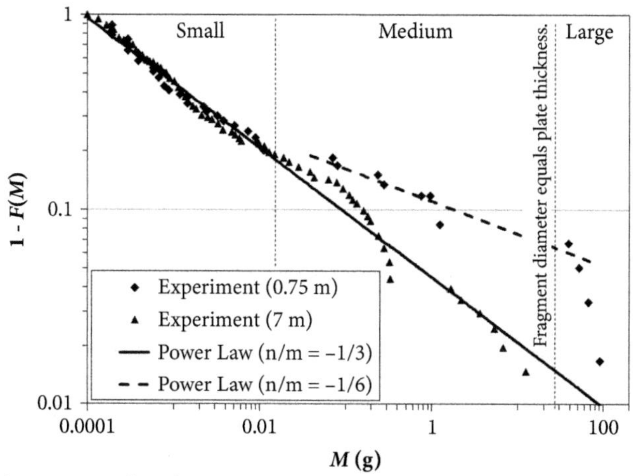

(b) Experimental results.

Figure 4.8 Computational (DEM) vs. experimental results for fragment size distributions produced by thin plates impacting solid surfaces.

Computational and experimental results taken from Linna et al. (2004) and Åström et al. (2004).

The examples given so far all involve ideal size distributions. Ideal size distributions are general, i.e., they occur in many different applications. The remainder of this section will give examples of non-ideal size distributions. Non-ideal size distributions are specific, i.e., they are unique to a given problem. To illustrate the basic principles, this discussion will begin with two examples involving experimental rather than computational results. As the first example, Figure 4.9 shows that solid spheres impacting non-responding flat surfaces often obtain non-ideal fragment size distributions. In such cases, the fragment sizes become increasingly uniform as the impact speed V increases. More specifically, the absolute value of the power law exponent increases roughly linearly as the specific kinetic energy $V^2/2$ increases.

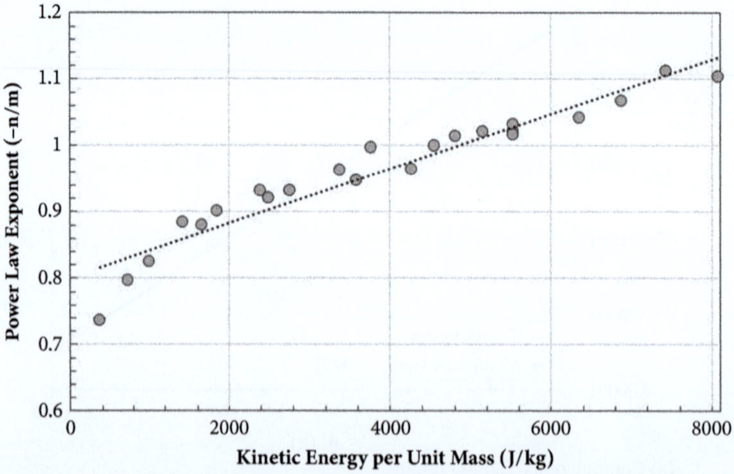

Figure 4.9 Experimental results for fragment size distributions as a function of kinetic energy for gun-launched solid spheres impacting a non-responding steel plate. The solid spheres are 18 mm-diameter and consist of Composition B (59.5% RDX, 39.5% TNT, 1% paraffin wax by weight). Fragment sizes determined based on physical collection in an unlined catch box. Results taken from Springer et al. (2014) and Atwood et al. (2014).

As the second example, Figure 4.10 shows that solid cubes impacting non-responding flat surfaces may obtain either ideal, quasi-ideal, or non-ideal fragment size distributions, depending on the orientation, kinetic energy, strength, and other properties of the cube at the moment of impact. More specifically, in ideal fragmentation, the power law exponent does not depend on the kinetic energy. Instead, it obtains universal rational values such as 1/2 or 5/6. This occurs when the cube lands face-on, resulting in uniform loading conditions. In *quasi-ideal fragmentation*, the power law exponent again does not depend on the kinetic energy. Instead, it jumps between non-universal real values depending on the orientation of the cube during impact. Quasi-ideal size distributions occur once a cube absorbs the maximum possible amount of kinetic energy for a particular impact configuration, after accounting for rebound, splitting, and other effects. In other words, quasi-ideal size distributions are due to energy *saturation*. In this example, quasi-ideal size distributions tend to occur once the kinetic energy of a cube exceeds 10^6 J/kg. Finally, in non-ideal fragmentation, the power law exponent increases continuously with the kinetic energy. In this example, this tends to occur when the cube lands corner-on or side-on with a kinetic energy below 10^6 J/kg.

Figure 4.11 shows the results of various different computations for impact fragmentation. Notice that these results agree qualitatively with the experimental results given earlier in Figures 4.9 and 4.10. In particular, Ma et al. (2018a, b) found that the power law exponent transitions gradually from linear non-ideal to a constant quasi-ideal asymptote as the specific kinetic energy increases. Similarly, Carmona et al. (2008) and Timár et al. (2010, 2012) found quasi-ideal asymptotes. On the other hand, when using uniform external energy input to simulate ultrasound-assisted impact fragmentation, Wittel (2010)

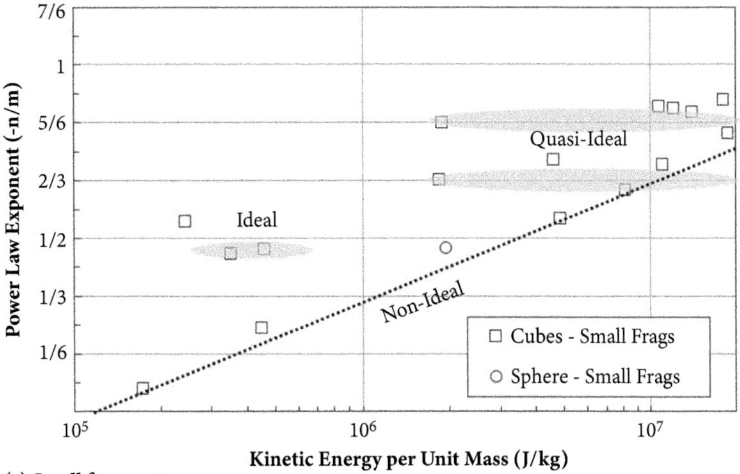

(a) Small fragments.

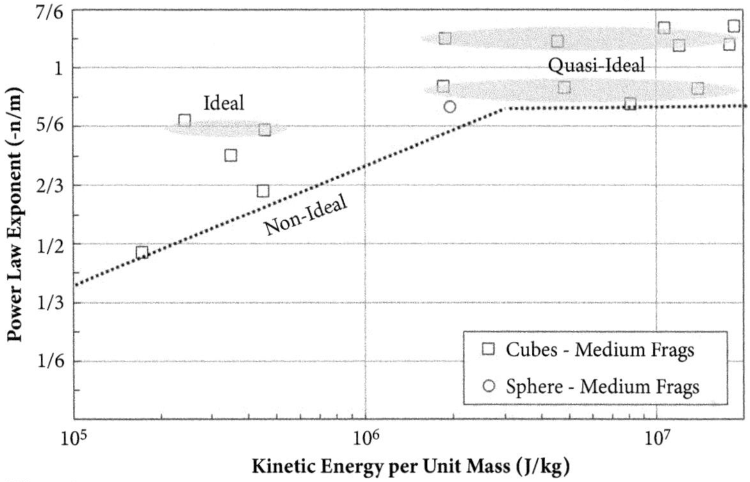

(b) Medium Fragments.

Figure 4.10 Experimental results for fragment size distributions as a function of kinetic energy for 1.5 mm cubes and spheres of basalt and shale impacting a non-responding steel plate protected by a 50 μm-thick layer of water and plastic. Fragment sizes based on physical collection in an unlined steel tray with a cover.
Results taken from Wickham-Eade et al. (2018).

found an ideal asymptote. While ideal and quasi-ideal fragmentation may seem similar, they have different root causes. Ideal fragment size distributions result from homogenous, isotropic stresses and strains during fragment formation, while quasi-ideal size distributions result from energy saturation. Ideal size distributions are generic and occur in many different problems while quasi-ideal size distributions are unique to a given problem. Examples 4.4 and 4.5 give additional details on some of the calculations shown in Figure 4.11.

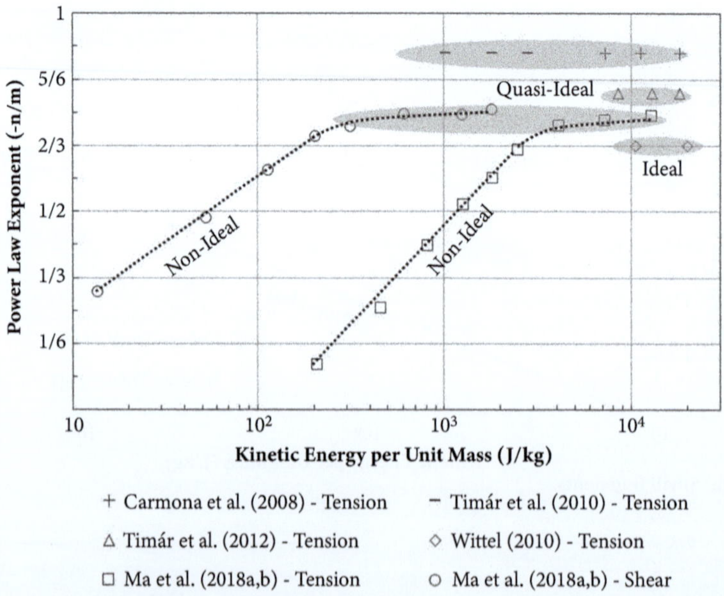

Figure 4.11 Computational results for fragment size distributions as a function of kinetic energy for small spheres impacting rigid flat surfaces.

All results were obtained using DEM, except for those from Ma et al. (2018a, b), who used hybrid DEM-FEM. All results concern medium-sized fragments.

Example 4.4 Non-ideal impact fragmentation

Ma et al. (2018a, b) used a "combined finite and discrete element method (FDEM) coupled with a rate-dependent cohesive zone fracture model" to simulate fragmentation of heterogeneous brittle spheres due to impact with non-responding flat surfaces. Here, the term "heterogeneous" refers to "cemented granular materials" such as sandstone where "sand grains are fused together through interfacial bonds" or concrete where "aggregates with different gradings [are] joined by cementitious materials."

Ma et al. (2018a, b) say that FDEM uses FEM to treat "the deformation and stress of each discrete body" and DEM to treat "the contact and motion of discrete bodies." They note that "FDEM is one of the few methods can handle the interaction between discrete bodies as well as resolving the strain and stress field within each discrete body." They further clarify that "element deletion [a.k.a. erosion] is not used ... meaning that even if a fragment is completely separated from the main body, its position, velocity, internal stress and strain are still tracked and updated. This allows fragments to interact with each other after the first impact and ... possible secondary fracturing can be fully accounted [for]."

Ma et al. (2018a, b) used a material model that requires over a dozen different input parameters including the Young's modulus, tensile strength, cohesion, fracture energy, and the "strength threshold of the failure criterion." To give the calculations a stochastic quality, Ma et al. (2018a, b) randomly varied two of these properties—namely, the tensile strength and the cohesion—to mimic pre-existing micro-flaws. Ma et al. (2018a, b) found that the fragment mass distribution can be well-characterized by a power law. More specifically, as seen in Figure 4.11, the power law exponent increased with impact velocity, up to some critical velocity, after which it "converged to a relatively constant value." Figure 4.12 shows a typical result.

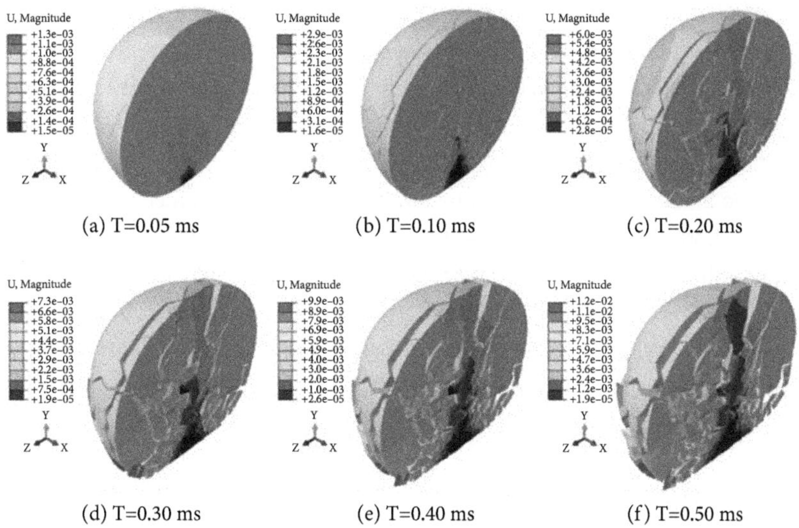

Figure 4.12 Computational (FEM-DEM) results for the response of a heterogeneous solid sphere impacting a rigid flat surface. The flat surface, while not shown, is at the bottom of the sphere.

Reprinted from Ma et al. (2018a) with permission from Elsevier.

Example 4.5 Quasi-ideal impact fragmentation

Carmona et al. (2008) and Timár et al. (2010) used a DEM model that treated spherical polymer projectiles as an array of spherical particles "connected via beam-truss elements that deform by elongation, shear, bending, and torsion." They allowed beams to break due to bending and stretching. In addition, in some cases, "broken particle contacts are able to reconnect when compressed against each other, leading eventually to the healing of cracks." In Carmona et al. (2008), the projectiles were 16 mm plastic spheres with velocities between 100 and 160 m/s. In Timár et al. (2010), the projectiles were 4 mm plastic spheres with velocities between 45 and 75 m/s.

As described by Carmona et al. (2008), the DEM calculations used a material model that depends on more than a dozen different parameters including tensile strength and bending thresholds. The parameters were chosen to reflect the mechanical properties of polymers such as Polymethyl methacrylate (PMMA), Polyamide (PA), and Polypropylene (PP). In terms of loading, "the specimen is placed close to the [target] plate with an [uniform] impact velocity in the negative z direction, assigned to all its composing elements," meaning that "highly hydrostatic compressive stresses are induced near the contact point."

To give the calculations a stochastic character, Carmona et al. (2008) "averaged over 36 simulations, changing the random breaking thresholds and randomly rotating the sample to obtain different impact points." In addition, they assigned the DEM particles random initial velocities and sizes, choosing between two different sizes with "more or less equal fractions." Timár et al. (2010) used a similar approach, except that instead of only two different particle sizes they used a "random packing of spheres with a narrow size distribution."

When the breakup was dominated by tensile stresses, Carmona et al. (2008) and Timár et al. (2010) both found that mid-sized fragments obtained quasi-ideal power law size distributions with $n/m = -0.9 \pm 0.2$, regardless of the impact velocity, projectile diameter, projectile composition, and other properties. However, when the breakup was dominated by bending and healing, Timár et al. (2010) found that mid-sized fragments obtained quasi-ideal power law size distributions with $n/m = -0.25 \pm 0.06$. To validate the latter result, Timár et al. (2010) conducted a series of experiments as follows:

> **A device was used which accelerates particles one-by-one by centrifugal force in a rotor [so that they] hit ...[a] hard wall at a rectangular angle in an evacuated environment ... Spherical polypropylene particles of diameter 4 mm were fragmented at different impact velocities in the range 30 m/s –180 m/s ... At each impact velocity, 400 particles were fragmented.**

The resulting fragments were collected in an unlined chamber and measured using digital photography. The as-measured fragments obtained power law size distributions with $n/m = -0.2 \pm 0.06$. In both the experiments and computations, below the critical velocity of 60 m/s, the projectiles broke into only few pieces with many smooth surfaces inherited from the original sphere, which implies $m = 2$. Above the critical velocity of 60 m/s, the particles broke into many pieces that were mostly rough on all sides, which implies $m = 3$.

Table 4.5 shows a number of examples of quasi-ideal size distributions obtained from simplified computational models of impact fragmentation. In all cases, the projectile is a two-dimensional disk or three-dimensional sphere and the target is a non-responding flat surface. Notice that, where error bounds are given, they generally exclude universal rational values of n/m such as $-1/3$, $-1/2$, or -1. Also notice that, because they are quasi-ideal, the values of n/m are essentially unique to each given problem. Compare this with the ideal size distributions given in Tables 4.1 to 4.4, where the same universal rational values of n/m occur repeatedly, regardless of the exact details.

Table 4.5 Examples of late-time quasi-ideal size distributions obtained using simplified computational methods for impact fragmentation of 2D disks and 3D spheres.

a. Weibull size distributions. Large fragments only.

$r = n/m$	m	d	Details	References
-1.9 ± 0.1^1	3	3	16 mm polymer spheres, 120–150 m/s, impact a flat surface. Fragments form mainly due to tension.	Carmona et al. (2008)

b. Power law size distributions. Medium fragments only.

$r = n/m$	m	d	Details	References
-0.25 ± 0.06^1	2, 3	3	4 mm polypropylene spheres, 45–75 m/s, impact a flat surface. Fragments form mainly due to bending.	Timár et al. (2010)
-0.6 ± 0.1^1	3	3	12 mm two-phase crystalline-amorphous spheres, 194J, impact a flat surface.	Carmona et al. (2015)
-0.74 ± 0.03^2	3	3	60 mm concrete spheres, 25–90 m/s, impact a flat surface. Fragments form due to tension or shear.	Ma et al. (2018a, b)
-0.8 ± 0.05^1	3	3	1 mm brittle spheres, 130–190 m/s, impact a flat surface. Fragments form due mainly to tension.	Timár et al. (2012)
-0.9 ± 0.2^1	3	3	16 mm polymer spheres, 120–150 m/s, impact a flat surface. Fragments form mainly due to tension.	Carmona et al. (2008)
-0.9 ± 0.1^1	2, 3	3	4 mm polypropylene spheres, 45–75 m/s, impact a flat surface. Fragments form mainly due to tension.	Timár et al. (2010)
$-1.1^{1,*}$	2	2	Thin round brittle disks, traveling 0.0005 times the speed of sound of the disk, impact a flat surface.	Potapov & Campbell (1997)
-1.27 ± 0.05^1	2	2	Two thin round granular solid disks, 2–35 m/s, collide with each other.	Kun & Herrmann (1999)
$-1.4^{1,*}$	2	2	Thin round brittle disks, traveling 0.0098 times the speed of sound of the disk, impact a flat surface.	Potapov & Campbell (1997)

(1) DEM
(2) Combination of FEM and DEM, a.k.a., FDEM
(*) The original paper lacks error bounds.

4.9 Conclusions

This chapter introduced five computational methods for aggregation and four computational methods for fragmentation. In Tables 4.6 and 4.7, each method is classified according to the four properties introduced at the beginning of the chapter. As these tables show: (1) all nine methods are Lagrangian or Arbitrary Lagrangian–Eulerian (ALE); (2) two of the methods are mesh-based and seven are meshless; (3) two of the methods are at least partly non-deterministic and seven of the methods are fully non-deterministic; (4) four of the methods are high-fidelity and five are simplified.

At one extreme, some computational approaches for aggregation and fragmentation emphasize realistic first-principles physics modeling, and deemphasize chaos and randomness, e.g., FEM. At the other extreme, some computational approaches for aggregation and

Table 4.6 Selected Computational Methods for Aggregation.

Name	Typical Properties	Description
1. Molecular Dynamics (MD)	- Lagrangian - Meshless - Chaotic/Non-Deterministic - High-Fidelity	Particles interact with each other through *short-range* body forces, resulting in curved trajectories with acceleration and deceleration.
2. N-Body (NB) methods	- Lagrangian - Meshless - Chaotic/Non-Deterministic - High-Fidelity	Particles interact with each other through *long-range* body forces, resulting in curved trajectories with acceleration and deceleration.
3. Brownian Dynamics (BD)	- Lagrangian - Meshless - Random/Non-Deterministic - Simplified	Particles obtain random piecewise-linear paths. Along each line segment, particles may decelerate but do not accelerate.
4. Monte-Carlo (MC) methods	- Lagrangian - Meshless - Random/Non-Deterministic - Simplified	Particles obtain random piecewise-linear paths. Along each line segment, particles neither accelerate nor decelerate.
5. Event-Driven Molecular Dynamics (EDMD)	- Lagrangian - Meshless - Chaotic/Non-Deterministic - Simplified	Particles obtain seemingly random piecewise-linear paths due to binary collisions. Along each line segment, particles neither accelerate nor decelerate.

Table 4.7 Selected Computational Methods for Fragmentation.

Name	Typical Properties	Description
1. Finite Element Methods (FEMs)	- Lagrangian or ALE[1] - Mesh-Based - Partly Non-Deterministic[2] - High-Fidelity	Highly detailed, highly realistic material models are typically used to estimate properties such as tensile strength and yield strength.
2. Smoothed Particle Hydrodynamics (SPH)	- Lagrangian - Meshless - Chaotic/Non-Deterministic - High-Fidelity	Moderately detailed, moderately realistic material models are typically used to estimate properties such as tensile strength and yield strength.
3. Finite Volume Methods (FVMs)	- ALE[1] - Mesh-Based - Partly Non-Deterministic[2] - Simplified	Analytic or semi-empirical models are used for subgrid scales, as discussed in the next chapter. As a result, FVMs obtain high computational efficiency.
4. Discrete Element Methods (DEMs)	- Lagrangian - Meshless - Chaotic/Non-Deterministic - Simplified	Notional material models are typically used, such as ones based on breakable damped springs and beams.

(1) Arbitrary Lagrangian–Eulerian.
(2) While traditionally considered to be deterministic, these methods can be modified to be partly non-deterministic, e.g., by randomly varying the material strength, by reducing the numerical dissipation, etc.

fragmentation emphasize chaos and randomness, and deemphasize realistic first-principles physics modeling, e.g., EDMD and DEM. The examples given in this chapter show that either extreme may obtain similar results. However, if something has to be sacrificed, it counterintuitively seems it should be the first-principles physics. In other words, even the most realistic physics models may not produce accurate results unless they adequately reflect the randomness or seeming-randomness present in the actual problem. However, highly unrealistic physics models—such as those that replace binding molecular forces by damped springs—may still produce accurate results provided they adequately reflect the randomness or seeming-randomness present in the actual problem.

Generating chaos with computational methods generally requires simulating very large numbers of very small events. On the one hand, numerical rounding errors may act like an additional source of chaos. In fact, Lorenz (1963, 1993) discovered chaos partly because the round off errors in 1960s-era computers were so large. On the other hand, dissipative numerical errors—sometimes called *artificial dissipation* or, for fluids, *artificial viscosity*—may act like a sink of chaos. To counteract this, researchers often find it helpful to deliberately introduce artificial randomness into computational simulations of aggregation and fragmentation, either in the initial conditions or throughout the calculation. In addition, researchers often find it helpful to reduce or eliminate both numerical *and physical* dissipation, which would otherwise erroneously damp out chaos.

The remainder of this book tends to focus on analytic models for aggregation and fragmentation. As one reason, if simplified models obtain results that are just as good as complex high-fidelity models, at least for the universal aspects of the problem, why not simplify as much as possible? As another reason, many applications-oriented computational methods depend on subgrid-scale models—often based in part or in whole on analytic methods—for modeling aggregation, fragmentation, and turbulence. Anything that improves the analytic components of these subgrid-scale models has the potential to improve these applications-oriented computational codes. That having been said, computational results will appear throughout the book, e.g., in comparisons between computational results, analytic results, and experimental results.

References

S. J. Aarseth, Dynamical Evolution of Clusters of Galaxies I, *Monthly Notices of the Royal Astronomical Society*, 126(3), 223–255, 1963, https://doi.org/10.1093/mnras/126.3.223

S. J. Aarseth, *Gravitational N-Body Simulations*, Cambridge University Press, 2003, https://doi.org/10.1017/CBO9780511535246.001

B. J. Alder & T. E. Wainwright, Phase Transition for a Hard Sphere System, *The Journal of Chemical Physics*, 27(5), 1208–1209, 1957, https://doi.org/10.1063/1.1743957

B. J. Alder & T. E. Wainwright, Studies in Molecular Dynamics. I. General Method, *The Journal of Chemical Physics*, 31(2), 459–466, 1959, https://doi.org/10.1063/1.1730376

W. Arnold & E. Rottenkolber, Fragment Mass Distribution of Metal Cased Explosive Charges, *International Journal of Impact Engineering*, 35(12), 1393–1398, 2008, https://doi.org/10.1016/j.ijimpeng.2008.07.049

J. A. Åström, R. P. Linna, J. Timonen, P. F. Møller, & L. Oddershede, Exponential and Power-Law Mass Distributions in Brittle Fragmentation, *Physical Review E*, 70, 026104, 2004, https://doi.org/10.1103/PhysRevE.70.026104

A. I. Atwood, K. P. Ford, I. Purifoy, C. J. Wheeler, & E. Woods, Composition B Damage Studies, Naval Air Warfare Center Weapons Division Technical Memorandum NAWCWD TM 8740, March 2014, http://apps.dtic.mil/sti/citations/ADA603031

S. Bastea, Aggregation Kinetics in a Model Colloidal Suspension, *Physical Review Letters*, 028305, 2006, https://doi.org/10.1103/PhysRevLett.96.028305

Z. P. Bažant, Mechanics of Distributed Cracking, *Applied Mechanics Review*, 39(5), 675–705, 1986, https://doi.org/10.1115/1.3143724

B. Behera, F. Kun, S. McNamara, & H. J. Herrmann, Fragmentation of a Circular Disc by Impact on a Frictionless Plate, *Journal of Physics: Condensed Matter*, 17(24), S2439, 2005, https://doi.org/10.1088/0953-8984/17/24/005

B. Bewick, G. Rolater, K. Marchand, A. Ziemba, M. Sanai, & C. Laney, *Fragmentation of Solid Materials Using Shock Tubes. Part 2: First Test Series in a Large Diameter Shock Tube*, Defense Threat Reduction Agency Technical Report DTRA-TR-17-50, December 2017, http://apps.dtic.mil/sti/citations/AD1043784

B. Bewick, G. Rolater, K. Marchand, A. Ziemba, M. Sanai, & C. Laney, Fragmentation of Solid Materials Using Shock Tubes. Part 4: Second Test Series in a Large Diameter Shock Tube, Defense Threat Reduction Agency Technical Report, Defense Threat Reduction Agency Technical Report DTRA-TR-19-5, January 2019, http://apps.dtic.mil/sti/citations/AD1065932

S. Braun, L. Wieth, S. Holz, T. F. Dauch, M. C. Keller, G. Chaussonnet, et al., Numerical Prediction of Air-Assisted Primary Atomization Using Smoothed Particle Hydrodynamics, *International Journal of Multiphase Flow*, 114, 303–315, 2019, https://doi.org/10.1016/j.ijmultiphaseflow.2019.03.008

N. V. Brilliantov & T. Pöschel, *Kinetic Theory of Granular Gases*, Oxford University Press, 2004, https://doi.org/10.1093/acprof:oso/9780198530381.001.0001

Y. G. Bushuev, S. V. Davletbaeva, & O. I. Koifman, Molecular Dynamics Simulations of Aqueous Glycine Solutions, *CrystEngComm*, 19(47), 7197–7206, 2017, https://doi.org/10.1039/c7ce01271c

T. R. Butkovich, O. R. Walton, & F. E. Heuze, Insights in Cratering Phenomenology Provided by Discrete Element Modeling, in P. A. Cundall, A. M. Starfield, & R. L. Sterling, editors, *Key Questions in Rock Mechanics*, CRC Press, 1988

H. A. Carmona, F. K. Wittel, F. Kun, & H. J. Herrmann, Fragmentation Processes in Impact of Spheres, *Physical Review E*, 77(5), 051302, 2008, https://doi.org/10.1103/PhysRevE.77.051302

H. A. Carmona, F. K. Wittel, & F. Kun, From Fracture to Fragmentation with Discrete Element Modeling: Complexity of Crackling Noise and Fragmentation Phenomena Revealed by Discrete Element Simulations, *The European Physical Journal Special Topics*, 223, 2369–2382, 2014, https://doi.org/10.1140/epjst/e2014-02270-3

H. A. Carmona, A. V. Guimarães, J. S. Andrade, Jr., I. Nikolakopoulos, F. K. Wittel, & H. J. Herrmann, Fragmentation Processes in Two-Phase Materials, *Physical Review E*, 91(1), 012402, 2015, https://doi.org/10.1103/PhysRevE.91.012402

D. Cereceda, L. Graham-Brady, & N. Daphalapurkar, Modeling Dynamic Fragmentation of Heterogeneous Brittle Materials, *International Journal of Impact Engineering*, 99, 85–101, 2017, https://doi.org/10.1016/j.ijimpeng.2016.09.012

D. Cereceda, D. Kats, N. Daphalapurkar, & L. Graham-Brady, A Micro-Mechanical Modeling Approach for Dynamic Fragmentation in Brittle Multi-Phase Materials, *International Journal of Solids and Structures*, 134, 116–129, 2018, https://doi.org/10.1016/j.ijsolstr.2017.10.026

A. Chakrabarti, D. Fry, & C. M. Sorensen, Molecular Dynamics Simulation of the Transition from Dispersed to Solid Phase, *Physical Review E*, 69, 031408, 2004, https://doi.org/10.1103/PhysRevE.69.031408

J.-S. Chen, M. Hillman, & S.-W. Chi, Meshfree Methods: Progress Made After 20 Years, *Journal of Engineering Mechanics*, 143(4), 04017001, 2017, https://doi.org/10.1061/(ASCE)EM.1943-7889.0001176

R. W. Clough, The Finite Element Method in Plane Stress Analysis, Proceedings of the 2nd American Society of Civil Engineers (ASCE) Conference on Electronic Computation, 345–378, 1960

R. W. Clough, Original Formulation of the Finite Element Method, *Finite Elements in Analysis and Design*, 7(2), 89–101, 1990, https://doi.org/10.1016/0168-874X(90)90001-U

R. W. Clough, Early History of the Finite Element Method from the View Point of a Pioneer, *International Journal for Numerical Methods in Engineering*, 60(1), 283–287, 2004, https://doi.org/10.1002/nme.962

P. A. Cundall, A Computer Model for Simulating Progressive Large Scale Movements in Blocky Rock Systems, in *Rock Fracture: Proceedings of the International Symposium on Rock Mechanics*, International Society for Rock Mechanics, 1971

P. A. Cundall & O. D. L. Strack, A Discrete Numerical Model for Granular Assemblies, *Géotechnique*, 29(1), 47–65, 1979, https://doi.org/10.1680/geot.1979.29.1.47

N. P. Daphalapurkar, K. T. Ramesh, L. Graham-Brady, & J.-F. Molinari, Predicting Variability in the Dynamic Failure Strength of Brittle Materials Considering Pre-Existing Flaws, *Journal of the Mechanics and Physics of Solids*, 59(2), 297–319, 2011, https://doi.org/10.1016/j.jmps.2010.10.006

C. A. Felippa, A Historical Outline of Matrix Structural Analysis: A Play in Three Acts, *Computers & Structures*, 79(14), 1313–1324, 2001, https://doi.org/10.1016/S0045-7949(01)00025-6

C. A. Felippa, The Origins of the Finite Element Method, in *Introduction to Finite Element Methods*, Boulder: University of Colorado Press, 2004, https://vulcanhammernet.files.wordpress.com/2017/01/ifem.pdf

J. B. Gibson, A. N. Goland, M. Milgram, & G. H. Vineyard, Dynamics of Radiation Damage, *Physical Review*, 120(4), 1229–1253, 1960, https://doi.org/10.1103/PhysRev.136.A405

R. A. Gingold & J. J. Monaghan, Smoothed Particle Hydrodynamics: Theory and Application to Non-Spherical Stars, *Monthly Notices of the Royal Astronomical Society*, 181(3), 375–389, 1977, https://doi.org/10.1093/mnras/181.3.375

S. Godunov, A Difference Method for the Numerical Computation of Discontinuous Solutions of Hydrodynamics, *Matematicheskii Sbornik*, 47, 271–306, 1959, https://hal.archives-ouvertes.fr/hal-01620642/document

D. E. Grady & D. A. Benson, Fragmentation of Metal Rings by Electromagnetic Loading, *Experimental Mechanics*, 23(4), 393–400, 1983, https://doi.org/10.1007/BF02330054

F. F. Grinstein, editor, *Coarse Grained Simulation and Turbulent Mixing*, Cambridge University Press, 2016, https://doi.org/10.1017/CBO9781316480243

J. K. Holmen, J. Johnsen, O. S. Hopperstad, & T. Børvik, Influence of Fragmentation on the Capacity of Aluminum Alloy Plates Subjected to Ballistic Impact, *European Journal of Mechanics-A/Solids*, 55, 221–233, 2016, https://doi.org/10.1016/j.euromechsol.2015.09.009

P. H. Hünenberger, Thermostat Algorithms for Molecular Dynamics Simulations, in C. Holm & K. Kremer, *Advanced Computer Simulation for Soft Matter Sciences I: Advances in Polymer Science*, 173, Springer, 2005, https://doi.org/10.1007/b99427

S. Kempf, S. Pfalzner, & T. K. Henning, N-Particle-Simulations of Dust Growth I. Growth Driven by Brownian Motion, *Icarus*, 141(2), 388–398, 1999, https://doi.org/10.1006/icar.1999.6171

A. R. Khoei, *Extended Finite Element Method: Theory and Applications*, Wiley, 2015, https://doi.org/10.1002/9781118869673

G. W. King, Monte-Carlo Method for Solving Diffusion Problems, *Industrial and Engineering Chemistry*, 43(11), 2475–2478, 1951, https://doi.org/10.1021/ie50503a021

R. Koch, S. Braun, L. Wieth, G. Chaussonnet, T. Dauch, & H.-J Bauer, Prediction of Primary Atomization using Smoothed Particle Hydrodynamics, *European Journal of Mechanics B/Fluids*, 61(2), 271–278, 2017, https://doi.org/10.1016/j.euromechflu.2016.10.007

X. Kong, W. Wu, J. Li, F. Liu, P. Chen, & Y. Li, A Numerical Investigation on Explosive Fragmentation of Metal Casing using Smoothed Particle Hydrodynamic Method, *Materials and Design*, 51, 729–741, 2013, https://doi.org/10.1016/j.matdes.2013.04.041

X. Kong, Q. Fang, L. Chen, & H. Wu, A New Material Model for Concrete Subjected to Intense Dynamic Loadings, *International Journal of Impact Engineering*, 120, 60–78, 2018, https://doi.org/10.1016/j.ijimpeng.2018.05.006

F. Kun & H. J. Herrmann, Fragmentation of Colliding Discs, *International Journal of Modern Physics C*, 7(6), 837–855, 1996a, https://doi.org/10.1142/S0129183196000697

F. Kun & H. J. Herrmann, A Study of Fragmentation Processes using a Discrete Element Method, *Computer Methods in Applied Mechanics and Engineering*, 138(1-4), 3–18, 1996b, https://doi.org/10.1016/S0045-7825(96)01012-2

F. Kun & H. J. Herrmann, Transition from Damage to Fragmentation in Collision of Solids, *Physical Review E*, 59(3), 2623–2632, 1999, https://doi.org/10.1103/PhysRevE.59.2623

F. Kun, G. A. D'Addetta, H. J. Herrmann, & E. Ramm, Two-Dimensional Dynamic Simulation of Fracture and Fragmentation of Solids, *Computer Assisted Mechanics and Engineering Sciences*, 6, 385–402, 1999, https://www.researchgate.net/publication/268996928_Two-dimensional_dynamic_simulation_of_fracture_and_fragmentation_of_solids

C. B. Laney, *Computational Gasdynamics*, Cambridge University Press, 1998, http://doi.org/10.1017/CBO9780511605604

C. B. Laney, On the Relationships Between Liquid and Solid Fragmentation, Defense Threat Reduction Agency Technical Report DTRA-TR-16-19, January 2016, "http://apps.dtic.mil/sti/citations/AD1004358"

P. Langevin, Sur la Théorie du Mouvement Brownien, *Comptes Rendus de l'Académie des Sciences*, 146(10), 967–970, 1908, https://doi.org/10.1119/1.18725

A. H. Lefebvre & V. G. McDonell, *Atomization and Sprays*, 2nd edition, CRC Press, 2017, https://doi.org/10.1201/9781315120911

S. Levy & J. F. Molinari, Dynamic Fragmentation of Ceramics, Signature of Defects and Scaling of Fragment Sizes, *Journal of the Mechanics and Physics of Solids*, 58(1), 12–26, 2010, https://doi.org/10.1016/j.jmps.2009.09.002

S. Levy, J. F. Molinari, & R. Radovitzky, Dynamic Fragmentation of a Brittle Plate under Biaxial Loading: Strength or Toughness Controlled?, *International Journal of Fracture*, 174(2), 203–215, 2012, https://doi.org/10.1007/s10704-012-9691-5

R. P. Linna, J. A. Åström, & J. Timonen, Unconstrained Periodic Boundary Conditions for Solid State Elasticity, *Computer Physics Communications*, 158(1), 26–38, 2004, https://doi.org/10.1016/j.comphy.2003.10.001

E. N. Lorenz, Deterministic Nonperiodic Flow, *Journal of Atmospheric Sciences*, 20(2), 130–141, 1963, https://doi.org/10.1175/1520-0469(1963)020<0130:DNF>2.0.CO;2

E. N. Lorenz, *The Essence of Chaos*, University of Washington Press, 1993, https://doi.org/10.1201/9781482288988

L. J. Lorig & P. A. Cundall, Modeling of Reinforced Concrete Using the Distinct Element Method, in S. P. Shah & S. E. Swartz, editors, *Fracture of Concrete and Rock*, Springer, 1989, https://doi.org/10.1007/978-1-4612-3578-1_28

L. B. Lucy, A Numerical Approach to the Testing of the Fission Hypothesis, *The Astronomical Journal*, 82(12), 1013–1024, 1977, https://doi.org/10.1086/112164

G. W. Ma, X. J. Wang, & F. Ren, Numerical Simulation of Compressive Failure of Heterogeneous Rock-Like Materials using SPH Method, *International Journal of Rock Mechanics & Mining Science*, 48(3), 353–363, 2011, https://doi.org/10.1016/j.ijrmms.2011.02.001

G. Ma, W. Zhou, Y. Zhang, Q. Wang, & X. Chang, Fractal Behavior and Shape Characteristics of Fragments Produced by the Impact of Quasi-Brittle Spheres, *Powder Technology*, 325, 498–509, 2018a, https://doi.org/10.1016/j.powtec.2017.11.030

G. Ma, Y. Zhang, W. Zhou, T.-T. Ng, Q. Wang, & X. Chen, The Effect of Different Fracture Mechanisms on Impact Fragmentation of Brittle Heterogeneous Solid, *International Journal of Impact Engineering*, 113, 132–143, 2018b, https://doi.org/10.1016/j.ijimpeng.2017.11.016

R. W. MacCormack & A. J. Paullay, Computational Efficiency Achieved by Time Splitting of Finite Difference Operators, AIAA Paper 72–154, 1972, https://doi.org/10.2514/6.1972-154

G. M. Magnotti & C. L. Genzale, Recent Progress in Primary Atomization Model Developments for Diesel Engine Simulations, in K. Saha, A. K. Agarwal, K. Ghosh, & S. Som, editors, *Two-Phase Flow for Automotive and Power Generation Sectors*, Springer, 2019, https://doi.org/10.1007/978-981-13-3256-2_4

P. Manneville, *Instabilities, Chaos, and Turbulence*, 2nd edition, Imperial College Press, 2010, https://doi.org/10.1142/p642

S. Markutsya, S. Subramaniam, R. D. Vigil, & R. O. Fox, On Brownian Dynamics Simulation of Nanoparticle Aggregation, *Industrial & Engineering Chemical Research*, 47(10), 3338–3345, 2008, https://doi.org/10.1021/ie0711168

H.-G. Matuttis & J. Chen, *Understanding the Discrete Element Method: Simulation of Non-Spherical Particles for Granular and Multi-Body Systems*, Wiley, 2014, https://doi.org/10.1002/9781118567210

P. W. McDonald, The Computation of Transonic Flow through Three-Dimensional Gas Turbine Cascades, ASME Paper 71-GT-89, 1971, https://doi.org/10.1115/71-GT-89

N. Metropolis, A. W. Rosenbluth, M. N. Rosenbluth, A. H. Teller, & E. Teller, Equation of State Calculations by Fast Computing Machines, *The Journal of Chemical Physics*, 21(4), 1087–1092, 1953, https://doi.org/10.1063/1.1699114

M.-G. Mithun, P. Koukouvinis, & M. Gavaises, Numerical Simulation of Cavitation and Atomization using a Fully Compressible Three-Phase Model, *Physical Review Fluids*, 3, 064304, 2018, https://doi.org/10.1103/PhysRevFluids.3.064304

W. Mock & W. H. Holt, Fragmentation Behavior of Armco Iron and HF-1 Steel Explosive-Filled Cylinders, *Journal of Applied Physics*, 54(5), 2344–2351, 1983, https://doi.org/10.1063/1.332346

J. J. Monaghan, Smoothed Particle Hydrodynamics, *Reports on Progress in Physics*, 68(8), 1703–1759, 2005, https://doi.org/10.1088/0034-4885/68/8/R01

N. F. Mott & E. H. Linfoot, *A Theory of Fragmentation*, U. K. Ministry of Supply, AC3348, January 1943; see also section II.1 of Grady (2006), https://doi.org/10.1007/b138675

J. F. Moxnes, A. K. Prytz, Ø. Frøyland, S. Klokkehaug, S. Skriudalen, E. Friis, et al., Experimental and Numerical Study of the Fragmentation of Expanding Warhead Casings by Using Different Numerical Codes and Solution Techniques, *Defence Technology*, 10(2), 161–176, 2014, https://doi.org/10.1016/j.dt.2014.05.009

Y. D. Murray, Users Manual for LS-DYNA Concrete Material Model 159, U.S. Federal Highway Administration FHWA-HRT-05-062, May 2007, https://hwww.fhwa.dot.gov/publications/research/infrastructure/pavements/05062

N. N. Myagkov, Scaling Invariance of Spherical Projectile Fragmentation upon High-Velocity Impact on a Thin Continuous Shield, *Journal of Experimental and Theoretical Physics*, 124(1), 57–69, 2017, https://doi.org/10.1134/S1063776116150115

E. O'Hare, D. Stevens, & K. Marchand, *Concrete Debris Mass Distribution Metamodels*, Protection Engineering Consultants (PEC), Technical Report, 2019 (unpublished)

S. N. Pathak, Z. Jabeen, D. Das, & R. Rajesh, Energy Decay in Three-Dimensional Freely Cooling Granular Gas, *Physical Review Letters*, 112, 038001, 2014, https://doi.org/10.1103/PhysRevLett.112.038001

F. Pierce, A. Chakrabarti, D. Fry, & C. M. Sorensen, Computer Simulation of Selective Aggregation in Binary Colloids, *Langmuir*, 20(6), 2498–2502, 2004, https://doi.org/10.1021/la0356452

A. Potapov, C. S. Campbell, & M. A. Hopkins, A Two-Dimensional Dynamic Simulation of Solid Fracture Part I: Description of the Model, *International Journal of Modern Physics C*, 6(3), 371–398, 1995a, https://doi.org/10.1142/S0129183195000277

A. Potapov, C. S. Campbell, & M. A. Hopkins, A Two-Dimensional Dynamic Simulation of Solid Fracture Part II: Examples, *International Journal of Modern Physics C*, 6(3), 399–425, 1995b, https://doi.org/10.1142/S0129183195000289

A. V. Potapov & C. S. Campbell, The Two Mechanisms of Particle Impact Breakage and the Velocity Effect, Powder Technology, 93(1), 13–21, 1997, https://doi.org/10.1016/S0032-5910(97)03242-7

A. Rahman, Correlations in the Motion of Atoms in Liquid Argon, *Physical Review*, 136(2A), A405–A411, 1964, https://doi.org/10.1103/PhysRev.136.A405

P. W. Randles & L. D. Libersky, Smoothed Particle Hydrodynamics: Some Recent Improvements and Applications, *Computer Methods in Applied Mechanics and Engineering*, 139(1–4), 375–408, 1996, https://doi.org/10.1016/S0045-7825(96)01090-0

J. G. Rasico, C. A. Newman, & M. R. Jensen, Modelling Fragmentation of a 155 mm Artillery Shell IED in a Buried Mine Blast Event, *International Journal of Vehicle Performance*, 4(4), 323–346, 2018, https://doi.org/10.1504/IJVP.2018.095752

R. Ramani & S. Shkoller, A Multiscale Model for Rayleigh–Taylor and Richtmyer–Meshkov Instabilities, *Journal of Computational Physics*, 405, 109177, 2020 https://doi.org/10.1016/j.jcp.2019.109177

P. Ray, Statistical Physics Perspective of Fracture in Brittle and Quasi-Brittle Materials, *Philosophical Transactions of the Royal Society A: Mathematical, Physical and Engineering Sciences*, 377(2136), 20170396, 2018, https://doi.org/10.1098/rsta.2017.0396

J. N. Reddy, *An Introduction to Nonlinear Finite Element Analysis: With Applications to Heat Transfer, Fluid Mechanics, and Solid Mechanics*, 2nd edition, Oxford University Press, 2015, https://doi.org/10.1093/acprof:oso/9780199641758.001.0001

A. W. Rizzi & M. Inouye, Time-Split Finite-Volume Method for Three-Dimensional Blunt Body Flow, *AIAA Journal*, 11(11), 1478–1485, 1973, https://doi.org/10.2514/3.50614

M. Rottereau, J. C. Gimela, T. Nicolai, & D. Durand, Monte Carlo Simulation of Particle Aggregation and Gelation: I. Growth, Structure and Size Distribution of the Clusters, *The European Physical Journal E*, 15, 133–140, 2004, https://doi.org/10.1140/epje/i2004-10044-x

A. Satoh, *Introduction to Practice of Molecular Simulation: Molecular Dynamics, Monte Carlo, Brownian Dynamics, Lattice Boltzmann and Dissipative Particle Dynamics*, Elsevier, 2011, https://doi.org/10.1016/C2010-0-67006-4

M. S. Shadloo, G. Oger, & D. Le Touzé, Smoothed Particle Hydrodynamics Method for Fluid Flows, Towards Industrial Applications: Motivations, Current State, and Challenges, *Computers & Fluids*, 136, 11–34, 2016, https://doi.org/10.1016/j.compfluid.2016.05.029

J. Shinjo, Recent Advances in Computational Modeling of Primary Atomization of Liquid Fuel Sprays, *Energies*, 11(11), 2971, 2018, https://doi.org/10.3390/en11112971

C. Singh & M. G. Mazza, Early-Stage Aggregation in Three-Dimensional Charged Granular Gas, *Physical Review E*, 97, 022904, 2018, https://doi.org/10.1103/PhysRevE.97.022904

M. Spellings, R. L. Marson, J. A. Anderson, & S. C. Glotzer, GPU Accelerated Discrete Element Method (DEM) Molecular Dynamics for Conservative, Faceted Particle Simulations, *Journal of Computational Physics*, 334, 460–467, https://doi.org/10.1016/j.jcp.2017.01.014

H. K. Springer, A. I. Atwood, & K. P. Ford, *Analysis of Composition B Sphere Impact Fragmentation Experiments*, Lawrence Livermore National Laboratory Technical Report LLNL-TR-650866, February 2014, https://www.osti.gov/biblio/1124854

Y. Sun, Z. Guan, & K. Hooman, Cavitation in Diesel Fuel Injector Nozzles and its Influence on Atomization and Sprays, *Chemical Engineering & Technology*, 42(1), 6–29, 2019, https://doi.org/10.1002/ceat.201800323

E. B. Tadmor & R. E. Miller, *Modeling Materials: Continuum, Atomistic and Multiscale Techniques*, Cambridge University Press, 2011, https://doi.org/10.1017/CBO9781139003582

F. X. Tanner, Continuum Based Models for Sprays, in N. Ashgriz, editor, *Handbook of Atomization and Sprays: Theory and Applications*, Springer, 2011, https://doi.org/10.1007/978-1-4419-7264-4_19

B. Terranova, A. Whittaker, & L. Schwer, Simulation of Wind-Borne Missile Impact using Lagrangian and Smooth Particle Hydrodynamics Formulations, *International Journal of Impact Engineering*, 117, 1–12, 2018, https://doi.org/10.1016/j.ijimpeng.2018.02.010

G. Timár, J. Blömer, F. Kun, & H. J. Herrmann, New Universality Class for the Fragmentation of Plastic Materials, *Physical Review Letters*, 104(9), 095502, 2010, https://doi.org/10.1103/PhysRevLett.104.095502

G. Timár, F. Kun, H. A. Carmona, & H. J. Herrmann, Scaling Laws for Impact Fragmentation of Spherical Solids, *Physical Review E*, 86(1), 016113, 2012, https://doi.org/10.1103/PhysRevE.86.016113

B. C. Trent, L. G. Margolin, P. A. Cundall, & E. S. Gaffney, The Micro-Mechanics of Cemented Granular Material, in C. S. Desai, E. Krempl, P. D. Kiousis, & T. Kundu, editors, *Constitutive Laws for Engineering Materials: Theory and Applications*, Elsevier, 1987

E. Trizac & J.-P. Hansen, Dynamic Scaling Behavior of Ballistic Coalesence, *Physical Review Letters*, 74(21), 4414–4417, 1995, https://doi.org/10.1103/PhysRevLett.74.4114

E. Trizac & J.-P. Hansen, Dynamics and Growth of Particles Undergoing Ballistic Coalescence, *Journal of Statistical Physics*, 82(5–6), 1345–1370, 1996, https://doi.org/10.1007/BF02183386

M. J. Turner, R. W. Clough, H. C. Martin, & L. J. Topp, Stiffness and Deflection Analysis of Complex Structures, *Journal of the Aeronautical Sciences*, 23(9), 805–823, 1956, https://doi.org/10.2514/8.3664

S. von Hoerner, Die numerische Integration des n-Körper-Problemes für Sternhaufen. I. [The Numerical Integration of the N-body Problem for Star Clusters], *Zeitschrift für Astrophysik*, 50(3), 184–214, 1960, https://ui.adsabs.harvard.edu/abs/1960ZA ... 50 ... 184V

S. von Hoerner, Die numerische Integration des n-Körper-Problemes für Sternhaufen. II. [The Numerical Integration of the N-body Problem for Star Clusters], *Zeitschrift für Astrophysik*, 57(2), 47–82, 1963, https://ui.adsabs.harvard.edu/abs/1963ZA ... 57 ... 47V

S. C. Wallace & T. R. Quinn, N-Body Simulations of Terrestrial Planet Growth with Resonant Dynamical Friction, *Monthly Notices of the Royal Astronomical Society*, 489(2), 2159–2176, 2019, https://doi.org/10.1093/mnras/stz2284

Y.-F. Wang, Q.-X. Huang, F. Wang, Y. Chi, & J.-H. Yan, Brownian Dynamics Simulation of Soot Primary Particle Aggregation in Laminar Ethylene Diffusion Flames, *Physica A: Statistical Mechanics and its Applications*, 514, 936–947, 2019, https://doi.org/10.1016/j.physa.2018.09.119

J. E. Wickham-Eade, M. J. Burchell, M. C. Price, & K. H. Harriss, Hypervelocity Impact Fragmentation of Basalt and Shale Projectiles, *Icarus*, 311, 52–68, 2018, https://doi.org/10.1016/j.icarus.2018.03.017

F. A. Williams, Spray Combustion and Atomization, *The Physics of Fluids*, 1(6), 541–545, 1958, https://doi.org/10.1063/1.1724379

F. K. Wittel, F. Kun, H. J. Herrmann, & B. H. Kröplin, Fragmentation of Shells, *Physical Review Letters*, 93(3), 035504, 2004, https://doi.org/10.1103/PhysRevLett.93.035504

F. K. Wittel, F. Kun, H. J. Herrmann, & B. H. Kröplin, Breakup of Shells Under Explosion and Impact, *Physical Review E*, 71(1), 016108, 2005, https://doi.org/10.1103/PhysRevE.71.016108

F. K. Wittel, F. Kun, B. H. Kröplin, & H. J. Herrmann, Study on the Fragmentation of Shells, *International Journal of Fracture*, 140(1–4), 243–254, 2006, https://doi.org/10.1007/s10704-005-3995-7

F. K. Wittel, Single Particle Fragmentation in Ultrasound Assisted Impact Comminution, *Granular Matter*, 12(4), 447–455, 2010, https://doi.org/10.1007/s10035-010-0189-4

F. Zhou, J.-F. Molinari, & K. T. Ramesh, Characteristic Fragment Size Distributions in Dynamic Fragmentation, *Applied Physics Letters*, 88(26), 261918, 2006, https://doi.org/10.1063/1.2216892

A. Zubelewicz & Z. Mróz, Numerical Simulation of Rock Burst Processes Treated as Problems of Dynamic Instability, *Rock Mechanics and Rock Engineering*, 16(4), 253–274, 1983, https://doi.org/10.1007/BF01042360

A. Zubelewicz & Z. P. Bažant, Interface Element Modeling of Fracture in Aggregate Composites, *Journal of Engineering Mechanics*, 113(11), 1619–1630, 1987, https://doi.org/10.1061/(ASCE)0733-9399(1987)113:11(1619)

5
Analytic, Semi-Empirical, and Empirical Methods

This chapter surveys analytic, semi-empirical, and empirical models for the size distributions produced by aggregation and fragmentation. In many fields of study, computational methods, such as those discussed in the last chapter, have long since displaced analytic methods. However, this is not the case for large systems where chaos plays a critical role. While computational techniques can generate chaos, they often do so imperfectly and at great expense, challenging or exceeding the limits of modern high-performance computers. In such cases, the bulk of the computational effort may be spent capturing innumerable fine details specific to one problem. Because first-principles methods can only solve one problem at a time, it can be hard to tell for certain which aspects of the solution apply to other problems. In addition, because computational methods can only offer approximations, it can be hard to tell for certain if a given number is real or rational, e.g., does $n/m \cong 0.32155$ actually mean that $n/m = 1/3$?

Unlike computational techniques, analytic techniques can capture chaos in ways that are both straightforward and effective. In fact, these techniques typically rely on chaos, which tends to drive disparate problems towards a limited set of solutions with universal features. In cases with spatial homogeneity and isotropy, analytic methods can obtain general results, quickly and efficiently, with exact rational exponents, without becoming bogged down in the exact details of any specific problem. As their main weakness, the analytical techniques surveyed in this chapter only predict particle size distributions. Having said that, some of these same techniques may also apply to shapes, velocities, and kinetic energies as discussed in Chapters 6, 7, 8, respectively. In treating these particle properties separately, this book implicitly assumes that they are in some sense statistically independent of one another. The validity of this assumption will be explored in future chapters.

In addition to analytic models, this chapter also surveys semi-empirical and empirical models for aggregation and fragmentation. Semi-empirical and empirical models exploit data from observations, experiments, and/or computations. According to Diggle (2014), semi-empirical models "aim to encapsulate the underlying scientific processes that generated the data" while "empirical models aim . . . only to describe the pattern of a data set." More specifically, in this chapter, semi-empirical models are derived by using data to choose free parameters in analytic expressions, while empirical models are derived by using generic curve fitting procedures.

5.1 Mean-Field Theory

This section briefly describes how to apply a general analytic technique, known as *mean-field theory*, to aggregation and fragmentation. To simplify random or seemingly random problems, while still capturing their essential universal behaviors, mean-field theory replaces

Particle Dynamics with Aggregation and Fragmentation. Culbert B. Laney, Oxford University Press. © Culbert B. Laney (2025).
DOI: 10.1093/9780198920687.003.0005

selected fluctuating fields by average fields. Opper & Saad (2001) say mean-field theory is "one of the simplest and most promising approaches . . . when the number of random variables is large enough [that] exact calculation becomes infeasible and also Monte-Carlo sampling techniques . . . reach their limit." Täuber (2014) says that mean-field "approximations work best . . . in high dimensions when the interactions with many neighboring sites average out local fluctuations."

In general, particle systems can be characterized by five fields, namely, the position, velocity, rotation, size, and shape of each particle at a given time. If the particle system is spatially homogeneous and isotropic, the position field does not affect the statistics of the other four fields. Invoking mean-field theory, suppose the velocity, rotation, and shape fields are replaced by a single average quantity. More specifically, for aggregation, suppose their combined effects can be represented by a single size-dependent function K, which describes the average rate at which collisions between two particles cause them to merge. This leads to the following mean-field governing equation for aggregation in continuous particle systems:

$$\frac{\partial c(M,t)}{\partial t} = -\,c(M,t)\int_0^{\infty} K(M,M')\,c(M't)\,dM' + \frac{1}{2}\int_0^{M} K(M-M',M)\,c(M-M',t)\,c(M',t)\,dM' \tag{5.1}$$

where the first term on the right hand side represents the rate at which particles of mass M are lost due to aggregation, while the second term on the right hand side represents the rate at which particles of mass M are gained due to aggregation. Because such governing equations balance population gains against population losses in each particle size range, they are sometimes called *population balance equations*; see, e.g., Ramkrishna (2000) and Ramkrishna & Singh (2014). In general, such mean-field governing equations can be solved using simple, efficient numerical methods. However, in some cases, they can be solved analytically. For example, if $K(M, M') = K_0 = \text{const.}$, then one possible solution of Equation (5.1) is as follows:

$$c(M,t) = \frac{C_0 M_0}{M_{\text{avg}}^2(t)} \exp\left[-\frac{M}{M_{\text{avg}}(t)}\right] \tag{5.2}$$

where M_0 is the initial average particle mass, $M_{\text{avg}}(t) = M_0(1 + K_0 C_0 t/2)$ is the average particle mass at time t, C_0 is the initial total particle concentration, and $C(t) = C_0 M_0 / M_{\text{avg}}(t)$ is the total concentration at time t. This solution can be rewritten as follows:

$$f(M,t) = \frac{c(M,t)}{C(t)} = \frac{1}{M_{\text{avg}}(t)} \exp\left[-\frac{M}{M_{\text{avg}}(t)}\right]. \tag{5.3}$$

In either form, this is an exponential size distribution, as discussed earlier in Section 3.7.

As discussed in Example 5.1, mean-field theory has been used successfully to model aggregation and fragmentation for over 100 years. Aldous (1999) notes that, for many applications, "mean-field models [are] . . . regarded as old fashioned." However, this is not the case for aggregation and fragmentation. Indeed, papers based on mean-field theory for aggregation and fragmentation still appear often in the research literature; see, e.g., Akimkin et al. (2023), Bhoi et al. (2024), Breschi & Fontelos (2023), Diemer (2023), Eggers & Fontelos (2023), Wilkinson (2023).

Example 5.1 Historical overview of mean-field theory for aggregation and fragmentation

Smoluchowski (1916, 1917) first used mean-field theory to model colloid formation due to Brownian diffusion. Marian Smoluchowski was a Professor of Experimental Physics at the Jagiellonian University in Kraków, Poland. Chandraeskhar (1986) notes that "Smoluchowski is chiefly remembered as the originator, along with Einstein, of the theory of Brownian motion." Gudowska-Nowak et al. (2017) say that "after his seminal papers on [Brownian] diffusion, Einstein's mind soon focused on other fundamental topics . . . In contrast, Smoluchowski did not let go of the topic of diffusion. Two additional major achievements in the theory of stochastic motion bear his name . . . The second one, maybe underlining his genius best, is Smoluchowski's theory of the diffusion-limited coagulation of two colloidal particles. This work may be viewed as one of the fundamental cornerstones of molecular physical chemistry." Immediately after publishing this work, Seneta (2001) notes that Smoluchowski "died of dysentery (doubtless due to wartime conditions) in the summer of 1917 at the age of 45." While his premature death limited his scientific contributions, Chandraeskhar (1986) has described Smoluchowski as "the founder of the physics of stochastic phenomena."

While Smoluchowski first used mean-field theory for aggregation, Kolmogorov (1941a) first used mean-field theory for fragmentation. Arnold & Dresden (1989) say the following:

> **Andrei Nokolaevich Kolmogorov . . . was one of the greatest scientists in Russian history. His work in probability theory, turbulence (the "law of 2/3"), and dynamical systems ("chaos") was fundamental and is now regarded as classic . . . Though educated in abstract set-theoretical mathematics, Kolmogorov was . . . always interested in the natural sciences and other applications, in which he would put aside the shackles of mathematical rigor to obtain a concrete result. But after having guessed a result he always tried to formulate it rigorously as a mathematical theorem . . . [Thus] unlike most of the great mathematicians of the 20th century, A. N. Kolmogorov revolutionized both mathematics and physics, following in the line of Newton and Poincaré.**

Kolmogorov's work on fragmentation focused on applications involving milling and grinding (a.k.a., comminution). More specifically, King et al. (2012) say: "Kolmogorov proved that if a particle and its progeny are broken successively and if each breakage event produces a random number of fragments having random sizes, then, if there is no preferential selection of sizes for breakage, the distribution of particle sizes will tend to the log-normal distribution after many successive fracture events." However, King et al. (2012) caution that "this theoretical analysis makes assumptions that are violated in practical comminution operations." In later work, Filippov (1961) used mean-field theory to prove that fragmentation cascades generally obtain Rosin–Rammler rather than log normal size distributions. Aleksei Fedorovich Filippov was a colleague of Kolmogorov at Moscow State University. Filippov (1961) notes that "the topic of the paper was suggested by A. N. Kolmogorov, who [provided] . . . assistance and advice during its writing." In addition, he notes that "a lecture containing the results of this paper was given at a seminar on probability theory at the Moscow State University in Spring 1952." In other words, there was almost a ten year delay before this seminal fragmentation work was published.

5.2 Stochastic Geometry

This section briefly describes how to apply a general analytic technique, taken from the field of stochastic geometry, to aggregation and fragmentation problems. Coupier (2019) says that "*stochastic geometry* can be succinctly described as the study of random spatial patterns" such as "geometric random graphs, percolation, convex geometry, random fields, and spatial statistics." Møller (1998) says "the study of *random tessellations* is one of the major subjects in stochastic geometry . . . A tessellation is roughly speaking a subdivision of the space into sets called *cells*." As it turns out, particle size distributions due to aggregation and fragmentation may be inferred from random tessellations. This is true even when real events and random tessellations obtain completely different particle shapes.

First, consider discrete particle systems. As shown in Example 5.2, stochastic geometry takes a particularly simple form if all of the particles are monomers or chains of monomers, i.e., each monomer can support only two links. Researchers have suggested various ways to extend this simple example to cases where each monomer can support three or more links; see, e.g., Darivakis et al. (1990, 1994). This difficult geometry problem will not be discussed here.

Example 5.2 Stochastic geometry for discrete particle systems composed of chains, a.k.a., strings, strands, threads, and so forth

Suppose a set of identical monomers are arranged to form a long chain. In the simplest case, the chain is a straight line; however, this is not necessary. Let p be the probability that a pair of adjacent monomers in the chain is linked. Conversely, let $1 - p$ be the probability that a pair of adjacent monomers in the chain is *not* linked. If i monomers in a row are linked together, then $i - 1$ links form the interior of a chain segment and two non-links form the ends of a chain segment. This implies the following:

$$f_i = p^{i-1}(1-p)^2. \tag{5.4}$$

Using the relationship $p^{i-1} = \exp[(\ln p)(i-1)]$, this can be rewritten as follows:

$$f_i \cong A \exp(-bi) \tag{5.5}$$

for large enough i where $A = (1-p)^2$ and $b = -\ln p$. Notice that $A > 0$ and $b > 0$ because $0 < p < 1$. More generally, if p is a function of time t then $f_i(t) \cong A(t) \exp[-b(t)i]$. In either case, this is an exponential size distribution. Kuhn (1930) first obtained this result for fragmentation cascades such as those that occur in cellulose degradation; see also, e.g., Mark & Simha (1940). While Kuhn (1930) obtained this result for fragmentation cascades, Flory (1936) obtained the same result for aggregation cascades; see also section 3.6 of Kumar & Gupta (2018). Comparing his result to Kuhn's, Flory (1936) said it is "obvious . . . that the two distributions should be identical . . . when one considers the equivalence of the formation of molecules . . . through random formation of linkages and through random breaking of linkages already formed." In addition to aggregation and fragmentation cascades, the same result applies to unitary fragmentation. This is true because it does not matter if the "random breaking of linkages already formed" occurs in a series of steps or all at once.

Next, consider continuous particle systems. As shown in Example 5.3, stochastic geometry takes a particularly simple form provided that all of the particles are straight line segments.

Example 5.3 Stochastic geometry for continuous particle systems composed of straight line segments

Suppose a long straight line is broken at random points spaced apart by an average distance of L_{avg}. Then, according to the Poisson distribution, the probability of finding i break points along a length L of the line is as follows:

$$p(i, L) = \frac{1}{i!}\left(\frac{L}{L_{avg}}\right)^i \exp\left(-\frac{L}{L_{avg}}\right). \tag{5.6}$$

Starting from one break point, the probability of finding no other break points within a length L is as follows:

$$p(0, L) = \exp\left(-\frac{L}{L_{avg}}\right)$$

while the probability of finding one break point within an additional length dL is as follows:

$$p(1, dL) = \frac{dL}{L_{avg}}.$$

This implies the following probability density function:

$$f(L)dL = p(0, L)p(1, dL) = \frac{dL}{L_{avg}} \exp\left(-\frac{L}{L_{avg}}\right). \tag{5.7}$$

For one-dimensional objects, the particle mass M, diameter D, and length L are all the same to within a constant factor. Then, Equation (5.7) can be rewritten as follows:

$$f(M) = \frac{1}{M_{avg}} \exp\left(-\frac{M}{M_{avg}}\right) \tag{5.8}$$

which is an exponential size distribution. Lienau (1936a, b, c) first obtained this result for unitary fragmentation; see also section 2.1 of Grady (2006). However, the same result applies to fragmentation cascades. This is true because it does not matter if fragmentation occurs all at once or in a time-staggered series of steps. By the same token, the same result applies to aggregation cascades, assuming that all of the particles are straight line segments, which can be joined together at their endpoints.

Researchers have suggested various ways to extend Example 5.3 to realistic multidimensional particles. In the earliest such work, Mott & Linfoot (1943) suggested dividing a large rectangle into smaller rectangles using randomly spaced horizontal and vertical lines; see Figure 5.1. This approach yields something similar to an exponential size distribution provided that the number of fragments N is sufficiently large; see also Grady & Kipp (1985). Such multidimensional tessellations apply both to unitary fragmentation and, less obviously, to fragmentation cascades. This is true because, in many cases, it does not matter whether an object is subdivided all at once or gradually in stages. For example, an object can be subdivided into two random pieces, and then each of those pieces can be subdivided into two random pieces, and so forth. While this is a fragmentation cascade, the results may apply equally to unitary fragmentation; see, e.g., Parisio & Dias (2011).

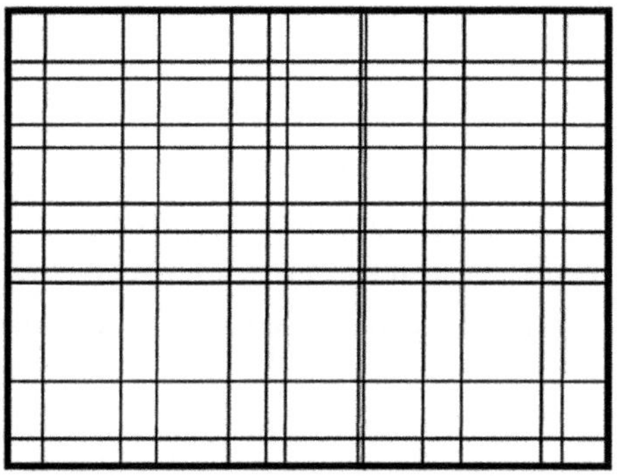

Figure 5.1 Simple tessellation of a rectangle using random horizontal and vertical lines.

In one-dimensional cases, random tessellations can be used to describe both aggregation and fragmentation. However, in multidimensions, random tessellations can be used to describe fragmentation but *not* aggregation. This is because, in multidimensions, particles do not necessarily aggregate into any particular shape such as a chain or line segment. By a similar token, in one-dimensional cases, random tessellations capture the actual particle shapes. However, in multidimensions, random tessellations generally replace the actual rough irregular particles by smooth geometric shapes, hopefully without altering their average shape or size distributions. For example, in Figure 5.1, the actual fragments—which could be, for example, jagged shards of glass from a plate glass window—have been replaced by perfect rectangles. Because it replaces random or seemingly random shape fields by averages, stochastic geometry is related to mean-field theory. As with mean-field theory, solutions can often be found analytically or, at worst, with simple numerical methods suitable for a personal computer. As discussed in Example 5.4, stochastic geometry has been used successfully to model fragmentation for over 90 years.

Example 5.4 Historical overview of stochastic geometry for fragmentation

As noted in Example 5.2, Kuhn (1930) was the first researcher to use stochastic geometry to describe particle fragmentation in one dimension. In this regard, Morris (1986) says the following:

> **In the early 1930s, attention turned to the physical chemistry of polymers [including] . . . the shape of polymers in solutions . . . and the kinetics of polymerization. Statistical mechanics was very useful in this research . . . A pioneering figure in this field was the Swiss physical chemical physicist Werner Kuhn, a brilliant polymath whose research also included the study of the earth's core and the air bladder of the fish . . . Kuhn first ventured into the field of polymer science in 1930 . . . He used statistics to calculate the molecular weight of degraded cellulose by assuming the molecule was broken up in a random fashion. This was probably the first application of statistical theory in polymer science.**

Mott & Linfoot (1943) were the first known researchers to use stochastic geometry to describe particle fragmentation in two dimensions. In this regard, Grady (2006) says the following:

> **For a brief period during the latter part of World War II, Nevill F. Mott led a theoretical group at Fort Halstead in the United Kingdom that tackled scientific issues related to pressing war-time concerns. Among later awards and honors, Mott was knighted and a recipient of the Nobel Prize. While at Fort Halstead, he undertook an effort to theoretically describe the statistical fragmentation of munitions subjected to intense explosive loading.**

More specifically, Davis (1998) says that Mott's research addressed questions such as "why German shells, which were made of steel with a high carbon content, fragmented into smaller pieces than the British shells." Davis (1997) describes the technical approach that Mott used in this and subsequent research efforts as follows:

> **His philosophy, which he never abandoned during the remainder of his life, was to have theoreticians and experimentalists working closely together. Although many today see the advantages of such a liaison, it has never been easy to implement such a policy. Mott's ability to draw theory and experiment together, to persuade practitioners on both**

sides to work on common problems—indeed to blur any distinction between the two approaches—was one of his hallmarks for which he will be remembered.

Because it was done for national defense, much of Mott's work on fragmentation originally appeared only in classified technical reports. Thus, its influence was limited for many years. Over 40 years later, Grady & Kipp (1985) summarized and extended Mott's work; see also Grady (2006). Since then, articles using Mott's approach, i.e., using stochastic geometry to model unitary fragmentation, have appeared on a semi-regular basis. However, for anything other than unitary fragmentation, mean-field theory is still a far more common approach.

5.3 Maximum Entropy Theory

This section briefly describes how to apply a general analytic technique, known as maximum entropy theory, to aggregation and fragmentation problems. As one possible definition, *maximum entropy theory* converts complex random or seemingly random problems into constrained optimization problems. Golan (2018) explains the general motivation as follows:

> The maximum entropy solution . . . can be thought of as the most uninformed solution out of all the solutions that satisfy the available information . . . The maximum entropy principle . . . eliminates all irrelevant details from the calculations . . . From the point of view of someone who wants to understand the broad macroscopic features of a system, most of the microscopic details turn out to be irrelevant, because most are unobserved . . . Any inferential method that manages to ignore irrelevant microscopic details will therefore have a huge gain in efficiency.

Maximum entropy theory treats aggregation and fragmentation in exactly the same way. Instead, the main distinction is discrete vs. continuous particle systems. First, consider discrete particle systems. In this case, the *partition function* Ω is defined as the number of ways in which N_{total} identical monomers can be grouped into sets. More specifically, let Ω be the number of ways in which N_{total} identical monomers can be formed into n_1 monomers, n_2 dimers (i.e., particles composed of two monomers), and n_i i-mers (i.e., particles composed of i monomers). Then, the *most probable size distribution* is obtained by maximizing Ω, subject to any constraints that might exist. Alternatively, similar to Boltzmann's equation from statistical mechanics, the *particle size entropy* may be defined as follows:

$$S_\Omega = k_\Omega \ln \Omega \tag{5.9}$$

where k_Ω is constant. Intuitively, anything that causes particle sizes to fluctuate in random or apparently random ways should cause the particle size entropy to increase. Then, the most probable size distribution is obtained by maximizing S_Ω, subject to any constraints that may exist. When no selection bias is present, maximizing S_Ω is equivalent to maximizing

Ω. However, when selection bias is present, Matsoukas (2014) warns that "the maximized quantity . . . is not the entropy [S_Ω] of the distribution as in statistical mechanics, but the microcanonical partition function [Ω] . . . It is the partition function that is of fundamental importance."

In the earliest known work, Flory (1942) used maximum entropy theory to derive an exponential size distribution. As noted in Example 5.2, Flory (1936) had already obtained the same result earlier using stochastic geometry. However, maximum entropy theory is simpler and avoids potentially limiting assumptions about the shapes of the resulting particles. Inspired by Flory, Stockmayer (1943) used maximum entropy theory to obtain the following size distribution:

$$f_i \cong A i^{-3/2} \exp(-bi) \tag{5.10}$$

for large enough i, where A and b are constants. This is a universal gamma size distribution with $n/m = -1/2$. Stockmayer's derivation is lengthy and relies on advanced combinatoric reasoning, including an analogy involving frames connected by bolts.

Next, consider continuous particle systems. Because such systems do not involve identical monomers, statistical mechanics does not apply. Instead, similar to Shannon's information theory, the particle size entropy may be defined as follow:

$$S_M = -k_M \int_0^\infty \ln[f(M)]\, f(M)\, dM \tag{5.11}$$

where k_M is a constant. In the first known work, Grady & Kipp (1985) maximized S_M subject to the following two constraints:

$$\int_0^\infty f(M)\, dM = 1$$

$$\int_0^\infty M f(M)\, dM = M_{avg}$$

which are both true by definition. This leads to the following exponential size distribution:

$$f(M) = \frac{1}{M_{avg}} \exp\left(-\frac{M}{M_{avg}}\right). \tag{5.12}$$

In this case and in general, maximum entropy theory relies on a classic technique, described in most calculus textbooks, known as *Lagrange multipliers*. Grady & Kipp (1985) showed that the same result can be obtained using stochastic geometry. However, maximum entropy theory is simpler and avoids potentially limiting assumptions about the shapes of the resulting particles.

In addition to Equation (5.11), there are many other possible definitions for particle size entropy. More generally, the particle size entropy can be defined as follows:

$$S_X = -k_X \int_0^{\infty} \ln[f(X)]\, f(X)\, dX \tag{5.13}$$

where k_X is a constant and X is any particle size metric such as diameter D, inverse diameter D^{-1}, mass M, inverse mass M^{-1}, and so forth.

In the current state-of-the-art, maximum entropy theory for discrete particle systems only leads to a few different analytic solutions, including gamma size distributions with n/m equal to 1, −1/2, and −3/2. By contrast, maximum entropy theory for continuous particle systems may lead to many different analytic solutions. In fact, almost any size distribution can be obtained through some choice of X and constraints on $f(X)$; see, e.g., Table 17.1 in Cover & Thomas (2006) and Table 5.1 in Park & Bera (2009). Thus, for continuous particle systems, extra steps must be taken to ensure that the resulting solutions have a physical basis. Despite such concerns, Example 5.5 shows that maximum entropy theory has been used with reasonable success to model aggregation and fragmentation for over 70 years.

Example 5.5 Historical overview of maximum entropy theory for aggregation and fragmentation

Flory (1942) was the first known researcher to use maximum entropy theory to describe aggregation in discrete particle systems. According to Chayut (1993), Flory used maximum entropy theory to show "that the number of chains present in a given polymer solution decrease exponentially with increasing length of the polymer chains. This is the simplest, the most probable, kind of distribution of species formed in polymerization." At that time, Paul J. Flory worked at the Esso Laboratories of the Standard Oil Development Company. Later, in 1948, he joined the faculty of Cornell University. Flory went on to produce over 500 publications and win the Nobel Prize for Chemistry in 1974. Patterson et al. (2016) called him "one of the greatest scientists of the twentieth century."

Stockmayer (1943) was the second researcher to use maximum entropy theory to describe aggregation in discrete particle systems. According to Kovac & Fixman (2017):

> **In 1941 Stockmayer moved to Columbia University, which had advertised for an instructor to teach evening classes for extension students. He was interested in Columbia because Joseph E. Mayer—perhaps the preeminent figure in statistical mechanics in the United States at that time—was a member of the chemistry department. Mayer and his wife Maria Goeppert Mayer had just published their textbook, *Statistical Mechanics*, on the subject... As he [Stockmayer] developed the theory, he had help from Maria Mayer in**

understanding the complicated combinatorics. He also wrote to Flory, who invited him to visit at the Esso Laboratory in Linden, NJ. They had a productive conversation, and Flory encouraged Stockmayer to write up his results . . . All this research was done on the side, because the first priority after teaching was war work.

Griffith (1943) was the first known researcher to use maximum entropy theory to describe fragmentation in discrete systems. In this work, he independently developed an approach similar to Flory (1942). In particular, like Flory, he derived an exponential size distribution. At that time, Laurence Griffith worked for the Research Department of the Hudson Bay Mining and Smelting Company in Manitoba, Canada.

Grady & Kipp (1985) were the first known researchers to use maximum entropy theory to describe unitary fragmentation in continuous solid systems. At that time, Dennis Grady and Martin Kipp worked at Sandia National Laboratory in Albuquerque, NM. Thadhani (2007) has cited Grady "for his pioneering contributions into the fundamental principles controlling dynamic failure and fragmentation," noting that he "has published over 300 technical papers and reports on a range of materials and applications issues in the intense shock environment."

Sellens & Brzustowski (1985) were the first known researchers to use maximum entropy theory to describe unitary fragmentation in continuous liquid systems. At that time, Richard Sellens and Thomas Brzustowski worked at the University of Waterloo in Ontario, Canada. According to Sweet (2020), Brzustowski published over 50 papers and received numerous honors and awards over his lifetime, including the Ontario Professional Engineering Association's Gold Medal in 2006.

5.4 Dimensional Analysis

This section briefly describes how to apply a general analytic technique, known as dimensional analysis, to aggregation and fragmentation problems. As a rough description, dimensional analysis involves forming algebraic expressions from critical parameters based on their units. According to Simon et al. (2017), "dimensional analysis removes unessential information . . . thus reducing the number of variables." This "unessential information" may include random or seemingly random fluctuations. As will be discussed later in Chapter 8, dimensional analysis has long been used to derive structure functions and kinetic energy spectra; see, e.g., Kolmogorov (1941b, c). However, relatively few researchers have used dimensional analysis to derive particle size distributions; examples include Friedlander (1960a, b), Hunt (1982), and Connaughton et al. (2004, 2006). As one limitation, dimensional analysis can only be used to derive power law size distributions—unlike the other analytic techniques discussed in this chapter, dimensional analysis cannot be used to derive exponential, gamma, or other types of size distributions. Example 5.6 discusses how dimensional analysis may be used to model particle aggregation.

Example 5.6 Dimensional analysis for temporally homogeneous aggregation in continuous particle systems

In general, aggregation causes particle sizes to increase over time. However, if there is a source of small particles that replaces the ones lost to aggregation, aggregation may be stationary, a.k.a., temporally homogeneous. In such cases, suppose the four critical variables are as follows:

1.	Particle mass	M	g
2.	Particle concentration	c	$g^{-1}cm^{-3}$
3.	Source mass flux per unit volume	S	$g\ cm^{-3}s^{-1}$
4.	Average collision rate times volume	K	cm^3s^{-1}

The cgs units given previously assume three-dimensional particles, i.e., $m = 3$. If S and K are constants, then dimensional analysis gives the following size distribution:

$$c(M) = A_1 (S/K)^{1/2} M^{-3/2} \tag{5.14}$$

where A_1 is a unitless constant and M is larger than some minimum M_{min} value imposed by the small-particle source. This is a power law size distribution with $n/m = -1/2$. Friedlander (1960a, b) first used dimensional analysis to derive this result. More recently, Connaughton et al. (2004) offered a more advanced proof that works for a wide variety of different K, not limited to constant values. According to Jeffrey (1981), dimensional analysis implicitly assumes that "collisions between particles of about the same size will make the most contribution . . . If the most important coagulation occurs between particles of very different size, then [dimensional analysis] . . . is inappropriate from the start." In addition, according to Jeffrey (1981) and White (1983), dimensional analysis only works if the source introduces particles with a single mass M_{min}. If the source introduces particles with a range of different masses, a parameter must be introduced to characterize the mass spread. In this case, White (1983) says there are two "physically independent quantities with the same units. Dimensional reasoning alone cannot, in consequence, yield an expression."

5.5 Semi-Empirical Models

The analytic methods discussed so far obtain complete solutions, i.e., the free parameters can be chosen based on initial conditions, boundary conditions, and so forth. However, the vast majority of analytic methods obtain incomplete solutions, i.e., some of the free parameters depend on the unknown solution. In such cases, the parameters that depend on the unknown solution must be estimated based on observations, experiments, or computations. The result is generally known as a *semi-empirical model*. As shown in Example 5.7, semi-empirical models tend to be highly specific to a single application. This book is focused on models that apply equally to a wide variety of different applications. Thus, this book will rarely discuss semi-empirical methods.

Example 5.7 Semi-empirical methods for instability-driven liquid fragmentation in automotive engines

In automotive engines, liquid fuel sprays are fragmented, a.k.a., *atomized*, to obtain efficient combustion. Magnotti & Genzale (2019) say that "the most widely employed . . . breakup model used within nearly all . . . [of the relevant finite-volume] codes, such as KIVA, CONVERGE, and OpenFOAM, assumes that the growth of hydrodynamic instabilities, formed due to the velocity difference at the liquid-gas interface of fuel spray, is the sole mechanism driving the primary breakup process." More specifically, this model assumes parent droplets with diameter D experience aerodynamic breakup due to Kelvin–Helmholtz (KH) surface instabilities. A linearized stability analysis obtains:

$$\frac{dD}{dt} = \frac{D - D_c}{\tau_{KH}}$$

where:

$$D_c = 2B_0\Lambda_{KH}$$

and:

$$\tau_{KH} = \frac{3.726B_1 a}{\Lambda_{KH}\Omega}.$$

In these expressions, τ_{KH} is a characteristic breakup time, Ω is the maximum growth rate of the most unstable surface wave, and Λ_{KH} is the wavelength of the most unstable surface wave. In addition, B_0 and B_1 are empirical constants chosen to "achieve agreement between modeled and measured liquid-phase penetration." There are multiple possible liquid fragmentation mechanisms including liquid turbulence, gas turbulence, cavitation, and Rayleigh–Taylor instabilities. Thus, Magnotti & Genzale (2019) say "it seems unlikely that a spray model assuming a single breakup mechanism would be capable of yielding reliable prediction . . . Although the calibrated spray model [as given previously] was found to capture many of the experimentally observed trends . . . results suggest that a spray breakup model premised on aerodynamic breakup, such as the KH model, is not fully predictive in representing the spray structure." For additional discussion of the KH breakup model given previously see, e.g., Chryssakis et al. (2011) and Bravo & Kweon (2014).

5.6 Empirical Models

This section briefly describes empirical models for aggregation and fragmentation problems. According to Pardo (2016) empirical modeling involves "making inferences about parameters of a probability distribution function. An inference is a generalization made from some specific data . . . The data are presumed to be a (relatively) small subset of values obtained, measured, or observed in some way from a larger population (sample space)." This section will consider one type of empirical model, namely, those based on curve fits to experimental or computational results. There are two main challenges with curve fitting: (a) determining the best possible functional form; and (b) determining the best possible values for the free parameters in that functional form. In terms of challenge (a), common choices for aggregation and fragmentation include power law, exponential, gamma, and Weibull size distributions, as discussed earlier in Section 3.7. In terms of challenge (b), if the particle system is spatially homogeneous and isotropic, curve fits may yield seemingly universal parameters. Of course, it is hard to be completely certain, since curve fits produce real numbers for the free parameters, generally with some error bounds, rather than rational numbers. However, if nearly the same value for a given parameter is found over and over again, under varying conditions, it may be reasonable to conclude that this is a universal value.

The difficulty of curve fitting increases as the number of free parameters in the chosen form increases. The simplest forms involve only one free parameter, e.g., power law and exponential size distributions. These forms may appear to involve two free parameters; however, one parameter is not actually free due to the constraint that the probability density function must have an integral of one. For power law size distributions, the free parameter is an exponent such as n/m. For exponential size distributions, the free parameter is a characteristic size, such as the mass mean diameter. As an obvious way to distinguish power laws from other possible parametric forms, power law size distributions should be linear when plotted in a log-log plane. Similarly, as an obvious way to distinguish exponentials from other possible parametric forms, exponential size distributions should be linear when plotted in a linear-log plane. For example, Figure 5.2 shows a possible power law curve fit in the log-log plane. In this figure, the smallest fragments appear to obtain a power law

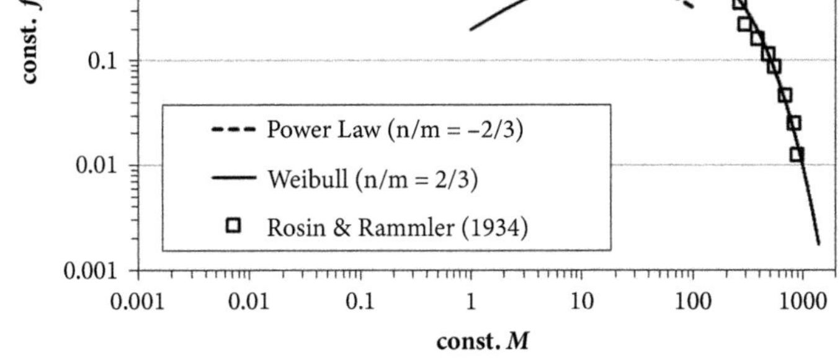

Figure 5.2 Empirical vs. experimental results for fragment size distributions from ball milling of iron. The notation "const." used in the axis labels refers to an arbitrary constant.
Experimental results taken from Rosin & Rammler (1934); see also Brown & Wohletz (1995).

size distribution with $n/m \cong -2/3$. However, Clauset et al. (2009) warns that "it is difficult to rule out alternative fits to such data, even when they are truly power-law distributed, and conversely the power-law form may appear to be a good fit even when the data are drawn from a non-power-law distribution." Furthermore, "commonly used methods for analyzing power-law data, such as least-squares fitting, can produce substantially inaccurate estimates of parameters for power-law distributions" due to issues such as "the large fluctuations that occur in the tail of the distribution—the part of the distribution representing large but rare events." As a possible alternative, Clauset et al. (2009) discuss maximum likelihood estimation (MLE). However, they say that "MLEs will give us no warning that our fits are wrong: they tell us only the best fit to the power-law form, not whether the power law is in fact a good model for the data." In addition, Cady (2017) says a "significant problem with MLE in general . . . is overfitting; especially if your dataset is small, you run the risk of getting parameters that predict your specific data very well, but generalize terribly." As result, most empirical aggregation and fragmentation models are still based on least-squares fitting. In fact, for model design, the number of different data sets is ultimately more important than the quality of the curve fitting done for any particular data set; see Example 5.8.

Example 5.8 Universal power law size distributions for mining engineering and other applications

In an early study of rock crushing, Gates (1915) found that rock fragments often obtained power law size distributions. This was later confirmed by Gaudin (1926) and Schuhmann (1940). Gates (1915) characterized fragment size distributions based on the "reciprocal of the diameter" D^{-1} and "the weight of the material passing through a screen of this diameter opening" $f_M(D)$; see also Example 3.3. The simplest possible size distribution involving these variables is $f_M(D)$= const. D^{-1}. In fact, according to Åström et al. (2004), "within the mining engineering community, an empirical scale-invariant fragment size distribution has long been known as ...the Gates–Gaudin–Schuhmann distribution. According to this distribution, the total mass of fragments . . . scales over a few orders of magnitude in the small-fragment limit as" $f_M(D)$ = const. D^{-1}. As shown in Appendix A.2, this is equivalent to $f(M)$ = const. $M^{n/m-1}$ where $n/m = -1$. In an attempt to validate and extend this result, Carpinteri & Pugno (2002a, b; 2004, 2005) reviewed a wide variety of data including early work done in the mining engineering community. Based on this review, they conclude that power laws such as $n/m = -2/3, -5/6$, and -1 are "universal laws" that occur in many different applications including drilling, impact fragmentation, and "biological cell growth processes . . . for small mammals."

The next simplest forms involve two free parameters. For Weibull and gamma size distributions, the two free parameters are an exponent and a characteristic particle size, as explained earlier in Section 3.7. For example, in Figure 5.2, the largest fragments appear to obtain a Weibull size distribution with $n/m \cong 2/3$. For another example, see Example 5.9.

Example 5.9 Empirical Weibull models for fragmentation

Mott & Linfoot (1943) analyzed a wide range of experimental results involving "a model bomb with a casing." Based on curve fits to the casing fragment masses, Mott & Linfoot (1943) suggested the following size distribution:

$$f(M) = \frac{A_1}{M_{01}}\left(\frac{M}{M_{01}}\right)^{-2/3} \exp\left[-b_1\left(\frac{M}{M_{01}}\right)^{1/3}\right] \tag{5.15}$$

where A_1 and b_1 are unitless constants and M_{01} is a characteristic mass. This is a Weibull size distribution with $n/m = 1/3$. In addition, they suggested the following size distribution:

$$f(M) = \frac{A_2}{M_{02}}\left(\frac{M}{M_{02}}\right)^{-1/2} \exp\left[-b_2\left(\frac{M}{M_{02}}\right)^{1/2}\right] \tag{5.16}$$

where A_2 and b_2 are unitless constants and M_{02} is a characteristic mass. This is a Weibull size distribution with $n/m = 1/2$. As discussed earlier in Example 5.4, Mott & Linfoot (1943) were the first known researchers to use stochastic geometry to describe particle fragmentation in two dimensions. However, their empirical models obtained significantly better agreement with experimental data.

Most experimental measurement techniques only produce results over a limited size range. For example, optical measurement techniques generally have a limited resolution (e.g., pixel size) which imposes a minimum particle size, and a limited field-of-view which imposes a maximum particle size. For Weibull and gamma size distributions, Yano et al. (2016) notes that if the range spans "only one decade of the particle diameter" or less—i.e., if there is a factor of ten or less between the maximum and minimum diameter—then different exponents n/m often obtain equally good fits to the experimental data. "To obtain more robust results, data for a much wider range [of] . . . sizes is required." In fact, ideally, particle diameters should span at least three orders-of-magnitude or, equivalently, particle masses should span at least one-and-a-half orders-of-magnitude. As one way to view the issue, suppose that one of the two free parameters is an average particle size. Any experimental technique that imposes an artificial minimum size will typically overestimate the average particle size. For Weibull and gamma size distributions, overestimating one free parameter (i.e., the average) leads to overestimating the other free parameter (i.e., the exponent n/m). Similar remarks apply to any experimental technique that imposes an artificial maximum size. Ideally, errors due to artificial maxima and minima will cancel each other out. However, in more typical cases, one dominates the other. For example, suppose improved optical techniques are used to reduce the minimum observable particle size without changing the maximum observable particle size. If the minimum decreases while the maximum stays the same, then the estimated n/m will tend to decrease—not because of any change in the physics of the experiment but only because of changes in how it is observed.

Like power laws and exponentials, Weibull and gamma fits may benefit from transforming them to planes in which they are linear. Example 5.10 discusses this in more detail.

Example 5.10 The classic approach to curve fitting experimental data from aggregation and fragmentation

Consider Weibull size distributions where:

$$\bar{F}(M) = 1 - F(M) = \exp\left[-b_0(M/M_{\text{char}})^{n/m}\right]. \tag{5.17}$$

Mock & Holt (1983) studied the "fragmentation response of explosive-filled cylinders of . . . HF-l steel." In one experiment, curve fitting in the log-log plane obtains $n/m \cong 1/2$. As seen in Figure 5.3, this curve fit provides a good agreement with experimental results. In another experiment, curve fitting in the log-log plane obtains $n/m \cong 7/12$. As seen in Figure 5.4, this curve fit provides only mediocre agreement with experimental results. As an alternative, suppose the curve fitting is done in a different plane where $\bar{F}(M)$ is linear. For example, if $n/m = 1/2$, then $\bar{F}(M)$ is linear in a square-root-log plane. This is true because:

$$\ln \bar{F}(M) = -\text{ const. } M^{1/2} + \text{const.} \tag{5.18}$$

where "const." refers to a constant. This approach was first suggested by Mott & Linfoot (1943). The resulting plots are sometimes called *Mott plots*. As seen in Figure 5.5, when viewed in this plane, a Weibull size distribution with $n/m = 1/2$ obtains good agreement with experiment results, but only for fragment masses less than about 100 grains (6.5 grams). Above 100 grains, the slope of the experimental data changes. This change in slope is not obvious in the log-log plane; however, it is unmistakable in the Mott plot.

If explosive forces overwhelm the cylinder strength, most of the fragments form in three-dimensions. This can be confirmed by examining the shape of the fragments. More specifically, fragments formed in three-dimensions are rough on all sides. In such cases, a single Weibull size distribution tends to obtain a good fit to test data regardless of the fragment size; see Figure 5.3. In less extreme cases, many of the larger fragments form in two-dimensions. Again, this can be confirmed by examining the shape of the fragments. More specifically, fragments formed in two-dimensions have smooth surfaces that were originally part of the inner and outer surfaces of the cylindrical shell. In cases that generate a mixture of two- and three-dimensional fragments, a single Weibull size distribution tends to obtain a mediocre fit to test data; see Figure 5.4. Instead, suppose one Weibull size distribution is used for the smaller three-dimensional fragments with diameters less than the shell thickness, while another Weibull size distribution is used for the larger two-dimensional fragments with diameters greater than the shell thickness. In such cases, the two Weibull size distributions together tend to obtain a good fit to test data; see Figure 5.5.

In general, transformation techniques such as Mott plots are a valuable tool for determining the range of validity of a given curve fit. However, transformation techniques do not address every issue with empirical curve fitting. In particular, as discussed in the main text, the ambiguity in the curve fit tends to increase as data range decreases. Unless a given trend spans a wide enough range of sizes, it may be impossible to characterize it accurately.

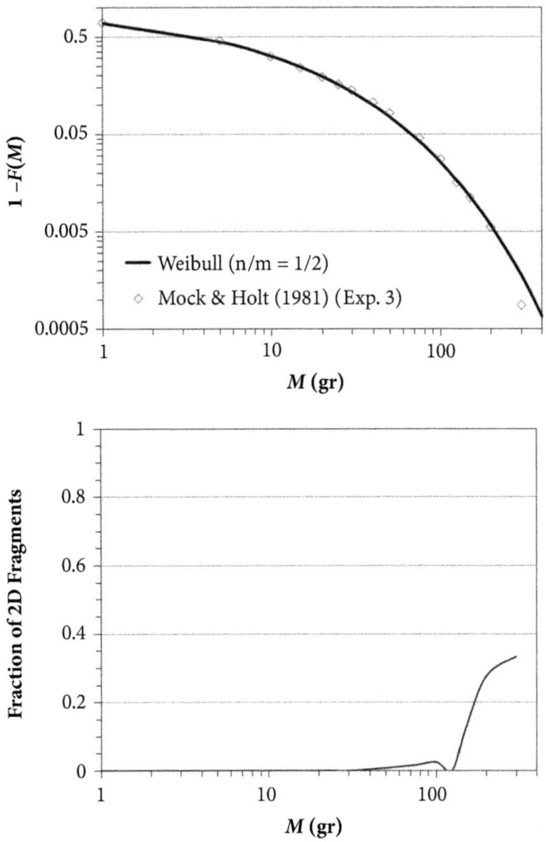

Figure 5.3 Empirical vs. experimental results for fragment size distributions from highly-loaded explosively-driven cylindrical steel shells. In this case, where most of the fragments form in three dimensions, a single Weibull size distribution obtains a good fit to test data.
Experimental results taken from Mock & Holt (1981).

Now consider forms involving three free parameters. This includes Rosin–Rammler size distributions; see Example 3.9. Babinsky & Sojka (2002) found that Rosin–Rammler size distributions with three free parameters often obtain substantially more accurate curve fits to specific sets of experimental data than Weibull or gamma size distributions with only two free parameters. In fact, "generally, it was found that the greater the number of adjustable parameters in the equation of the distribution function, the better the fit to experimental data. This is to be expected, given the greater degree of 'shape freedom' provided by distributions with a large number of parameters." In many cases, different combinations of the three free parameters in Rosin–Rammler size distributions obtain almost equally good fits to experimental data. Unlike forms with one or two parameters, this ambiguity cannot be resolved by transformations or by improved measurement techniques that capture a broader range of particle sizes. Instead, Babinsky & Sojka (2002) say that Rosin–Rammler size distributions have fundamental "problems with the stability of parameters." Similarly, Voutilainen et al. (2001) say "from the mathematical point of view, the determination of the

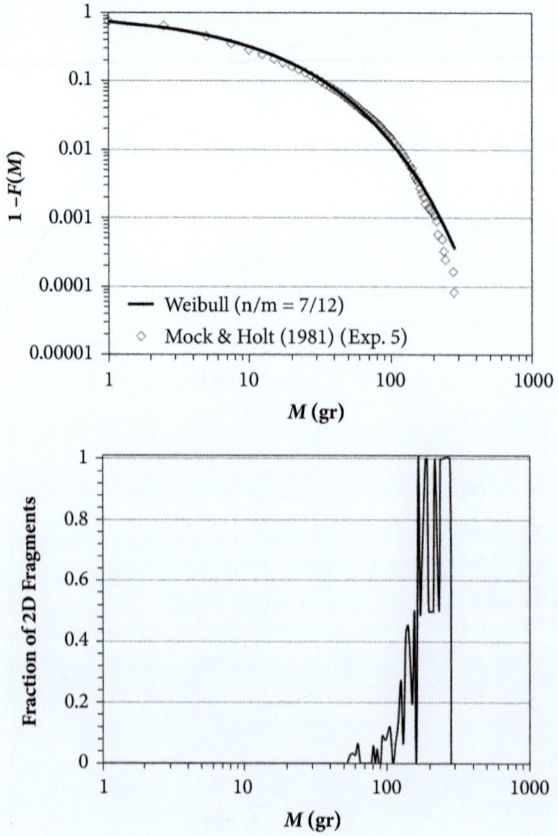

Figure 5.4 Empirical vs. experimental results for fragment size distributions for medium-loaded explosively-driven cylindrical steel shells. In this case, where fragments form both in two- and three-dimensions, a single Weibull size distribution obtains a mediocre fit to test data. Experimental results taken from Mock & Holt (1981).

[Rosin–Rammler] size distribution function is an ill-posed problem . . . Since the problem is usually severely underdetermined, the solution is not unique. In order to stabilize the problem some further prior information on the solution must be incorporated to the problem." As one possible approach, Voultainen et al. (2001) suggest using "Tikhonov-type regularization methods;" however, such techniques are rarely used in particle aggregation and fragmentation studies. As a more common approach, Yano et al. (2016) recommend combining experimental results with analytic theory, such as maximum entropy theory, to limit the range of possible outcomes.

So far, all of the empirical models discussed in this section concern ideal cases, i.e., cases that are spatially homogeneous and isotropic. Figure 5.6 shows an example of non-ideal fragmentation known as *bench blasting*. Example 5.11 describes a non-ideal empirical model developed for bench blasting. As this example shows, non-ideal empirical models tend to be highly specific to a single application. This book is focused on models that apply equally to a wide variety of different applications. Thus, this book will rarely discuss non-ideal empirical models.

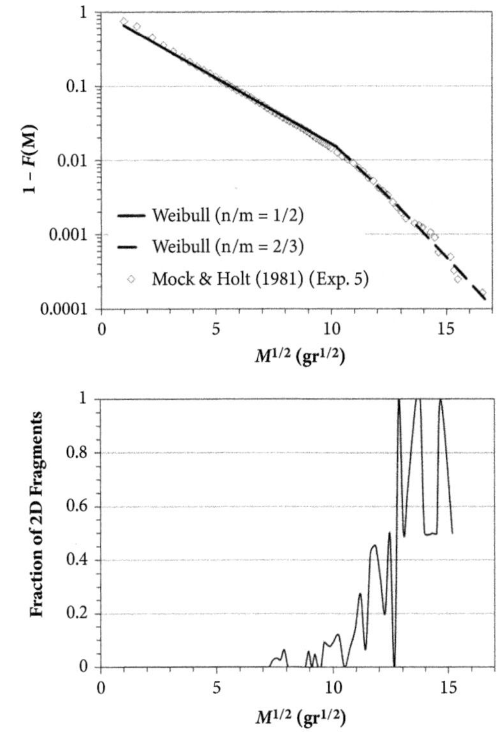

Figure 5.5 Empirical vs. experimental results for fragment size distributions for medium-loaded explosively-driven cylindrical steel shells. When fragments form in both two- and three-dimensions, using one Weibull size distribution for the smaller three-dimensional fragments, and a different Weibull size distribution for the larger two-dimensional fragments, obtains a good fit to test data.
Experimental results taken from Mock & Holt (1981).

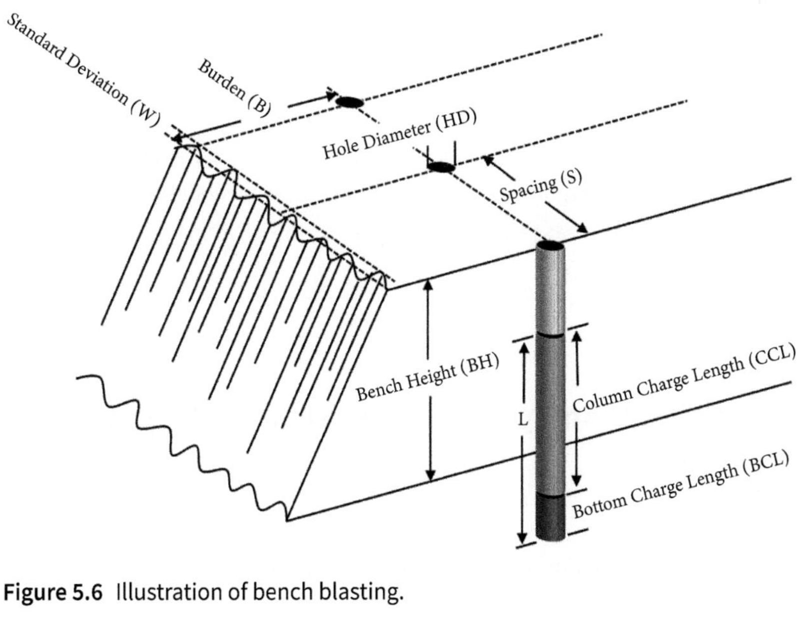

Figure 5.6 Illustration of bench blasting.

Example 5.11 Non-ideal empirical models for bench blasting

In *bench blasting*, a standard technique used in rock quarries, one or more rows of evenly spaced holes are drilled adjacent to the open face of a rock bench; see Figure 5.6. Each hole is filled with a two-part explosive, typically ammonium nitrate–fuel oil (ANFO) as a *column charge* and a dynamite booster as a *bottom charge*. According to Cunningham (1983, 1987, 2005), the resulting fragments typically obtain size distributions of the following form:

$$F_M(D) = 1 - \exp\left[-0.693\left(\frac{D}{D_m}\right)^n\right] \tag{5.19}$$

where D_m is a characteristic fragment diameter and n is an exponent. In the mining engineering community, distributions of this general form are called *Rosin–Rammler–Sperling–Bennett (RRSB) size distributions*; see Rosin, Rammler & Sperling (1933) and Appendix A.6. According to Kuznetsov (1973) and Cunningham (1983, 1987, 2005):

$$D_m = 0.06(RMD + RDI + HF)K^{-0.8}\ Q^{1/6}(115/RWS)^{19/20} \tag{5.20}$$

where RMD is a "rock mass description," RDI is a rock "density influence," HF is rock "hardness factor," K is the "kg of explosive per cubic meter of rock," Q is the "mass of explosive in the hole," and RWS and is the explosive "weight strength relative to ANFO." In addition, according to Cunningham (1983, 1987, 2005):

$$n = \left(2.2 - 14\frac{B}{HD}\right)\left(1 - \frac{W}{B}\right)\sqrt{\frac{1}{2} + \frac{S}{2B}}\left(0.1 + \left|\frac{BCL - CCL}{L}\right|\right)^{0.1}\frac{L}{BH} \tag{5.21}$$

where B, L, S, W, HD, BH, BCL, and CCL are defined in Figure 5.6. Notice that all lengths in Equation (5.21) are measured in meters except for HD, which is measured in millimeters. Because it combines an expression for the characteristic fragment diameter suggested by Kuznetsov (1973) with a fragment size distribution suggested by Rosin, Rammler & Sperling (1933), this is called the *Kuz–Ram model*. The Kuz–Ram model appears in most modern textbooks on rock blasting; see, e.g., Roy (2005), Rustan et al. (2010), Zhang (2016), and Zou (2017). Like all non-ideal fragmentation models, the Kuz–Ram model only applies to the specific problem for which it was designed—it implies little if anything about fragmentation in general.

5.7 Examples of Analytic and Empirical Model Results for Aggregation

This section gives examples of results from analytic and empirical methods for aggregation. If a particle system is spatially homogeneous and isotropic, analytic and empirical methods generally obtain ideal size distributions. In other words, as shown in Table 5.1, n/m is a

Table 5.1 Sizes distributions for late-time aggregation obtained with analytic and empirical methods.

a. Power law size distributions.

n/m	*m*	*d*	Field	References
−1/2[1]	3	3	Astrophysics	Field & Saslaw (1965)
−1/2[4]	3	3	Aerosol science	Friedlander (1960a, b)
−1[4]	3	3	Aerosol and hydrosol science	Hunt (1982)
−7/6[1]	3	3	Aerosol science	Klett (1975)

b. Weibull size distributions.

n/m	*m*	*D*	Field	References
−1/3[1]	3	3	Cloud physics	Horvai, Nazarenko & Stein (2008)
1/3[5]	3	3	Cloud physics	Marshall & Palmer (1948)
2/3[3]	3	3	Cloud physics	Liu et al. (1995); Liu (1995)
1[1]	3	3	Aerosol science	Schumann (1940)

c. Gamma size distributions.

n/m	*m*	*D*	Field	References
−1/2[1]	3	3	Astrophysics	Safronov (1962, 1969)
−1/2[1]	3	3	Cloud physics	Golovin (1963)
−1/2[3]	<3[a]	3	Polymer science	Stockmayer (1943)
−1/2[5]	<3[a]	3	Colloid science	Lin et al. (1989a, b, 1990)
−3/2[1]	3	3	Aerosol science	McLeod (1962, 1964)
−3/2[3]	<3[a]	3	Polymer science	Cohen & Benedek (1982); Spouge (1983)
1[1]	<3[a]	3	Colloid and aerosol science	Smoluchowski (1916, 1917)
1[2,3]	1	1	Polymer science	Flory (1936, 1942)

(1) Mean-field theory
(2) Stochastic geometry
(3) Maximum entropy theory
(4) Dimensional analysis
(5) Empirical
(a) Particles typically have fractal shapes. In such cases, experimental and computational results indicate that *m* is usually in the range 1.4–2.6; see, e.g., Lin et al. (1989a, b, 1990), Kempf et al. (1999), Jungblut (2019a, b).

universal rational number such as $\pm 1/3$, $\pm 1/2$, or ± 1 regardless of the exact details. According to Table 5.1, when aggregation is viewed through the prism of a single field, there appears to be only a few universal size distributions. However, when aggregation is viewed across a range of different fields, there are at least a dozen universal size distributions. Notice that Table 5.1 includes examples from the following five fields:

1. *Aerosol science.* These applications involve small particles, less than 100μm in diameter, suspended in air such as fog, haze, steam, smoke, and soot. A closely related field, *hydrosol science,* concerns small particles suspended in water such as marine sediments.
2. *Astrophysics.* These applications include collisional aggregation of molecular clouds, aggregation of dust and debris in planetary rings and circumstellar disks, and aggregation of dust leading to protoplanet formation.

3. *Cloud physics.* These applications include the formation and evolution of cloud droplets, rain droplets, ice crystals, snowflakes, and hail stones due to aggregation and fragmentation, including the effects of aerodynamic breakup.
4. *Colloid science.* These applications include the formation and destruction of clusters of atoms and molecules with diameters up to 1μm or so, bonded by van der Waals or other non-covalent forces, suspended in a fluid medium.
5. *Polymer science.* These applications include the formation and, in some cases, simultaneous destruction of long chains and cross-linked networks of covalently bonded atoms and molecules.

Table 5.2 Books that discuss standard size distributions for aggregation.

(a) Aerosol & colloid sciences: Smoluchowski (1916, 1917) ($n/m = 1$)
I. Agranovski, *Aerosols—Science and Technology*, Wiley, 2010
J. C. Berg, *Interfaces and Colloids: The Bridge to Nanoscience*, World Scientific, 2010
I. Colbeck & M. Lazaridis, *Aerosol Science: Technology and Applications*, Wiley, 2014
J. Gregory, *Particles in Water*, CRC Press, 2006
J. Lyklema, *Fundamentals of Interface and Colloid Science*, Elsevier, 2005
Z. Tan, *Air Pollution and Greenhouse Gases: From Basic Concepts to Engineering Applications for Air Emission Control*, Springer, 2014
(b) Cloud physics: Marshall–Palmer (1948) ($n/m = 1/3$)
R. A. Houze Jr., *Cloud Dynamics*, 2nd edition, Academic Press, 2014
A. P. Khain & M. Pinsky, *Physical Processes in Clouds and Cloud Modeling*, Cambridge University Press, 2018
V. I. Khvorostyanov & J. A. Curry, *Thermodynamics, Kinetics, and Microphysics of Clouds*, Cambridge University Press, 2014
D. Lamb & J. Verlinde, *Physics and Chemistry of Clouds*, Cambridge University Press, 2011
U. Lohmann, F. Lüönd, & F. Mahrt, *An Introduction to Clouds: From the Microscale to Climate*, Cambridge University Press, 2016
H. R. Pruppacher & J. D. Klett, *Microphysics of Clouds and Precipitation*, 2nd edition, Springer, 2010
J. H. Seinfeld & S. N. Pandis, *Atmospheric Chemistry and Physics*, 3rd edition, Wiley, 2016
P. K. Wang, *Physics and Dynamics of Clouds and Precipitation*, Cambridge University Press, 2013
D. S. Wilks, *Statistical Methods in the Atmospheric Sciences*, 3rd edition, Academic Press, 2011
(c) Colloid science: Lin et al. (1989, 1990) ($n/m = -1/2$)
F. Babick, *Suspensions of Colloidal Particles and Aggregates*, Springer, 2016
R. Hidalgo-Alvarez, *Structure and Functional Properties of Colloidal Systems*, CRC Press, 2017
P. Somasundaran, *Encyclopedia of Surface and Colloid Science*, 2nd edition, volume 2, CRC Press, 2006
G. Sposito, *The Surface Chemistry of Natural Particles*, Oxford University Press, 2004
(d) Polymer science: Flory (1936, 1941, 1942), a.k.a., Flory–Schulz ($n/m = 1$)
M. Alger, *Polymer Science Dictionary*, 3rd edition, Springer, 2017
J. Chalmers & R. Meier, *Molecular Characterization and Analysis of Polymers*, Elsevier, 2008
M. Chanda, *Introduction to Polymer Science and Chemistry*, 2nd edition, CRC Press, 2013
P. C. Hiemenz & T. P. Lodge, *Polymer Chemistry*, 2nd edition, CRC Press, 2007
G. Odain, *Principles of Polymerization*, 4th edition, Wiley, 2004
F. Rodriguez et al., *Principles of Polymer Systems*, 6th edition, CRC Press, 2014

Except for astrophysics, each of the previous fields has adopted a single *standard* size distribution for aggregation. Table 5.1 highlights these standard size distributions in gray. Table 5.2 lists a number of modern textbooks and handbooks that discuss these standard size distributions. These references indicate that these standards, despite being at least 30 years old, are still in widespread use today.

5.8 Examples of Analytic and Empirical Model Results for Fragmentation

This section gives examples of results from analytic and empirical methods for fragmentation. If a particle system is spatially homogeneous and isotropic, analytic and empirical methods generally obtain ideal size distributions. In other words, as shown in Table 5.3, n/m tends to be a universal rational number such as $\pm 1/3$, $\pm 1/2$, or ± 1 regardless of the exact details. According to Table 5.3, when fragmentation is viewed through the prism of a single field, there appears to be only a few universal size distributions. However, when fragmentation is viewed across a range of different fields, there are at least a dozen universal size distributions. Notice that Table 5.3 includes examples from the following six fields:

1. *Astrophysics.* These applications include collisional fragmentation of asteroids and collapse-driven fragmentation of molecular clouds, ultimately leading to the formation of protostars, stars, circumstellar disks, etc.
2. *Liquid sprays.* These applications include conversions of bulk liquids into droplets, e.g., by passing a thin stream of liquid through a nozzle. The resulting droplets usually form in two stages, with fragmentation in the second stage driven by aerodynamic forces.
3. *Mining engineering.* These applications include drilling and blasting of materials such a rocks, minerals, and concrete. In addition, these applications include crushing, grinding, milling, and other processes used to decrease size and increase uniformity.
4. *Small-scale explosions.* These applications include high-speed fragments from compact metal-cased high-explosives, including sympathetic detonations, and high-speed projectile impacts on metal, ceramic, or other types of armor.
5. *Large-scale explosions.* These applications include cratering and ejecta of geologic materials, as well as planar blast damage to building façade materials such as glass windows, brick masonry walls, clay tile roofs, and concrete spandrel panels.
6. *Space debris.* These applications include collisions between high-speed debris and space systems. Fragments from one space system may impact other space systems, conceivably leading to catastrophic cascades, as featured in the 2013 film *Gravity*.

With the exception of mining engineering, each of the previous fields has adopted a single *standard* size distribution for fragmentation. Table 5.3 highlights these standard size distributions in gray. Table 5.4 lists a number of textbooks and handbooks that discuss standard size distributions for fragmentation. These references indicate that the standard size distributions, despite being at least 30 years old, are still in widespread use today.

Table 5.3 Sizes distribution for late-time fragmentation obtained using analytical and empirical methods.

a. Power law size distributions.

n/*m*	*m*	*d*	Field	References
−2/3[1]	3	3	Astrophysics	Piotrowski (1952, 1953)
−2/3[2]	3	3	Space debris	Gilvarry (1961); Gilvarry & Bergstrom (1961)
−2/3[2]	3	3	Astrophysics	Inaoka & Takayasu (1996, 1999)
−2/3[4]	3	3	Astrophysics	Aschwanden et al. (2016)
−2/3[5]	3	3	Mining engineering	Åström et al. (2004)
−5/6[1]	3	3	Astrophysics	Dohnanyi (1967, 1969, 1970)
−5/6[5]	3	3	Large-scale explosions	Seebaugh (1976, 1977); Schoutens (1979)
−5/6[5]	3	3	Mining engineering	Carpinteri & Pugno (2002a, b, 2004, 2005)
−0.85[5]	2	2	Space debris	Bess (1975); Johnson et al. (2001)
−1[5]	3	3	Mining engineering	Carpinteri & Pugno (2002a, b, 2004, 2005)
−1.35[5]	3	3	Astrophysics	Salpeter (1955); Kroupa (2001, 2002)

b. Weibull size distributions.

n/*m*	*m*	*d*	Field	References
−1[1]	3	3	Liquid sprays, etc.	Krapivsky & Ben-Naim (2003)
1/3[5]	3	3	Small-scale explosions	Mott & Linfoot (1943)
1/2[5]	2	2	Small-scale explosions	Mott & Linfoot (1943); Mott (1943a, b)
1[2]	1	1	Small-scale explosions	Lienau (1936a, b, c)
1[3]	3	3	Small-scale explosions	Grady & Kipp (1985)
2[2]	1	1	Small-scale explosions	Harris et al. (1973); Zhou et al. (2006)

c. Gamma size distributions. Converted from diameter to mass; see Laney (2015b; 2016a)

n/*m*	*m*	*d*	Field	References
0.62[5]	3	3	Liquid sprays	Marmottant & Villermaux (2004a)
1[3]	3	3	Liquid sprays	Li & Tankin (1987)
1.03[5]	3	3	Liquid sprays	Marmottant & Villermaux (2004b)

d. Root normal size distributions. Converted from $\sigma_{M,D}$ to *n*/*m* as shown in Laney (2015a, b).

$\sigma_{M,D}$ (*n*/*m*)	*m*	*d*	Field	References
0.29 (1/3)[5]	3	3	Liquid sprays	Spielbauer et al. (1989)
0.24 (5/12)[5]	3	3	Liquid sprays	Simmons (1977a, b)
0.20 (1/2)[5]	3	3	Liquid sprays	Empie et al. (1993, 1995, 1997)

(1) Mean-field theory
(2) Stochastic geometry
(3) Maximum entropy theory
(4) Dimensional analysis
(5) Empirical

Table 5.4 Books that discuss standard size distributions for fragmentation.

(a) Astrophysics/Solids: Dohnanyi (1967, 1969, 1970) ($n/m = -5/6$)
L. T. Elkins-Tanton & B. P. Weiss, *Planetesimals: Early Differentiation and Consequences for Planets*, Cambridge University Press, 2017
J. A. Fernandez, *Comets: Nature, Dynamics, Origin, and their Cosmogonical Relevance*, Springer, 2005
E. Grün, B. A. S. Gustafson, S. Dermott, & H. Fechtig, *Interplanetary Dust*, Springer, 2011
R. Hawks, I. Mann, & P. Brown, *Modern Meteor Science: An Interdisciplinary View*, Springer, 2005
(b) Astrophysics/Gases and Plasmas: Salpeter (1955) ($n/m \approx -1.35 \approx -4/3$)
P. H. Bodenheimer, *Principles of Star Formation*, Springer-Verlag, 2011
K. R. Lang, *Essential Astrophysics*, Springer-Verlag, 2013
M. S. Longair, *High Energy Astrophysics*, 3rd edition, Cambridge University Press, 2011
M. Salaris & S. Cassisi, *Evolution of Stars and Stellar Populations*, Wiley, 2005
L. S. Sparke & J. S. Gallagher, *Galaxies in the Universe: An Introduction*, 2nd edition, Cambridge University Press, 2007
D. Ward-Thompson & A. P. Whitworth, *An Introduction to Star Formation*, Cambridge University Press, 2011
J. R. Webb, *Extragalactic Astrophysics*, 2nd edition, IOP Publishing, 2022
(c) Liquid Sprays: Simmons (1977a, b) ($n/m \approx 5/12$)
N. Ashgriz, *Handbook of Atomization and Sprays: Theory and Applications*, Springer, 2011
C. E. Brennen, *Fundamentals of Multiphase Flows*, Cambridge University Press, 2005
E. Michaelides et al., *Multiphase Flow Handbook*, 2nd edition, CRC Press, 2016
A. H. Lefebvre & V. G. McDonell, *Atomization and Sprays*, 2nd edition, CRC Press, 2017
(d) Small-scale explosions: Mott–Linfoot (1943) ($n/m \approx 1/2$)
A. M. J. Bull et al., editors, *Blast Injury Science and Engineering*, 2nd edition, Springer, 2022
D. E. Carlucci & S. S. Jacobson, *Ballistics*, 3rd edition, CRC Press, 2018
D. Grady, *Fragmentation of Rings and Shells: The Legacy of N. F. Mott*, Springer, 2006
D. Grady, *Physics of Shock and Impact*, Volume 1, IOP Publishing, 2017
C. E. Needham, *Blast Waves*, 2nd edition, Springer, 2018
(e) Large-scale explosions: Seebaugh–Schoutens ($n/m \approx -5/6$)
R. Botet & M. Ploszajczak, *Universal Fluctuations: The Phenomenology of Hadronic Matter*, World Scientific, 2002
J. Dubois, *Non-Linear Dynamics in Geophysics*, Wiley, 1998
S. Hergarten, *Self-Organized Criticality in Earth Systems*, Springer, 2002
I. Rodríguez-Iturbe & A. Rinaldo, *Fractal River Basins: Chance and Self-Organization*, Cambridge University Press, 2001
D. L. Turcotte, *Fractals and Chaos in Geology and Geophysics*, 2nd edition, Cambridge University Press, 1997
(f) Space debris: NASA Standard Breakup Model ($n/m \approx -0.85 \approx -5/6$ when $m \approx 2$)
H. Klinkrad, *Space Debris: Models and Risk Analysis*, Springer, 2006
J. N. Pelton & F. Allahdadi, *Handbook of Cosmic Hazards and Planetary Defense*, Springer, 2015
N. N. Smirnov, *Space Debris: Hazard Evaluation and Debris*, CRC Press, 2001
A. Tan & R. C. Reynolds, *Theory of Satellite Fragmentation in Orbit*, World Scientific, 2020

The following gives some additional background information on these standard size distributions:

- *Dohnanyi size distribution.* While the other standards concern unitary fragmentation, this standard concerns cascade fragmentation. In other words, this standard applies to fragments formed through repetitive multi-stage breakup as discussed in Section 1.1.
- *Salpeter–Kroupa size distribution.* In the original paper, Salpeter (1955) found $n/m \approx -1.35 \approx -4/3$ regardless of the particle size. However, in more recent work, Kroupa (2001, 2002) and Bastian et al. (2010) found $n/m \approx 1.3 \approx -4/3$ for objects larger than half a solar mass and $n/m \approx -0.3 \approx -1/3$ for objects smaller than half a solar mass.
- *Simmons size distribution.* Instead of being specified by free parameters such as $\sigma_{M,D}$, this size distribution is usually specified by MMD/SMD = 1.21, where MMD is the mass median diameter and SMD is the Sauter mean diameter, as defined in Example 3.6.
- *Mott–Linfoot size distribution.* In the original paper, Mott & Linfoot (1943) suggested letting $n/m = 1/2$ for 2D fragments and $1/3$ for 3D fragments, respectively. However, in the end, $n/m = 1/2$ became the community standard for both 2D and 3D fragments.
- *Seebaugh–Schoutens size distribution.* While most of the other standards concern unitary fragmentation after primary breakup, this one concerns unitary fragmentation after secondary breakup. In other words, this standard applies to fragments formed through two-stage breakup as discussed in Section 1.5.
- *National Aeronautics and Space Administration (NASA) Standard Breakup Model.* According to this standard, $n \cong -1.71$. If fragments have $m = 2$, then $n/m \cong -0.85 \cong -5/6$. Bess (1975) and Johnson et al. (2001) first suggested this standard; see also Stansbery et al. (2015) and Olivieri et al. (2022, 2023).

5.9 Examples of Analytic Model Results for Aggregation and Fragmentation

Consider cases where aggregation and fragmentation occur simultaneously. As one analytic approach, maximum entropy theory treats aggregation, fragmentation, and various combinations of aggregation and fragmentation in exactly the same way. At the other extreme, mean-field theory treats aggregation, fragmentation, and various combinations of aggregation of fragmentation in different ways customized to each given problem. For example, suppose that J describes the average rate at which particles fragment into children with a number distribution m and suppose that K describes the average rate at which collisions between two particles cause them to merge. This leads to the following mean-field governing equation for aggregation and fragmentation in continuous particle systems:

$$\frac{\partial c(M,t)}{\partial t} = -c(M,t)\int_0^\infty K(M,M')\,c(M't)\,dM' - J(M)c(M,t)$$
$$+\frac{1}{2}\int_0^M K(M-M',M)\,c(M-M',t)\,c(M',t)\,dM' + \int_M^\infty m(M,M')\,J(M')\,c(M',t)\,dM' \tag{5.22}$$

where the first two terms on the right hand side represent the rate at which particles of mass M are lost due to aggregation and fragmentation, while the second two terms on the right hand side represent the rate at which particles of a given mass M are gained due to aggregation and fragmentation. Examples 5.12 and 5.13 give some possible mean-field solutions for particle systems that experience both aggregation and fragmentation.

Example 5.12 Mean-field theory for combined aggregation and fragmentation

Many aggregating particle systems also experience some degree of fragmentation. In the simplest case, where each parent creates exactly two children, fragmentation is often called *reverse aggregation*. In earliest known work, Becker & Döring (1935) studied discrete particle systems where each aggregation and fragmentation event either added or subtracted a single monomer. Table 5.5 lists six possible mean-field solutions for combined aggregation and fragmentation. The first two rows involve simple reverse aggregation. The remaining four rows involve more complex forms of fragmentation.

Table 5.5 Gamma size distributions for late-time aggregation and fragmentation obtained using mean-field theory. The first two rows involve reverse aggregation.

n/m	m	d	Field	References
1	3	3	Aerosol science	Becker & Döring (1935) [see also Wattis (2009)]
1	3	3	Polymer science	Blatz & Tobolsky (1945)
1	3	3	Astrophysics	Barrow (1981); Chervanyov & Heinrich (2013)
1	3	3	Colloid science	Patil & Andrews (1998); McCoy & Madras (1998)
−1/2	3	3	Astrophysics	Brilliantov et al. (2009, 2015); Bodrova et al. (2019)
−1/2	3	3	Mathematics	Laurençot (2019)

Example 5.13 Mean-field theory for self-organized groups of living organisms

As discussed earlier in Section 2.2, the earliest work on self-organization focused on the relationships between living and non-living systems. In this regard, Ginot et al. (2018) note that "self-organization, the spontaneous emergence of spatial structures, is a generic phenomenon occurring from atomic to macroscopic length scales, both in inert and living matter" to include "swarming bacteria, birds, or fishes." From one point of view, a living organism may be considered to be a monomer, and groups of living organisms may be considered to be unbonded clusters of monomers. A number of researchers have found that mean-field theory models of combined aggregation and fragmentation accurately describe real-life animal groupings such as schools of fish, flocks of geese, herds of buffalo, and so forth. This is true even though mean-field theory was originally used to describe bonded rather than unbonded particles. Table 5.6 lists some proposed mean-field solutions for aggregating and fragmenting systems composed of living organisms and, in one case, lifelike self-propelled colloids.

Table 5.6 Gamma size distributions for self-organized groups of living organisms and life-like self-propelled colloids obtained using mean-field theory for aggregation and fragmentation.

n/m	m	d	Field	References
−1	3	3	Biology	Ma et al. (2011)
$-0.850 \approx -5/6$ [1]	3	3	Colloids	Moran & Posner (2017); Ginot et al. (2018)
$-0.698 \approx -2/3$ [2]	3	3	Biology	Bonabeau et al. (1998, 1999)
−0.580 [3]	3	3	Biology	Hayakawa & Furuhashi (2012)
−1/2 [4,5]	3	3	Biology	Bonabeau & Dagorn (1995); Degond et al. (2017)
0 [5]	3	3	Biology	Niwa (2003); Ma et al. (2011)
1/3 [5]	3	3	Biology	Degond et al. (2017); Pego (2017)
1 [6]	3	3	Biology	Gueron & Levin (1995); Bonabeau et al. (1999)
2	3	3	Biology	Gueron & Levin (1995)

(1) Validated using experimental results for clusters of self-propelled Janus microswimmers.
(2) Validated using "tuna fish schools caught in the vicinity of a fish-aggregating device."
(3) Validated using observations of "skeins of Greater White-fronted geese" in Northern Japan.
(4) Validated using "data from commercial fisheries in the vicinity of Cape Lopez over 7 years."
(5) Validated using observations taken in six studies of pelagic fish school sizes.
(6) Validated using observations of "herd size distribution for the African buffalo."

The combination of aggregation and fragmentation often produces exactly the same size distributions as aggregation by itself or fragmentation by itself. In other words, if a particle system is spatially homogeneous and isotropic, n/m tends to be a universal rational number such as ±1/3, ±1/2, or ±1 regardless of the exact details; c.f., Tables 5.1 to 5.4. For simplicity, this book focuses on aggregation by itself and fragmentation by itself. For more information on mean-field theory for the combination of aggregation and fragmentation see, e.g., Banasiak et al. (2019a, b).

5.10 Experimental Calibration

Knowing that a given aggregation or fragmentation problem is ideal substantially narrows down the possible range of values for n/m. However, for each specific problem, additional information must be introduced to obtain a unique value for n/m. As seen in the last chapter, high-fidelity computational methods can account for all of the relevant information about a given problem and thus can generally obtain unique solutions. By contrast, analytical and empirical methods generally can *not* account for all the relevant information about a given problem and thus can *not* generally obtain unique solutions.

One possible approach for obtaining a unique solution is to use past experience. For example, Laney (2016b) surveyed over 170 solid fragmentation experiments including those

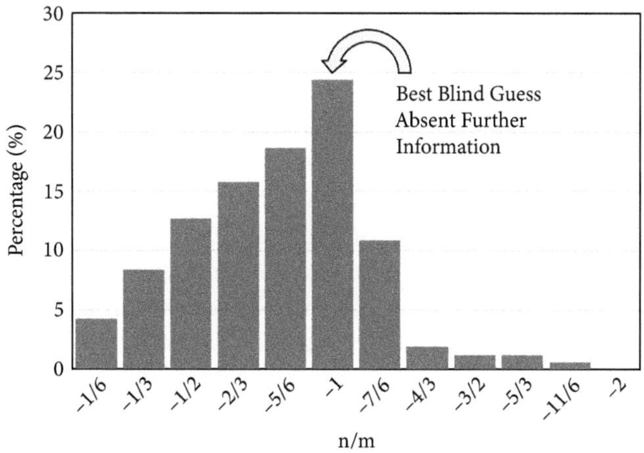

Figure 5.7 Results of a survey of power law curve fits to experimental results for the size distributions of solid fragments.
Based on Figure 11 in Laney (2016b).

found in Hartmann (1969), Turcotte (1986), Kaminski & Jaupart (1998) and Yanovsky et al. (2010). As shown in Figure 5.7, these experiments all obtained power law size distributions where n/m had a mean of about −5/6 and a mode of −1. For ideal problems that resemble those included in this particular dataset, $n/m = -1$ has about a 25% chance of being exactly correct and about a 55% chance of being approximately correct to within ±1/6.

For another example, Laney (2015a, b) surveyed liquid fragmentation experiments dating back to the 1940s including those found in Paloposki (1994) and Pimentel et al. (2010). While the original results were fitted with variety of different parametric forms, including root normal size distributions, Laney (2015a, b) suggested alternative fits using a single parametric form, namely, a Weibull size distribution. As a result, he found the following:

- Simmons (1977a, b) obtained a mean value for n/m of about 5/12.
- Li & Tankin (1987) obtained a modal value for n/m of about 2/3.
- Pimentel et al. (2010) obtained a modal value for n/m of about 3/2.

As these results show, newer spray systems tend to obtain higher absolute values of n/m, corresponding to more uniform droplet size distributions. This presumably reflects progressive improvements in spray technology. Thus, in this application, the best estimate for n/m depends, as much as anything, on the vintage of the spray system.

As these examples illustrate, the traditional approach for mapping a specific problem into a single analytic or empirical solution involves using standard size distributions optimized for different applications; see also Tables 5.2 and 5.4. Alternatively, in recent years, researchers have suggested using neural networks and machine learning techniques; see, e.g., Guo et al. (2020), Xie et al. (2021), Wang et al. (2022), Larsen et al. (2023). Given large enough training sets, these rapidly developing technologies should offer ways of associating problems with solutions that are clearly superior to traditional approaches.

5.11 Conclusions

For aggregation and fragmentation in spatially homogeneous and isotropic particle systems, the computational methods surveyed in the last chapter often obtained power law, Weibull or gamma size distributions where n/m appeared to be a rational number such as $\pm 1/3$, $\pm 2/3$, or ± 1 for sufficiently late times, regardless of the exact details; see Tables 4.1 to 4.4. Similarly, for aggregation and fragmentation in spatially homogeneous and isotropic particle systems, the analytic and empirical methods surveyed in this chapter often obtained power law, Weibull or gamma size distributions where n/m is a rational number such as $\pm 1/3$, $\pm 2/3$, or ± 1 for sufficiently late times, regardless of the exact details; see Tables 5.1 to 5.6. As defined in this book, these particle size distributions are *ideal* or *universal*.

As discussed in the last chapter, simplified computational methods make some of the same assumptions as analytic methods, e.g., they may ignore particle rotations, they may assume rather than model particle shapes and velocities, and so forth. As a result, simplified computational methods may obtain multiple different solutions, depending on the exact assumptions, just like analytic methods do. Even high-fidelity computational methods may obtain multiple different solutions. This is because high-fidelity computational solutions typically vary markedly depending on the assumed amount of chaos and randomness. In addition, there are often uncertainties which can only be accounted for by considering a range of possible alternatives. For example, for aggregation leading to crystal formation, Bushuev et al. (2017) tried three different atomic charge sets in a molecular dynamics (MD) method, which resulted in three different particle size distributions; see the earlier discussion in Section 4.2.

This chapter focused on ideal cases. However, of course, aggregation and fragmentation problems are not always ideal. Many practical problems are quasi-ideal or completely non-ideal; see Examples 4.4, 4.5, and 5.11. In addition, many practical problems are piecewise ideal; see Examples 4.2, 4.3, and 5.10. In piecewise-ideal cases, the physics of the problem depends on the particle size range, resulting in different size distributions for small, medium, and large particles. For example, for colloids, the smallest particles may experience Reaction-Limited Cluster-Cluster Aggregation (RLCA), the medium-sized particles may experience Diffusion-Limited Cluster-Cluster Aggregation (DLCA), and the largest particles may experience gravity-driven sedimentation; see, e.g., Lazzari et al. (2016). For another example, in unitary fragmentation, the change from small to medium sizes is typically due to a size-related reduction in dissipation. In addition, the change from medium to large sizes is typically due to a change in the fragment dimension m. Borrowing terminology from the fluid turbulence community, the small fragments may be called the *dissipative range*, the medium fragments may be called the *semi-inertial range*, and the large fragments may be called the *energy-containing range*.

In many applications, the particle size distribution is the most important part of the problem. This chapter showed that it is often possible to solve for the particle size distribution in isolation, without simultaneously solving for the particle shape distribution, particle velocity distribution, and so forth. Having said this, analytic techniques must account for particle shapes, particle velocities, and the other particle properties in some way, however hidden and opaque that way may be. For example, mean-field theory accounts for particle shapes, particle velocities, and the other particle properties through the choice of a mean aggregation or fragmentation rate. For another example, maximum entropy theory accounts for particle shapes, particle velocities, and the other particle properties through the choice of constraints. In essence, each analytic, semi-empirical, or empirical solution for particle size distributions implicitly corresponds to certain assumptions about the particle shape distribution, the particle velocity distribution, and so forth. The next two chapters will have more to say about shape and velocity distributions and their influence on size distributions.

References

V. Akimkin, A. V. Ivlev, P. Caselli, M. Gong, & K. Silsbee, Coagulation–Fragmentation Equilibrium for Charged Dust: Abundance of Submicron Grains Increases Dramatically in Protoplanetary Disks, *The Astrophysical Journal*, 953(1), 2023, https://doi.org/10.3847/1538-4357/ace2c5

D. J. Aldous, Deterministic and Stochastic Models for Coalescence (Aggregation and Coagulation): A Review of the Mean-Field Theory for Probabilists, *Bernoulli*, 5(1), 3–48, 1999, https://doi.org/10.2307/3318611

V. I. Arnold & M. Dresden, A. M. Kolmogorov, *Physics Today*, 42(10), 148–150, 1989, https://doi.org/10.1063/1.2811198

M. J. Aschwanden, N. B. Crosby, M. Dimitropoulou, M. K. Georgoulis, S. Hergarten, J. McAteer, et al., 25 Years of Self-Organized Criticality: Solar and Astrophysics, *Space Science Reviews*, 198(1), 47–166, 2016, https://doi.org/10.1007/s11214-014-0054-6

J. A. Åström, R. P. Linna, J. Timonen, P. F. Møller, & L. Oddershede, Exponential and Power-Law Mass Distributions in Brittle Fragmentation, *Physical Review E*, 70, 026104, 2004, https://doi.org/10.1103/PhysRevE.70.026104

E. Babinsky & P. E. Sojka, Modeling Drop Size Distributions, *Progress in Energy and Combustion Science*, 28(4), 303–329, 2002, https://doi.org/10.1016/S0360-1285(02)00004-7

J. Banasiak, W. Lamb, & P. Laurencot, *Analytic Methods for Coagulation-Fragmentation Models, Volume I*, CRC, 2019a, https://doi.org/10.1201/9781315154428

J. Banasiak, W. Lamb, & P. Laurencot, *Analytic Methods for Coagulation-Fragmentation Models, Volume II*, CRC, 2019b, https://doi.org/10.1201/9780429280320

J. D. Barrow, Coagulation with Fragmentation, *Journal of Physics A: Mathematical and General*, 14(3), 729–733, 1981, https://doi.org/10.1088/0305-4470/14/3/019

N. Bastian, K. R. Covey, and M. R. Meyer, Universal Stellar Initial Mass Function? A Critical Look at Variations, *Annual Review of Astronomy and Astrophysics*, 48, 339–389, 2010, https://doi.org/10.1146/annurev-astro-082708-101642

R. Becker & W. Döring, Kinetische Behandlung der Keimbildung in Übersättigten Dämpfen, *Annalen der Physik*, 416(8), 719–752, 1935, https://doi.org/10.1002/andp.19354160806

T. D. Bess, Mass Distribution of Orbiting Man-Made Space Debris, NASA Technical Note TN D-8108, December 1975, https://ntrs.nasa.gov/search.jsp?R=19760007896

S. Bhoi, S. R. Kolan, A. Bück, & E. Tsotsas, Population Balance Modeling of Formation and Breakage of Nanoparticle Agglomerates in a Spouted Bed, *Powder Technology*, 433, 119271, 2024, https://doi.org/10.1016/j.powtec.2023.119271

P. J. Blatz & A. V. Tobolsky, Note on the Kinetics of Systems Manifesting Simultaneous Polymerization-Depolymerization Phenomena, *Journal of Physical Chemistry*, 49(2), 77–80, 1945, https://doi.org/10.1021/j150440a004

A. S. Bodrova, V. Stadnichuk, P. L. Krapivsky, J. Schmidt, & N. V. Brilliantov, Kinetic Regimes in Aggregating Systems with Spontaneous and Collisional Fragmentation, *Journal of Physics*

A: Mathematical and Theoretical, 52(20), 205001, 2019, https://doi.org/10.1088/1751-8121/ab1616

E. Bonabeau & L. Dagorn. Possible Universality in the Size Distribution of Fish Schools, *Physical Review E*, 51(6), R5220, 1995, https://doi.org/10.1103/PhysRevE.51.R5220

E. Bonabeau, L. Dagorn, & P. Fréon, Space Dimension and Scaling in Fish School-Size Distributions, *Journal of Physics A: Mathematical and General*, 31(44), L731, 1998, https://doi.org/10.1088/0305-4470/31/44/001

E. Bonabeau, L. Dagorn, & P. Fréon, Scaling in Animal Group-Size Distributions, *Proceedings of the National Academy of Sciences*, 96(8), 4472–4477, 1999, https://doi.org/10.1073/pnas.96.8.4472

L. Bravo & C.-B. Kweon, A Review on Liquid Spray Models for Diesel Engine Computational Analysis, Army Research Laboratory Technical Report ARL-TR-6932, May 2014, http://apps.dtic.mil/sti/citations/ADA603658

G. Breschi & M. A. Fontelos, On Global in Time Self-Similar Solutions of Smoluchowski Equation with Multiplicative Kernel, *IMA Journal of Applied Mathematics*, 88(2), 405–428, 2023, https://doi.org/10.1093/imamat/hxad012

N. V. Brilliantov, A. S. Bodrova, & P. L. Krapivsky, A Model of Ballistic Aggregation and Fragmentation, *Journal of Statistical Mechanics: Theory and Experiment*, P06011, 2009, https://doi.org/10.1088/1742-5468/2009/06/P06011

N. V. Brilliantov, P. L. Krapivsky, A. Bodrova, F. Spahn, H. Hayakawa, V. Stadnichuk, & J. Schmidt, Size Distribution of Particles in Saturn's Rings from Aggregation and Fragmentation, *Proceedings of the National Academy of Sciences*, 112(31), 9536–9541, 2015, https://doi.org/10.1073/pnas.1503957112

W. K. Brown & K. H. Wohletz, Derivation of the Weibull Distribution Based on Physical Principles and its Connection to the Rosin–Rammler and Logormal Distributions, *Journal of Applied Physics*, 78(4), 2758–2763, 1995, https://doi.org/10.1063/1.360073

Y. G. Bushuev, S. V. Davletbaeva, & O. I. Koifman, Molecular Dynamics Simulations of Aqueous Glycine Solutions, *CrystEngComm*, 19(47), 7197–7206, 2017, https://doi.org/10.1039/c7ce01271c

F. Cady, *The Data Science Handbook*, Wiley, 2017, https://doi.org/10.1002/9781119092919

A. Carpinteri & N. Pugno, One, Two, and Three-Dimensional Universal Laws for Fragmentation due to Impact and Explosion, *Transactions of the ASME*, 69(6), 854–856, 2002a, https://doi.org/10.1115/1.1488937

A. Carpinteri & N. Pugno, A Fractal Comminution Approach to Evaluate the Drilling Energy Dissipation, *International Journal for Numerical and Analytical Methods in Geomechanics*, 26(5), 499–513, 2002b, https://doi.org/10.1002/nag.209.

A. Carpinteri & N. Pugno, Scale-Effects on Mean and Standard Deviation of the Mechanical Properties of Condensed Matter: An Energy-Based Unified Approach, *International Journal of Fracture*, 128(1–4), 253–261, 2004, https://doi.org/10.1023/B:FRAC.0000040988.61253.05

A. Carpinteri & N. Pugno, Are Scaling Laws on Strength of Solids Related to Mechanics or Geometry?, *Nature Materials*, 4(6), 421–423, 2005, https://doi.org/10.1038/nmat1408

S. Chandraeskhar, Marian Smoluchowski as the Founder of the Physics of Stochastic Phenomena, in R. S. Ingarden, editor, *Marian Smoluchowski: His Life and Scientific Work*, Polish Scientific Publishers, 1986; see also S. Chandreasekhar, *Selected Papers, Volume 3: Stochastic, Statistical and Hydromagnetic Problems in Physics and Astronomy*, The University of Chicago Press, 1989

M. Chayut, New Sites for Scientific Change: Paul Flory's Initiation into Polymer Chemistry, *Historical Studies in the Physical and Biological Science*, 23(2), 193–218, 1993, https://doi.org/10.2307/27757698

A. I. Chervanyov & G. Heinrich, Coagulation–Fragmentation with Mass Conservation: Self-Consistent Mean Rate Model, *Physica B*, 421, 117–121, 2013, https://doi.org/10.1016/j.physb.2013.04.017

C. A. Chryssakis, D. N. Assanis, & F. X. Tanner, Atomization Models, in N. Ashgriz, editor, *Handbook of Atomization and Sprays: Theory and Application*, Springer, 2011, https://doi.org/10.1007/978-1-4419-7264-4_9

A. Clauset, C. R. Shalizi, & M. E. J. Newman, Power-Law Distributions in Empirical Data, *SIAM Review*, 51(4), 661–703, 2009, https://doi.org/10.1137/070710111

R. J. Cohen & G. B. Benedek, Equilibrium and Kinetic Theory of Polymerization and the Sol-Gel Transition, *The Journal of Physical Chemistry*, 86(19), 3696–3714, 1982, https://doi.org/10.1021/j100216a005

C. Connaughton, R. Rajesh, & O. Zaboronski, Stationary Kolmogorov Solutions of the Smoluchowski Aggregation Equation with a Source Term, *Physical Review E*, 69(6), 061114, 2004, https://doi.org/10.1103/PhysRevE.69.061114

C. Connaughton, R. Rajesh, & O. Zaboronski, Cluster-Cluster Aggregation as an Analogue of a Turbulent Cascade: Kolmogorov Phenomenology, Scaling Laws and the Breakdown of Self-Similarity, *Physica D: Nonlinear Phenomena*, 222(1–2), 97–115, 2006, https://doi.org/10.1016/j.physd.2006.08.005

D. Coupier, editor, Preface to *Stochastic Geometry: Modern Research Frontiers*, Springer, 2019, https://doi.org/10.1007/978-3-030-13547-8

T. M. Cover & J. A. Thomas, *Elements of Information Theory*, 2nd edition, Wiley, 2006, https://doi.org/10.1002/047174882X

C. V. B. Cunningham, The Kuz-Ram Model for Prediction of Fragmentation from Blasting, in A. Holmberg & A. Rustan, editors, *Proceedings of the 1st International Symposium on Rock Fragmentation by Blasting*, Luleå University of Technology, 1983

C. V. B. Cunningham, Fragmentation Estimations and the Kuz-Ram Model – Four Years On, in W. L. Fourney, & R. D. Dick, editors, *Proceedings of the 2nd International Symposium of Rock Fragmentation by Blasting*, Society for Experimental Mechanics, 1987

C. V. B. Cunningham, The Kuz-Ram Fragmentation Model—20 Years On, in R. Holmberg, editor, *Proceedings of the 3rd EFEE World Conference on Explosives and Blasting Technique*, EFEE, 2005

G. S. Darivakis, W. A. Peters, & J. B Howard, Rationalization for the Molecular Weight Distributions of Coal Pyrolysis Liquids, *AIChE Journal*, 36(8), 1189–1199, 1990, https://doi.org/10.1002/aic.690360808

G. S. Darivakis, J. B. Howard, & W. A. Peters, A Rationale for Heating Rate and Coal Type Effects on Liquids Yields and Substrate Morphology Changes during Rapid Pyrolysis, *Energy & Fuels*, 8(5), 1024–1032, 1994, https://doi.org/10.1021/ef00047a004

E. A. Davis, Professor Sir Nevill Francis Mott, *Europhysics News*, 28(3), 84–88, 1997, https://doi.org/10.1007/s00770-997-0084-7

E. A. Davis, editor, *Nevill Mott: Reminiscences and Appreciations*, Taylor & Francis, 1998

P. Degond, J.-G. Liu, & R. L. Pego, Coagulation–Fragmentation Model for Animal Group-Size Statistics, *Journal of Nonlinear Science*, 27(2), 379–424, 2017, https://doi.org/10.1007/s00332-016-9336-3

R. B. Diemer Jr., Simplifying Population Balance Models to Promote Broader Use in Industry, *Chemical Engineering Research and Design*, 190, 91–105, 2023, https://doi.org/10.1016/j.cherd.2022.12.002

P. J. Diggle, *Statistical Analysis of Spatial and Spatio-Temporal Point Patterns*, 3rd edition, CRC Press, 2014, https://doi.org/10.1201/b15326

J. S. Dohnanyi, Collisional Model of Meteoroids, in J. L. Weinberg, editor, *The Zodiacal Light and the Interplanetary Medium*, NASA SP-150, 1967, https://ntrs.nasa.gov/search.jsp?R=19680018003

J. S. Dohnanyi, Collisional Model of Asteroids and their Debris, *Journal of Geophysical Research*, 74(10), 2531–2554, 1969, https://doi.org/10.1029/JB074i010p02531

J. S. Dohnanyi, On the Origin and Distribution of Meteoroids, *Journal of Geophysical Research*, 75(17), 3468–3493, 1970, https://doi.org/10.1029/JB075i017p03468

J. Eggers & M. Fontelos, Anomalous Dimensions of the Smoluchowski Coagulation Equation, *Physical Review E*, 108, 064110, 2023, https://doi.org/10.1103/PhysRevE.108.064110

H. J. Empie, S. J. Lien, W. Yang, & D. B. Samuels, *Kraft Black Liquor Delivery Systems Report Number 4*, Institute of Paper Science and Technology Progress Report, January 1993, http://hdl.handle.net/1853/535

H. J. Empie, S. J. Lien, W. Yang, & T. N. Adams, Spraying Characteristics of Commercial Black Liquor Nozzles, *TAPPI Journal*, 78(1), 121–128, 1995, https://imisrise.tappi.org/TAPPI/Products/95/JAN/95JAN121.aspx

H. J. Empie, S. J. Lien, & W. Yang, Drop Size Modification in Black Liquor Sprays From Commercial Nozzles Using Pulsed Flow, *Atomization and Sprays*, 7(5), 457–466, 1997, https://doi.org/10.1615/AtomizSpr.v7.i5.10

G. B. Field & W. C. Saslaw, A Statistical Model of the Formation of Stars and Interstellar Clouds, *The Astrophysical Journal*, 142, 568–583, 1965, https://doi.org/10.1086/148318

A. F. Filippov, On the Distribution of the Sizes of Particles which Undergo Splitting, *Theory of Probability & Its Applications*, 6(3), 275–294, 1961, https://doi.org/10.1137/1106036

P. J. Flory, Molecular Size Distribution in Linear Condensation Polymers, *Journal of the American Chemical Society*, 58(10), 1877–1885, 1936, https://doi.org/10.1021/ja01301a016

P. J. Flory, Molecular Size Distribution in Three Dimensional Polymers. I. Gelation, *Journal of the American Chemical Society*, 63(11), 3083–3090, 1941, https://doi.org/10.1021/ja01856a061

P. J. Flory, Random Reorganization of Molecular Weight Distribution in Linear Condensation Polymers, *Journal of the American Chemical Society*, 64 (9), 2205–2212, 1942, https://doi.org/10.1021/ja01261a049

S. K. Friedlander, On the Particle-Size Spectrum of Atmospheric Aerosols, *Journal of Meteorology*, 17(3), 373–374, 1960a, https://doi.org/10.1175/1520-0469(1960)017<0373:OTPSSO>2.0.CO;2

S. K. Friedlander, Similarity Considerations for the Particle-Size Spectrum of a Coagulating, Sedimenting Aerosol, *Journal of Meteorology*, 17(5), 479–483, 1960b, https://doi.org/10.1175/1520-0469(1960)017<0479:SCFTPS>2.0.CO;2

A. O. Gates, Kick vs. Rittinger: An Experimental Investigation in Rock Crushing, Performed at Purdue University, *Transactions of the AIME*, 52, 875–909, 1915, http://www.onemine.org/document/abstract.cfm?docid=14056

A. M. Gaudin, An Investigation of Crushing Phenomena, *Transactions of the AIME*, 73, 253–310, 1926, http://www.onemine.org/document/abstract.cfm?docid=164880

J. J. Gilvarry, Fracture of Brittle Solids. I. Distribution Function for Fragment Size in Single Fracture (Theoretical), *Journal of Applied Physics*, 32(3), 391–399, 1961, https://doi.org/10.1063/1.1736016

J. J. Gilvarry & B. H. Bergstrom, Fracture of Brittle Solids. II. Distribution Function for Fragment Size in Single Fracture (Experimental), *Journal of Applied Physics*, 32(3), 400–410, 1961, https://doi.org/10.1063/1.1736017

A. Golan, *Foundations of Info-Metrics: Modeling, Inference, and Imperfect Information*, Oxford University Press, 2018, https://10.1093/oso/9780199349524.001.0001

A. M. Golovin, The Solution of the Coagulation Equation for Cloud Droplets in a Rising Air Current, *Bulletin of the Academy of Sciences of the USSR, Geophysics Series*, 5, 482–487, 1963

F. Ginot, I. Theurkauff, F. Detcheverry, C. Ybert, & C. Cottin-Bizonne, Aggregation-Fragmentation and Individual Dynamics of Active Clusters, *Nature Communications*, 9, 696, 2018, https://doi.org/10.1038/s41467-017-02625-7

D. E. Grady & M. E. Kipp, Geometric Statistics and Dynamic Fragmentation, *Journal of Applied Physics*, 58(3), 1210–1222, 1985, https://doi.org/10.1063/1.336139

D. E. Grady, *Fragmentation of Rings and Shells: The Legacy of N. F. Mott*, Springer, 2006, https://doi.org/10.1007/b138675

L. Griffith, A Theory of the Size Distribution of Particles in a Comminuted System, *Canadian Journal of Research*, 21A(6), 57–64, 1943, https://doi.org/10.1139/cjr43a-005

E. Gudowska-Nowak, K. Lindenberg, & R. Metzler, Preface: Marian Smoluchowski's 1916 Paper—A Century of Inspiration, *Journal of Physics A: Mathematical and Theoretical*, 50(38), 380301, 2017, https://doi.org/10.1088/1751-8121/aa8529

S. Gueron & S. A. Levin, The Dynamics of Group Formation, *Mathematical Biosciences*, 128(1–2), 243–264, 1995, https://doi.org/10.1016/0025-5564(94)00074-A

H. Guo, J. Zhou, F. Liu, Y. He, H. Huang, & H. Wang, Application of Machine Learning Method to Quantitatively Evaluate the Droplet Size and Deposition Distribution of the UAV Spray Nozzle, *Applied Sciences*, 10(5), 1759, 2020, https://doi.org/10.3390/app10051759

C. C. Harris, B. V. Patil, & B. P. Jalan, One-Dimensional Fracture, *Powder Technology*, 8(5-6), 253–260, 1973, https://doi.org/10.1016/0032-5910(73)80090-7

W. K. Hartmann, Terrestrial, Lunar, and Interplanetary Rock Fragmentation, *Icarus*, 10(2), 201–213, 1969, https://doi.org/10.1016/0019-1035(69)90022-0

Y. Hayakawa & S. Furuhashi, Group-Size Distribution of Skeins of Wild Geese, *Physical Review E*, 86, 031924, 2012, https://doi.org/10.1103/PhysRevE.86.031924

P. Horvai, S. V. Nazarenko, & T. H. M. Stein, Coalescence of Particles by Differential Sedimentation, *Journal of Statistical Physics*, 130, 1177–1195, 2008, https://doi.org/10.1007/s10955-007-9466-y

J. R. Hunt, Self-Similar Particle-Size Distributions during Coagulation: Theory and Experimental Verification, *Journal of Fluid Mechanics*, 122, 169–185, 1982, https://doi.org/10.1017/S0022112082002158

H. Inaoka & H. Takayasu, Universal Fragment Size Distribution in a Numerical Model of Impact Fracture, *Physica A: Statistical Mechanics and its Applications*, 229(1), 5–25, 1996, https://doi.org/10.1016/0378-4371(96)00048-9

H. Inaoka & H. Takayasu, Application of Statistical Physics to Impact Fragmentation, *Physica A: Statistical Mechanics and its Applications*, 274(1), 300–309, 1999, https://doi.org/10.1016/S0378-4371(99)00426-4

D. J. Jeffrey, Quasi-Stationary Approximations for the Size of Distribution of Aerosols, *Journal of the Atmospheric Sciences*, 38, 2440–2443, 1981, https://doi.org/10.1175/1520-0469(1981)038<2440:QSAFTS>2.0.CO;2

N. L. Johnson, P. H. Krisko, J.-C. Lieu, & P. D. Am-Meador, NASA's New Breakup Model of Evolve 4.0, *Advances in Space Research*, 28(9), 1377–1384, 2001, https://doi.org/10.1016/S0273-1177(01)00423-9

S. Jungblut, J.-O. Joswig, & A. Eychmüller, Diffusion-Limited Cluster Aggregation: Impact of Rotational Diffusion, *The Journal of Physical Chemistry C*, 123(1), 950–954, 2019a, https://doi.org/10.1021/acs.jpcc.8b10805

S. Jungblut, J.-O. Joswig, & A. Eychmüller, Diffusion- and Reaction-Limited Cluster Aggregation Revisited, *Physical Chemistry Chemical Physics*, 21, 5723–5729, 2019b, https://doi.org/10.1039/c9cp00549h

E. Kaminski & C. Jaupart, The Size Distribution of Pyroclasts and the Fragmentation Sequence in Explosive Volcanic Eruptions, *Journal of Geophysical Research*, 103(B12), 29759–29779, 1998, https://doi.org/10.1029/98JB02795

S. Kempf, S. Pfalzner, & T. K. Henning, N-Particle-Simulations of Dust Growth I. Growth Driven by Brownian Motion, *Icarus*, 141(2), 388–398, 1999, https://doi.org/10.1006/icar.1999.6171

R. P. King, C. L. Schneider, & E. A. King, *Modeling and Simulation of Mineral Processing Systems*, 2nd edition, Society for Mining, Metallurgy, and Exploration (SME), 2012

J. D. Klett, A Class of Solutions to the Steady-State, Source-Enhanced Kinetic Coagulation Equation, *Journal of Atmospheric Sciences*, 32(2), 380–389, 1975, https://doi.org/10.1175/1520-0469(1975)032<0380:ACOSTT>2.0.CO;2

A. N. Kolmogorov, O Logarifmicheski Normal'nom Zakone Raspredeleniya Razmerov Chastis pri Droblenii [On the Logarithmic Normal Distribution of Particle Sizes under Grinding], *Doklady Akademii Nauk SSSR*, 31(2), 99–101, 1941a; see also The Logarithmically Normal Law of Distribution of Dimensions of Particles when Broken into Small Parts, NASA Technical Translation TT F-12, 287, 1969, https://archive.org/details/nasa_techdoc_19690019884

A. N. Kolmogorov, The Local Structure of Turbulence in Incompressible Viscous Fluids for Very Large Reynolds Numbers, *Doklady Akademii Nauk SSSR*, 30(4), 299–303, 1941b; see also *Proceedings of the Royal Society of London A*, 434, 9–13, 1991, https://doi.org/10.1098/rspa.1991.0075

A. N. Kolmogorov, Dissipation of Energy in Locally Isotropic Turbulence, *Doklady Akademii Nauk SSSR*, 32(1), 16–18, 1941c; see also *Proceedings of the Royal Society of London A*, 434, 15-17, 1991, https://doi.org/10.1098/rspa.1991.0076

J. Kovac & M. Fixman, Walter Hugo Stockmayer, *Biographical Memoirs of the National Academy of Science*, 2017, http://www.nasonline.org/publications/biographical-memoirs/memoir-pdfs/stockmayer-walter.pdf

P. L. Krapivsky & E. Ben-Naim, Shattering Transitions in Collision-Induced Fragmentation, *Physical Review E*, 68(2), 021102, 2003, https://doi.org/10.1103/PhysRevE.68.021102

P. Kroupa, On the Variation of the Initial Mass Function, *Monthly Notices of the Royal Astronomical Society*, 322(2), 231–246, 2001, https://doi.org/10.1046/j.1365-8711.2001.04022.x

P. Kroupa, The Initial Mass Function of Stars: Evidence for Uniformity in Variable Systems, *Science*, 295(5552), 82–91, 2002, https://doi.org/10.1126/science.1067524

W. Kuhn, Über die Kinetik des Abbaues hochmolekularer Ketten [The Kinetics of the Degradation of High Molecular Weight Chains], *Chemische Berichte*, 63(6), 1503–1509, 1930, https://doi.org/10.1002/cber.19300630631

A. Kumar & R. K. Gupta, *Fundamentals of Polymer Engineering*, 3rd edition, CRC Press, 2018,

V. M. Kuznetsov, The Mean Diameter of the Fragments Formed by Blasting Rock, *Soviet Mining Science*, 9(2), 144–148, 1973, https://doi.org/10.1007/BF02506177

C. B. Laney, *Improved Root Normal Size Distributions for Liquid Atomization*, Defense Threat Reduction Agency Technical Report DTRA-TR-16-03, November 2015a, http://apps.dtic.mil/sti/citations/ADA624919

C. B. Laney, *Transformation and Self-Similarity Properties of Gamma and Weibull Fragment Size Distributions*, Defense Threat Reduction Agency Technical Report DTRA-TR-16-06, December 2015b, http://apps.dtic.mil/sti/citations/ADA624878

C. B. Laney, *On the Relationships Between Liquid and Solid Fragmentation*, Defense Threat Reduction Agency Technical Report DTRA-TR-16-19, January 2016a, http://apps.dtic.mil/sti/citations/AD1004358

C. B. Laney, *Effects of Mixtures on Liquid and Solid Fragment Size Distributions*, Defense Threat Reduction Agency Technical Report DTRA-TR-16-52, May 2016b, http://apps.dtic.mil/sti/citations/AD1008708

K. E. Larsen, R. Bevilacqua, O. S. Mulekar, E. L. Jerome, & T. J. Hatch-Aguilar, Predicting Dynamic Fragmentation Characteristics from High-Impact Energy Events Utilizing Terrestrial Static Arena Test Data and Machine Learning, *Acta Astronautica*, 209, 67–81, 2023, https://doi.org/10.1016/j.actaastro.2023.04.036

P. Laurençot, Stationary Solutions to Coagulation-Fragmentation Equations, *Annales de l'Institut Henri Poincaré C, Analyse non Linéaire*, 2019, https://doi.org/10.1016/j.anihpc.2019.06.003

S. Lazzari, L. Nicoud, B. Jaquet, M. Lattuada, & M. Morbidelli, Fractal-Like Structures in Colloid Science, *Advances in Colloid and Interface Science*, 235, 1–13, 2016, https://doi.org/10.1016/j.cis.2016.05.002

X. Li & R. S. Tankin, Droplet Size Distribution: A Derivation of a Nukiyama–Tanasawa Type Distribution Function, *Combustion Science and Technology*, 56 (1–3), 65–76, 1987, https://doi.org/10.1080/00102208708947081

C. C. Lienau, Random Fracture of Brittle Solids, *Journal of the Franklin Institute*, 221(4), 485–494, 1936a, https://doi.org/10.1016/S0016-0032(36)90309-5

C. C. Lienau, Random Fracture of Brittle Solids (Continued), *Journal of the Franklin Institute*, 221(5), 673–686, 769–780, 1936b, https://doi.org/10.1016/S0016-0032(36)91108-0

C. C. Lienau, Random Fracture of Brittle Solids (Concluded), *Journal of the Franklin Institute*, 221(6), 769–780, 1936c, https://doi.org/10.1016/S0016-0032(36)90526-4

M. Y. Lin, H. M. Lindsay, D. A. Weitz, R. C. Ball, R. Klein, & P. Meakin, Universality in Colloid Aggregation, *Nature*, 339, 360–362, 1989a, https://doi.org/10.1038/339360a0

M. Y. Lin, H. M. Lindsay, D. A. Weitz, R. C. Ball, R, Klein, & P. Meakin, Universality of Fractal Aggregates as Probed by Light Scattering, *Proceedings of the Royal Society A: Mathematical and Physical Sciences*, 423(1864), 71–87, 1989b, https://doi.org/10.1098/rspa.1989.0042

M. Y. Lin, M. H. Lindsay, D. A. Weitz, R. C. Ball, R. Klein, & P. Meakin, Universal Reaction-Limited Colloid Aggregation, *Physical Review A*, 41(4), 2005–2020, 1990, https://doi.org/10.1103/PhysRevA.41.2005

Y. Liu, L. You, W. Yang, & F. Liu, On the Size Distribution of Cloud Droplets, *Atmospheric Research*, 35(2–4), 201–216, 1995, https://doi.org/10.1016/0169-8095(94)00019-A

Y. Liu, On the Generalized Theory of Atmospheric Particle Systems, *Advances in Atmospheric Sciences*, 12(4), 419–438, 1995, https://doi.org/10.1007/BF02657003

Q. Ma, A. Johansson, &D. J. T. Sumpter, A First Principles Derivation of Animal Group Size Distributions, *Journal of Theoretical Biology*, 283, 35–43, 2011, https://doi.org/10.1016/j.jtbi.2011.04.031

G. M. Magnotti & C. L. Genzale, Recent Progress in Primary Atomization Model Developments for Diesel Engine Simulations, in K. Saha, A. K. Agarwal, K. Ghosh, & S. Som, editors, *Two-Phase Flow for Automotive and Power Generation Sectors*, Springer, 2019, https://doi.org/10.1007/978-981-13-3256-2_4

H. Mark & R. Simha, Degradation of Long Chain Molecules, *Transactions of the Faraday Society*, 35, 611–618, 1940, https://doi.org/10.1039/TF9403500611

P. Marmottant & E. Villermaux, On Spray Formation, *Journal of Fluid Mechanics*, 498, 73–111, 2004a, https://doi.org/10.1017/S0022112003006529

P. Marmottant & E. Villermaux, Fragmentation of Stretched Liquid Ligaments, *Physics of Fluids*, 16(8), 2732–2741, 2004b, https://doi.org/10.1063/1.1756030

J. S. Marshall & W. M. Palmer, The Distribution of Raindrops with Size, *Journal of Meteorology*, 5(4), 165–166, 1948, https://doi.org/10.1175/1520-0469(1948)005<0165:TDORWS>2.0.CO;2

T. Matsoukas, Statistical Thermodynamics of Clustered Populations, *Physical Review E*, 90, 022113, 2014, https://doi.org/10.1103/PhysRevE.90.022113

B. J. McCoy & G. Madras, Evolution to Similarity Solutions for Fragmentation and Aggregation, *Journal of Colloid and Interface Science*, 201(2), 200–209, 1998, https://doi.org/10.1006/jcis.1998.5434

J. B. McLeod, On an Infinite Set of Nonlinear Differential Equations, *The Quarterly Journal of Mathematics*, 13(1), 119–128, 1962, https://doi.org/10.1093/qmath/13.1.119

J. B. McLeod, On the Scalar Transport Equation, *Proceedings of the London Mathematical Society*, S3, 14(3), 445–458, 1964, https://doi.org/10.1112/plms/s3-14.3.445

W. Mock & W. H. Holt, Fragmentation Behavior of Armco Iron and HF-1 Steel Explosive-Filled Cylinders, *Journal of Applied Physics*, 54(5), 2344–2351, 1983, https://doi.org/10.1063/1.332346

J. Møller, Topics in Voronoi and Johnson-Mehl tessellations, in O. E. Barndorff-Nielsen,W. S. Kendall, & M. N. M. van Lieshout, editors, *Stochastic Geometry: Likelihood and Computation*, Taylor & Francis, 1998, https://doi.org/10.1201/9780203738276

J. L. Moran & J. D. Posner, Phoretic Self-Propulsion, *Annual Review of Fluid Mechanics*, 49, 511–540, 2017, https://doi.org/10.1146/annurev-fluid-122414-034456

P. J. T. Morris, *Polymer Pioneers: A Popular History of the Science and Technology of Large Molecules*, Beckman Center for the History of Chemistry, 1986

N. F. Mott, *Fragmentation of H.E. Shells: A Theoretical Formula for the Distribution of Weights of Fragments*, U. K. Ministry of Supply, AC3642, March 1943a; see also section II.2 of Grady (2006), https://doi.org/10.1007/b138675

N. F. Mott, *A Theory of the Fragmentation of Shells and Bombs*, U. K. Ministry of Supply, AC4035, May 1943b; see also section II.3 of Grady (2006), https://doi.org/10.1007/b138675

N. F. Mott & E. H. Linfoot, *A Theory of Fragmentation*, U. K. Ministry of Supply, AC3348, January 1943; see also section II.1 of Grady (2006), https://doi.org/10.1007/b138675

H.-S. Niwa, Power-Law versus Exponential Distributions of Animal Group Sizes, *Journal of Theoretical Biology*, 224(4), 451–457, 2003, https://doi.org/10.1016/S0022-5193(03)00192-9

L. Olivieri, P. A. Smocovich, C. Giacomuzzo, & A. Francesconi, Characterization of the Fragments Generated by a Picosatellite Impact Experiment, *International Journal of Impact Engineering*, 168, 104313, 2022, https://doi.org/10.1016/j.ijimpeng.2022.104313

L. Olivieri, C. Giacomuzzo, & A. Francesconi, Analysis of Fragments Larger than 2 mm Generated by a Picosatellite Fragmentation Experiment, *Acta Astronautica*, 204, 418–424, 2023, https://doi.org/10.1016/j.actaastro.2023.01.020

M. Opper & D. Saad, editors, *Advanced Mean Field Methods: Theory and Practice*, MIT Press, 2001, https://doi.org/10.7551/mitpress/1100.001.0001

T. Paloposki, Drop Size Distributions in Liquid Sprays, *Acta Polytechnica Scandinavica, Mechanical Engineering Series*, 114, 1994; see also *Proceedings of the Sixth International Conference on Liquid Atomization and Spray Systems (ICLASS 94)*, Begell House, 1994, https://doi.org/10.1615/ICLASS-94.1160

S. A. Pardo, *Empirical Modeling and Data Analysis for Engineers and Applied Scientists*, Springer, 2016, https://doi.org/10.1007/978-3-319-32768-6

F. Parisio & L. Dias, Minimal Fragmentation Problem, *Physical Review E*, 84(3), 035101(R), 2011, https://doi.org/10.1103/PhysRevE.84.035101

S. Y. Park & A. K. Bera, Maximum Entropy Autoregressive Conditional Heteroskedasticity Model, *Journal of Econometrics*, 150, 219–230, 2009, https://doi.org/10.1016/j.jeconom.2008.12.014

D. P. Patil & J. R. G. Andrews, An Analytical Solution to Continuous Population Balance Model Describing Floc Coalescence and Breakage—A Special Case, *Chemical Engineering Science*, 53(3), 599–601, 1998, https://doi.org/10.1016/S0009-2509(97)00314-X

G. D. Patterson, J. E. Mark, J. R. Fried, & D. Y. Yoon, *Paul John Flory: A Life of Science and Friends*, CRC Press, 2016, https://doi.org/10.1201/b18755

R. L. Pego, How to Count Fish Using Mathematics, *SIAM News*, 50(3), 9–11, 2017, https://sinews.siam.org/Details-Page/how-to-count-fish-using-mathematics

R. Pimentel, R. Stowe, P. Harris, A. DeChamplain, & D. Kretschmer, Spray Characterization Based on the Pearson System of Frequency Curves, *Atomization and Sprays*, 20(5), 365–386, 2010, https://doi.org/10.1615/AtomizSpr.v20.i5.10

S. L. Piotrowski, The Collisions of Asteroids, *The Astronomical Journal*, 57(1), 23–24, 1952, https://doi.org/10.1086/106692

S. L. Piotrowski, The Collisions of Asteroids, *Acta Astronomica, Series A*, 5, 115–138, 1953, https://ui.adsabs.harvard.edu/abs/1952AJ . . . . . 57 . . . 23P/abstract

D. Ramkrishna, *Population Balances: Theory and Applications to Particulate Systems in Engineering*, Elsevier, 2000, https://doi.org/10.1016/B978-0-12-576970-9.X5000-0

D. Ramkrishna & M. R. Singh, Population Balance Modeling: Current Status and Future Prospects, *Annual Review of Chemical and Biomolecular Engineering*, 5, 123–146, 2014, https://doi.org/10.1146/annurev-chembioeng-060713-040241

P. Rosin, E. Rammler, & K. Sperling, *Korngrößenprobleme des Kohlenstaubes und ihre Bedeutung für die Vermahlung* [Particle Size Problems of Coal Dust and their Importance for Grinding], Bericht C 52 des Reichskohlenrates [Report C52 of the National Coal Council], Verlag: VDI-Verlag, 1933

P. Rosin & E. Rammler, Die Kornzusammensetzung des Mahlgutes im Lichte der Wahrscheinlichkeitslehre [Grain Composition of Ground Material in Light of Probability Theory], *Kolloid-Zeitschrift*, 67(1), 16–26, 1934, https://doi.org/10.1007/BF01439773

P. P. Roy, *Rock Blasting: Effects and Operations*, CRC Press, 2005

A. Rustan et al., editors, *Mining and Rock Construction Technology Desk Reference: Rock Mechanics, Drilling and Blasting*, CRC Press, 2010, https://doi.org/10.1201/b10543

V. S. Safronov, Chastnyi Sluchai Resheniya Uravnenii Koagulyatsii [A Particular Solution of the Coagulation Equations], *Doklady Akademii Nauk SSSR*, 147(1), 64, 1962; see also V. S. Safronov, A Restricted Solution of the Accretion Equation, *Soviet Physics Doklady*, 7(11), 967–969, 1963

V. S. Safronov, Evolyutsiya Doplanetnogo Oblaka i Obrazovanie Zemli i Planet [Evolution of the Protoplanetary Cloud and Formation of the Earth and its Planets], *Akademii Nauk SSSR*, 1969; see also NASA TT F-677, 1972, https://archive.org/details/nasa_techdoc_19720019068

E. E. Salpeter, The Luminosity Function and Stellar Evolution, *The Astrophysical Journal*, 121, 161–167, 1955, https://doi.org/10.1086/145971

J. E. Schoutens, editor, *Empirical Analysis of Nuclear and High-Explosive Cratering and Ejecta*, in *Nuclear Geoplosics Sourcebook, Volume IV, Part II, Empirical Analysis of Nuclear and High-Explosive Cratering and Ejecta*, Defense Nuclear Agency 6501H-4-2, March 1979, http://apps.dtic.mil/sti/citations/ADA095097

E. V. Schuhmann, Principles of Communition, Size Distribution and Surface Calculations, *AIME Technical Publication*, 1189, 1–11, 1940, http://www.onemine.org/document/abstract.cfm?docid=9197

T. E. W. Schumann, Theoretical Aspects of the Size Distribution of Fog Particles, *Quarterly Journal of the Royal Meteorological Society*, 66(285), 195–208, 1940, https://doi.org/10.1002/qj.49706628508

W. R. Seebaugh, *A Dynamic Crater Ejecta Model*, Defense Nuclear Agency DNA 4144F, October 1976, http://apps.dtic.mil/sti/citations/ADA042546

W. R. Seebaugh, A Dynamic Crater Ejecta Model, in D. J. Roddy, R. O. Pepin, & R. B. Merrill, editors, *Impact and Explosion Cratering*, Pergammon Press, 1977, http://adsabs.harvard.edu/full/1977iecp.symp.1043S

R. W. Sellens & T. A. Brzustowski, A Prediction of the Drop-Size Distribution in a Spray from First Principles, *Atomisation and Spray Technology*, 1(2), 89–102, 1985

E. Seneta, Marian Smoluchowski, in C. C. Heyde, E. Seneta, P. Crépel, S. E. Fienberg, & J. Gani, editors, *Statisticians of the Century*, Springer, 2001, https://doi.org/10.1007/978-1-4613-0179-0_64

H. C. Simmons, The Correlation of Drop-Size Distributions in Fuel Nozzle Sprays – Part I: The Drop-Size/Volume-Fraction Distribution, *Journal of Engineering for Power*, 99(3), 309–314, 1977a, https://doi.org/10.1115/1.3446488

H. C. Simmons, The Correlation of Drop-Size Distributions in Fuel Nozzle Sprays—Part II: The Drop-Size/Number Distribution, *Journal of Engineering for Power*, 99(3), 315–319, 1977b, https://doi.org/10.1115/1.3446489

V. Simon, V. Weigand, & H. Gomaa, *Dimensional Analysis for Engineers*, Springer, 2017, https://doi.org/10.1007/978-3-319-52028-5

M. V. Smoluchowski, Drei Vorträge über Diffusion, Brownsche Molekularbewegung und Koagulation von Kolloidteilche [Three Lectures on Diffusion, Brownian Motion and Coagulation of Colloid Particles], *Physikalische Zeitschrift*, 17, 557–571, 587–599, 1916, https://eudml.org/doc/215805

M. V. Smoluchowski, Versuch einer mathematischen Theorie der Koagulationskinetik kolloider Lösunger [Mathematical Theory of the Kinetics of the Coagulation of Colloidal Solutions], *Zeitschrift für Physikalische Chemie*, 92, 129–168, 1917, https://doi.org/10.1515/zpch-1918-9209

T. M. Spielbauer, T. N. Adams, J. E. Monacelli, & R. T. Bailey, Droplet Size Distribution of Black Liquor Sprays, in *Proceedings of the National Chemical Recovery Conference*, Ottawa: TAPPI Press, 1989, http://hdl.handle.net/1853/2595

J. L. Spouge, Equilibrium Polymer Size Distributions, *Macromolecules*, 16(1), 121–127, 1983, https://doi.org/10.1021/ma00235a024

E. G. Stansbery, et al., *NASA Orbital Debris Engineering Model ORDEM 3.0 – Verification and Validation*, NASA Technical Publication TP-2015-218592, 2015.

W. H. Stockmayer, Theory of Molecular Size Distribution and Gel Formation in Branched-Chain Polymers, *The Journal of Chemical Physics*, 11(2), 45–55, 1943, https://doi.org/10.1063/1.1723803

B. Sweet, Remembering Tom Brzustowski, Waterloo's First Provost, *Waterloo Stories*, June 24, 2020, https://uwaterloo.ca/stories/remembering-tom-brzustowski-waterloos-first-provost

U. C. Täuber, *Critical Dynamics: A Field Theory Approach to Equilibrium and Non-Equilibirum Scaling Behavior*, Cambridge University Press, 2014, https://doi.org/10.1017/CBO9781139046213.001

N. Thadhani, Chair, Selection Committee, George E. Duvall Shock Compression Science Award, https://www.aps.org/programs/honors/prizes/shock.cfm; APS Announces 2007 Award Winners, *Physics Today*, 60(2), 76–78, 2007, https://doi.org/10.1063/1.2711645

D. L. Turcotte, Fractals and Fragmentation, *Journal of Geophysical Research*, 91(B2), 1921–1926, 1986, https://doi.org/10.1029/JB091iB02p01921

A. Voutilainen, V. Kolehmainen, F. Stratmann, & J. P. Kaipio, Computational Methods for the Estimation of Aerosol Size Distributions, in L. A. Uvarova & A. V. Latyshev, editors, *Mathematical Modeling. Problems, Methods, Applications*, Springer, 2001, https://doi.org/10.1007/978-1-4757-3397-6_22

J. L. Wang, J. H. Curtis, N. Riemer, & M. West, Learning Coagulation Processes with Combinatorial Neural Networks, *Journal of Advances in Modeling Earth Systems*, 14(12), 3252, 2022, https://doi.org/10.1029/2022MS003252

J. A. D. Wattis, The Modified Becker–Döring System with Aggregation-Dominated Rate Coefficients, *Journal of Physics A: Mathematical and Theoretical*, 42(4), 045002, 2009, https://doi.org/10.1088/1751-8113/42/4/045002

W. H. White, On the Form of Stationary Size Distributions Established by Coagulation and Sedimentation, *Journal of the Atmospheric Sciences*, 40(7), 1855–1858, 1983, https://doi.org/10.1175/1520-0469(1983)040<1855:OTFOSS>2.0.CO;2

M. Wilkinson, Quantifying the Lucky Droplet Model for Rainfall, *Journal of Statistical Physics*, 190, 37, 2023, https://doi.org/10.1007/s10955-022-03056-5

C. Xie, H. Nguyen, X.-N. Bui, Y. Cho, J. Zhou, & T. Nguyen-Trang, Predicting Rock Size Distribution in Mine Blasting using Various Novel Soft Computing Models Based on Meta-Heuristics and Machine Learning Algorithms, *Geoscience Frontiers*, 12(3), 101108, 2021, https://doi.org/10.1016/j.gsf.2020.11.005

J.-I. Yano, A. J. Heymsfield, & V. T. J. Phillips, Size Distributions of Hydrometeors: Analysis with the Maximum Entropy Principle, *Journal of the Atmospheric Sciences*, 73(1), 95–108, 2016, https://doi.org/10.1175/JAS-D-15-0097.1

V. V. Yanovsky, A. V. Tur, & O. V. Kuklina, Universal Elements of Fragmentation, *Journal of Experimental and Theoretical Physics*, 110(5), 863–876, 2010, https://doi.org/10.1134/S1063776110050183

Z.-X. Zhang, *Rock Fracture and Blasting: Theory and Applications*, Elsevier, 2016, https://doi.org/10.1016/C2014-0-01408-6

F. Zhou, J.-F. Molinari, & K. T. Ramesh, Characteristic Fragment Size Distributions in Dynamic Fragmentation, *Applied Physics Letters*, 88(26), 261918, 2006, https://doi.org/10.1063/1.2216892

D. Zou, *Theory and Technology of Rock Excavation for Civil Engineering*, Springer, 2017, https://doi.org/10.1007/978-981-10-1989-0

6

Particle Shape Distributions

This chapter concerns particle shapes in aggregating and fragmenting systems. In general, at least three parameters are required to specify the size and shape of a three-dimensional particle. As one possible approach, suppose those parameters are the dimensions a, b, and c of a rectangular box that bounds the particle with the minimum possible volume; see Figure 6.1. Alternatively, suppose those parameters are the axes a, b, and c of an ellipsoid that bounds the particle with the minimum possible volume; see Figure 6.2. Rectangular boxes are more common for fragmentation while ellipsoids are more common for aggregation. But, as a practical matter, the choice is not especially critical. In either case, as a notational convention, this chapter will assume that:

$$a \leq b \leq c \tag{6.1}$$

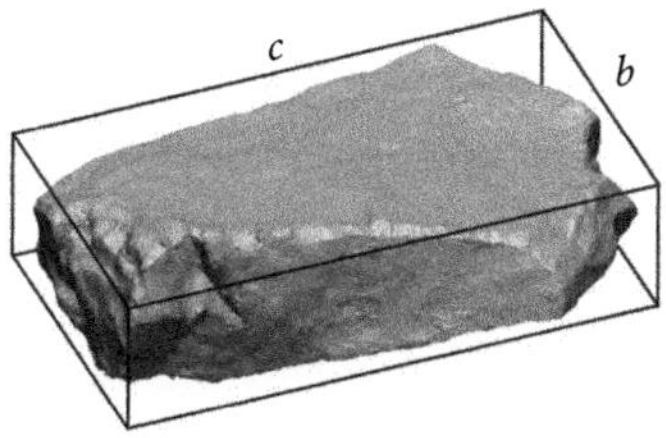

Figure 6.1 Particle sizes and shapes can be measured by the dimensions of the smallest rectangular box in which the particle fits.
[The notation in this figure has been changed to match the notation used in this book.]
Reprinted with permission from Domokos et al. (2015).

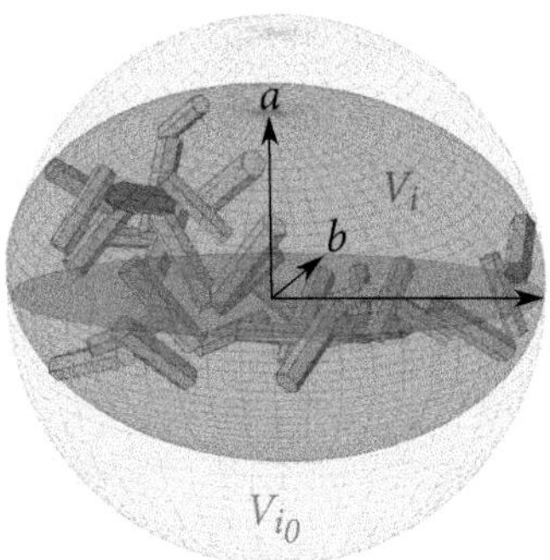

Figure 6.2 Particle sizes and shapes can be measured by the dimensions of the smallest ellipsoid in which the particle fits.
[The notation in this figure has been changed to match the notation used in this book.]
Reprinted from Westbrook et al. (2008) and Dunnavan et al. (2019). Copyright 2008 and 2019 American Meteorological Society. Used with permission.

Particle Dynamics with Aggregation and Fragmentation. Culbert B. Laney, Oxford University Press. © Culbert B. Laney (2025).
DOI: 10.1093/9780198920687.003.0006

For either boxes or ellipsoids, one of the three dimensions a, b, or c can be arbitrarily chosen to describe the size of the particle. Then the other two dimensions, divided by the size, describe the shape of the particle. For example, if a describes the size then the ratios b/a and c/a describe the shape. These are known as *aspect ratios*. Aspect ratios are often expressed as $1: b/a: c/a$.

As another possible way to describe the size and shape of a compact particle, suppose the three parameters are the diameter D, the surface area $\mathscr{A}$, and the mass M. The surface area and mass are well-defined physical quantities. However, the diameter is, generally speaking, only a concept. As before, one of the three quantities D, $\mathscr{A}$, or M can be arbitrarily chosen to describe the size of the particle. Then the other two quantities describe the shape of the particle. For example, if D describes the size then $\mathscr{A}$ and M describe the shape.

There are three cases commonly seen in the research literature: all particles have roughly the same shape; different particles have the same shape on average; and different particles have completely different shapes. This chapter considers each of these cases in turn.

6.1 Different Particles Have the Same Shape

First, consider the case where all particles have roughly the same shape. As one possible approach, if a describes the size, then the aspect ratios $b/a = \text{const}$ and $c/a = \text{const.}$ describe the shape. These expressions use the notation "const." as described earlier in Section 1.3. As another possible approach, if D describes the size, then $\mathscr{A}$ and M describe the shape where:

$$\mathscr{A} = \text{const.}\, D^s \tag{6.2}$$

$$M = \text{const.}\, \rho D^m \tag{6.3}$$

In these expressions $\max(1, m-1) \le s \le m$ is the particle *surface dimension*, $1 \le m \le d$ is the particle *mass dimension*, $d = 1, 2,$ or 3 is the *spatial dimension*, ρ is the particle density, and "const." refers to constants. For smooth regular particles:

$$s = \max(1, m - 1).$$

However, for rough regular particles, it is possible that $s = m$; see, e.g., Schaefer (1988). The most important thing is that, if all the particles have roughly the same shape, then s and m are constants. Examples 6.1, 6.2, and 6.3 give specific examples of Equations (6.2) and (6.3).

Example 6.1 Spherical particles

In many cases, fragmentation of liquids yields particles that are roughly spherical. Even when the particles are not initially spherical, they often gradually become spherical due to the effects of surface tension. For some particle shapes, the diameter D is only a concept. However, for spherical particles, the diameter D is a true physical quantity just like the surface area $\mathscr{A}$ and mass M. Notice that:

$$\mathscr{A} = \pi D^2$$

$$M = \rho \pi D^3/6.$$

Thus, in this case, $s = 2$, $m = 3$, and $d = 3$.

Example 6.2 Three-dimensional rectangular particles

In some cases, fragmentation of three-dimensional solids yields particles that are roughly three-dimensional rectangular boxes; see Figure 6.1. If the rectangular boxes have dimensions a, b, and c then:

$$\mathcal{A} = 2(ab + ac + bc)$$

$$M = \rho abc.$$

In addition, D may be set equal to a, b, c, or combinations such as $\sqrt{(a^2 + b^2 + c^2)/3}$. Suppose the boxes all have the same aspect ratios, i.e., b/a = const. and c/a = const. In addition, suppose $D = a$. Then:

$$\mathcal{A} = 2[b/a + c/a + (b/a)(c/a)]a^2 = \text{const.}\, D^2$$

$$M = \rho(b/a)(c/a)a^3 = \text{const.}\, \rho D^3.$$

In this case, $s = 2$, $m = 3$, and $d = 3$.

Example 6.3 Two-dimensional rectangular particles

In some cases, fragmentation of two-dimensional solids (e.g., thin plates) yields particles that are roughly two-dimensional rectangles. If the rectangles have dimensions a and b then:

$$\mathcal{A} = 2(a + b)$$

$$M = \rho ab.$$

In addition, D may be set equal to $a, b, \sqrt{(a^2 + b^2)/2}$, etc. Suppose that the rectangles all have the same aspect ratio, i.e., b/a = const. In addition, suppose $D = a$. Then:

$$\mathcal{A} = 2[b/a + 1]a = \text{const.}\, D$$

$$M = \rho(b/a)a^2 = \text{const.}\, \rho D^2.$$

In this case, $s = 1$, $m = 2$, and $d = 2$.

6.2 Different Particles Have the Same Shape on Average

If different particles have different shapes, they may still have the same shape on average. For example, in a series of impact fragmentation tests, Fujiwara et al. (1978) found that the average aspect ratios were $1{:}\sqrt{2}{:}\ 2$. These same aspect ratios were subsequently observed in a number of other impact fragmentation tests; see, e.g., Lange & Ahrens (1981), Matsui et al. (1982), Bianchi et al. (1984) and Capaccioni et al. (1984, 1986). For another example, Domokos et al. (2015) found that "both the mass and shape of fragments have an astonishing universality . . . irrespective of material details and loading conditions." More specifically, they found that "above a characteristic size the overall shape of all fragments can be well approximated by rectangles of dimensions 1:1.56:2.32." This result was based on seven datasets ranging from "weathering . . . [that] gradually removes pieces from the surface of rock walls under the action of gravity" to "explosions ... performed in stone mines."

As a third example, Mock & Holt (1981, 1983) conducted experiments on explosively-driven steel and iron shells. As described in Example 6.4, their results indicate that the largest fragments obtained universal aspect ratios of around $1{:}\sqrt{2.5}{:}\ 2.5$. While the largest fragments obtained this universal shape, the smallest fragments did not. In other words, the fragment shapes tended to converge gradually to a universal as the fragment size increased.

More generally, La Spina & Paolicchi (1996) found that universal fragment shapes obtain:

$$b/a = \sqrt{c/a}. \tag{6.4}$$

Notice that the three examples given previously exactly or approximately obey this rule:

$1 : \sqrt{2} : 2$	Fujiwara et al. (1978).
1:1.56:2.32	Domokos et al. (2015) [notice that $\sqrt{2.32} \approx 1.52$].
$1 : \sqrt{2.5} : 2.5$	Mock & Holt (1981,1983) [see Example 6.4].

Figure 6.3 shows results from eight different experimental series. In this figure, about two thirds of the results obey Equation (6.4) to within ±10%, including the three universal shapes listed previously.

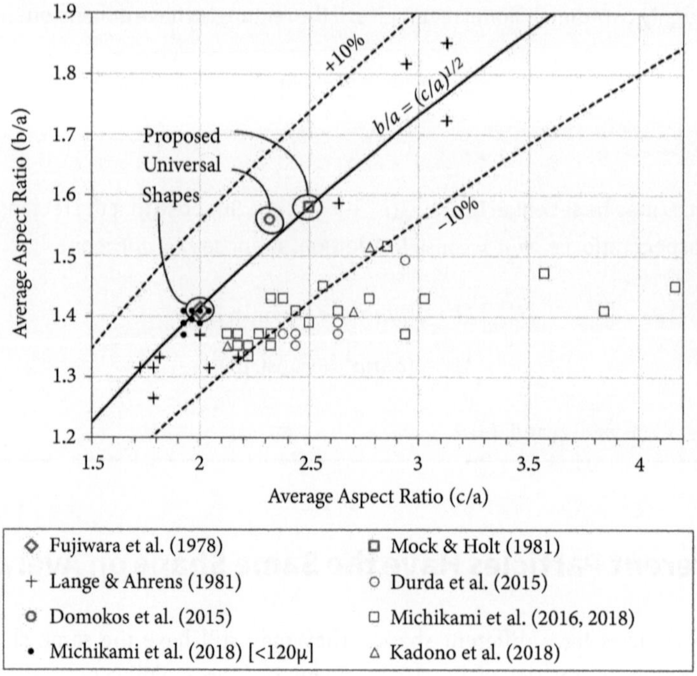

Figure 6.3 Experimental results for average aspect ratios in unitary fragmentation. Except for Mock & Holt (1981) and Domokos et al. (2015), these tests involve targets impacted by small high-speed projectiles.

Example 6.4 Average shapes produced by shell fragmentation—Part 1

Consider explosively loaded cylindrical shells as shown in Figure 6.4. Mock & Holt (1981, 1983) categorized the resulting shell fragments as follows:

Type 1. Fragments retain two smooth surfaces from the inner and outer shell surfaces.
Type 2. Fragments retain one smooth surface from the outer shell surface.
Type 3. Fragments retain one smooth surface from the inner shell surface.
Type 4. Fragments are rough on all sides, e.g., they were formed entirely by fracture.

Generally speaking, in shell fragmentation, all of the largest fragments are Type 1, all of the smallest fragments are Type 4, while the intermediate fragments are a mixture of types. Mock & Holt (1981) measured the widths, lengths, and thicknesses of selected Type 1 fragments obtained from explosively-driven steel and iron shells. For example, the fragment length was defined as the maximum distance along a smooth surface in the direction parallel to the original cylindrical axis. For another example, the fragment thickness was defined as the minimum distance between the two smooth surfaces. Mock and Holt (1981) found that the fragment thickness varied randomly, due to residual strain, meaning it was 10% to 60% less than the original wall thickness. For the largest steel fragments, the width corresponds to a, the thickness corresponds to b, and the length corresponds to c. More generally, these correspondences depend on the fragment material and size. In Figure 6.5, length/thickness seems to be almost statistically independent of thickness/width. However, clear patterns emerge if the aspect ratios are plotted as functions of fragment mass. In fact, as seen in Figures 6.6 and 6.7, steel shells obtain aspect ratios of:

$$a/a : b/a : c/a \;=\; 1{:}1.58{:}2.5$$

on average for large enough fragments. According to Equation (6.4), this is a universal shape. On the other hand, as seen in Figure 6.8, iron shells obtain aspect ratios of:

$$a/a : b/a : c/a \;=\; 1{:}1{:}8$$

on average for large enough fragments. This means that iron cylinders produce long thin strips. According to Equation (6.4), this is *not* a universal shape. Iron produces larger aspects ratios partly because iron is more brittle than steel. For ductile materials, Curran (1997), Goto et al. (2008), and Hiroe et al. (2008) report aspect ratios in the range of one to six. For brittle materials at early times, Kun et al. (2006) report aspect ratios as large as 15 to 35.

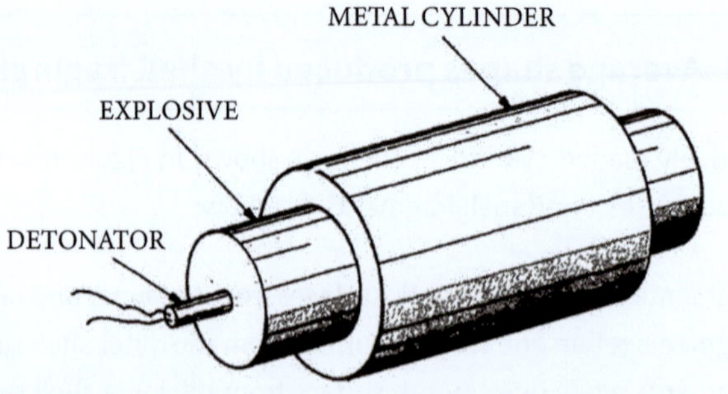

Figure 6.4 Diagram showing an explosively-loaded cylindrical metal shell.
Reprinted from Mock & Holt (1983) with the permission of AIP Publishing.

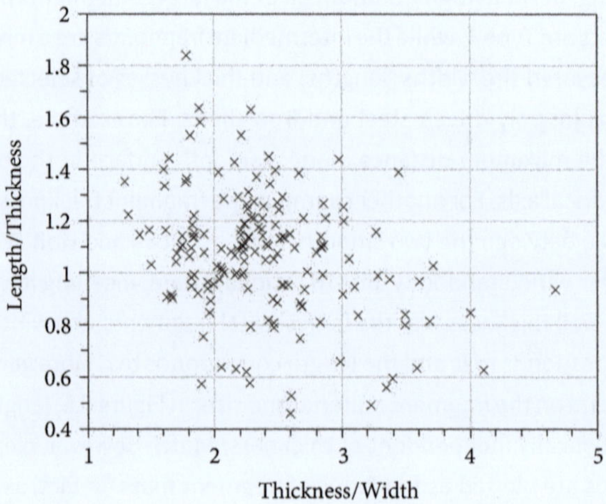

Figure 6.5 For explosively-driven steel shells, when fragment sizes are not accounted for, the two fragment aspect ratios appear to be uncorrelated. Aspect ratios for 162 Type 1 fragments collected in four tests conducted by Mock & Holt (1981).

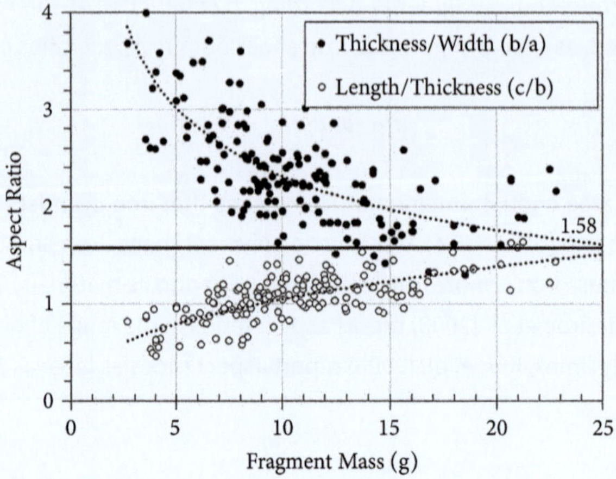

Figure 6.6 For explosively-driven steel shells, the two fragment aspect ratios converge to 1.58 on average as the fragment mass increases. Aspect ratios for 162 Type 1 fragments collected in four tests conducted by Mock & Holt (1981).

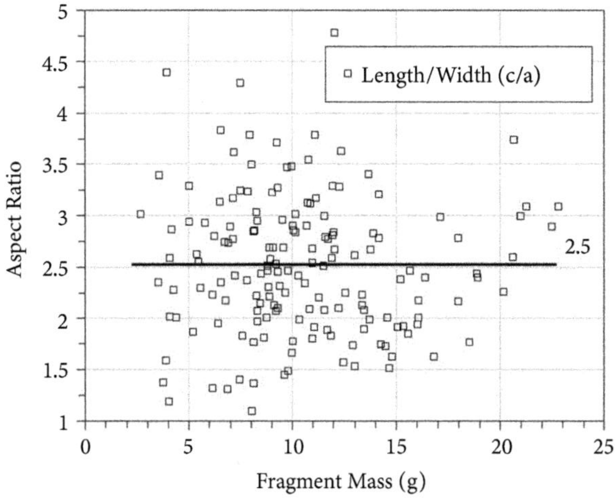

Figure 6.7 For explosively-driven steel shells, one fragment aspect ratio has an average of 2.5 regardless of fragment mass. Aspect ratios for 162 Type 1 fragments collected in four tests conducted by Mock & Holt (1981).

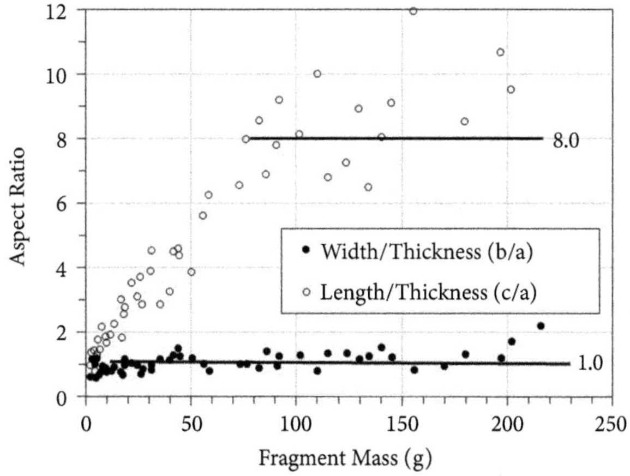

Figure 6.8 For explosively-driven iron shells, one fragment aspect ratio has an average of 8 for large enough fragments while another fragment aspect ratio has an average of 1 regardless of fragment mass. Aspect ratios for 50 Type 1 fragments collected in two tests conducted by Mock & Holt (1981).

Instead of aspect ratios, particle shapes may also be described by surface area and mass; see Example 6.5.

Example 6.5 Average shapes produced by shell fragmentation—Part 2

Kun et al. (2006) and Wittel et al. (2006) studied explosively-driven fragmentation of eggshells and round glass shells. In these tests, the shell fragments obtained a constant thickness. In measuring these fragments, the fragment diameter D was defined to be twice the radius of gyration, i.e., twice the root-mean-square-average distance between the points on the fragment surface and the center of mass. In addition, as seen in Figure 6.9, $\mathscr{A}$ was defined as being the fragment circumference times the constant thickness and M was defined as being the fragment cross-sectional area times the constant thickness and density. With these definitions, Kun et al. (2006) and Wittel et al. (2006) found that:

$$M_{\mathrm{avg}} \cong \frac{1}{N}\sum_{k=1}^{N} M_k \cong \text{const.}\, \rho D^m \tag{6.5}$$

where M_{avg} is the average mass of N particles that have different masses M_k but the same diameter D. More specifically, they found that $m = 2 \pm 0.05$ for compact eggshell fragments and $m \cong 1.5 \pm 0.05$ for jagged glass fragments. They say that $m < 2$ "implies that the fragments have a self-affine character, i.e., the larger they are, the more elongated they get."

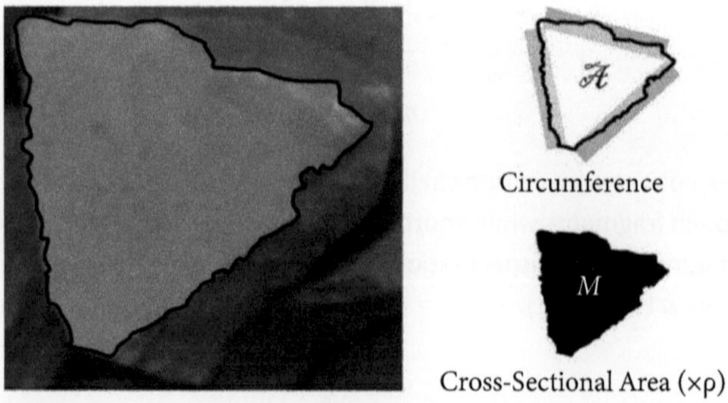

Figure 6.9 Possible definitions for the surface area $\mathscr{A}$ and mass M of a fragment with a constant thickness.

So far, this section has focused on fragmentation. However, most if not all of the same principles apply to aggregation. In particular, just like fragmentation, experimental results suggest that aggregation often produces universal shapes, at least on average. For example, Westbrook et al. (2004a, b) found substantial evidence of a universal aggregate shape based on "simulations and experimental observation ... and theoretical arguments." According to Example 6.6, their universal shape is 1: 1.58: 2.5. As discussed earlier, the same universal shape occurs for fragmentation.

Example 6.6 Average shapes produced by aggregation of ice crystals

Westbrook et al. (2004a, b) studied aggregation of "ice crystals in Cirrus clouds ... [where] small pristine ice particles are formed at the top of the cloud, and proceed to fall through it, colliding with one another and sticking to produce aggregates." Such aggregates may eventually evolve into snowflakes.

Westbrook et al. (2004a, b) defined an aspect ratio based on two-dimensional "projections in an arbitrary plane." In each plane, the aspect ratio was defined as the "longest span divided into the span perpendicular to the longest." This is roughly equivalent to drawing a bounding rectangle with the minimum area around a two-dimensional side view of the aggregate. For most shapes, the as-measured aspect ratio will vary substantially depending on the viewing plane. In such cases, Dunnavan et al. (2019) note that the 2D "projection process itself ruins any attempt at inferring 3D geometry from single view aggregate imaging." However, suppose the aggregate obeys Equation (6.4). If the "longest span" observed is c, then the as-measured aspect ratio will be b/c on average. On the other hand, if the "longest span" observed is b, then the as-measured aspect ratio will be a/b on average. But according to Equation (6.4):

$$a/b = b/c. \tag{6.6}$$

More generally, when the longest span observed is somewhere between b and c, the exact choice of the two-dimensional viewing plane does not affect the as-measured aspect ratio, on average. Instead, as shown in Figure 6.10, Westbrook et al. (2004a, b) found that the average as-measured aspect ratio converges to about 0.63 as the aggregate size increases. This was true for experimental results, computational results in which the monomers were rosettes, and computational results in which the monomers were rods. If $a/b = b/c \approx 0.63$, then $1 : b/a : c/a \approx 1 : 1.58 : 2.5$, which agrees with the universal shape found in Example 6.4.

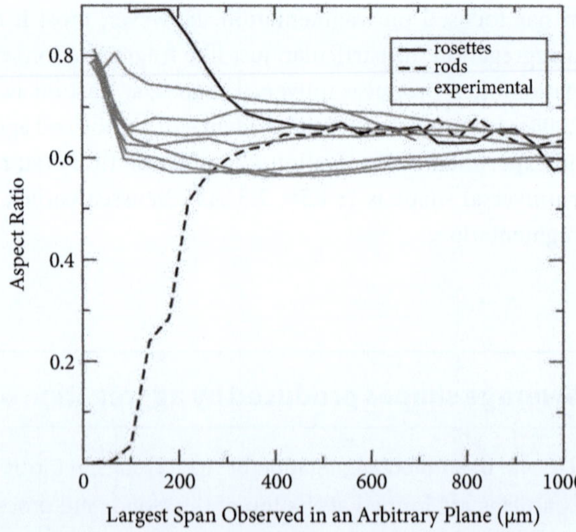

Figure 6.10 For aggregation of ice crystals, the as-measured aspect ratio converges to approximately 0.63 as the aggregate size increases.
[The notation in this figure has been changed to match the notation used in this book.]
Reprinted with permission from Westbrook et al. (2004a). Copyright 2004 American Geophysical Union.

Most aggregation studies report only a single aspect ratio. As seen in Example 6.6, it may be possible to convert that single aspect ratio into two aspect ratios—in other words, to estimate both b/a and c/a—assuming that Equation (6.4) applies. Alternatively, instead of making assumptions, b/a and c/a can be measured directly. This is difficult to do in experiments; see, e.g., Jiang et al. (2019). However, it is relatively straightforward to do in computations. In fact, as seen in Figure 6.11, three computational studies all obtained:

$$b/a \cong \sqrt{c/a}$$

to within ±10%, where larger aspect ratios are generally associated with larger aggregation rates. In Figure 6.11, Warren & Ball (1989) and Fry et al. (2004) studied six types of colloids including those formed by Diffusion-Limited (DL), Reaction-Limited (RL), Ballistic-Limited (BL), Cluster-Monomer (A), and Cluster-Cluster (CA) aggregation. Recall that these terms were defined earlier in Section 1.3. In addition, in Figure 6.11, Dunnavan et al. (2019) studied aggregation of airborne ice crystals of various shapes and sizes. These computations account for the effects of ballistics, gravity, and shape-dependent drag while generally assuming that clusters may merge with monomers but not with other clusters. While Figure 6.11 offers evidence that Equation (6.4) applies to aggregation, it does not offer evidence that true universal shapes occur, i.e., no specific value for b/a or c/a occurs substantially more frequently than any other.

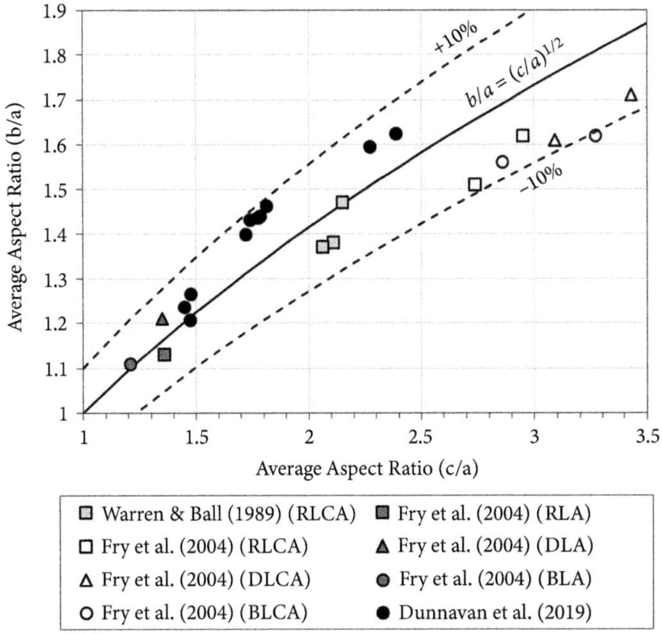

Figure 6.11 Computational results for average aspect ratios in aggregation.

Instead of aspect ratios, aggregate shapes may also be described by surface area and mass; see Example 6.7.

Example 6.7 Fractal particle shapes in colloidal aggregation

In aggregating systems, different particles may have the same shape on average, even though the exact details such as the branched structures vary in seemingly random ways. For example, Schaefer (1988) used light and neutron scattering experiments to study various kinds of colloids including those formed through DL, RL, and BL aggregation. Recall that these terms were defined earlier in Section 1.3. In general, Schaefer (1988) found that:

$$\mathscr{A} = \text{const. } D^s$$

$$M = \text{const. } \rho D^m$$

apply on average, where s and m are constant real numbers. Schaefer's technique measured $2m + s$ rather than s and m separately. However, by combining experimental results with computational results, Schaefer surmised that m varied between 1.8 and 3, and that s varied between $m - 1$ and m.

Since this pioneering study, numerous other researchers have reported a wide range of experimental and computational values for m. For example, Jungblut et al. (2019a, b) used Molecular Dynamics (MD) to study DL and RL colloidal aggregation. Recall that MD was discussed earlier in Section 4.1. In general, Jungblut et al. (2019a, b) found that $M = \text{const. } \rho D^m$ where D is twice the radius of gyration, defined as the root mean square average of the three "eigenvaules of the gyration tensor extracted from the positions of all particles in the aggregation." When the clusters were allowed to rotate, the aggregating particle system obtained $m = 1.4$ to 2.1. In general, Jungblut et al. (2019a, b) found that their values of m were significantly lower than those reported in most earlier computational studies, which did not allow clusters to rotate.

6.3 Different Particles Have Different Shapes

As seen in the last section, even when different particles have the same shape on average, the following is generally true:

- Only particles above a minimum size or in a certain size range obtain this average.
- There may be considerable scatter about the average; see, e.g., Figures 6.6, 6.7, and 6.8.

Thus, rather than trying to identify and exploit average shapes, it may be better to simply assume that different particles have completely different shapes. As seen earlier in Examples 6.2 and 6.3, if different particles have the same shape on average, the exact definition of the diameter does not matter very much. On the other hand, as seen in Example 6.8, if different particles have completely different shapes, the definition of the particle diameter matters a great deal. As this example shows, while analytic and computational approaches are free to adopt any definition of diameter they like, observational and experimental approaches may be forced to use a specific definition of diameter, due to the measurement technique and the type of commercial equipment involved.

Example 6.8 Effects of particle shapes on measurements of particle diameters

Percival & Lindsay (1997) note that different size measurement techniques often yield different results. This is due to "the definition of the size being determined: projected area size by image analysis, intermediate diameter size by sieve, volume size by electronic sensor counter, or quartz equivalent spherical sedimentation diameter by pipette method."

To provide further insight into this issue, Cepuritis et al. (2017) measured the diameter distributions for 3 to 200μm rock particles "produced from 10 different blasted and crushed rocks with an original size range of about 4 to 22 mm." As seen in Figure 6.12, these particle samples included a wide range of different particle shapes. Cepuritis et al. (2017) compared the diameter distributions obtained using four different commercial systems:

- *Malvern Mastersizer 2000S*. Uses laser diffraction to estimate the volume-equivalent spherical diameters of particles suspended in a liquid medium.
- *Micromeritics SediGraph 5100*. Uses x-rays to estimate gravitational settling rates and mass-equivalent spherical diameters of particles suspended in a liquid medium.
- *Zeiss Xradia 500 Versa*. Uses x-ray microcomputed tomography to estimate the volume-equivalent spherical diameters of particles suspended in a solid medium.

- *AnaTec Fine Particle Analyzer.* Uses high-speed photography to estimate the cross-sectional area-equivalent circular diameters of particles suspended in air.

Figure 6.13 compares the cumulative distribution function $F(D)$ obtained using these four commercial systems. Because each system uses a substantially different definition for the particle diameter, each system obtained a substantially different result for the particle diameter distribution. For more information on the effects of particle shapes on the measurements of particle diameters, see the classic review by Matthews (1991); see also the more recent work by Califice et al. (2013), Jia & Garboczi (2016), Garboczi et al. (2017), and Neumann et al. (2018).

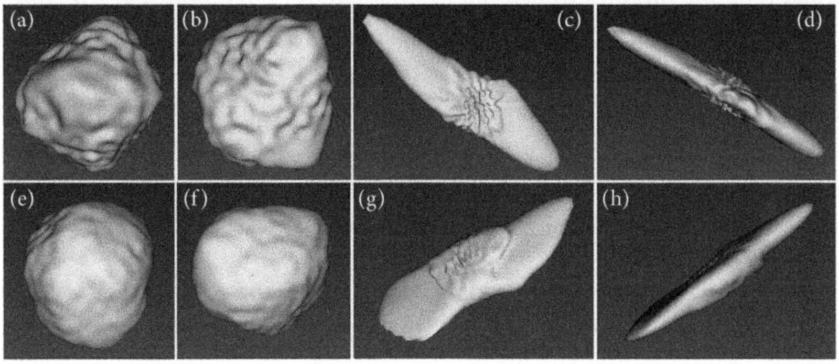

Figure 6.12 Selected particle shapes present in samples of aplite dust.
Reprinted from Cepuritis et al. (2017) with permission from Elsevier.

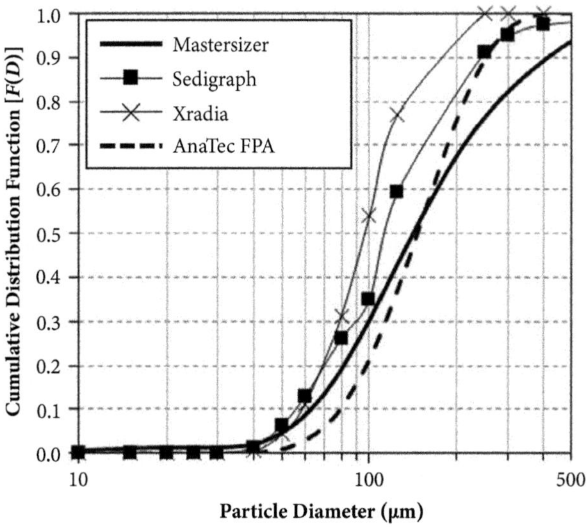

Figure 6.13 Experimental results for the size distributions of seemingly randomly shaped regular particles, as found using four different commercial measuring systems, each of which uses a different definition of particle diameter.
[The notation in this figure has been changed to match the notation used in this book.]
Reprinted from Cepuritis et al. (2017) with permission from Elsevier.

This book usually uses the notation n/m to express particle size distributions. This notation assumes that $M = c_1\rho D^m$. This is always true if different particles have roughly the same shape. However, if different particles have different shapes, it is necessary to choose between two options. The first option leaves the definition of m alone and changes the definition of the diameter D to ensure that $M = c_1\rho D^m$. As the main advantage of this option, m remains constant; see, e.g., Example 6.9. This book adopts the first option for regular particles. The second option leaves the definition of the diameter D alone and changes the definition of m to ensure that $M = c_1\rho D^m$. In this option, m varies from particle to particle such that:

$$m = \frac{\ln(M/c_1\rho)}{\ln D}.$$

This book adopts the second option for fractal particles. For fractal particles, experimental techniques typically produce something like the radius of gyration; see, e.g., Bushell et al. (2002). Thus, the particle diameter is typically chosen to be twice the radius of gyration.

To summarize, if all of the particles have the same shape, most of the obvious size metrics are essentially interchangeable. However, if different particles have different shapes, the size metrics are distinct. In theory, the problem itself dictates the best choice. As described in Examples 6.10 and 6.11, the best choice could be some type of particle diameter; alternatively, it could be the particle mass or surface area. As the default, this book generally uses particle masses to describe particle sizes. This is because, unlike the particle diameter, there is always a unique physical definition for the particle mass. In addition, this is because, unlike the particle surface area, the particle mass is relatively easy to measure and is widely reported.

Example 6.9 Possible definitions of diameter when different particles have different shapes

Suppose the particles are all rectangular boxes with sides of length a, b, and c such that $a < b < c$. Assume that different boxes have different shapes, i.e., $b/a \neq$ const. and $c/a \neq$ const. Four possible definitions of particle diameter are as follows:

1. *Physical dimension*. Arbitrarily choose a, b, or c. For example, suppose that:

$$D = a.$$

If $b/a \neq$ const. and $c/a \neq$ const. then $\mathcal{A} = 2[b/a + c/a + (b/a)(c/a)]D^2 \neq \text{const.}\, D^2$ and $M = \rho(b/a)(c/a)D^3 \neq \text{const.}\, \rho D^3$. Thus, with this definition of D, Equations (6.2) and (6.3) do not apply. The same conclusion applies if $D = b$ or c.

2. *Radius of gyration*. Let the diameter be twice the radius of gyration, i.e., twice the root-mean-square average of the distance from the center of mass. According to Domokos et al. (2015):

$$D \sim \sqrt{(a^2 + b^2 + c^2)/3}$$

meaning that the diameter is proportional to the *root mean square average* of a, b, and c. If $b/a \neq$ const. and $c/a \neq$ const. then $\mathcal{A} = 2(ab + ac + bc) \neq \text{const.}\, D^2$ and

$M = \rho abc \neq \text{const.}\, \rho D^3$. Thus, with this definition of D, Equations (6.2) and (6.3) do not apply.

3. *Volume-to-area ratio*. Let the diameter be proportional to $M/\rho \mathcal{A}$. Then:

$$\frac{1}{D} = \frac{1}{3}\left(\frac{1}{a} + \frac{1}{b} + \frac{1}{c}\right)$$

which is the *harmonic mean* of a, b, and c. If $b/a \neq \text{const.}$ and $c/a \neq \text{const.}$ then $\mathcal{A} = 2(ab + ac + bc) \neq \text{const.}\, D^2$ and $M = \rho abc \neq \text{const.}\, \rho D^3$. Thus, with this definition of D, Equations (6.2) and (6.3) do not apply.

4. *Spherical volume equivalent*. Suppose that boxes and spheres with the same diameters also have the same volumes. Then:

$$D = (6abc/\pi)^{1/3}$$

meaning that the diameter is proportional to the *geometric mean* of a, b, and c. If $b/a \neq \text{const.}$ and $c/a \neq \text{const.}$ then $M = \rho abc = \pi\rho D^3/6$ and $\mathcal{A} = 2(ab + ac + bc) \neq \text{const.}\, D^2$. Thus, with this definition of D, Equation (6.3) applies but not Equation (6.2).

To summarize, if the particles are rectangular boxes with varying aspect ratios, only one of the four proposed definitions of D makes Equation (6.3) true, and none makes Equation (6.2) true. According to Figure 6.1, the size and shape of any particle can be characterized by bounding rectangular boxes. Thus, the same conclusions apply in general, i.e., Equation (6.3) but not Equation (6.2) applies if D is defined as the geometric mean of the linear dimensions.

Example 6.10 Shape and size distributions produced by shell fragmentation—Part 1

Kun et al. (2006) and Wittel et al. (2006) studied explosively-driven fragmentation of ellipsoidal eggshells and spherical glass shells. In general, at least two parameters are required to specify the shape and size of a two-dimensional fragment. As one possible approach, suppose those parameters are the dimensions a and b of a rectangle that bounds each fragment with the minimum possible area where $a \leq b$. As another possible approach, as discussed earlier in Example 6.5, suppose D is twice the radius of gyration, $\mathcal{A}$ is the circumference, and M is the cross-sectional area time density. These two approaches can be related as follows:

$$D \sim \sqrt{a^2 + b^2}; \quad \mathcal{A} \sim 2a + 2b; \quad M \sim \rho ab$$

where the tilde refers to a shape-dependent factor. Instead of choosing a, b, D, $\mathcal{A}$, or M to characterize the fragment shapes, Kun et al. (2006) and Wittel et al. (2006) chose the following dimensionless parameter:

$$S = \frac{\mathscr{A}D}{M}. \tag{6.7}$$

Notice that if $\mathscr{A} = \text{const.}\, D^{m-1}$ and $M = \text{const.}\, \rho D^{m}$, then S is constant. However, because these two relationships do *not* apply, S is *not* constant. Instead:

$$S \sim \frac{2(a+b)}{\rho ab}\sqrt{a^2+b^2} = \frac{2}{\rho}(1+b/a)\sqrt{(a/b)^2+1}.$$

Then $S \sim \text{const.}$ for isotropic fragments with small aspect ratios $b/a \approx 1$. On the other hand:

$$S \sim b/a \tag{6.8}$$

for anisotropic fragments with large aspect ratios b/a. Based on curve fits to experimental results, Kun et al. (2006) and Wittel et al. (2006) found that the shape distribution $f(S)$ obtained a power law for sufficiently large $S \sim b/a$. However, as seen in Figure 6.14, this power law is non-ideal, in the sense that the exponent depends on material composition and other factors.

In addition to fragment shape, Kun et al. (2006) and Wittel et al (2006) considered two possible definitions of fragment size. First, they considered $\mathscr{A}/M$. From one point of view, $\mathscr{A}/M$ is an inverse diameter. When they curve fitted the experimental results to estimate $f(\mathscr{A}/M)$, they found the fit was sensitive to the exact conditions and applied only over a limited range of fragment shapes with $b/a \approx 1$. Next, they considered mass M. When they curve fitted the experimental results to estimate $f(M)$, they found that a good fit could be obtained with a power law size distribution over a wide range of sufficiently large M. This power law size distribution was ideal in the sense that it involved a "universal exponent" $n/m \approx -1/3$ that remained the same under "widely varying conditions." This implies that the inverse diameter $\mathscr{A}/M$ is not an appropriate way to measure fragment size. Instead, the mass M is the appropriate way to measure fragment size, at least in this specific case.

Example 6.11 Shape and size distributions produced by shell fragmentation—Part 2

Curran (1997) studied explosively-driven steel shells including "fragment aspect ratio data from 1/50th and 1/6th scale BLU-109 experiments, and 1/5th scale AN-Mk 1 AP 1600 bomb experiments." To characterize the fragment shapes, Curran (1997) chose the aspect ratio b/a where a and b are the dimensions of a rectangle that bounds each 2D fragment with the minimum possible area such that $a \leq b$. Based on curve fits to experimental results, Curran (1997) found that $f(b/a)$ obtained a power law form. However, this power law was non-ideal, in the sense that the exponent was "a material dependent geometrical factor."

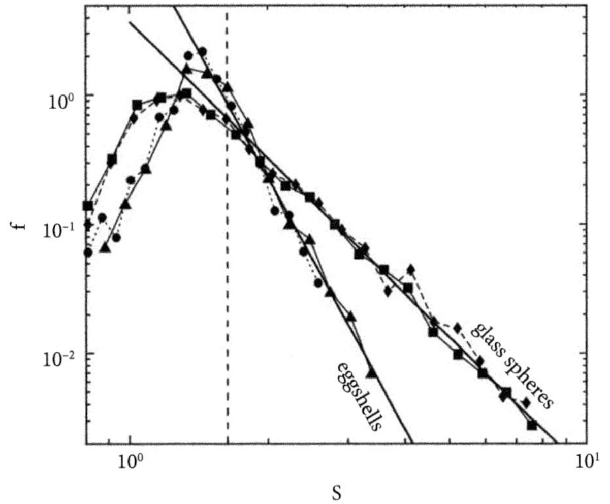

Figure 6.14 Experimental results for fragment shape distributions for explosively-driven eggshells and glass spheres. On the horizontal axis, S equals the fragment area times its diameter divided by its mass. On the vertical axis, f is the probability density function.
Reprinted with permission from Kun et al. (2006). Copyright 2016 American Physical Society.

In addition to fragment shape, Curran (1997) considered two possible definitions of fragment size. First, he considered width a. From one point of view, a is a diameter. When he curve fitted the experimental results to estimate $f(a)$, he found that a good fit could be obtained with an exponential size distribution. This size distribution was ideal in the sense it did not depend on material type or other factors. Curran noted that the distributions for size a and shape b/a were "completely decoupled." Second, he considered mass M. In this case, Curran (1997) found the following:

If we convert the ... size distribution to a mass distribution, the simplicity disappears because longer fragments with the same widths have larger masses ... One obtains complex mixed mass and shape distributions ... If our model of fragment formation is correct, the size and shape distributions are naturally separable.

This implies that mass is not an appropriate way to measure fragment size. Rather, the diameter a is the appropriate way to measure fragment size, at least in this specific case. Thus, even though this example is closely related to the previous example, the opposite conclusion about mass vs. diameter was reached.

As seen in Examples 6.10 and 6.11, curve fits to two-dimensional fragmentation data indicate that $f(b/a)$ tends to approach a power law distribution for large enough b/a. Similarly, as seen in Example 6.12, curve fits to three-dimensional aggregation data indicate that $f(b/a)$ and $f(c/a)$ tend to approach power laws for large enough b/a and c/a, respectively. In this sense, aggregation and fragmentation tend to be similar.

Example 6.12 Shape distributions produced by aggregation of ice crystals

Dunnavan et al. (2019) note that, during snowfall, "each aggregate represents a connected ensemble of individual ice particles, each called a 'monomer', that have collided and stuck to one another while falling." Dunnavan et al. (2019) studied two sets of snowflake data, one from the North Slope of Alaska and the other from Alta, Utah. In both cases, they found that a good fit could be obtained with the following:

$$f(y) = \text{const.}\, y^{\alpha-1}(1-y)^{\beta-1} \tag{6.9}$$

where $y = a/b$ or a/c and where $\alpha > 1$ and $\beta > 1$. This implies that $f(a/b)$ and $f(a/c)$ approach power laws for small enough a/b and a/c, respectively. However, this book usually works with ratios $x = b/a$ and c/a. As it turns out, $f(b/a)$ and $f(c/a)$ approach power laws for large enough b/a and c/a, respectively. To prove this, notice that:

$$f(x) = \left|\frac{dy}{dx}\right| f(y). \tag{6.10}$$

If $y = 1/x$, i.e., if $x = b/a$ or c/a, then Equation (6.9) can be rewritten as follows:

$$f(x) = \text{const.}\, x^{-\alpha-1}\left(1-\frac{1}{x}\right)^{\beta-1}. \tag{6.11}$$

Then:

$$f(x) \cong \text{const.}\, x^{-\alpha-1} \tag{6.12}$$

for large enough x. In other words, $f(b/a)$ and $f(c/a)$ tend to approach power laws for large enough b/a and c/a, respectively.

Dunnavan et al. (2019) report that "both observations and numerical studies support claims of geometric universality." However, this universality does not seem to apply to α or β. For example, for b/a, Dunnavan et al. (2019) report that $\alpha \cong 4.17$, $\beta \cong 1.91$ for snowflakes from the North Slope of Alaska, while $\alpha \cong 6.12$, $\beta \cong 2.55$ for snowflakes from Alta, Utah. This raises "questions [about] the universality of some mass-dimensional power-law[s] ...."

In addition to aggregate shapes, Dunnavan et al. (2019) briefly considered aggregate sizes. As one possible definition, suppose the aggregate size is its maximum ellipsoidal dimension $2c$. Then Dunnavan et al. (2019) say that for large enough snowflakes formed through "many different stages of aggregation ... the maximum dimension and shape can be considered separate and can be specified using different [distribution] functions ...."

6.4 Conclusions

This chapter discussed the shapes of particles produced by aggregating and fragmenting systems. This discussion was mainly limited to compact particles, i.e., those whose aspect ratios are not particularly large. By limiting this discussion to compact shapes, which can be well-represented by bounding rectangular boxes or ellipsoids, this chapter deliberately excluded a number of important applications. For example, according to Koltzenburg et al. (2017) polymers "can be divided into different groups according to their architecture as linear, branched, or cross-linked structures. In addition, linear polymers, whose chains are longer than a certain critical length tend to form entanglements." Along these same lines, Stuart (2002) says "some of the more unusual polymer structures include star polymers, which contain three or more polymer chains connected to a central unit, ladder polymers, which consist of repeating ring structures, and dendrimers, which show a star-like structure with branching." Thus, in general, polymer particles are not compact and cannot be well-represented by bounding rectangular boxes or ellipsoids. For another example, as discussed in Chapter 8, turbulent fluid flows produce self-organized recirculating structures through a fragmentation-like process. In many cases, Davidson (2013) says "the intermediate to small scales in a turbulence flow can be pictured as a sparse tangle of spaghetti-like vortex tubes." Thus, in general, turbulent structures are not compact and cannot be well-represented by bounding rectangular boxes or ellipsoids. As usual, this book focuses on results that apply for many different applications, which excludes specific discussions of polymer shapes, turbulent vortex shapes, and so forth.

For compact particles, this chapter has shown that there are two common ways of representing particle shapes. In the first approach, sometimes known as the *Euclidean description*, particle shapes are represented by aspect ratios such as b/a or c/a, where a, b, and c are the dimensions of a bounding rectangular box or ellipsoid. A number of studies have found that that shape distributions such as $f(b/a)$ and $f(c/a)$ tend to approach power laws for large enough b/a and c/a, respectively. In addition, if different particles have the same shape on average, a number of studies have found that:

$$b/a = \sqrt{c/a}$$

on average. This includes so-called universal shapes such as $1{:}\sqrt{2}{:}\ 2$ and $1{:}\sqrt{2.5}{:}\ 2.5$. As noted by La Spina & Paolicchi (1996), the former universal shape is more common when particles "suffer secondary and artificial breakups, taking place preferentially along the longest axis, [which decreases] ... the mean values of measured [aspect] ratios." Conversely, the latter universal is more common when particles do *not* suffer secondary or artificial breakups.

In the second approach, sometimes known as the *fractal description*, particle shapes are represented by power law relationships between size metrics such as:

$$M = \text{const.}\ \rho D^m.$$

For regular particles, this book assumes that m is a constant or piecewise-constant integer, namely, 1, 2, or 3. This true regardless of whether the particles have the same shapes on average on not. On the other hand, for fractal particles, this book assumes that m is a real number between 1 and 3. If fractal particles have the same shape on average, then m may be the same for different particles. However, if fractal particles have different shapes, m may be different for different particles. These are all, to a large extent, just convenient assumptions. Other assumptions are possible. For example, regular fragments sometimes form in somewhere between two and three dimensions; e.g., Type 2 and 3 fragments as discussed in Example 6.4. In such cases, it would be reasonable to choose a real value of m between 2 and 3. However, lacking a well-established approach for doing so, this book rounds to the nearest integer, i.e., m is set equal to either 2 or 3.

The results given in this chapter are empirical, i.e., they are entirely based on curve fits to specific sets of observational, experimental, or computational data. As a result, many of the conclusions are somewhat tentative. As noted in the last chapter, where they exist, analytic methods can often obtain firm general results, quickly and efficiently, without becoming bogged down in the exact details of any specific problem, subject only to the validity of the underlying assumptions. Unfortunately, there are no well-established analytic approaches for modeling particle shapes. This chapter has focused on just a few specific applications, all of which involve solid particles. While similar results probably apply to liquid particles, this is difficult to prove because liquid particles tend to either have simple spherical shapes in static cases or complex rapidly evolving shapes in dynamic cases; see, e.g., Stevenin et al. (2016) and Pham et al. (2018).

If different particles in a given system have different shapes, each problem seems to uniquely dictate the proper definition of particle size, to within a constant factor. More specifically, *when the particle size definition is properly chosen, particle size is statistically independent of the particle shape.* This size definition can only be determined by considering particle size and shape together in a unified fashion on a problem-by-problem basis. Unfortunately, at present, there is no good analytic theory for doing this—it must be done after-the-fact using observations, experimental results, or computational results; see, e.g., Examples 6.10 and 6.11. In most cases, this book defaults to using mass to describe particle sizes, while somewhat hedging its bets by using the notation n/m to allow for easy conversion from mass to diameter.

Unless stated otherwise, the remainder of this book assumes that particles formed through aggregation and fragmentation experience rotation. In many cases, rotating particles with complex shapes can be treated like non-rotating particles with simpler shapes. For example, a circular coin flipped end-over-end may behave like a non-rotating sphere of the same radius. In general, rotating particle shapes can be well-represented by mass and diameter, where the latter is generally defined as twice the radius of gyration or the spherical volume equivalent, depending on the axes of rotation.

References

R. Bianchi, F. Capaccioni, P. Cerroni, M. Coradini, E. Flamini, P. Hurren, et al., Experimental Simulation of Asteroidal Fragmentation by Macroscopic Hypervelocity Impacts Against Free Falling Bodies, *Astronomy and Astrophysics*, 139(1), 1–6, 1984, http://adsabs.harvard.edu/abs/1984A&A...139....1B

G. C. Bushell, Y. D. Yan, D. Woodfield, J. Raper, & R. Amal, On Techniques for the Measurement of the Mass Fractal Dimension of Aggregates, *Advances in Colloid and Interface Science*, 95(1), 1–50, 2002, https://doi.org/10.1016/S0001-8686(00)00078-6

A. Califice, F. Michel, G. Dislaire, & E. Pirard, Influence of Particle Shape on Size Distribution Measurements by 3D and 2D Image Analyses and Laser Diffraction, *Powder Technology*, 237, 67–75, 2013, https://doi.org/10.1016/j.powtec.2013.01.003

F. Capaccioni, P. Cerroni, M. Coradini, P. Farinella, E. Flamini, G. Martelli, et al., Shapes of Asteroids Compared with Fragments from Hypervelocity Impact Experiments, *Nature*, 308, 832–834, 1984, https://doi.org/10.1038/308832a0

F. Capaccioni, P. Cerroni, M. Coradini, M. Di Martino, P. Farinella, E. Flamini, et al., Asteroidal Catastrophic Collisions Simulated by Hypervelocity Impact Experiments, *Icarus*, 66(3), 487–514, 1986, https://doi.org/10.1016/0019-1035(86)90087-4

R. Cepuritis, E. J. Garboczi, C. F. Ferraris, S. Jacobsen, & B. E. Sørensen, Measurement of Particle Size Distribution and Specific Surface Area for Crushed Concrete Aggregate Fines, *Advanced Powder Technology*, 28(3), 706–720, 2017, https://doi.org/10.1016/j.apt.2016.11.018

D. R. Curran, Simple Fragment Size and Shape Distribution Formulae for Explosively Fragmenting Munitions, *International Journal of Impact Engineering*, 20(1–5), 197–208, 1997, https://doi.org/10.1016/S0734-743X(97)87493-3

P. A. Davidson, *Turbulence in Rotating, Stratified, and Electrically Conducting Fluids*, Cambridge University Press, 2013, https://doi.org/10.1017/CBO9781139208673

G. Domokos, F. Kun, A. Á. Sipos, & T. Szabó, Universality of Fragment Shapes, *Scientific Reports*, 5, 9147, 2015, https://doi.org/10.1038/srep09147

E. L. Dunnavan, Z. Jiang, J. Y. Harrington, J. Verlinde, K. Fitch, & T. J. Garrett, The Shape and Density Evolution of Snow Aggregates, *Journal of the Atmospheric Sciences*, 76(12), 3919–3940, 2019, https://doi.org/10.1175/JAS-D-19-0066.1

D. D. Durda, A. Campo-Bagatin, R. A. Alemañ, G. J. Flynn, M. M. Strait, A. N. Clayton, & E. B. Patmore, The Shapes of Fragments from Catastrophic Disruption Events: Effects of Target Shape and Impact Speed, *Planetary and Space Science*, 107, 77–83, 2015, https://doi.org/10.1016/j.pss.2014.10.006

D. Fry, A. Mohammad, A. Chakrabarti, & C. M. Sorensen, Cluster Shape Anisotropy in Irreversibly Aggregating Particulate Systems, *Langmuir*, 20, 7871–7879, 2004, https://doi.org/10.1021/la0494369

A. Fujiwara, G. Kamimoto, & A. Tsukamoto, Expected Shape Distribution of Asteroids Obtained from Laboratory Impact Experiments, *Nature*, 272, 602–603, 1978, https://doi.org/10.1038/272602a0

E. J. Garboczi, K. A. Riding, & M. Mirzahosseini, Particle Shape Effects on Particle Size Measurement for Crushed Waste Glass, *Advanced Powder Technology*, 28(2), 648–657, 2017, https://doi.org/10.1016/j.apt.2016.12.002

D. M. Goto, R. Becker, T. J. Orzechowski, H. K. Springer, A. J. Sunwoo, & C. K. Syn, Investigation of the Fracture and Fragmentation of Explosively Driven Rings and Cylinders, *International Journal of Impact Engineering*, 35(12), 1547–1556, 2008, https://doi.org/10.1016/j.ijimpeng.2008.07.081

T. Hiroe, K. Fujiwara, H. Hata, & H. Takahashi, Deformation and Fragmentation Behaviour of Exploded Metal Cylinders and the Effects of Wall Materials, Configuration, Explosive Energy and Initiated Locations, *International Journal of Impact Engineering*, 35(12), 1578–1586, 2008, https://doi.org/10.1016/j.ijimpeng.2008.07.002

X. Jia & E. J. Garboczi, Advances in Shape Measurement in the Digital World, *Particuology*, 26, 19–31, 2016, https://doi.org/10.1016/j.partic.2015.12.005

S. Jungblut, J.-O. Joswig, & A. Eychmüller, Diffusion-Limited Cluster Aggregation: Impact of Rotational Diffusion, *The Journal of Physical Chemistry C*, 123(1), 950–954, 2019a, https://doi.org/10.1021/acs.jpcc.8b10805

S. Jungblut, J.-O. Joswig, & A. Eychmüller, Diffusion- and Reaction-Limited Cluster Aggregation Revisited, *Physical Chemistry Chemical Physics*, 21(10), 5723–5729, 2019b, https://doi.org/10.1039/C9CP00549H

T. Kadono, T. Tanigawa, K. Kurosawa, T. Okamoto, T. Matsui, & H. Mizutani, Correlation Between Fragment Shape and Mass Distributions in Impact Disruption, *Icarus*, 309, 260–264, 2018, https://doi.org/10.1016/j.icarus.2018.03.014

S. Koltzenburg, M. Maskos, & O. Nuyken, *Polymer Chemistry*, Springer, 2017, https://doi.org/10.1007/978-3-662-49279-6

F. Kun, F. K. Wittel, H. J. Herrmann, B. H. Kröplin, & K. J. Måløy, Scaling Behavior of Fragment Shapes, *Physical Review Letters*, 96(2), 025504, 2006, https://doi.org/10.1103/PhysRevLett.96.025504

M. A. Lange & T. J. Ahrens, Fragmentation of Ice by Low Velocity Impact, In *Proceedings of the 12th Lunar and Planetary Science Conference, Volume 2*, Pergamon Press, 1981

A. La Spina & P. Paolicchi, Catastrophic Fragmentation as a Stochastic Process: Sizes and Shapes of Fragments, *Planetary and Space Science*, 44(12), 1563–1578, 1996, https://doi.org/10.1016/S0032-0633(96)00064-5

T. Matsui, T. Waza, K. Kani, & S. Suzuki, Laboratory Simulation of Planetesimal Collision, *Journal of Geophysical Research*, 87(B13), 10968–10982, 1982, https://doi.org/10.1029/JB087iB13p10968

M. D. Matthews, The Effect of Grain Shape and Density on Size Measurement, in J. P. M. Syvitski, editor, *Principles, Methods, and Applications of Particle Size Analysis*, Cambridge University Press, 1991, https://doi.org/10.1017/CBO9780511626142.004

T. Michikami, A. Hagermann, T. Kadokawa, A. Yoshida, A. Shimada, S. Hasegawa, et al., Fragment Shapes in Impact Experiments Ranging from Cratering to Catastrophic Disruption, *Icarus*, 264, 316–330, 2016, https://doi.org/10.1016/j.icarus.2015.09.038

T. Michikami, T. Kadokawa, A. Tsuchiyama, A. Hagermann, T. Nakano, K. Uesugi, et al., Influence of Petrographic Textures on the Shapes of Impact Experiment Fine Fragments Measuring Several Tens of Microns: Comparison with Itokawa Regolith Particles, *Icarus*, 302, 109–125, 2018, https://doi.org/10.1016/j.icarus.2017.10.040

W. Mock & W. H. Holt, *Computational and Experimental Determination of Fragmentation for Naturally Fragmenting Warheads*, U.S. Naval Surface Warfare Center Technical Report, NSWC TR 80-238, May 1981, http://apps.dtic.mil/sti/citations/ADA110444

W. Mock & W. H. Holt, Fragmentation Behavior of Armco Iron and HF-1 Steel Explosive-Filled Cylinders, *Journal of Applied Physics*, 54(5), 2344–2351, 1983, https://doi.org/10.1063/1.332346

J. Neumann, J.-W. Simon, & S. Reese, Digital Sieving of Irregular 3D Particles—A Study Using XRCT and Statistically Similar Synthetic Data, *Powder Technology*, 338, 1001–1015, 2018, https://doi.org/10.1016/j.powtec.2018.07.002

J. B. Percival and P. J. Lindsay, Measurement of Physical Properties of Sediments, in A. Mudroch, J. M. Azcue, and P. Mudroch, editors, *Manual of Physico-Chemical Analysis of Aquatic Sediments*, CRC Press, 1997, https://doi.org/10.1201/9780203748176

P. X. Pham, A. Kourmatzis, & A. R. Masri, Local Characteristics of Fragments in Atomizing Sprays, *Experimental Thermal and Fluid Science*, 95, 44–51, 2018, https://doi.org/10.1016/j.expthermflusci.2018.01.024

D. W. Schaefer, Fractal Models and the Structure of Materials, *MRS Bulletin*, 13(2), 22–27, 1988, https://doi.org/10.1557/S088376940006632X

C. Stevenin, S. Tomas, A. Vallet, M. Amielh, & F. Anselmet, Flow Characteristics of a Large-Size Pressure-Atomized Spray using DTV, *International Journal of Multiphase Flow*, 84, 264–278, 2016, https://doi.org/10.1016/j.ijmultiphaseflow.2016.05.004

B. H. Stuart, *Analytical Techniques in the Sciences: Polymer Analysis*, Wiley, 2002, https://doi.org/10.1002/9780470511343

P. B. Warren & R. C. Ball, Anisotropy and the Approach to Scaling in Monodisperse Reaction-Limited Cluster-Cluster Aggregation, *Journal of Physics A: Mathematical and General*, 22(9), 1405–1413, 1989, https://doi.org/10.1088/0305-4470/22/9/027

C. D. Westbrook, R. C. Ball, P. R. Field, & A. J. Heymsfield, Universality in Snowflake Aggregation, *Geophysical Research Letters*, 31(15), L15104, 2004a, https://doi.org/10.1029/2004GL020363

C. D. Westbrook, R. C. Ball, P. R. Field, & A. J. Heymsfield, Theory of Growth by Differential Sedimentation, With Application to Snowflake Formation, *Physical Review E*, 70, 021403, 2004b, https://doi.org/10.1103/PhysRevE.70.021403

C. D Westbrook, R. J. Hogan, & A. J. Illingworth, The Capacitance of Pristine Ice Crystals and Aggregate Snowflakes, *Journal of the Atmospheric Sciences*, 65, 206–219, 2008, https://doi.org/10.1175/2007JAS2315.1

F. K. Wittel, F. Kun, B. H. Kröplin, & H. J. Herrmann, Study on the Fragmentation of Shells, *International Journal of Fracture*, 140(1–4), 243–254, 2006, https://doi.org/10.1007/s10704-005-3995-7

7

Particle Velocity Distributions

This chapter concerns particle speeds and velocities in aggregating and fragmenting systems. Because there are only a few general results available in the research literature, this chapter is relatively short. This chapter assumes that the components of velocity are statistically independent of one another. In addition, this chapter assumes that the components of velocity are statistically independent of size, shape, rotation, and other particle properties.

7.1 Joint Probability Density Functions

Chapter 2 showed that joint probability density functions can be used to describe the same quantity at different points in time and space. Similarly, joint probability density functions can be used to describe different quantities at the same point in time and space. However, if a particle system is temporally and spatially homogenous, there is no need to specify the time or location. Let X_1 and X_2 be two particle properties. For example, X_1 could be mass and X_2 could be speed. For another example, X_1 could be the x-component of velocity and X_2 could be the y-component of velocity. Previous chapters assumed $0 < X_i < \infty$. However, to allow for negative components of velocity, this chapter assumes $-\infty < X_i < \infty$. Then the *joint probability density function* $f(x_1, x_2)$ is defined as follows:

$$\int_{x_{1\,\min}}^{x_{1\,\max}} \int_{x_{2\,\min}}^{x_{2\,\max}} f(x_1, x_2)\, dx_1 dx_2 = P[x_{1\,\min} \le X_1 < x_{1\,\max} \text{ and } x_{2\,\min} \le X_2 < x_{2\,\max}] \tag{7.1}$$

for any $x_{1\,\min}, x_{1\,\max}, x_{2\,\min}, x_{2\,\max}$ where P refers to probability. Notice that:

$$\int_{-\infty}^{\infty} \int_{-\infty}^{\infty} f(x_1, x_2)\, dx_1 dx_2 = 1. \tag{7.2}$$

The *marginal probability density functions* for X_1 and X_2 are defined as follows:

$$f_1(x_1) = \int_{-\infty}^{\infty} f(x_1, x_2)\, dx_2 \tag{7.3}$$

$$f_2(x_2) = \int_{-\infty}^{\infty} f(x_1, x_2)\, dx_1 \tag{7.4}$$

Particle Dynamics with Aggregation and Fragmentation. Culbert B. Laney, Oxford University Press.
DOI: 10.1093/9780198920687.003.0007

where

$$\int_{-\infty}^{\infty} f_1(x_1)\,dx_1 = \int_{-\infty}^{\infty} f_2(x_2)\,dx_2 = 1. \tag{7.5}$$

By definition, X_1 and X_2 are statistically independent if:

$$f(x_1, x_2) = f_1(x_1) f_2(x_2). \tag{7.6}$$

More generally, $X_1, X_2, \ldots X_d$ are statistically independent if:

$$f(\boldsymbol{x}) = f(x_1, x_2, \ldots, x_d) = f_1(x_1) f_2(x_2) \ldots f_d(x_d) \tag{7.7}$$

where $f_1, f_2, \ldots, f_d$ can be different functions or the same function.

If X_1 and X_2 are not statistically independent, the *conditional probability density function* for X_1 for a given fixed value of X_2 is defined follows:

$$f(x_1|x_2) = \frac{f(x_1, x_2)}{f_2(x_2)}. \tag{7.8}$$

Similarly, the conditional probability density function for X_2 for a given fixed value of X_1 is defined as follows:

$$f(x_2|x_1) = \frac{f(x_1, x_2)}{f_1(x_1)}. \tag{7.9}$$

Notice that:

$$\int_{-\infty}^{\infty} f(x_1|x_2)\,dx_1 = \int_{-\infty}^{\infty} f(x_2|x_1)\,dx_2 = 1. \tag{7.10}$$

The *expected value* X_1 of for a given value of X_2 is defined as follows:

$$E(X_1|X_2) = \int_{-\infty}^{\infty} f(x_1|x_2)\,x_1\,dx_1. \tag{7.11}$$

This book will use the following notation:

$$X_{1\,\mathrm{avg}}(X_2) = E(X_1|X_2). \tag{7.12}$$

For example, if X_1 is the particle speed V and X_2 is the particle mass M, then the average particle speed V_{avg} for a given particle mass M is as follows:

$$V_{\mathrm{avg}}(M) = E(V|M). \tag{7.13}$$

For more information on the material given in this section, readers may consult any basic textbook on probability theory such as, e.g., Ross (2018) or Rossi (2018).

7.2 Velocity Distributions for Aggregation

Suppose an aggregating particle system obtains molecular chaos. As described in Chapter 2, molecular chaos occurs when all of the components of the velocity vector are statistically independent, both for one particle and for pairs of adjacent particles. In such cases, there may be simple algebraic expressions for average particle speeds, speed distributions, and velocity distributions; see Examples 7.1, 7.2, and 7.3.

Example 7.1 Average speeds for ballistic aggregation

Consider ballistic aggregation as discussed in Section 4.3. Paul & Das (2017, 2018) used Event-Driven Molecular Dynamics (EDMD) to show that if ballistic aggregation obtains molecular chaos then:

$$V_{\mathrm{rms}} \cong \text{const.}\, M_{\mathrm{avg}}^{-1/2}. \tag{7.14}$$

where:

$$V_{\mathrm{rms}}^2 = \left(V^2\right)_{\mathrm{avg}} \cong \frac{1}{N}\sum_{k=1}^{N} V_k^2 \quad \text{and} \quad M_{\mathrm{avg}} \cong \frac{1}{N}\sum_{k=1}^{N} M_k.$$

In these expressions, V_k and M_k are samples of particle speed and mass, respectively, and "rms" stands for "root mean square." In earlier work, Carnevale et al. (1990) obtained the same result analytically; however, their proof was limited to one dimension.

Example 7.2 Speed distributions for granular gas aggregation. Part 1

As described earlier in Example 2.1, molecular gases are composed of identical hard spheres that interact through instantaneous, binary, perfectly elastic (i.e., frictionless) collisions. If the collisions are frequent enough to obtain molecular chaos, the spheres in molecular gases obtain the *Maxwell–Boltzmann* (MB) *speed distribution*. According to Bolton (2010), the MB speed distribution can be written as follows:

$$f(V) = 4\pi\left(\frac{M_1}{2\pi k_B T}\right)^{3/2} V^2 \exp\left(-\frac{M_1 V^2}{2k_B T}\right) \tag{7.15}$$

where T is the temperature, k_B is the Boltzmann constant, and M_1 is the monomer mass. Notice that this is a type of Rosin–Rammmler distribution; see Section 3.7. As described earlier in Examples 2.2 and 2.6, granular gases are composed of identical hard spheres interacting through instantaneous, binary, partially elastic (i.e., frictional) collisions. As it turns out, the MB speed distribution, or something similar, often occurs in granular gases. For example, Das et al. (2016) studied the "formation and evolution of plugs or clusters in dense granular flows." Using Event-Driven Molecular Dynamics (EDMD), at early times, Das et al. (2016) found that "an arbitrary initial velocity distribution rapidly evolves . . . to the MB distribution." However, "once clustering sets in . . . at later times, deviations from the MB distribution are seen in the tail region."

Example 7.3 Speed distributions for granular gas aggregation. Part 2

As discussed in the previous example, granular gases are composed of identical hard spheres interacting through instantaneous, binary, partially elastic collisions. For large enough times and speeds, particle speed distributions tend obtain self-similar forms such as the following:

$$f(V,t) = \left[\frac{V}{V_{\text{char}}(t)}\right]^{d/2} \exp\left\{-\left[\frac{V}{V_{\text{char}}(t)}\right]^{\beta}\right\} \tag{7.16}$$

where V_{char} is a characteristic speed, d is the particle velocity dimension, and β is an exponent. Notice the Equation (7.16) is a type of Rosin–Rammler distribution. If energy is neither lost nor gained, mean-field theory leads to the MB velocity distribution where:

$$\beta = 2. \tag{7.17}$$

This result agrees with Equation (7.15). In this case, V_{char} is the most probable speed where $V_{\text{char}} = \sqrt{2/3}\,V_{\text{rms}} = \sqrt{3\pi/8}\,V_{\text{avg}} = \sqrt{2k_B T(t)/M_1}$ and where $T(t)$ is the temperature, k_B is the Boltzmann constant, and M_1 is the monomer mass. If energy lost to dissipation is fully replaced, Goldshtein & Shapiro (1995) used mean-field theory to derive the following:

$$\beta = 3/2 \tag{7.18}$$

for large enough times and speeds; see also van Noije & Ernst (1998). This case is sometimes called *driven*. Alternatively, if energy lost to dissipation is not replaced at all, Esipov & Pöschel (1997) used mean-field theory to derive the following:

$$\beta = 1 \tag{7.19}$$

for large enough times and speeds; see also van Noije & Ernst (1998). This case is sometimes called *freely cooling* or *freely decaying*. According to Pöschel et al. (2006), Equations (7.17), (7.18), and (7.19) are examples of *universal speed distributions*. Researchers are still debating the conditions under which such universals occur. In driven cases, Biswas et al. (2020) say the following:

Though determining the value of the exponent β has been the subject of many studies, the results are still not conclusive . . . A number of experiments and simulations conclude that the exponent β is approximately 1.5, and universal in the sense that it does not differ for a wide range of densities and driving parameters . . . At the same time . . . other experiments [have found] . . . a wide range of values for β instead of a universal form.

In freely cooling cases, Pöschel et al. (2006) report that Monte-Carlo calculations obtained $\beta \cong 2$ for $V/V_{char} < 1.5$ and $\beta \cong 1$ for $V/V_{char} > 1.5$. Similarly, Yu et al. (2020) report that microgravity tests obtained $\beta \cong 2$ for $V/V_{char} < 1.5$ and $\beta \cong 1$ for $V/V_{char} > 1.5$, i.e., "a Maxwell–Boltzmann distribution [except that the] high energy tail decays exponentially."

7.3 Velocity Distributions for Fragmentation

In molecular chaos, adjacent particles obtain uniformly random trajectories. At the other extreme, in unitary fragmentation, suppose adjacent particles obtain nearly identical trajectories. In such cases, there may be simple algebraic expressions for the average particle speeds, average particle trajectories, and speed distributions; see Examples 7.4, 7.5, and 7.6.

Example 7.4 Average speeds for explosive fragmentation of cylindrical shells

Using conservation of energy, Gurney (1943) showed that explosively-driven cylindrical metal shells often produce fragments with nearly uniform initial speeds:

$$V_{avg} = \frac{G}{\sqrt{M_{cyl}/M_{H.E.} + 1/2}} \tag{7.20}$$

where G is the Gurney constant, M_{cyl} is the mass of the cylindrical shell, and $M_{H.E.}$ is the mass of the high explosive. This expression assumes that the explosive completely fills the cylindrical shell and detonates symmetrically about the axis of the cylindrical shell. It also assumes a long cylindrical shell, so that non-cylindrical end effects can be ignored. The resulting fragments obtain nearly the same trajectory angles. If the trajectory angle is measured in the direction of the detonation wave and relative to the surface normal then:

$$\theta_{avg} = \sin^{-1}\left(V_{avg}/2U_{avg}\right) \tag{7.21}$$

where U_{avg} is the average high-explosive detonation velocity; see Figure 7.1. For more information on Equations (7.20) and (7.21) see, e.g., Kennedy (1998), Huang et al. (2015), Li et al. (2015, 2017), and Grady (2017).

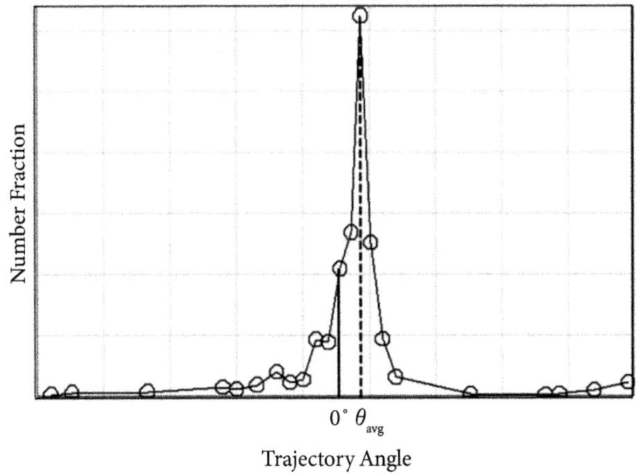

Figure 7.1 Cartoon showing how the number fraction of fragments might vary as a function of trajectory angle for an explosively-driven cylindrical metal shell.

Example 7.5 Average speeds for impact fragmentation of plates

Consider impact fragmentation, as discussed earlier in Section 1.5, where the impactor is a small high-speed projectile and the target is a flat plate. Based on a survey of impact fragmentation experiments dating back to the 1960s, Giblin, Davis & Ryan (2004) report that "a power-law relationship between ... [target fragment] mass and velocity ... holds, at least statistically." In other words:

$$V_{\text{avg}}(M) = \text{const.}\, M^{\alpha}. \tag{7.22}$$

where $V_{\text{avg}}(M)$ the average plate fragment speed as defined as in Equation (7.13), M is the plate fragment mass, and α is a negative rational exponent. According Giblin et al. (2004), Sachse et al. (2015), and Leliwa-Kopystyński et al. (2016), typical experiments obtain values such as $\alpha \cong -1/6, -1/3$, or $-1/2$. On the other hand, Giblin et al. (2004) note that many "other laboratory experimental studies have indicated little or no correlation between fragment velocity and mass for essentially similar experimental parameters."

As a general rule, optical measurement techniques are most likely to capture the largest, slowest fragments and, conversely, least likely to capture the smallest, fastest fragments. This bias could, by itself, explain the experimentally observed inverse correlations between fragment sizes and speeds. In other words, in theory, Equation (7.22) could be entirely an artifact of the measurement techniques used in typical experiments. Giblin et al. (2004) refer to this as "the substantial problem of selection effects."

To avoid experimental measurement errors, Pál et al. (2014) and Pál & Kun (2016) used a computational approach. More specifically, they used the Discrete Element Method (DEM), as discussed earlier in Section 4.7, to model impact fragmentation of plates. In these studies:

... The [target] sample is represented as a random packing of spherical particles with a uniform distribution of the diameter in a narrow interval around the average ... Cohesive interaction is provided by beams which connect the particles ... The total deformation of a beam is calculated as the superposition of elongation, torsion, ... bending and shearing. Cracks are formed when overstressed beams break according to a physical breaking rule [that] ... takes into account the stretching and bending of beams.

Pál & Kun (2016) say that fragment "mass-velocity correlation is only [found] ... for thin plates, while it disappears for three-dimensional bulk samples." More specifically, for thin plates, they found that $V_{\text{avg}}(M) = \text{const.}\ M^{\alpha}$ where $\alpha \approx -1/3$ for the smallest fragments and $\alpha \approx -1/6$ for the largest fragments. However, as plate thickness or impactor velocity increases, "the power law correlation gradually disappears so that the average velocity attains a constant value independent of the mass of fragments."

Example 7.6 Speed distributions for blast fragmentation of panels

Bewick et al. (2017) used a used a small-diameter explosively-driven shock tube to shatter simply supported plates with high-pressure, long-duration blast loads. These tests included a *fragment stripper*, consisting of a 10-ft tall by 1-ft wide slit located 5-ft from downstream from the plates. The fragment stripper excluded fragments traveling at angles greater than 5° relative to a vertical plane passing through the midsection of the plates. As a result, the surviving fragments obtained nearly parallel trajectories. As its main benefit, the fragment stripper made it possible to accurately estimate fragment sizes and speeds using a single side-view high-speed video camera. This is because the fragment stripper ensured that the fragments maintained a known fixed distance from the camera lens. Fragment strippers are a standard experimental technique; see also, e.g., Onose & Fujiwara (2004).

As seen in Figure 7.2, Bewick et al. (2017) found that the fragment speed was statistically independent of the fragment mass, i.e., $f(V, M) \cong f_1(V) f_2(M)$. In addition, Bewick et al. (2017) found that the fragment speeds obtained Weibull distributions where:

$$f_1(V) \cong \frac{A_0}{V_{\text{avg}}}\left(\frac{V}{V_{\text{avg}}}\right)^{r-1} \exp\left[-b_0\left(\frac{V}{V_{\text{avg}}}\right)^{r}\right] \text{ or } F_1(V) \cong 1 - B_0 \exp\left[-b_0\left(\frac{V}{V_{\text{avg}}}\right)^{r}\right]. \quad (7.23)$$

In this expression, Bewick et al. (2017) found that r ranged between 1.30 and 2.29, depending on the exact experimental conditions. In fact, as seen in Figure 7.3, about half of the thirteen tests obtained roughly the same Weibull speed distribution with $r = 5/3 \pm 0.1$. This was true despite significant differences in the test conditions including differences in the plate material, plate thickness, peak shock overpressure, and so forth. This suggests that, in some cases, fragment speed distributions may be universal. For more examples of Weibull fragment speed distributions see, e.g., Stanton & Rutland (1995), Pan & Render (1996), Moreira et al. (2010), Moreira & Panão (2011) and Guégan et al. (2011).

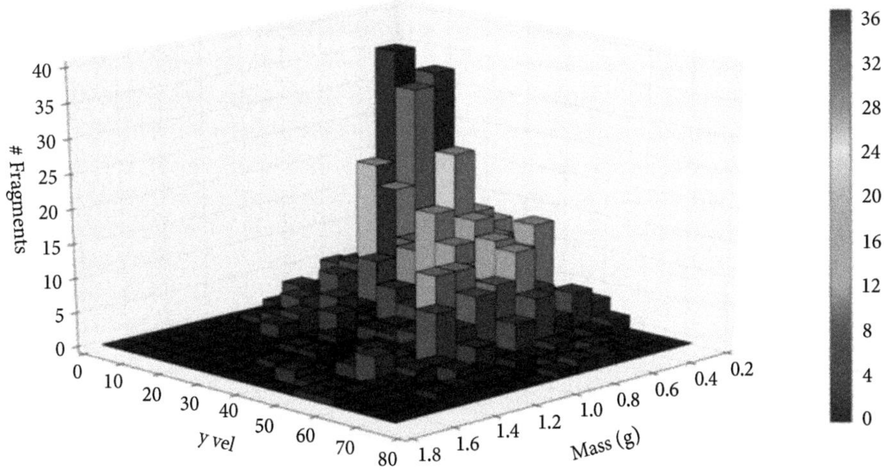

Figure 7.2 Experimental results for number of fragments vs. fragment speed ("y vel") vs. fragment mass for shock-loaded plates.
Reprinted with permission from Bewick et al. (2017).

7.4 Conclusions

There is a wealth of experimental and computational data on particle velocities in aggregating and fragmenting systems to be found in the research literature. However, the vast majority of these results are specific to a single situation. This chapter has summarized some of the few results that potentially apply to a broad range of different problems. Two tentative conclusions reached in this chapter are as follows: (1.) the average particle speed may be a simple algebraic function of the particle mass or the average particle mass; and (2.) the particle speed and velocity distributions may obtain universal forms. The results given in this chapter are mostly empirical, i.e., they are mainly based on curve fits to specific sets of experimental or computational data. As a result, the previous two conclusions are somewhat tentative.

As noted in Chapter 5, where they exist, analytic methods can often obtain general results, quickly and efficiently, without becoming bogged down in the exact details of any specific problem, subject only to the validity of the underlying assumptions. For example, consider mean-field theory. When applied to velocity, mean-field theory requires assuming that particles all have the same sizes, shapes, and so forth, at least on average. This assumption makes sense in certain cases—namely, ideal or granular gases composed of identical monomers—but not in most cases relevant to this book. For particle mass, there are dozens of possible mean-field governing equations, which lead to dozens of possible universal solutions, depending partly on the sources and sinks of mass. However, for particle velocity, there are only a few possible mean-field governing equations—such as the Boltzmann equation, the Enskog–Boltzmann equation, and the Boltzmann–Fokker–Planck equation—which obtain only three known universal solutions, depending partly on the sources and sinks of kinetic energy. As another possible analytic approach, Chapter 19 will discuss maximum entropy theory, which can lead directly to a wide variety of Rosin–Rammler speed distributions, such as Weibull speed distributions.

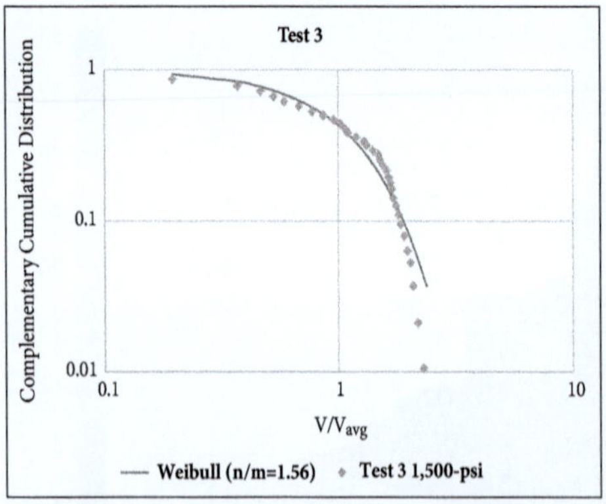

(a) Masonry plate at ~1,500 psi obtains $r = 1.56$.

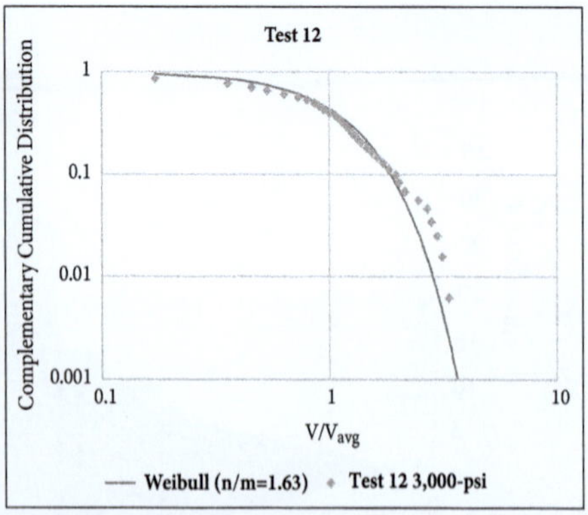

(b) Masonry plate at ~3,000 psi obtains $r = 1.63$.

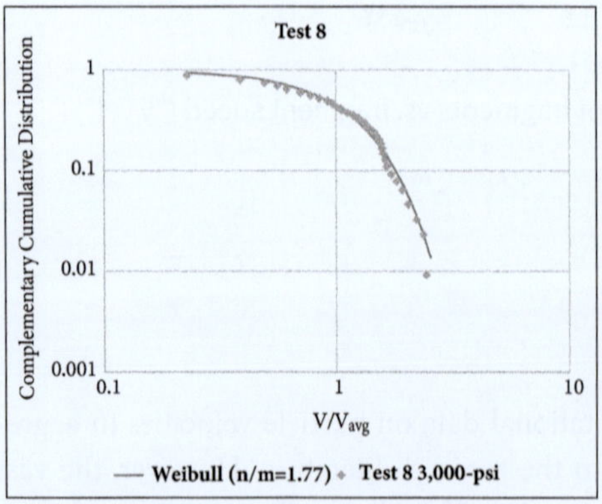

(c) Concrete plate at ~3,000 psi obtains $r = 1.77$.

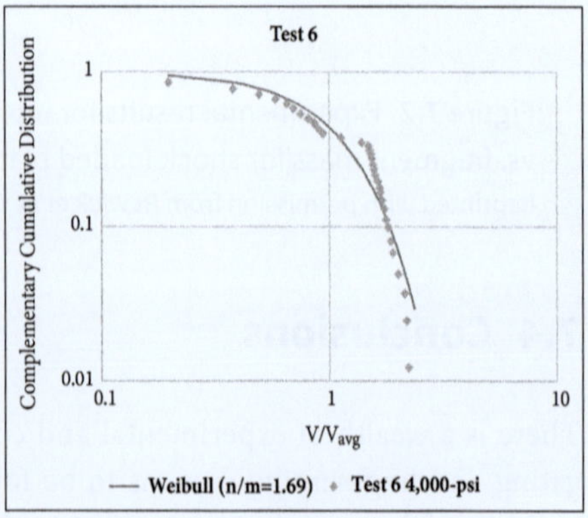

(d) Concrete plate at ~4,000 psi obtains $r = 1.69$.

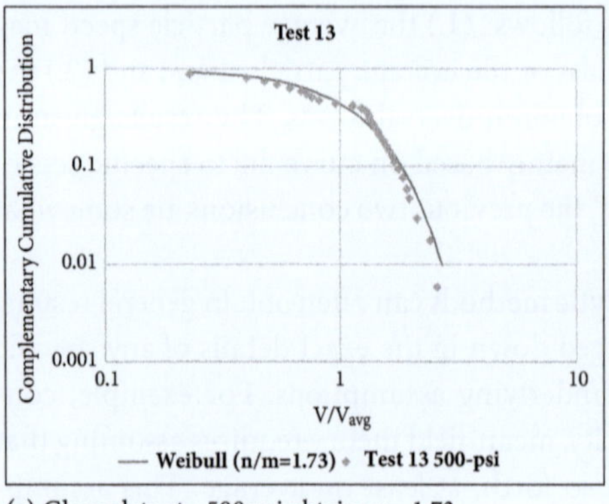

(e) Glass plate at ~500 psi obtains $r = 1.73$.

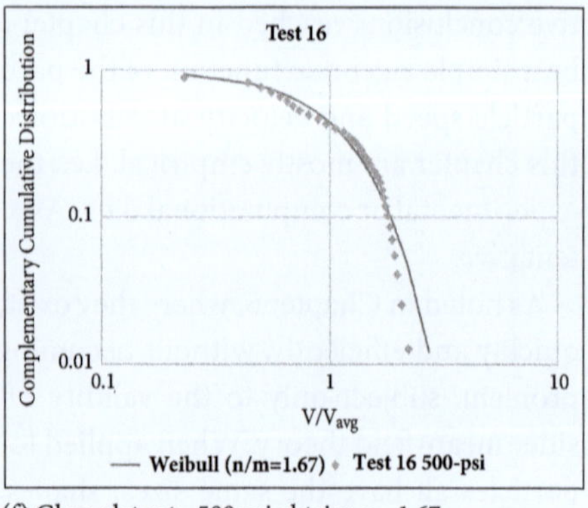

(f) Glass plate at ~500 psi obtains $r = 1.67$

Figure 7.3 Experimental (symbols) vs. analytic (curves) results for speed distributions $F(V/V_{avg})$ for shock-loaded plates.

Reprinted with permission from Bewick et al. (2017).

References

B. Bewick, G. Rolater, J. Bui, M. Barsotti, A. Ziemba, M. Sanai, & C. Laney, *Fragmentation of Solid Materials Using Shock Tubes. Part 1: First Test Series in a Small Diameter Shock Tube*, Defense Threat Reduction Agency Technical Report DTRA-TR-17-21, January 2017, http://apps.dtic.mil/sti/citations/AD1025099

A. Biswas, V. V. Prasad, & R. Rajesh, Asymptotic Velocity Distribution of a Driven One-Dimensional Binary Granular Maxwell Gas, *Journal of Statistical Mechanics: Theory and Experiment*, 013202, 2020, https://doi.org/10.1088/1742-5468/ab6095

J. Bolton, *Classical Physics of Matter*, CRC Press, 2010, https://doi.org/10.1201/9780429187780

G. F. Carnevale, Y. Pomeau, & W. R. Young, Statistics of Ballistic Agglomeration, *Physical Review Letters*, 64(24), 2913–2916, 1990, https://doi.org/10.1103/PhysRevLett.64.2913

P. Das, S. Puri, & M. Schwartz, Clustering and Velocity Distributions in Granular Gases Cooling by Solid Friction, *Physical Review E*, 94, 032907, 2016, https://doi.org/10.1103/PhysRevE.94.032907

S. E. Esipov & T. Pöschel, The Granular Phase Diagram, *Journal of Statistical Physics*, 86, 1385–1395, 1997, https://doi.org/10.1007/BF02183630

I. Giblin, D. R. Davis, & E. V. Ryan, On the Collisional Disruption of Porous Icy Targets Simulating Kuiper Belt Objects, *Icarus*, 171, 487–505, 2004, https://doi.org/10.1016/j.icarus.2004.03.006

A. Goldshtein & M. Shapiro, Mechanics of Collisional Motion of Granular Materials. Part 1. General Hydrodynamic Equations, *Journal of Fluid Mechanics*, 282, 75–114, 1995, https://doi.org/10.1017/S0022112095000048

D. E. Grady, *Physics of Shock and Impact, Volume 1: Fundamentals and Dynamic Failure*, IOP Publishing, 2017, https://doi.org/10.1088/978-0-7503-1254-7

P. Guégan, R. Othman, D. Lebreton, F. Pasco, P. Villedieu, J. Meyssonnier, & S. Wintenberger, Experimental Investigation of the Kinematics of Post-Impact Ice Fragments, *International Journal of Impact Engineering*, 38(10), 786–795, 2011, https://doi.org/10.1016/j.ijimpeng.2011.05.003

R. W. Gurney, *The Initial Velocities of Fragments from Bombs, Shells, and Grenades*, Ballistic Research Laboratories Report Number 405, 1943, http://apps.dtic.mil/sti/citations/ADA289704

G.-Y. Huang, W. Li, & S.-S. Feng, Axial Distribution of Fragment Velocities from Cylindrical Casing under Explosive Loading, *International Journal of Impact Engineering*, 76, 20–27, 2015, https://doi.org/10.1016/j.ijimpeng.2014.08.007

J. E. Kennedy, The Gurney Model for Explosive Output for Driving Metal, in J. A. Zukas & W. P. Walters, editors, *Explosive Effects and Applications*, Springer, 1998, https://doi.org/10.1007/978-1-4612-0589-0_7

J. Leliwa-Kopystyński, I. Włodarczyk, & M. J. Burchell, Analytical Model of Impact Disruption of Satellites and Asteroids, *Icarus*, 268, 266–280, 2016, https://doi.org/10.1016/j.icarus.2015.12.023

W. Li, G.-Y. Huang, & S.-S. Feng, Effect of Eccentric Edge Initiation on the Fragment Velocity Distribution of a Cylindrical Casing Filled with Charge, *International Journal of Impact Engineering*, 80, 107–115, 2015, https://doi.org/10.1016/j.ijimpeng.2015.01.007

Y. Li, Y.-H. Li, & Y.-Q. Wen, Radial Distribution of Fragment Velocity of Asymmetrically Initiated Warhead, *International Journal of Impact Engineering*, 99, 39–47, 2017, https://doi.org/10.1016/j.ijimpeng.2016.09.007

A. L. N. Moreira, A. S. Moita, & M.R. Panão, Advances and Challenges in Explaining Fuel Spray Impingement: How Much of Single Droplet Impact Research is Useful?, *Progress in Energy and Combustion Science*, 36(5), 554–580, 2010, https://doi.org/10.1016/j.pecs.2010.01.002

A. L. Moreira & M. R. O. Panão, Spray-Wall Impact, in N. Ashgriz, editor, *Handbook of Atomization and Sprays: Theory and Applications*, Springer, 2011, https://doi.org/10.1007/978-1-4419-7264-4

N. Onose & A. Fujiwara, Mass-Velocity Distributions of Fragments in Oblique Impact Cratering on Gypsum, *Meteoritics & Planetary Science*, 39(2), 321–331, 2004, https://doi.org/10.1111/j.1945-5100.2004.tb00343.x

G. Pál, I. Varga, & F. Kun, Emergence of Energy Dependence in the Fragmentation of Heterogeneous Materials, *Physical Review E*, 90(6), 062811, 2014, https://doi.org/10.1103/PhysRevE.90.062811

G. Pál & F. Kun, Mass-Velocity Correlation in Impact Induced Fragmentation of Heterogeneous Solids, *Granular Matter*, 18, 74, 2016, https://doi.org/10.1007/s10035-016-0670-9

H. Pan & P. M. Render, Impact Characteristics of Hailstones Simulating Ingestion by Turbofan Aeroengines, *Journal of Propulsion and Power*, 12(3), 457–462, 1996, https://doi.org/10.2514/3.24057

S. Paul & S. K. Das, Ballistic Aggregation in Systems of Inelastic Particles: Cluster Growth, Structure, and Aging, *Physical Review E*, 96, 012105, 2017, https://doi.org/10.1103/PhysRevE.96.012105

S. Paul & S. K. Das, Dimension Dependence of Clustering Dynamics in Models of Ballistic Aggregation and Freely Cooling Granular Gas, *Physical Review E*, 97, 032902, 2018, https://doi.org/10.1103/PhysRevE.97.032902

T. Pöschel, N. V. Brilliantov, & A. Formella, Impact of High-Energy Tails on Granular Gas Properties, *Physical Review E*, 74, 041302, 2006, https://doi.org/10.1103/PhysRevE.74.041302

S. Ross, *A First Course in Probability*, 10th edition, Pearson, 2018

R. J. Rossi, *Mathematical Statistics: An Introduction to Likelihood Based Inference*, Wiley, 2018, https://doi.org/10.1002/9781118771075

M. Sachse, J. Schmidt, S. Kempf, & F. Spahn, Correlation between Speed and Size for Ejecta from Hypervelocity Impacts, *Journal of Geophysical Research: Planets*, 120(11), 1847–1858, 2015, https://doi.org/10.1002/2015JE004844

D. W. Stanton & C. J. Rutland, Modeling Fuel Film Formation and Wall Interaction in Diesel Engines, *SAE Transactions*, 103, 808–824, 1996, https://doi.org/10.4271/960628

T. P. C. van Noije & M. H. Ernst, Velocity Distributions in Homogeneous Granular Fluids: The Free and the Heated Case, *Granular Matter*, 1, 57–64, 1998 https://doi.org/10.1007/s100350050009

P. Yu, M. Schröter, & M. Sperl, Velocity Distribution of a Homogeneously Cooling Granular Gas, *Physical Review Letters*, 124, 208007, 2020, https://doi.org/10.1103/PhysRevLett.124.208007

8

Particle Energy Distributions

This chapter concerns the kinetic energy spectra of the fluctuations associated with randomness and seeming-randomness in aggregating and fragmenting particle systems. The fluctuating kinetic energy spectrum is an example of a structure factor. This chapter will also discuss other examples of structure factors. Most of the relevant material found in the research literature concerns self-organized systems as discussed earlier in Section 2.2. This chapter will consider two specific types of self-organized systems, namely, granular gases and turbulent flows. As part of the discussion of granular gases, this chapter will discuss four possible ways of defining self-organized clusters, including one deterministic definition and three statistical definitions.

8.1 Variances and Covariances

This section defines variances and covariances. These definitions will be used later in the chapter and elsewhere in the book. Most of these definitions can be found in basic textbooks on probability theory; see, e.g., Ross (2018) and Rossi (2018). While introductory textbook treatments are usually cast in terms of *expected values*, this treatment is cast in terms of *ensemble averages* that depend on the position $\boldsymbol{r}$ and time t. Before continuing, readers may wish to review the material on probability theory given earlier in Sections 2.4, 2.5, and 3.5. As in those sections, let $X(\boldsymbol{r}, t)$ refer to a random or seemingly random non-negative scalar property such as the monomer density ρ in a granular gas. The ensemble average of $X(\boldsymbol{r}, t)$ is defined as follows:

$$\langle X \rangle(\boldsymbol{r}, t) = \int_0^\infty x f(x; \boldsymbol{r}, t) dx \tag{8.1}$$

where $f(x; \boldsymbol{r}, t)$ is the one-point, one-time probability density function. Then the *fluctuation* from the ensemble average is defined as follows:

$$X(\boldsymbol{r}, t) - \langle X \rangle(\boldsymbol{r}, t). \tag{8.2}$$

The *variance* is defined as the ensemble average of the square of the fluctuation. In other words:

$$\mathrm{Var}[X(\boldsymbol{r}, t)] = \int_0^\infty [x - \langle X \rangle(\boldsymbol{r}, t)]^2 f(x; \boldsymbol{r}, t) dx. \tag{8.3}$$

Particle Dynamics with Aggregation and Fragmentation. Culbert B. Laney, Oxford University Press. © Culbert B. Laney (2025).
DOI: 10.1093/9780198920687.003.0008

Notice that:

$$\begin{aligned}\mathrm{Var}[X(\boldsymbol{r},t)] &= \langle [X(\boldsymbol{r},t) - \langle X\rangle(\boldsymbol{r},t)]^2\rangle \\ &= \langle X(\boldsymbol{r},t)^2\rangle - [\langle X\rangle(\boldsymbol{r},t)]^2.\end{aligned} \tag{8.4}$$

For two times t_1 and t_2, the *temporal covariance* or *autocovariance* may be defined as follows:

$$\mathrm{Cov}\,[X(\boldsymbol{r},t_1),X(\boldsymbol{r},t_2)]$$

$$= \int_0^{\infty}\int_0^{\infty} [x_1 - \langle X\rangle(\boldsymbol{r},t_1)]\,[x_2 - \langle X\rangle(\boldsymbol{r},t_2)]\,f(x_1,x_2;\ \boldsymbol{r},t_1,t_2)\,dx_1dx_2. \tag{8.5}$$

where $f(x_1, x_2; \boldsymbol{r}, t_1, t_2)$ is the one-point, two-time probability density function. As it turns out:

$$\begin{aligned}\mathrm{Cov}[X(\boldsymbol{r},t_1),X(\boldsymbol{r},t_2)] &= \langle [X(\boldsymbol{r},t_1) - \langle X(\boldsymbol{r},t_1)\rangle][X(\boldsymbol{r},t_2) - \langle X(\boldsymbol{r},t_2)\rangle]\rangle \\ &= \langle X(\boldsymbol{r},t_1)\,X(\boldsymbol{r},t_2)\rangle - \langle X(\boldsymbol{r},t_1)\rangle\langle X(\boldsymbol{r},t_2)\rangle.\end{aligned} \tag{8.6}$$

The temporal covariance is a measure of *temporal correlation.* As another measure, the *coefficient of temporal correlation,* a.k.a., *autocorrelation,* is defined as follows:

$$\frac{\mathrm{Cov}\,[X(\boldsymbol{r},t_1),X(\boldsymbol{r},t_2)]}{\sqrt{\mathrm{Var}\,[X(\boldsymbol{r},t_1)]\,\mathrm{Var}\,[X(\boldsymbol{r},t_2)]}}. \tag{8.7}$$

Notice that the coefficient of temporal correlation always lies between −1 and 1 where −1 corresponds to perfect anticorrelation, 1 corresponds to perfect correlation, and 0 corresponds to no correlation. If t_1 and t_2 are far enough apart, then $X(\boldsymbol{r}, t_1)$ and $X(\boldsymbol{r}, t_2)$ tend to be *statistically independent.* In other words:

$$f(x_1,x_2;\boldsymbol{r},t_1,t_2) = f(x_1;\boldsymbol{r},t_1)\,f(x_2;\boldsymbol{r},t_2). \tag{8.8}$$

If $X(\boldsymbol{r}, t_1)$ and $X(\boldsymbol{r}, t_2)$ are statistically independent then they are *uncorrelated.* In other words:

$$\mathrm{Cov}\,[X(\boldsymbol{r},t_1),X(\boldsymbol{r},t_2)] = 0. \tag{8.9}$$

For stationary systems, $\mathrm{Cov}[X(\boldsymbol{r}, t), X(\boldsymbol{r}, t + \Delta t)]$ depends on Δt and $\boldsymbol{r}$ but not on t. Thus, a convenient notation is as follows:

$$C_X(\boldsymbol{r},\Delta t) = \mathrm{Cov}[X(\boldsymbol{r},t),X(\boldsymbol{r},t+\Delta t)]. \tag{8.10}$$

For two positions $\boldsymbol{r}_1$ and $\boldsymbol{r}_2$, the *spatial covariance* may be defined as follows:

$$\mathrm{Cov}\,[X(\boldsymbol{r}_1,t),X(\boldsymbol{r}_2,t)]$$

$$= \int_0^{\infty}\int_0^{\infty} [x_1 - \langle X(\boldsymbol{r}_1,t)\rangle][x_2 - \langle X(\boldsymbol{r}_2,t)\rangle]\,f(x_1,x_2;\boldsymbol{r}_1,\boldsymbol{r}_2,t)dx_1dx_2. \tag{8.11}$$

where $f(x_1, x_2; \boldsymbol{r}_1, \boldsymbol{r}_2, t)$ is the two-point, one-time probability density function.

Notice that:

$$\begin{aligned}\mathrm{Cov}[X(\boldsymbol{r}_1,t),X(\boldsymbol{r}_2,t)] &= \langle[X(\boldsymbol{r}_1,t)-\langle X(\boldsymbol{r}_1,t)\rangle][X(\boldsymbol{r}_2,t)-\langle X(\boldsymbol{r}_2,t)\rangle]\rangle\\ &= \langle X(\boldsymbol{r}_1,t)X(\boldsymbol{r}_2,t)\rangle-\langle X(\boldsymbol{r}_1,t)\rangle\langle X(\boldsymbol{r}_2,t)\rangle. \end{aligned} \tag{8.12}$$

The spatial covariance is a measure of *spatial correlation*. As another measure, the *coefficient of spatial correlation* is defined as follows:

$$\frac{\mathrm{Cov}[X(\boldsymbol{r}_1,t),X(\boldsymbol{r}_2,t)]}{\sqrt{\mathrm{Var}[X(\boldsymbol{r}_1,t)]\mathrm{Var}[X(\boldsymbol{r}_2,t)]}}. \tag{8.13}$$

Notice that the coefficient of spatial correlation always lies between −1 and 1 where −1 corresponds to perfect anticorrelation, 1 corresponds to perfect correlation, and 0 corresponds to no correlation. If $\boldsymbol{r}_1$ and $\boldsymbol{r}_2$ are far enough apart, then $X(\boldsymbol{r}_1, t)$ and $X(\boldsymbol{r}_2, t)$ tend to be *statistically independent*. In other words:

$$f(x_1,x_2;\boldsymbol{r}_1,\boldsymbol{r}_2,t) = f(x_1;\boldsymbol{r}_1,t)\,f(x_2;\boldsymbol{r}_2,t). \tag{8.14}$$

If $X(\boldsymbol{r}_1, t)$ and $X(\boldsymbol{r}_2, t)$ are statistically independent then they are *uncorrelated*. In other words:

$$\mathrm{Cov}[X(\boldsymbol{r}_1,t),X(\boldsymbol{r}_2,t)] = 0. \tag{8.15}$$

For spatially homogeneous and isotropic systems, $\mathrm{Cov}[X(\boldsymbol{r}, t), X(\boldsymbol{r} + \Delta\boldsymbol{r}, t)]$ depends on the vector magnitude $\Delta r = |\Delta\boldsymbol{r}|$ and t but not on $\boldsymbol{r}$. Thus, a convenient notation is as follows:

$$C_X(\Delta r,t) = \mathrm{Cov}[X(\boldsymbol{r},t),X(\boldsymbol{r}+\Delta\boldsymbol{r},t)]. \tag{8.16}$$

Rather than using fluctuations such as $X(\boldsymbol{r}_1,t)-\langle X(\boldsymbol{r}_1,t)\rangle$ to determine covariances, Das et al. (2016) suggest using $\mathrm{sign}[X(\boldsymbol{r}_1,t) - \langle X(\boldsymbol{r}_1,t)\rangle]$ where $\mathrm{sign}(x) = 1$ if $x > 0$ and $\mathrm{sign}(x) = -1$ if $x < 0$. They refer to this alteration as *hardening*.

So far, this section has considered random scalar processes such as the monomer density in a granular gas. The remainder of this section considers random vector processes $\boldsymbol{V}(\boldsymbol{r}, t)$ such as the monomer velocity in a granular gas. Let the components be $V_x(\boldsymbol{r}, t)$, $V_y(\boldsymbol{r}, t)$, $V_z(\boldsymbol{r}, t)$. The ensemble average of the vector component $V_i(\boldsymbol{r}, t)$ where $i = x, y,$ or z is defined as follows:

$$\langle V_i\rangle(\mathbf{r},t) = \int_{-\infty}^{\infty} v_i f(v_i;\mathbf{r},t)\,dv_i. \tag{8.17}$$

where $f(v_i;\ \boldsymbol{r},\ t)$ is the one-point, one-time probability density function. In this case, a *fluctuation* is defined as follows:

$$V_i(\boldsymbol{r},t) - \langle V_i\rangle(\boldsymbol{r},t). \tag{8.18}$$

The *variance* is defined as the ensemble average of the square of the fluctuation:

$$\mathrm{Var}[V_i(\boldsymbol{r},t)] = \int_{-\infty}^{\infty} [v_i - \langle v_i\rangle(\boldsymbol{r},t)]^2 f(v_i;\boldsymbol{r},t)dv_i. \tag{8.19}$$

Notice that:

$$\begin{aligned}\mathrm{Var}[V_i(\boldsymbol{r},t)] &= \langle [V_i(\boldsymbol{r},t) - \langle V_i\rangle (\boldsymbol{r},t)]^2\rangle \\ &= \langle V_i(\boldsymbol{r},t)^2\rangle - \langle V_i\rangle(\boldsymbol{r},t)^2.\end{aligned} \tag{8.20}$$

The *temporal covariance* or *autocovariance* is defined as follows:

$$\begin{aligned}&\mathrm{Cov}[V_i(\boldsymbol{r},t_1), V_j(\boldsymbol{r},t_2)] \\ &= \int_{-\infty}^{\infty}\int_{-\infty}^{\infty} [v_i - \langle v_i\rangle (\boldsymbol{r},t_1)]\,[v_j - \langle v_j\rangle (\boldsymbol{r},t_2)]\,f(v_i,v_j;\boldsymbol{r},t_1,t_2)\,dv_i dv_j.\end{aligned} \tag{8.21}$$

where $f(v_i, v_j; \boldsymbol{r}, t_1, t_2)$ is the one-point, two-time joint probability density function. Notice that the temporal covariance of the velocity is a tensor. Also notice that:

$$\begin{aligned}\mathrm{Cov}[V_i(\boldsymbol{r},t_1), V_j(\boldsymbol{r},t_2)] &= \langle [V_i(\boldsymbol{r},t_1) - \langle V_i(\boldsymbol{r},t_1)\rangle][V_j(\boldsymbol{r},t_2) - \langle V_j(\boldsymbol{r},t_2)\rangle]\rangle \\ &= \langle V_i(\boldsymbol{r},t_1)V_j(\boldsymbol{r},t_2)\rangle - \langle V_i(\boldsymbol{r},t_1)\rangle\langle V_j(\boldsymbol{r},t_2)\rangle.\end{aligned} \tag{8.22}$$

For stationary systems, $\mathrm{Cov}[V_i(\boldsymbol{r}, t), V_j(\boldsymbol{r}, t + \Delta t)]$ depends on Δt and $\boldsymbol{r}$ but not on t. Thus, a convenient notation is as follows:

$$(C_V)_{ij}(\boldsymbol{r},\Delta t) = \mathrm{Cov}[V_i(\boldsymbol{r},t), V_j(\boldsymbol{r},t+\Delta t)]. \tag{8.23}$$

The *spatial covariance* is defined as follows:

$$\begin{aligned}&\mathrm{Cov}[V_i(\boldsymbol{r}_1,t), V_j(\boldsymbol{r}_2,t)] \\ &= \int_{-\infty}^{\infty}\int_{\infty}^{\infty} [v_i - \langle v_i\rangle(\boldsymbol{r}_1,t)][v_j - \langle v_j\rangle(\boldsymbol{r}_2,t)]\,f(v_i,v_j;\boldsymbol{r}_1,\boldsymbol{r}_2,t)dv_i dv_j\end{aligned} \tag{8.24}$$

where $f(v_i, v_j; \boldsymbol{r}_1, \boldsymbol{r}_2, t)$ is the two-point, one-time probability density function. Notice that the spatial covariance of the velocity is a tensor. Also notice that:

$$\begin{aligned}\mathrm{Cov}[V_i(\boldsymbol{r}_1,t), V_j(\boldsymbol{r}_2,t)] &= \langle [V_i(\boldsymbol{r}_1,t) - \langle V_i(\boldsymbol{r}_1,t)\rangle][V_j(\boldsymbol{r}_2,t) - \langle V_j(\boldsymbol{r}_2,t)\rangle]\rangle \\ &= \langle V_i(\boldsymbol{r}_1,t)V_j(\boldsymbol{r}_2,t)\rangle - \langle V_i(\boldsymbol{r}_1,t)\rangle\langle V_j(\boldsymbol{r}_2,t)\rangle.\end{aligned} \tag{8.25}$$

For spatially homogeneous and isotropic systems, $\mathrm{Cov}[V_i(\boldsymbol{r}, t), V_j(\boldsymbol{r} + \Delta\, \boldsymbol{r}, t)]$ depends on the vector magnitude $\Delta r = |\Delta\boldsymbol{r}|$ and t but not on $\boldsymbol{r}$. Thus, a convenient notation is as follows:

$$(C_V)_{ij}(\Delta r,t) = \mathrm{Cov}[V_i(\boldsymbol{r},t), V_j(\boldsymbol{r}+\Delta\boldsymbol{r},t)]. \tag{8.26}$$

Instead of this tensor quantity, consider instead the following scalar quantity:

$$C_V(\Delta r,t) = \sum_i \mathrm{Cov}\,[V_i(\boldsymbol{r},t), V_i(\boldsymbol{r}+\Delta\boldsymbol{r},t)] \tag{8.27}$$

where the sum is taken over the x, y, and z components. Notice that:

$$\begin{aligned} C_V(\Delta r, t) &= \langle [\boldsymbol{V}(\boldsymbol{r}, t) - \langle \boldsymbol{V}(\boldsymbol{r}, t)\rangle] \cdot [\boldsymbol{V}(\boldsymbol{r} + \Delta \boldsymbol{r}, t) - \langle \boldsymbol{V}(\boldsymbol{r} + \Delta \boldsymbol{r}, t)\rangle]\rangle \\ &= \langle \boldsymbol{V}(\boldsymbol{r}, t) \cdot \boldsymbol{V}(\boldsymbol{r} + \Delta \boldsymbol{r}, t)\rangle - \langle \boldsymbol{V}(\boldsymbol{r}, t)\rangle \cdot \langle \boldsymbol{V}(\boldsymbol{r} + \Delta \boldsymbol{r}, t)\rangle. \end{aligned} \tag{8.28}$$

8.2 Structure Factors

Instead of raw covariances, most of the applications relevant to this book use *Fourier transforms* of the covariances. First, let $X(\boldsymbol{r}, t)$ refer to a random non-negative scalar process such as the monomer density ρ in a granular gas. Define $S_X(\omega, \boldsymbol{r})$ to be the temporal Fourier transform of $C_X(\boldsymbol{r}, \Delta t)$:

$$S_X(\omega, \boldsymbol{r}) = A_1 \int C_X(\boldsymbol{r}, \Delta t) \exp(\iota \omega \Delta t) d(\Delta t) \tag{8.29}$$

where $\iota = \sqrt{-1}$, A_1 is a multiplicative constant, ω is the *frequency*, and $C_X(\boldsymbol{r}, \Delta t)$ is defined in Equation (8.10). Similarly, define $S_X(\boldsymbol{\kappa}, t)$ to be the spatial Fourier transform of $C_X(\Delta r, t)$:

$$S_X(\boldsymbol{\kappa}, t) = A_2 \iiint C_X(\Delta r, t) \exp(\iota \boldsymbol{\kappa} \cdot \Delta \boldsymbol{r}) d(\Delta \boldsymbol{r}) \tag{8.30}$$

where $\iota = \sqrt{-1}$, A_2 is a multiplicative constant, $\boldsymbol{\kappa}$ is the *wave vector*, and $C_X(\Delta r, t)$ is defined in Equation (8.16). Alternatively, to remove the directional information, let:

$$S_X(\kappa, t) = \oiint S_X(\boldsymbol{k}, t) d\boldsymbol{k} = \iiint S_X(\boldsymbol{k}, t) \delta(|\boldsymbol{k}| - \kappa) d\boldsymbol{k} \tag{8.31}$$

where $\kappa = |\boldsymbol{\kappa}|$ is the wave number, δ is the Dirac delta function, and the surface integral is taken over a sphere of radius κ. In this expression, S_X is called a *structure factor* or a *fluctuation spectrum*. The former term is borrowed from solid state physics and crystallography, where it is used to describe scattering of x-rays, neutrons, and electrons.

Next, let $\boldsymbol{V}(\boldsymbol{r}, t)$ refer to a random vector process with components $V_i(\boldsymbol{r}, t)$ such as the monomer velocity in a granular gas. Define $(S_V)_{ij}$ to be the spatial Fourier transform of $(C_V)_{ij}$:

$$(S_V)_{ij}(\boldsymbol{\kappa}, t) = B_1 \iiint (C_V)_{ij}(\Delta r, t) \exp(\iota \boldsymbol{\kappa} \cdot \Delta \boldsymbol{r}) d(\Delta \boldsymbol{r}) \tag{8.32}$$

where $\iota = \sqrt{-1}$, B_1 is a multiplicative constant, $\boldsymbol{\kappa}$ is the wave vector, and $(C_V)_{ij}$ is defined in Equation (8.26). Alternatively, to remove the directional information, let:

$$(S_V)_{ij}(\kappa, t) = \oiint (S_V)_{ij}(\boldsymbol{k}, t) d\boldsymbol{k} = \iiint (S_V)_{ij}(\boldsymbol{k}, t) \delta(|\boldsymbol{k}| - \kappa) d\boldsymbol{k} \tag{8.33}$$

where $\kappa = |\boldsymbol{\kappa}|$ is the wave number, δ is the Dirac delta function, and the surface integral is taken over a sphere of radius κ. In these expressions, the tensor $(S_V)_{ij}$ is sometimes called the *velocity spectrum*. This term is borrowed from the fluid dynamics community, where it is used to describe turbulent flows. Similarly, define S_V to be the spatial Fourier transform of C_V:

$$S_V(\boldsymbol{\kappa}, t) = B_2 \iiint C_V(\Delta \boldsymbol{r}, t) \exp(\iota \boldsymbol{\kappa} \cdot \Delta \boldsymbol{r}) d(\Delta \boldsymbol{r}) \tag{8.34}$$

where $\iota = \sqrt{-1}$, B_2 is a multiplicative constant, $\boldsymbol{\kappa}$ is the wave vector, and C_V is defined in Equation (8.27). Alternatively, to remove the directional information let:

$$S_V(\kappa, t) = \oiint S_V(\boldsymbol{k}, t) d\boldsymbol{k} = \iiint S_V(\boldsymbol{k}, t)\delta(|\boldsymbol{k}| - \kappa) d\boldsymbol{k}. \tag{8.35}$$

In this expression, the scalar S_V is generally called either the *velocity structure factor* or the *fluctuating kinetic energy spectrum*, where the former term is drawn from solid-state physics and crystallography, and the latter term is drawn from fluid dynamics. As a standard result in fluid dynamics, if $B_2 = (2\pi)^{-3}$ then the kinetic energy per unit mass of the fluctuations is as follows:

$$\int_0^\infty S_V(\kappa, t) d\kappa = \frac{1}{2} \langle [\boldsymbol{V}(t) - \langle \boldsymbol{V} \rangle(t)] \cdot [\boldsymbol{V}(t) - \langle \boldsymbol{V} \rangle(t)] \rangle. \tag{8.36}$$

8.3 Structure Functions

The last two sections characterized random and seemingly random particle systems based on deviations from the average. In another classic approach, this section characterizes random and seemingly random particle systems based on the differences between two points. More specifically, the *second-order structure function* is defined as follows:

$$\mathfrak{D}_{ij}(\boldsymbol{r}_1, \boldsymbol{r}_2, t) = \langle [V_i(\boldsymbol{r}_1, t) - V_i(\boldsymbol{r}_2, t)][V_j(\boldsymbol{r}_1, t) - V_j(\boldsymbol{r}_2, t)] \rangle. \tag{8.37}$$

Pope (2000) describes this as "the covariance of the difference in velocity between two points." For spatially homogeneous and isotropic systems, $\mathfrak{D}_{ij}(\boldsymbol{r}, \boldsymbol{r} + \Delta \boldsymbol{r}, t)$ depends on the vector magnitude $\Delta r = |\Delta \boldsymbol{r}|$ and t but not on $\boldsymbol{r}$. Thus, a convenient notation is as follows:

$$\mathfrak{D}_{ij}(\Delta r, t) = \langle [V_i(\boldsymbol{r}, t) - V_i(\boldsymbol{r} + \Delta \boldsymbol{r}, t)][V_j(\boldsymbol{r}, t) - V_j(\boldsymbol{r} + \Delta \boldsymbol{r}, t)] \rangle. \tag{8.38}$$

Instead of this tensor quantity, consider instead the following scalar quantity:

$$\mathfrak{D}(\Delta r, t) = \sum_i \mathfrak{D}_{ii}(\boldsymbol{r}, \boldsymbol{r} + \Delta \boldsymbol{r}, t) \tag{8.39}$$

where the sum is taken over the x, y, and z components. Notice that the structure *factors* defined in the previous section involve Fourier transforms of covariances of one type, while the structure *functions* defined in this section involve covariances of another type. The next two sections consider applications involving structure factors; these are quite common. Following that, Section 8.6 considers applications involving structure functions; these are intriguing but relatively rare.

8.4 Granular Gases

Consider a granular gas composed of identical compact monomers (e.g., spheres, disks, short cylinders) interacting through instantaneous, binary, partially elastic collisions. Here, the term *partially elastic* means that the collisions are dissipative. For example, Burton et al. (2013) considered a granular gas composed of small 8×10 cm cylinders of dry ice sliding on a flat cast aluminum plate. They report that "even the most elastic collisions between dry-ice particles result in a loss of $\approx$40% of the kinetic energy." As shown in Examples 2.2 and 2.6, granular gases are chaotic and self-organized, provided the monomer concentrations and collision rates are high enough. As a result, according to Pöschel et al. (2005), "the most striking phenomenon which distinguishes granular gases from molecular gases is the self-organized formation of spatio-temporal structures such as clusters and vortices." This section will use the terms *clusters* or *particles* to refer to self-organized groupings of monomers, including but not limited to rotating *vortices.*

Prasad et al. (2017) notes that "granular gases are of two kinds: *freely cooling* in which there is no input of energy; or *driven*, in which energy is injected at a constant rate." In freely cooling granular gases, the temperature—i.e., the average kinetic energy of the particles—decreases monotonically with time. By contrast, in driven granular gases, the temperature often remains constant with time. Examples of driven granular gases include "coal or gravel flowing down a chute or snow (or rocks) avalanching down a mountain slope" per Goldhirsch (1999) and "a collection of grains inside a box being brought into a gas-like state by vertical shaking [with] vibrations . . . so intense that gravity can be neglected" per Andreotti et al. (2013).

As one possibility, clusters in granular gases can be defined using deterministic cluster-finding algorithms; see Example 8.1. This approach makes sense for nearly stationary clusters in which the monomers have small speeds and random directions, due to repeated back-and-forth collisions.

Example 8.1 Event-driven molecular dynamics for nearly stationary clusters in granular gases

Pathak et al. (2014) used Event-Driven Molecular Dynamics (EDMD) to model freely cooling granular gases with $m = d = 3$ dominated by nearly stationary clusters. In these simulations, the monomers consisted of up to 8 million identical hard spheres. The monomer volume fractions varied between 0.052 and 0.208. To give the calculation a stochastic character, the monomers were "distributed uniformly within a periodic three-dimensional box . . . with initial velocities chosen from a normal distribution." In these simulations, self-organized clusters began forming almost immediately. Cluster shapes and sizes were judged strictly based on proximity. More specifically, Pathak et al. (2014) subdivided the computational domain into square boxes, with box sides equal in length to the monomer diameter. A box was *occupied* if it contained the center of a monomer; otherwise, a box was *vacant*. Following this, Pathak et al. (2014) used a standard cluster-finding algorithm, of the sort developed for percolation and computer science applications, to decide which monomers belonged to which clusters; see, e.g., Figure 8.1. The results were dominated by aggregation, i.e., the average and maximum cluster sizes continuously increased throughout the simulation. In fact, by the time the simulation ended, a giant cluster had formed that contained ~75% of the monomers in the entire system. This behavior is sometimes called *gelation*. Omitting the one giant cluster, Pathak et al. (2014) obtained a power law size distribution with $n/m \approx -1.7 \approx -5/3$. To put this result in perspective, Pathak et al. (2014) used a similar approach to model ballistic aggregation. In one such calculation, Pathak et al. (2014) obtained a power law size distribution with $n/m \approx 0.8 \approx 5/6$. Since size spread is inversely proportional to $|n|/m$, this means that the ordinary bonded particles obtained a much broader size distribution than the self-organized unbonded particles. Figure 1.5 compares the shapes of self-organized unbonded vs. ordinary bonded particles. Notice that, unlike ordinary bonded particles, self-organized particles have nebulous, indistinct boundaries.

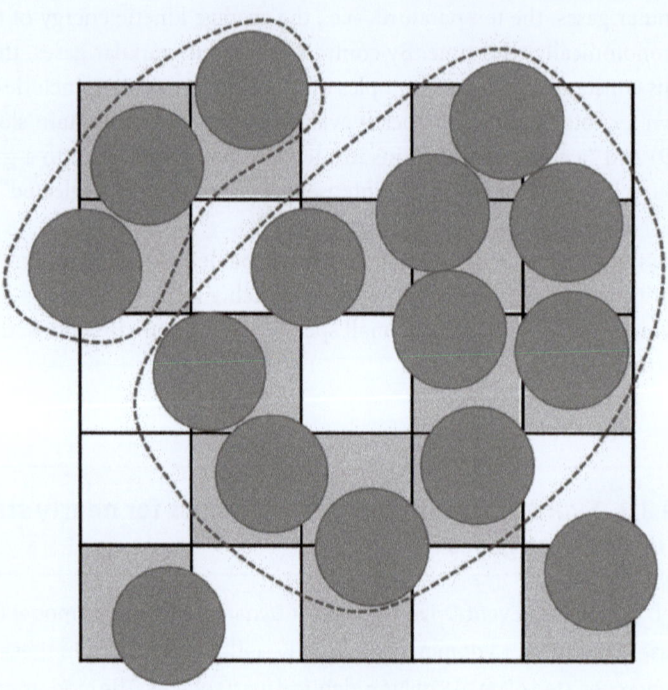

Figure 8.1 The Hoshen–Kopelman cluster-finding algorithm can be used to determine which monomers are part of which clusters. Monomer positions are overlaid on a square grid. Grey indicates an occupied grid cell while white indicates a vacant grid cell.
This figure draws on an example given in Bunde & Havlin (1996).

As another possibility, clusters in granular gases can be defined using probability theory. More specifically, clusters are commonly defined based on one of the following three properties:

(a) Spatial correlations between monomer positions or densities. This choice makes the most sense for clusters that experience little or no translation or rotation; see Figure 8.2.

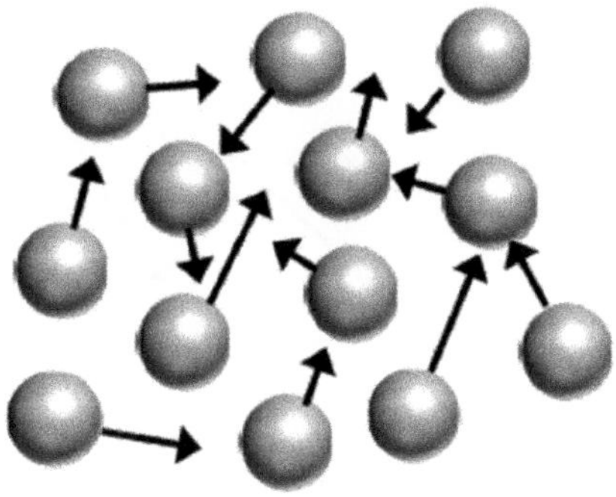

Figure 8.2 Clusters that experience no significant translation or rotation are best defined by correlations between monomer positions or density.

(b) Spatial correlations between the *same* component of monomer velocity. This makes the most sense for clusters that experience translation but not rotation; see Figure 8.3.

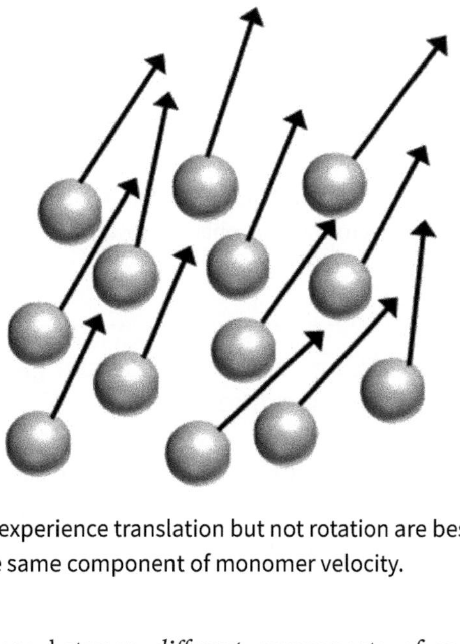

Figure 8.3 Clusters that experience translation but not rotation are best defined by correlations between the same component of monomer velocity.

(c) Spatial correlations between *different* components of monomer velocity. This makes the most sense for clusters that experience rotation but not translation; see Figure 8.4.

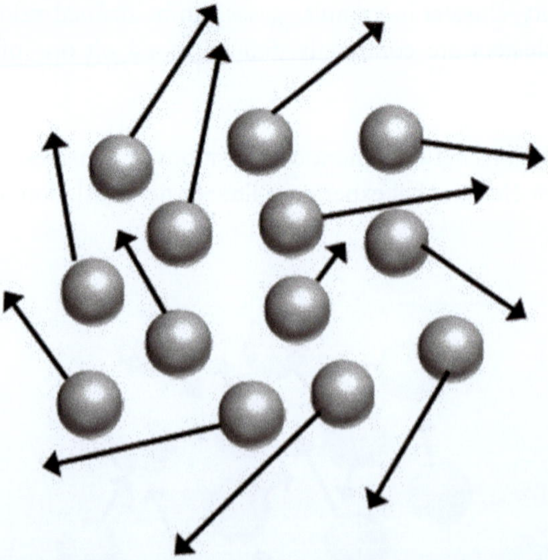

Figure 8.4 Clusters that experience rotation but not translation are best defined by correlations between different components of monomer velocity.

First, consider choice (a). As noted earlier, this choice makes the most sense for clusters that experience little or no translation or rotation, e.g., when frequent back-and-forth collisions within each cluster largely cancel each other out. In this case, monomer collision rates tend to be higher inside than outside of clusters; thus, the system loses energy more rapidly once clusters form. Goldhirsch (1999) explained the formation of such clusters as follows: "since the [monomer] collisions are inelastic, the granular temperature in the dense region [clusters] decreases faster than in neighboring regions, causing the pressures in this region to be lower than in neighboring regions. The resulting pressure gradient leads to a migration of ... [monomers and small clusters] into the dense region [large cluster], thus further increasing its density and decreasing its pressure." According to Van Zon & Mackintosh (2005), "as the [large] cluster grows in size, it can no longer be destroyed by the impact of high velocity [monomers] ... and [thus] the cluster remains stable." Instead, the granular gas consists of "coexisting 'cool' liquid-like clusters surrounded by a 'hot' gaseous state." Example 8.2 shows how to define self-organized particles in terms of spatial correlations between monomer densities. In this example, the structure factor obtains power law distributions with universal exponents such as −3/2 or −3.

Example 8.2 Characterizing clusters in granular gases based on spatial correlations for monomer density

Das et al. (2016) used EDMD to simulate freely cooling granular gases with $m = d = 2$. In these simulations, the monomers consisted of up to 250,000 identical circular disks interacting in a square domain with periodic boundary conditions. The monomer areal fraction, a.k.a., packing fraction, was either 0.157 or 0.236. To give the calculations a stochastic character, "the system is initialized by randomly placing... [monomers] in the simulation box...[with] no overlap" such that the monomers "have the same speed but the velocity vector points in random directions." The monomers were allowed to interact through perfectly elastic collisions, until they reached a thermal equilibrium with a Maxwell–Boltzmann velocity, as discussed earlier in Example 7.2. This equilibrium state, with its seemingly random monomer positions and velocities, provided the initial conditions for the main calculation. To reduce scatter, "all statistical quantities... [were] obtained as averages over ten independent runs."

At early times, Das et al. (2016) found that "particles in the high-density regions lose more energy than those in the low-density regions. Particles entering randomly from the adjacent low-density surrounding regions have a finite probability to be trapped within the high-density region. That probability increases with the number of particles in the [high-density] region." This is typical of nearly stationary clusters. To characterize the "morphology of clustering in the density field" at early times, Das et al. (2016) estimated the structure factor S_ρ. In particular, based on curve fits to computational results, Das et al. (2016) found that:

$$S_\rho(\kappa, t) \sim \kappa^{-3}$$

for middling values of κ. Borrowing a term from solid-state physics and crystallography, Das et al. (2016) say this power-law structure factor obeys *Porod's law* "which characterizes scattering off... sharp interfaces." For comparison, if a granular gas with $m = d = 2$ is randomly driven rather than freely cooling, Peng & Ohta (1998) found that $S_\rho(\kappa, t) \sim \kappa^{-3/2}$.

Now, consider choice (b). As noted earlier, this choice makes the most sense for clusters that experience translation but not rotation. Van Zon & Mackintosh (2005) say that such "clusters are highly dynamic and assemble and disassemble as they move around." In fact, the faster they move, the faster such clusters tend to appear and disappear. In this case, monomer collision rates tend to be lower inside than outside the clusters; thus, the system loses energy less rapidly once clusters form. In addition, collisions between two clusters may cause either aggregation or fragmentation, depending on their relative size and speed. During cluster collisions, the monomer collision rates tend to suddenly spike up. For example, Burton et al. (2013) conducted experiments involving collisions between two large self-organized clusters, each composed of 50 to 100 dry ice cylinders floating on a flat aluminum plate. They found that "a large fraction of the system's kinetic energy is lost over a short time scale [immediately following a collision] as a jamming front traverses the [monomers]... in each cluster. The energy decays as a power law in time . . . to 5–20% of its initial value." For another example, see Example 8.3.

Example 8.3 Characterizing clusters in granular gases based on spatial correlations between the *same* component of monomer velocity

Isobe (2017) used EDMD to simulate freely cooling granular gases with $m = d = 2$. In these simulations, the monomers consisted of 16.7 million identical circular disks interacting in a square domain with periodic boundary conditions. Isobe considered a relatively dense case, namely a monomer areal fraction of 0.6. This is sometimes known as the *packing fraction*. To give the calculation a stochastic character, the monomers were allowed to interact through perfectly elastic collisions until they reached thermal equilibrium. This equilibrium state, with its seemingly random monomer positions and velocities, provided the initial conditions for the main calculation. Isobe (2017) used a thermostat algorithm, as mentioned earlier in Section 4.1, to rescale the monomer velocities "so as to keep the energy of the whole system . . . constant." Because this was a preliminary study, Isobe characterized the cluster sizes qualitatively but not quantitatively, i.e., he did not report any structure factors. Isobe (2017) observed an early-time *clustering regime* in which selected groups of monomers "moved collectively as a cluster." In addition, he observed a "novel" late-time *cluster impact regime* in which "collisions between clusters spontaneously" occurred. The early-time clustering regime was characterized by rapidly decreasing monomer collision rates and, conversely, the late-time cluster impact regime was characterized by rapidly increasing monomer collision rates. Isobe (2017) suggests that the late-time "cluster impact regime can be detected only in [relatively dense] systems [with] more than several million [monomers]."

Finally, consider choice (c). As noted earlier, this choice makes the most sense for clusters that experience rotation but not translation. In this context, the term "rotation" refers to recirculating motions of all kinds, including but not limited to classic rigid-body rotation. There are two common types of rotating clusters: compact formations that rotate about a single center point; and extended *vortex tubes* or *vortex filaments* that rotate about an axial line or curve; see, e.g., Oyama et al. (2019). Example 8.4 shows how to define self-organized particles in terms of correlations between different components of monomer velocity. In this example, the structure factor for velocity, a.k.a. the fluctuating kinetic energy spectrum, obtains a power law distribution with the universal exponent −4.

Example 8.4 Characterizing clusters in granular gases based on spatial correlations between *different* components of monomer velocity

As described earlier in Example 8.2, Das et al. (2016) used EDMD to simulate freely cooling granular gases with $m = d = 2$. At early times, they found that most of the monomer clusters were nearly stationary. However, at later time, they found that "the evolution of the velocity field is characterized by the emergence and diffusive coarsening of vortices . . . Since there is no relative velocity among . . . [monomers] in a given plug, particles do not lose energy." Based on curve fits to numerical results, Das et al. (2016) found that:

$$S_V(\kappa, t) \sim \kappa^{-4}$$

for middling values of κ. Das et al. (2016) say this structure factor obeys *generalized Porod's law*. Baldassarri et al. (2002) observes that generalized Porod's law "is the signature of the presence of vortices . . . Vortices form spontaneously and represent the boundaries between regions that selected different orientations of the velocities . . . [When] random initial conditions [are] adopted, vortices are born at the smallest scales and subsequently grow in size by pair annihilation . . ."

At any given time, some clusters may be nearly stationary, some clusters may experience translation but not rotation, some clusters may experience rotation but not translation, and some clusters may experience both rotation and translation. The most common choice is to characterize the overall result using the velocity structure factor, a.k.a. the fluctuating kinetic energy spectrum. Table 8.1 gives a number of examples of computational results for the fluctuating kinetic energy spectra. As this table shows, when granular gas is spatially homogeneous and isotropic, the fluctuating kinetic energy spectrum generally obtains power law distributions with universal exponents such as −1 and −2. For comparison with granular gases, Table 8.1 includes two results for dense granular flows. Dense granular flows were discussed earlier in Section 2.3.

Table 8.1 Fluctuating kinetic energy spectra for spatially homogeneous and isotropic granular gases and dense granular flows. All results were obtained with simplified computational methods at late times, except where noted. All results take the form $S_V(\kappa, t) \sim \kappa^{r-1}$ for intermediate values of κ.

r	m	d	Details	References
-5[1]	3	3	Freely cooling	Das et al. (2017)
-3[1]	2	2	Freely cooling	Das & Puri (2003), Das et al. (2016)
-3[1]	2	2	Freely cooling, constant density	Baldassarri et al. (2002, 2003)
-2[1]	2	2	Freely cooling, thermostat, early time	Isobe (2003, 2012)
-1[1]	2	2	Randomly driven, stationary	van Noije et. al. (1999)
-2/3[2,3]	3	3	Randomly driven, stationary	Taguchi (1993, 1995)
-2/3[1]	3	3	Freely cooling, thermostat, early time	Isobe (2012)
3/2[2]	3	3	Shear driven, stationary	Oyama et. al. (2019)

(1) Granular gases modeled with Event-Driven Molecular Dynamics (EDMD).
(2) Dense granular materials modeled with the Discrete Element Method (DEM).
(3) Kellay et al. (2007) obtained similar results in experiments with vibrating glass beads.

8.5 Fluid Turbulence

In many cases, fluid flows spontaneously form numerous recirculating features called *eddies*. In such cases, the flow is said to be *turbulent*. According to Mathieu & Scott (2000), "most flows occurring in nature and in industrial applications are said to be turbulent." As discussed in Examples 8.5 and 8.6, turbulence is a manifestation of chaos and self-organization in fluids. However, turbulence is not the only possible manifestation, e.g., Reetz et al. (2019) describes cases where fluids form "alternating turbulent and laminar regions forming a stripe pattern."

Example 8.5 Turbulent fluid flows are chaotic

Davidson (2015) says "turbulence is, in effect, a hierarchy of tangled vortex tubes and ribbons of various sizes, advecting each other in a chaotic fashion." According to the definition of chaos given in Section 2.1:

1. *Sensitive to Initial Conditions*	Davidson (2015) says "turbulence is extremely sensitive to its initial conditions and it is for this reason that no two realizations of a turbulent experiment are ever exactly the same." Berera & Ho (2018) found that the rate of exponential "divergence of two initially close trajectories in phase space" increases with Reynolds number.
2. *Seemingly Random*	Davidson (2015) says that "the vorticity field is [seemingly] random in both space and time ... While V is apparently random in any one realization, the ensemble average $\langle \boldsymbol{V} \cdot \boldsymbol{V} \rangle$ is a smooth function of time ... That is, although $\boldsymbol{V}$ is chaotic, the average rate of loss of [kinetic] energy is a smooth, repeatable function of time."
3. *Forgets Initial Conditions*	Mathieu & Scott (2000) say that "although steady turbulent flow is nonunique in detail, its statistical properties do seem to be uniquely determined by the boundary conditions alone, that is, the flow forgets its initial conditions." Put another way, turbulent flow properties gradually decorrelate as time increases.
4. *Molecular Chaos (Optional)*	Babić & Rotach (2018) note that, for locally isotropic turbulence, spatial "correlations between velocity components ... vanish." However, not all turbulence is locally isotropic; see, e.g., Bernard & Erinen (2018). In other words, turbulent flows may or may not obtain molecular chaos, depending on the exact circumstances.

Example 8.6 Turbulent fluid flows are self-organized

Marov & Kolesnichenko (2013) say that fluid turbulence can "be interpreted as an unbounded sequence of a self-organization process." According to the definition of self-organization given in Section 2.2:

1. *Ordered*	Fluid flows spontaneously form eddies, independent of their initial conditions, provided the Reynolds number is large enough.
2. *Dissipation*	Turbulence only occurs in viscous fluids; however, even the slightest amount of viscous dissipation is sufficient to allow turbulence.
3. *Chaos or Randomness*	Smol'yakov & Tkachenko (1983) say that "turbulent flow ... is an exceedingly complicated, tortuous and chaotic phenomenon [in which] ... all ... parameters fluctuate in an irregular fashion ... Turbulence is commonly studied by statistical, probabilistic methods because of the [seemingly] random nature of turbulent fluctuations."

Fluid turbulence is important to this book for three reasons. First, particles are often suspended in a gas or liquid medium. In such cases, fluid turbulence in the medium may cause the particles to collide, leading to aggregation or fragmentation. Second, the particles are often liquid. For liquid particles traveling at sufficiently high speeds relative to a fluid medium, drag may induce turbulence in the particles, leading to breakup. Third, there are interesting analogies between fluid turbulence, particle aggregation, and particle fragmentation. Thus, some of the results developed for fluid turbulence also apply to particle aggregation and fragmentation.

Turbulence can be classified as either compressible (variable density) or incompressible (constant density). This section concerns only incompressible turbulence. In addition, turbulence can be classified as either freely decaying or driven. Sagaut & Cambon (2018) say that "in the former case, no source of turbulent kinetic energy is present, and the turbulent kinetic energy is a monotonically decaying function of time, while in the latter a kinetic energy source is used to reach a statistically steady state." Common turbulence drivers include buoyancy and shear near solid boundaries. Turbulent eddies can generally be classified as either: compact formations that rotate or circulate about a single center point; and extended *vortex tubes* or *vortex filaments* that rotate or circulate about an axial line or curve. Regarding the former, Davidson (2013) says they may be "rather flat (pancake-like) and . . . sit in a plane." Regarding the latter, which are more typical, Davidson (2013) says "the intermediate to small scales in a turbulence flow can be pictured as a sparse tangle of spaghetti-like vortex tubes."

Davidson (2015) notes that "the lifespan of a typical eddy is rather short, of the order of its so-called turn-over time." As a result, turbulent eddies participate in an *energy cascade*. Davidson (2015) says "the largest eddies, which are created by instabilities in the mean flow, are themselves subject to internal instabilities and rapidly break-up or evolve into yet smaller vortices . . . The smaller eddies are themselves unstable and they, in turn, pass their energy onto even smaller structures and so on." From one point of view, the instabilities that cause turbulent eddies to break up are due to *vortex stretching*. Davidson (2015) says "one popular cartoon of homogeneous turbulence [involves] . . . vortex tubes. These tubes . . . undergo continual stretching as they are washed around by the chaotic velocity field, contracting until their diameter reaches the Kolmogorov scale. In this cartoon, the energy cascade is fueled . . . by the continual stretching and folding of vortex tubes."

In general, turbulent cascades result in three different size regimes: the largest scales are called the *energy-containing* or *production range*; the intermediate scales are called the *inertial range*; and the smallest scales are called the *dissipative range*. As this terminology implies, the majority of the turbulent energy originates at the largest scales while the majority of the turbulent energy is dissipated at the smallest scales. The largest eddy size l, and the matching velocity scale u, are called the *integral scales*. The smallest eddy size η, and the matching velocity scale v, are called the *Kolmogorov microscales*.

Even the slightest amount of dissipation may lead to turbulence. However, too much dissipation can damp out turbulence. In fact, turbulence only occurs when the fluid viscosity is substantially smaller than the flow inertia. The *Reynolds number* (*Re*) is defined as the non-dimensional ratio between inertial forces and viscosity forces. More specifically:

$$Re = \rho LV/\mu = LV/\nu \tag{8.40}$$

where L is an inertial length scale, V is an inertial velocity scale, ρ is the density, μ is the *dynamic viscosity*, and $\nu = \mu/\rho$ is the *kinematic viscosity*. Then fluid turbulence only occurs for sufficiently large Reynolds numbers. The Reynolds number increases as the kinematic viscosity ν decreases. Davidson (2015) says "as we let ν become progressively smaller we might expect that the dissipation becomes less, but it does not. Instead we find that finer and finer structures appear in the fluid, and these fine-scale structures have a thickness which is just sufficient to ensure that the rate of viscous dissipation per unit mass remains finite and equal to $\sim u^3/l$ . . . Viscosity plays a rather passive role in the so-called energy cascade . . . This entire process is driven by inertia and viscosity plays a role only when the eddy size reaches the dissipation scale η. In short, viscosity provides a dustbin for energy at the end of the cascade but does not (cannot) influence the cascade itself."

As described in Examples 8.7, 8.8, and 8.9, turbulent fluids and particle systems may exhibit analogous behaviors. The previous section discussed various different ways to define clusters in granular gases. However, only one of those ways applies to eddies in turbulent fluids, namely, structure factors for velocity. In this context, these are better known as *turbulent kinetic energy spectra*. Turbulent kinetic energy spectra tend to obtain Rosin–Rammler distributions of the following form:

$$S_V(\kappa, t) \sim \frac{A}{\kappa_{\text{ref}}}\left(\frac{\kappa}{\kappa_{\text{ref}}}\right)^{r-1} \exp\left[-b\left(\frac{\kappa}{\kappa_{\text{ref}}}\right)^q\right] \tag{8.41}$$

where κ_{ref} is a reference wavenumber; see, e.g., Martínez et al. (1997) and Khurshid et al. (2018). As one possibility, $\kappa_{\text{ref}} = \kappa_{\max} = 2\pi/\eta$. Common special cases include power law distributions with $q = 0$ and gamma distributions with $q = 1$. As noted earlier, turbulent kinetic energy spectra often obtain three different size ranges. In general, turbulent kinetic energy spectra obtain different distributions in each range.

Example 8.7 The analogy between granular gases and turbulent fluid flows

Granular gases and turbulent fluid flows have many things in common. These include the following: (1.) Granular gas clusters and turbulent eddies may both experience forward cascades with fragmentation-like behaviors, i.e., large structures transfer mass, momentum, and/or energy to smaller structures; and these structures transfer mass, momentum, and/or energy to even smaller structures; and so on down to a small size limit. (2.) Granular gas clusters and turbulent eddies may both experience reverse cascades with aggregation-like behaviors, wherein smaller-scale structures transfer mass, momentum, and/or energy to larger-scale structures which, in turn, transfer mass, momentum, and/or energy to even larger-scale structures. (3.) Granular gas clusters and turbulent eddies both have a minimum size: the monomer size in the former and the Kolmogorov microscale in the latter. (4.) Granular gases and turbulent fluid flows both dissipate energy mainly on the smallest scales. (5.) Granular gases and turbulent fluid flows are both chaotic and self-organized. (6.) Clustering in granular gases and eddies in turbulent fluids are both caused by even the slightest amount of dissipation. (7.) Clusters in granular gases and eddies in turbulent fluid flows both may take forms such as compact disks or extended vortex tubes. (8.) Clusters in granular gases and eddies in turbulent fluid flows both tend to be nebulous, indistinct and short-lived. (9.) Granular gases and turbulent fluid flows may both be either driven or freely decaying. (10.) Granular gases and turbulent fluid flows both obtain the same sorts of fluctuating kinetic energy spectra. As one difference, while particles in compressible granular gases can be defined by their density as seen in Figure 8.1, eddies in incompressible turbulence cannot. For more information on the relationships between granular gases and turbulent flows see, e.g., Taguchi (1993, 1995) and Isobe (2003, 2012).

Example 8.8 The analogy between dense granular flows and turbulent fluid flows

Dense granular flows and incompressible turbulent fluid flows have many things in common. In fact, most of the ten items listed in the previous example for granular gases apply equally to dense granular flows. One exception is item (4.). While turbulent fluid flows dissipate energy mainly on the smallest scales, this is not true for dense granular flows. Instead, Bak et al. (1988) say that "energy is dissipated on all length scales" and thus dense granular flows are only "a toy model of generalized turbulence." Another exception is item (8.). While turbulent eddies tend to be short-lived, vortices in dense granular flows tend to be even shorter-lived. In either case, "short" must be defined relative to the time scales of the main flow. In this regard, Radjai & Roux (2002) note that while "eddy-like structures appear quite frequently" in dense granular flows, they only "survive for... short times... This behavior is different from turbulence where the eddies survive long enough to undergo significant distortions due to fluid motion." Such distortions include the vortex stretching that drives energy cascades, as discussed in the main text. As the final difference, dense granular flows can store and sporadically release energy; however, turbulent flows cannot; see the earlier discussion in Section 2.3. For more information on the relationships between dense granular flows and turbulent fluid flows see, e.g., Radjai & Roux (2002) and Oyama et al. (2019).

Example 8.9 The analogy between aggregation and fragmentation in particle systems and turbulent fluid flows

In this analogy, most of the focus is on forward and reverse cascades. More specifically, forward cascades in turbulent fluids resemble fragmentation cascades in ordinary bonded particle systems; see, e.g., Saveliev & Gorokhovski (2012) and Bažant & Caner (2014). In both cases, a flux of mass, momentum, and/or energy flows from large scales to small scales. Similarly, reverse cascades in turbulent fluids resemble aggregation cascades in regular bonded particle systems; see, e.g., Friedlander (1960a, b), Hunt (1982), Pushkin & Aref (2002) and Connaughton et al. (2004, 2006). In both cases, a flux of mass, momentum, and/or energy flows from small scales to large scales. In fact, all three problems—turbulence cascades, aggregation cascades, and fragmentation cascades—can be modeled using dimensional analysis. As discussed in Example 5.6, dimensional analysis for ordinary bonded particle systems assumes that most aggregation events involve particles of similar sizes. By the same token, dimensional analysis for turbulent flows assumes that eddies interact primarily with other eddies of similar sizes; see, e.g., Kolmogorov (1941a, b). Having said this, Horvai et al. (2008) note that ordinary bonded particle systems are often dominated by aggregation between particles of substantially different sizes. Similarly, Elsinga & Marusic (2016) say that "the underlying assumptions in Kolmogorov's [turbulence] theory appear not to hold, with most experimental and numerical data showing evidence of . . . significant direct energy transfer between the large- and the small-scales." Despite such objections, dimensional analysis often agrees well with real-world experiments and observations. In all three problems, there may be three different size ranges. For turbulent flows, these are generally called the energy-containing range, the inertial range, and the dissipative range. As discussed in Section 5.11, similar terminology can be used for aggregation and fragmentation in ordinary bonded particle systems. In all three problems, the size and kinetic energy distributions tend to be different in each range. Example 8.10 briefly compares the histories of fragmentation and turbulence research.

Example 8.10 The history of fragmentation research vs. the history of turbulence research

Schmitt (2017) says that the earliest widely cited work on fluid turbulence is "Boussinesq's eddy-viscosity closure . . . introduced in 1872." For comparison, Lynch & Rowland (2005) say that the earliest widely cited work on solid fragmentation is "Peter von Rittinger's energy-size reduction theory" published in 1867. Thus, the solid fragmentation and fluid turbulence problems are about equally old and, arguably, about equally difficult. Jiménez (2004) says that two "papers published by Kolmogorov (1941a, b) . . . are generally considered to be the origin of modern turbulence theory." During the same period, Kolmogorov (1941c, 1949) also published two key papers on particle fragmentation; see the earlier discussion of Andrei Kolmogorov in Example 5.1.

As one famous result, for spatially homogenous and isotropic turbulence, Kolmogorov (1941a, b) used dimensional analysis to argue that:

$$\mathfrak{D}(\Delta r, t) \sim \varepsilon^{2/3} \Delta r^{2/3} \tag{8.42}$$

in the inertial range, where $\varepsilon \sim u^3/l$ is the dissipation. This is sometimes called *Kolmogorov's 2/3 law*.

As a closely related expression:

$$S_V(\kappa, t) \sim \varepsilon^{2/3}\kappa^{-5/3}. \tag{8.43}$$

This is sometimes called *Kolmogorov's 5/3 law.* McComb (2014) notes that Kolmogorov's 2/3 law typically only applies for very large Reynolds numbers on the order of 10^3 to 10^6. By contrast, Kolmogorov's 5/3 law is "almost embarrassingly successful, with the 5/3 spectrum emerging at . . . Reynolds numbers as low as . . . 28." Table 8.2 summarizes this and some of the other better-known turbulent kinetic energy spectra; see also the earlier survey by Alexakis & Biferale (2018). Table 8.2 highlights Kolmogorov's result in gray to indicate that it is

Table 8.2 Late-time turbulent kinetic energy spectra for spatially homogeneous and isotropic turbulent flows. All results were obtained with analytic or empirical methods.

(a) Power law distributions for the energy-containing range (large scales). In these cases, the energy spectrum obtains a power law $S_V(\kappa, t) \sim \kappa^{r-1}$ for small values of κ.

r	m	d	Details	References
5	3	3	Freely decaying	Batchelor & Proudman (1956)[1]
3	3	3	Freely decaying	Saffman (1967)[1]
0	3	3	Near-wall flow	Tchen (1954), Katul & Chu (1998)[2]
0	3	3	-	Tatsumi (1960, 1980)[3]

(b) Power law distributions for the inertial range (middling scales). In these cases, the energy spectrum obtains a power law $S_V(\kappa, t) \sim \kappa^{r-1}$ for middling values of κ.

r	m	d	Details	References
−1/2	3	3	Magnetohydrodynamics	Iroshnikov (1964), Kraichnan (1965)[4]
−2/3	3	3	-	Kolomogorov (1941a, b)
−1	3	3	-	Tatsumi (1960, 1980)[3]
−1	3	3	Channel flows	Long (2003)[5]
−2	2	2	-	Kraichnan (1967), Batchelor (1969)[6]

(c) Gamma distributions for the dissipative range (small scales). In these cases, the energy spectrum obtains a gamma distribution $S_V(\kappa, t) \sim \kappa^{r-1}\exp(-\text{const.}\,\kappa)$ for large values of κ.

r	m	d	Details	References
4, 3	3	3	-	Kraichnan (1959)
−2/3	3	3	-	Kraichnan (1959)[7]
−1	3	3	-	Domaradzki (1992)
−2	2	2	-	Kida et al. (1988)

(1) See also the textbooks by Mathieu & Scott (2000), McComb (2014) and Davidson (2013, 2015).
(2) See also the textbooks by Davidson (2013, 2015) and Sagaut & Cambon (2018).
(3) See also the textbooks by Chorin (1994) and Lesieur (2008).
(4) See also the textbooks by Lesieur (2008) and Davidson (2013).
(5) See also the textbooks by Thorpe (2005), Cushman-Roisin & Beckers (2011), Simpson & Sharples (2012) and Tsinober (2019).
(6) See also the textbooks by Wyngaard (2010), Davidson (2015) and Sagaut & Cambon (2018).
(7) See also the textbooks by Pope (2000) and Davidson (2015).

a community standard. Notice that Table 8.2 for the fluctuating kinetic energy in turbulent flows resembles Tables 5.1 and 5.3 for the size distributions in aggregating and fragmentation particle systems. This is an aspect of the analogy between turbulence, aggregation, and fragmentation.

The results given in Table 8.2 are *ideal* or *universal*, i.e., they involve rational exponents r that do not depend on the exact conditions. In this regard Sagaut & Cambon (2018) say "the spectrum shape relaxes towards a universal shape at small scales provided that the Reynolds number is high enough to allow for the existence of the inertial range." They attribute this to the combined influence of forward and backward cascades. Researchers are still working to determine the conditions under which the various universal turbulent kinetic energy spectra apply. However, such work is relatively rare, partly because Kolmogorov's 5/3 law works so well for so many different problems. In this regard, McComb (2014) says "it is possible to encounter active turbulence researchers . . . who have never heard of any disagreement and . . . just assume that Kolmogorov was correct. This situation is not helped (and indeed is presumably to some extent caused) by the heterogeneous nature of the turbulence community, which consists of engineers (themselves subdivided into mechanical, chemical, aeronautical, and civil), mathematicians, meteorologists, climatologists, oceanographers, and physicists. Nor is it helped by the relatively low level of effort put into fundamentals [compared to] . . . urgent, practical applications." Similar observations apply to particle aggregation and fragmentation research. However, while the different communities involved in turbulence research have all adopted the same standard for the fluctuating kinetic energy spectrum, different communities involved in aggregation and fragmentation research have adopted different standards for the particle size distribution; see the earlier discussion in Sections 5.7 and 5.8.

8.6 Relationships Between Kinetic Energy and Size Distributions

This chapter has described two basic approaches for characterizing self-organized particle systems. In the first approach, deterministic cluster-finding algorithms are used to estimate particle shapes and sizes. This approach obtains probability density functions for particle diameter D or mass M. In the second approach, self-organized unbonded particles are characterized by statistical correlations between density or velocity. Here, the term "particle" may be interpreted broadly to include the eddies observed in turbulent flows. This approach yields structure functions and structure factors for the wave number κ. What is the relationship between these two approaches? For simplicity, in answering this question, this section assumes temporal homogeneity so that time dependencies can be largely ignored.

As the most common way of relating these two approaches, the particle diameter D is considered to be equivalent to a wavelength of $2\pi/\kappa$. In fact, for turbulent flows, Davidson (2015) says that "the conventional interpretation of $S_V(\kappa)$ [is] that it represents the energy in eddies of size $2\pi/\kappa$. [This] is convenient and works well for many purposes." On the other hand, he says "a random array of simple eddies of fixed size D has an energy spectrum which grows as κ^4 (or else κ^2) for small κ, peaks around $2\pi/D$, and then declines exponentially. The key point is that eddies of a given size contribute to $S_V(\kappa)$ across a full range of wavenumbers."

Davidson (2015) describes three different alternatives to the conventional interpretation of $S_V(\kappa)$. Each of these alternatives potentially offers a more accurate way of converting statistical quantities to particle size distributions in the following form:

$$f_E(D) \tag{8.44}$$

where E refers to the kinetic energy of the chaotic fluctuations and f_E refers to the fraction of E contained in turbulent eddies with diameters between D and $D + dD$. Notice that Equation (8.44) uses the notation $f_Y(x)$ introduced in Section 3.3 where $x = D$ and $Y = E$. Also notice that the very existence of $f_E(D)$ implies that the size distribution depends on the kinetic energy E of the fluctuations or, equivalently, on the velocity V of the fluctuations. By contrast, the approaches used for ordinary bonded particles in earlier chapters generally assumed that distributions of diameter D depend only on D, distributions of the velocity V only depend on V, and so forth.

In the first alternative suggested by Davidson (2015), the velocity structure factor S_V is reinterpreted as a probability density function. After a change of variables, this alternative yields the following:

$$f_E(D) = -\text{const.}\,\frac{d}{dD}\left(\frac{2\pi}{D}\right) S_V\left(\frac{2\pi}{D}\right) = \text{const.}\,D^{-2} S_V\left(\frac{2\pi}{D}\right). \tag{8.45}$$

This result assumes that $f_E(D) = |dX/dD|\,f_E(X)$ where $X = 2\pi/D$.

In the second alternative suggested by Davidson (2015), the second-order structure function $\mathfrak{D}$ is reinterpreted as a cumulative distribution function. This was first suggested by Townsend (1956). As usual, differentiation converts a cumulative distribution function to a probability distribution function. Thus, this alternative yields the following:

$$f_E(D) = \text{const.}\,\frac{d\mathfrak{D}}{dD}. \tag{8.46}$$

In the third alternative suggested by Davidson (2015), an attempt is made to correct the errors inherent in the first two alternatives. This was first suggested by Davidson & Pearson (2005). This alternative yields the following:

$$f_E(D) = -\frac{D^2}{2}\frac{\partial}{\partial D}\left(\frac{1}{D}\frac{d\mathfrak{D}}{dD}\right). \tag{8.47}$$

For example, for typical turbulent flows in the inertial range with $-2 \le r < 0$, equation (G.23) in Pope (2000) says that:

$$S_V(\kappa) \approx C_1 \kappa^{r-1} \tag{8.48}$$

for sufficiently large κ where C_1 is a constant that depends on r. In addition, equation (G.24) in Pope (2000) says that

$$\mathfrak{D}(D) \approx C_2 D^{-r} \tag{8.49}$$

for sufficiently small D where C_2 is a constant that depends on r.

In this case, all three of Davidson's alternatives give the following result:

$$f_E(D) \approx C_3 D^{-r-1} \tag{8.50}$$

where, in ideal cases, r is an even multiple of 1/6 as seen earlier in Table 8.2.b. While the three alternatives given previously were originally developed for fluid turbulence, they should apply to granular gases and other self-organized unbonded particle systems, provided that $-2 \leq r < 0$.

To summarize, there are two main approaches for characterizing particle sizes. The first approach obtains probability density functions such as $f(D)$ or $f(M)$. In ideal three-dimensional cases, these probability density functions obtain Rosin–Rammler forms where the diameter exponents n are even multiples of 1/2 or, equivalently, the mass exponents $r = n/m$ are even multiples of 1/6 when $m = 3$. This approach assumes the size distribution is statistically independent of the velocity distribution. This approach is generally the best choice when rotation and circulation are not important for characterizing particle sizes; see Figures 8.2 and 8.3. By contrast, the second approach obtains fluctuating kinetic energy spectra or, equivalently, probability density functions such as $f_E(D)$. In ideal cases, these probability density functions obtain Rosin–Rammler forms where the diameter exponents r are even multiples of 1/6. This assumes that the size distribution is statistically dependent on the velocity distribution. The second approach is generally the best choice when rotation and circulation are important for characterizing particle sizes.

8.7 Conclusions

This chapter described fluctuating kinetic energy spectra for self-organized particle systems. While irrotational elements such as normal bonded particles are best described by their size distributions as described in earlier chapters, rotational elements such as self-organized vortex tubes are best described by their fluctuating kinetic energy spectra. In the former case, when the systems are spatially homogeneous and isotropic, particle sizes tend to obtain power law, gamma, and other Rosin–Rammler distributions where the exponents are universal rational numbers such as −1/2, −2/3, and −1. Similarly, in the latter case, when the systems are spatially homogeneous and isotropic, the kinetic energy spectra tend to obtain power law, gamma, and other Rosin–Rammler distributions where the exponents are universal rational numbers such as −1/2, −2/3, and −1. This chapter has focused on two self-organized systems, namely, granular gases and turbulent fluid flows. Notice that Table 8.1 for granular gases includes many of the same universal kinetic energy spectra as Table 8.2 for turbulent fluid flows. For example, Taguchi (1995) found that three-dimensional granular gases—more specifically, "monodisperse glass beads shaken by a loudspeaker"—obtained results "analogous to Kolmogorov's 5/3 energy spectrum" with $r = -2/3$. For another example, Isobe (2003) found that two-dimensional granular gases at early to mid-times obtained "energy spectra . . . close to the expectation of Kraichnan-Batchelor theory" for turbulence with $r = 2$.

Improved understanding of turbulent flows may lead to improved understanding of aggregation and fragmentation. This is especially true for aggregation and fragmentation in granular gases, dense granular flows, and ordinary bonded particle systems that are bathed in a fluid medium. Regarding turbulent flows, Falkovich & Sreenivasan (2006) say the following;

> It is trite to regard turbulence as the last unsolved problem in classical physics and to cite many books and authorities to justify the opinion. It is likewise a cliché to list great physicists and mathematicians, such as Werner Heisenberg, Richard Feynman, and Andrei Kolmogorov, who "failed" to solve the problem despite much effort. Horace Lamb and others have been credited with wishing to seek heavenly wisdom on the subject when they arrived in heaven. With such lists and stories, youngsters are cautioned, directly and indirectly, that turbulence is beyond reasonable grasp. [However] one need not apologize for or despair over the difficulty of turbulence.

Similar remarks could be made about particle aggregation and fragmentation. In particular, just like many well-known scientists and mathematicians have contributed to the understanding of turbulence, so have many well-known scientists and mathematicians contributed to the understanding of aggregation and fragmentation. As discussed in Chapter 5, they include Marian Smoluchowski, Nevill Mott, Paul Flory, and Andrei Kolmogorov. From one point of view, the main challenge that aggregation, fragmentation, and turbulence share in common is associating specific problems with specific solutions in the face of unknowns; see the earlier discussion in Section 5.10. In all cases, while computational, analytic, semi-empirical, and empirical methods tend to obtain completely legitimate solutions—provided they properly account for randomness, seeming-randomness, and dissipation—they may not obtain the desired solution that matches properly with a given real-world problem. A common approach is to use community standards to estimate key solution elements. This approach addresses unknowns by drawing on past experience with similar problems.

References

A. Alexakis & L. Biferale, Cascades and Transitions in Turbulent Flows, *Physics Reports*, 767–769, 1–101, 2018, https://doi.org/10.1016/j.physrep.2018.08.001

B. Andreotti, Y. Forterre, & O. Pouliquen, *Granular Media: Between Fluid and Solid*, Cambridge University Press, 2013, https://doi.org/10.1017/CBO9781139541008.001

K. Babić & M. W. Rotach, Turbulence Kinetic Energy Budget in the Stable Boundary Layer Over a Heterogeneous Surface, *Quarterly Journal of the Royal Meteorological Society*, 144(713), 1045–1062, 2018, https://doi.org/10.1002/qj.3274

P. Bak, C. Tang, & K. Wiesenfeld, Self-Organized Criticality, *Physical Review A*, 38(1), 364–374, 1988, https://doi.org/10.1103/PhysRevA.38.364

A. Baldassarri, U. M. B. Marconi, & A. Puglisi, Cooling of a Lattice Granular Fluid as an Ordering Process, *Physical Review E*, 65, 051301, 2002, https://doi.org/10.1103/PhysRevE.65.051301

A. Baldassarri, U. M. B. Marconi, & A. Puglisi, Velocity Fluctuations in Cooling Granular Gases, in T. Pöschel & N. Brilliantov, editors, *Granular Gas Dynamics*, Springer, 2003, https://doi.org/10.1007/978-3-540-39843-1_4

G. K. Batchelor & I. Proudman, The Large-Scale Structure of Homogeneous Turbulence, *Philosophical Transactions of the Royal Society of London A: Mathematical and Physical Sciences*, 248(949), 369–405, 1956, https://doi.org/10.1098/rsta.1956.0002

G. K. Batchelor, Computation of the Energy Spectrum in Homogeneous Two-Dimensional Turbulence, *The Physics of Fluids*, 12(12), II-233 to II-239, 1969, https://doi.org/10.1063/1.1692443

Z. P. Bažant & F. C. Caner, Impact Comminution of Solids due to Local Kinetic Energy of High Shear Strain Rate: I. Continuum Theory and Turbulence Analogy, *Journal of the Mechanics and Physics of Solids*, 64, 223–235, 2014, https://doi.org/10.1016/j.jmps.2013.11.008

A. Berera & R. D. J. G. Ho, Chaotic Properties of a Turbulent Isotropic Fluid, *Physical Review Letters*, 102, 024101, 2018, https://doi.org/10.1103/PhysRevLett.120.024101

P. S. Bernard & M. A. Erinen, Fluid Particle Dynamics and the Non-Local Origin of the Reynolds Shear Stress, *Journal of Fluid Mechanics*, 847, 520–551, 2018, https://doi.org/10.1017/jfm.2018.333

A. Bunde & S. Havlin, Percolation, in A. Bunde & S. Havlin, editors, *Fractals and Disordered Systems*, Springer, 1996 [see section 2.7.1], https://doi.org/10.1007/978-3-642-84868-1_2

J. C. Burton, P. Y. Lu, & S. R. Nagel, Collision Dynamics of Particle Clusters in a Two-Dimensional Granular Gas, *Physical Review E*, 88, 062204, 2013, https://doi.org/10.1103/PhysRevE.88.062204

A. J. Chorin, *Vorticity and Turbulence*, Springer, 1994, https://doi.org/10.1007/978-1-4419-8728-0

C. Connaughton, R. Rajesh, & O. Zaboronski, Stationary Kolmogorov Solutions of the Smoluchowski Aggregation Equation with a Source Term, *Physical Review E*, 69(6), 061114, 2004, https://doi.org/10.1103/PhysRevE.69.061114

C. Connaughton, R. Rajesh, & O. Zaboronski, Cluster-Cluster Aggregation as an Analogue of a Turbulent Cascade: Kolmogorov Phenomenology, Scaling Laws and the Breakdown of Self-Similarity, *Physica D: Nonlinear Phenomena*, 222(1–2), 97–115, 2006, https://doi.org/10.1016/j.physd.2006.08.005

B. Cushman-Roisin & J.-M. Beckers, *Introduction to Geophysical Fluid Dynamics: Physical and Numerical Aspects*, 2nd edition, Academic Press, 2011 [see chapter 5.1], https://doi.org/10.1016/B978-0-12-088759-0.00001-8

S. K. Das & S. Puri, Kinetics of Inhomogeneous Cooling in Granular Gases, *Physical Review E*, 68, 011302, 2003, https://doi.org/10.1103/PhysRevE.68.011302

P. Das, S. Puri, & M. Schwartz, Clustering and Velocity Distributions in Granular Gases by Cooling by Solid Friction, *Physical Review E*, 94, 032907, 2016, https://doi.org/10.1103/PhysRevE.94.032907

P. Das, M. Schwartz, & S. Puri, Frictional Cooling of Granular Gases: A Molecular Dynamics Study, *Journal of Physics: Conference Series*, 905(1), 012035, 2017, https://doi.org/10.1088/1742-6596/905/1/012035

P. A. Davidson & B. R. Pearson, Identifying Turbulent Energy Distributions in Real, Rather than Fourier, Space, *Physical Review Letters*, 95, 214501, 2005, https://doi.org/10.1103/PhysRevLett.95.214501

P. A. Davidson, *Turbulence in Rotating, Stratified, and Electrically Conducting Fluids*, Cambridge University Press, 2013, https://doi.org/10.1017/CBO9781139208673

P. A. Davidson, *Turbulence: An Introduction for Scientists and Engineers*, 2nd edition, Oxford University Press, 2015, https://doi.org/10.1093/acprof:oso/9780198722588.001.0001

J. A. Domaradzki, Nonlocal Triad Interactions and the Dissipation Range of Isotropic Turbulence, *Physics of Fluids A: Fluid Dynamics*, 4(9), 2037–2045, 1992, https://doi.org/10.1063/1.858373

G. E. Elsinga & I. Marusic, The Anisotropic Structure of Turbulence and its Energy Spectrum, *Physics of Fluids*, 28(1), 011701, 2016, https://doi.org/10.1063/1.4939471

G. Falkovich & K. Sreenivasan, Lessons from Hydrodynamic Turbulence, *Physics Today*, 59(4), 43–49, 2006, https://doi.org/10.1063/1.2207037

S. K. Friedlander, On the Particle-Size Spectrum of Atmospheric Aerosols, *Journal of Meteorology*, 17(3), 373–374, 1960a, https://doi.org/10.1175/1520-0469(1960)017<0373:OTPSSO>2.0.CO;2

S. K. Friedlander, Similarity Considerations for the Particle-Size Spectrum of a Coagulating, Sedimenting Aerosol, *Journal of Meteorology*, 17(5), 479–483, 1960b, https://doi.org/10.1175/1520-0469(1960)017<0479:SCFTPS>2.0.CO;2

I. Goldhirsch, Scales and Kinetics of Granular Flows, *Chaos*, 9(3), 659–672, 1999, https://doi.org/10.1063/1.166440

P. Horvai, S. V. Nazarenko & T. H. M. Stein, Coalescence of Particles by Differential Sedimentation, *Journal of Statistical Physics*, 130(6), 1177–1195, 2008, https://doi.org/10.1007/s10955-007-9466-y

J. R. Hunt, Self-Similar Particle-Size Distributions during Coagulation: Theory and Experimental Verification, *Journal of Fluid Mechanics*, 122, 169–185, 1982, https://doi.org/10.1017/S0022112082002158

P. S. Iroshnikov, Turbulence of a Conducting Fluid in a Strong Magnetic Field, *Soviet Astronomy*, 7(4), 566–571, 1964, http://adsabs.harvard.edu/abs/1964SvA...7...566I

M. Isobe, Velocity Statistics in Two-Dimensional Granular Turbulence, *Physical Review E*, 68, 040301(R), 2003, https://doi.org/10.1103/PhysRevE.68.040301

M. Isobe, Statistical Law of Turbulence in Granular Gas, *Journal of Physics: Conference Series*, 402(1), 012041, 2012, https://doi.org/10.1088/1742-6596/402/1/012041

M. Isobe, Clustering Impact Regime with Shocks in Freely Evolving Granular Gas, in F. Radjai, S. Nezamabadi, S. Luding, & J. Y. Delenne, editors, *Powders and Grains: 8th International Conference on Micromechanics on Granular Media, EPJ Web of Conferences*, 140, 04006, 2017, https://doi.org/10.1051/epjconf/201714004006

J. Jiménez, The Contributions of A. N. Kolmogorov to the Theory of Turbulence, *Arbor*, 178(704), 589–606, 2004, https://doi.org/10.3989/arbor.2004.i704.550

H. Kellay, Y. Amarouchene, & J. F. Boudet, Intermittency of the Velocity Fluctuations in a Granular Surface Flow, *Physics of Fluids*, 19, 078104, 2007, https://doi.org/10.1063/1.2750679

S. Khurshid, D. A. Donzis, & K. R. Sreenivasan, Energy Spectrum in the Dissipation Range, *Physical Review Fluids*, 3, 082601(R), 2018, https://doi.org/10.1103/PhysRevFluids.3.082601

S. Kida, M. Yamada & K. Ohkitani, The Energy Spectrum in the Universal Range of Two-Dimensional Turbulence, *Fluid Dynamics Research*, 4(4), 271–301, 1988, https://doi.org/10.1016/0169-5983(88)90029-9

A. N. Kolmogorov, The Local Structure of Turbulence in Incompressible Viscous Fluids for Very Large Reynolds Numbers, *Doklady Akademii Nauk SSSR*, 30(4), 299–303, 1941a; see also

Proceedings of the Royal Society of London A, 434, 9–13, 1991, https://doi.org/10.1098/rspa.1991.0075

A. N. Kolmogorov, Dissipation of Energy in Locally Isotropic Turbulence. *Doklady Akademii Nauk SSSR*, 32(1), 16–18, 1941b; see also *Proceedings of the Royal Society of London A*, 434, 15–17, 1991, https://doi.org/10.1098/rspa.1991.0076

A. N. Kolmogorov, On the Log Normal Distribution of Particle Sizes under Grinding, *Doklady Akademii Nauk SSSR*, 31(2), 99–101, 1941c; see also *The Logarithmically Normal Law of Distribution of Dimensions of Particles when Broken into Small Parts*, NASA Technical Translation TT F-12, 287, 1969, https://archive.org/details/nasa_techdoc_19690019884

A. N. Kolmogorov, On the Breakage of Drops in a Turbulent Flow, *Doklady Akademii Nauk SSSR*, 66(5), 825–828, 1949; see also V. M. Tikhomirov, editor, *Selected Works of A. N. Kolmogorov, Volume 1: Mathematics and Mechanics*, Springer, 1991

G. Katul & C.-R. Chu, A Theoretical and Experimental Investigation of Energy-Containing Scales in the Dynamic Sublayer of Boundary-Layer Flows, *Boundary-Layer Meteorology*, 86, 279–312, 1998, https://doi.org/10.1023/A:1000657014845

R. H. Kraichnan, The Structure of Isotropic Turbulence at Very High Reynolds Number, *Journal of Fluid Mechanics*, 5(4), 497–543, 1959, https://doi.org/10.1017/S0022112059000362

R. H. Kraichnan, Inertial-Range Spectrum of Hydromagnetic Turbulence, *The Physics of Fluids*, 8(7), 1385–1387, 1965, https://doi.org/10.1063/1.1761412

R. H. Kraichnan, Inertial Ranges in Two-Dimensional Turbulence, *The Physics of Fluids*, 10(7), 1417–1423, 1967, https://doi.org/10.1063/1.1762301

M. Lesieur, *Turbulence in Fluids*, 4th edition, Springer, 2008, https://doi.org/10.1007/978-1-4020-6435-7

R. R. Long, Do Tidal Channel Turbulence Measurements Support $k^{-5/3}$?, *Environmental Fluid Mechanics*, 3(2), 109–127, 2003, https://doi.org/10.1023/A:1022086815714

A. J. Lynch & C. A. Rowland, *The History of Grinding*, Society for Mining, Metallurgy, and Exploration, 2005

M. Y. Marov & A. V. Kolesnichenko, *Turbulence and Self-Organization: Modeling Astrophysical Objects*, Springer, 2013, https://doi.org/10.1007/978-1-4614-5155-6

D. O. Martínez, S. Chen, G. D. Doolen, R. H. Kraichnan, L.-P. Wang, & Y. Zhou, Energy Spectrum in the Dissipation Range of Fluid Turbulence, *Journal of Plasma Physics*, 57(1), 195–201, 1997, https://doi.org/10.1017/S0022377896005338

J. Mathieu & J. Scott, *An Introduction to Turbulent Flow*, Cambridge University Press, 2000, https://doi.org/10.1017/CBO9781316529850

W. D. McComb, *Homogeneous, Isotropic Turbulence: Phenomenology, Renormalization, and Statistical Closures*, Oxford University Press, 2014, https://doi.org/10.1093/acprof:oso/9780199689385.001.0001

N. Oyama, H. Mizuno, & K. Saitoh, Avalanche Interpretation of the Power-Law Energy Spectrum in Three-Dimensional Dense Granular Flow, *Physical Review Letters*, 122, 188004, 2019, https://doi.org/10.1103/PhysRevLett.122.188004

S. N. Pathak, Z. Jabeen, D. Das, & R. Rajesh, Energy Decay in Three-Dimensional Freely Cooling Granular Gas, *Physical Review Letters*, 112, 038001, 2014, https://doi.org/10.1103/PhysRevLett.112.038001

G. Peng & T. Ohta, Steady State Properties of a Driven Granular Medium, *Physical Review E*, 58(1), 4737–4746, 1998, https://doi.org/10.1103/PhysRevE.58.4737

S. B. Pope, *Turbulent Flows*, Cambridge University Press, 2000, https://doi.org/10.1017/CBO9780511840531.001

T. Pöschel, N. V. Brilliantov, & T. Schwager, Transient Clusters in Granular Gases, *Journal of Physics: Condensed Matter*, 17(24), S2705–S2713, 2005, https://doi.org/10.1088/0953-8984/17/24/021

V. V. Prasad, D. Das, S. Sabhapandit, & R. Rajesh, Velocity Distribution of a Driven Inelastic One-Component Maxwell Gas, *Physical Review E*, 95, 032909, 2017, https://doi.org/10.1103/PhysRevE.95.032909

D. O. Pushkin & H. Aref, Self-Similarity Theory of Stationary Coagulation, *Physics of Fluids*, 14(2), 694–703, 2002, https://doi.org/10.1063/1.1430440

F. Radjai & S. Roux, Turbulentlike Fluctuations in Quasistatic Flow of Granular Media, *Physical Review Letters*, 89(6), 064302, 2002, https://doi.org/10.1103/PhysRevLett.89.064302

F. Reetz, T. Kreilos, & T. M. Schneider, Exact Invariant Solution Reveals the Origin of Self-Organized Oblique Turbulent-Laminar Stripes, *Nature Communications*, 10, 2227, 2019, https://doi.org/10.1038/s41467-019-10208-x

S. Ross, *A First Course in Probability*, 10th edition, Pearson, 2018.

R. J. Rossi, *Mathematical Statistics: An Introduction to Likelihood Based Inference*, Wiley, 2018, https://doi.org/10.1002/9781118771075

P. G. Saffman, The Large-Scale Structure of Homogeneous Turbulence, *Journal of Fluid Mechanics*, 27(3), 581–593, 1967, https://doi.org/10.1017/S0022112067000552

P. Sagaut & C. Cambon, *Homogeneous Turbulence Dynamics*, 2nd edition, Springer, 2018, https://doi.org/10.1007/978-3-319-73162-9

V. L. Saveliev & M. A. Gorokhovski, Renormalization of the Fragmentation Equation: Exact Self-Similar Solutions and Turbulent Cascades, *Physical Review E*, 86(6), 061112, 2012, https://doi.org/10.1103/PhysRevE.86.061112

F. G. Schmitt, Turbulence from 1870 to 1920: The Birth of a Noun and a Concept, *Comptes Rendus Mécanique*, 345(9), 620–626, 2017, https://doi.org/10.1016/j.crme.2017.06.003

J. H. Simpson & J. Sharples, *Introduction to the Physical and Biological Oceanography of Shelf Seas*, Cambridge University Press, 2012 [see chapter 4], https://doi.org/10.1017/CBO9781139034098

A. V. Smol'yakov & V. M. Tkachenko, *The Measurement of Turbulent Fluctuations: An Introduction to Hot-Wire Anemometry and Related Transducers*, Springer, 1983, https://doi.org/10.1007/978-3-642-81983-4

Y.-H. Taguchi, $k^{-5/3}$ Power Spectrum in Powder-Turbulent Flow in a Vibrated Bed: Numerical Results, *Europhysics Letters*, 24(3), 203–209, 1993, https://doi.org/10.1209/0295-5075/24/3/008

Y.-H. Taguchi, Numerical Study of Granular Turbulence and the Appearance of the $k^{-5/3}$ Energy Spectrum without Flow, *Physica D: Nonlinear Phenomena*, 80(1–2), 61–71, 1995, https://doi.org/10.1016/0167-2789(95)90059-4

T. Tatsumi, Energy Spectra in Magneto-Fluid Dynamic Turbulence, *Reviews of Modern Physics*, 32(4), 807–812, 1960, https://doi.org/10.1103/RevModPhys.32.807

T. Tatsumi, Theory of Homogenous Turbulence, *Advances in Applied Mechanics*, 20, 39–133, 1980, https://doi.org/10.1016/S0065-2156(08)70232-1

C.-M. Tchen, Transport Processes as Foundations of the Heisenberg and Obukhoff Theories of Turbulence, *Physical Review*, 93(1), 4–14, 1954, https://doi.org/10.1103/PhysRev.93.4

S. A. Thorpe, *The Turbulent Ocean*, Cambridge University Press, 2005 [see section 1.7], https://doi.org/10.1017/CBO9780511819933

A. A. Townsend, *The Structure of Turbulent Shear Flow*, Cambridge University Press, 1956, https://doi.org/10.1017/S0022112056210366

A. Tsinober, *The Essence of Turbulence as a Physical Phenomenon: With Emphasis on Issues of Paradigmatic Nature*, 2nd edition, Springer, 2019, https://doi.org/10.1007/978-3-319-99531-1

T. P. C. van Noije, M. H. Ernst, E. Trizac, & I. Pagonabarraga, Randomly Driven Granular Fluids: Large-Scale Structure, *Physical Review E*, 59(4), 4236–4341, 1999, https://doi.org/10.1103/PhysRevE.59.4326

J. S. van Zon & F. C. MacKintosh, Velocity Distributions in Dilute Granular Systems, *Physical Review E*, 72, 051301, 2005, https://doi.org/10.1103/PhysRevE.72.051301

J. C. Wyngaard, *Turbulence in the Atmosphere*, Cambridge University Press, 2010, https://doi.org/10.1017/CBO9780511840524

9

Aggregation Cascades Part 1: Mean-Field Governing Equations

This is the first of three chapters on mean-field theory for aggregation cascades. In particular, this chapter describes the mean-field governing equations for aggregation cascades. This chapter assumes that binary collisions cause two particles to form a single particle. The probability of such collisions depends on five fields, namely, the particle position, size, velocity, rotation rate, and shape fields. If the particle system is spatially homogeneous and isotropic, the position field does not affect the statistics of the other fields. In that case, suppose the velocity, rotation, and shape fields can be replaced by an average. More specifically, suppose they can be combined into a single size-dependent aggregation rate. This rate is commonly known as the *collision kernel*, *collection kernel*, or *collision frequency function*. Among other things, the collision kernel captures the average effects of randomness and chaos. As a result, mean-field theory represents complex chaotic particle systems by relatively simple deterministic governing equations, where the independent variables are the particle size and time, and the dependent variable is the particle concentration.

9.1 Discrete Governing Equation

In the simplest type of discrete systems, particles are composed of identical *monomers* with mass M_1. Then the particle concentration $c_i(t)$ is defined as the number of particles per unit volume composed of i monomers at time t. In addition, the *collision kernel* K_{ij} is defined such that:

> $K_{ij}c_i(t)c_j(t)$ is the average rate per unit volume at which particles of size i and j collide to form particles of size $i + j$.

Notice that K_{ij} has units of volume divided by time. As shown by Smoluchowksi (1916, 1917), a discrete mean-field governing equation can be written as follows:

$$\frac{dc_i(t)}{dt} = -c_i(t)\sum_{j=1}^{\infty} K_{ij}c_j(t) + \frac{1}{2}\sum_{j=1}^{i-1} K_{i-j,j}c_{i-j}(t)c_j(t). \tag{9.1}$$

This is called *Smoluchowski's coagulation equation*. This equation can be interpreted as follows:

$$\begin{pmatrix}\textit{rate of change}\\ \textit{in particles of}\\ \textit{size } i\end{pmatrix} = -\begin{pmatrix}\textit{rate at which particles of}\\ \textit{size } i \textit{ merge to form particles}\\ \textit{larger than size } i\end{pmatrix} + \begin{pmatrix}\textit{rate at which particles}\\ \textit{smaller than size } i \textit{ merge}\\ \textit{to form particles of size } i\end{pmatrix}.$$

Particle Dynamics with Aggregation and Fragmentation. Culbert B. Laney, Oxford University Press. © Culbert B. Laney (2025).
DOI: 10.1093/9780198920687.003.0009

Equations of this general sort are sometimes called *population balance equations*. If sources and sinks are present, Equation (9.1) becomes as follows:

$$\frac{dc_i(t)}{dt} = -c_i(t)\sum_{j=1}^{\infty} K_{ij}c_j(t) + \frac{1}{2}\sum_{j=1}^{i-1} K_{i-j,\,j}c_{i-j}(t)c_j(t) + a_i(t) + b_i(t)c_i(t). \tag{9.2}$$

In this equation, $a_i(t)$ and $b_i(t)c_i(t)$ are sources when they are positive and sinks when they are negative. As defined here, *sources* introduce new particles at uniformly random locations throughout the system. If the particles being introduced have the same or similar sizes, the source is called *hard*. If the particles being introduced have a wide range of different sizes, the source is called *soft*. Similarly, *sinks* remove existing particles at uniformly random locations throughout the system. If the particles being removed all have sizes above or below a specific threshold, the sink is called *hard*. Otherwise, if the effect gradually increases with size, the sink is called *soft*. For example, White (1982) suggested a hard source and a soft sink as follows:

$$a_i(t) = a_0\delta_{i1}$$
$$b_i(t) = -b_0 i^{\gamma}$$

where $\gamma > 0$, $a_0 > 0$, $b_0 > 0$ are constants and δ is the Kronecker delta function, i.e., $\delta_{i1} = 1$ when $i = 1$ and $\delta_{i1} = 0$ when $i \neq 1$. In this example, the source implies a hard minimum cutoff, i.e., it ensures that there are always particles as small as monomers, while the sink imposes a soft maximum cutoff, i.e., it gradually tapers the number of particles as size increases.

9.2 Continuous Governing Equation

In continuous systems, the concentration $c(x, t)$ of particles of mass x at time t is defined as discussed in Section 3.2. In this case, the collision kernel $K(x, y)$ is defined such that:

> $K(x, y)c(x, t)c(y, t)dy$ is the average rate per unit mass and volume at which parent particles of masses x and $y \pm dy$ collide to form particles of mass $x + y$.

Notice that $K(x, y)$ has units of volume divided by time. As shown by Schumann (1940), a continuous mean-field governing equation can be written as follows:

$$\frac{\partial c(x,t)}{\partial t} = -c(x,t)\int_0^{\infty} K(x,y)c(y,t)dy + \frac{1}{2}\int_0^{x} K(x-y,x)c(x-y,t)c(y,t)dy. \tag{9.3}$$

The first term on the right hand side represents the rate at which particles of a given size x are lost due to aggregation, while the second term on the right hand side represents the rate at which particles of a given size x are gained due to aggregation. As noted before, equations of this general sort are sometimes called *population balance equations*; see, e.g., Ramkrishna (2000).

If sources and sinks are present, Equation (9.3) becomes as follows:

$$\frac{\partial c(x,t)}{\partial t} + \frac{\partial[\alpha(x,t)c(x,t)]}{\partial x} = -c(x,t)\int_0^\infty K(x,y)c(y,t)dy$$
$$+\frac{1}{2}\int_0^x K(x-y,x)c(x-y,t)c(y,t)dy + a(x,t) + b(x,t)c(x,t). \tag{9.4}$$

In this equation, $a(x, t)$, $b(x, t)c(x, t)$, and $\partial[\alpha(x, t)c(x, t)]/\partial x$ are sources when they are positive and sinks when they are negative. The earlier definitions of hard and soft sources and sinks apply again here. However, while $a(x, t)$ and $b(x, t)c(x, t)$ may be either hard or soft, $\partial[\alpha(x, t)c(x, t)]/\partial x$ is always soft. For example, Ramabhadran et al. (1976) suggested a hard (monodisperse) source and a soft (power law) sink:

$$a(x,t) = a_0\delta(M - M_0)$$
$$\alpha(x,t) = -\alpha_0 x^\gamma$$

where $0 \le \gamma \le 1$, $a_0 > 0$, $\alpha_0 > 0$ are constants and δ is the Dirac delta, i.e., $\delta(x) = 1$ when $x = 0$ and $\delta(x) = 0$ when $x \neq 0$.

In Equation (9.4), $a(x, t)$ and $b(x, t)$ $c(x, t)$ refer to sources and sinks that add or subtract particles in their entirety. By contrast, $\partial[\alpha(x, t)c(x, t)]/\partial x$ refers to sources and sinks that add to or subtract from existing particles gradually, e.g., due to condensation and evaporation. For example, Singh & Rodgers (1996) say that when $a(x, t)$ or $b(x, t)c(x, t)$ are "employed to remove mass from a system ...[they typically] remove mass by removing complete particles." However, when $\partial[\alpha(x, t)c(x, t)]/\partial x$ is "employed to remove mass from a system ... mass disappears from the particles themselves, yet the particles ... still remain behind, albeit in diminished form, in the system." Similarly, Abulwafa et al. (2006) says "the exposed surface of the particle recedes continuously eventually leading to a total loss of the mass of the particle."

Hounslow (1990) says that "solutions of the population balance for systems involving aggregation are notoriously difficult to obtain." As an alternative to Equation (9.3), he suggested the following:

$$\frac{c(x,t) - c_1(x)}{t} = -c(x,t)\int_0^\infty K(x,y)c(y,t)dy + \frac{1}{2}\int_0^x K(x-y,x)c(x-y,t)c(y,t)dy \tag{9.5}$$

where $c_1(x)/t$ is a source. Notice that Equation (9.3) is an *integro-differential* equation while Equation (9.5) is a pure *integral* equation. As one possibility, if t is the ordinary time, then Equation (9.5) is sometimes called an unsteady *relaxation time approximation* to Equation (9.3). Killingbeck & Cole (1971) warn that "an approximation of this form will not be rigorous." As another possibility, if t is a fixed time scale characteristic of the particle system, then Equation (9.5) is sometimes called a *steady-state population balance equation.* For example, Hounslow (1990) let $t = Q/\mathcal{V}$ where $\mathcal{V}$ is the system volume and Q is the volumetric flow rate through the system driven by an exponential source.

In addition, Hounslow (1990) set the term $c_1(x)$ on the left hand side of Equation (9.5) as follows:

$$c_1(x) = \frac{C_1}{M_1}\exp\left(-\frac{x}{M_1}\right)$$

where $C_1 = \int_0^\infty c_1(x)dx$ is the total source concentration and $M_1 = \left(\int_0^\infty xc_1(x)dx\right)/C_1$ is the average source mass.

9.3 Kernels

For simplicity, the remainder of this chapter will consider only continuous particle systems. Horvai et al. (2008) defined two types of aggregation: (a) *local* or *forced locality* where binary collisions preferentially involve particles of similar sizes, shapes, and rotations; and (b) *non-local* or *free merging* where binary collisions preferentially involve particles of different sizes, shapes, and rotations. To use terms familiar from statistical mechanics, local cases involve pairs of *indistinguishable* particles, while non-local cases involve pairs of *distinguishable particles*; see, e.g., chapter 6 of Stowe (2007). Local and non-local aggregation both allow for useful approximations. More specifically, for local cases where $x \approx y$ it is often true that:

$$K(x,y) \approx K_0 x^\lambda \approx K_0 y^\lambda \tag{9.6}$$

where K_0 is a constant and λ is a power law exponent. Similarly, for non-local cases where $x \gg y$ or $x \ll y$, it is often true that:

$$K(x,y) \approx K_0 \max\left(x^\lambda, y^\lambda\right). \tag{9.7}$$

Even when neither of these approximations apply, it is often true that:

$$K(ax, ay) = a^\lambda K(x,y) \tag{9.8a}$$

where a is any constant. In this case, the collision kernel is called *homogeneous* and the exponent λ is called the *degree of homogeneity*. Equivalent definitions include the following:

$$K(x,y) = x^\lambda K_1(y/x) \tag{9.8b}$$
$$K(x,y) = y^\lambda K_2(x/y). \tag{9.8c}$$

The degree of homogeneity plays a critical role in mean-field solutions. As one observation, if $\lambda < 0$, small particles tend to grow faster than large particles. Conversely, if $\lambda > 0$, large particles tend to grow faster than small particles. As another observation, if $\lambda < 1$, no single particle grows dramatically faster than any other particle. However, if $\lambda > 1$, a single giant particle may form, which vastly exceeds the size of all other particles. As a third observation, λ helps determine the late-time size distributions.

For example, for sufficiently large times, mean-field solutions may obtain gamma size distributions:

$$c(M,t) = \frac{A_0}{M_{\text{avg}}(t)}\left[\frac{M}{M_{\text{avg}}(t)}\right]^{\frac{n}{m}-1} \exp\left[-\frac{b_0 M}{M_{\text{avg}}(t)}\right]$$

where:

$$\frac{n}{m} = -\lambda + 1.$$

In ideal cases, the degree of homogeneity is typically a rational number such that:

$$\lambda = \frac{j_1 + j_2}{k} + 1 \tag{9.9}$$

where k is the number of *degrees of freedom* while j_1 and j_2 may be defined as follows:

j_1 ±1 *times the integer dimension of the phase space for particle* 1 $(-k \leq j_1 \leq k)$

j_2 ±1 *times the integer dimension of the phase space for particle* 2 $(-k \leq j_2 \leq k)$.

The situation in phase space is similar to the situation in physical space, where d is the upper bound on the physical dimension m of each particle; see Figure 1.8. Equation (9.9) exploits the fact that chaotic systems, such as those described by collision kernels, inherently involve a limited number of degrees of freedom, as discussed earlier in Section 2.1. More specifically, the degrees of freedom capture the net effect of particle positions, velocities, rotations, and shapes. As a simplifying assumption, most mean-field theory treatments of aggregation assume non-rotating spherical particles. For example, Michaelides et al. (2017) assume non-rotating "perfect, rigid spheres undergoing instantaneous collisions that also yield spheres." While this may often be true for liquids, due to the effects of surface tension, this is rarely true for solids. However, rotating non-spherical particles may behave, on average, like non-rotating spherical particles. As a simple example, a circular coin flipped end-over-end may behave, on average, like a non-rotating sphere of the same radius. Thus, the remainder of this discussion assumes that all particles are non-rotating spheres. In this case, in Equation (9.9), j_1, j_2, and k capture the net effect of particle positions and velocities.

For local cases with $d = 3$, when aggregation is dominated by collisions between pairs of particles of similar sizes:

$$k = 3 \tag{9.10}$$

corresponding to the three components of the relative velocity needed to describe two indistinguishable particles. In this case, because the particles are indistinguishable, their relative position makes no difference. For example, Chakraborty (1991) says that if two "particles are assumed to be rigid and spherically symmetric point masses ... the colliding particle have only three degrees of freedom due to translational motion."

For non-local cases with $d = 3$, when aggregation is dominated by collisions between pairs of particles of different sizes:

$$k = 6 \tag{9.11}$$

corresponding to a maximum of three components of relative velocity and three components of relative position needed to describe two distinguishable particles. Example 9.1 describes an analogy between particle energies and sizes. As this example shows, the concept of degrees of freedom works equally well in either case, provided the particle system is chaotic. For more information on the analogies between particle energies and sizes, see Chapter 8 and 18.

Example 9.1 Degrees of freedom for molecular gases

According to de Regt (2017), the concept of degrees of freedom was first introduced by Maxwell (1872) and Watson (1876). This concept ultimately led to the following standard formula for the ratio of specific heat capacities γ in an ideal molecular gas:

$$\gamma = \frac{2}{k} + 1 \tag{9.12}$$

where k is the number of degrees of freedom. For ideal monoatomic gases:

$$k = 3 \tag{9.13}$$

corresponding to three components of velocity. Similarly, for ideal diatomic gases:

$$k = 6 \tag{9.14}$$

corresponding to three components of velocity, two components of rotation, and one component of vibration, all taken relative to the molecular center-of-mass. In general, for an ideal gas with N atoms per molecule, $k = 3N$ corresponding to three components of velocity, either two or three components of rotation, and the rest components of vibration; see, e.g., McQuarrie (2008) and Burrows et al. (2021). Notice that, in practice, the number of degrees of freedom varies as a function of temperature; for example, for diatomic gases such as hydrogen, $k = 3$ at low temperatures, $k = 5$ at normal temperatures, and $k = 7$ at high temperatures; see, e.g., Schwabl (2006) and Giancoli (2014). Equations (9.12), (9.13), (9.14) concern particle *energies* while Equations (9.9), (9.10), (9.11) concern particle *sizes*. Otherwise, these two sets of equations are identical, at least in cases where $j_1 + j_2 = 2$. This is because molecular gases and particle aggregation both involve chaos as seen in Chapter 2, and because chaos affects particle energies and sizes in analogous ways as seen in Chapter 8.

For local cases, Equations (9.9) and (9.10) imply the following:

$$\lambda \text{ is an integer multiple of } 1/3 \text{ between } -1 \text{ and } 3. \tag{9.15}$$

Similarly, for non-local cases, Equations (9.9) and (9.11) imply the following:

$$\lambda \text{ is an integer multiple of } 1/6 \text{ between } -1 \text{ and } 3. \tag{9.16}$$

Table 9.1 gives some examples of collision kernels proposed in the research literature for small particles with diameters roughly between 10^{-10} and 10^{-4}m. To obtain the simplest possible expressions, $K(x, y)$ is divided by an application-specific constant K_0. Notice that, in all cases, λ is an integer multiple of 1/3 or 1/6 between -1 and 3. Table 9.1 is based partly

Table 9.1 Examples of homogeneous collision kernels for small particles. In these expressions, x and y refer to particle masses and K_0 refers to an application-specific constant.

λ	$K(x,y)/K_0$	Application	References
$-2/3$	$x^{-2/3}+y^{-2/3}$	Diffusion-controlled growth of supported metal crystallites	da Costa (1998, 2015)
$-1/3$	$(x^{1/3}+y^{1/3})\times(x^{-2/3}+y^{-2/3})$	Brownian coagulation between continuum and free molecular	Williams & Loyalka (1991), eq. (4.46)
$-1/6$	$(x^{1/3}+y^{1/3})\times(xy)^{1/2}(x+y)^{-3/2}$	Gas kinetic theory	Schumann (1940)
0	1	Condensation polymerization, linear chains	Dostal & Raff (1936); Ziff (1980)
0	$(x^{1/3}+y^{1/3})^2\times x^{-1/3}y^{-1/3}$	Brownian coagulation in the continuum regime ("diffusion")	Smoluchowski (1916, 1917)
0	$(x^{1/3}+y^{1/3})\times(x^{-1/3}+y^{-1/3})$	Brownian coagulation in the continuum regime ("diffusion")	Smoluchowski (1917)
1/6	$(x^{1/3}+y^{1/3})^2\times(x^{-1}+y^{-1})^{1/2}$	Brownian coagulation in the free molecular regime ("diffusion")	Fuchs (1964); Brock & Hidy (1965)
1/3	$x^{1/3}+y^{1/3}$	Coagulation due to Brownian motion and gravitational settling	Williams & Loyalka (1991), eq. (4.325)
2/3	$x^{2/3}+y^{2/3}$	Sintering controlled catalyst aging	Ruckenstein & Pulvermacher (1973)
2/3	$(x^{1/3}+y^{1/3})^2$	Gas kinetic theory	Fuchs (1964); Abrahamson (1975)
5/6	$\|x^{1/6}-y^{1/6}\|\times(x^{1/3}+y^{1/3})^2$	Collection by gravitational settling for particle radii greater than 100 µ	Camp & Stein (1943); Golovin (1963)
1	$x+y$	Colliding water droplets	Golovin (1963)
1	$x+y$	Branched polymers with AB-bonds	Ziff (1980)
1	$\|x^{1/3}-y^{1/3}\|\times(x^{1/3}+y^{1/3})^2$	Coagulation by gravitational settling for particle radii of 50–100 µ	Berry (1967)
1	$(x^{1/3}+y^{1/3})^3$	Coagulation in linear shear flow	Smoluchowski (1917)
1	$(x^{1/3}+y^{1/3})^3$	Coagulation due to diffusion in isotropic fluid turbulence	Saffman & Turner (1956)
1	$(x^{1/3}+y^{1/3})^3$	Coagulation due to pure strain flow	Zeichner & Schowalter (1977)
4/3	$\|x^{2/3}-y^{2/3}\|\times(x^{1/3}+y^{1/3})^2$	Coagulation due to inertia in isotropic fluid turbulence	Saffman & Turner (1956); Levich (1962)
4/3	$\|x^{2/3}-y^{2/3}\|\times(x^{1/3}+y^{1/3})^2$	Coagulation by gravitational settling for particle radii less than 50 µ	Findeisen (1939); Enukashvili (1964)
2	x^2+y^2	*(same as the above)*	Long (1974)
2	xy	Condensation polymerization, branched chains	Stockmayer (1943); Ziff (1980)

on previous surveys by Drake (1972), White (1982), Pearson et al. (1984), Ernst (1986), Williams & Loyalka (1991), Smit et al. (1994), da Costa (1998), Camacho (2001), da Costa (2015) and Jeldres et al. (2018). Similarly, Table 9.2 gives some examples of collision kernels proposed in the research literature for large particles with diameters roughly between 10^6 and 10^{11}m. As before, in all cases, λ is an even multiple of 1/3 or 1/6 between −1 and 3.

As a typical example, Friedlander (2000) gives a complex two-page proof showing that Brownian diffusion of small spherical aerosol particles suspended in air leads to:

$$K(x,y) = K_0\left(\frac{1}{x^{1/3}} + \frac{1}{y^{1/3}}\right)(x^{1/3} + y^{1/3})$$

where $K_0 = 2k_B T/3\mu$, k_B is the Boltzmann constant, T is the temperature of the ambient air, and μ is the dynamic viscosity of the ambient air. While typical proofs are application-specific, with a focus on the most-probable velocity vectors, the resulting homogeneous collision kernels may apply to a broad range of different problems, except for the application-specific constant K_0. For example, according to Table 9.1, the following homogenous collision kernel:

$$K(x,y) = K_0\left(x^{1/3} + y^{1/3}\right)^3$$

applies to three different problems. In this sense, homogeneous kernels may be *universal*. By contrast, non-homogeneous kernels apply only to certain specific problems; see Example 9.2.

Table 9.2 Examples of homogeneous collision kernels for large particles. In these expressions, x and y refer to particle masses and K_0 refers to an application-specific constant.

λ	$K(x,y)/K_0$	Application	Reference
0	1	Protoplanet formation	Trubnikov (1971); Tanaka & Nakazawa (1994)
1/3	$x^{1/3} + y^{1/3}$	Planetesimal formation	Zvyagina & Safronov (1972)
2/3	$(x^{1/3} + y^{1/3})^2$	Bodies that experience little or no gravity	Safronov (1962, 1969, p. 92)
2/3	$x^{2/3} + y^{2/3}$	Mathematical model of protoplanet formation	Zvyagina & Safronov (1972)
1	$x + y$	Bodies that experience moderate gravity	Safronov (1962, 1969, p. 92) Trubnikov (1971)
4/3	$x^{4/3} + y^{4/3}$	Protoplanet formation	Zvyagina & Safronov (1972)
4/3	$(x^{2/3} + y^{2/3})^2$	Protoplanet formation	Zvyagina & Safronov (1972)
4/3	$(x^{1/3} + y^{1/3})(x + y)$	Bodies that experience significant gravity	Safronov (1962, 1969, p. 92)
2	xy	Protoplanet formation	Trubnikov (1971); Tanaka & Nakazawa (1994)
7/3	$\min(x,y)\times (x^{1/3} + y^{1/3})(x + y)$	"Planetesimal growth in the early runaway stage"	Makino et al. (1998); Inaba et al. (1999)

Example 9.2 Non-homogeneous collision kernels for bubble aggregation

Consider bubble coalescence in mildly agitated gas-liquid dispersions where ε is the turbulent dissipation rate and t_c is a characteristic bubble coalescence time. Lee et al. (1987) suggested the following non-homogeneous collision kernel:

$$K(x,y) = K_0\varepsilon^{1/3}(x+y)^2\left(x^{2/3}+y^{2/3}\right)^{1/2}\exp\left[K_1\varepsilon^{1/3}t_c(x+y)^{-2/3}\right]$$

where x and y are bubble diameters and K_0 and K_1 are constants. This expression only applies under a very specific set of circumstances, namely, cases where liquid turbulence causes "the approach of two bubbles to within a distance of 10–100 μm," followed by gradual "thinning of the liquid layer between the bubbles to 0.01–0.1 μm," followed by "rupture of the thin liquid layer via an instability mechanism." For additional examples of non-homogeneous collision kernels for droplet and bubble aggregation see, e.g., Liao & Lucas (2009, 2010).

As Example 9.2 shows, even when when the kernel itself is not universal, the exponents that appear in the kernel—which generally have geometric values such as 1/3—may be universals. As noted earlier, this is especially true for the degree of homogeneity λ which is, in essence, the grand total of the exponents appearing in a given kernel.

9.4 Initial Conditions

In general, solutions of mean-field governing equations for aggregation depend on the initial and boundary conditions. This section concerns initial conditions while the next section concerns boundary conditions. For aggregation, the most frequently studied initial conditions are the *monodisperse* and *exponential* size distributions. In self-similar form, monodisperse initial conditions may be defined as follows:

$$c(M,0) = \frac{C_0}{M_0}\delta\left(\frac{M}{M_0}-1\right) \tag{9.17}$$

where δ is the Dirac delta, M_0 is the initial mass of each particle, and $C_0 = \int_0^\infty c(M,0)dM$ is the initial total particle concentration. In self-similar form, exponential initial conditions may be defined as follows:

$$c(M,0) = \frac{C_0}{M_0}\exp\left(-\frac{M}{M_0}\right) \tag{9.18}$$

where $M_0 = \int_0^\infty Mc(M,0)dM/C_0$ is the initial average particle mass and $C_0 = \int_0^\infty c(M,0)dM$ is the initial total particle concentration.

If the initial conditions are monodisperse, the particle sizes will be discrete for all times, i.e., all particles masses will be integer multiples of M_0. Thus, most treatments of aggregation with monodisperse initial conditions solve a discrete form of the governing equations such as Equation (9.1) or (9.2). If the initial conditions are exponential, the particle sizes will be continuous for all times. Thus, all treatments of aggregation with continuous exponential initial conditions must solve a continuous form of the governing equations such as Equation (9.3), (9.4), or (9.5).

Notice that monodisperse initial conditions have, by any metric, the smallest possible size spread. Conversely, exponential initial conditions have, by some metrics, the largest possible size spread. As first observed by Schumann (1940), for constant collision kernels, these two bounding cases—monodisperse and exponential initial conditions—typically lead to the same late-time solution. More specifically, he says "there should be a tendency for the same distribution to be approached asymptotically after a sufficiently long time, no matter what the initial distribution."

As a general rule, deterministic (non-random, non-chaotic) systems tend to preserve monodisperse initial conditions. In other words, if a deterministic system starts with particles that are all the same size, it tends to produce particles that are all the same size regardless of time, assuming there are no sources of new particles.[1] In addition, if a deterministic system starts with particles that have different sizes, it tends to produce particles that are increasingly similar in size as time progresses. Thus, deterministic systems tend to remember their initial conditions and progressively narrow their size distributions over time. By contrast, random and chaotic systems tend to preserve or even, by some metrics, broaden exponential initial conditions. In other words, if a chaotic system starts with particles that have different sizes, it tends to evolve particles that have increasingly different sizes as time progresses. Thus, random and chaotic systems tend to forget their initial conditions and progressively widen their size distributions over time.

9.5 Boundary Conditions

Unlike many problems, which involve boundaries in *space*, aggregation involves boundaries in *size*, e.g., maximum and minimum particle sizes. These boundary conditions involve sources and sinks, which may or may not take the form of additional terms in the governing equations. As noted earlier in Section 1.1, sources transfer mass, momentum, and energy from the surroundings to the particle system and, conversely, sinks transfer mass, momentum, and energy from the particle system to the surroundings. Examples of boundary conditions involving sources and sinks for aggregating particle systems include the following:

- *Free-field boundary conditions.* In this case, there are no sources or sinks. In other words, this case allows an infinite range of particle sizes.
- *Single-source boundary conditions.* In this case, a source replenishes the small-size ranges depleted by aggregation.
- *Single-sink boundary conditions.* In this case, a sink depletes the large-size ranges replenished by aggregation.
- *Periodic boundary conditions.* In this case, there is a source of small particles and an equal-but-opposite sink of large particles; see Example 9.3.

[1] For example, this occurs in when airborne droplets grow through regular condensation; see Example 11.1.

Example 9.3 Example of periodic boundary conditions for aggregation of interstellar clouds

In a classic paper on star formation from interstellar clouds, Field & Saslaw (1965) say:

> [Each interstellar] cloud starts its life cycle ... as a small region of high density compressed by surrounding hot gas ... It undergoes inelastic collisions with other clouds in the vicinity, engendering new clouds of large mass through coalescence. Through repeated coalescence processes, the mass of the cloud increases to the point that self-gravitation becomes significant and ultimately dominant: At this point the cloud becomes gravitationally unstable, undergoes rapid collapse, and become subject to gravitational fragmentation. The final result is a group of young stars, the brightest of which ionize gas remaining in the cloud ... and thereby regenerate small clouds as before. These small clouds are then available for a new cycle of activity.

One way to model this system is to include both aggregation and fragmentation terms in the mean-field governing equations; see Section 5.9. Another way is to include only aggregation terms in the mean-field governing equations and to use periodic boundary conditions to account for fragmentation. More specifically, in Equation (9.2), Field & Saslaw (1965) used a hard sink that removes existing large clouds such that $c_i(t) = 0$ when $i > i_{\max}$. In addition, in Equation (9.2), Field & Saslaw (1965) used a hard source that introduces new small clouds where $a_i(t) = a(t)\delta_{i1}$. In constructing these boundary conditions, Field & Saslaw (1965) chose $a(t)$ so that the mass lost by removing large existing clouds exactly equaled the mass gained by adding new small clouds. Figure 9.1 illustrates the effects of this source-sink pair. As their main result, Field & Saslaw (1965) found stationary power law size distributions with $n/m = -1/2$. While this is a somewhat exotic example, the same principle applies in general—aggregation with periodic boundary conditions may be equivalent to aggregation-fragmentation with free-field boundary conditions—provided that only the largest particles experience fragmentation.

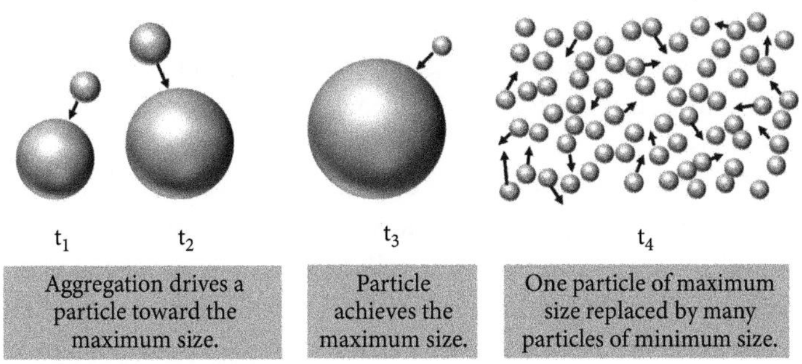

Figure 9.1 Example of periodic boundary conditions for aggregation. In this figure, a particle grows over time by aggregating with other particles until it reaches a maximum size. An instant later, the single particle of the maximum size is replaced by many particles of the minimum size. In this figure, $t_1 < t_2 < t_3 < t_4$ refer to time.

With free-field boundary conditions, aggregation always results in progressively larger particle sizes; such systems are never stationary. On the other hand, with periodic boundary conditions, the particle sizes tend to remain within a finite range; such systems may become stationary (recall that this term was defined earlier in Section 2.4). In fact, as seen as in Example 9.4, periodic boundary conditions can be modeled without introducing an explicit source-sink pair simply by assuming the existence of a stationary solution.

Example 9.4 Example of periodic boundary conditions for aggregation of atmospheric aerosols

Brock (1972) suggested that aerosols in the continental troposphere grow through "binary accretion" governed by Equation (9.3). In solving Equation (9.3), Brock (1972) assumed free merging, i.e., that all aggregating collisions involve particles with masses x and y where $x \gg y$ or $y \ll x$. To obtain stationary solutions, Brock (1972) assumed that new small "particles are continually supplied by condensational growth from a reservoir of particles which is itself supplied by nucleation, coagulation, and condensational growth." At the same time, an equal mass of large "particles are also lost ... by processes such as mixing and dispersion, sedimentation, impaction, and other mechanisms." Unlike the previous example, Brock did not introduce periodic boundary conditions via explicit terms in the governing equations. Rather, he introduced them implicitly by simply assuming the existence of a stationary state. As the main conclusion, Brock (1972) found that the particle addition and "removal processes cause any arbitrary initial distribution to be 'forgotten' and the aerosol to assume under certain conditions a 'characteristic' ... long-time distribution." More specifically, Brock (1972) found stationary Weibull size distributions with $n/m = 1/3$ or $2/3$; see also Example 11.4.

In general, because the original material in a particle system depends on the initial conditions, while the added material does not, sources and sinks help particle systems "forget" their initial conditions.[2] This mimics a key behavior of chaotic and stochastic systems. By constantly replenishing the small-size ranges depleted by aggregation, sources help particle systems obtain a broader range of particle sizes. This mimics another key behavior of chaotic and stochastic systems, as discussed in the last section. By contrast, by constantly depleting the large-size ranges replenished by aggregation, sinks cause particle systems to obtain a narrower range of particle sizes which, in the extreme, sometimes leads to physically unrealistic mean-field solutions.

As noted earlier in Section 9.3, locality occurs when aggregation preferentially involves particles of similar sizes. Because sinks reduce particle size spreads, especially among the largest particles that are the most likely to collide, sinks are associated with locality. Conversely, non-locality (free merging) occurs when aggregation preferentially involves particles of different sizes. Because sources increase particle size spreads, sources are associated with free merging.

[2] In this context, the term "original material" refers to particles that were initially present in the system, while the term "added material" refers to particles introduced to the system at some later time by a source.

9.6 Conservation of Mass

In deterministic systems, the governing equations are typically expressions of conservation of mass, momentum, and energy. In such cases, exact or approximate solutions must rigorously ensure conservation of mass, momentum, and energy, see, e.g., Laney (1998). By contrast, in random or chaotic systems, the governing mean-field equations are not necessarily expressions of conservation of mass, momentum, or energy. In fact, most exact or approximate mean-field solutions violate conservation of mass due to particle sources and sinks. Even when sources and sinks are not deliberately introduced, mean-field aggregation as governed by Equations (9.1) or (9.3) may spontaneously develop a large-particle sink after some finite time if $\lambda > 1$. From the point of view of the universe that includes both the particle systems and its surroundings, sources and sinks obey conservation of mass, momentum, and energy. However, from the point of view of the particle system alone—which is the point of view assumed by the mean-field governing equations—sources and sinks violate conservation of mass, momentum, and energy.

Having said this, there are useful relations for particle systems that obey conservation of mass. Consider finite particle systems, i.e., systems in which $0 < N(t) < \infty$ and $0 < \mathcal{V}^o < \infty$ where $N(t)$ is the total number of particles in the system and $\mathcal{V}^o = \text{const.}$ is the total volume of the system including the space between particles. If conservation of mass applies then:

$$N(t)M_{\text{avg}}(t) = M_{\text{total}}(t) = \text{const.} \tag{9.19}$$

where $N(t)$ is the total number of particles, $M_{\text{avg}}(t)$ is the average mass of particles as defined in Table 3.2, and $M_{\text{total}}(t)$ is the total mass of particles in the system. Similarly:

$$C(t)M_{\text{avg}}(t) = \rho_{\text{total}}(t) = \text{const.} \tag{9.20}$$

where $C(t) = N(t)/\mathcal{V}^o$ is the total concentration and $\rho_{\text{total}}(t)$ is the total density of particles in the system. For infinite particle systems, i.e., systems where $N(t) = \infty$ and $\mathcal{V}^o = \infty$ such that $0 < C(t) < \infty$, Equation (9.19) no longer applies but Equation (9.20) still applies.

9.7 Uniqueness

As already noted, mean-field aggregation may or may not obtain conservation of mass. If it obtains conservation of mass, mean-field theory tends to produce unique solutions; see, e.g., Dubovskii (1994) and Leyvraz (2003). However, if it does not obtain conservation of mass, mean-field theory may produce infinitely many different solutions, each corresponding to a specific set of sources and sinks. For example, Menon & Pego (2004) show that Equation (9.3) with $K(x, y) = K_0$, $K(x, y) = K_0(x + y)$, or $K(x, y) = K_0xy$ allows an infinite one-parameter set of analytic solutions, each in the form of an infinite series. For these kernels, all but one of the multiple solutions violates conservation of mass.

9.8 Conclusions

As discussed in Chapter 4, computational models can treat particle positions, velocities, rotations, shapes, and sizes. In many applications, such as diffusion-limited aggregation, this leads to extremely expensive calculations that push the limits of today's fastest supercomputers. Furthermore, without the explicit introduction of a sufficient amount of chaos or randomness, first-principle computational models may still produce unphysical results.

As a common alternative to computational methods, mean-field theory assumes that the average particle positions, velocities, and shapes lead to an average size-dependent aggregation rate. In principle, this average aggregation rate can be predicted in advance; however, in practice, this can be difficult. As Pumir & Wilkinson (2016) say "in principle, the calculation of the collision rate appears to be a straightforward exercise in elementary kinetic theory, but in practice it is a surprisingly complex problem." Assuming the size-dependent aggregation rate is known, the single remaining unknown—namely, the particle size distribution—can be found by solving a relatively simple mean-field governing equation. Physically relevant collision rates usually lead to physically relevant particle size distributions, without the need to explicitly introduce chaos or randomness. In summary, this chapter has shown that the three key ingredients of mean-field aggregation theory are as follows:

(a) the collision kernel, i.e., the average particle collision rate;
(b) the initial conditions; and
(c) the boundary conditions, i.e., sources of small particles and sinks of large particles.

Regarding collision kernels, the most important property is their degree of homogeneity, e.g., the power-law exponent observed in extreme cases such as free merging and forced locality. Regarding initial conditions, their influence tends to fade over time due to chaos, as discussed earlier in Section 2.1.

References

J. Abrahamson, Collision Rates of Small Particles in a Vigorously Turbulent Fluid, *Chemical Engineering Science*, 30(11), 1371–1379, 30, 1975, https://doi.org/10.1016/0009-2509(75)85067-6

E. M. Abulwafa, M. A. Abdou, & A. A. Mahmoud, The Solution of Nonlinear Coagulation Problem with Mass Loss, *Chaos, Solitons, & Fractals*, 29(2), 313–330, 2006, https://doi.org/10.1016/j.chaos.2005.08.044

E. X. Berry, Cloud Drop Growth by Collection, *Journal of Atmospheric Sciences*, 24(6), 688–701, 1967, https://doi.org/10.1175/1520-0469(1967)024<0688:CDGBC>2.0.CO;2

J. R. Brock & G. M. Hidy, Collision-Rate Theory and the Coagulation of Free-Molecule Aerosols, *Journal of Applied Physics*, 36(6), 1857–1862, 1965, https://doi.org/10.1063/1.1714366

J. R. Brock, Condensational Growth of Atmospheric Aerosols, *Journal of Colloidal and Interface Sciences*, 39(1), 32–36, 1972, https://doi.org/10.1016/0021-9797(72)90139-7

A. Burrows, J. Holman, A. Parsons, G. Pilling, & G. Price, *Chemistry*^{3}*: Introducing Inorganic, Organic and Physical Chemistry*, 4th edition, Oxford University Press, 2021 [see section 10.5]

J. Camacho, Scaling in Steady-State Aggregation with Injection, *Physical Review E*, 63(4), 046112, 2001, https://doi.org/10.1103/PhysRevE.63.046112

T. R. Camp & P. C. Stein, Velocity Gradients and Internal Work in Fluid Motion. *Journal of Boston Society of Civil Engineering*, 30(4), 219–237, 1943

B. Chakraborty, *Principles of Plasma Mechanics*, 2nd edition, Wiley, 1991 [see also *Principles of Plasma Mechanics*, 4th edition, New Age International Publishers, 2003]

F. P. da Costa, A Finite-Dimensional Dynamical Model for Gelation in Coagulation Processes, *Journal of Nonlinear Science*, 8(6), 619–653, 1998, https://doi.org/10.1007/s003329900061

F. P. da Costa, Mathematical Aspects of Coagulation-Fragmentation Equations, in J. P. Bourguignon, R. Jeltsch, A. Pinto, & M. Viana, editors, *Mathematics of Energy and Climate Change, CIM Series in Mathematical Sciences, Volume2*, Springer, 2015, https://doi.org/10.1007/978-3-319-16121-1_5

H. W. de Regt, *Understanding Scientific Understanding*, Oxford University Press, 2017 [see section 6.2], https://doi.org/10.1093/oso/9780190652913.001.0001

H. Dostal & R. Raff, Über die Kinetik einer thermischen Polykondensation, *Zeitschrift für Physikalische Chemie*, 32B(1), 117, 1936, https://doi.org/10.1515/zpch-1936-3211

R. L. Drake, A General Mathematical Survey of the Coagulation Equation, in G. M. Hidy & J. R. Brock, editors, *Topics in Current Aerosol Research, Part 2: International Reviews in Aerosol Physics and Chemistry, Volume3*, 201–376, Pergamon Press, 1972, https://doi.org/10.1016/B978-0-08-016809-8.50012-7

P. B. Dubovskii, *Mathematical Theory of Coagulation*, Seoul National University, Research Institute of Mathematics, Global Analysis Research Center, 1994

I. M. Enukashvili, On the Solution of the Kinetic Coagulation Equation, *Bulletin of the Academy of Sciences of the USSR, Geophysics Series*, 10, 944–948, 1964

M. H. Ernst, Kinetics of Clustering in Irreversible Aggregation, in L. Pietronero & E. Tosatti, editors, *Fractals in Physics*, North-Holland, 1986, https://doi.org/10.1016/B978-0-444-86995-1.50057-3

G. B. Field & W. C. Saslaw, A Statistical Model of the Formation of Stars and Interstellar Clouds, *The Astrophysical Journal*, 142(2), 568–583, 1965, https://doi.org/10.1086/148318

W. Findeisen, Zur Frage der Regentropfenbildung in reinen Wasserwolken, *Meteorologische Zeitschrift*, 56, 365–368, 1939

S. K. Friedlander, *Smoke, Dust and Haze*, 2nd edition, Oxford University Press, 2000

N. A. Fuchs, *The Mechanics of Aerosols*, Pergamon Press, 1964

D. C. Giancoli, *Physics for Scientists and Engineers*, 4th edition, Pearson Prentice Hall, 2014 [see section 19.8]

A. M. Golovin, The Solution of the Coagulation Equation for Cloud Droplets in a Rising Air Current, *Bulletin of the Academy of Sciences of the USSR, Geophysics Series, 5*, 482–487, 1963

P. Horvai, S. V. Nazarenko & T. H. M. Stein, Coalescence of Particles by Differential Sedimentation, *Journal of Statistical Physics*, 130(6), 1177–1195, 2008, https://doi.org/10.1007/s10955-007-9466-y

M. J. Hounslow, A Discretized Population Balance for Continuous Systems at Steady State, *AIChE Journal*, 36(1), 106–116, 1990, https://doi.org/10.1002/aic.690360113

S. Inaba, H. Tanaka, K. Ohtsuki, & K. Nakazawa, High-Accuracy Statistical Simulation of Planetary Accretion: I. Test of the Accuracy by Comparison with the Solution to the Stochastic Equation, *Earth, Planets and Space*, 51, 205–217, 1999, https://doi.org/10.1186/BF03352224

R. I. Jeldres, P. D. Fawell, & B. J. Florio, Population Balance Modelling to Describe the Particle Aggregation Process: A Review, *Powder Technology*, 326, 190–207, 2018, https://doi.org/10.1016/j.powtec.2017.12.033

J. P. Killingbeck and G. H. A. Cole, *Mathematical Techniques and Physical Applications*, Academic Press, 1971, https://doi.org/10.1016/B978-0-12-406850-6.50014-2

C. B. Laney, *Computational Gasdynamics*, Cambridge University Press, 1998, https://doi.org/10.1017/CBO9780511605604

C.-H. Lee, Bubble Breakup and Coalescence in Turbulent Gas-Liquid Dispersions, *Chemical Engineering Communications*, 59, 65–84, 1987, https://doi.org/10.1080/00986448708911986

B. G. Levich, *Physicochemical Hydrodynamics*, Prentice Hall, 1962

F. Leyvraz, Scaling Theory and Exactly Solved Models in the Kinetics of Irreversible Aggregation, *Physics Reports*, 383(2–3), 95–212, 2003, https://doi.org/10.1016/S0370-1573(03)00241-2

Y. Liao & D. Lucas, A Literature Review of Theoretical Models for Drop and Bubble Breakup in Turbulent Dispersions, *Chemical Engineering Science*, 64(15), 3389–3406, 2009, https://doi.org/10.1016/j.ces.2009.04.026

Y. Liao & D. Lucas, A Literature Review on Mechanisms and Models for the Coalescence Process of Fluid Particles, *Chemical Engineering Science*, 65(10), 2851–2864, 2010, https://doi.org/10.1016/j.ces.2010.02.020

A. B. Long, Solutions to the Droplet Collection Equation for Polynomial Kernels, *Journal of the Atmospheric Sciences*, 31(4), 1040–1052, 1974, https://doi.org/10.1175/1520-0469(1974)031%3C1040:STTDCE%3E2.0.CO;2

J. Makino, T. Fukushige, Y. Funato, & E. Kokubo, On the Mass Distribution of Planetesimals in the Early Runaway Stage, *New Astronomy*, 3(7), 411–417, 1998, https://doi.org/10.1016/S1384-1076(98)00021-9

J. C. Maxwell, On Reciprocal Figures, Frames, and Diagrams of Forces: (a.) *Transactions of the Royal Society of Edinburgh*, 26(1), 1–40, 1872, https://doi.org/10.1017/S0080456800026351; (b.) W. D. Niven, editor, *The Scientific Papers of James Clerk Maxwell*, Volume 2, Cambridge University Press, 1890; Dover Books, 1965; https://doi.org/10.1017/CBO9780511710377.014

D. McQuarrie, *Quantum Chemistry*, 2nd edition, University Science Book, 2008 [see section 5.11]

G. Menon & R. L. Pego, Approach to Self-Similarity in Smoluchowski's Coagulation Equations, *Communications on Pure and Applied Mathematics*, 57(9), 1197–1232, 2004, https://doi.org/10.1002/cpa.3048

E. Michaelides, C. T. Crowe, & J. D. Schwarzkopf, editors, *Multiphase Flow Handbook*, 2nd edition, CRC Press, 2017 [see section 20.3], https://doi.org/10.1201/9781315371924

H. J. Pearson, I. A. Valioulis, & E. J. List, Monte-Carlo Simulation of Coagulation in Discrete Particle-Size Distributions. Part 1. Brownian Motion and Fluid Shearing, *Journal of Fluid Mechanics*, 143, 367–385, 1984, https://doi.org/10.1017/S0022112084001397

A. Pumir & M. Wilkinson, Collisional Aggregation Due to Turbulence, *Annual Review of Condensed Matter Physics*, 7, 141–170, 2016, https://doi.org/10.1146/annurev-conmatphys-031115-011538

T. E. Ramabhadran, T. W. Peterson, & J. H. Seinfeld, Dynamics of Aerosol Coagulation and Condensation, *AIChE Journal*, 22(5), 840–851, 1976, https://doi.org/10.1002/aic.690220505

D. Ramkrishna, *Population Balances: Theory and Applications to Particulate Systems in Engineering*, Elsevier, 2000, https://doi.org/10.1016/B978-0-12-576970-9.X5000-0

E. Ruckenstein & B. Pulvermacher, Kinetics of Crystallite Sintering During Heat Treatment of Supported Metal Catalysts, *AIChE Journal*, 19(2), 356–364, 1973, https://doi.org/10.1002/aic.690190222

P. G. Saffman & J. S. Turner, On the Collision of Drops in Turbulent Clouds, *Journal of Fluid Mechanics*, 1(1), 16–30, 1956, https://doi.org/10.1017/S0022112056000020

V. S. Safronov, Chastnyi Sluchai Resheniya Uravnenii Koagulyatsii [A Particular Solution of the Coagulation Equations], *Doklady Akademii Nauk SSSR*, 147(1), 64, 1962; see also V. S. Safronov, A Restricted Solution of the Accretion Equation, *Soviet Physics Doklady*, 7(11), 967–969, 1963

V. S. Safronov, *Evolyutsiya Doplanetnogo Oblaka i Obrazovanie Zemli i Planet (Evolution of the Protoplanetary Cloud and Formation of the Earth and its Planets)*, Akademii Nauk SSSR, 1969; see also NASA TT F-677, 1972, https://archive.org/details/nasa_techdoc_19720019068

T. E. W. Schumann, Theoretical Aspects of the Size Distribution of Fog Particles, *Quarterly Journal of the Royal Meteorological Society*, 66(285), 195–208, 1940, https://doi.org/10.1002/qj.49706628508

F. Schwabl, *Statistical Mechanics.* 2nd edition, Springer-Verlag, 2006, https://doi.org/10.1007/3-540-36217-7

P. Singh & G. J. Rodgers, Coagulation Processes with Mass Loss, *Journal of Physics A: Mathematical and General*, 29(2), 437–450, 1996, https://doi.org/10.1088/0305-4470/29/2/021

M. V. Smoluchowski, Drei Vorträge über Diffusion, Brownsche Molekularbewegung und Koagulation von Kolloidteilche [Three Lectures on Diffusion, Brownian Motion and Coagulation of Colloid Particles], *Physikalische Zeitschrift*, 17, 557–571, 587–599, 1916, https://eudml.org/doc/215805

M. V. Smoluchowski, Versuch einer mathematischen Theorie der Koagulationskinetik kolloider Lösunger [Mathematical Theory of the Kinetics of the Coagulation of Colloidal Solutions], *Zeitschrift für Physikalische Chemie*, 92, 129–168, 1917, https://doi.org/10.1515/zpch-1918-9209

D. J. Smit, M. J. Hounslow, & W. R. Paterson, Aggregation and Gelation – I. Analytical Solutions for CST and Batch Operation, *Chemical Engineering Science*, 49(7), 1025–1035, 1994, https://doi.org/10.1016/0009-2509(94)80009-X

W. H. Stockmayer, Theory of Molecular Size Distribution and Gel Formation in Branched-Chain Polymers, *The Journal of Chemical Physics*, 11(2), 45–55, 1943, https://doi.org/10.1063/1.1723803

K. Stowe, *An Introduction to Thermodynamics and Statistical Mechanics*, 2nd edition, Cambridge University Press, 2007, https://doi.org/10.1017/CBO9780511801570

H. Tanaka & K. Nakazawa, Validity of the Statistical Coagulation Equation and Runaway Growth of Protoplanets, *Icarus*, 107(2), 404–412, 1994, https://doi.org/10.1006/icar.1994.1032

B. A. Trubnikov, Solution of the Coagulation Equations in the Case of a Bilinear Coefficient of Adhesion of Particles, *Soviet Physics Doklady*, 16(2), 124–125, 1971

H. W. Watson, *A Treatise on the Kinetic Theory of Gases*, Oxford University Press, 1876

W. H. White, On the Form of Steady-State Solutions to the Coagulation Equations, *Journal of Colloid and Interface Science*, 87(1), 204–208, 1982, https://doi.org/10.1016/0021-9797(82)90382-4

M. M. R. Williams & S. K. Loyalka, *Aerosol Science: Theory and Practice, with Special Applications to the Nuclear Industry*, Pergamon, 1991

G. R. Zeichner & W. R. Schowalter, Use of Trajectory Analysis to Study Stability of Colloidal Dispersions in Flow Fields, *AIChE Journal*, 23(3), 243–254, 1977, https://doi.org/10.1002/aic.690230306

R. M. Ziff, Kinetics of Polymerization, *Journal of Statistical Physics*, 23(2), 241–263, 1980, https://doi.org/10.1007/BF01012594

E. V. Zvyagina & V. S. Safronov, Mass Distribution of Protoplanetary Bodies, *Soviet Astronomy*, 15(5), 810–817,1972, http://adsabs.harvard.edu/abs/1972SvA...15...810Z

10

Aggregation Cascades Part 2: Full Mean-Field Solutions

This chapter concerns non-stationary mean-field solutions for aggregation. In particular, this chapter describes discrete particle systems governed by Equations (9.1) and (9.2) where:

$K_{ij} = K_0$	Constant kernel	$\lambda = 0$
$K_{ij} = K_0\,(i + j)$	Sum kernel	$\lambda = 1$
$K_{ij} = K_0\,ij$	Product kernel	$\lambda = 2$

As it turns out, these three kernels—and only these three kernels—allow for closed-form time-dependent solutions of Equations (9.1) and (9.2). In most cases, these solutions assume free-field boundary conditions and *monodisperse* initial conditions, i.e., the particles all start off with the same size. In addition, this chapter describes continuous particle systems governed by Equations (9.3), (9.4), and (9.5) where:

$K(x, y) = K_0$	Constant Kernel	$\lambda = 0$
$K(x, y) = K_0\,(x + y)$	Sum Kernel	$\lambda = 1$
$K(x, y) = K_0 xy$	Product Kernel	$\lambda = 2$

As it turns out, these three kernels—and only these three kernels—allow for closed-form time-dependent solutions of Equations (9.3), (9.4), and (9.5). In most cases, these solutions assume free-field boundary conditions and gamma distribution initial conditions. At large enough times, continuous systems generally obtain about the same size distributions as discrete systems. However, while the particle sizes may be the same, the particle shapes are generally different. In particular, as discussed earlier in Section 1.3, discrete systems often obtain fractal *non-integer* values of m while continuous systems always obtain *integer* values of m.

This chapter assumes that aggregating systems consist of an infinite number of particles homogeneously and isotopically distributed in an infinite spatial volume with a finite concentration C. In other words, $0 < C = N/\mathcal{V} < \infty$ in the limit $N \to \infty$ and $\mathcal{V} \to \infty$. This is sometimes called the *thermodynamic limit*; see, e.g., Hendriks (1984a) and Lushnikov (2006). Because there are an infinite number of particles, aggregation never ends but continues forever. In addition, in most cases, this chapter assumes that there is no fixed maximum particle size. As a result, aggregation always produces progressively larger particles. Obviously, neither of these two assumptions is realistic—aggregation must end at some finite time and with some finite maximum particle size—but they may be approximately true in many cases.

This chapter focuses on basic phenomenology rather than on mathematical proofs. Readers interested in proofs may consult the original research papers cited below, review articles such as Aldous (1999) and Leyvraz (2003), and textbooks such as Bertoin (2006),

Particle Dynamics with Aggregation and Fragmentation. Culbert B. Laney, Oxford University Press. © Culbert B. Laney (2025).
DOI: 10.1093/9780198920687.003.0010

Krapivsky et al. (2010) and Banasiak et al. (2019a, b). Instead of providing proofs, this chapter includes most of the mathematical expressions required to verify that the solutions given here do, in fact, satisfy the mean-field governing equations for aggregation as claimed.

10.1 Moments

As discussed in Chapter 3, size distributions may be expressed in terms of volumetric concentration c or number fraction f. The *zero-th moments* of $c(M, t)$ and $f(M, t)$ are as follows:

$$\int_0^\infty c(M,t)\,dM = C(t) \tag{10.1}$$

$$\int_0^\infty f(M,t)\,dM = 1. \tag{10.2}$$

The *first moments* of $c(M, t)$ and $f(M, t)$ are as follows:

$$\int_0^\infty Mc(M,t)\,dM = C(t)M_{\text{avg}}(t) \tag{10.3}$$

$$\int_0^\infty Mf(M,t)\,dM = M_{\text{avg}}(t). \tag{10.4}$$

More generally, the *k-th moments* of $c(M, t)$ and $f(M, t)$ are as follows:

$$\int_0^\infty M^k c(M,t)\,dM \tag{10.5}$$

$$\int_0^\infty M^k f(M,t)\,dM. \tag{10.6}$$

For convenience, the following notation will be used throughout the rest of the chapter:

$$C_0 = C(0) = \int_0^\infty c(M,0)\,dM \tag{10.7}$$

$$M_0 = M_{\text{avg}}(0) = \int_0^\infty Mf(M,0)\,dM. \tag{10.8}$$

According to Equation (9.20), if conservation of mass applies then:

$$C(t)\,M_{avg}(t) = C_0 M_0 = \text{const.} \tag{10.9}$$

10.2 Common Size Distributions

The results found in this chapter often become gamma size distributions for sufficiently large times and particle sizes. In regular form, gamma size distributions can be written as follows:

$$c(M,t) = C(t) f(M,t) = C(t) A(t) M^{\frac{n}{m}-1} \exp[-b(t) M]. \tag{10.10}$$

In self-similar form, gamma size distributions can be written as follows:

$$c(M,t) = \frac{C(t)A_0}{M_{char}(t)} \left[\frac{M}{M_{char}(t)}\right]^{\frac{n}{m}-1} \exp\left[-\frac{b_0 M}{M_{char}(t)}\right] \tag{10.11}$$

where $M_{char}(t)$ is a characteristic particle size. For constant kernels, a possible solution is a gamma size distribution with $n/m = 1$:

$$c(M,t) \approx \frac{C_0 M_0}{M_{avg}^2(t)} \exp\left[-\frac{M}{M_{avg}(t)}\right] \tag{10.12}$$

for sufficiently large times and particle sizes where $C(t) = C_0 M_0 / M_{avg}(t)$ and:

$$M_{avg}(t)/M_0 \cong \frac{1}{2} K_0 C_0 t. \tag{10.13}$$

For sum kernels, a possible solution is a gamma size distribution with $n/m = -1/2$:

$$c(M,t) \approx \frac{C(t)A_0}{M_{char}(t)} \left[\frac{M}{M_{char}(t)}\right]^{-3/2} \exp\left[-\frac{b_0 M}{M_{char}(t)}\right] \tag{10.14}$$

for sufficiently large times and particle sizes. In this case, $M_{char}(t)$ is related to $M_{avg}(t)$ where:

$$M_{avg}(t)/M_0 \cong \exp(\text{const.}t). \tag{10.15}$$

For product kernels, a possible solution is a gamma size distribution with $n/m = -3/2$:

$$c(M,t) \approx \frac{C(t)A_0}{M_{char}(t)} \left[\frac{M}{M_{char}(t)}\right]^{-5/2} \exp\left[-\frac{b_0 M}{M_{char}(t)}\right] \tag{10.16}$$

for sufficiently large times and particle sizes. In this case, $M_{char}(t)$ is related to $M_{avg}(t)$ where:

$$M_{avg}(t)/M_0 \cong \left(1 - t/t_g\right)^{-1} \tag{10.17}$$

for $t < t_g$ where t_g is the *gel time*. As this chapter will show, while self-similar solutions are often observed in practice, the available analytic solutions for product and sum kernels are almost but not quite self-similar.

10.3 Gelation—Part 1

For free-field boundary conditions, if $\lambda < 1$, all of the particles tend to grow at roughly the same rate, i.e., no single particle is dramatically larger than any other particle. In this case, the particle system obeys conservation of mass regardless of time. On the other hand, if $\lambda = 1$, a relatively small number of particles may grow much faster than any of the other particles. These rare but extremely large particles occasionally merge to form even larger particles. However, the particle system still obeys conservation of mass regardless of time. Finally, if $\lambda > 1$, then a single infinite[1] particle tends to form at some finite time. Being infinitely large, this particle vastly exceeds the size of all other particles. Before the infinite particle forms, the system obeys conservation of mass. However, after the infinite particle forms, the system violates conservation of mass. In standard terminology, borrowed from the polymer science community, the formation of a single infinite particle is called *gelation*, the single infinite particle is called the *gel*, and the remaining finite particles are called the *sol*; see, e.g., Leyvraz & Tschudi (1981) and Lushnikov (2006). As another possible terminology, borrowed from the astrophysics community, the formation of a single infinite particle is called *runaway growth*; see, e.g., Alfvén & Arrhenius (1976) and Wetherill (1976). As a third possible terminology, borrowed from the cloud physics community, the single infinite particle is called a *lucky particle*; see, e.g., Kostinski & Shaw (2005) and Dziekan & Pawlowska (2017).

From the mean-field point of view, any sol particles that join the gel essentially disappear. Lushnikov (2006) says "a single giant object (a gel) invisible in the thermodynamic limit arises after a critical time and begins to consume the mass of the smaller but higher populated fraction (the sol)." In essence, the gel acts like a sink, which preferentially removes the largest sol particles from system. Of course, because there are an infinite number of sol particles, the gel can never completely absorb them; thus, the gel and sol continue to evolve together indefinitely. Once the system violates conservation of mass due to gelation, there is no longer a unique solution. Instead, there are multiple possible solutions, depending on how hard or soft the resulting sink is presumed to be. For example, for a product kernel, a hard sink leads to a power law size distribution while a soft sink leads to a gamma size distribution; see, e.g., Leyvraz & Lushnikov (2015).

In short, for free-field boundary conditions, the degree of homogeneity λ determines whether or not gelation occurs. More specifically:

$\lambda < 1$	Gelation tends not to not occur	e.g., $\lambda = 0$ for $K(x,y) = K_0$
$\lambda = 1$	Gelation tends to occur at $t_g = \infty$	e.g., $\lambda = 1$ for $K(x,y) = K_0(x+y)$
$\lambda > 1$	Gelation tends to occur at $0 \leq t_g < \infty$	e.g., $\lambda = 2$ for $K(x,y) = K_0 xy$

Leyvraz & Lushnikov (2015) argue that the gel time t_g depends on the initial conditions. More specifically, for the product kernel, they argue that "the gel time is inversely proportional to the initial second moment... If the first moment is finite, but not the second, then we deal with instantaneous gelation ... that is, mass is lost from the very start ... If, on the other hand, the second moment is finite but not the third, then gelation occurs at finite [non-zero] time."

[1] Infinite particle sizes occur in the thermodynamic limit. Of course, in reality, particle systems can contain only a finite number of finite-size particles. The differences between gelation in infinite particle systems and in more-realistic finite particle systems will be discussed later in Section 10.7.

10.4 Non-Stationary Closed-Form Solutions for Constant Kernels

This section gives four examples of exact closed-form solutions for constant kernels.

Example 10.1 Time-dependent closed-form solutions for constant kernels with monodisperse initial conditions and free-field boundary conditions

For $K_{ij} = K_0 = \text{const.}$, Smoluchowski (1916, 1917) obtained the following exact solution to Equation (9.1):

$$c_i(t) = C_0 \frac{\tau(t)^{i-1}}{(1+\tau(t))^{i+1}} \tag{10.18}$$

where:

$$\tau(t) = \frac{1}{2} K_0 C_0 t \tag{10.19}$$

is a scaled non-dimensional time. Notice that Equation (10.18) is a geometric series. For geometric series, $\sum_{i=0}^{\infty} r^i = 1/(1-r)$. Then $C(t) = \sum_{i=1}^{\infty} c_i(t) = C_0/(1+\tau(t))$ and:

$$f_i(t) = \frac{c_i(t)}{C(t)} = \frac{\tau(t)^{i-1}}{(1+\tau(t))^i} \tag{10.20}$$

and:

$$M_{\text{avg}}(t) = \sum_{i-1}^{\infty} M_i f_i = M_0(1+\tau(t)). \tag{10.21}$$

Notice that $\sum_{i=1}^{\infty} M_i c_i(t) = C(t) M_{\text{avg}}(t) = C_0 M_0 = \text{const.}$ In other words, conservation of mass applies. Also notice that:

$$c_i(t) \cong C_0 \delta_{i1} \tag{10.22}$$

for small enough times, i.e., the initial condition is monodisperse. Finally, notice that:

$$c_i(t) \cong \frac{C_0 M_0}{\text{M}^2_{\text{avg}}(t)} \exp\left[-\frac{M_0 i}{M_{\text{avg}}(t)}\right] \tag{10.23}$$

for large enough times and particle sizes, i.e., the size distribution is a self-similar exponential. In summary, this example obtains a hybrid solution. The hybrid is a Dirac delta for small times and particle sizes. This describes unreacted particles, i.e., particles that have never participated in aggregation. Conversely, the hybrid becomes an exponential size distribution for large times and particle sizes. This describes fully reacted particles, i.e., particles formed through multiple aggregation events that have "forgotten" their initial conditions.

Example 10.2 Time-dependent closed-form solutions for constant kernels with exponential initial conditions and free-field boundary conditions

For $K(x, y) = K_0 = \text{const.}$, Schumann (1940) obtained the following exact solution to Equation (9.3):

$$c(M,t) = \frac{C_0}{M_0(1+\tau(t))^2} \exp\left(-\frac{1}{1+\tau(t)}\frac{M}{M_0}\right) \tag{10.24}$$

where:

$$\tau(t) = \frac{1}{2}K_0 C_0 t \tag{10.25}$$

is a scaled non-dimensional time. Notice that $C(t) = \int_0^\infty c(M,t)\,dM = C_0/(1+\tau(t))$. Then:

$$f(M,t) = \frac{c(M,t)}{C(t)} = \frac{1}{M_0(1+\tau(t))} \exp\left(-\frac{1}{1+\tau(t)}\frac{M}{M_0}\right) \tag{10.26}$$

and:

$$M_{\text{avg}}(t) = \int_0^\infty Mf(M,t)\,dM = M_0(1+\tau(t)) \cong \frac{1}{2}K_0 C_0 M_0 t. \tag{10.27}$$

Notice that $\int_0^\infty Mc(M,t)\,dM = C(t)\,M_{\text{avg}}(t) = C_0 M_0 = \text{const.}$ In other words, conservation of mass applies. Also notice that Equations (10.24) and (10.26) can be rewritten as follows:

$$c(M,t) = \frac{C_0 M_0}{M_{\text{avg}}^2(t)} \exp\left[-\frac{M}{M_{\text{avg}}(t)}\right] \tag{10.28}$$

$$f(M,t) = \frac{1}{M_{\text{avg}}(t)} \exp\left[-\frac{M}{M_{\text{avg}}(t)}\right]. \tag{10.29}$$

In summary, the solution is always a self-similar exponential size distribution, regardless of time or particle size. As one way to understand this example, if the initial distribution obtains the broadest possible size spread, according to the metric associated with K, aggregation will preserve that initial distribution for all time. In this case, reacted and unreacted particles have the same size distribution, i.e., particles formed through zero, one, or multiple aggregation events all have an exponential size distribution. However, if the initial distribution obtains anything less than the broadest possible size spread, aggregation will progressively drive the distribution towards the broadest possible size spread, as seen in the previous example. As an important point, this example typically concerns nearly-spherical liquid droplets with $m = 3$, such as those found in fog. However, the last example typically concerns fractal particles with m between 1.4 and 2.6, such as those found in colloidal systems. Thus, even when the particle size distributions are the same, the particle shapes may be completely different.

Example 10.3 Time-dependent closed-form solutions for constant kernels with gamma initial conditions and free-field boundary conditions

For $K(x,y) = K_0 = \text{const.}$, Scott (1968) and Ramabhadran et al. (1976) obtained the following exact solution to Equation (9.3):

$$c(M,t) = \frac{2C_0}{M_0}\frac{[1-\tau(t)]^2}{\sqrt{\tau(t)}}\exp\left(-2\frac{M}{M_0}\right)\sinh\left(2\sqrt{\tau(t)}\frac{M}{M_0}\right) \tag{10.30}$$

where $\sinh(x) = [\exp(x) - \exp(-x)]/2$ is the hyperbolic sine function and where:

$$\tau(t) = \frac{\frac{1}{2}K_0C_0t}{1+\frac{1}{2}K_0C_0t}. \tag{10.31}$$

Notice that $C(t) = \int_0^\infty c(M,t)\,dM = C_0/(1+K_0C_0t/2)$. Then:

$$f(M,t) = \frac{c(M,t)}{C(t)} = \frac{2}{M_0}\frac{1-\tau(t)}{\sqrt{\tau(t)}}\exp\left(-2\frac{M}{M_0}\right)\sinh\left(2\sqrt{\tau(t)}\frac{M}{M_0}\right). \tag{10.32}$$

and:

$$M_{\text{avg}}(t) = \int_0^\infty Mf(M,t)\,dM = M_0(1+K_0C_0t/2). \tag{10.33}$$

Notice that $\int_0^\infty Mc(M,t)\,dM = C(t)\,M_{\text{avg}}(t) = C_0M_0 = \text{const.}$ In other words, conservation of mass applies. Also, notice that:

$$c(M,t) \cong \frac{4C_0M}{M_0^2}\exp\left(-2\frac{M}{M_0}\right) \tag{10.34}$$

for small enough times, which is true because $\sinh(x) = x$ as $x \to 0$. In addition, notice that:

$$c(M,t) \cong [1-\tau(t)]^2\frac{C_0}{M_0}\exp\left\{-[1-\tau(t)]\frac{M}{M_0}\right\} = \frac{C_0M_0}{[M_{\text{avg}}(t)]^2}\exp\left[-\frac{M}{M_{\text{avg}}(t)}\right] \tag{10.35}$$

for large enough times and particle sizes, which is true because $\sinh(x) \cong \exp(x)/2$ when $x \to \infty$. In summary, this example obtains a hybrid solution that is a self-similar gamma size distribution with $n/m = 2$ for small times or particle sizes. This describes unreacted particles, i.e., particles that have never participated in aggregation. Conversely, the hybrid becomes a self-similar exponential size distribution for large times and particle sizes. This describes fully-reacted particles, i.e., particles formed through multiple aggregation events that have 'forgotten' their initial conditions.

Most of the remaining examples in this chapter involve modified Bessel functions of the first kind. By definition, *modified Bessel functions of the first kind* $I_\alpha(x)$ are solutions to the following ordinary differential equation:

$$x^2 \frac{d^2}{dx^2} I_\alpha(x) + x \frac{d}{dx} I_\alpha(x) + \left(x^2 - \alpha^2\right) I_\alpha(x) = 0$$

where α is the *order*. However, this definition does not really explain modified Bessel functions of the first kind. As one way to understand them, they are closely related to a number of other common transcendental functions. For example:

$$I_{1/2}(x) = \sqrt{2\pi/x}\ \sinh(x) \tag{10.36}$$

$$I_{-1/2}(x) = \sqrt{2\pi/x}\ \cosh(x) \tag{10.37}$$

where $\sinh(x)$ and $\cosh(x)$ are the hyperbolic sine and cosine, respectively. For another example:

$$I_{\pm 1/3}(x) = \frac{1}{2}\sqrt{\frac{3}{z}}\left[\mp \mathrm{Ai}(z) + \mathrm{Bi}(z)\right] \tag{10.38}$$

where $z = (3x/2)^{2/3}$ and $\mathrm{Ai}(z)$ and $\mathrm{Bi}(z)$ are *Airy functions*. As a final example:

$$I_0(x) - I_1(x) = \exp(-x)\ {}_1F_1(1/2, 2, 2x) \tag{10.39}$$

where ${}_1F_1(x, y, z)$ is *Kummer's confluent hypergeometric function*. Kummer's confluent hypergeometric function will be discussed in more detail in Chapter 13. As another way to understand modified Bessel functions of the first kind, consider the following limits:

$$I_\alpha(x) \cong \frac{1}{\Gamma(\alpha + 1)}\left(\frac{x}{2}\right)^\alpha \tag{10.40}$$

as $x \to 0$ and:

$$I_\alpha(x) \cong \frac{1}{\sqrt{2\pi x}} \exp(x) \tag{10.41}$$

as $x \to \infty$. In addition, in many cases, $I_\alpha(x)$ is monotone increasing or decreasing for $x > 0$. In our application, analytic solutions typically obtain the first limit, Equation (10.40), for sufficiently small times or small particles. This describes *unreacted* particles, i.e., particles that have not participated in aggregation and thus still have their initial conditions. Conversely, analytic solutions obtain typically obtain the second limit, Equation (10.41), for sufficiently large times or large particles. This describes *fully-reacted* particles, i.e., particles formed through enough aggregation events that they have "forgotten" their initial conditions.

As a third way to understand modified Bessel functions of the first kind, consider integral expressions such as the following:

$$\int_0^\infty x \exp\left(-\frac{a+b}{2}x\right) I_0\left(\frac{b-a}{2}x\right) dx = \frac{1}{2}\left(\frac{1}{a^{3/2}b^{1/2}} + \frac{1}{a^{1/2}b^{3/2}}\right) \tag{10.42}$$

$$\int_0^\infty x \exp\left(-\frac{a+b}{2}x\right) I_1\left(\frac{b-a}{2}x\right) dx = \frac{1}{2}\left(\frac{1}{a^{3/2}b^{1/2}} - \frac{1}{a^{1/2}b^{3/2}}\right). \tag{10.43}$$

In addition:

$$\int_0^\infty \exp\left[-\left(y^2+1\right)x\right] I_1(2xy)\, dx = \begin{cases} \dfrac{y}{(1-y^2)} & 0 \le y < 1 \\ \dfrac{1}{y(y^2-1)} & y > 1 \end{cases} \tag{10.44}$$

$$\int_0^\infty x \exp\left[-\left(y^2+1\right)x\right] I_1(2xy)\, dx = \frac{2y}{\left(y^2-1\right)^3} \tag{10.45}$$

$$\int_0^\infty x^{-1} \exp\left[-\left(y^2+1\right)x\right] I_1(2xy)\, dx = \begin{cases} y & 0 \le y < 1 \\ \dfrac{1}{y} & y \ge 1 \end{cases} \tag{10.46}$$

Such integral expressions mean that most solutions involving modified Bessel functions of the first kind obtain reasonably simple algebraic expressions for their moments such as $C(t)$.

Modified Bessel functions of the first kind arise naturally in a number of different applications. As one early example, based on empirical curve fits to various types of biometric data, McKay (1932) suggested the following distribution:

$$f(x) = \frac{\sqrt{\pi}\left(c^2-1\right)^{\alpha+1/2}}{\Gamma(\alpha+1/2)\,2^\alpha d^{\alpha+1}} x^\alpha \exp\left(-\frac{c}{d}x\right) I_\alpha\left(\frac{x}{d}\right)$$

where x is a random variable and α, c, d are free parameters. If $a = (c-1)/d$ and $b = (c+1)/d$ are functions of time, and if $x = M$, this expression becomes:

$$f(M,t) = \frac{\sqrt{\pi}}{\Gamma(\alpha+1/2)} \frac{[a(t)\,b(t)]^{\alpha+1/2}}{[b(t)-a(t)]^\alpha} M^\alpha \exp\left[-\frac{a(t)+b(t)}{2}M\right] I_\alpha\left[\frac{b(t)-a(t)}{2}M\right]. \tag{10.47}$$

While distributions such as this one involving Bessel functions were originally derived empirically, they have since been derived rigorously using mean-field theory for particle aggregation; see, e.g., Example 10.4.

Example 10.4 Time-dependent closed-form solutions for relaxation-time approximations with constant kernels, exponential initial conditions, free-field boundary conditions, and exponential sources

Suppose the particle system is governed by the relaxation-time approximation given by Equation (9.5) where $K(x, y) = K_0 = \text{const.}$ and:

$$c_1(x) = (C_0/M_0)\exp(-x/M_0). \tag{10.48}$$

In this case, Hounslow (1990) obtained the following exact solution:

$$c(M,t) = \frac{C_0}{M_0}\frac{1}{\sqrt{1+2\tau(t)}}\exp\left(-\frac{1+\tau(t)}{1+2\tau(t)}\frac{M}{M_0}\right)\left[I_0\left(-\frac{\tau(t)}{1+2\tau(t)}\frac{M}{M_0}\right) - I_1\left(-\frac{\tau(t)}{1+2\tau(t)}\frac{M}{M_0}\right)\right]$$

where $I_0(x)$ and $I_1(x)$ are zeroth- and first-order modified Bessel function of the first kind and where:

$$\tau(t) = K_0 C_0 t. \tag{10.49}$$

Notice that $I_0(x) - I_1(x) = \exp(-x)\,{}_1F_1(1/2, 2, 2x)$ where ${}_1F_1(x, y, z)$ is Kummer's confluent hypergeometric function, as described in the main text. Then:

$$c(M,t) = \frac{C_0}{M_0}\frac{1}{\sqrt{1+2\tau(t)}}\exp\left(-\frac{M}{M_0}\right){}_1F_1\left(\frac{1}{2}, 2, -\frac{2\tau(t)}{1+2\tau(t)}\frac{M}{M_0}\right). \tag{10.50}$$

Also notice that $C(t) = C_0\left(\sqrt{1+2\tau(t)} - 1\right)/\tau(t) = 2C_0/\left(\sqrt{1+2\tau(t)} + 1\right)$. Then:

$$f(M,t) = \frac{c(M,t)}{C(t)} = \frac{1+\sqrt{1+2\tau(t)}}{2M_0\sqrt{1+2\tau(t)}}\exp\left(-\frac{M}{M_0}\right){}_1F_1\left(\frac{1}{2}, 2, -\frac{2\tau(t)}{1+2\tau(t)}\frac{M}{M_0}\right). \tag{10.51}$$

In addition, notice that $M_{\text{avg}}(t) = M_0\left(\sqrt{1+2\tau(t)} + 1\right)/2$ and $\int_0^\infty Mc(M,t)\,dM = C(t)M_{\text{avg}}(t) = C_0M_0 = \text{const.}$ In other words, conservation of mass applies. Finally, notice that:

$$c(M,t) \cong \frac{C_0}{M_0}\exp\left(-\frac{M}{M_0}\right) \text{ as } t \to 0 \tag{10.52}$$

$$c(M,t) \cong \frac{C_0}{M_0}\frac{1}{\sqrt{2\pi\tau(t)}}\left(1+\frac{1}{2\tau(t)}\right)\left(\frac{M}{M_0}\right)^{-3/2}\exp\left[-\frac{1}{1+2\tau(t)}\frac{M}{M_0}\right] \text{ as } M \to \infty. \tag{10.53}$$

In summary, the solution obtains a self-similar gamma size distribution with $n/m = 1$ for small enough times and particle sizes and a regular gamma size distribution with $n/m = -1/2$ for large enough times and particle sizes. For more information see, e.g., Dubovski (1994, equation 8.11), Nicmanis & Hounslow (1998) and Roussos et al. (2005).

10.5 Non-Stationary Closed-Form Solutions for Sum Kernels

Examples 10.5 gives an exact closed-form solution of Equation (9.1) for sum kernels. Similarly, Example 10.6 gives an exact closed-form solution of Equation (9.3) for sum kernels.

Example 10.5 Time-dependent closed-form solutions for sum kernels with monodisperse initial conditions and free-field boundary conditions

For $K_{ij} = K_0(i + j)$, Golovin (1963) obtained the following exact solution to Equation (9.1):

$$c_i(t) = C_0[1 - \tau(t)]\frac{1}{i!}[\tau(t)\,i]^{i-1}\exp[-\tau(t)\,i] \tag{10.54}$$

where:

$$\tau(t) = 1 - \exp(-K_0C_0t). \tag{10.55}$$

Notice that $C(t) = \sum_{i=1}^{\infty} c_i(t) = C_0[1 - \tau(t)]$. Then:

$$f_i(t) = \frac{c_i(t)}{C(t)} = \frac{1}{i!}[\tau(t)\,i]^{i-1}\exp[-\tau(t)\,i] \tag{10.56}$$

and:

$$M_{\text{avg}}(t) = \sum_{i-1}^{\infty} M_i f_i = \frac{M_0}{1 - \tau(t)} = M_0\exp(K_0C_0t). \tag{10.57}$$

Notice that $\sum_{i=1}^{\infty} M_i c_i(t) = C(t)M_{\text{avg}}(t) = C_0M_0 = \text{const}$. In other words, conservation of mass applies. In addition, notice that:

$$c_i(t) \cong C_0\delta_{i1} \text{ as } t \to 0 \tag{10.58}$$

$$c_i(t) \cong \sqrt{\frac{M_0}{2\pi M_{\text{avg}}(t)}\frac{C_0M_0^2}{M_{\text{avg}}^2(t)}}\left[\frac{M_0 i}{M_{\text{avg}}(t)}\right]^{-3/2}\exp\left[-\frac{M_0 i}{M_{\text{avg}}(t)}\right] \text{ as } i \to \infty. \tag{10.59}$$

due to Stirling's approximation, i.e., $i! \cong \sqrt{2\pi}i^{i+1/2}\exp(-i)$ as $i \to \infty$. In summary, this example obtains a hybrid solution that is monodisperse for small times and particle sizes. This describes unreacted particles, i.e., particles that have never participated in aggregation. Conversely, the hybrid becomes a regular gamma size distribution with $n/m = -1/2$ for large times and particle sizes. This describes fully-reacted particles, i.e., particles formed through multiple aggregation events that have "forgotten" their initial conditions. For more information on this example see, e.g., Leyvraz (2003) and Krapivsky et al. (2010). This example concerns discrete particle systems, which often have real values of m. By contrast, the next example (Example 10.6) concerns continuous particle systems with integer m.

Example 10.6 Time-dependent closed-form solutions for sum kernels with exponential initial conditions and free-field boundary conditions

For $K(x,y) = K_0(x+y)$, Safronov (1962, 1969) obtained the following exact solution to Equation (9.3):

$$c(M,t) = C_0 \frac{1-\tau(t)}{\sqrt{\tau(t)}} \frac{1}{M} \exp\left[-(\tau(t)+1)\frac{M}{M_0}\right] I_1\left[2\sqrt{\tau(t)}\frac{M}{M_0}\right] \tag{10.60}$$

where $I_1(x)$ is a first-order modified Bessel function of the first kind and:

$$\tau(t) = 1 - \exp(-K_0 C_0 M_0 t). \tag{10.61}$$

Notice that $C(t) = C_0 \exp(-K_0 C_0 M_0 t) = C_0[1-\tau(t)]$. Then:

$$f(M,t) = \frac{c(M,t)}{C(t)} = \frac{1}{\sqrt{\tau(t)}} \frac{1}{M} \exp\left[-(\tau(t)+1)\frac{M}{M_0}\right] I_1\left[2\sqrt{\tau(t)}\frac{M}{M_0}\right] \tag{10.62}$$

and:

$$M_{avg}(t) = \int_0^\infty M f(M,t)\, dM = \frac{M_0}{1-\tau(t)} = M_0 \exp(K_0 C_0 M_0 t). \tag{10.63}$$

In self-similar form, the small-time limit is as follows:

$$c(M,t) \cong \frac{C_0}{M_0} \exp\left(-\frac{M}{M_0}\right) \tag{10.64}$$

which is true because $I_1(x) \cong x/2$ as $x \to 0$. In regular form, the large-size limit is as follows:

$$c(M,t) \cong \frac{C_0}{M_0} \frac{1-\tau(t)}{2\sqrt{\pi}} \left[\sqrt{\tau(t)}\frac{M}{M_0}\right]^{-\frac{3}{2}} \exp\left[-\left(1-\sqrt{\tau(t)}\right)^2 \frac{M}{M_0}\right] \tag{10.65}$$

which is true because $I_1(x) \cong \exp(x)/\sqrt{2\pi x}$ as $x \to \infty$. In summary, this example obtains a hybrid solution that is a self-similar exponential distribution for small times and particle sizes. This describes unreacted particles, i.e., particles that have never participated in aggregation. Conversely, the hybrid becomes a regular gamma distribution with $n/m = -1/2$ for large times and particle sizes. This describes fully-reacted particles, i.e., particles formed through multiple aggregation events that have "forgotten" their initial conditions. For more information on this example see e.g., Scott (1968), Ramabhadran et al. (1976), Williams & Loyalka (1991, Equation 5.23, p. 247) and Aldous (1999, Table 10.2). This example has different early-time conditions than the previous example. However, it still obtains the same late-time size distribution. This is because aggregation tends to drive all solutions towards the broadest possible size distribution, where the metric for "broadest" depends on K. While the late-time size distributions may be the same, the particle shapes are generally different.

10.6 Non-Stationary Closed-Form Solutions for Product Kernels

This section gives three examples (Examples 10.7, 10.8, and 10.9) of exact closed-form solutions for product kernels. In these examples, according to the particle size distribution, the chances of observing a large particle decreases monotonically as size increases. Thus, at any given time, there is only an infinitesimal chance of observing an infinitely large particle. However, according to the average particle size, after enough time passes, a single infinitely large particle develops instantaneously, seemingly out of nowhere. The next section discusses this behavior in more detail.

Example 10.7 Time-dependent closed-form solutions for product kernels with monodisperse initial conditions and free-field boundary conditions

For $K_{ij} = K_0 ij$, McLeod (1962) obtained the following exact solution to Equation (9.1):

$$c_i(t) = \frac{1}{i!} C_0 \tau(t)^{i-1} i^{i-2} \exp[-\tau(t)\, i] \tag{10.66}$$

$$\tau(t) = C_0 K_0 t. \tag{10.67}$$

Notice that $C(t) = C_0[1 - \tau(t)/2]$ for $0 \leq \tau(t) < 2$. Then $f_i(t) = c_i(t)/C(t)$ and:

$$M_{\text{avg}}(t) = \frac{M_0}{1 - \tau(t)/2} \tag{10.68}$$

for $0 \leq \tau(t) < 2$. Notice that $\sum_{i=1}^{\infty} M_i c_i(t) = C(t) M_{\text{avg}}(t) = \text{const.}$ for $0 \leq \tau(t) \leq 2$ but not for $\tau(t) > 2$. In other words, conservation of mass applies until gelation occurs at $\tau(t) = 2$. Notice that the initial conditions are monodisperse. In other words:

$$\lim_{t \to 0} c_i(t) = A\delta_{i1} \tag{10.69}$$

where A is constant. In addition, notice that for large enough particle sizes prior to gelation, the size distribution is a gamma size distribution with $n/m = -3/2$. If $\tau(t) > 2$, the particle system violates conservation of mass, and thus it may admit multiple solutions. As one possible solution for $\tau(t) > 2$, Leyvraz & Tschudi (1982) suggest:

$$c_i(t) \cong \frac{A}{\tau(t)} \frac{i^{i-2}}{i!} \exp(-i) \tag{10.70}$$

which becomes a power law size distribution with $n/m = -3/2$ for sufficiently large times and particle sizes. Levyraz & Lushnikov (2015) say this case applies when "the interactions between sol and gel are discarded." As another possible solution for $\tau(t) > 2$, Levyraz

& Lushnikov (2015) suggest a regular gamma size distribution with $n/m = -3/2$. In this case, the same size distribution occurs both before and after gelation. Levyraz & Lushnikov (2015) say this case applies when the "the gel is viewed as just another large cluster, and its interactions with the finite clusters are fully taken into account." From another point of view, the gel acts like a hard sink (power law case) or a soft sink (gamma case) of large particles.

Example 10.8 Time-dependent closed-form solutions for product kernels with gamma initial conditions and free-field boundary conditions

For $K(x,y) = K_0xy$, McLeod (1964) obtained the following exact solution to Equation (9.3):

$$c(M,t) = \frac{A}{M_0}\frac{1}{\sqrt{\tau(t)}}\left(\frac{M}{M_0}\right)^{-2} \exp\left[-(\tau(t)+1)\frac{M}{M_0}\right] I_1\left[2\sqrt{\tau(t)}\frac{M}{M_0}\right] \tag{10.71}$$

for $0 \leq \tau(t) < 2$ and $M \geq M_{\min}$ where A is a constant that depends on $M_{\min}$, $I_1(x)$ is a first-order modified Bessel function of the first kind, and:

$$\tau(t) = K_0AM_0^2t. \tag{10.72}$$

Notice that $C(t) = C_0[1 - \tau(t)/2]$ for $0 \leq \tau(t) < 2$. Then $f_i(t) = c_i(t)/C(t)$ and:

$$M_{\text{avg}}(t) = \frac{M_0}{1 - \tau(t)/2} \tag{10.73}$$

for $0 \leq \tau(t) < 2$. Notice that $\int_0^\infty Mc(M,t)\,dM = C(t)\,M_{\text{avg}}(t) = \text{const.}$ for $0 \leq \tau(t) \leq 2$ but not for $\tau(t) > 2$. In other words, conservation of mass applies until gelation occurs at $\tau(t) = 2$. In self-similar form, the small-time limit is as follows:

$$c(M,t) \cong \frac{A}{M_0}\left(\frac{M}{M_0}\right)^{-1} \exp\left(-\frac{M}{M_0}\right) \tag{10.74}$$

which is true because $I_1(x) \cong x/2$ as $x \to 0$. Thus, the initial condition is a gamma size distribution with $n/m = 0$. In regular form, the large size limit is as follows:

$$c(M,t) \cong \frac{1}{2\sqrt{\pi}}\frac{A}{M_0}\sqrt{\tau(t)}\left(\sqrt{\tau(t)}\frac{M}{M_0}\right)^{-\frac{5}{2}} \exp\left[-\left(\sqrt{\tau(t)}-1\right)^2\frac{M}{M_0}\right] \tag{10.75}$$

which is true because $I_1(x) \cong \exp(x)/\sqrt{2\pi x}$ as $x \to \infty$. Thus, for large enough particle sizes prior to gelation, the size distribution is a gamma size distribution with $n/m = -3/2$. Gelation occurs when $\tau(t) > 2$, i.e., when $t > t_g = 2/(K_0AM_0^2)$. Then, because the particle system violates conservation of mass, it admits multiple solutions. As one possible solution,

Ernst et al. (1984) suggest a power law size distribution with $n/m = -3/2$. In this case, Ernst et al. (1984) show that:

$$\int_0^{\infty} Mc(M,t)\,dM = \begin{cases} C_0 M_0 = \text{const.} & \tau(t) \leq 2 \\ \dfrac{C_0 M_0}{\sqrt{\tau(t)/2}} \neq \text{const.} & \tau(t) > 2 \end{cases} \tag{10.76}$$

which confirms that the solution obeys conservation of mass for $\tau(t) \leq 2$ but not for $\tau(t) > 2$.

Example 10.9 Time-dependent closed-form solution for product kernels with gamma initial conditions and large-particle sinks

Consider a particle system governed by Equation (9.4) with $K(x,y) = K_0 xy$ and a single sink as follows:

$$\alpha(x,t) = -\alpha_0 \tag{10.77.a}$$

$$a(x,t) = b(x,t) = 0 \tag{10.77.b}$$

where $\alpha_0 > 0$ is a constant. Suppose A and M_0 are defined as in the previous example. If $K_0 A M_0^2/2 > \alpha_0$, Singh & Rodgers (1996) say "the mass removal from the system is strong enough to prevent the formation of an infinite gel cluster." In this case, Equation (9.4) obtains a gamma size distribution with $n/m = -3/2$ for sufficiently large times and particle sizes. On the other hand, if $0 < \alpha_0 < K_0 A M_0^2/2$, Singh & Rodgers (1996) say "the mass removal is insufficient to prevent the formation of an infinite gel cluster, which forms in a finite time ... However, the mass removal does slow down the process of gel formation." In this case, Singh & Rodgers (1996) and Abulwafa et al. (2006) obtained the following exact solution to Equation (9.4):

$$c(M,t) = \frac{A}{M_0}\frac{1}{\sqrt{\tau(t)}}\left(\frac{M}{M_0}\right)^{-2} \exp\left\{-\left[\tau(t)\exp(-\alpha_0 t) + \exp(\alpha_0 t)\right]\frac{M}{M_0}\right\} I_1\left[2\sqrt{\tau(t)}\frac{M}{M_0}\right] \tag{10.78}$$

where $0 \leq \tau(t) \leq 2$, $M \geq M_{\min}$, A is a constant that depends on $M_{\min}$, $I_1(x)$ is a first-order modified Bessel function of the first kind and:

$$\tau(t) = K_0 A M_0^2 \exp(\alpha_0 t)\,\frac{\sinh(\alpha_0 t)}{\alpha_0}. \tag{10.79}$$

Singh & Rodgers show that Equation (10.78) is a self-similar gamma size distribution with $n/m = 0$ for sufficiently small particle sizes and times; this is true because $I_1(x) \cong x/2$ as $x \to 0$. In addition, Singh & Rodgers (1996) show that Equation (10.78) is a regular gamma size distribution with $n/m = -3/2$ for sufficiently large particle sizes and times prior to gelation; this is true because $I_1(x) \cong \exp(x)/\sqrt{2\pi x}$ as $x \to \infty$. Finally, Singh & Rodgers (1996) suggest that, following gelation, the solution is a power law size distribution with $n/m = -3/2$. In this case, Singh & Rodgers (1996) show that:

$$\int_0^\infty Mc(M,t)\,dM = \begin{cases} C_0 M_0 \exp(-\alpha_0 t) & \tau(t) \leq 2 \\ \dfrac{C_0 M_0 \exp(-\alpha_0 t)}{\sqrt{\tau(t)/2}} & \tau(t) > 2 \end{cases} \tag{10.80}$$

Thus, the particle system continuously loses mass, regardless of time, due to the initial sink with strength α_0. The rate of mass loss increases once gelation occurs, due to the spontaneous formation of a second sink. Notice that the results for this example are the same as those for the previous example in the limit $\alpha_0 \to 0$. For other time-dependent solutions of Equation (9.4) see, e.g., Ramabhadran et al. (1976) and Alexopoulos et al. (2004).

10.7 Computational Validation

Figure 10.1 compares mean-field theory to computational results. In this figure, the dashed lines are mean-field solutions for an infinite particle system with a product kernel, as given in Example 10.7, assuming that a gamma size distribution applies both before and after gelation. The circles are computational results for a finite particle system with a product kernel. The computational results were obtained using a Monte-Carlo method, as discussed in Section 4.3, averaged over 5,000 trials. In this case, the Monte-Carlo results assume that N_{total} monomers aggregate to form N particles, where N_{total} is fixed at 40. Notice that N decreases from 33 to 6 as time progresses and gelation occurs when $N = 22$. The mean-field theory results agree with the computational results except for the largest clusters. At early times, the mean-field solution predicts more large clusters than the computational solution. However, after gelation occurs, the mean-field solution predicts fewer large clusters than the computational solution. This is because the mean-field solution violates conservation of mass—the infinitely large gel acts like a sink that removes other large clusters from the system—while the computational solution does not.

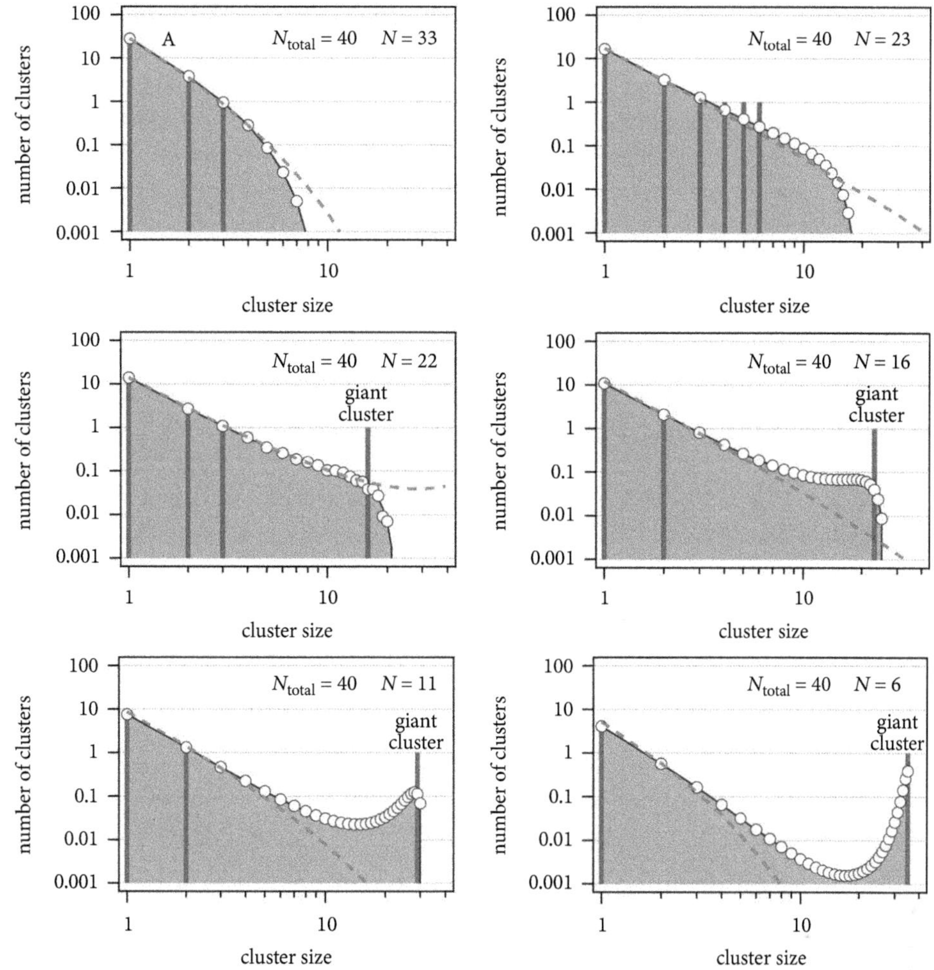

Figure 10.1 Gelation with an infinite number of particles (dashed line) vs. gelation with a finite number of particles (circles). Time increases from left to right and top to bottom. The results for an infinite number of particles were obtained using mean-field theory with a product kernel; see Example 10.7. The results for a finite number N of particles were found with a Monte-Carlo method.

Figure reprinted with permission from Matsoukas (2015).

10.8 Experimental Validation

This section gives experimental evidence for the analytic solutions presented earlier in this chapter. In evaluating experimental results, it is important to realize that particle systems may experience several collision kernels at once. This is because: (a.) aggregation causes the range of particle sizes to increase with time and (b.) the aggregation mechanism tends to depend on the particle size. For example, in colloidal systems, after enough time has passed, it may be true that the smallest particles experience Reaction-Limited Cluster-Cluster Aggregation (RLCA) with $\lambda = 1$, the medium-sized particles experience Diffusion-Limited Cluster-Cluster Aggregation (DLCA) with $\lambda = 0$, and the largest particles experience gravity-driven sedimentation with $\lambda = 4/3$; see, e.g., Lazzari et al. (2016). Thus, while most available collision kernels only account for a *single* type of aggregation, many real-world particle systems actually experience *multiple* types of aggregation at once. To avoid this issue, the validation examples (Examples 10.10 and 10.11) given below were chosen specifically so that the results are dominated by a single type of aggregation.

Example 10.10 Experimental vs. mean-field results for colloidal aggregation of polystyrene spheres

Broide & Cohen (1990, 1992) studied "an aqueous suspension of surfactant-free charged polystyrene microspheres, 0.258μm in radius, which are induced to irreversibly aggregate by the addition of salt ($MgCl_2$)." To ensure monodisperse initial conditions, they "sonicated suspensions of microspheres in an ultrasonic bath in order to break up any initial aggregates: this pretreatment enabled [them] to obtain suspensions which were 99.8% unaggregated." The subsequent colloidal aggregation was studied for a 52-hour period. To induce Diffusion-Limited Aggregation (DLA) with $\lambda = 0$, Broide & Cohen (1990) added "high concentrations of salt [so that] the electrostatic repulsion between spheres is screened and aggregation proceeds at a rate near the Brownian collision rate." Alternatively, to induce Reaction-Limited Aggregation (RLA) with $\lambda = 1$, Broide & Cohen (1990) used "lower concentrations of salt [so that] the repulsion between spheres is only partially screened and aggregation proceeds at a rate slower than the Brownian collision rate."

As seen in Figure 10.2, the experimental results for average particle sizes generally agree with mean-field theory. More specifically, for DLA with $\lambda = 0$, Broide & Cohen (1992) say that the experimental average "grows linearly in time with a slope near unity" for $K_0C_0t/2 < 15$, consistent with Equations (10.13), (10.21), and (10.27). In addition, for RLA with $\lambda = 1$, Broide & Cohen (1992) say that the experimental averages "grow approximately exponentially in time" for $K_0C_0t/2 < 13$. At very early times, the slope in the linear-log plane is about 2, which is consistent with monodisperse initial conditions; see Equation (10.57). However, after that, the slope in the linear-log plane suddenly drops to about 0.4, which is more consistent with exponential initial conditions; see Equation (10.63).

As seen in Figure 10.3, the experimental results for particle size distributions generally agree with mean-field theory. More specifically, for DLA with $\lambda = 0$, the experiments obtain an exponential size distribution for sufficiently large particles, consistent with Equations (10.12), (10.23), and (10.28). In addition, for RLA with $\lambda = 1$, the experiments obtain a gamma size distribution with $n/m = -1/2$ regardless of particle size, consistent with Equation (10.14). In Figure 10.3, the size distributions are normalized by M_{char}, which is chosen to be "directly proportional to weight-average mean cluster size" $M_{M,\text{avg}}$.

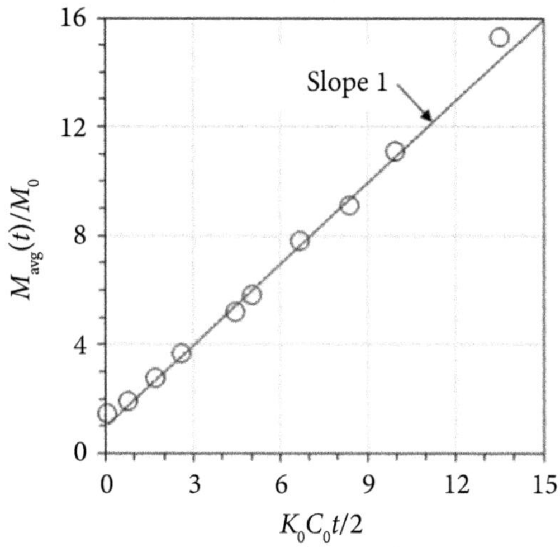

(a) Average particle mass for DLCA with $\lambda = 0$.

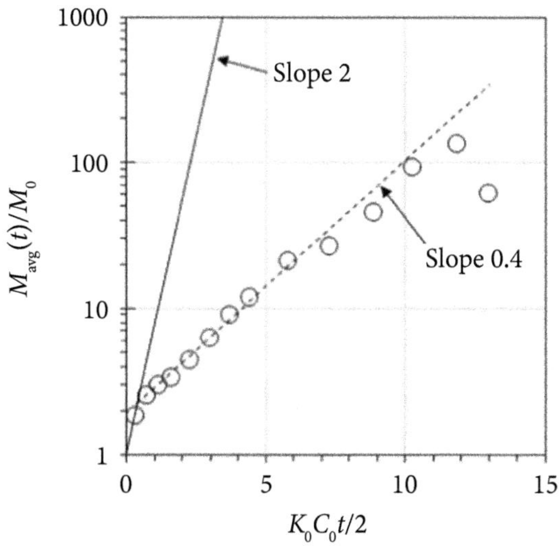

(b) Average particle mass for RLCA with $\lambda = 1$.

Figure 10.2 Analytical (mean-field theory) vs. experimental results for average particle diameters due to colloidal aggregation of polystyrene microspheres. The circles refer to experimental data as reported by Broide & Cohen (1990, 1992) while the lines refer to mean-field theory.

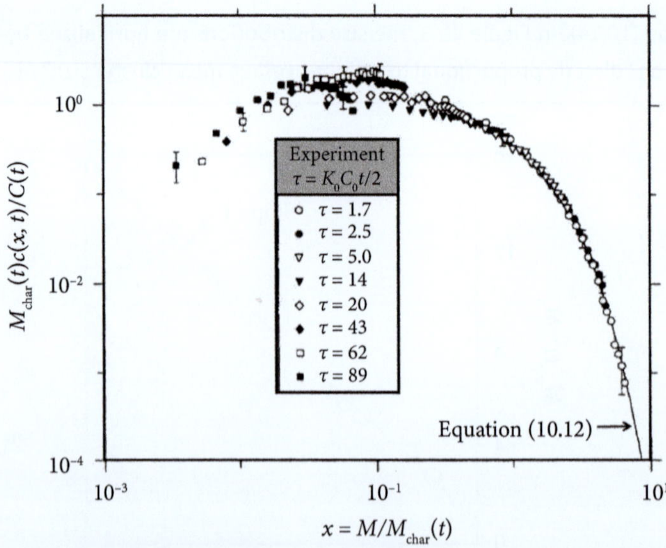

(a) Particle size distribution for DLCA with $\lambda = 0$.

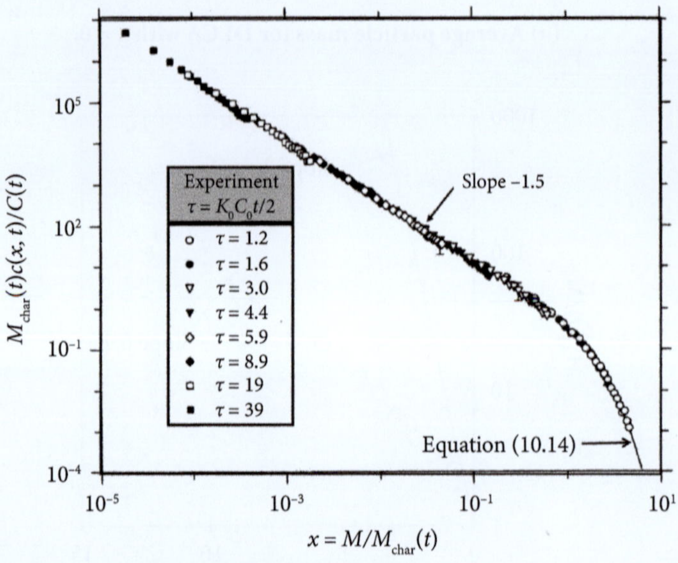

(b) Particle size distribution for RLCA with $\lambda = 1$.

Figure 10.3 Analytical (mean-field theory) vs. experimental results for particle size distributions due to colloidal aggregation of polystyrene microspheres. The symbols refer to experimental data while the solid lines refer to mean-field theory.

[The notation in this figure has been changed to match the notation used in this book.]

Example 10.11 Experimental vs. mean-field results for colloidal aggregation of gold, silica, and polystyrene latex particles

Lin et al. (1989, 1990a, b) studied aggregation of three different materials in charge-stabilized aqueous solutions: gold particles of radius 7.5nm; silica particles of radius 3.5nm; and polystyrene latex particles of radius 19nm. For each material, one chemical was used to retard aggregation while another chemical was used to accelerate aggregation. By adjusting the retardant-to-accelerant ratio, per the following table, either DLCA or RLCA was achieved:

	Retardant	Accelerant: DLCA	Accelerant: RLCA
Gold	"Citrate ions adsorbed on the surface"	9×10^{-3} M pyridine	9×10^{-6} M pyridine
Silica	"SiO^- groups on the surface"	1.7 M salt [NaCl]	0.6 M salt [NaCl]
Polystyrene Latex	"Charged carboxylic acid groups on the surface"	1.2 M acid [HCl]	0.2 M salt [NaCl]

The resulting aggregation was studied over a 10 hour period. To avoid gravitational settling of the largest particles, the particle system was inverted every 15 minutes. To avoid secondary aggregation, a.k.a. restructuring, the system was "not subjected to any mechanical disturbances." Finally, they used dynamic light scattering to "extract a true hydrodynamic radius" and "to help determine the shape of the cluster mass distribution."

As seen in Figure 10.4, the experimental results for average particle sizes generally agree with mean-field theory. In Figure 10.4.a, for DLCA with $\lambda = 0$, Lin et al. (1990a) note that the experimental results are linear in the log-log plane with a slope of 0.54 ± 0.02. Then:

$$M_{\text{avg}}(t) = \text{const.}\, D_{\text{avg}}(t)^m = \text{const.}\left(t^{0.54}\right)^m \approx \text{const.}\, t$$

where the as-measured fractal mass dimension is $m = 1.86 \pm 0.01$. Thus "the characteristic mass grows linearly in time," which is consistent with Equations (10.13), (10.21), and (10.27). In Figure 10.4.b for RLCA with $\lambda = 1$, Lin et al. (1990b) note that the average particle size obtained "exponential growth as indicated by the linear behavior in the semi-logarithmic plot," which is consistent with Equations (10.15), (10.57), and (10.63). Notice that, because the results have not been normalized by a characteristic particle size, the slope in the linear-log plane depends on the exact conditions including the particle material type.

While no figure is given, the experimental results for particle size distributions generally agree with mean-field theory. For DLCA with $\lambda = 0$, Lin et al. (1990a) say that, at any given time, the particle size distribution is "relatively flat up to some characteristic mass ... after which it decreases exponentially," which is consistent with Equation (10.12), (10.23), and (10.28). In addition, for RLCA with $\lambda = 1$, Lin et al. (1990b) obtained a gamma size distribution with $n/m = -1/2$, which is consistent with Equation (10.14), (10.59), and (10.65). Lin et al. (1989) say that these results "provide convincing experimental evidence that the two regimes of aggregation [diffusion- and reaction-limited] are indeed universal."

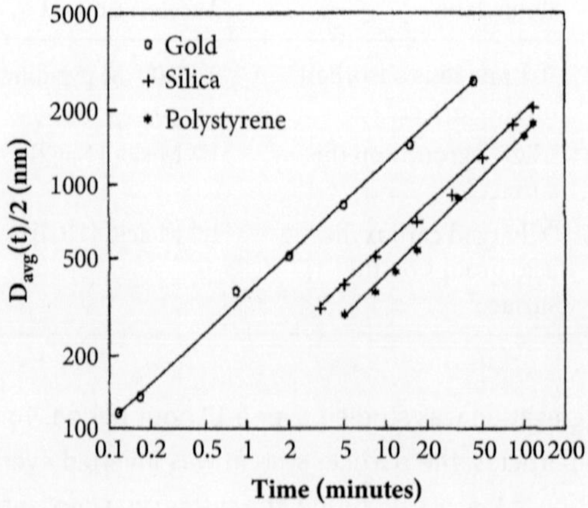

(a) Average particle diameters for DLCA with λ = 0. Reprinted from Lin et al. (1990a). Copyright IOP Publishing. [The notation in this figure has been changed to match the notation used in this book.]

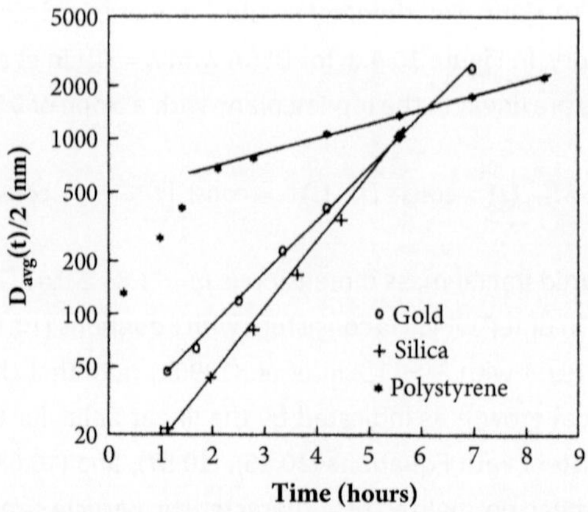

(b) Average particle diameters for RLCA with λ = 1. Reprinted with permission from Lin et al. (1990b). Copyright 1990 American Physical Society. [The notation in this figure has been changed to match the notation used in this book.]

Figure 10.4 Analytical (mean-field theory) vs. experimental results for the average particle diameters due to colloidal aggregation of gold, silica, and polystyrene microparticles. The symbols refer to experimental data while the solid lines refer to mean-field theory.

Example 10.12 Experimental vs. mean-field results for polymer aggregation

Mori & Barth (1999) says that "by far the most popular and convenient method of determining the average molecular weight and molecular weight distribution of a polymer is *size exclusion chromatography (SEC)* ... SEC separates on the basis of molecular hydrodynamic volume or size ... In SEC, the polymer is dissolved in an appropriate solvent and is injected into a column packed with porous particles of fairly defined pore size ... High-molecular weight material elutes first from an SEC column, following by low-molecular weight components."

Lusignan et al. (1995, 1999) used size exclusion chromatography to measure "mass distributions for a series of randomly branched polyester samples with long linear chain sections." More specifically, they "prepared stable randomly branched polyesters, by reacting all the acid groups in an excess of hydroxyl groups, thereby halting the chemical reaction by stoichiometric imbalance." In simple terms, aggregation ended when one of the reactants ran out. By design, this occurred just before the gel point. This led to long, strong, cross-linked, entangled chains with roughly 900 monomers between branch points. This aggregation process required on the order of 36 hours. However, instead of using time, the polymer science community often prefers to use the *extent of reaction* ε where:

$$\varepsilon(t) = 1 - p(t)/p(t_g) \tag{10.81}$$

where $p(t)$ is the number of cross-links at time t and t_g is the gel point. According to Lusignan (1995), the reaction time t is proportional to $1-\varepsilon$ In other words:

$$t/t_g = 1 - \varepsilon(t). \tag{10.82}$$

First, consider the average particle size. According to equation 6 and table I in Lusignan et al. (1999):

$$M_{M,\,\mathrm{avg}}(t)/M_{M,\,\mathrm{avg}}(0) = \varepsilon(t)^{-1} \tag{10.83}$$

where $M_{M,\,\mathrm{avg}}(t)$ was defined earlier in Table 3.2. Laney (2015) showed that:

$$\frac{M_{M,\,\mathrm{avg}}(t)}{M_{\mathrm{avg}}(t)} = \frac{D^m_{M,\,\mathrm{avg}}(t)}{D^m_{\mathrm{avg}}(t)} = \left\{\frac{\Gamma[(l+1)/n]\,\Gamma[(l+m+2)/n]}{\Gamma[(l+2)/n]\,\Gamma[(l+m+1)/n]}\right\}^m = \mathrm{const.} \tag{10.84}$$

for Rosin-Rammler size distributions. Equations (10.82), (10.83), and (10.84) imply that:

$$M_{\mathrm{avg}}(t)/M_{\mathrm{avg}}(0) \cong \left(1 - t/t_g\right)^{-1} \tag{10.85}$$

which agrees with Equation (10.17) for product kernels with $\lambda = 2$. Now, consider the particle size distributions. As seen in Figure 10.5, Lusignan et al. (1999) obtained a gamma size distribution with $n/m = -3/2$, which agrees with Equation (10.16) for product kernels with $\lambda = 2$. In Figure 10.5, Lusignan et al. (1999) normalized the size distribution by M_{char}, defined as the "characteristic largest molar mass ... determined by fitting the molar mass distribution to the ... Stockmayer (1943) theory;" see the earlier discussion in Section 5.3.

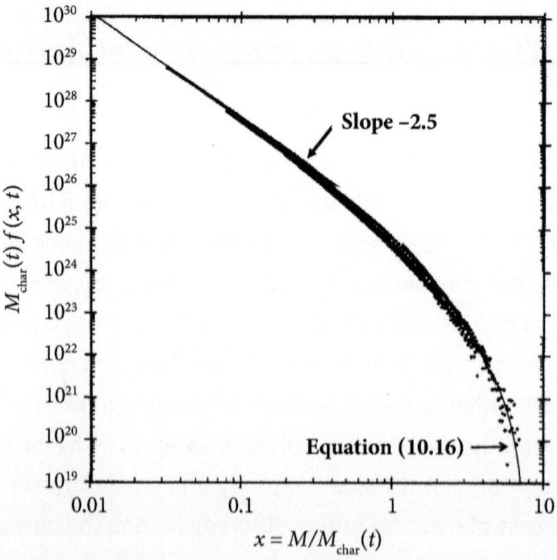

Figure 10.5 Analytical (mean-field theory) vs. experimental results for the size distribution of polyester particles. The symbols refer to experimental data while the solid line refers to mean-field theory with $\lambda = 2$.

[The notation in this figure has been changed to match the notation used in this book.]
Figure reprinted with permission from Lusignan et al. (1999). Copyright 1999 American Physical Society.

10.9 Gelation—Part 2

As discussed earlier, when $\lambda > 1$, particle aggregation typically leads to gelation. When there is a finite number of particles, gelation is relatively straightforward—one particle gradually becomes larger than any of the other particles as time progresses. However, when there is an infinite number of particles, gelation may be more complex—an infinitely large particle can appear instantaneously. In fact, there are at least three possible ways of describing gelation in the thermodynamic limit. As one possible description, gelation is a phase transition. More specifically, Matsoukas (2014, 2016) says "the emergence of a giant component from a population of finite-size clusters ... is a phase transition that is governed by equilibrium criteria analogous to those in molecular systems." In some cases, this phase transition may be analogous to freezing. For example, for colloids with large enough volume fractions, Lattuada et al. (2004) say the following:

> The process of forming a gel may be divided into two steps. The first step is the aggregation process that leads to the formation of clusters up to the arrest time, which we define as the time where the clusters, being in close proximity to each other are hindered from free movement and further growth. This close proximity is quantified as the time when their cumulative occupied volume being equal to 50% of the total available volume, in analogy to the freezing condition of hard sphere colloids. The second step is the interconnection among the clusters leading to the formation of the infinite network characteristic of the gel.

As another possible description, gelation involves the spontaneous formation of discontinuities, similar to those observed in hyperbolic partial differential equations. For example, for inviscid fluid flows, conservation of mass, momentum, and energy lead to a set of five partial differential equations known as the Euler equations. If the Mach number exceeds one, the Euler equations admit discontinuous flow features called *shocks*; see, e.g., Laney (1998). While gels occur only at infinity, shocks may occur anywhere, after which they may progressively propagate towards infinity.

As a third possible description, gelation is a manifestation of a classic chaotic behavior known as the *butterfly effect*. In a famous talk, Lorenz (1972) described the butterfly effect as follows:

> If the atmosphere is unstable with respect to small perturbations [then] . . . a single flap of a butterfly's wings can be instrumental in generating a tornado ... We do not know exactly how many butterflies there are, nor where they are located, let alone which ones are flapping their wings at any instant . . . Miniscule disturbances neither increase nor decrease the frequency of occurrence of various weather events such as tornadoes; the most that they may do is to modify the sequence in which these events occur.

Put another way, the chances that the flapping of any given butterfly's wings will lead to a tornado is infinitesimal. However, under the right circumstances, the chances that the flapping of *some* butterfly's wing—or any other small disturbance—will lead to a tornado is 100%. From the point of view of the butterfly effect, gelation is a small fluctuation amplified by instability. However, mean-field theory specifically averages out such fluctuations; see Section 5.1. As a result, the chances that any given particle will experience runaway growth is infinitesimal, as reflected in the mean-field size distribution. However, the chance that *some* particle will experience runaway by growth is 100%, as reflected in the infinite mean-field average size. Mean-field theory is a simple mathematically-idealized way of representing chaotic fluctuations. Thus, not surprisingly, mean-field theory captures an important feature of aggregation in chaotic particle systems—gelation—in a simple mathematically-idealized fashion.

10.10 Conclusions

In a classic paper, Schumann (1940) "surmised that there should be a tendency for the same universal distribution to be approached asymptotically after a sufficiently long time, no matter what the initial distribution might be," while warning that "a formal and direct proof of this supposition appears to be very difficult." Since then, researchers have advanced various proofs; see, e.g., Friedlander & Wang (1966), Lushnikov (1974) and Mullholland et al. (1977). While this chapter did not attempt a general proof, it gave a substantial amount of evidence in three specific cases where full time-dependent closed-form mean-field solutions are available, namely, a constant kernel with $\lambda = 0$, a sum kernel with $\lambda = 1$, and a product kernel with $\lambda = 2$. These solutions showed that, when there are an infinite number of particles, there will always be some particles that have never participated in aggregation and thus retain their initial sizes. In such cases, for spatially homogeneous and isotropic particle systems, the full time-dependent solution is a hybrid of two different size distributions—an arbitrary *non-universal* initial size distribution for unreacted particles and a second *universal* size distribution for fully-reacted particles. The balance progressively shifts from one to the other as the time and particle size increase.

As summarized in Table 10.1, the late-time universal size distributions for the constant kernel with $\lambda = 0$, the sum kernel with $\lambda = 1$, and the product kernel with $\lambda = 2$ are gamma size distributions with $n/m = 1$, $1/2$, and $3/2$, respectively. As seen in Example 10.4, changing the exact mean-field governing equation may change the exact correspondence between λ and n/m. However, there are still only the same three collision kernels and the same three analytic late-time universal size distributions.

Of course, in reality, there are more than just three collision kernels and more than just three late-time universal size distributions. As seen in the next chapter, a full range of results can be derived by simply ignoring the early-time dependency and going straight to the late-time universal size distribution.

This chapter gave two examples of experimental validation of mean-field theory where $\lambda = 0$ and 1, depending on the exact conditions. These examples were taken from the colloid science community, which has undertaken extensive efforts specifically dedicated to such validation. For additional examples see, e.g., von Schulthess et al. (1980), Hoekstra et al. (1992), Sherman & Cohen (1995), Fernández-Barbero et al. (1996a, b), Axford (1997), Tirado-Miranda et al. (1999), Schmitt et al. (2000), Sandkühler et al. (2003), Odriozola et al.

Table 10.1 Examples of gamma size distributions obtained as late-time analytic solutions of the non-stationary mean-field governing equations for aggregation.

λ	n/m	Field	References
0	1	Colloid science	Smoluchowski (1916, 1917)[1]
0	1	Aerosol science	Schumann (1940)[3]
0	1	Cloud physics	Scott (1968)[3]
0	1	Fire science	Mulholland & Baum (1980)[3]
0	−1/2	Crystallization	Hounslow (1990)[5]
1	−1/2	Astrophysics	Safronov (1962, 1969)[3]
1	−1/2	Cloud physics	Golovin (1963)[1]
2	−3/2	Applied mathematics	McLeod (1962)[1,a]
2	−3/2	Applied mathematics	McLeod (1964); Leyvraz & Lushnikov (2015)[3,a]
2	−3/2	Polymer science	Hendriks (1984b)[2,a]
2	−3/2	Applied physics	Singh & Rodgers (1996); Abulwafa et al. (2006)[4,a]

(1) Solution to Equation (9.1).
(2) Solution to Equation (9.2).
(3) Solution to Equation (9.3).
(4) Solution to Equation (9.4).
(5) Solution to Equation (9.5).
(a) Gelation may occur.

(2004), Runkana et al. (2005), Lattuada et al. (2003, 2004, 2006) and Gauer et al. (2009). As a general caveat about aggregation experiments, even when the degree of homogeneity λ is clear, the collision kernel K is difficult to measure directly and thus may not be clear. While constant kernels obtain $\lambda = 0$, so do other kernels. Similarly, while product kernels obtain $\lambda = 1$, so do other kernels.

This chapter gave one example of experimental validation of mean-field theory for $\lambda = 2$. This example was taken from the polymer science community, which has as undertaken extensive efforts specifically dedicated to such validation. In addition, this chapter gave one example of computational validation of mean-field theory for $\lambda = 2$. For additional examples, see, e.g., Laurenzi & Diamond (1999), Friesen & Dabros (2003), Alfonso et al. (2008), Drążkowska et al. (2013), Matsoukas (2015, 2019) and Wei (2016). The advantage of computations over experiments is that both the degree of homogeneity λ and the collision kernel K are generally completely clear. In fact, all of the papers just cited specifically concern the product kernel.

References

E. M. Abulwafa, M. A. Abdou, & A. A. Mahmoud, The Solution of Nonlinear Coagulation Problem with Mass Loss, *Chaos, Solitons & Fractals*, 29(2), 313–330, 2006, https://doi.org/10.1016/j.chaos.2005.08.044

D. J. Aldous, Deterministic and Stochastic Models for Coalescence (Aggregation and Coagulation): A Review of the Mean-Field Theory for Probabilists, *Bernoulli*, 5(1), 3–48, 1999, https://projecteuclid.org/euclid.bj/1173707093

A. H. Alexopoulos, A. I. Roussos, & C. Kiparissides, Part I: Dynamic Evolution of the Particle Size Distribution in Particulate Processes Undergoing Combined Particle Growth and Aggregation, *Chemical Engineering Science*, 59(24), 5751–5769, 2004, https://doi.org/10.1016/j.ces.2004.06.021

H. Alfvén & G. Arrhenius, *Evolution of the Solar System*, NASA SP-345, 1976, https://ntrs.nasa.gov/search.jsp?R=19770006016

L. Alfonso, G. B. Raga, & D. Baumgardner, The Validity of the Kinetic Collection Equation Revisited, *Atmospheric Chemistry and Physics*, 8, 969–982, 2008, https://doi.org/10.5194/acp-8-969-2008

S. D. T. Axford, Aggregation of Colloidal Silica: Reaction-Limited Kernel, Stability Ratio and Distribution Moments, *Journal of the Chemical Society, Faraday Transactions*, 93(2), 303–311, 1997, https://doi.org/10.1039/A606195H

J. Banasiak, W. Lamb, & P. Laurencot, *Analytic Methods for Coagulation-Fragmentation Models, Volume I*, CRC, 2019a, https://doi.org/10.1201/9781315154428

J. Banasiak, W. Lamb, & P. Laurencot, *Analytic Methods for Coagulation-Fragmentation Models, Volume II*, CRC, 2019b, https://doi.org/10.1201/9780429280320

J. Bertoin, *Random Fragmentation and Coagulation Processes*, Cambridge University Press, 2006, https://doi.org/10.1017/CBO9780511617768

M. L. Broide & R. J. Cohen, Experimental Evidence of Dynamic Scaling in Colloidal Aggregation, *Physical Review Letters*, 64(17), 1990, https://doi.org/10.1103/PhysRevLett.64.2026

M. L. Broide & R. J. Cohen, Measurements of Cluster-Size Distributions Arising in Salt-Induced Aggregation of Polystyrene Microspheres, *Journal of Colloid and Interface Science*, 153(2), 493–508, 1992, https://doi.org/10.1016/0021-9797(92)90340-R

J. Drążkowska, F. Windmark, & C. P. Dullemond, Planetesimal Formation via Sweep-Up Growth at the Inner Edge of Dead Zones, *Astronomy & Astrophysics*, 556, A37, 2013, https://doi.org/10.1051/0004-6361/201321566

P. B. Dubovski, *Mathematical Theory of Coagulation*, Seoul National University, Research Institute of Mathematics, Global Analysis Research Center, 1994, https://www.researchgate.net/publication/265038723_Mathematical_Theory_of_Coagulation_1

P. Dziekan & H. Pawlowska, Stochastic Coalescence in Lagrangian Cloud Microphysics, *Atmospheric Chemistry and Physics*, 17, 13509–13520, 2017, https://doi.org/10.5194/acp-17-13509-2017

M. H. Ernst, R. M Ziff, & E. M. Hendriks, Coagulation Processes with a Phase Transition, *Journal of Colloid and Interface Science*, 97(1), 266–277, 1984, https://doi.org/10.1016/0021-9797(84)90292-3

A. Fernández-Barbero, M. Cabrerizo-Vílchez, R. Martínez-García, & R. Hidalgo-Álvarez, Effect of the Particle Surface Charge Density on the Colloidal Aggregation Mechanism *Physical Review E*, 53(5), 4981–4989, 1996a, https://doi.org/10.1103/PhysRevE.53.4981

A. Fernández-Barbero, A. Schmitt, M. Cabrerizo-Vílchez, & R. Martínez-García, Cluster-Size Distribution in Colloidal Aggregation Monitored by Single-Cluster Light Scattering, *Physica A: Statistical Mechanics and its Applications*, 230(1–2), 53–74, 1996b, https://doi.org/10.1016/0378-4371(96)00049-0

S. K. Friedlander & C. S. Wang, The Self-Preserving Particle Size Distribution for Coagulation by Brownian Motion, *Journal of Colloid and Interface Science*, 22(2), 126–132, 1966, https://doi.org/10.1016/0021-9797(66)90073-7

W. I. Friesen & T. Dabros, Constant-Number Monte Carlo Simulation of Aggregating and Fragmentation Particles, *The Journal of Chemical Physics*, 119(5), 2825–2838, 2003, https://doi.org/10.1063/1.1586693

C. Gauer, H. Wu, & M. Morbidelli, Effect of Surface Properties of Elastomer Colloids on Their Coalescence and Aggregation Kinetics, *Langmuir*, 25(20), 12073–12083, 2009, https://doi.org/10.1021/la900963f

A. M. Golovin, The Solution of the Coagulation Equation for Cloud Droplets in a Rising Air Current, *Bulletin of the Academy of Sciences of the USSR, Geophysics Series*, 5, 482–487, 1963

E. M. Hendriks, Cluster Size Distributions in Equilibrium, *Zeitschrift für Physik B – Condensed Matter*, 57(4), 307–314, 1984a, https://doi.org/10.1007/BF01470421

E. M. Hendriks, Exact Solution of a Coagulation Equation with a Removal Term, *Journal of Physics A: Mathematical and General*, 17(11), 2299–2303, 1984b, https://doi.org/10.1088/0305-4470/17/11/024

L. L. Hoekstra, R. Vreeker, & W. G. M. Agterof, Aggregation of Colloidal Nickel Hydroxycarbonate Studied by Light Scattering, *Journal of Colloid and Interface Science*, 151(1), 17–25, 1992, https://doi.org/10.1016/0021-9797(92)90234-D

M. J. Hounslow, A Discretized Population Balance for Continuous Systems at Steady State, *AIChE Journal*, 36(1), 106–116, 1990, https://doi.org/10.1002/aic.690360113

A. B. Kostinski & R. A. Shaw, Fluctuations and Luck in Droplet Growth by Coalescence, *Bulletin of the American Meteorological Society*, 86(2), 235–244, 2005, https://doi.org/10.1175/BAMS-86-2-235

P. L. Krapivsky, S. Redner, & E. Ben-Naim, *A Kinetic View of Statistical Physics*, Cambridge University Press, 2010, https://doi.org/10.1017/CBO9780511780516 [see Chapter 5: Aggregation]

C. B. Laney, *Computational Gasdynamics*, Cambridge University Press, 1998, https://doi.org/10.1017/CBO9780511605604

C. B. Laney, *Transformation and Self-Similarity Properties of Gamma and Weibull Fragment Size Distributions*, Defense Threat Reduction Agency Technical Report DTRA-TR-16-06, December 2015, http://apps.dtic.mil/sti/citations/ADA624878

M. Lattuada, P. Sandkühler, H. Wu, J. Sefcik, & M. Morbidelli, Aggregation Kinetics of Polymer Colloids in Reaction Limited Regime: Experiments and Simulations, *Advances in Colloid and Interface Science*, 103(1), 33–56, 2003, https://doi.org/10.1016/S0001-8686(02)00082

M. Lattuada, H. Wu, P. Sandkühler, J. Sefcik, & M. Morbidelli, Modelling of Aggregation Kinetics of Colloidal Systems and its Validation by Light Scattering Measurements, *Chemical Engineering Science*, 59(8–9), 1783–1798, 2004, https://doi.org/10.1016/j.ces.2004.01.033

M. Lattuada, H. Wu, J. Sefcik, & M. Morbidelli, Detailed Model of the Aggregation Event between Two Fractal Clusters, *The Journal of Physical Chemistry B*, 110(13), 6574–6586, 2006, https://doi.org/10.1021/jp056538e-9

I. J. Laurenzi & S. L. Diamond, Monte Carlo Simulation of the Heterotypic Aggregation Kinetics of Platelets and Neutrophils, *Biophysical Journal*, 77(3), 1733–1746, 1999, https://doi.org/10.1016/S0006-3495(99)77019-0

S. Lazzari, L. Nicoud, B. Jaquet, M. Lattuada, & M. Morbidelli, Fractal-Like Structures in Colloid Science, *Advances in Colloid and Interface Science*, 235, 1–13, 2016, https://doi.org/10.1016/j.cis.2016.05.002

F. Leyvraz & H. R. Tshudi, Singularities in the Kinetics of Coagulation Processes, *Journal of Physics A: Mathematical and General*, 14(12), 3389–3405, 1981, https://doi.org/10.1088/0305-4470/14/12/030

F. Leyvraz & H. R. Tschudi, Critical Kinetics near Gelation, *Journal of Physics A: Mathematical and General*, 15(6), 1951–1964, 1982, https://doi.org/10.1088/0305-4470/15/6/033

F. Leyvraz, Scaling Theory and Exactly Solved Models in the Kinetics of Irreversible Aggregation, *Physics Reports*, 383(2–3), 95–212, 2003, https://doi.org/10.1016/S0370-1573(03)00241-2

F. Leyvraz & A. A. Lushnikov, Scaling Anomalies in the Sol–Gel Transition, *Journal of Physics A: Mathematical and Theoretical*, 48(20), 205002, 2015, https://doi.org/10.1088/1751-8113/48/20/205002

M. Y. Lin, H. M. Lindsay, D. A. Weitz, R. C. Ball, R. Klein, & P. Meakin, Universality in Colloid Aggregation, *Nature*, 339, 360–362, 1989, https://doi.org/10.1038/339360a0

M. Y. Lin, H. M. Lindsay, D. A. Weitz, R. Klein, R. C. Ball & P. Meakin, Universal Diffusion-Limited Colloid Aggregation, *Journal of Physics: Condensed Matter*, 2(13), 3093–3113, 1990a, https://doi.org/10.1088/0953-8984/2/13/019

M. Y. Lin, M. H. Lindsay, D. A. Weitz, R. C. Ball, R. Klein, & P. Meakin, Universal Reaction-Limited Colloid Aggregation, *Physical Review A*, 41(4), 2005–2020, 1990b, https://doi.org/10.1103/PhysRevA.41.2005

E. Lorenz, The Butterfly Effect, 139th Meeting of the American Association for the Advancement of Science, Washington DC, 1972, in R. Abraham & Y. Ueda, editors, *The Chaos Avant-Garde: Memories of the Early Days of Chaos Theory*, World Scientific, 2000, https://doi.org/10.1142/4510

A. A. Lushnikov, Evolution of Coagulating Systems. II. Asymptotic Size Distributions and Analytical Properties of Generating Functions, *Journal of Colloid and Interface Science*, 48(3), 400–409, 1974, https://doi.org/10.1016/0021-9797(74)90183-0

A. A. Lushnikov, Gelation in Coagulating Systems, *Physica D: Nonlinear Phenomena*, 222(1-2), 37–53, 2006, https://doi.org/10.1016/j.physd.2006.08.002

C. P. Lusignan, T. H. Mourey, J. C. Wilson, & R. H. Colby, Viscoelasticity of Randomly Branched Polymers in the Critical Percolation Class, *Physical Review E*, 52, 6271–6280, 1995, https://doi.org/10.1103/PhysRevE.52.6271

C. P. Lusignan, T. H. Mourey, J. C. Wilson, & R. H. Colby, Viscoelasticity of Randomly Branched Polymers in the Vulcanization Class, *Physical Review E*, 60(5), 5657–5669, 1999, https://doi.org/10.1103/PhysRevE.60.5657

T. Matsoukas, Statistical Thermodynamics of Clustered Populations, *Physical Review E*, 022113, 2014, https://doi.org/10.1103/PhysRevE.90.022113

T. Matsoukas, Statistical Thermodynamics of Irreversible Aggregation: The Sol-Gel Transition, *Scientific Reports*, 5, 8855, 2015, https://doi.org/10.1038/srep08855

T. Matsoukas, Thermodynamics of Irreversible Aggregation, in M. Papadrakakis, V. Papadopoulos, G. Stefanou, and V. Plevris, editors, *7th European Congress on Computational Methods in Applied Sciences and Engineering, ECCOMAS Congress 2016*, National Technical University of Athens, 2016

T. Matsoukas, Kinetic Gelation, in *Generalized Statistical Thermodynamics: Thermodynamics of Probability Distributions and Stochastic Processes*, Springer, 2019, https://doi.org/10.1007/978-3-030-04149-6_9

A. T. McKay, A Bessel Function Distribution, *Biometrika*, 24(1-2) 39–44, 1932, https://doi.org/10.1093/biomet/24.1-2.39

J. B. McLeod, On an Infinite Set of Nonlinear Differential Equations, *The Quarterly Journal of Mathematics*, 13(1), 119–128, 1962, https://doi.org/10.1093/qmath/13.1.119

J. B. McLeod, On the Scalar Transport Equation, *Proceedings of the London Mathematical Society*, S3, 14(3), 445–458, 1964, https://doi.org/10.1112/plms/s3-14.3.445

S. Mori & H. G. Barth, *Size Exclusion Chromatography*, Springer, 1999, https://doi.org/10.1007/978-3-662-03910-6

G. W. Mulholland, T. G. Lee, & H. R. Baum, The Coagulation of Aerosols with Broad Initial Size Distributions, *Journal of Colloid and Interface Science*, 62(3), 406–420, 1977, https://doi.org/10.1016/0021-9797(77)90092-3

G. W. Mulholland & H. R. Baum, Effect of Initial Size Distribution on Aerosol Coagulation, *Physical Review Letters*, 45(9), 761–763, 1980, https://doi.org/10.1103/PhysRevLett.45.761

M. Nicmanis & M. J. Hounslow, Finite-Element Methods for Steady-State Population Balance Equations, *AIChE Journal*, 44(10), 2258–2272, 1998, https://doi.org/10.1002/aic.690441015

G. Odriozola, R. Leone, A. Schmitt, J. Callejas-Fernández, R. Martínez-García, & R. Hidalgo-Álvarez, Irreversible versus Reversible Aggregation: Mean Field Theory and Experiments, *The Journal of Chemical Physics*, 121(11), 5468–5481, 2004, https://doi.org/10.1063/1.1779571

T. E. Ramabhadran, T. W. Peterson, & J. H. Seinfeld, Dynamics of Aerosol Coagulation and Condensation, *AIChE Journal*, 22(5), 840–851, 1976, https://doi.org/10.1002/aic.690220505

V. Runkana, P. Somasundaran, & P. C. Kapur, Reaction-Limited Aggregation in Presence of Short-Range Structural Forces, *AIChE Journal*, 51(4), 1233–1245, 2005, https://doi.org/10.1002/aic.10375

A. I. Roussos, A. H. Alexopoulos, & C. Kiparissides, Part III: Dynamic Evolution of the Particle Size Distribution in Batch and Continuous Particulate Processes: A Galerkin on Finite Elements Approach, *Chemical Engineering Science*, 60(24), 6998–7010, 2005, https://doi.org/10.1016/j.ces.2005.06.021

V. S. Safronov, Chastnyi Sluchai Resheniya Uravnenii Koagulyatsii [A Particular Solution of the Coagulation Equations], *Doklady Akademii Nauk SSSR*, 147(1), 64, 1962; see also V. S. Safronov, A Restricted Solution of the Accretion Equation, *Soviet Physics Doklady*, 7(11), 967–969, 1963

V. S. Safronov, *Evolyutsiya Doplanetnogo Oblaka i Obrazovanie Zemli i Planet (Evolution of the Protoplanetary Cloud and Formation of the Earth and its Planets)*, Akademii Nauk SSSR, 1969; see also NASA TT F-677, 1972, https://archive.org/details/nasa_techdoc_19720019068

P. Sandkühler, J. Sefcik, M. Lattuada, H. Wu, & M. Morbidelli, Modeling Structure Effects on Aggregation Kinetics in Colloidal Dispersions, *AIChE Journal*, 49(6), 1542–1555, 2003, https://doi.org/10.1002/aic.690490618

A. Schmitt, A. Fernández-Barbero, M. A. Cabrerizo-Vílchez, & R Hidalgo-Álvarez, Dynamic Scaling in Surface-Controlled Colloidal Aggregation, *Journal of Physics: Condensed Matter*, 12(8A), A281–A286, 2000, https://doi.org/10.1088/0953-8984/12/8A/336

T. E. W. Schumann, Theoretical Aspects of the Size Distribution of Fog Particles, *Quarterly Journal of the Royal Meteorological Society*, 66(285), 195–208, 1940, https://doi.org/10.1002/qj.49706628508

W. T. Scott, Analytic Studies of Cloud Droplet Coalescence I, *Journal of the Atmospheric Sciences*, 25(1), 54–65, 1968, https://doi.org/10.1175/1520-0469(1968)025<0054:ASOCDC&> 2.0.CO;2

D. A. Sherman & R. J. Cohen, Kinetic and Equilibrium Cluster-Size Distributions of Antibody-Antigen-Induced Colloidal Aggregation, in F. Family, P. Meakin, B. Sapoval, & R. Wool, editors, *Fractal Aspects of Materials*, Materials Research Society, 1995, https://doi.org/10.1557/PROC-367-435

W. T. Scott, Analytic Studies of Cloud Droplet Coalescence I, *Journal of the Atmospheric Sciences*, 25(1), 54–65, 1968, https://doi.org/10.1175/1520-0469(1968)025<0054:ASOCDC>2.0.CO;2

P. Singh & G. J. Rodgers, Coagulation Processes with Mass Loss, *Journal of Physics A: Mathematical and General*, 29(2), 437–450, 1996, https://doi.org/10.1088/0305-4470/29/2/021

M. V. Smoluchowski, Drei Vorträge über Diffusion, Brownsche Molekularbewegung und Koagulation von Kolloidteilche [Three Lectures on Diffusion, Brownian Motion and Coagulation of Colloid Particles], *Physikalische Zeitschrift*, 17, 557–571, 587–599, 1916, https://eudml.org/doc/215805

M. V. Smoluchowski, Versuch einer mathematischen Theorie der Koagulationskinetik kolloider Lösunger [Mathematical Theory of the Kinetics of the Coagulation of Colloidal Solutions],

Zeitschrift für Physikalische Chemie, 92, 129–168, 1917, https://doi.org/10.1515/zpch-1918-9209

W. H. Stockmayer, Theory of Molecular Size Distribution and Gel Formation in Branched-Chain Polymers, *The Journal of Chemical Physics*, 11(2), 45–55, 1943, https://doi.org/10.1063/1.1723803

M. Tirado-Miranda, A. Schmitt, J. Callejas-Fernández, & A. Fernández-Barbero, Colloidal Clusters with Finite Binding Energies: Fractal Structure and Growth Mechanism, *Langmuir*, 15(10), 3437–3444, 1999, https://doi.org/10.1021/la981233v

G. von Schulthess, G. B. Benedek, & R. De Blois, Measurement of the Cluster Size Distributions for High Functionality Antigens Cross-Linked by Antibody, *Macromolecules*, 13(4), 939–945, 1980, https://doi.org/10.1021/ma60076a033

J. Wei, A Monte Carlo Simulation for Particle Aggregation Containing a Sol–Gel Phase Transition, *Journal of Sol-Gel Science and Technology*, 78(2), 270–278, 2016, https://doi.org/10.1007/s10971-015-3947-2

G. W. Wetherill, The Role of Large Bodies in the Formation of the Earth and Moon, *Proceedings of the 7th Lunar Science Conference*, Volume 3, Pergamon Press, 1976, http://adsabs.harvard.edu/full/1976LPSC.... 7.3245W

M. M. R. Williams & S. K. Loyalka, *Aerosol Science: Theory and Practice, with Special Applications to the Nuclear Industry*, Pergamon Press, 1991

11

Aggregation Cascades Part 3: Late-Time Mean-Field Solutions

As seen in the last chapter, for spatially homogeneous and isotropic particle systems, mean-field solutions for particle aggregation gradually transition from arbitrary *non-universal* size distributions to *universal* size distributions as time and particle size increase. The former depends only on the initial conditions while the later depends only on the collision kernel and boundary conditions. As it turns out, as seen in the last chapter, full time-dependent analytic solutions are only possible for three specific collision kernels, namely, the constant, sum, and product kernels. For each of these kernels, closed-form time-dependent analytic solutions are only possible for certain initial conditions and for free-field boundary conditions.

Alternatively, suppose that only the late-time universal size distribution is of interest. In fact, for many practical applications, the final state is more important than intermediate states. When full time-dependency can be sacrificed, mean-field theory can lead directly to universal size distributions. More specifically, universal size distributions often occur in two limiting cases:

1. **Systems where aggregating particles tend to have different sizes.** This can be called *non-local* or *free merging*; see Horvai et al. (2008), Ball et al. (2011) and Connaughton et al. (2018). Alternatively, Blum & Wurm (2008) call this *Group B* if one of the colliding particles is always a monomer or *Group C* if one of the colliding particles is *not* always a monomer. This case tends to occur when sources continuously add small particles to the system, thus increasing the range of particle sizes. In fact, with or without sources, aggregation naturally tends to increase the range of particle sizes as time increases. As a result, collisions between unequal-sized particles tend to become increasingly common, just through random chance, unless sinks or other factors are present; see Example 11.1.
2. **Systems where aggregating particles tend to have the same or similar sizes.** This can be called *local* or *forced locality*; see Horvai et al. (2008), Ball et al. (2011), and Connaughton et al. (2018). Alternatively, Blum & Wurm (2008) call this *Group A*. This case tends to occur when sinks remove large particles from the system, thus reducing the range of particle sizes. In addition, this case tends to occur when:
   - Collisions between particles of different sizes obtain poor reaction efficiency, e.g., smaller particles tend to glance harmlessly off of larger particles.
   - Gravity causes large particles to settle, leaving only the smaller particles in suspension; then aggregation may be dominated by collisions between small particles.
   - Small particles entrained in a fluid medium follow streamlines around large particles; only particles large enough to avoid entrainment in the fluid medium can collide.

Particle Dynamics with Aggregation and Fragmentation. Culbert B. Laney, Oxford University Press.
DOI: 10.1093/9780198920687.003.0011

While mean-field theory can account for sinks, it cannot easily account for these other phenomena. Thus, this chapter considers only forced locality due to sinks of large particles, including but not limited to gel-type sinks that form spontaneously when $\lambda > 1$.

Of course, aggregation may be somewhere in between these two extremes. In any event, these two extremes tend to bound the possible outcomes.

Example 11.1 Condensation

As defined here, *condensation* is dominated by collisions between vapor molecules and liquid droplets. There are two main types of condensation:

Regular or *Maxwellian condensation*. By definition, regular condensation is deterministic. As discussed in Section 9.4, deterministic systems typically remember their initial conditions and progressively narrow their size distributions over time. Regular condensation tends to occur when the vapor concentration is high, the droplet concentration is low, and no seemingly random elements are present. In such cases, each liquid droplet continuously accretes large numbers of vapor molecules at a predictable rate. Droplets rarely, if ever, collide or merge with other droplets.

Stochastic condensation. As the term implies, stochastic condensation is either genuinely or apparently random. As discussed earlier in Section 9.4, such systems typically forget their initial conditions and progressively broaden their size distributions over time. Stochastic condensation tends to occur when the vapor concentration is low, the droplet concentration is high, and seemingly random elements are present. For example, vapor molecules may sporadically collide and merge with liquid droplets in a seemingly random fashion. Less commonly, droplets may collide with other droplets in a seemingly random fashion. The apparent randomness may be enhanced by other factors such as atmospheric turbulence.

Stochastic condensation is a classic example of free merging aggregation. As it turns out, clouds in the troposphere have all the hallmarks of stochastic condensation. More specifically, Khvorostyanov & Curry (2014) say the following:

> **Understanding the formation of the size spectra and their parameterizations have been hampered by the contradiction between the theory of Maxwellian condensation and observations of size spectra in clouds . . . The theory of regular condensation in a constant updraft has been shown to narrow the droplet spectrum with time . . . Since broad spectra are observed in clouds, and are required to initiate collision and coalescence growth of precipitation, numerous attempts have been made to remove this contradiction . . .**
>
> **The theory of stochastic condensation attributes the presence of broad drop size spectra in clouds to the occurrence of condensation in a turbulent medium . . . In applying the**

kinetic equation for stochastic condensation . . . , we are faced with the following dilemmas: . . . Analytical solutions to stochastic kinetic equations in the low-frequency regime have been obtained of the Weibull distribution . . . , which does not agree very well with observations that are typically in better agreement with a gamma distribution . . . This suggests there are deficiencies in the formulation of the kinetic equations for stochastic condensation.

This chapter describes a wide range of analytic solutions, including those that involve gamma rather than Weibull size distributions, which may help to address this concern.

11.1 Boundary Conditions

Building on the earlier discussion in Section 9.5, this chapter will focus on the following two types of boundary conditions:

1. *Free-field boundary conditions for late-time non-stationary aggregation.* In this case, there are no sources or sinks in the initial or boundary conditions. However, sinks of large particles may develop spontaneously at some finite time due to gelation.
2. *Periodic boundary conditions for late-time stationary aggregation.* In this case, there is a source of small particles and an equal-but-opposite sink of large particles. These can be either explicit or implicit, i.e., they can appear in the governing equations or not.

As discussed earlier in Chapter 9, sinks and sources are either hard or soft. More specifically, for purposes of this discussion, *hard sources* introduce small particles of a fixed size x_{min} while *hard sinks* remove large particles of a fixed size x_{max} or larger. By contrast, *soft sources* introduce a wide range of small particles while *soft sinks* gradually remove a greater percentage of the particles as their size increases.

11.2 Common Size Distributions

This section briefly reviews the size distributions commonly observed in aggregating particle systems. Here and in the rest of the chapter, the variable x refers to particle masses. The effects of the initial conditions gradually fade over time. Thus, two other two key factors—the collision kernel (i.e., λ) and the boundary conditions (i.e., sources and sinks)—jointly determine the late-time size distributions. In this chapter, all such size distributions are Rosin-Rammler size distributions. In regular form, these may be written as follows:

$$c(x,t) = C(t)f(x,t) = C(t)A(t)x^{n/m-1}\exp\left[-b(t)x^{l/m}\right]. \tag{11.1}$$

As one subset, gamma size distributions have $l = 1$ or, alternatively, $l/m = 1$:

$$c(x,t) = C(t)f(x,t) = C(t)A(t)x^{n/m-1}\exp\left[-b(t)x\right]. \tag{11.2}$$

As another subset, Weibull size distributions have $l = n$ or, equivalently, $l/m = n/m$:

$$c(x,t) = C(t)f(x,t) = C(t)A(t)x^{n/m-1}\exp\left[-b(t)x^{n/m}\right]. \quad (11.3)$$

As a third subset, power law size distributions have $l = 0$ or, equivalently, $l/m = 0$:

$$c(x,t) = C(t)f(x,t) = C(t)A(t)x^{n/m-1}. \quad (11.4)$$

If $n/m < 0$, power law and gamma size distributions obtain infinite $C(t)$ unless there is a lower size limit $x_{\min}$, which naturally occurs at late times if there is a hard source of small particles of size $x_{\min}$. By contrast, Weibull size distributions obtain finite $C(t)$ by gradually tapering the frequency of small particles relative to power law size distributions, which naturally occurs at late times for certain soft sources of small particles. Similar observations apply for large particles for $n/m > 0$.

11.3 Size Distributions for Non-Local Aggregation

By definition, non-local particle systems are dominated by collisions between particles of unequal masses x and y. In other words, for most collisions that lead to aggregation, $x \gg y$ or $x \ll y$. Then, for most homogeneous collision kernels:

$$K(x,y)/K_0 \cong \begin{cases} x^{\nu}y^{\mu} & x \gg y \\ x^{\mu}y^{\nu} & x \ll y \end{cases} \quad (11.5a)$$

or equivalently:

$$K(x,y)/K_0 \cong x^{\nu}y^{\mu} + x^{\mu}y^{\nu} \text{ for } x \gg y \text{ or } x \ll y. \quad (11.5b)$$

Notice that μ is associated with the *smaller* of the two particles, ν is associated with the *larger* of the two particles, and $\lambda = \mu + \nu$. Table 11.1 gives some examples of Equation (11.5). As this table shows, K often depends only on the larger of the two particles, i.e., $\mu = 0$ and $\lambda = \nu$.

First, consider non-stationary cases with free-field boundary conditions. According to the best available theory, as reviewed in the following examples, non-local aggregation in such cases leads to gamma size distributions where:

$$n/m = -\lambda + 1 \text{ for } \lambda < 1 \quad (11.6)$$

for large enough times and particle sizes. Equation (11.6) first appeared in Lushnikov (1973). If $\lambda > 1$, gelation occurs. If gelation occurs, aggregation tends to be local rather than non-local, as discussed earlier. Thus, Equation (11.6) specifies $\lambda < 1$ so that gelation does not occur. According to Equation (9.16), λ is an integer multiple of 1/6 between −1 and 3. If $\lambda < 1$ then:

$$n/m \text{ is an integer multiple of } 1/6 \text{ between } -2 \text{ and } 0. \quad (11.7)$$

In essence, Equations (11.6) and (11.7) generalize the results of Section 10.4 from $\lambda = 0$ to other values of λ. As the main compromise, while the results of Section 10.4 apply

Table 11.1 Examples of approximate collision kernels suitable for free merging cases. All examples are drawn from Tables 9.1 and 9.2.

λ	$K(x,y)/K_0$	$\cong K(x,y)/K_0$ for $x \gg y$	$\cong K(x,y)/K_0$ for $y \gg x$	μ	ν
−2/3	$x^{-2/3}+y^{-2/3}$	$y^{-2/3}$	$x^{-2/3}$	−2/3	0
−1/3	$(x^{1/3}+y^{1/3})(x^{-2/3}+y^{-2/3})$	$x^{1/3}y^{-2/3}$	$x^{-2/3}y^{1/3}$	−2/3	1/3
−1/6	$(x^{1/3}+y^{1/3})(xy)^{1/2}(x+y)^{-3/2}$	$x^{-2/3}y^{1/2}$	$x^{1/2}y^{-2/3}$	1/2	−2/3
0	1	1	1	0	0
0	$(x^{1/3}+y^{1/3})^2x^{-1/3}y^{-1/3}$	$x^{1/3}y^{-1/3}$	$x^{-1/3}y^{1/3}$	−1/3	1/3
0	$(x^{1/3}+y^{1/3})(x^{-1/3}+y^{-1/3})$	$x^{1/3}y^{-1/3}$	$x^{-1/3}y^{1/3}$	−1/3	1/3
1/6	$(x^{1/3}+y^{1/3})^2(x^{-1}+y^{-1})^{1/2}$	$x^{2/3}y^{-1/2}$	$x^{-1/2}y^{2/3}$	−1/2	2/3
1/3	$x^{1/3}+y^{1/3}$	$x^{1/3}$	$y^{1/3}$	0	1/3
2/3	$x^{2/3}+y^{2/3}$	$x^{2/3}$	$y^{2/3}$	0	2/3
2/3	$(x^{1/3}+y^{1/3})^2$	$x^{2/3}$	$y^{2/3}$	0	2/3
5/6	$\lvert x^{1/6}-y^{1/6}\rvert(x^{1/3}+y^{1/3})^2$	$x^{5/6}$	$y^{5/6}$	0	5/6
1	$x+y$	x	y	0	1
1	$\lvert x^{1/3}-y^{1/3}\rvert(x^{1/3}+y^{1/3})^2$	x	y	0	1
1	$(x^{1/3}+y^{1/3})^3$	x	y	0	1
4/3	$x^{4/3}+y^{4/3}$	$x^{4/3}$	$y^{4/3}$	0	4/3
4/3	$(x^{2/3}+y^{2/3})^2$	$x^{4/3}$	$y^{4/3}$	0	4/3
4/3	$\lvert x^{2/3}-y^{2/3}\rvert(x^{1/3}+y^{1/3})^2$	$x^{4/3}$	$y^{4/3}$	0	4/3
4/3	$(x^{1/3}+y^{1/3})(x+y)$	$x^{4/3}$	$y^{4/3}$	0	4/3
2	x^2+y^2	x^2	y^2	0	2
2	xy	xy	xy	1	1
7/3	$\min(x,y)(x^{1/3}+y^{1/3})(x+y)$	$x^{4/3}y$	$xy^{4/3}$	1	4/3

regardless of time and particle size, the results given here apply only for large enough times and particle sizes.

Next, consider stationary cases with periodic boundary conditions. According to the best available theory, as reviewed in the following examples, non-local aggregation in such cases leads to Weibull size distributions where:

$$n/m = -\nu + 1 \tag{11.8}$$

for large enough times and particle sizes. Equation (11.8) first appeared in Brock (1972). This equation assumes that the sink in the periodic boundary condition prevents gelation regardless of λ or ν. According to Table 11.1, ν is an integer multiple of 1/6 between −1 and 3. Then:

$$n/m \text{ is an integer multiple of } 1/6 \text{ between } -2 \text{ and } 2. \tag{11.9}$$

The remainder of this section discusses possible proofs of Equation (11.6) and (11.8). For simplicity, this section only gives outlines rather than full proofs. Readers who are interested in full proofs can consult the references cited in the examples. Readers who are not interested in proofs at all can skip ahead to the next section.

Examples 11.2 and 11.3 concern Equation (11.6), i.e. cases where late-time non-stationary aggregation with free-field boundary conditions leads to gamma size distributions. In these examples, gelation does not occur because $\lambda < 1$. Examples 11.2 and 11.3 use the following type of proof:

(1) derive a simplified governing equation for large particles;
(2) guess a parametric form for the solution of the simplified governing equation;
(3) show that the assumed parametric form satisfies the simplified governing equations.

Examples 11.2 and 11.3 assume that gelation does not occur. If gelation does, in fact, occur then aggregation is likely to be local, as discussed in the next section, rather than non-local.

Example 11.2 Late-time non-stationary non-local aggregation with $\lambda < 1$. Part 1

Consider aggregation with free-field boundary conditions and $\lambda < 1$. As discussed in the last chapter, $\lambda < 1$ ensures that gelation does not occur. In other words, in this example, there are no sinks initially and sinks do not spontaneously form at some later time. Without sinks to artificially limit the range of particle sizes, aggregation progressively increases particle size spreads. Thus, free merging aggregation between large and small particles should become increasingly common. In this case, van Dongen & Ernst (1985a) argue that:

$$K(x,y)/K_0 \cong \begin{cases} x^{\nu}y^{\mu} & x \gg y \\ x^{\mu}y^{\nu} & x \ll y \end{cases}$$

where:

$$\nu < 1. \tag{11.10}$$

As seen in Table 11.1, for any of the usual collision kernels, assuming $\nu < 1$ is essentially equivalent to assuming that:

$$\lambda = \mu + \nu < 1. \tag{11.11}$$

In other words, assuming $\nu < 1$ is essentially equivalent to assuming that gelation does not occur. As seen in the last chapter, the non-stationary solutions for $K(x,y) = K_0$ are self-similar gamma size distributions for sufficiently large times and particle sizes. Thus, it reasonable to assume that the non-stationary solutions for $\nu < 1$ and $\lambda < 1$ are also self-similar gamma size distributions for sufficiently large times and particle sizes. In other words:

$$c(x,t) \cong \frac{C(t)A_0}{M_{\text{char}}(t)}\left[\frac{x}{M_{\text{char}}(t)}\right]^{-n/m} \exp\left[-\frac{b_0 x}{M_{\text{char}}(t)}\right] \tag{11.12}$$

for sufficiently large times and particle sizes where A_0 and b_0 are constants and M_{char} is a characteristic particle mass. In fact, van Dongen & Ernst (1985a) proved that Equation (11.12) is an approximate solution to Equation (9.3) for sufficiently large times and particle sizes provided that:

$$n/m = -\lambda + 1 \tag{11.13}$$

which agrees with Equation (11.6). To complete the solution, it can be shown that:

$$M_{\text{char}}(t) \cong a_1 t^{1/(n/m)} = a_1 t^{1/(-\lambda+1)} \tag{11.14}$$

for sufficiently large times where a_1 is a constant; see Section 11.5.

Example 11.3 Late-time non-stationary non-local aggregation with $\lambda < 1$. Part 2

Consider aggregation with free-field boundary conditions and $\lambda < 1$. This example is similar to the previous example. As one difference, in the first step, van Dongen (1987) simplifies the governing equation by assuming that:

$$K(x,y)/y \to 0 \text{ as } y \to \infty. \tag{11.15}$$

As seen in Table 11.1, for any of the usual collision kernels, Equation (11.15) is essentially equivalent to assuming that $K(x,y)/K_0 \cong x^{\mu}y^{\nu}$ for $x \ll y$ where $\nu < 1$. As the second step, instead of guessing gamma size distributions as in the last example, van Dongen (1987) guessed Rosin-Rammler size distributions; see Equation (11.1). As the third step, Van Dongen (1987) showed that Rosin-Rammler size distributions are approximate solutions of Equation (9.3) for large enough times and sizes if $n/m = -\lambda + 1$ and $l/m = 1$. Because $l/m = 1$, this solution corresponds to a gamma size distribution. So far, the results of this example seem to be identical to those of the previous example. However, it turns out that assuming Rosin-Rammler rather than gamma size distributions yields a significant amount of additional information. First, in order for Equation (11.6) to be valid, von Dongen (1987) found that the initial conditions must be bounded by an exponential size distribution. In other words:

$$c(x,0) < A_{\max}\exp(-b_{\max}x) \tag{11.16}$$

for some constants $A_{\max}$ and $b_{\max}$. Second, in order for Equation (11.6) to be valid, van Dongen (1987) found that "$K(x, 1-x)$ must be continuous and strictly positive . . . in order to exclude kernels with vanishing diagonal elements, such as $K(x,y) = |x-y|$, for which a monodisperse distribution is stationary." As seen in Table 11.1, this condition excludes several common collision kernels, e.g., the kernel with $\lambda = 5/6$. Third, von Dongen (1987) found that the solution evolves in three distinct stages. At the earliest times, starting from monodisperse initial conditions, van Dongen (1987) showed that:

$$c(x,t) \cong A(t)\,x^{-\lambda} \tag{11.17}$$

for sufficiently large particles; see also Lushnikov (1973). Notice that Equation (11.17) is a power law size distribution with $n/m = -\lambda + 1$. At middling times, van Dongen (1987) showed that:

$$c(x,t) \cong A(t)\,x^{-\lambda}\exp\left[-b(t)x\right] \tag{11.18}$$

for sufficiently large particles. Notice that Equation (11.18) is a regular gamma size distribution with $n/m = -\lambda + 1$. Finally, at late times, van Dongen (1987) showed that:

$$c(x,t) \cong \frac{C(t)A_0}{M_{\text{char}}(t)}\left[\frac{x}{M_{\text{char}}(t)}\right]^{-\lambda}\exp\left[-\frac{b_0 x}{M_{\text{char}}(t)}\right] \tag{11.19}$$

for sufficiently large particles, where M_{char} is a characteristic mass. Notice that Equation (11.19) is a self-similar gamma size distribution with $n/m = -\lambda + 1$, just as in the previous example. For more information on this example, see also van Dongen & Ernst (1986, 1987).

Now consider late-time stationary aggregation with periodic boundary conditions. In the simplest case, one of the colliding particles is always a monomer. If the particle system consists of liquid droplets and vapor molecules, this may be called *stochastic condensation*; see Example 11.1. More generally, as noted earlier, this may be called *Group B* aggregation; see Blum & Wurm (2008). In this case, the governing equation can be written as follows:

$$\frac{\partial c(x,t)}{\partial t} + \alpha_0\frac{\partial\left[x^{\nu}c(x,t)\right]}{\partial x} = a(x,t) + b(x,t)\,c(x,t) \tag{11.20}$$

where $\alpha_0 > 0$; see, e.g., Ramabhadran et al. (1976), Peterson et al. (1978), Seinfeld & Pandis (2016). Notice that that Equation (11.20) is the same as Equation (9.4) without the integrals. As discussed in Section 9.2, when they are positive, $a(x, t)$ and $b(x, t)c(x, t)$ are sources that add entirely new particles. By contrast, when α_0 is positive, $\alpha_0\partial[x^{\nu}c(x,t)]/\partial x$ is a source that gradually increases the sizes of existing particles. In addition to the sources, assume that there is an equal-but-opposite sink. Then Equation (11.20) admits stationary solutions $c(x)$, which depend on the sources and sinks but not the initial conditions. As one possibility, suppose that:

$$a(x,t) = a_0\delta(x - x_1),\quad b(x,t) = 0 \tag{11.21}$$

where a_0 is a constant. Here, a hard source injects monomers of size x_1, which replace those lost to mergers with larger particles. Then Equation (11.20) admits the following stationary solution:

$$c(x) = A_0 x^{-\nu}\ \text{ for }\ x > x_1 \tag{11.22}$$

where A_0 is a constant; see, e.g., Smith et al. (2016). Notice that this is a power law size distribution where $n/m = -\nu + 1$. As another possibility, suppose that:

$$a(x,t) = 0, \; b(x,t) = b_0 c(x,t) \tag{11.23}$$

where b_0 is a constant. Then Equation (11.20) admits the following stationary solution:

$$c(x) = A_0\, x^{-\nu} \exp\left[-\frac{b_0}{\alpha_0(-\nu+1)} x^{-\nu+1}\right] \tag{11.24}$$

where A_0 is a constant; see, e.g., Brock (1972), Horvai et al. (2008) and Ball et al. (2011). Notice that this is a Weibull size distribution where $n/m = -\nu + 1$.

In general, the smaller particle in each aggregation event may or may not be a monomer. As noted earlier, this can be called *Group C* aggregation; see Blum & Wurm (2008). As it turns out, Group C aggregation can often be approximated by Group B aggregation, provided that gelation does not occur. In other words, the late-time stationary results are the same, just so long as one particle in each aggregation event is much smaller than the other, regardless of whether or not the smaller particle is a monomer. Examples 11.4 and 11.5 outline the required proofs for $\lambda < 1$ and $\lambda > 1$, respectively. Notice that Example 11.5 assumes that $\mu = 0$, i.e., the collision kernel does not depend on the size of the smaller particle. These examples show that stationary non-local aggregation tends to obtain Weibull size distributions, regardless of λ and ν.

Example 11.4 Late-time stationary non-local aggregation with $\nu < 1$ and $\lambda < 1$

Assume periodic boundary conditions with a balanced source and sink; see Example 9.4. If the source and sink are sufficiently soft, the particle sizes may be anywhere between 0 and ∞. Assume that small particles $x < x_{mid}$ preferentially collide with large particles $x > x_{mid}$ where x_{mid} is a notional dividing point. According to Brock (1972), Equation (9.3) becomes:

$$\frac{\partial c(x,t)}{\partial t} \cong -c(x,t)\int_0^{x_{mid}} K(x,y)\, c(y,t)\, dy + \frac{1}{2}\int_0^{x_{mid}} K(x-y,x)\, c(x-y,t)\, c(y,t)\, dy. \tag{11.25}$$

for $x > x_{mid}$. Using a Taylor series, Equation (11.25) can be approximated by the following second-order partial differential equation:

$$\frac{\partial c(x,t)}{\partial t} = \frac{\partial^2}{\partial x^2}\left[\beta(x)\, c(x,t)\right] - \frac{\partial}{\partial x}\left[\alpha(x)\, c(x,t)\right] + b_0 c(x,t) \tag{11.26}$$

where b_0 is a "measure of the lifetime of the particles" and where:

$$\alpha(x) = \lim_{t\to\infty} \int_0^{x_{\mathrm{mid}}} K(x,y)\, y c(y,t)\, dy = \lim_{t\to\infty} \left[K_0 \int_0^{x_{\mathrm{mid}}} y^{\mu+1} c(y,t)\, dy \right] x^{\nu} \equiv \alpha_0 x^{\nu} \tag{11.27}$$

$$\beta(x) = \lim_{t\to\infty} \frac{1}{2} \int_0^{x_{\mathrm{mid}}} K(x,y)\, y^2 c(y,t)\, dy = \lim_{t\to\infty} \left[\frac{1}{2} K_0 \int_0^{x_{\mathrm{mid}}} y^{\mu+2} c(y,t)\, dy \right] x^{\nu} \equiv \beta_0 x^{\nu} \tag{11.28}$$

which assumes that $K(x,y)/K_0 \cong x^{\nu} y^{\mu}$ with $\nu < 1$ for $y \ll x$. Equations such as (11.26) have been studied by the fluid dynamics community for over 100 years; see, e.g., Bonkile et al. (2018). By analogy with fluid dynamics, the term involving $\alpha(x)$ is called *convective* while the term involving $\beta(x)$ is called *diffusive* or *dissipative*. As a "severe simplification," Brock (1972) says that the diffusive term can be neglected provided that the small particles with $x < x_{\mathrm{mid}}$ have "masses not differing greatly among themselves and vanishingly small." In revisiting Brock (1972), Matsoukas & Lin (2006) found that this assumption was unnecessary. Instead, they found that "dropping the diffusive term ... is generally acceptable ... [for] coagulation or other processes that result in substantial broadening of the [size] distribution." In this case, Equation (11.26) becomes:

$$\frac{\partial c(x,t)}{\partial t} + \alpha_0 \frac{\partial}{\partial x} \left[x^{\nu} c(x,t) \right] = b_0 c(x,t). \tag{11.29}$$

for $x > x_{\mathrm{mid}}$. Equation (11.29) admits Equation (11.24) as stationary solution. In other words:

$$c(x) = A_0\, x^{-\nu} \exp\left[-\frac{b_0}{\alpha_0(-\nu+1)} x^{-\nu+1} \right].$$

If $\nu < 1$, this is a Weibull size distribution with $n/m = -\nu + 1 > 0$.

Example 11.5 Late-time stationary non-local aggregation with $\nu > 1$ and $\lambda > 1$

From the previous chapter, recall that collision kernels with $\lambda > 1$ tend to experience gelation, i.e., an infinite particle forms after some finite time. To prevent this, Horvai et al. (2008) and Ball et al. (2011) introduced a hard sink "which removes [particles] ... from the system when their mass exceeds a specified cutoff size" x_{max}. To ensure that free merging occurs despite the presence of the sink, and to ensure stationary solutions, Horvai et al. (2008) and Ball et al. (2011) introduced a hard source which adds particles of a fixed size $x_{\mathrm{min}} \ll x_{\mathrm{max}}$. Then, assuming free merging with a kernel $K(x,y)/K_0 = \max(x^{\nu}, y^{\nu})$, Horvai et al. (2008) and Ball et al. (2011) showed that Equation (9.3) can be approximated as follows:

$$\frac{\partial c(x,t)}{\partial t} + \alpha(t) \frac{\partial}{\partial x} \left[x^{\nu} c(x,t) \right] = b(t)\, c(x,t) \tag{11.30}$$

where $\alpha(t)$ and $b(t)$ are K_0 times the first and ν-th moments of $c(x, t)$, respectively. As in the last example, the proof requires splitting $\partial c/\partial t$ into two parts and using Taylor series. As the main difference, this example assumes finite lower bounds on the small particles $x_{\min}$ instead of finite upper bounds on the small particles x_{mid}; and finite upper bounds on the large particles $x_{\max}$ instead of finite lower bounds on the large particles x_{mid}. Let:

$$\alpha_0 = \lim_{t\to\infty} \alpha(t) = K_0 \lim_{t\to\infty} \int_{x_{\min}}^{x_{\max}} yc(y,t)\,dy \tag{11.31}$$

$$b_0 = \lim_{t\to\infty} b(t) = K_0 \lim_{t\to\infty} \int_{x_{\min}}^{x_{\max}} y^{\nu} c(y,t)\,dy. \tag{11.32}$$

Then Equation (11.30) admits Equation (11.24) as a stationary solution. In other words:

$$c(x) = A_0\, x^{-\nu} \exp\left[-\frac{b_0}{\alpha_0(-\nu+1)} x^{-\nu+1}\right].$$

If $\nu > 1$, this is a Weibull size distribution with $n/m = -\nu + 1 < 0$. Notice that:

$$c(x) \cong A_0\, x^{-\nu}$$

for large enough x, which is a power law size distribution. Based on computations, Ball et al. (2011) found: "When monomers are injected into the system, until the cutoff is felt, the mass in the system just increases linearly. However, as soon as large particles are generated, their absorption of the monomers is a runaway process which very rapidly converts the monomers which have accumulated in the system into large particles which are immediately removed by the cutoff. This then resets the system close to its initial state in which there are almost no particles in the system. The dynamics then repeat." This results in long-lived periodic oscillations in the total particle concentration, meaning such systems may approach a stationary state extremely slowly. This is a consequence of using a hard instead of a soft sink.

11.4 Size Distributions for Local Aggregation

By definition, local particle systems are dominated by collisions between particles of nearly equal masses x and y. In other words, for most collisions $x \approx y$. In this case, one possible approximation of the collision kernel is as follows:

$$K(x,y)/K_0 \cong \text{const.}\, x^{\lambda} \cong \text{const.}\, y^{\lambda} \quad \text{for } x \approx y. \tag{11.33}$$

Another possible approximation is as follows:

$$K(x,y)/K_0 \cong \text{const.}\, x^{\nu} y^{\mu} \quad \text{for } x \approx y \tag{11.34}$$

where μ and ν are given in Table 11.1.

Equation (11.34) has several advantages over Equation (11.33). As one advantage, Equation (11.34) is consistent with Equation (11.5). In other words, the same approximation applies for $x \approx y$ and $y \ll x$, except for the multiplicative constant. As a second advantage, Equation (11.34) includes the product kernel, $K(x, y)/K_0 = xy$, which played a major role in the last chapter. As a third advantage, if $K(x, y)/K_0$ is exactly or approximately equal to const. $x^{\nu}y^{\mu}$ for any reason, then forced locality occurs provided that:

$$|\mu - \nu| < 1. \tag{11.35}$$

In other words, according to Connaughton et al. (2018), Equations (11.34) and (11.35) lead to "coalescence being dominated by collisions between similar mass particles" for sufficiently late times and sufficiently large particles. Thus, it is not necessary to assume forced locality in advance; rather, it occurs naturally. In the research literature, Equation (11.35) appears in various different forms. For example, Camacho (2001) says $\nu < (1 + \lambda)/2$ while Connaughton et al. (2004) say $\mu - \nu + 1 > 0$. These conditions are both equivalent to Equation (11.35) provided that $\nu \geq \mu$. Notice that $\nu \geq \mu$ is true for all but one of the kernels listed in Table 11.1.

First, consider non-stationary cases with free-field boundary conditions. According to the best available theory, as reviewed in the following examples, local aggregation in such cases leads to power law or gamma size distributions where:

$$\frac{n}{m} = -\frac{\lambda + 1}{2} \quad \text{for } \lambda > 1 \tag{11.36a}$$

for large enough times and particle sizes. Equation (11.36a) first appeared in Lushnikov & Smirnov (1975) and Lushnikov & Piskunov (1976). According to Equation (9.15), λ is an integer multiple of 1/3 between −1 and 3. However, according to Equation (11.36a), $\lambda > 1$. Taking the intersection, λ is an integer multiple of 1/3 between 1 and 3. Then:

$$n/m \text{ is an integer multiple of } 1/6 \text{ between } -2 \text{ and } -1. \tag{11.37a}$$

In essence, Equations (11.36a) and (11.37a) generalize the results of Section 10.6 from $\lambda = 2$ to other values of λ. As the main compromise, the results of Section 10.6 apply regardless of time and particle size, while the results give here apply only for large enough times and particle sizes.

Next, consider stationary cases with periodic boundary conditions. According to the best available theory, as reviewed in the following examples, local aggregation in such cases leads to power law or gamma size distributions where:

$$\frac{n}{m} = -\frac{\lambda + 1}{2} \tag{11.36b}$$

for large enough times and particle sizes. Equation (11.36b) first appeared in Klett (1975). According to Equation (9.15), λ is an integer multiple of 1/3 between −1 and 3. Then:

$$n/m \text{ is an integer multiple of } 1/6 \text{ between } -2 \text{ and } 0. \tag{11.37b}$$

To summarize, mean-field theory for late-time aggregation obtains power law, gamma, and Weibull size distributions where n/m is an integer multiple of 1/6. Local vs. non-local aggregation may affect the possible range of n/m but, in either case, n/m is an integer multiple of 1/6. Similarly, free-field vs. periodic boundary conditions may affect the possible range of n/m but, in either case, n/m is an integer multiple of 1/6. This conclusion is consistent with a wide body of computational and experimental evidence; see, e.g., Tables 4.1, 4.2, and 5.1.

The remainder of this section discusses possible proofs of Equations (11.36a) and (11.36b). For simplicity, this section only gives outlines rather than full proofs. Readers who are interested in seeing full proofs can consult the references given below. Readers who are not interested in proofs at all can skip ahead to the next section. Examples 11.7 and 11.8 concern Equation (11.36a), i.e., they show that late-time non-stationary aggregation with free-field boundary conditions leads to power law size distributions if there is a hard sink of large particles. In these examples, gelation occurs because $\lambda > 1$. Similarly, Examples 11.9 and 11.10 concern Equations (11.36b), i.e., they show that late-time stationary aggregation with periodic boundary conditions leads to a power law or gamma size distribution, depending on whether there is a hard or soft sink of large particles, respectively. In these examples, gelation does not occur either because $\lambda < 1$ or because the sink in the periodic boundary conditions absorbs any incipient gel particles. The proofs given in Examples 11.7 and 11.10 are based on *generating functions*, which are a standard part of probability theory; see e.g., Ross (2018) and Rossi (2018). Krapivsky et al. (2010) say:

> A powerful approach for solving for the cluster mass distribution is the generating function method. This technique is ideally suited for aggregation because the master equations are in the form of a discrete convolution . . . Notice that the generating function is simply the discrete analog of the Laplace transform.

As shown in Examples 11.7 and 11.10, rewriting the governing equations in terms of appropriately chosen generating functions typically results in a simple quadratic equation. However, it is generally difficult to convert the resulting solution back into the original physical variables. While proofs based on generating functions are powerful and general, Examples 11.8 and 11.9 show that simpler alternatives may be used successfully in many cases.

Example 11.7 Late-time non-stationary local aggregation with $|\mu - \nu| < 1$ and $\lambda > 1$. Part 1

Consider aggregation with free-field boundary conditions and $\lambda > 1$. As discussed in the last chapter, $\lambda > 1$ implies that gelation occurs, i.e., a sink forms spontaneously after some finite time. Assume that this sink leads to forced locality such that $K_{ij} \cong i^{\nu} j^{\mu}$ where $\lambda = \mu + \nu > 1$, $|\mu - \nu| < 1$, and $\nu > 0$. Then Equation (9.1) can be approximated as follows:

$$\frac{dc_k(t)}{dt} = \sum_{i+j=k} i^{\nu} j^{\mu} c_i c_j - k^{\nu} c_k M_{\mu} - k^{\mu} c_k M_{\nu} \tag{11.38}$$

where M_n is the n-th moment:

$$M_n = \sum_{k=1}^{\infty} k^n c_k(t).$$

Hendriks et al. (1983) introduced the following three *generating functions*:

$$G_\nu(y,t) = \sum_{k=1}^{\infty} k^\nu c_k(t)\exp(ky) \cong M_\nu + a_\nu(-y)^{\alpha_\nu} \tag{11.39}$$

$$G_\mu(y,t) = \sum_{k=1}^{\infty} k^\mu c_k(t)\exp(ky) \cong M_\mu + a_\mu(-y)^{\alpha_\mu} \tag{11.40}$$

$$G_0(y,t) = \sum_{k=1}^{\infty} c_k(t)\exp(ky) \tag{11.41}$$

Rewriting Equation (11.38) in terms of these three generating functions, Hendriks et al. (1983) obtained:

$$\frac{\partial G_0}{\partial t} + M_\nu M_\mu = (G_\nu - M_\nu)(G_\mu - M_\mu). \tag{11.42}$$

Using a singularity analysis, which will not be discussed here, Hendriks et al. (1983) concluded that:

$$c_k(t) \cong \left[\frac{E(t)}{2\pi}\right]^{1/2} k^{-(\lambda+3)/2} \tag{11.43}$$

for large enough k where $E(t) = dM_1(t)/dt$. This is a power law size distribution with $n/m = -(\lambda+1)/2$. This result assumes that the gel acts like a hard sink for large particles. If the gel acts like a soft sink, a gamma size distribution will occur; see, e.g., Example 10.7. For more information on this example see, e.g., Ziff et al. (1982) and Hayakawa (1987).

Example 11.8 Late-time non-stationary local aggregation with $|\mu - \nu| < 1$ and $\lambda > 1$. Part 2

Consider aggregation with free-field boundary conditions and $\lambda > 1$. As in the previous example, assume that gelation leads to forced locality such that $K_{ij} \cong i^\nu j^\mu$ where $\lambda = \mu + \nu > 1$, $|\mu - \nu| < 1$, and $\nu > 0$. As an alternative to using generating functions, this example uses the following three step approach:

(1) Assume the mass flux transferred from sol to gel is greater than 0 but less than ∞.
(2) Based on the first assumption, guess a general parametric form for the solution.
(3) Show that this form ensures that the mass flux is greater than 0 but less than ∞.

Regarding the first step, according to Leyvraz & Tschudi (1982), Equation (9.1) implies that:

$$\dot{M}^{(\infty)}(t) = -\lim_{k\to\infty} \sum_{i=1}^{k} \sum_{j=k-i+1}^{\infty} iK_{ij}c_i(t)c_j(t) \tag{11.44}$$

where $\dot{M}^{(\infty)}(t)$ is the mass flux transferred from sol to gel at time $t > t_g$.

Notice that:

$$0 < \dot{M}^{(\infty)}(t) < \infty. \tag{11.45}$$

Regarding the second step, according to van Dongen & Ernst (1985b), Equation (11.45) will be true "if $c_i(t)$ has a sufficiently slow (algebraic) decay at large k." In other words:

$$c_k(t) \cong A(t)\, k^{-\tau} \tag{11.46}$$

as $k \to \infty$. Substituting Equation (11.46) into (11.44):

$$\dot{M}^{(\infty)}(t) = -A(t)^2 \left(\lim_{k\to\infty} k^{3+\lambda-2\tau}\right) \int_0^1 dx \int_{1-x}^{\infty} dy\, x\, K(x,y)\,(xy)^{-\tau}.$$

Then $0 < \dot{M}^{(\infty)}(t) < \infty$ if and only if:

$$3 + \lambda - 2\tau = 0.$$

Solving for τ:

$$\tau = \frac{\lambda + 3}{2}$$

Thus, just like the last example, this example obtains a power law size distribution with $n/m = -\tau + 1 = -(\lambda + 1)/2$. Just like the last example, this result assumes that the gel acts like a hard sink for large particles.

Example 11.9 Late-time stationary local aggregation with a hard sink

Assume there is a hard sink of large particles of size $x_{\max}$ or greater and an equal-but-opposite hard source of small particles of size $x_{\min}$. This will result in a steady average mass flux E from small to large particles. In stationary cases, Equation (9.3) becomes as follows:

$$\frac{1}{2}\int_{x_{\min}}^{x} K(x-y,x)\,c(x-y)\,c(y)\,dy = c(x)\int_{x_{\min}}^{x_{\max}} K(x,y)\,c(y)\,dy \tag{11.47}$$

for $x_{\min} \le x < x_{\max}$. Jeffrey (1981) says "we consider the integral on the right-hand side... and estimate it by assuming that collisions between particles of about the same size will make the most contribution to it." More specifically, if $K(x,y) \approx K(x,x)$ and $c(y) \approx c(x)$ when:

$$(1 - \text{const.})\,x \le y \le (1 + \text{const.})\,x.$$

If this inequality is true, then the right hand side of Equation (11.47) becomes:

$$\text{const.}\, K(x,x)\, c^2(x)\, x.$$

Multiply by const. x^2 to convert from a number flux to a mass flux:

$$E = \text{const.}\, K(x,x)\, c^2(x)\, x^3.$$

Invert to find:

$$c(x) = \text{const.}\left[\frac{E}{K(x,x)}\right]^{1/2} x^{-3/2}. \tag{11.48}$$

Assuming that Equation (11.33) applies, $K(x,x) \cong \text{const.}\, x^{\lambda}$. Then the final result is as follows:

$$c(x) = \text{const.}\, E^{1/2} x^{-(\lambda+3)/2} \tag{11.49}$$

which is a power law size distribution with $n/m = -(\lambda+1)/2$. While this proof used mean-field theory, the same result can be derived using dimensional analysis; see, e.g., Connaughton et al. (2004, 2006). According to Jeffrey (1981), dimensional analysis implicitly assumes that "collisions between particles of about the same size will make the most contribution . . . If the most important coagulation occurs between particles of very different size, then [dimensional analysis] . . . is inappropriate from the start." In addition, Jeffrey (1981) shows that dimensional analysis applies for hard but not soft sinks such as those due to differential sedimentation, a.k.a., gravitational settling. In essence, soft sinks involve a length scale proportional the range of particle sizes affected. Then, according to White (1983), the particle diameter and the sink's length scale constitute two "physically independent quantities with the same units. Dimensional reasoning alone cannot, in consequence, yield an expression."

Example 11.10 Late-time stationary local aggregation with a soft sink

As in the previous example, consider stationary aggregation on a finite periodic domain. Then, according to Equation (9.2):

$$0 = -c_i \sum_{j=1}^{i_{\max}} K_{ij} c_j + \frac{1}{2} \sum_{j=1}^{i-1} K_{i-j,j} c_{i-j} c_j + a_i + b_i c_i \tag{11.50}$$

for $i \le i_{\max}$. Assume that:

$$K_{ij} = K_0 i^{\lambda/2} j^{\lambda/2} \tag{11.51}$$

$$a_i = a\delta_{i1} \tag{11.52}$$

$$b_i = b i^{\lambda/2} \tag{11.53}$$

where $\delta_{i1} = 1$ when $i = 1$ and $\delta_{i1} = 0$ when $i \neq 1$.

Notice that the source is hard, i.e., it involves a single particle size. By contrast, notice that the sink is soft, i.e., its effect gradually increases as a function of particle size. To solve this problem, White (1982) used the following generating function:

$$G(y) = \sum_i i^{\lambda/2} c_i y^2. \tag{11.54}$$

Rewriting Equations (11.51) to (11.54) in terms of $G(y)$:

$$\frac{1}{2} K_0 G^2(y) - [K_0 G(1) + b]\, bG(y) + ay = 0 \tag{11.55}$$

which can be solved using the quadratic equation. By matching terms in a power series, White (1982) showed the following:

$$c_i = \left(\frac{2a}{K_0 w}\right)^{1/2} \left| \begin{pmatrix} 1/2 \\ i \end{pmatrix} \right| i^{-\lambda/2} w^i \cong \left(\frac{a}{2\pi K_0 w}\right)^{1/2} i^{-(\lambda+3)/2} \exp(\ln w\, i) \tag{11.56}$$

where the approximation on the right hand side is based on Stirling's formula and where:

$$w = \frac{2aK_0}{2aK_0 + b^2}.$$

Notice that $w < 1$ and therefore $\ln w < 0$. Then Equation (11.56) is a gamma size distribution with $n/m = -(\lambda+1)/2$. White (1982, 1983) argues that the same result applies for any collision kernel and sink provided that $K_{ij} \leq K_0 i^{\lambda/2} j^{\lambda/2}$ and $b_i \geq b i^{\lambda/2}$ so that "the rate coefficient for physical removal is at least as strong as . . . [the] coagulation kernel."

11.5 Average Particle Sizes

In the last chapter, mean-field theory provided expressions for both the particle size distribution and the average particle size. However, that is not the case in this chapter. Instead, so far, mean-field theory has only obtained expressions for the particle size distribution. To obtain a general expression for the average particle size, suppose that the size distribution is self-similar for large enough times and particle sizes. In fact, Leyvraz (2003) says that "every solution starting from a 'narrow' . . . size distribution evolves for large times into a self-similar solution . . . [and these are] in fact the only relevant solutions at large times, as they are approached from essentially any initial conditions." This section will show that self-similarity occurs under certain conditions. As part of the proof, this section will obtain expressions for the average particle size, which complement the earlier expressions for the particle size distribution. From Section 3.6, recall that self-similar solutions have the following form:

$$f(M,t) = \frac{c(M,t)}{C(t)} = \frac{1}{M_{\text{char}}(t)} \tilde{f}\left[\frac{M}{M_{\text{char}}(t)}\right] \tag{11.57}$$

for sufficiently large times and particle sizes.

To prove that self-similarity occurs, consider the following two step approach:

(1) Guess a general expression for $M_{\text{char}}(t)$ and substitute the guess into Equation (11.57).
(2) Show that Equation (11.57) with the guess for $M_{\text{char}}(t)$ approximately satisfies the mean-field governing equation.

Regarding the first step, it seems reasonable to guess that:

$$M_{\text{char}}(t) = M_{\text{avg}}(t) \tag{11.58}$$

where:

$$M_{\text{avg}}(t) = \int_0^\infty M f(M,t)\, dM.$$

For a given size distribution, other average masses such as $M'_{\text{avg}}(t)$ and $M_{M,\,\text{avg}}(t)$ are equal to $M_{\text{avg}}(t)$ to within in a constant factor, provided that all of the averages in question are finite; see, e.g., Laney (2015). When this is true, the choice of $M_{\text{char}}(t)$ does not matter. However, letting $M_{\text{char}}(t) = M_{\text{avg}}(t)$ simplifies the proof. If conservation of mass applies then:

$$C(t)\, M_{\text{avg}}(t) = C_0 M_0 = \text{const.} \tag{11.59}$$

per Equation (10.9). Then Equation (11.57) can be written as follows:

$$c(M,t) = \frac{C_0 M_0}{M^2_{\text{avg}}(t)}\, \tilde{c}\left[\frac{M}{M_{\text{avg}}(t)}\right]. \tag{11.60}$$

Regarding the second step, substitute Equation (11.60) into Equation (9.3). Assume free-field boundary conditions. Also assume that K_0 is defined as in Sections 9.3, 11.3, and 11.4. Then van Dongen & Ernst (1985a, eq. 5) obtain the following approximate differential equation:

$$\frac{dM_{\text{avg}}(t)}{dt} = a_0 K_0 C_0 M_0 M^{\lambda}_{\text{avg}}(t) \tag{11.61}$$

where a_0 is a unitless positive constant; see also Leyvraz (2003, eq. 3.31) and Krapivsky et al. (2010, eq. 5.59). Equation (11.61) has the following solution:

$$M_{\text{avg}}(t) = M_0\left[(1-\lambda)\, a_0 K_0 C_0 M_0^{\lambda} t + 1\right]^{\frac{1}{1-\lambda}} \tag{11.62}$$

provided that $\lambda \neq 1$, which is easy to verify through back substitution. If $\lambda > 1$, then Equation (11.62) can be rewritten as follows:

$$M_{\text{avg}}(t) = M_0\left[(t_g - t)/t_g\right]^{\frac{1}{1-\lambda}} \tag{11.63}$$

where:

$$t_g = \frac{1}{(\lambda-1)\, a_0 K_0 C_0 M_0^{\lambda}} \tag{11.64}$$

is called the *gel time*. Notice that Equation (11.63) applies only at sufficiently late times prior to gelation, i.e., for $t < t_g$.

To summarize, the mean-field governing equation for aggregation with free-field boundary conditions admits a conservative self-similar solution if:

$$M_{avg}(t) = \begin{cases} a_1 t^{1/(1-\lambda)} & \lambda < 1,\ t \to \infty \\ M_0 \exp(-a_0 K_0 C_0 M_0 t) & \lambda = 1 \\ a_2 (t_g - t)^{1/(1-\lambda)} & \lambda > 1,\ t \to t_g \end{cases} \tag{11.65}$$

where a_1 and a_2 are constants. Notice that, in all cases, $M_{avg}(t)$ is monotone increasing with time, as expected. This result only applies when conservation of mass applies, i.e., when there are no sources or sinks. The average particle size may be completely different if there are sources or sinks due to boundary conditions or gelation. Example 11.11 discusses three specific examples of Equation (11.65).

Example 11.11 Average particle sizes for constant, sum, and product kernels

If $K(x, y) = K_0$, Scott (1968) showed that integrating Equation (9.3) obtains the following:

$$M_{avg}(t) = M_0 (1 + K_0 C_0 t/2) \cong K_0 C_0 M_0 t/2 \tag{11.66a}$$

$$C(t) = C_0 M_0 / M_{avg}(t) = \frac{C_0}{1 + K_0 C_0 t/2}. \tag{11.66b}$$

This result agrees with Example 10.2. If $K(x, y) = K_0(x+y)$, Scott (1968) showed that integrating Equation (9.3) obtains the following:

$$M_{avg}(t) = M_0 \exp(K_0 C_0 M_0 t) \tag{11.67a}$$

$$C(t) = C_0 M_0 / M_{avg}(t) = C_0 \exp(-K_0 C_0 M_0 t). \tag{11.67b}$$

This result agrees with Example 10.6. Finally, if $K(x, y) = K_0 xy$, then Scott (1968) showed that integrating Equation (9.3) obtains the following:

$$M_{avg}(t) = \frac{M_0}{1 - K_0 C_0 M_0^2 t/2} \tag{11.68a}$$

$$C(t) = C_0 M_0 / M_{avg}(t) = C_0 \left(1 - K_0 C_0 M_0^2 t/2\right). \tag{11.68b}$$

This result agrees with Example 10.8. Notice that Equations (11.66), (11.67), and (11.68) all agree with Equation (11.65).

While the research literature has devoted a great deal of attention to self-similarity, it turns out to be surprisingly difficult to write most specific analytic solutions in self-similar forms. In the last chapter, Examples 10.1 and 10.2 naturally yielded self-similar forms. However, most of the other examples did not. While self-similarity may be hard to observe in theory, it is easy to observe in practice; see, e.g., the experimental results given in Figures 10.3 and 10.5.

11.6 Computational Validation

Tables 4.1 and 4.2 give examples of computational results for particle aggregation and, in most cases, lesser amounts of particle fragmentation. In all cases, the results are either power law or gamma size distributions where the exponents n/m appear to be even multiples of 1/6 to within the limits of computational accuracy. However, because these computational studies did not report values of λ, no conclusions can be reached about the relationship between n/m and λ. Instead, consider numerical solutions of the mean-field governing equations. In one such study, Lee (2001) found that aggregation just prior to gelation obtained a power law size distribution where $n/m \approx -(\lambda + 1)/2$ to within about 7%. In another such study, Leyvraz (2006) found that the difference between the numerical and analytic mean-field solution decreased as the sink strength increased, i.e., as λ increased. For example:

- Numerical results for $\lambda = 4/3$ obtained $n/m \approx -(\lambda + 1)/2 = -7/6$ to within 8%.
- Numerical results for $\lambda = 3/2$ obtained $n/m \approx -(\lambda + 1)/2 = -5/2$ to within 2%.
- Numerical results for $\lambda = 2$ obtained $n/m \approx -(\lambda + 1)/2 = -3/2$ to within 1%.

More recently, Connaughton et al. (2018) showed that using two sinks together instead of only one may cause n/m to change from, say, $-\lambda/2$ to $-(\lambda + 1)/2$. In other words, a combination of two sinks may produce forced locality, even when neither sink by itself can.

11.7 Experimental Validation

This section gives experimental evidence for some of the analytic solutions presented earlier in the chapter. This section builds on the experimental evidence given earlier in Section 10.8. However, while Section 10.8 was limited to $\lambda = 0$, 1, and 2, this section considers a broad range of different λ. Examples 11.12 and 11.13 concern colloidal aggregation without and with gelation, respectively. Example 11.14 concerns polymer aggregation with gelation. Finally, Example 11.15 shows that gelation may help explain why rain showers develop so quickly.

Example 11.12 Experiment vs. mean-field theory for colloid aggregation without gelation

Table 11.2 shows selected experimental results for late-time colloidal aggregation. As in Example 10.11, aggregation was controlled by adding various concentrations of salt, pyridine, or polyvinyl pyridine. In all cases, the researchers observed either power law or gamma size distributions. In addition, in almost all cases, the exponents n/m appear to be even multiples of 1/6, to within the limits of experimental accuracy. Most of the experiments listed in Table 11.2 tried to measure the degree of homogeneity λ using Equation (11.65). For example, suppose that an experiment obtains:

$$M_{avg}(t) = \text{const.}\, t^z \tag{11.69}$$

for sufficiently large times where z is a constant exponent less than one.

Then Equation (11.65) implies that:

$$\lambda = 1 - \frac{1}{z}. \tag{11.70}$$

For another example, suppose that an experiment obtains:

$$M_{\text{avg}}(t) = \text{const. } \exp(-\text{const. } t) \tag{11.71}$$

for sufficiently large times. Then Equation (11.65) implies that:

$$\lambda = 1. \tag{11.72}$$

Recall that Equation (11.65) only applies for free-field boundary conditions at sufficiently late times. However, it appears that these conditions were only met in 5 of the 14 experiments listed in Table 11.2. In some cases, the experiments involved sinks of large particles due to gravitational settling, a.k.a., differential sedimentation. In other cases, the experiments only captured relatively early times; this was especially common for slow aggregation, a.k.a., reaction-limited cluster-cluster aggregation (RLCA). The fourth column in Table 11.2 only reports results for λ for the five experiments that clearly obtained free-field boundary conditions and sufficiently late times. Three of the five results for λ are consistent with free-merging, i.e., a power law or gamma distribution with $n/m = -\lambda + 1$. Two of the five results are consistent with something in between free merging and forced locality, i.e., an exponential size distribution.

Table 11.2 Selected experimental results for late-time colloidal aggregation without gelation.

Size Distribution			λ	Monomers		
n/m	m	Type		Material	Diameter	Reference
−1.0	–	Power Law[1, 2]	2[3]	Polystyrene Latex	0.9 μm	Pefferkorn & Varoqui (1989)
−1.0 to −0.85	2.05[a] ± 0.06	Power Law[1]	–	Silica	7 nm	Martin (1987)
−0.86	1.84[b] ± 0.08	Power Law[1]	–	Silica	13.5 nm	Martin et al. (1986)
−0.80 ± 0.08	2.65[a] ± 0.05	Power Law[1]	–	Polystyrene	44 nm	Rarity et al. (1989)
−0.66 to −0.40	–	Power Law[1]	–	Polystyrene Latex	0.9 μm	Elaissari & Pefferkorn (1991)
−0.66	2[a]	Power Law	–	Petroleum Asphaltene	–	Fenistein & Barré (2001)
−0.50	2.1[a] ± 0.05	Gamma	1[4]	Gold	7.5 nm	Lin et al. (1989, 1990b)
−0.50	2.10[a]	Gamma	1[5]	Polystyrene Latex	258 nm	Broide & Cohen (1990, 1992)
−0.30[6] ± 0.12	2.13[a] ± 0.06	Gamma	–	α-crystallin (eye lens protein)	34 nm	Bassi et al. (1995)

Continued

Table 11.2 *Continued*

Size Distribution			λ	Monomers		
n/m	m	Type		Material	Diameter	Reference
0	2.02[a] ± 0.02	Gamma	–	β-lactoglobulin (globular protein)	3.6 nm	Gimel et al. (1994)
0.31	1.75[b]	Power Law[1]	–	Gold	7.5 nm	Wilcoxon et al. (1989)
0.5	2.05[a]	Power Law[1]	–	Gold	9.5 nm to 15 nm	Olivier & Sorensen (1990)
1	1.86[b] ± 0.01	Gamma	0[4]	Gold	7.5 nm	Lin et al. (1989, 1990a)
1	1.75[b]	Gamma	0[5]	Polystyrene Latex	258 nm	Broide & Cohen (1990, 1992)

(a) Slow aggregation, i.e., RLCA.
(b) Fast aggregation, i.e., DLCA.
(1) Described as a power law times an optional exponential cutoff factor for large particles. This allows for possibility that the experimental results did not capture large enough particles to reliably distinguish between (untapered) power law and (tapered) gamma size distributions.
(2) Described as "the envelope of the size distribution at different times . . . which corresponds roughly to the peak of the distributions." The individual size distributions appear to be Weibull.
(3) ". . . These particular experiments were performed in a mixture . . . having the same density as the latex; no drift due to gravitation occurs and no macroscopic sedimentation was observed."
(4) "The cell was . . . inverted every 15 min to achieve zero gravity on average, and thus eliminate any effects of sedimentation."
(5) "The near equality of the densities of the latex spheres (1.055 g/cm^3) and water prevents appreciable gravitational settling of clusters during the initial stages of aggregation."
(6) The best of several possible proposed fits to the experimental data.

Example 11.13 Experimental vs. mean-field results for colloidal aggregation with gelation

In colloidal systems, gravity both creates and destroys gels. To explain, gravity causes the largest particles to settle. The largest particles accrete smaller particles as they settle; thus their growth rate accelerates. However, the largest particles may be extremely porous, weak, and unstable. Thus, gravity—and opposing forces such as aerodynamic drag and buoyancy—may cause the largest particles to experience secondary fragmentation and secondary aggregation, a.k.a., restructuring or collapse. In this regard, Manley et al. (2005) says "gel formation is constrained by gravity, which can impose a size-limiting strain on the aggregates, or drive sedimentation before a network can form. Moreover, even after gels form, they can be mechanically unstable . . . For weak interparticle attractions they collapse to form a loose sediment . . . Network collapse is ubiquitous; it is observed for gels made from a variety of materials, and for a wide range of [interparticle forces]. . ."

To avoid this, colloid researchers have sometimes resorted to microgravity environments. For example, Manley et al. (2004) describe colloidal aggregation experiments conducted on the International Space Station. In these tests, the microgravity environment led to larger particles than terrestrial gravity environments. Unfortunately, these tests ran for only 16 days rather than the estimated 90 days required to form a true gel. As a faster alternative, Ivlev et al. (2002), Konopka et al. (2005) and Morfill & Ivlev (2009) describe charged-particle aggregation experiments conducted on the International Space Station. In these experiments, "up to ~10^6 melamine-formaldehyde particles of 3.4 μm size were injected into a chamber filled with Argon gas at a pressure of 0.7 mbar." By design, each of the clusters was "charged, positively or negatively... [to] at least a few thousand electron charges." This caused aggregation to occur "at a rate ~10^4–10^6 times faster than that usually observed with uncharged grains." As a result, in each experiment, "a single, very large (runaway) aggregate was observed to form" starting at about 0.1s. After several seconds, "these aggregates contained up to ~10^5 microparticles, or up to ~10% of the total particles injected." They note that the aggregates were several millimeters in diameter, had a "loose fluffy structure," and obtained fractal dimensions of $m = 1.6 - 2$.

Based on the intensity of reflected laser light, Ivlev et al. (2002), Konopka et al. (2005) and Morfill & Ivlev (2009) estimated that the typical aggregates obtained a power law size distribution with $n/m = -1.5 \pm 0.5$, consistent with forced locality with $\lambda = 2$. However, instead of assuming $\lambda = 2$, Ivlev et al. (2002) proposed a kernel of the following form:

$$K(x, y; q, r) = x^{\mu} y^{\nu} |r|^{\varepsilon} + y^{\mu} x^{\nu} |q|^{\varepsilon} \tag{11.73}$$

where x, y are the masses, r, q are the charges of two colliding particles, $\lambda = \mu + \nu$ is the ordinary *geometric* degree of homogeneity, and ε is the *dielectric* degree of homogeneity. Ivlev et al. (2002) suggest that ε always obtains integer values such as 0, 1, or 2. This kernel assumes that the mass and charge of each cluster varies independently of each other in a seemingly random fashion. Konopka et al. (2005) conclude that "our measurements exhibit all the features peculiar to gelation: along with the formation of a single large ('infinite') aggregate observed for large amounts of injected particles... The spectrum of smaller ('finite') clusters in those cases is not bounded at the high-mass end but has a power law tail... [with a slope] within the range necessary for runaway growth."

Example 11.14 Experimental vs. mean-field results for polymers with gelation

Table 11.3 surveys experimental results for polymer formation. The size distributions were typically measured using size exclusion chromatography, as described in Example 10.12. In all but one case, the researchers observed either power law or gamma size distributions. In addition, in almost all cases, the exponents n/m appear to be even multiples of $1/6$ to within the limits of experimental accuracy. Unlike the colloid experiments reported in Table 11.2, the polymer experiments in Table 11.3 generally made no attempt to measure the degree of homogeneity. However, in one case, Lusignan et al. (1999) provided enough information to infer that $\lambda = 2$ and $n/m \sim -(\lambda + 1)/2 = -3/2$ consistent with forced locality; see Example 10.12.

In Table 11.3, most of the results were obtained near but somewhat before the gel point. However, in one case, Colby et al. (1992, 1993) obtained results after the gel point. Trappe et al. (1992) say the following: "Sol-gel transitions can . . . be observed in small amplitude oscillatory shear experiments. In the pre-gel region the material behaves as a viscoelastic liquid, in the post-gel region it exhibits properties of a viscoelastic solid. At the gel point the relaxation modulus follows a power law." In other words, polymer gelation can be determined based on mechanical properties rather than particle size measurements.

Table 11.3 Selected experimental size distributions for polymers with gelation.

Size Distribution					
n/m	m	Type	λ (Gel)	Material	Reference
−1.47 ± 0.05	2.07 ± 0.02	Gamma	2[1] (Pre)	"Branched polyesters with long linear chain sections"	Lusignan et al. (1999)
−1.35 ± 0.03	1.88 ± 0.09	Power Law[2]	– (Post)	"Branched polyesters beyond the gel point"	Colby et al. (1992, 1993)
−1.3 ± 0.04	–	Gamma	– (Pre)	Branched polyesters "up to the gel point"	Patton et al. (1989)
−1.3 ± 0.1	–	Gamma	– (Pre)	Polystyrene chains in solution "crosslinked by γ irradiation"	Leibler & Schosseler (1985)
−1.27 ± 0.04	–	Gamma	– (Pre)	"Polydimethylsiloxane (PDMS) chains in the melt"	Lapp et al. (1989)
−1.20 ± 0.04	2.5 ± 0.06	Power Law[2]	– (Pre)	Clusters of "deuterated and hydrogenated polyurethane"	Bouchaud et al. (1986); Adam et al. (1991)
−1.17 ± 0.08	–	(3)	– (Pre)	"Branched polyester samples below their gel point"	Lusignan et al. (1995)
−1.17 ± 0.03	2.1 ± 0.2	Power Law[2]	– (Pre)	Crosslinked chains of "anhydride cured epoxies"	Trappe et al. (1992)
−0.53 ± 0.05	–	Gamma	– (Pre)	"AB/AB_2 hyperbranched polyesters"	Kunamaneni et al. (2003)
−0.35	–	Gamma	– (Pre)	Short-chain hyperbranched polyesters	Suneel et al. (2002)

(1) Not directly stated but inferred as described in Example 10.12.
(2) A power law times an optional "exponential cutoff" for large particles. This allows for possibility that the observed result is actually the lower end of a gamma size distribution.
(3) "Fitted to a power law times a shifted Gaussian cutoff function;" see Equation (11.79).

Example 11.15 Experimental vs. mean-field results for rain initiation

Rain initiation is arguably an unsolved problem. For example, Grabowski & Wang (2013) say "it is difficult to explain the rapid growth of cloud droplets in the size range 15-40 μm in radius for which neither the diffusional mechanism [i.e., regular condensation] nor the gravitational collision-coalescence [i.e., aggregation] mechanism is effective . . . The question is what drives droplet growth through this bottleneck size range?" Similarly, Pumir & Wilkinson (2016) say that "the challenge is to explain the growth of droplets from 10 μm in size to approximately 50 μm, and in particular, to identify the physical process leading to the rapid onset of rain showers, which can develop in less than half an hour . . . Rain falls as droplets ~1 mm in size."

According to Grabowski & Wang (2013), two possible explanations for rain initiation include: (1) condensation accelerated by "turbulent fluctuations of the water-vapor supersaturation" and (2.) aggregation including "turbulent collision-coalescence" and "the effects of giant aerosol particles." Regarding the first explanation, Li et al. (2019) found that turbulent-driven "supersaturation-fluctuation-facilitated condensation may easily overcome the bottleneck." Regarding the second explanation, Pumir & Wilkinson (2016) found that "turbulence-mediated collisions may be an important ingredient to explain the runaway growth of rain drops in warm cumulus clouds." More specifically, Pumir & Wilkson (2016) focused on the following mean-field collision kernel:

$$K(x,y) = K_0 \left|x^{2/3} - y^{2/3}\right| \left(x^{1/3} + y^{1/3}\right)^2 \tag{11.74}$$

which is designed to model gravity-driven sedimentation of particles between 10 and 50 μm. They note that gelation is possible because this kernel obtains $\lambda = 4/3$, which is significantly larger than the gelation threshold value of 1. Having said this, Pumir & Wilkinson (2016) question whether gelation is an actual physical phenomenon or simply a mathematical artifact of mean-field theory. For instance, they note "under the mean-field approximation the local depletion of particles due to the formation of a large cluster is not taken into account." In addition, they observe that "the time required for the gelation transition may be strictly equal to zero. This instantaneous gelation is clearly unphysical; it implies that mean-field descriptions based on the Smoluchowski approach have to be applied with caution." These kinds of concerns, common among practitioners, are addressed at the end of Section 11.9.

In another study of rain initiation, Dziekan & Pawlowska (2017) note that "some droplets undergo a series of unlikely collisions and grow much faster than an average droplet. . . These few lucky droplets are argued to be responsible for droplet spectra broadening and rain forming quicker than predicted . . . [The] single luckiest droplet out of 1,000 grows five times faster than average and the luckiest out of 100,000, 11 times faster." In comparing the results of Monte-Carlo calculations to mean-field theory, they found that "lucky droplets . . . cannot be [fully] described using the Smoluchowski equation," as discussed earlier in Section 10.7.

However they form, the largest raindrops have two possible fates: (a.) they exit the parent cloud, cease growing, and fall unchanged to the ground, in which case the resulting raindrops are formed by aggregation; (b.) they break up due to aerodynamic drag, in which case the resulting raindrops near the ground are formed by fragmentation. Researchers are still debating which of these fates is most typical; see, e.g., Villermaux & Bossa (2009).

11.8 Aggregation Combined with Fragmentation

Rather than pure aggregation, many real-world particle systems experience some combination of aggregation and fragmentation. There is a hard and an easy way to model this combination. In the hard way, terms representing fragmentation are explicitly included in the governing equations; see the earlier discussion in Section 5.9. In the easy way, in a discrete system, fragmentation is assumed to cause the largest particles to decompose into their constituent monomers. In other words, combined aggregation-fragmentation is treated like pure aggregation with periodic boundary conditions, i.e., a sink of large particles and an equal source of monomers; see Example 9.3. In this approach, the main challenge is to choose which particles to fragment, using selection rules simple enough to allow for analytic solutions. For example, in Brilliantov et al. (2015), colliding particles either aggregate or fragment depending on their relative kinetic energies, such that the ratio of aggregating-to-fragmenting collisions is a constant. For another example, in Connauhgton et al. (2018), colliding particles either aggregate or fragment with a uniformly random probability. For a generalized product kernel of the form K_{ij} = const. $(ij)^{\lambda/2}$, Brilliantov et al. (2015) and Connauhgton et al. (2018) both obtained a gamma size distribution with $n/m = -(\lambda + 1)/2$, consistent with forced locality with a soft sink. This illustrates a general principle: *the combination of aggregation and fragmentation generally produces exactly the same types of size distributions as aggregation by itself or fragmentation by itself.* In addition to analytic results, this principle has been observed repeatedly in computational studies. For example, Table 4.1 gives computational results for combined aggregation and fragmentation; in all cases, the results appear to be gamma or power law size where n/m is even multiple of 1/6. In other words, the results are exactly the same as if only aggregation occurred.

11.9 Conclusions

This chapter has shown that, according to mean-field theory, late-time aggregation often obtains power law, gamma, or Weibull size distributions. Furthermore, the exponent n/m in these size distributions is typically an integer multiple of 1/6, provided the particle system is spatially homogeneous and isotropic. As discussed earlier in Chapters 1 and 5, such particle size distributions may be called *ideal* or *universal.* While this chapter focused on local and non-local aggregation, universal size distributions may also occur in other cases. For example, late-time aggregation governed by the sum kernel is neither local nor non-local. As shown in Section 10.5, if the particle system is spatially homogeneous and isotropic, late-time aggregation with a sum kernel obtains a universal gamma size distribution with $n/m = -1/2$.

For free-field boundary conditions, mean-field theory provides not just the exponent n/m but the entire late-time particle size distribution. To summarize:

$$f(M,t) = \frac{c(M,t)}{C(t)} \cong \frac{A_0}{M_{\text{avg}}(t)}\left[\frac{M}{M_{\text{avg}}(t)}\right]^{n/m-1} \exp\left[-b_0\frac{M}{M_{\text{avg}}(t)}\right] \tag{11.75}$$

for sufficiently large particle sizes times and sufficiently large times prior to gelation. In this expression, $C(t) = C_0M_0/M_{\text{avg}}(t)$ and:

$$M_{\text{avg}}(t) \cong \begin{cases} \text{const. } t^{1/(1-\lambda)} & \lambda < 1 \\ \text{const. } \exp(-\text{const. } t) & \lambda = 1 \\ \text{const. } (t_g - t)^{-1/(\lambda-1)} & \lambda > 1 \end{cases} \tag{11.76}$$

If $\lambda < 1$, late-time aggregation is dominated by collisions between particles of substantially different sizes. In this case, the exponent n/m is related to λ as follows:

$$n/m = -\lambda + 1. \tag{11.77}$$

If $\lambda > 1$, late-time aggregation is dominated by collisions between particles of substantially similar sizes. In this case, the exponent n/m is related to λ as follows:

$$n/m = -(\lambda + 1)/2. \tag{11.78}$$

If $\lambda = 1$, late-time aggregation evolves through collisions between particles of both similar and dissimilar sizes, without either type of collision dominating the other. In this case the exponent n/m is related to λ as follows:

$$n/m = -1/2.$$

Figure 11.1 summarizes the results of the preceding three equations. Instead of a sudden jump at $\lambda = 1$, Figure 11.1 shows a continuous transition. While notional, this continuous transition is arguably more realistic than a discontinuous transition.

Mean-field theory typically predicts power law, gamma, or Weibull size distributions. However, based on curve fits to experiments, researchers have suggested a number of other possibilities. For example, Lusignan et al. (1995) suggested "a power law times a shifted Gaussian cutoff function" of the following form:

$$c(M) = A\left(\frac{M}{M_{\text{char}}}\right)^{-n/m-1} \exp\left\{-\left[0.681 - \left(\frac{M}{M_{\text{char}}}\right)^{\sigma}\right]^2\right\} \tag{11.79}$$

where σ is a constant parameter. For another example, Friedlander & Wang (1966) suggested a log-normal size distribution of the following form:

$$c(M) = \frac{A}{\sqrt{2\pi\sigma^2}}\frac{1}{M}\exp\left\{-\frac{1}{2\sigma^2}[\ln M - a]^2\right\} \tag{11.80}$$

where $\sigma \cong 1.32$ is a universal constant; see also, e.g., Cohen & Vaughan (1971), Lee (1983), Olivier et al. (1992) and Park et al. (2000).

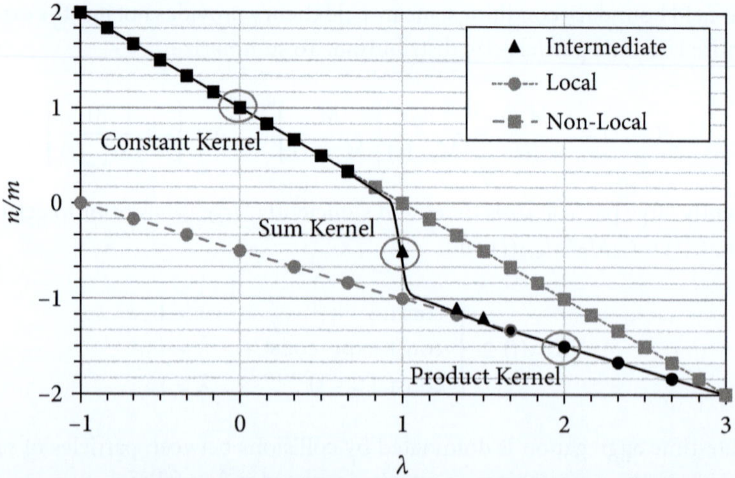

Figure 11.1 Mean-field theory for aggregation with free-field boundary conditions predicts that the exponent n/m in the gamma size distribution depends on the degree of homogeneity λ for sufficiently large particle sizes and times. The circles represent local cases where λ is an integer multiple of 1/3 and n/m is an integer multiple of 1/6. The squares represent non-local cases where λ and n/m are both integer multiples of 1/6. The triangles represent a notional transition between local and non-local.

As it turns out, shifted Gaussian, log normal, gamma, Weibull, and other types of size distributions can mimic each other; see, e.g., Villermaux (2007) and Wu & McFarquhar (2018). In particular, a log normal size distribution with $\sigma \cong 1.32$ strongly resembles an exponential size distribution with $n/m = 1$. Thus, a log normal size distribution with $\sigma \cong 1.32$ strongly resembles the classic Smoluchowski (1917) universal size distribution for:

$$K(x,y) = K_0\left(x^{1/3} + y^{1/3}\right)\left(x^{-1/3} + y^{-1/3}\right).$$

This highlights the value of using analytic approaches such as mean-field theory. Only analytic approaches—as opposed to experimental or computational approaches which always have some inherent degree of uncertainty—can definitively distinguish between the actual size distribution and various possible approximations.

Using mean-field theory for guidance, it is generally possible to accurately infer particle size distributions from computational or experimental results; see also the earlier discussion in Section 5.6. In fact, for spatially homogeneous and isotropic particle systems, there is a substantial amount of computational and experiment evidence to support the notion that late-time aggregation produces power law, gamma, and Weibull size distributions where n/m is an integer multiple of 1/6; see, e.g., Sections 4.2, 4.3, 5.7, 10.7, 10.8, 11.6, and 11.7. By contrast, it is much harder to infer mean-field quantities such as λ, μ, ν from computational or experimental results. In some computations, a collision kernel with a known λ, μ, ν may be imposed. However, this is non-trivial and relatively rare; see, e.g., Sections 10.7 and 11.6. In some experiments, λ may be estimated using Equation (11.76). However, again, this is non-trivial and relatively rare; see, e.g., Example 11.12. Thus, while there is substantial computational and experimental evidence suggesting that n/m is an integer multiple of 1/6, there is much less computational and experimental evidence showing that

λ, μ, ν are integer multiples of 1/3 or 1/6. As a result, there is only limited computational and experimental evidence for relationships such as $n/m = -\lambda + 1$, $n/m = -\nu + 1$, and $n/m = -(\lambda + 1)/2$. Having said that, such mean-field relationships have been used successfully for over 50 years, starting with Brock (1972), Lushnikov (1973), Lushnikov & Smirnov (1975) and Klett (1975).

The last chapter spent a substantial amount of time discussing gelation for product kernels. In particular, the last chapter argued that, for free-field boundary conditions, gelation occurs whenever $\lambda > 1$. Intuitively, gelation means that extremely large particles develop extremely rapidly. An especially intriguing example is rain initiation; however, the interpretation of raindrops as gels is still controversial; see Example 11.15. While mean-field theory predicts that gelation occurs, mean-field theory cannot predict the size distribution of gels. This is because, when gelation occurs, mean-field theory predicts that an infinite particle system produces a single infinite gel particle instantaneously. In reality, each large (but not infinite) spatial subregion may produce a single large (but not infinite) gel particle over a relatively short (but not zero) period of time. Furthermore, this process may occur repeatedly (not just once) every time that a significant concentration of smaller particles—which had been evolving with $\lambda \leq 1$–increases in size to the point that $\lambda > 1$ due to the influence of gravity, electromagnetism, or some other effect.

Gelation essentially introduces a sink of large particles. When a sink of large particles is present, due to gelation or for any other reason, some rules of thumb are as follows:

- Locality may occur regardless of λ and boundary conditions. Generally speaking, locality leads to $n/m = -(\lambda + 1)/2$ for large enough times.
- If the large-particle sink is hard, the size distribution tends to become a power law as the particle size increases.
- If the large-particle sink is soft, the size distribution tends to become a tapered power law as the particle size increases, e.g., a positive Gamma or positive Weibull size distribution.

On the other hand, when a source of small particles is present, some rules of thumb are as follows:

- Non-locality may occur regardless of λ and boundary conditions. Generally speaking, non-locality leads either to $n/m = -\lambda + 1$ or $n/m = -\nu + 1$ for large enough times.
- If the small-particle source is hard, the size distribution tends to become a power law as the particle size decreases.
- If the small-particle source is soft, the size distribution tends to become a tapered power law as the particle size decreases, e.g., a negative Weibull size distribution.

When both a source and sink are present, as in periodic boundary conditions, the outcome may be either local or non-local, and the extremes of the size distribution may be either power laws or tapered power laws. In some cases, particle systems can be divided into multiple size ranges, each dominated by a different physical phenomenon such as condensation, Brownian motion, shear, or differential sedimentation. In such cases, each size range may be delimited by its own source and sink; and each size range may experience locality or non-locality separately; see, e.g., Friedlander (1960) and Jeffrey (1981).

References

M. Adam, D. Lairez, F. Boué, J. P. Busnel, D. Durand, & T. Nicolai, Verification of the Hyperscaling Law on Polymer Clusters by Neutron-Scattering Experiments, *Physical Review Letters*, 67(24), 3456–3458, 1991, https://doi.org/10.1103/PhysRevLett.67.3456

R. C. Ball, C. Connaughton, T. H. M. Stein, & O. Zaboronski, Instantaneous Gelation in Smoluchowski's Coagulation Equation Revisited, *Physical Review E*, 84(1), 011111, 2011, https://doi.org/10.1103/PhysRevE.84.011111

F. A. Bassi, G. Arcovito, M. De Spirito, A. Mordente, & G. E. Martorana, Self-Similarity Properties of α-Crystallin Supramolecular Aggregates, *Biophysical Journal*, 69(6), 2720–2727, 1995, https://doi.org/10.1016/S0006-3495(95)80143-8

J. Blum & G. Wurm, The Growth Mechanisms of Macroscopic Bodies in Protoplanetary Disks, *Annual Review of Astronomy and Astrophysics*, 46, 21–56, 2008, https://doi.org/10.1146/annurev.astro.46.060407.145152

M. P. Bonkile, A. Awasthi, C. A. Lakshmi, V. Mukundan, & V. S. Aswin, Systematic Literature Review of Burgers' Equation with Recent Advances, *Pramana—Journal of Physics*, 90(69), 2018, https://doi.org/10.1007/s12043-018-1559-4

E. Bouchaud, M. Delsanti, M. Adam, M. Daoud, & D. Durand, Gelation and Percolation: Swelling Effect, *Journal de Physique*, 47(8), 1273–1277, 1986, https://doi.org/10.1051/jphys:019860047080127300

N. Brilliantov, P. L. Krapivsky, A. Bodrova, F. Spahn, H. Hayakawa, V. Stadnichuk, et al., Size Distribution of Particles in Saturn's Rings from Aggregation and Fragmentation, *Proceedings of the National Academy of Sciences*, 112(31), 9536–9541, 2015, https://doi.org/10.1073/pnas.1503957112

J. R. Brock, Condensational Growth of Atmospheric Aerosols, *Journal of Colloidal and Interface Sciences*, 39(1), 32–36, 1972, https://doi.org/10.1016/0021-9797(72)90139-7

M. L. Broide & R. J. Cohen, Experimental Evidence of Dynamic Scaling in Colloidal Aggregation, *Physical Review Letters*, 64(17), 1990, https://doi.org/10.1103/PhysRevLett.64.2026

M. L. Broide & R. J. Cohen, Measurements of Cluster-Size Distributions Arising in Salt-Induced Aggregation of Polystyrene Microspheres, *Journal of Colloid and Interface Science*, 153(2), 493–508, 1992, https://doi.org/10.1016/0021-9797(92)90340-R

J. Camacho, Scaling in Steady-State Aggregation with Injection, *Physical Review E*, 63(4), 046112, 2001, https://doi.org/10.1103/PhysRevE.63.046112

E. R. Cohen & E. U. Vaughan, Approximate Solution of the Equations for Aerosol Agglomeration, *Journal of Colloid and Interface Science*, 35(4), 612–623, 1971, https://doi.org/10.1016/0021-9797(71)90219-0

R. H. Colby, M. Rubinstein, J. R. Gillmor, & T. H. Mourey, Scaling Properties of Branched Polyesters. 2. Static Scaling above the Gel Point, *Macromolecules*, 25(26), 7180–7187, 1992, https://doi.org/10.1021/ma00052a017

R. H. Colby, J. R. Gillmor, & M. Rubinstein, Dynamics of Near-Critical Polymer Gels, *Physical Review E*, 48(5), 3712–3716, 1993, https://doi.org/10.1103/PhysRevE.48.3712

C. Connaughton, R. Rajesh, & O. Zaboronski, Stationary Kolmogorov Solutions of the Smoluchowski Aggregation Equation with a Source Term, *Physical Review E*, 69(6), 061114, 2004, https://doi.org/10.1103/PhysRevE.69.061114

C. Connaughton, R. Rajesh, & O. Zaboronski, Cluster–Cluster Aggregation as an Analogue of a Turbulent Cascade: Kolmogorov Phenomenology, Scaling Laws and the Breakdown of Self-Similarity, *Physica D: Nonlinear Phenomena*, 222(1–2), 97–115, 2006, https://doi.org/10.1016/j.physd.2006.08.005

C. Connaughton, A. Dutta, R. Rajesh, N. Siddharth, & O. Zaboronski, Stationary Mass Distribution and Nonlocality in Models of Coalescence and Shattering, *Physical Review E*, 97(2), 022137, 2018, https://doi.org/10.1103/PhysRevE.97.022137

P. Dziekan & H. Pawlowska, Stochastic Coalescence in Lagrangian Cloud Microphysics, *Atmospheric Chemistry and Physics*, 17, 13509–13520, 2017, https://doi.org/10.5194/acp-17-13509-2017

A. Elaissari & E. Pefferkorn, Polyelectrolyte Induced Aggregation of Latex Particles: Influence of the Structural Relaxation of Adsorbed Macromolecules on the Colloid Aggregation Mode, *Journal of Colloid and Interface Science*, 141(2), 522–533, 1991, https://doi.org/10.1016/0021-9797(91)90349-D

D. Fenistein & L. Barré, Experimental Measurement of the Mass Distribution of Petroleum Asphaltene Aggregates using Ultracentrifugation and Small-Angle X-ray Scattering, *Fuel*, 80(2), 283–287, 2001, https://doi.org/10.1016/S0016-2361(00)00072-7

S. K. Friedlander, On the Particle-Size Spectrum of Atmospheric Aerosols, *Journal of Meteorology*, 17(3), 373–374, 1960, https://doi.org/10.1175/1520-0469(1960)017<0373:OTPSSO>2.0.CO;2

S. K. Friedlander & C. S. Wang, The Self-Preserving Particle Size Distribution for Coagulation by Brownian Motion, *Journal of Colloid and Interface Science*, 22(2), 126–132, 1966, https://doi.org/10.1016/0021-9797(66)90073-7

J. C. Gimel, D. Durand, & T. Nicolai, Structure and Distribution of Aggregates Formed After Heat-Induced Denaturation of Globular Proteins, *Macromolecules*, 27(2), 583–589, 1994, https://doi.org/10.1021/ma00080a037

W. W. Grabowski & L.-P. Wang, Growth of Cloud Droplets in a Turbulent Environment, *Annual Review of Fluid Mechanics*, 45, 293–324, 2013, https://doi.org/10.1146/annurev-fluid-011212-140750

H. Hayakawa, Irreversible Kinetic Coagulations in the Presence of a Source, *Journal of Physics A: Mathematical and General*, 20(12), L801–L805, 1987, https://doi.org/10.1088/0305-4470/20/12/009

E. M. Hendriks, M. H. Ernst, & R. M. Ziff, Coagulation Equations with Gelation, *Journal of Statistical Physics*, 31(3), 519–563, 1983, https://doi.org/10.1007/BF01019497

P. Horvai, S. V. Nazarenko & T. H. M. Stein, Coalescence of Particles by Differential Sedimentation, *Journal of Statistical Physics*, 130, 1177–1195, 2008, https://doi.org/10.1007/s10955-007-9466-y

A. V. Ivlev, G. E. Morfill, & U. Konopka, Coagulation of Charged Microparticles in Neutral Gas and Charge-Induced Gel Transitions, *Physical Review Letters*, 89(19), 195502, 2002, https://doi.org/10.1103/PhysRevLett.89.195502

D. J. Jeffrey, Quasi-Stationary Approximations for the Size of Distribution of Aerosols, *Journal of the Atmospheric Sciences*, 38, 2440–2443, 1981, https://doi.org/10.1175/1520-0469(1981)038<2440:QSAFTS>2.0.CO;2

V. I. Khvorostyanov & J. A. Curry, *Thermodynamics, Kinetics, and Microphysics of Clouds*, Cambridge University Press, 2014, https://doi.org/10.1017/CBO9781139060004

J. D. Klett, A Class of Solutions to the Steady-State, Source-Enhanced Kinetic Coagulation Equation, *Journal of Atmospheric Sciences*, 32(2), 380–389, 1975, https://doi.org/10.1175/1520-0469(1975)032<0380:ACOSTT>2.0.CO;2

U. Konopka, F. Mokler, A. V. Ivlev, M. Kretschmer, G. E Morfill, H. M. Thomas, et al., Charge-Induced Gelation of Microparticles, *New Journal of Physics*, 7, 227, 2005, https://10.1088/1367-2630/7/1/227

P. L. Krapivsky, S. Redner, & E. Ben-Naim, *A Kinetic View of Statistics*, Cambridge University Press, 2010, https://doi.org/10.1017/CBO9780511780516

S. Kunamaneni, D. M. A. Buzza, D. Parker, & W. J. Feast, Size Exclusion Chromatography and Rheology of AB/AB_2 Hyperbranched Polymers, *Journal of Materials Chemistry*, 13, 2749–2755, 2003, https://doi.org/10.1039/B303473A

A. Lapp, L. Leibler, F. Schosseler, & C. Strazielle, Scaling Behavior of Pregel Sols obtained by End-Linking of Linear Chains, *Macromolecules*, 22(6), 2871–2873, 1989, https://doi.org/10.1021/ma00196a064

C. B. Laney, *Transformation and Self-Similarity Properties of Gamma and Weibull Fragment Size Distributions*, Defense Threat Reduction Agency Technical Report DTRA-TR-16-06, December 2015, http://apps.dtic.mil/sti/citations/ADA624878

K. W. Lee, Change of Particle Size Distribution during Brownian Coagulation, *Journal of Colloid and Interface Science*, 92(2), 315–325, 1983, https://doi.org/10.1016/0021-9797(83)90153-4

M. H. Lee, A Survey of Numerical Solutions to the Coagulation Equation, *Journal of Physics A: Mathematical and General*, 34(47), 10219–10241, 2001, https://doi.org/10.1088/0305-4470/34/47/323

L. Leibler & F. Schosseler, Gelation of Polymer Solutions: An Experimental Verification of the Scaling Behavior of the Size Distribution Function, *Physical Review Letters*, 55(10), 1110–1113, 1985, https://doi.org/10.1103/PhysRevLett.55.1110

F. Leyvraz, Scaling Theory and Exactly Solved Models in the Kinetics of Irreversible Aggregation, *Physics Reports*, 383(2–3), 95–212, 2003, https://doi.org/10.1016/S0370-1573(03)00241-2

F. Leyvraz, Scaling Theory for Gelling Systems: Work in Progress, *Physica D: Nonlinear Phenomena*, 222(1-2), 21–28, 2006, https://doi.org/10.1016/j.physd.2006.08.011

F. Leyvraz & H. R. Tschudi, Critical Kinetics near Gelation, *Journal of Physics A: Mathematical and General*, 15(6), 1951–1964, 1982, https://doi.org/10.1088/0305-4470/15/6/033

X.-Y. Li, G. Svensson, A. Brandenburg, & N. E. L. Haugen, Cloud-Droplet Growth due to Supersaturation Fluctuations in Stratiform Clouds, *Atmospheric Chemistry and Physics*, 19, 639–648, 2019, https://doi.org/10.5194/acp-19-639

M. Y. Lin, H. M. Lindsay, D. A. Weitz, R. C. Ball, R. Klein, & P. Meakin, Universality in Colloid Aggregation, *Nature*, 339, 360–362, 1989, https://doi.org/10.1038/339360a0

M. Y. Lin, H. M. Lindsay, D. A. Weitz, R. Klein, R. C. Ball, & P. Meakin, Universal Diffusion-Limited Colloid Aggregation, *Journal of Physics: Condensed Matter*, 2(13), 3093–3113, 1990a, https://doi.org/10.1088/0953-8984/2/13/019

M. Y. Lin, M. H. Lindsay, D. A. Weitz, R. C. Ball, R. Klein, & P. Meakin, Universal Reaction-Limited Colloid Aggregation, *Physical Review A*, 41(4), 2005–2020, 1990b, https://doi.org/10.1103/PhysRevA.41.2005

A. A. Lushnikov, Evolution of Coagulating Systems, *Journal of Colloid and Interface Science*, 45(3), 549–556, 1973, https://doi.org/10.1016/0021-9797(73)90171-9

A. A. Lushnikov & V. I. Smirnov, Steady-State Coagulation and Formation of Particle Size Distributions in the Atmosphere, *Izvestiya, Akademii Nauk SSSR, Fizika, Atmosfery i Okeana (Academy of Sciences, USSR, Physics of Atmosphere and Ocean)*, 11(2), 139–152, 1975

A. A. Lushnikov & V. N. Piskunov, Coagulation in the Presence of External Sources, *Doklady Akademii Nauk SSSR*, 231, 1403–1406, 1976

C. P. Lusignan, T. H. Mourey, J. C. Wilson, & R. H. Colby, Viscoelasticity of Randomly Branched Polymers in the Critical Percolation Class, *Physical Review E*, 52, 6271–6280, 1995, https://doi.org/10.1103/PhysRevE.52.6271

C. P. Lusignan, T. H. Mourey, J. C. Wilson, & R. H. Colby, Viscoelasticity of Randomly Branched Polymers in the Vulcanization Class, *Physical Review E*, 60(5), 5657–5669, 1999, https://doi.org/10.1103/PhysRevE.60.5657

S. Manley, L. Cipelletti, V. Trappe, A. E. Bailey, R. J. Christianson, U. Gasser, V. Prasad, et al., Limits to Gelation in Colloidal Aggregation, *Physical Review Letters*, 93(10), 108302, 2004, https://doi.org/10.1103/PhysRevLett.93.108302

S. Manley, J. M. Skotheim, L. Mahadevan, & D. A. Weitz, Gravitational Collapse of Colloidal Gels, *Physical Review Letters*, 94(21), 218302, 2005, https://doi.org/10.1103/PhysRevLett.94.218302

J. E. Martin, D. W. Schaefer, & A. J. Hurd, Fractal Geometry of Vapor-Phase Aggregates, *Physical Review A*, 33(5), 3540–3543, 1986, https://doi.org/10.1103/PhysRevA.33.3540

J. E. Martin, Slow Aggregation of Colloidal Silica, *Physical Review A*, 36(7), 3415–3426, 1987, https://doi.org/10.1103/PhysRevA.36.3415

T. Matsoukas & Y. Lin, Fokker–Planck Equation for Particle Growth by Monomer Attachment, *Physical Review E.*, 74, 031122, 2006, https://doi.org/10.1103/PhysRevE.74.031122

G. E. Morfill & A. V. Ivlev, Complex Plasmas: An Interdisciplinary Research Field, *Reviews of Modern Physics*, 81(4), 1353–1404, 2009, https://doi.org/10.1103/RevModPhys.81.1353

B. J. Olivier & C. M. Sorensen, Evolution of the Cluster Size Distribution during Slow Colloid Aggregation, *Journal of Colloid and Interface Science*, 134(1), 139–146, 1990, https://doi.org/10.1016/0021-9797(90)90259-Q

B. J. Olivier, C. M. Sorensen, & T. W. Taylor, Scaling Dynamics of Aerosol Coagulation, *Physical Review A*, 45(8), 5614–5623, 1992, https://doi.org/10.1103/PhysRevA.45.5614

S. H. Park, R. Xiang, & K. W. Lee, Brownian Coagulation of Fractal Agglomerates: Analytical Solution Using the Log-Normal Size Distribution Assumption, *Journal of Colloid and Interface Science*, 231(1), 129–135, 2000, https://doi.org/10.1006/jcis.2000.7102

E. V. Patton, J. A. Wesson, M. Rubinstein, J. C. Wilson, & L. E. Oppenheimer, Scaling Properties of Branched Molecules, *Macromolecules*, 22(4), 1946–1959, 1989, https://doi.org/10.1021/ma00194a072

E. Pefferkorn & R. Varoqui, Dynamics of Latex Aggregation. Modes of Cluster Growth, *The Journal of Chemical Physics*, 91(9), 5679–5686, 1989, https://doi.org/10.1063/1.457521

T. W. Peterson, F. Gelbard, & J. H. Seinfeld, Dynamics of Source-Reinforced, Coagulating, and Condensing Aerosols, *Journal of Colloid and Interface Science*, 63(3), 426–445, 1978, https://doi.org/10.1016/S0021-9797(78)80004-6

A. Pumir & M. Wilkinson, Collisional Aggregation Due to Turbulence, *Annual Review of Condensed Matter Physics*, 7, 141–170, 2016, https://doi.org/10.1146/annurev-conmatphys-031115-011538

T. E. Ramabhadran, T. W. Peterson, & J. H. Seinfeld, Dynamics of Aerosol Coagulation and Condensation, *AIChE Journal*, 22(5), 840–851, 1976, https://doi.org/10.1002/aic.690220505

J. G. Rarity, R. N. Seabrook, & R. J. G. Carr, Light-Scattering Studies of Aggregation, *Proceedings of the Royal Society of London A: Mathematical, Physical and Engineering Sciences*, 423(1864), 89–102, 1989, https://doi.org/10.1098/rspa.1989.0043

S. Ross, *A First Course in Probability*, 10th edition, Pearson, 2018

R. J. Rossi, *Mathematical Statistics: An Introduction to Likelihood Based Inference*, Wiley, 2018, https://doi.org/10.1002/9781118771075

W. T. Scott, Analytic Studies of Cloud Droplet Coalescence I, *Journal of the Atmospheric Sciences*, 25(1), 54–65, 1968, https://doi.org/10.1175/1520-0469(1968)025<0054:ASOCDC> 2.0.CO;2

J. H. Seinfeld & S. N. Pandis, *Atmospheric Chemistry: From Air Pollution to Climate Change*, Wiley, 2016

N. Smith, N. J. Shaviv, & H. Svensmark, Approximate Analytical Solutions to the Condensation-Coagulation Equation of Aerosols, *Aerosol Science and Technology*, 50(6), 578–590, 2016, https://doi.org/10.1080/02786826.2016.1168921

M. V. Smoluchowski, Versuch einer mathematischen Theorie der Koagulationskinetik kolloider Lösunger [Mathematical Theory of the Kinetics of the Coagulation of Colloidal Solutions], *Zeitschrift für Physikalische Chemie*, 92, 129–168, 1917, https://doi.org/10.1515/zpch-1918-9209

Suneel, D. M. A. Buzza, D. J. Groves, T. C. B. McLeish, D. Parker, A. J. Keeney & W. J. Feast, Rheology and Molecular Weight Distribution of Hyperbranched Polymers, *Macromolecules*, 35(25), 9605–9612, 2002, https://doi.org/10.1021/ma020820r

V. Trappe, W. Richtering & W. Burchard, Critical Behavior of Anhydride Cured Epoxies, *Journal de Physique II*, 2(7), 1453–1463, 1992, https://doi.org/10.1051/jp2:1992212

P. G. J. van Dongen & M. H. Ernst, Dynamic Scaling in the Kinetics of Clustering, *Physical Review Letters*, 54(13), 1396–1399, 1985a, https://doi.org/10.1103/PhysRevLett.54.1396

P. G. J. van Dongen & M. H. Ernst, Cluster Size Distribution in Irreversible Aggregation at Large Times, *Journal of Physics A: Mathematical and General*, 18(14), 2779–2793, 1985b, https://doi.org/10.1088/0305-4470/18/14/028

P. G. J. van Dongen & M. H. Ernst, Tail Distribution for Large Clusters in Irreversible Aggregation, in L. Pietro Nero & E. Tosatti, editors, *Fractals in Physics*, Elsevier, 1986, https://doi.org/10.1016/B978-0-444-86995-1.50058-5

P. G. J. van Dongen & M. H. Ernst, Tail Distribution of Large Clusters from the Coagulation Equation, *Journal of Colloid and Interface Science*, 115(1), 27–35, 1987, https://doi.org/10.1016/0021-9797(87)90005-1

P. G. J. Van Dongen, Solutions of Smoluchowski's Coagulation Equation at Large Cluster Sizes, *Physica A: Statistical Mechanics and its Applications*, 145(1–2), 15–66, 1987, https://doi.org/10.1016/0378-4371(87)90240-8

E. Villermaux, Fragmentation, *Annual Review of Fluid Mechanics*, 39, 419–446, 2007, https://doi.org/10.1146/annurev.fluid.39.050905.110214

E. Villermaux & B. Bossa, Single-Drop Fragmentation Determines Size Distribution of Raindrops, *Nature Physics*, 5, 697–702, 2009, https://doi.org/10.1038/nphys1340

W. H. White, On the Form of Steady-State Solutions to the Coagulation Equations, *Journal of Colloid and Interface Science*, 87(1), 204–208, 1982, https://doi.org/10.1016/0021-9797(82)90382-4

W. H. White, On the Form of Stationary Size Distributions Established by Coagulation and Sedimentation, *Journal of the Atmospheric Sciences*, 40(7), 1855–1858, 1983, https://doi.org/10.1175/1520-0469(1983)040<1855:OTFOSS>2.0.CO;2

J. P. Wilcoxon, J. E. Martin, & D. W. Schaefer, Aggregation in Colloidal Gold, *Physical Review A*, 39(5), 2675–2688, 1989, https://doi.org/10.1103/PhysRevA.39.2675

W. Wu & G. M. McFarquhar, Statistical Theory on the Functional Form of Cloud Particle Size Distributions, *Journal of the Atmospheric Sciences*, 75(8), 2801–2814, 2018, https://doi.org/10.1175/JAS-D-17-0164.1

R. M. Ziff, E. M. Hendriks, & M. H. Ernst, Critical Properties for Gelation: A Kinetic Approach, *Physical Review Letters*, 49(8), 593–595, 1982, https://doi.org/10.1103/PhysRevLett.49.593

12

Fragmentation Cascades Part 1: Mean-Field Theory for Binary Fragmentation

This is the first of three chapters on mean-field theory for fragmentation cascades. In fragmentation cascades, particles instantaneously divide into multiple pieces, over and over again. As standard terminology, the particles participating in each fragmentation event are called *parents* or *mothers* and the products of each fragmentation event are called *children*, *daughters*, or *progeny*. As seen in Table 12.1, there are three common types of fragmentation cascades. This chapter discusses the first type, binary fragmentation, where each fragmentation event creates exactly two children. Binary fragmentation is sometimes also called *fission*, *mitosis*, *scission*, or *splitting*. Example 12.1 describes a common application in which binary fragmentation occurs.

Table 12.1 Three common types of fragmentation cascades.

Type	Description	Typical Applications
Binary Fragmentation	Randomly selected particles fragment into exactly two pieces with independently random sizes.	Thermal and oxidative degradation of polymers
Random Fragmentation	Randomly selected particles fragment into two or more pieces with independently random sizes.	Grinding of minerals, ores, and chemicals
Collisional Fragmentation	Random binary collisions cause one or both of the participating particles to fragment into two or more pieces with independently random sizes.	Evolution of objects in the asteroid belt and the Kuiper belt

Example 12.1 Binary fragmentation in polymer degradation

Polymers degrade due to heat, oxidation, and other factors. According to Billingham (2000), "thermal degradation of polymers may be defined as the collection of chemical reactions that result in deterioration of properties [once] ... the chemical bonds of the polymer acquire enough thermal energy to be able to break spontaneously at significant rates." According to Ray & Cooney (2018), thermal degradation typically occurs alongside "physical processes like glass transition ..., melting, sublimation, evaporation, crystallization, cross-linking, charring, and water absorption ... Thus, the material may undergo changes in color, mass, density, and shape, and cracks and voids can also develop." According to Ray & Cooney (2018) and Alger (2017), polymer degradation typically involves *scission*, i.e., "breaking of

Particle Dynamics with Aggregation and Fragmentation. Culbert B. Laney, Oxford University Press. © Culbert B. Laney (2025).
DOI: 10.1093/9780198920687.003.0012

an in-chain chemical bond in a polymer molecule, producing two new molecules of lower molecular mass." According to McCoy & Madras (1997) scission may occur "(a) at any bond in the backbone chain (random-chain scission), (b) at the chain midpoint, or (c) at the end of the chain yielding a monomer (chain-end scission) ... [Random- and] chain-end scission occur in thermal decomposition ... Random-chain scission is characteristic of oxidative degradation reactions. Midpoint-chain scission dominates in mechanical degradation, for example, by ultrasonic radiation." According to Pielichowski et al. (2022), "polymers that do not depolymerize, like polyethylene, generally decompose into fragments that break again into smaller fragments and so on ... If such random scission events are repeated successively in a polymer and its degradation products, the result is initially a decrease in molecular weight and ultimately weight loss, as degraded products, with a broad range of carbon numbers, become small enough to evaporate without further cleavage." In summary, polymer degradation tends to involve binary fragmentation cascades.

12.1 Governing Equations

12.1.1 Introduction

The state of any fragmenting particle system can be described by random or seemingly random fields such as position, mass, momentum, energy, shape, and rotation. If the particle system is spatially homogeneous and isotropic, the position field does not affect the other fields. Suppose the effects of the other fields on the particle size can be combined into a single fragmentation rate. This rate, known as the binary fragmentation kernel, captures the average effects of randomness and chaos in the context of a simple deterministic governing equation.

12.1.2 Discrete Governing Equations

If $M_i = iM_1$, then the *binary fragmentation kernel* I_{ij} is defined as follows:

> I_{ij} is the average rate at which children of sizes i and j appear from parents of size $i + j$ or, equivalently, twice the average rate at which parents of size $i + j$ disappear.

Notice that I_{ij} has units of inverse time. In this case, Jellinek & White (1951) obtained the following mean-field governing equation:

$$\frac{dc_i(t)}{dt} = -\frac{1}{2}c_i(t)\sum_{j=1}^{i-1} I_{i-j,j} + \sum_{j=1}^{\infty} I_{ij}c_{i+j}(t). \tag{12.1}$$

See also Ziff & McGrady (1986). Equation (12.1) may be interpreted as follows:

$$\begin{pmatrix} \textit{rate of change} \\ \textit{in particles of} \\ \textit{size } i \end{pmatrix} = - \begin{pmatrix} \textit{rate at which particles of} \\ \textit{size } i \textit{ split to form particles} \\ \textit{smaller than size } i \end{pmatrix} + \begin{pmatrix} \textit{rate at which particles} \\ \textit{larger than size } x \textit{ split to} \\ \textit{form particles of size } i \end{pmatrix}.$$

As a special case, suppose that the fragmentation kernel only depends on size of the parent $i + j$ and not on the size of the children i and j. In other words, suppose that:

$$I_{ij} = I_{i+j}. \tag{12.2}$$

In this case, Equation (12.1) becomes:

$$\frac{dc_i(t)}{dt} = -\frac{1}{2}(i-1)I_i c_i(t) + \sum_{j=i+1}^{\infty} I_j c_j(t). \tag{12.3}$$

Binary fragmentation is the exact inverse of aggregation. Each aggregation event converts two smaller particles with seemingly random properties into one larger particle and, conversely, each binary fragmentation event converts one larger particle into two smaller particles with seemingly random properties. Thus, binary fragmentation is sometimes called *reverse aggregation*. As it turns out, the earliest studies of binary fragmentation often concerned the combination with aggregation. In addition, the earliest studies of binary fragmentation often assumed constant rate kernels. Examples 12.2, 12.3, and 12.4 summarize some of this early work.

Example 12.2 Combined aggregation and binary fragmentation where particles can only gain or lose a single monomer at a time

In this case, Becker & Döring (1935) obtained the following mean-field governing equation:

$$\frac{dc_i(t)}{dt} = -K_{i1}c_i(t)c_1(t) + K_{i-1,1}c_{i-1}(t)c_1(t) - I_{i1}c_i(t) + I_{i+1,1}c_{i+1}(t). \tag{12.4}$$

This equation combines Equation (9.1) for aggregation with Equation (12.1) for binary fragmentation. Despite its seeming simplicity, this governing equation poses substantial challenges; see, e.g., Ball et al. (1986), Ball & Carr (1988, 1990), King & Wattis (2002), Wattis (2009), da Costa (2015). This is true even if I_{i1} and K_{i1} are constants; see, e.g., Wattis & King (1998). Equation (12.4) assumes deterministic rather than random fragment sizes, i.e., one of the children from each event is always a single monomer. From now on, this chapter will only consider cases with random rather than deterministic fragment sizes, e.g., for polymer degradation, this chapter concerns random-chain instead of end-chain scission.

Example 12.3 Binary fragmentation for linear chains consisting of i monomers connected by $i-1$ linkages

In this case, Simha (1941) suggested the following two mean-field governing equations:

$$\frac{dc_i(t)}{dt} = -\frac{1}{2} I_0 (i-1) c_i(t) + I_0 \sum_{j=i+1}^{i_{\max}} c_j(t) \tag{12.5}$$

$$\frac{dc_i(t)}{dt} = -\frac{1}{2} I_0 c_i(t) + I_0 \sum_{j=i+1}^{i_{\max}} \frac{1}{j-1} c_j(t). \tag{12.6}$$

These equations assume monodisperse initial conditions, i.e., all particles initially consist of exactly $i_{\max}$ monomers. Simha (1941) says Equation (12.5) occurs when "all the linkages are broken at random." Simha (1941, Eqn. 2a′) found that Equation (12.5) obtains an exponential size distribution for sufficiently large times. Simha (1941) says Equation (12.6) occurs when the "chance of decomposing is roughly equal for all molecules independently of their size. The probability for a particular bond in a particular chain being cut is then inversely proportional to the number of linkages in that chain." Ziff (1992) says that Equation (12.6) has "an explicit solution [with] ... coefficients written in terms of multiple summations that are unwieldy for large systems ... In the continuum limit, [this is] ... at the borderline of the shattering transition, in which a finite fraction of the mass breaks up to zero-size particles."

Example 12.4 Combined aggregation and binary fragmentation for linear chains consisting of i monomers connected by i–1 linkages

In this case, Blatz & Tobolsky (1945) suggested a combination of Equation (9.1) for aggregation with $I_{ij} = I_0 = \text{const.}$ and Equation (12.1) for binary fragmentation with $K_{ij} = K_0 = \text{const.}$ Aizenman & Bak (1979) say that this occurs "when all the bonds break independently and with the same rate and ... the probability for two molecules to combine is independent of their lengths." In this case, the combination of aggregation and binary fragmentation obtains an exponential size distribution for sufficiently large times, just like aggregation alone or binary fragmentation alone; see also, e.g., Diemer & Olson (2002) and Vigil (2009).

12.1.3 Continuous Governing Equations

Let x and y be particle masses. Define the *binary fragmentation kernel* $I(x, y)$ as follows:

> $I(x, y)c(x + y, t)dy$ is the average rate per unit mass and volume at which children of sizes x and y form due to binary fragmentation of parents of size $x + y \pm dy$.

Notice that $I(x, y)$ has units of inverse mass and time.

In this case, Saito (1958) obtained the following mean-field governing equation:

$$\frac{\partial c(x,t)}{\partial t} = -\frac{1}{2} c(x,t) \int_0^x I(x-y,x)\, dy + \int_0^\infty I(x,y)\, c(x+y,t)\, dy. \tag{12.7}$$

See also Ziff & McGrady (1985, 1986), Redner (1990), and Section 6.1 of Krapivsky et al. (2010). Equation (12.7) may be interpreted as follows:

$$\begin{pmatrix} \textit{rate of change} \\ \textit{in particles of} \\ \textit{size } x \end{pmatrix} = -\begin{pmatrix} \textit{rate at which particles of} \\ \textit{size } x \textit{ split to form particles} \\ \textit{smaller than size } x \end{pmatrix} + \begin{pmatrix} \textit{rate at which particles} \\ \textit{larger than size } x \textit{ split to} \\ \textit{form particles of size } x \end{pmatrix}.$$

Equations of this general sort are sometimes called *population balance equations*; see, e.g., Ramkrishna (2000). In general, $I(x, y)$ depends on the sizes of the children x and y and, thus, also on the size of the parent $x + y$. As a special case, suppose the fragmentation kernel only depends on the parent size $x + y$ and not on the children sizes x, y. In other words, suppose that:

$$I(x,y) = I(x+y). \tag{12.8}$$

In this case, Equation (12.7) becomes:

$$\frac{\partial c(x,t)}{\partial t} = -\frac{1}{2} x I(x) c(x,t) + \int_x^\infty I(y) c(y,t)\, dy. \tag{12.9}$$

See also Ziff & McGrady (1985, 1986) and Section 6.1 of Krapivsky et al. (2010). As another special case, suppose the binary fragmentation kernel does not depend on either the sizes of the parents or the children. In other words, suppose that:

$$I(x,y) = I(x+y) = I_0 = \text{const.} \tag{12.10}$$

In this case, Equation (12.9) becomes:

$$\frac{\partial c(x,t)}{\partial t} = -\frac{1}{2} I_0 x c(x,t) + I_0 \int_x^\infty c(y,t)\, dy. \tag{12.11}$$

12.2 Kernels

For simplicity, the remainder of this chapter will consider only continuous rather than discrete particle systems, except where noted. In addition, the remainder of this chapter will consider only kernels of the form $I(x)$. By definition, the kernel $I(x)$ is *homogeneous* if it is a power law. More specifically:

$$I(x) = I_0 x^{\kappa-1} \tag{12.12}$$

where I_0 and κ are constants. In this case, κ is called the *degree of homogeneity*. In ideal cases with $m = d = 3$, researchers generally assume the following:

$$\kappa \text{ is an integer multiple of } 1/6 \text{ between } -2 \text{ and } 2. \tag{12.13}$$

Table 12.2 give examples of degrees of homogeneity for binary fragmentation taken from the research literature. Except for the last row, these examples obey Equation (12.13). The next chapter gives additional details on Equation (12.13), including an informal proof.

Table 12.2 Examples of ideal homogeneous kernels for binary fragmentation. In the first column, all values of κ are rational and were obtained analytically. In the second column, x refers to particle mass and I_0 refers to an application-specific constant.

κ	$I(x)/I_0$	Application	References
−2	x^{-3}	Polymer degradation — 1st model for shattering	Ziff & McGrady (1985)[1]
−1	x^{-2}	Polymer degradation — 2nd model for shattering	Ziff & McGrady (1985)[1]
0	x^{-1}	Polymer degradation — transitional	Ziff & McGrady (1985)[1,2]
1/3	$x^{-2/3}$	"Typical of fragmentation in shear flow"	Piskunov (2012)
1/2	$x^{-1/2}$	Polymers, colloids, or any "treelike" cluster	Family et al. (1986)
1/2	$x^{-1/2}$	Polymer degradation — 1st model for non-shattering	Ziff & McGrady (1985)[1]
2/3	$x^{-1/3}$	Polymer degradation — 2nd model for non-shattering	Ziff & McGrady (1985)[1]
1	1	Polymer degradation — 3rd model for non-shattering	Ziff & McGrady (1985)[1,2]
2	x	Polymer degradation — 4th model for non-shattering	Ziff & McGrady (1985)[1]
3	x^2	Polymer degradation — 5th model for non-shattering	Ziff & McGrady (1986)

(1) See also Ziff & McGrady (1986) and Meakin & Ernst (1988).
(2) See also Simha (1941).

12.3 Shattering

This section concerns *shattering*, i.e., the spontaneous formation of sinks of small particles. For free-field boundary conditions, shattering tends to occur when $\kappa < 0$. Shattering in binary fragmentation is related to gelation in aggregation; see Section 10.3. As one difference, gelation in aggregation involves sinks of *large* particles that form after some finite amount of time has passed, while shattering in binary fragmentation involves sinks of *small* particles that form right away. As another difference, gelation in aggregation occurs due to nonlinear interactions between particles, while shattering in binary fragmentation occurs because the fragmentation rate goes to infinity as the particle size goes to zero. For discrete systems, Matsoukas (2022) says shattering involves "the emergence of a ghost fraction of particles ('dust') ... [that] manifests itself as a breakdown of mass conservation ... Binary breakage with power-law rate ... leads to shattering when the exponent is negative ... From a kinetic perspective, shattering is the result of the accelerated fragmentation rate with decreasing size that produces an accumulation at the low end of the distribution ... [From a] statistical mechanics [perspective] ... shattering is a phase transition mathematically symmetric to gelation..." This will be discussed further in Chapter 13.

12.4 Size Distributions and Moments

This section provides a brief review of size distributions and moments. For a longer discussion, see Chapter 3. First, consider moments. The *zero-th moments* of $c(M, t)$ and $f(M, t)$ are as follows:

$$\int_0^\infty c(M,t)\,dM = C(t) \tag{12.14}$$

$$\int_0^\infty f(M,t)\,dM = 1. \tag{12.15}$$

The *first moments* of $c(M, t)$ and $f(M, t)$ are as follows:

$$\int_0^\infty Mc(M,t)\,dM = C(t)M_{\text{avg}}(t) \tag{12.16}$$

$$\int_0^\infty Mf(M,t)\,dM = M_{\text{avg}}(t). \tag{12.17}$$

For convenience, the following notation will be used throughout the rest of the chapter:

$$C_0 = C(0) \tag{12.18}$$

$$M_0 = M_{\text{avg}}(0). \tag{12.19}$$

As described in Example 12.5, if conservation of mass applies then:

$$C(t)\,M_{\text{avg}}(t) = C_0M_0 = \text{const.} \tag{12.20}$$

Example 12.5 Conservation of mass for binary fragmentation

In deterministic systems, the governing equations are typically expressions of conservation of mass, momentum, and energy. In such cases, exact or approximate solutions must rigorously ensure conservation of mass, momentum, and energy; see, e.g., Laney (1998). By contrast, in random or chaotic systems, the governing mean-field equations are not necessarily expressions of conservation of mass, momentum, or energy. Indeed, most mean-field solutions violate conservation of mass due to particle sources and sinks. Even when sources and sinks are not deliberately introduced, mean-field fragmentation may spontaneously develop a small-particle sink due to shattering. From the point of view of the universe that includes both the particle systems and its surroundings, sources and sinks obey conservation of mass, momentum, and energy. However, from the point of view of the particle system alone—which is the point of view assumed by the mean-field governing equations—sources and sinks violate conservation of mass, momentum, and energy.

Having said this, there are useful relations for particle systems that obey conservation of mass. Consider finite particle systems, i.e., systems in which $0 < N(t) < \infty$ and $0 < \mathcal{V}^o < \infty$ where $N(t)$ is the total number of particles in the system and $\mathcal{V}^o = \text{const.}$ is the total volume of the system including the spaces between particles. If conservation of mass applies then:

$$N(t)\,M_{\text{avg}}(t) = M_{\text{total}}(t) = \text{const.} \tag{12.22}$$

where $N(t)$ and $M_{\text{total}}(t)$ are the total number and mass of particles in the system, respectively. Similarly:

$$C(t)\,M_{\text{avg}}(t) = \rho_{\text{total}}(t) = \text{const.} \tag{12.23}$$

where $C(t) = N(t)/\mathcal{V}^o$ is the total concentration and $\rho_{\text{total}}(t) = M_{\text{total}}(t)/\mathcal{V}^o$ is the density of particles in the system. For infinite particle systems, i.e., systems where $N(t) = \infty$ and $\mathcal{V}^o = \infty$ such that $0 < C(t) < \infty$, Equation (12.22) no longer applies but Equation (12.23) still applies.

If $\kappa \neq 0$, binary fragmentation usually obtains Rosin–Rammler distributions for large enough times and small enough particles. In regular form, these distributions may be written as follows:

$$c(M,t) = C(t)f(M,t) = C(t)A(t)M^{n/m-1}\exp\left[b(t)M^{l/m}\right]. \tag{12.21}$$

If $\kappa = 0$, binary fragmentation usually obtains something related to a log normal distribution.

12.5 Full Mean-Field Solutions

This section gives seven examples of closed-form time-dependent solutions of Equation (12.9). More specifically, Examples 12.6 to 12.10 concern cases where $\kappa = 0$,

1, and 2. Because $\varkappa \geq 0$, shattering does not occur. In addition, Examples 12.11 and 12.12 concern cases where $\varkappa = -1$ and -2, respectively. Because $\varkappa < 0$, shattering occurs. To avoid extraneous factors of 1/2, and for consistency with the next chapter, these examples use the following notation:

$$J_0 = I_0/2. \tag{12.24}$$

In all seven examples, the full time-dependent solution is a hybrid of two different size distributions—an arbitrary *non-universal* initial size distribution for unreacted particles and a second *universal* size distribution for fully-reacted particles. The balance progressively shifts from the former to the latter as the time increases and particle size decreases.

Example 12.6 Time-dependent closed-form solutions for $\varkappa = 0$ with monodisperse initial conditions and free-field boundary conditions

For $I(x) = 2J_0/x^{-1}$, McGrady & Ziff (1987) obtained the following time-dependent solution to Equation (12.9):

$$c(M,t) = \frac{C_0}{M_0}\exp(-J_0M_0t)\,\delta\left(\frac{M}{M_0}-1\right) + C_0\exp(-J_0t)\sqrt{2J_0t}\,\frac{I_1\left[2\sqrt{2J_0t\ln(M_0/M)}\right]}{M_0\sqrt{\ln(M_0/M)}} \tag{12.25}$$

where C_0 is the initial total concentration, M_0 is the initial mass of each particle, and $I_1(x)$ is a first-order modified Bessel function of the first kind, as discussed earlier in Section 10.4. Notice that the zero-th moment $C(t)$ of $c(M,t)$ is $C_0\exp(J_0t)$, which is monotone increasing. In addition, the first moment C_0M_0 of $c(M,t)$ is constant, i.e., conservation of mass applies. Then according to Equation (12.20), $M_{avg}(t) = (C_0M_0)/C(t) = M_0\exp(-J_0t)$. For sufficiently small times and large particle sizes, Equation (12.25) becomes $c(M,t) \cong (C_0/M_0)\delta(M/M_0 - 1)$, which is a monodisperse size distribution. Conversely, for sufficiently large times and small particle sizes, Equation (12.25) becomes:

$$c(M,t) \cong \frac{C_0\exp(-J_0t)(2J_0t)^{1/4}\exp\left\{2[2J_0t\ln(M_0/M)]^{1/2}\right\}}{(4\pi)^{1/2}M_0[\ln(M_0/M)]^{3/4}}. \tag{12.26}$$

Instead of a continuous particle system with $I(x) = I_0/x$, consider a discrete particle system with $I_i = I_0/(i-1)$ Then, as noted in Example 12.3, instead of a nice closed-form solution like Equation (12.25), Simha (1941) obtained a solution involving multiple summations. As an alternative, Ziff (1992) considered $I_i = I_0/(i+1)$. Instead of multiple summations, Ziff (1992) obtained a solution involving a single summation "in a compact form." This solution trades physicality for simplicity. Although discrete and continuous particle systems have similar mean-field governing equations, after making the obvious transliteration between summation and integration, they can clearly obtain different solutions.

In many cases, binary fragmentation operates on linkages rather than on particles. In other words, binary fragmentation affects randomly selected linkages instead of randomly selected particles. Some researchers, including Simha (1941) and Ziff (1992), avoid this distinction by assuming chains, i.e., every monomer has exactly two linkages, except for the ones at the chain ends. More generally, researchers avoid this distinction by assuming that most monomers have a fixed number of linkages k. However, suppose that different monomers support different numbers of linkages. Then a particle composed of i monomers could have anywhere between $i - 1$ and $(i - 1)i/2$ linkages. In this case, two particles with the same mass may experience different probabilities of binary fragmentation, depending on how many linkages they each have. There is nothing entirely analogous to this in continuous particle systems. The closest analog is shape. For example, if each monomer can only support two linkages, discrete particle systems allow only linear chains. Similarly, certain continuous particle systems may only support linear shapes, e.g., binary fragmentation of dry spaghetti. This book tries to avoid getting bogged down in application-specific issues. Thus, this book focuses on cases where discrete and continuous particle systems obtain the same or similar solutions.

Example 12.7 Time-dependent closed-form solutions for $\varkappa = 1$ with monodisperse initial conditions and free-field boundary conditions

For $I(x) = I_0 = 2J_0 = \text{const.}$, Ziff & McGrady (1985) obtained the following time-dependent solution to Equation (12.11):

$$c(M,t) = \frac{C_0}{M_0} \exp(-J_0 M_0 t)\, \delta\!\left(\frac{M}{M_0} - 1\right) + C_0 J_0 t \exp(-J_0 M t)\,[2 + J_0 (M_0 - M)\, t] \qquad (12.27)$$

where C_0 is the initial total concentration and M_0 is the initial mass of each particle. Notice that $M \le M_0$ for all times so that:

$$C(t) = \int_0^{M_0} c(x,t)\, dx = C_0(1 + J_0 M_0 t). \qquad (12.28)$$

and $f(M,t) = c(M,t)/C(t)$ so that:

$$M_{\text{avg}}(t) = \int_0^{M_0} M f(M,t)\, dM = \frac{M_0}{1 + J_0 M_0 t}. \qquad (12.29)$$

Notice that $\int_0^{M_0} M c(M,t)\, dM = C(t) M_{\text{avg}}(t) = C_0 M_0 = \text{const.}$ In other words, conservation of mass applies. For sufficiently small times and large particle sizes, Equation (12.27) becomes:

$$c(M,t) \cong (C_0/M_0)\; \delta(M/M_0 - 1) \qquad (12.30)$$

which is a monodisperse size distribution.

Conversely, for sufficiently large times and small particle sizes, Equation (12.27) becomes:

$$c(M,t) \cong C_0 M_0 (J_0 t)^2 \exp(-J_0 M t) \tag{12.31}$$

or equivalently:

$$c(M,t) \cong \frac{C_0 M_0}{[M_{avg}(t)]^2} \exp\left[-\frac{M}{M_{avg}(t)}\right] \tag{12.32}$$

where $M_{avg}(t) \cong 1/(J_0 t)$ as $t \to \infty$. Notice that this is an exponential size distribution. While this example concerns continuous particle systems, discrete particle systems with $I_i = I_0$ also produce exponential distributions.

As seen in Examples 10.1, 10.2, and 10.3, aggregation with constant kernels produces exponential size distributions. In addition, as noted in Example 12.4, combined aggregation and binary fragmentation with constants kernels produces exponential size distributions. Because exponential size distributions tend to occur so often in so many different types of aggregation and fragmentation problems, they are sometimes called *universals*. By contrast, in the last example, binary fragmentation obtained a rare result, which does not occur except for binary aggregation in a continuous particle system (and perhaps, as discussed in Example 12.15, not even there). Such unusual or unique results are generally not considered to be universals.

Example 12.8 Time-dependent closed-form solutions for $\kappa = 1$ with exponential initial conditions and free-field boundary conditions

For $I(x) = I_0 = 2J_0 = \text{const.}$, Ziff & McGrady (1985) obtained the following time-dependent solution to Equation (12.11):

$$c(M,t) = (C_0/M_0)(1 + J_0 M_0 t)^2 \exp\left[-(1 + J_0 M_0 t) M/M_0\right] \tag{12.33}$$

where C_0 is the initial total concentration and M_0 is the initial average particle mass. Notice that:

$$C(t) = \int_0^\infty c(x,t)\,dx = C_0(1 + J_0 M_0 t) \tag{12.34}$$

and $f(M,t) = c(M,t)/C(t)$ so that:

$$M_{avg}(t) = \int_0^\infty M f(M,t)\,dM = \frac{M_0}{1 + J_0 M_0 t}. \tag{12.35}$$

Notice that $\int_0^{M_0} M c(M,t)\,dM = C(t) M_{avg}(t) = C_0 M_0 = \text{const.}$ In other words, conservation of mass applies.

For small times and large particle sizes, Equation (12.33) becomes:

$$c(M,t) \cong (C_0/M_0)\exp(-M/M_0) \tag{12.36}$$

which is an exponential size distribution. Conversely, for large times and small particle sizes, Equation (12.33) becomes:

$$c(M,t) \cong C_0 M_0 (J_0 t)^2 \exp(-J_0 M t) \tag{12.37}$$

which is also an exponential size distribution. Finally, notice that Equation (12.37) can be rewritten in self-similar form as follows:

$$c(M,t) \cong \frac{C_0 M_0}{[M_{avg}(t)]^2}\exp\left[-\frac{M}{M_{avg}(t)}\right] \tag{12.38}$$

or equivalently:

$$f(M,t) \cong \frac{1}{M_{avg}(t)}\exp\left[-\frac{M}{M_{avg}(t)}\right]. \tag{12.39}$$

where $M_{avg}(t) \cong 1/(J_0 t)$. In other words, in this case, the size distribution is exponential regardless of time. The discussion given earlier in Example 10.2 applies again here. In particular, if the initial size distribution obtains the broadest possible size spread—in this case, an exponential size distribution according to the metric associated with I—fragmentation will preserve that initial distribution for all time. Otherwise, fragmentation will progressively drive the distribution towards the broadest possible size spread, as seen in the last example.

Example 12.9 Time-dependent closed-form solutions for $\kappa = 2$ with monodisperse initial conditions and free-field boundary conditions

For $I(x) = I_0 x = 2J_0 x$, Ziff & McGrady (1985) obtained the following time-dependent solution to Equation (12.9):

$$c(M,t) = (C_0/M_0)\exp\left(-J_0 M_0^2 t\right)\delta(M/M_0 - 1) + 2C_0 J_0 M_0 t \exp\left(-J_0 M^2 t\right) \tag{12.40}$$

where C_0 is the initial total concentration and M_0 is the initial mass of each particle. Notice that $M \le M_0$ for all times so that:

$$C(t) = \int_0^{M_0} c(x,t)\,dx = C_0 \exp\left(-J_0 M_0^2 t\right) + C_0\sqrt{\pi J_0 M_0^2 t}\,\mathrm{erf}\left(\sqrt{J_0 M_0^2 t}\right) \tag{12.41}$$

where erf(x) is the *error function*.

Then $f(M,t) = c(M,t)/C(t)$ so that:

$$M_{avg}(t) = \int_0^{M_0} Mf(M,t)\,dM = \frac{M_0}{\exp\left(-J_0 M_0^2 t\right) + \sqrt{\pi J_0 M_0^2 t}\,\mathrm{erf}\left(\sqrt{J_0 M_0^2 t}\right)}. \tag{12.42}$$

Notice that $\int_0^{M_0} Mc(M,t)\,dM = C(t)\,M_{avg}(t) = C_0 M_0 = \text{const.}$ In other words, conservation of mass applies. For sufficiently small times and large particle sizes, Equation (12.40) becomes:

$$c(M,t) \cong (C_0/M_0)\,\delta(M/M_0 - 1). \tag{12.43}$$

which is a monodisperse size distribution. Conversely, for sufficiently large times and small particle sizes, Equation (12.40) becomes:

$$c(x,t) \cong 2C_0 J_0 M_0 t \exp\left(-J_0 M^2 t\right) \tag{12.44}$$

which is a Rosin–Rammler size distribution with $n/m = 1$ and $l/m = 2$. Notice that Equation (12.44) can be rewritten in a self-similar form as follows:

$$c(M,t) \cong \frac{2C_0 M_0}{\pi\left[M_{avg}(t)\right]^2} \exp\left\{-\frac{1}{\pi}\left(\frac{M}{M_{avg}(t)}\right)^2\right\} \tag{12.45}$$

or equivalently:

$$f(M,t) \cong \frac{2}{\pi M_{avg}(t)} \exp\left\{-\frac{1}{\pi}\left(\frac{M}{M_{avg}(t)}\right)^2\right\} \tag{12.46}$$

where $M_{avg}(t) \cong 1/\sqrt{\pi J_0 t}$ as $t \to \infty$. This is true because $\mathrm{erf}(x) \to 1$ and $\mathrm{erf}(-x) \to 0$ as $x \to \infty$. In self-similar form, it becomes clear that this is a *normal distribution* with a standard deviation of $\sqrt{\pi/2}\,M_{avg}(t)$. While normal distributions are common in other applications, they are rare in aggregation and fragmentation, and thus arguably should not be called universals.

Example 12.10 Time-dependent closed-form solutions for $\kappa = 2$ with exponential initial conditions and free-field boundary conditions

For $I(x) = I_0 x = 2J_0 x$, Ziff & McGrady (1985) obtained the following time-dependent solution to Equation (12.9):

$$c(M,t) = (C_0/M_0)\left[1 + 2J_0 M_0 (M + M_0)\,t\right] \exp\left[-(1 + J_0 M_0 M t)\,M/M_0\right] \tag{12.47}$$

where C_0 is the initial total concentration and M_0 is the initial average particle mass.

Notice that:

$$C(t) = \int_0^\infty c(x,t)\,dx = C_0 + C_0\sqrt{\pi J_0 M_0^2 t}\,\exp\left(\frac{1}{4J_0M_0^2t}\right)\mathrm{erfc}\left(\frac{1}{\sqrt{4J_0M_0^2t}}\right) \tag{12.48}$$

where $\mathrm{erfc}(x)$ is the *complementary error function*. Then $f(M,t) = c(M,t)/C(t)$ so that:

$$M_{\mathrm{avg}}(t) = \int_0^\infty Mf(M,t)\,dM = \frac{C_0M_0}{C(t)}. \tag{12.49}$$

Notice that $\int_0^\infty Mc(M,t)\,dM = C(t)\,M_{\mathrm{avg}}(t) = C_0M_0 = \mathrm{const.}$ In other words, conservation of mass applies. For sufficiently small times and large particle sizes, Equation (12.47) becomes:

$$c(M,t) \cong (C_0/M_0)\exp(-M/M_0) \tag{12.50}$$

which is an exponential size distribution. Conversely, for sufficiently large times and small particle sizes, Equation (12.47) becomes:

$$c(x,t) \cong 2C_0J_0M_0t\exp\left(-J_0M^2t\right). \tag{12.51}$$

which is a normal size distribution, just as in the previous example. Notice that Equation (12.51) can be rewritten in self-similar form as follows:

$$c(M,t) \cong \frac{2C_0M_0}{\pi\left[M_{\mathrm{avg}}(t)\right]^2}\exp\left\{-\frac{1}{\pi}\left(\frac{M}{M_{\mathrm{avg}}(t)}\right)^2\right\} \tag{12.52}$$

or equivalently:

$$f(M,t) \cong \frac{2}{\pi M_{\mathrm{avg}}(t)}\exp\left\{-\frac{1}{\pi}\left(\frac{M}{M_{\mathrm{avg}}(t)}\right)^2\right\} \tag{12.53}$$

where $M_{\mathrm{avg}}(t) \cong 1/\sqrt{\pi J_0 t}$ as $t \to \infty$. This is true because $\mathrm{erfc}(1/x) \to 1$ and $\exp(1/x) \to 1$ as $x \to \infty$. Notice that Equations (12.52) and (12.53) agree with Equations (12.45) and (12.46), respectively. While this example approaches the same normal size distribution as the previous example, it approaches it more rapidly.

Example 12.11 Time-dependent closed-form solutions for $\kappa = -1$ with monodisperse initial conditions and free-field boundary conditions

For $I(x) = 2J_0/x^{-2}$, Ziff & McGrady (1986) obtained the following time-dependent solution to Equation (12.9):

$$c(M,t) = \frac{C_0}{M_0}\exp\left(-\frac{J_0 t}{M_0}\right)\delta\left(\frac{M}{M_0} - 1\right) + 2C_0\exp\left(-\frac{J_0 t}{M_0}\right)\frac{J_0 t}{M_0^2}\left[1 + \frac{J_0 t}{2M_0}\left(\frac{M_0}{M} - 1\right)\right] \quad (12.54)$$

where C_0 is the initial total concentration and M_0 is the initial mass of each particle. Notice that $M \le M_0$ for all times so that:

$$C(t) = \int_0^{M_0} c(M,t)\,dM = \infty \quad (12.55)$$

for $t > 0$. Alternatively, suppose a hard sink removes particles with masses less than M_{min} where M_{min} approaches but does not equal 0. Then $M_{\text{min}} \le M \le M_0$ for all times so that:

$$C(t) = \int_{M_{\text{min}}}^{M_0} c(M,t)\,dM \quad (12.56)$$

is finite. In either case, if M_{min} is exactly equal to zero or only approaches zero:

$$\int_{M_{\text{min}}}^{M_0} Mc(M,t)\,dM \cong \int_0^{M_0} Mc(M,t)\,dM = C_0 M_0 \exp\left(-\frac{J_0 t}{M_0}\right)\left[1 + \frac{J_0 t}{M_0} + \frac{1}{2}\left(\frac{J_0 t}{M_0}\right)^2\right] \quad (12.57)$$

for $t > 0$. In other words, conservation of mass does not apply. Instead, the total particle mass is monotone decreasing due to shattering. For sufficiently small times and large particle sizes, Equation (12.54) becomes:

$$c(M,t) \cong \frac{C_0}{M_0}\,\delta\left(\frac{M}{M_0} - 1\right) \quad (12.58)$$

which is a monodisperse size distribution. Conversely, for sufficiently large times and small particle sizes, Equation (12.54) becomes:

$$c(M,t) \cong C_0 \exp\left(-\frac{J_0 t}{M_0}\right)\left(\frac{J_0 t}{M_0}\right)^2 \frac{1}{M} \quad (12.59)$$

which is a power law size distribution with $n/m = 0$. As discussed in the last chapter, aggregation also produces power law size distributions, where n/m obtains rational values such as 0, $\pm 1/2$, ± 1, and so forth. Because they occur so often for both aggregation and fragmentation, in both discrete and continuous particle systems, power law size distributions with rational exponents may fairly be called universals.

Example 12.12 Time-dependent closed-form solutions for $\varkappa = -2$ with monodisperse initial conditions and free-field boundary conditions

For $I(x) = 2J_0/x^{-3}$, Ziff & McGrady (1986) obtained the following time-dependent solution to Equation (12.9):

$$c(M,t) = \frac{C_0}{M_0}\exp\left(-\frac{J_0 t}{M_0^2}\right)\delta\left(\frac{M}{M_0}-1\right) + 2\frac{C_0}{M_0}\exp\left(-\frac{J_0 t}{M_0^2}\right)\frac{J_0 t}{M_0^2}. \tag{12.60}$$

where C_0 is the initial total concentration and M_0 is the initial mass of each particle. Notice that $M \leq M_0$ for all times so that:

$$C(t) = \int_0^{M_0} c(M,t)\,dM = C_0\exp\left(-\frac{J_0 t}{M_0^2}\right)\left(1+\frac{2J_0 t}{M_0^2}\right). \tag{12.61}$$

Then:

$$f(M,t) = \frac{c(M,t)}{C(t)} = \frac{1}{M_0}\frac{\delta(M/M_0-1)+2J_0 t/M_0^2}{1+2J_0 t/M_0^2}. \tag{12.62}$$

Notice that:

$$\int_0^{M_0} Mc(M,t)\,dM = C_0 M_0\exp\left(-\frac{J_0 t}{M_0^2}\right)\left(1+\frac{J_0 t}{M_0^2}\right). \tag{12.63}$$

In other words, conservation of mass does not apply. Instead, the total particle mass is monotone decreasing for all $t > 0$ due to shattering. In addition, notice that:

$$M_{\text{avg}}(t) = \int_0^{M_0} Mf(M,t)\,dM = \frac{1}{C(t)}\int_0^{M_0} Mc(M,t)\,dM = M_0\frac{1+J_0 t/M_0^2}{1+2J_0 t/M_0^2}. \tag{12.64}$$

For sufficiently small times and large particles, Equation (12.60) becomes:

$$c(M,t) \cong \frac{C_0}{M_0}\delta\left(\frac{M}{M_0}-1\right) \tag{12.65}$$

which is a monodisperse size distribution. Conversely, for sufficiently large times and small particle sizes, Equation (12.60) becomes:

$$c(M,t) \cong 2\frac{C_0}{M_0}\exp\left(-\frac{J_0 t}{M_0^2}\right)\frac{J_0 t}{M_0^2} \tag{12.66}$$

which is a *uniform size distribution*, i.e., it does not depend on M. Alternatively, this can be viewed as a power law size distribution with $n/m = 1$. As discussed in the last example, such power law size distributions are may fairly be called universals.

12.6 Late-Time Mean-Field Solutions

For many practical applications, the final state is more important than any intermediate state. Suppose the boundary conditions are free-field. In addition, suppose the initial conditions are monodisperse or Rosin–Rammler. If $\kappa > 0$ so that shattering does not occur, mean-field theory obtains a Rosin–Rammler size distribution with $n/m = 1$ and $l/m = \kappa$. More specifically:

$$c(M,t) \cong \frac{\kappa b^{1/\kappa}}{\Gamma(1/\kappa)} \frac{C_0 M_0}{[M_{avg}(t)]^2} \exp\left\{-b\left[\frac{M}{M_{avg}(t)}\right]^{\kappa}\right\} \tag{12.67}$$

where:

$$M_{avg}(t) \cong (\kappa a_0 J_0 t)^{-1/\kappa} \tag{12.68}$$

for small enough particles and large enough times. In these expressions, a_0 is a positive unitless constant and:

$$b = \left[\frac{\Gamma(2/\kappa)}{\Gamma(1/\kappa)}\right]^{\kappa}.$$

If $\kappa = 0$, mean-field theory obtains a log-normal distribution, or something reminiscent of a log-normal distribution, where:

$$M_{avg}(t) = M_0 \exp(-a_0 J_0 t) \tag{12.69}$$

and a_0 is a positive unitless constant. Finally, if $\kappa < 0$ so that shattering occurs, mean-field theory obtains a power law size distribution with $n/m = |\kappa| - 1$. More specifically:

$$c(M,t) \cong A(\kappa) \frac{C_0 M_0}{\left[M'_{avg}(t)\right]^2} \left[\frac{M}{M'_{avg}(t)}\right]^{|\kappa|-2} \tag{12.70}$$

where:

$$M'_{avg}(t) \cong M_0 \exp\left(J_0 M_0^{\kappa} t / |\kappa|\right) \left(J_0 M_0^{\kappa} t\right)^{-2/\kappa^2} \tag{12.71}$$

for small enough particles and large enough times. Equation (12.70) applies regardless of the strength of the sink formed due to shattering. However, Equation (12.71) only applies for relatively weak sinks. The next chapter will give additional expressions, including those for stronger sinks, for both stationary and non-stationary particle systems. To summarize, for sufficiently late times and small particles, binary fragmentation produces power law or Rosin–Rammler size distributions where, according to Equation (12.13), the exponents are integer multiples of 1/6. The power law size distributions occur in many different aggregation and fragmentation problems and thus can fairly be called universals. On the other hand, many of the Rosin–Rammler distributions are unusual and thus arguably should not be called universals.

12.7 Experimental Validation

As shown in Table 12.3, most of the extant experimental results concern $\kappa = 0$ or $\kappa = 1$. However, there is also evidence for other values. For example, Madras et al. (1997) obtained evidence for $\kappa = 2$, Rangarajan et al. (1998) obtained evidence for $\kappa = 5/6$, and Kruse et al. (2003) obtained evidence for $\kappa = -1$. However, on the whole, this evidence is substantially weaker than that given in Table 12.3. Examples 12.13, 12.14, and 12.15 give additional details on the experimental validations listed in the three shaded rows in Table 12.3.

Table 12.3 Experimental results for polymer degradation due to random-chain scission. Except for Rangarajan et al. (1998), the polymers were dissolved in a solution prior to degradation.

κ	Polymer	Initial Molecular Weight	Final Time (M_0/M_{avg})	Reference
0[1]	Poly(styrene-allyl alcohol)	Small (1,100)	Small (< 1.2)	Wang, Smith, & McCoy (1995)
0[2]	Polystyrene	Large (95,000)	Small (< 2)	Madras et al. (1997a)
0[2]	Polystyrene	Medium (12,000)	Small (< 2)	Madras et al. (1997b)
0[2]	Poly(vinyl acetate) Poly(ε-caprolactone)	Small to Large (15,400–80,000)	Large (~50)	Sivalingam & Madras (2004)
1[3]	Polystyrene	Large (95,000)	Small–Large (1.2–20)	Madras & McCoy (1997)
1[3]	Polystyrene	Large (50,000)	Large	Kim & McCoy (2000)
1[3]	Polystyrene	Large (500,000)	Medium	Sterling, Walline, & McCoy (2001)
1[2]	Low Density Polyethylene	Large (200,000)	Med.–Large (3–28)	Kumar, Kumar & Madras (2002)
1[2]	Polypropylene	Large (< 125,000)	Medium (2.5–12)	Kumar & Madras (2003)
1[2]	Polycarbonate	Medium (26,000)	Large	Sivalingam et al. (2003)
1[1]	Poly(alkyl methacrylate)	Large (90,000–340,000)	Medium (2.3–10)	Marimuthu & Madras (2008)
1[2]	Poly(acrylic acid)	Large (240,000)	Med.–Large (3–28)	Shukla, Daraboina, & Madras (2009)

(1) The corresponding size distribution is described as a gamma size distribution with $n/m > 1$.
(2) The corresponding size distribution is not described.
(3) The corresponding size distribution is described as an exponential size distribution, i.e., a gamma distribution with $n/m = 1$.

Example 12.13 Experimental results for thermal degradation of polystyrene vs. mean-field results for binary fragmentation with $\kappa = 1$

As one way to induce thermal degradation, solid polymers may be heated above their melting point. As noted by Sterling et al. (2001), this results in a complex "multiphase process ... [with] random degradation, chain-end degradation, vaporization, diffusion, and repeated initiation/recombination." As an alternative, solid polymers may be dissolved in a solution prior to heating. As noted by Sterling et al. (2001), "the advantage of polymer degradation studies conducted in solution is the elimination of some of these [complex] effects, and the ability to distinguish among others." More specifically, "during pyrolysis of polystyrene melt, chain-end products of degradation (oligomers) are significant, whereas in solution ... the generation rate of such products is negligible compared to products of random scission."

Sterling et al. (2001) used size exclusion chromatography (SEC), as described earlier in Example 10.12, to characterize thermal degradation of polystyrene dissolved in mineral oil; see also Madras et al. (1997a, b). In this case, the experimental results agree with mean-field theory results for binary fragmentation with $\kappa = 1$. More specifically, as seen in Examples 12.7 and 12.8, mean-field theory with monodisperse or exponential initial conditions obtains:

$$c(M,t) \cong \frac{C_0 M_0}{[M_{avg}(t)]^2} \exp\left[-\frac{M}{M_{avg}(t)}\right] \tag{12.72}$$

where:

$$M_{avg}(t) \cong \frac{M_0}{1 + J_0 M_0 t} \cong \frac{1}{J_0 t} \tag{12.73}$$

for sufficiently large times. Figure 12.1a compares Equation (12.72) to experimental results. According to Sterling et al. (2001), "the polymer molar distribution was observed to maintain the exponential distribution form throughout the experiments," as witnessed by "the linearity of the semi-log data plot." Figure 12.1b compares Equation (12.73) to the experimental results. According to Sterling et al. (2001), the "data linearity [in this figure] supports the assumption of negligible polymer mass loss, and shows the linear increase in polymer moles," where the term "moles" refers to the number of particles. More specifically, a mole is $6.02214076 \times 10^{23}$ particles.

According to Sterling et al. (2001), "weak links in the polymer backbone arise during polymerization when monomers attach head-to-head, or when oxygenated structures are embedded in the backbone. These bonds are weaker than the normal head-to-tail bonds causing an initial period of rapid degradation." Thus, Equations (12.72) and (12.73) only apply once enough time has passed, so that all the weak links are broken and only the strong links survive. As a rule of thumb, this is true when:

$$M_0/M_{avg}(t) < 2.$$

Prior to this, the experimental results tend to agree with mean-field theory for binary fragmentation with $\kappa = 0$; see, e.g., Madras et al. (1997a, b).

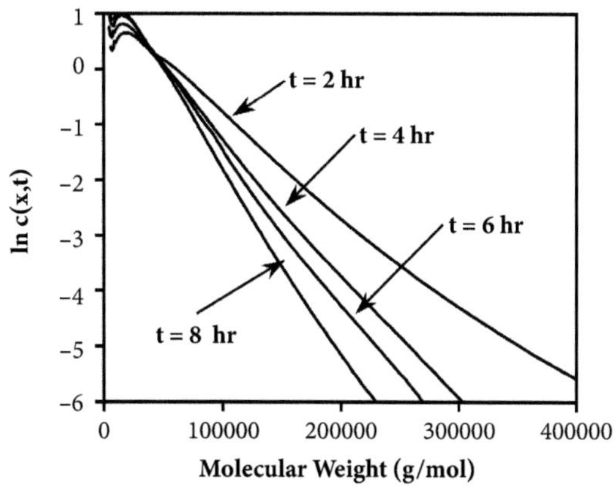

(a) Particle size distribution. Experimental results are shown as curves. Notice that these results are not normalized by average fragment sizes; thus the curves do not lie on top of one another. However, the fact that these curves are nearly linear in a linear-log plot indicates that they are exponentials, as predicted by mean-field theory for binary fragmentation with $\kappa = 1$.

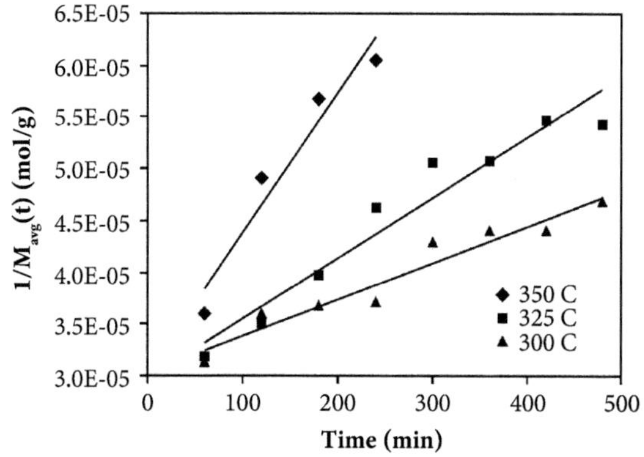

(b) Average particle size. Experimental results shown as symbols. Mean-field theory results for binary fragmentation with $\kappa = 1$ are shown as lines.

Figure 12.1 Analytical (mean-field theory) vs. experimental results for thermal degradation of polystyrene dissolved in mineral oil.

[The notation in these figures has been changed to match the notation used in this book.]
Figures reprinted from Sterling et al. (2001) with permission from Elsevier.

Example 12.14 Experimental results for oxidative degradation of polystyrene vs. mean-field results for binary fragmentation with $\varkappa = 1$

As noted in the previous example, thermal degradation typically involves both random- *and* end-chain scission. By contrast, oxidative degradation typically involves *only* random-chain scission. This makes such experiments especially valuable for validating the models given in this chapter, which concern only random- and not end-chain scission. However, oxidative degradation requires solving two governing equations—one for the polymer chain concentration and the other for the oxidizer concentration. More specifically, Madras & McCoy (1997) give the following governing equations:

$$\frac{\partial c_{\text{poly}}(x,t)}{\partial t} = -J_{\text{poly}}(x)\, c_{\text{poly}}(x,t) + \int_x^\infty m(x,y) J_{\text{poly}}(y)\, c_{\text{poly}}(y,t)\, dy \tag{12.74}$$

$$\frac{dc_{\text{oxid}}(t)}{dt} = -\int_0^\infty J_{\text{oxid}}(y)\, c_{\text{oxid}}(t)\, dy - c_{\text{oxid}}(t)\int_0^\infty J_{\text{poly}}(y)\, c_{\text{poly}}(x,t)\, dy \tag{12.75}$$

where $c_{\text{poly}}(x,t)$ is the polymer concentration, $c_{\text{oxid}}(t)$ is the oxidizer concentration, $J_{\text{poly}}(y)$ is the rate coefficient for polymer chain loss, $J_{\text{oxid}}(y)$ is the rate coefficient for oxidizer loss, and $m(x,y) = 1/y$ consistent with binary fragmentation. Assume that:

$$J_{\text{poly}}(y) = J_{0,\text{poly}}\, y \tag{12.76}$$

$$J_{\text{oxid}}(y) = J_{0,\text{oxid}}\, y \tag{12.77}$$

where $J_{0,\text{poly}}$ and $J_{0,\text{oxid}}$ are constants. Consider the following change of variables:

$$d\tau = J_{0,\text{poly}}\, c_{\text{oxid}}(t)\, dt. \tag{12.78}$$

For exponential initial conditions, Madras & McCoy (1997) obtained the following time-dependent solution to Equations (12.74) to (12.77):

$$c(M,t) = \frac{C_0 M_0}{[M_{\text{avg}}(t)]^2} \exp\left[-\frac{M}{M_{\text{avg}}(t)}\right] \tag{12.79}$$

where $M_{\text{avg}}(t) = M_{\text{avg}}(\tau(t))$ and:

$$M_{\text{avg}}(\tau) = \frac{M_0}{1 + M_0\tau} \cong \frac{1}{\tau} \tag{12.80}$$

consistent with the expressions found in Example 12.8 for binary fragmentation with $\varkappa = 1$. Madras & McCoy (1997) used SEC to characterize "oxidative degradation of polystyrene in tricholorobenzene solution at 428K at various concentrations of polymer and of peroxide." Figures 12.2.a and 12.2.b compare Equation (12.79) and (12.80), respectively, to experimental results; see also Kim & McCoy (2000).

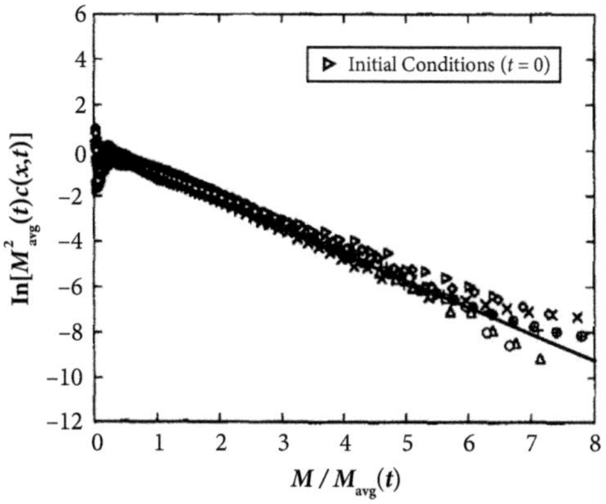

(a) Particle size distribution. Experimental results shown as symbols. Mean-field theory results for binary fragmentation with $\varkappa = 1$ shown as a line.

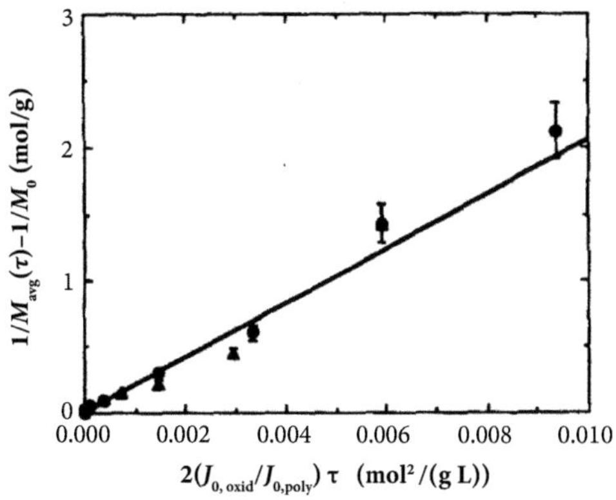

(b) Average particle size in the limit of large times. Experimental results shown as symbols. Mean-field theory results for binary fragmentation with $\varkappa = 1$ shown as a line.

Figure 12.2 Analytical (mean-field theory) vs. experimental results for oxidative degradation of polystyrene dissolved in tricholorobenzene.

[The notation in these figures has been changed to match the notation used in this book.]
Reprinted from Madras and McCoy (1997) with permission from Elsevier.

Example 12.15 Experimental results for thermal degradation of polycaprolactone vs. mean-field results for binary fragmentation with $\varkappa = 0$

Sivalingam & Madras (2004) used SEC to characterize thermal degradation of polycaprolactone (PCL)—also known as poly(ε-caprolactone)—dissolved in an inert solvent. In this case, they found that "the thermal degradation ... occurs by random chain scission." Their experimental results seem to agree with mean-field theory results for binary fragmentation with $\varkappa = 0$. As seen in Example 12.6, if $\varkappa = 0$, mean-field theory with monodisperse initial conditions obtains:

$$c(M,t) \cong \frac{C_0}{M_0} \frac{\exp(-J_0 t)(2J_0 t)^{1/4}}{(4\pi)^{1/2}[\ln(M_0/M)]^{3/4}} \exp\left\{2[2J_0 t \ln(M_0/M)]^{1/2}\right\} \tag{12.81}$$

for large enough times and small enough particles where:

$$\ln\left[M_0/M_{\text{avg}}(t)\right] = \ln\left[C(t)/C_0\right] = J_0 t \tag{12.82}$$

and where J_0 is a positive constant. Figure 12.3 compares Equation (12.82) to experimental results. Unfortunately, Sivalingam & Madras (2004) do not report any experimental results for size distributions; thus, Equation (12.81) cannot be compared to experimental results. In fact, a literature search did not uncover any experimental evidence for Equation (12.81). However, there is some limited experimental evidence for its cousin, the log-normal size distribution; see, e.g., Rangarajan et al. (1998) and Kruse et al. (2003).

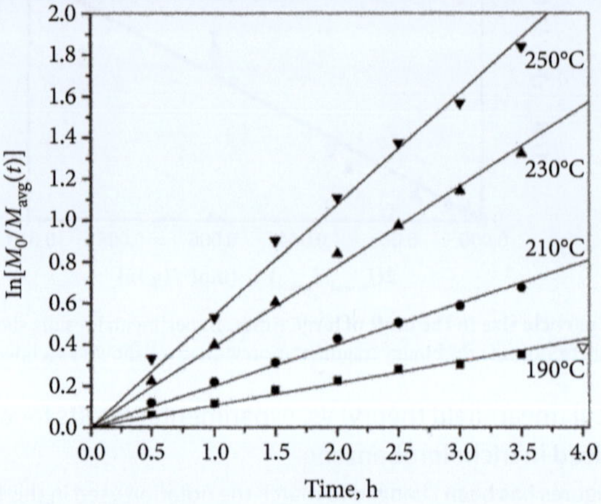

Figure 12.3 Analytical (mean-field theory) vs. experimental results for thermal degradation of PCL dissolved in diphenyl ether. Experimental results shown as symbols. Mean-field theory results for binary fragmentation with $\varkappa = 0$ shown as lines.
[The notation in this figure has been changed to match the notation used in this book.]

12.8 Conclusions

This chapter discussed mean-field theory for binary fragmentation. This chapter assumed that there are an infinite number of particles homogeneously and istotropically distributed in an infinite spatial volume with a finite concentration C. In other words, $0 < C = N/\mathcal{V} < \infty$ in the limit $N \to \infty$ and $\mathcal{V} \to \infty$. This is sometimes called the *thermodynamic limit*. Because there are an infinite number of particles, binary fragmentation never ends but continues forever. In order to maintain spatial homogeneity and isotropy, this chapter assumed that the children produced in each fragmentation event are instantaneously and evenly distributed over the entire spatial domain. Obviously, such assumptions are not realistic, but they may be approximately true in many cases.

The best-known real-world example of binary fragmentation is polymer degradation. More specifically, polymer degradation involves *random-chain scission* if both children in a given fragmentation event have random sizes. Alternatively, polymer degradation involves *end-chain scission* if one child is always a monomer, dimer (two monomers), trimer (three monomers), or has some other fixed size. Binary fragmentation, as treated in this chapter, includes random-chain scission but not end-chain scission. In other words, binary fragmentation, as treated in this chapter, assumes that each fragmentation event produces a random rather than a deterministic particle size distribution. In general, polymer degradation experiments provide reasonable support for mean-field theory. However, most of this evidence is limited to $\kappa = 1$ and, to a lesser extent, $\kappa = 0$. Even in these cases, polymer degradation experiments sometimes find evidence for gamma size distributions, even though mean-field theory predicts something else; see, e.g., Wang, Smith, & McCoy (1995) and Marimuthu & Madras (2008). As discussed in Chapters 10 and 11, materials formed through aggregation—which include polymers—often obtain gamma size distributions. Fragmentation replaces such initial conditions gradually as time progresses. Thus, when the initial conditions are gamma size distributions, they may survive at small and medium times, albeit in a modified form and over a limited range; see, e.g., Table 12.1 in Wang, Smith, & McCoy (1995).

As noted earlier, binary fragmentation is the exact inverse of aggregation. However, while aggregation kernels always depend on two particle sizes, binary fragmentation kernels may depend on only one particle size. As a result, mean-field theory for binary aggregation is substantially simpler than mean-field theory for aggregation. For example, for aggregation, mean-field theory produces full time-dependent closed-form solutions for only three different kernels with λ equal to 0, 1, or 2. By contrast, for binary aggregation, mean-field theory produces full time-dependent closed solutions for at least five different kernels with κ equal to −2, −1, 0, 1, and 2. (In fact, as seen in the next chapter, closed-form solutions are possible for any real value of κ between −2 and 2.) Although mean-field theory itself may be simpler for binary fragmentation, experimental validation of mean-field theory is simpler for aggregation. As a result, mean-field theory for binary fragmentation is only well-validated for $\kappa = 1$ and, to a lesser extent, $\kappa = 0$; see, e.g., Table 12.3. By contrast, mean-field theory for aggregation is well-validated for many different values of λ; see, e.g., Tables 11.2 and 11.3. This difference is perhaps partly due to the fact that aggregation is a more common problem and thus has received more attention than binary fragmentation.

References

M. Aizenman & T. A. Bak, Convergence to Equilibrium in a System of Reacting Polymers, *Communications in Mathematical Physics*, 65, 203–230, 1979, https://doi.org/10.1007/BF01197880

M. Alger, *Polymer Science Dictionary*, 3rd edition, Springer, 2017, https://doi.org./10.1007/978-94-024-0893-5

J. M. Ball, J. Carr, & O. Penrose, The Becker–Döring Cluster Equations: Basic Properties and Asymptotic Behaviour of Solutions, *Communications in Mathematical Physics*, 104, 657–692, 1986, https://doi.org/10.1007/BF01211070

J. M. Ball & J. Carr, Asymptotic Behaviour of Solutions to the Becker–Döring Equations for Arbitrary Initial Data, *Proceedings of the Royal Society of Edinburgh Section A: Mathematics*, 108(1–2), 109–116, 1988, https://doi.org/10.1017/S0308210500026561

J. M. Ball & J. Carr, The Discrete Coagulation-Fragmentation Equations: Existence, Uniqueness, and Density Conservation, *Journal of Statistical Physics*, 61(1–2), 203–234, 1990, https://doi.org/10.1007/BF01013961

R. Becker & W. Döring, Kinetische Behandlung der Keimbildung in Übersättigten Dämpfen, *Annalen der Physik*, 416(8), 719–752, 1935, https://doi.org/10.1002/andp.19354160806

N. C. Billingham, Degradation and Stabilization of Polymers, in M. Schütze, editor, *Corrosion and Environmental Degradation, Volume 2*, Wiley, 2000, https://doi.org/10.1002/9783527603978.mst0412

P. J. Blatz and A. V. Tobolsky, Note on the Kinetics of Systems Manifesting Simultaneous Polymerization-Depolymerization Phenomena, *Journal of Physical Chemistry*, 49(2), 77–80, 1945, https://doi.org/10.1021/j150440a004

F. P. da Costa, Mathematical Aspects of Coagulation-Fragmentation Equations, in J. P. Bourguignon, R. Jeltsch, A. Pinto, & M. Viana, editors, *Mathematics of Energy and Climate Change, CIM Series in Mathematical Sciences, Volume 2*, Springer, 2015, https://doi.org/10.1007/978-3-319-16121-1_5

R. B. Diemer & J. H. Olson, A Moment Methodology for Coagulation and Breakage Problems Part 1—Analytical Solution of the Steady-State Population Balance, *Chemical Engineering Science*, 57(12), 2193–2209, 2002, https://doi.org/10.1016/S0009-2509(02)00111-2:

F. Family, P. Meakin, & J. M. Deutch, Kinetics of Coagulation with Fragmentation: Scaling Behavior and Fluctuations, *Physical Review Letters*, 57(6), 727, 1986, https://doi.org/10.1103/PhysRevLett.57.727

H. H. G. Jellinek & G. White, The Degradation of Long-Chain Molecules by Ultrasonic Waves. I. Theoretical, *Journal of Polymer Science*, 6(6), 745–756, 1951, https://doi.org/10.1002/pol.1951.120060609

Y. C. Kim & B. J. McCoy, Degradation Kinetics Enhancement of Polystyrene by Peroxide Addition, *Industrial & Engineering Chemistry Research*, 39(8), 2811–2816, 2000, https://doi.org/10.1021/ie0000562

J. R. King & J. A. D. Wattis, Asymptotic Solutions of the Becker–Döring Equations with Size-Dependent Rate Constants, *Journal of Physics A: Mathematical and General*, 35(6), 1357–1380, 2002, https://doi.org/10.1088/0305-4470/35/6/303

P. L. Krapivsky, S. Redner, & E. Ben-Naim, *A Kinetic View of Statistical Physics*, Cambridge University Press, 2010, https://doi.org/10.1017/CBO9780511780516

T. M. Kruse, H.-W. Wong, & L. J. Broadbelt, Modeling the Evolution of the Full Polystyrene Molecular Weight Distribution during Polystyrene Pyrolysis, *Industrial & Engineering Chemistry Research*, 42(12), 2722–2735, 2003, https://doi.org/10.1021/ie020657o

G. S. Kumar, V. R. Kumar, & G. Madras, Continuous Distribution Kinetics for the Thermal Degradation of LDPE in Solution, *Journal of Applied Polymer Sciences*, 84(4), 681–690, 2002, https://doi.org/10.1002/app.2344

R. Kumar & G. Madras, Thermal Degradation Kinetics of Isotactic and Atactic Polypropylene, *Journal of Applied Polymer Science*, 90(8), 2206–2213, 2003, https://doi.org/10.1002/app.12880

C. B. Laney, *Computational Gasdynamics*, Cambridge University Press, 1998, https://doi.org/10.1017/CBO9780511605604

G. Madras & B. J. McCoy, Oxidative Degradation Kinetics of Polystyrene in Solution, *Chemical Engineering Science*, 52(16), 2707–2713, 1997, https://doi.org/10.1016/S0009-2509(97)00091-2

G. Madras, J. M. Smith, & B. J. McCoy, Thermal Degradation Kinetics of Polystyrene in Solution, *Polymer Degradation and Stability*, 58(1–2), 131–138, 1997a, https://doi.org/10.1016/S0141-3910(97)00036-0

G. Madras, G. Y. Chung, J. M. Smith, & B. J. McCoy, Molecular Weight Effect on the Dynamics of Polystyrene Degradation, *Industrial & Engineering Chemistry Research*, 36(6), 2019–2024, 1997b, https://doi.org/10.1021/ie9607513

A. Marimuthu & G. Madras, Continuous Distribution Kinetics for Microwave-Assisted Oxidative Degradation of Poly(alkyl methacrylates), *AIChE Journal*, 54(8), 2164–2173, 2008, https://doi.org/10.1002/aic.11548

T. Matsoukas, Stochastic Theory of Discrete Binary Fragmentation—Kinetics and Thermodynamics, *Entropy*, 24(2), 229, 2022, https://doi.org/10.3390/e24020229

B. J. McCoy & G. Madras, Degradation Kinetics of Polymers in Solution: Dynamics of Molecular Weight Distributions, *AIChE Journal*, 43(3), 802–810, 1997, https://doi.org/10.1002/aic.690430325

E. D. McGrady & R. M. Ziff, "Shattering" Transition in Fragmentation, *Physical Review Letters*, 58(9), 892–895, 1987, https://doi.org/10.1103/PhysRevLett.58.892

P. Meakin & M. H. Ernst, Scaling in Aggregation with Breakup Simulations and Mean-Field Theory, *Physical Review Letters*, 60(24), 2503–2506, 1988, https://doi.org/10.1103/PhysRevLett.60.2503

K. Pielichowski, J. Njuguna, & T. M. Majka, *Thermal Degradation of Polymeric Materials*, 2nd edition, Elsevier, 2022, https://doi.org/10.1016/C2019-0-04932-1

V. N. Piskunov, Analytical and Numerical Results in the Kinetics of Particle Coagulation and Fragmentation Processes, *Journal of Applied Mathematics and Mechanics*, 76(8), 688–705, 2012, https://doi.org/10.1016/j.jappmathmech.2013.02.006

D. Ramkrishna, *Population Balances: Theory and Applications to Particulate Systems in Engineering*, Elsevier, 2000, https://doi.org/10.1016/B978-0-12-576970-9.X5000-0

P. Rangarajan, D. Bhatttacharyya, & E. Grulke, HDPE Liquefaction: Random Chain Scission Model, *Journal of Applied Polymer Science*, 70(6), 1239–1251, 1998, https://doi.org/10.1002/(SICI)1097-4628(19981107)70:6%3C1239:AID-APP20%3E3.0.CO;2-T

S. Ray & R. P. Cooney, Thermal Degradation of Polymer and Polymer Composites, in M. Kutz, editor, *Handbook of Environmental Degradation of Materials*, 3rd edition, Elsevier, 2018, https://doi.org/10.1016/B978-0-323-52472-8.00009-5

S. Redner, Fragmentation, in J. Herrmann & S. Roux, editors, *Statistical Models for the Fracture of Disordered Media*, Elsevier, 1990, https://doi.org/10.1016/B978-0-444-88551-7.50021-5

O. Saito, On the Effect of High Energy Radiation to Polymers I. Cross-Linking and Degradation, *Journal of the Physical Society of Japan*, 13(2), 1980–206, 1958, https://doi.org/10.1143/JPSJ.13.198

N. B. Shukla, N. Daraboina, & G. Madras, Oxidative and Photooxidative Degradation of Poly(Acrylic Acid), *Polymer Degradation and Stability*, 94(8), 1238–1244, 2009, https://doi.org/10.1016/j.polymdegradstab.2009.04.020

R. Simha, Kinetics of Degradation and Size Distribution of Long Chain Polymers, *Journal of Applied Physics*, 12(7), 569–578, 1941, https://doi.org/10.1063/1.1712939

G. Sivalingam, K. Nagaveni, G. Madras, & M. S. Hegde, Kinetics of Catalytic Degradation of Polycarbonate in Benzene, *Industrial & Engineering Chemistry Research*,42(4), 687–691, 2003, https://doi.org/10.1021/ie020582o

G. Sivalingam & G. Madras, Thermal Degradation of Poly(vinyl acetate) and Poly(ε-caprolactone) and Their Mixtures in Solution, *Industrial & Engineering Chemistry Research*, 43(7), 1561–1567, 2004, https://doi.org/10.1021/ie034115y

W. J. Sterling, K. S. Walline, & B. J. McCoy, Experimental Study of Polystyrene Thermolysis to Moderate Conversion, *Polymer Degradation and Stability*, 73(1), 75–82, 2001, https://doi.org/10.1016/S0141-3910(01)00069-6

R. D. Vigil, On Equilibrium Solutions of Aggregation–Fragmentation Problems, *Journal of Colloid and Interface Science*, 336(2), 642–647, 2009, https://doi.org/10.1016/j.jcis.2009.04.061

M. Wang, J. M. Smith, & B. J. McCoy, Continuous Kinetics for Thermal Degradation of Polymer in Solution, *AIChE Journal*, 1521–1533, 41(6), 1995, https://doi.org/10.1002/aic.690410616

J. A. D. Wattis & J. R. King, Asymptotic Solutions of the Becker–Döring Equations, *Journal of Physics A: Mathematical and General*, 31(34), 7169–7189, 1998, https://doi.org/10.1088/0305-4470/31/34/018

J. A. D. Wattis, The Modified Becker–Döring System with Aggregation-Dominated Rate Coefficients, *Journal of Physics A: Mathematical and General*, 42(4), 045002, 2009, https://doi.org/10.1088/1751-8113/42/4/045002

R. M. Ziff & E. D. McGrady, The Kinetics of Cluster Fragmentation and Depolymerisation, *Journal of Physics A: Mathematical and General*, 18(15), 3027–3037, 1985, https://doi.org/10.1088/0305-4470/18/15/026

R. M. Ziff & E. D. McGrady, Kinetics of Polymer Degradation, *Macromolecules*, 19(10), 2513–2519, 1986, https://doi.org/10.1021/ma00164a010

R. M. Ziff, An Explicit Solution to a Discrete Fragmentation Equation, *Journal of Physics A: Mathematical and General*, 25(9), 2569–2576, 1992, https://doi.org/10.1088/0305-4470/25/9/027

13

Fragmentation Cascades Part 2: Mean-Field Theory for Random Fragmentation

This is the second of three chapters on mean-field theory for fragmentation cascades. This chapter discusses *random fragmentation*, a.k.a., *linear* or *spontaneous fragmentation*. As seen in Table 13.1, random fragmentation as discussed in this chapter is a generalization of binary fragmentation as discussed in the last chapter. While each event in binary fragmentation creates exactly two children, each event in random fragmentation may create any number of children. Example 13.1 describes a common application in which random fragmentation occurs.

Table 13.1 Three common types of fragmentation cascades.

Type	Description	Typical Applications
Binary Fragmentation	Randomly selected particles fragment into exactly two pieces with independently random sizes.	Thermal and oxidative degradation of polymers
Random Fragmentation	Randomly selected particles fragment into two or more pieces with independently random sizes.	Grinding of minerals, ores, and chemicals
Collisional Fragmentation	Random binary collisions cause one or both of the participating particles to fragment into two or more pieces with independently random sizes.	Evolution of objects in the asteroid belt and the Kuiper belt.

13.1 Governing Equations

13.1.1 Introduction

The state of any fragmenting particle system can be described by random or seemingly random fields such as position, mass, momentum, energy, shape, and rotation. If the particle system is spatially homogeneous and isotropic, the position field does not affect the other fields. Suppose that the effects of the other fields on the particle size can be combined into a single fragmentation rate. This rate, known as the random fragmentation kernel, captures the average effects of randomness and chaos in the context of a simple deterministic governing equation. This mean-field description relies on spatial homogeneity and isotropy. However, each fragmentation event creates new particles, whose properties vary markedly from the older particles surrounding them. In other words, each fragmentation event temporarily disrupts spatial homogeneity and isotropy. However, as a practical matter, spatial

Particle Dynamics with Aggregation and Fragmentation. Culbert B. Laney, Oxford University Press. © Culbert B. Laney (2025).
DOI: 10.1093/9780198920687.003.0013

homogeneity still applies provided the mixing rates are large enough, relative to the fragmentation rates. Most of the applications in this book involve particle systems with low volumetric fractions, which tend to produce relatively high mixing rates. However, many of the mechanisms that cause random fragmentation only occur in particle systems with high volumetric fractions, which tend to produce relatively low mixing rates. As it turns out, dense particle systems tend to trap children next to a fixed set of neighboring particles for extended periods of time. In addition to trapping mass, dense particle systems also trap momentum and energy, releasing it sporadically in a series of avalanches or catastrophes; see Section 2.3. Such momentum and energy releases tend to drive fragmentation rates up, at least locally and temporarily. To use mean-field theory on such dense systems, the usual definitions of spatially homogeneous and isotropic must be relaxed, so that they apply only in some average sense over sufficiently large spatial and temporal regions.

Example 13.1 Random fragmentation in ball mills

A ball mill is used to grind materials like minerals, ores, and chemicals. A ball mill consists of a horizontal rotating cylinder that contains steel or ceramic balls along with the particles to be ground. According to Lynch & Rowland (2005), "the mass moves up the wall of the cylinder as it rotates and falls back into the 'toe' of the mill when the force of gravity exceeds friction and centrifugal forces. Particles are broken in the toe of the mill when caught in the collision between the grinding media [i.e., the balls] and mill wall ... The grinding media and particles acquire potential energy that becomes kinetic energy as the mass falls from the rotating shell." According to Tavares & De Carvalho (2009), "if a particle is stressed with a sufficiently high energy to cause body fracture (cleavage or shattering, depending on the stressing energy) it will break, producing progeny fragments. However, if it is stressed with an insufficient amount of energy, then it will become damaged and its strength will be reduced, suffering only surface breakage [or] ... surface abrasion." In *batch* mills, a fixed set of material is ground until it obtains the desired size; such systems are *non-stationary*. This case amounts to a random fragmentation cascade with no sources or sinks. For *continuously operated* mills, new coarse material is added while fully ground fine material is removed; such systems are *stationary*. This case amounts to a random fragmentation cascade with a large-particle source and a small-particle sink. In general, ball mills involve dense particle systems, i.e., the volumetric fraction is relatively high. Thus, like all dense particle systems, ball mills continuously trap and sporadically release energy; see the earlier discussion in Section 2.3. The energy trapping phase occurs as the material rises along the wall of the rotating cylinder. The energy release phase occurs as the material avalanches from near the top to the "toe" of the mill. In addition, like all dense systems, ball mills tend to trap mass. If the volumetric fraction is low, children can travel freely, interacting and mixing with neighbors. However, if the volumetric fraction is high, children tend to be hemmed in on all sides, which reduces their mixing rate.

In this book, the mean-field governing equations are usually written in terms of the concentration c_i or $c(M)$. This makes perfect sense for aggregation and collisional fragmentation, where particle systems evolve solely due to things, such as binary collisions, which depend on concentration. This also generally makes sense for binary fragmentation, which is the inverse of aggregation, and is often assumed to occur alongside aggregation. However, this does not always make sense for random fragmentation, where particle systems generally evolve due to multiple different mechanisms, some of which may not depend on concentration. As another issue, measuring concentration requires sizing and counting. With a great deal of patience, researchers can size and count tens of thousands of particles by hand. For example, Weimer & Rogers (1979) report manually sizing and counting almost 80,000 solid fragments. With modern software and automated techniques, such as those described in Figure 3.1, researchers can size and count hundreds of thousands or even millions of particles. However, for dense particle systems, sizing and counting may be practically impossible. For example, according to Tavares (2017), "a 30cm slice of [a typical ball] mill could have more than two billion elements [particles] with sizes above 0.5mm." While it is hard to count this many particles, it is easy to weigh them. As seen in Examples 3.2 and 3.3, measuring $f_{M, i}$ or $f_M(D)$ is as simple as passing particles through a series of increasingly finer sieves. To allow for easy comparisons with such experimental results, the mean-field governing equations are generally written in terms of $f_{M, i}$ or $f_M(D)$. In order to accurately reflect the research literature, this chapter will write the mean-field governing equations both in terms of number-weighted distributions c_i or $c(M)$ and mass-weighted distributions $f_{M, i}$ or $f_M(D)$.

13.1.2 Discrete Governing Equations

Define the *random fragmentation kernel* (a.k.a. *breakage kernel*) J_{ij} as follows:

J_{ij} is the average rate at which children of size i are produced by parents of size j.

Notice that J_{ij} has units of inverse time. In this case, Spouge (1984, 1985) obtained the following mean-field governing equation:

$$\frac{dc_i(t)}{dt} = -\frac{c_i(t)}{i}\sum_{j=1}^{i-1} jJ_{ji} + \sum_{j=i+1}^{\infty} J_{ij}c_j(t). \tag{13.1}$$

Equation (13.1) may be interpreted as follows:

$$\begin{pmatrix}\textit{rate of change}\\ \textit{in particles of}\\ \textit{size i}\end{pmatrix} = -\begin{pmatrix}\textit{rate at which particles of}\\ \textit{size i break into particles}\\ \textit{smaller than size i}\end{pmatrix} + \begin{pmatrix}\textit{rate at which particles}\\ \textit{larger than size i break to}\\ \textit{form particles of size i}\end{pmatrix}$$

where the term "size" refers to mass. Equations of this general sort are sometimes called *population balance equations*. As a special case, suppose that J_{ij} can be written as follows:

$$J_{ij} = m_{ij}\, J_j \tag{13.2}$$

where:

m_{ij} is the average number of children of size i produced by parents of size j.
J_j is the rate at which particles of size j fragment.

In this special case, conservation of mass says the following:

$$\sum_{j=1}^{i-1} j m_{ji} = i. \tag{13.3}$$

If $J_{ij} = m_{ij} J_j$, Spicer & Pratsinis (1996) obtained the following mean-field governing equation:

$$\frac{dc_i(t)}{dt} = -J_i c_i(t) + \sum_{j=i+1}^{\infty} m_{ij} J_j c_j(t). \tag{13.4}$$

Comparing Equations (12.3) and (13.4), random fragmentation becomes binary fragmentation if:

$$I_i = m_{ji} J_i \tag{13.5}$$

and

$$m_{ji} = 2/(i-1). \tag{13.6}$$

13.1.3 Continuous Governing Equations

Define the *random fragmentation kernel* (a.k.a. *breakage kernel*) $J(x, y)$ as follows:

> $J(x, y)\, c(y, t)\, dy$ is the average rate per unit mass and volume at which children of size x are produced by parents of size $y \pm dy$.

Notice that $J(x, y)$ has units of inverse mass and inverse time. In this case, Melzak (1957) obtained the following mean-field governing equation:

$$\frac{\partial c(x,t)}{\partial t} = -\frac{c(x,t)}{x} \int_0^x y J(y,x)\, dy + \int_x^{\infty} J(x,y)\, c(y,t)\, dy. \tag{13.7}$$

See also Dubovskii (2000). Equation (13.7) may be interpreted as follows:

$$\begin{pmatrix}\textit{rate of change}\\ \textit{in particles of}\\ \textit{size } x\end{pmatrix} = -\begin{pmatrix}\textit{rate at which particles of}\\ \textit{size } x \textit{ break to form particles}\\ \textit{smaller than size } x\end{pmatrix} + \begin{pmatrix}\textit{rate at which particles}\\ \textit{larger than size } x \textit{ break to}\\ \textit{form particles of size } x\end{pmatrix}$$

where the term "size" refers to mass. As noted earlier, this may be called a *population balance equation*. As a special case, suppose that $J(x, y)$ can be written as follows:

$$J(x,y) = m(x,y) J(y) \tag{13.8}$$

where on average:

$m(x, y)\, dy$	is the number of children of size x produced by parents of sizes $y \pm dy$
$J(y)$	is the rate at which particles of size y fragment to produce children of size $m(x, y)$.

Notice that $m(x, y)$ has units of inverse mass and $J(y)$ has units of inverse time. If x and y refer to particle mass, then conservation of mass says the following:

$$\int_0^x ym(y,x)dy = x. \tag{13.9}$$

In this special case, Equation (13.7) becomes as follows:

$$\frac{\partial c(x,t)}{\partial t} = -J(x)\,c(x,t) + \int_x^\infty m(x,y)\,J(y)\,c(y,t)\,dy. \tag{13.10}$$

This governing equation was discovered independently by different researchers working in different fields. For example: (a.) Meyer et al. (1966) and Goren (1968) developed a similar equation for mechanical breakage and thermal degradation of polymers; (b.) Valentas et al. (1966) developed a similar governing equation for droplet breakup in dispersed systems due to agitation; see also Ramkrishna (1974) and Pandya & Spielman (1982); and (c.) Hellyer (1970) developed a similar governing equation for the evolution of asteroids and meteors; see also Bishop & Searle (1983). As these examples show, Equation (13.10) is typically used for particle systems with relatively low volumetric fractions. Examples 13.2 and 13.3 describe variants of this mean-field governing equation that are commonly used for particle systems with relatively high volumetric fractions, such as ball mills.

Example 13.2 A common alternative form of the continuous governing equation

Define the *selection function* $S(x)$ as follows:

> $S(x)dt$ is the mass-weighted probability that a particle of size x fragments during the time interval dt.

In addition, define the *breakage function* $g_M(x, y)$ as follows:

> $g_M(x, y)dy$ is the mass-weighted fraction of children of size x produced by parents of size $y \pm dy$.

In other words, $g_M(x, y)$ is the mass-weighted probability density function for the children produced in each fragmentation event. Using these definitions, Bass (1954) obtained the following mean-field governing equation:

$$\frac{\partial f_M(x,t)}{\partial t} = -S(x)f_M(x,t) + \int_x^\infty g_M(x,y)S(y)f_M(y,t)\,dy \tag{13.11}$$

where $f_M(x, t)$ is the mass-weighted probability density function for the particle system at time t.

While not absolutely necessary, x is almost always chosen to be the particle diameter. Notice that the logic used to derive Equation (13.11) is the same as the logic used to derive Equation (13.10). To be more specific, Equation (13.11) says the following:

$$\begin{pmatrix} \textit{rate of change in} \\ \textit{particles of size } x \end{pmatrix} = -\begin{pmatrix} \textit{rate at which particles of} \\ \textit{size } x \textit{ break to form particles} \\ \textit{smaller than size } x \end{pmatrix} + \begin{pmatrix} \textit{rate at which particles larger} \\ \textit{than size } x \textit{ break to form} \\ \textit{particles of size } x \end{pmatrix}$$

where the term "size" refers to diameter. Put another way, Equations (13.10) and (13.11) are the same, except for notation. As a result, solutions of Equation (13.11) can be converted to solutions of Equation (13.10) using a simple substitution of variables, namely, by substituting c for f_M, J for S, m for g_M, and M for D. Section 13.6 gives a series of examples of closed-form time-dependent mean-field solutions for random fragmentation. In most cases, these were originally given as solutions to Equation (13.11). In other words, the original research papers use a notation equivalent to $f_M(D, t)$ and $g_M(D)$. However, Section 13.6 converts these solutions to ones involving $c(M, t)$ and $m(M)$. This is done for consistency with the rest of the book, and for consistency with key research papers such as McGrady & Ziff (1987).

Example 13.3 The batch grinding equation

Suppose that a dense particle system, such as a batch grinding mill, is characterized using physical collection and sieving as described in Examples 3.2 and 3.3. While mean-field model results can be converted to match the experimental results, or vice versa, it is more convenient if the mean-field model directly matches the experimental results. Then define $f_{M,i}$ as follows:

$f_{M,i}$ is the mass fraction of particles captured in sieve i, i.e., the mass fraction of particles with diameters between the screen sizes x_{i-1} and x_i (where $x_0 = \infty$).

In other words, $f_{M,i}$ is the mass of the particles captured in sieve i divided by the mass of the particles captured in all of the sieves. Notice that, in the rest of this book, integer indices refer to discrete particle systems. However, in this example, integer indices refer to size classes in continuous particle systems. Also notice that, in the rest of this book, the integer indices go from smallest to largest. However, in this example, the integer indices go from largest to smallest. This reverse numbering scheme makes perfect sense for sieving. Consider a set of sieves numbered 1, 2, ..., N_{sieve} going from coarsest to finest. In other words, sieve 1 has the coarsest mesh, sieve 2 has the next coarsest mesh, and N_{sieve} has the finest mesh. Referring to Figure 3.1a, sieve 1 appears at the top of the stack. The particles to be measured are placed into sieve 1. Sieve 1 retains the largest particles and passes the rest to sieve 2 immediately below it, which retains the largest particles it receives from sieve 1 and passes the rest to sieve 3 immediately below it, and so forth. With this background in mind, Sedlatschek & Bass (1953) obtained the following mean-field governing equation:

$$\frac{df_{M,i}(t)}{dt} = -S_i f_{M,i}(t) + \sum_{j=1}^{i-1} g_{M,ij} S_j f_{M,j}(t) \tag{13.12}$$

where:

$S_i\,dt$ is the average mass fraction of particles in size class i that fragment during the time interval dt.

$g_{M,ij}$ is the average mass fraction of children in size class i produced by parents of size class j.

Equation (13.12) is known as the *batch grinding equation*. While it may look different, Equation (13.12) is exactly the same as Equation (13.11), except that the particles have been grouped into size classes. Thus, the same terminology applies as before, i.e., S_i is called the *selection function* and $g_{M,ij}$ is called the *breakage function*. Developing the batch grinding equation—along with the associated procedures for choosing the selection function, the breakage function, and the numerical solution techniques—was a major international effort that included contributions from Germany, Australia, the U.S., the U.K., and other countries; see, e.g., Sedlatschek & Bass (1953), Broadbent & Callcott (1956), Reid (1965) and Austin (1971, 1999).

13.2 Breakage Functions

For simplicity, unless stated otherwise, the remainder of this chapter will focus on Equation (13.10). This section concerns the size distributions produced by individual fragmentation events, i.e., $m(x, y)$. Chapters 15, 16, and 17 discuss individual fragmentation events in detail. This section provides just a few basics. As the simplest possibility, suppose that $m(x, y)$ is a power law size distribution. In other words, suppose that $m(x, y)$ has the following form:

$$m(x, y) = (\eta + 1)\, x^{\eta-1}/y^{\eta} \tag{13.13}$$

where $-1 < \eta \le 1$. For example, suppose that $\eta = 1$. Then

$$m(x, y) = 2/y. \tag{13.14}$$

In addition, suppose that:

$$J(x) = xI(x)/2. \tag{13.15}$$

Then Equation (13.10) for random fragmentation is the same as Equation (12.9) for binary fragmentation. More generally, as noted in Example 13.4, choosing a fixed value for η corresponds to choosing a fixed value for the average number of children produced in each fragmentation event. For example:

$\eta = 1$ corresponds to two children per fragmentation event (on average)
$\eta = 2/3$ corresponds to 2.5 children per fragmentation event (on average)
$\eta = 1/2$ corresponds to three children per fragmentation event (on average)

and so forth. Just like m in Equation (13.10) is often chosen to be a power law, the breakage function $g_{M},_{ij}$ in Equation (13.12) is often chosen to be a power law. Alternatively, in many cases, $g_{M},_{ij}$ in Equation (13.12) is chosen to be a linear hybrid of two power laws; see Example 13.5.

Of course, there are many other possibilities for m besides power laws; see Chapters 5 and 17. However, in general, all such alternatives obtain the following property:

$$m(ax, ay) = a^{-1} m(x, y) \tag{13.16a}$$

for any real positive number a. In this case, m is called *homogeneous* and the exponent -1 is called the *degree of homogeneity*. Equivalently:

$$m(x, y) = y^{-1}\, \tilde{m}(x/y). \tag{13.16b}$$

Most closed-form mean-field results are derived assuming that m is a power law. However, a few closed-form mean-field results can be derived using the more general assumption that m is homogeneous.

Example 13.4 Power law size expressions for $m(x, y)$

If children have power law size distributions then, according to conservation of mass, $m(x, y)$ may be written as follows:

$$m(x, y) = \frac{(\eta + 1)y}{x_{\max}^{\eta+1}(y) - x_{\min}^{\eta+1}(y)} x^{\eta-1} \tag{13.17}$$

where $x_{\min}(y)$ and $x_{\max}(y)$ are the minimum and maximum child masses and η is a real exponent. Notice that the total mass of children $\int_{x_{\min}(y)}^{x_{\max}(y)} x m(x, y) dx$ from parents of mass y equals y. Similarly, notice that the total number of children from parents of mass y equals:

$$N(y) = \int_{x_{\min}(y)}^{x_{\max}(y)} m(x, y) dx = \frac{\eta + 1}{\eta} \frac{x_{\max}^{\eta}(y) - x_{\min}^{\eta}(y)}{x_{\max}^{\eta+1}(y) - x_{\min}^{\eta+1}(y)} y. \tag{13.18}$$

The maximum possible range of child masses is $x_{\max}(y) = y$ and $x_{\min}(y) = 0$. This means that a child can be almost as large as its parent, in which case all but that one child's mass is almost zero. For the maximum possible range, Equation (13.17) becomes:

$$m(x,y) = (\eta + 1)\frac{x^{\eta-1}}{y^{\eta}}. \tag{13.19}$$

Similarly, for the maximum possible range, Equation (13.18) becomes:

$$N = \int_0^y m(x,y)\,dx = \begin{cases} \dfrac{\eta+1}{\eta} & \eta > 0 \\ \infty & -1 < \eta \le 0 \end{cases} \tag{13.20}$$

Notice that, according to Equation (13.20), N does not depend on the parent size y. Instead, N depends strictly on the children sizes via η. Also notice that, according to Equation (13.20), $\eta > 1$ implies $N < 2$. However, this is physically impossible. In reality, each parent must give rise to at least two children. Thus, Equation (13.20) implies the following:

$$-1 < \eta \le 1. \tag{13.21}$$

Notice, according to Equation (13.20), choosing a fixed value for η amounts to choosing a fixed value for N. In particular, if $\eta = 1/k$ where $k = 1, 2, 3$, then $N = k + 1$. For example:

$\eta = 1$ corresponds to exactly 2 children per fragmentation event;
$\eta = 1/2$ corresponds to exactly 3 children per fragmentation event;
$\eta = 1/3$ corresponds to exactly 4 children per fragmentation event,

and so forth.

Example 13.5 Power law size expressions for $g_{M,ij}$

It is common to assume that the breakage function $g_{M,ij}$ is a power law. Alternatively, it is common to assume that the cumulative breakage function $G_{M,ij}$ is a power law. By definition, the breakage function and cumulative breakage function are related as follows:

$$G_{M,ij} = \sum_{k=i}^{N_{\text{sieve}}} g_{M,kj} \tag{13.22a}$$

$$g_{M,ij} = G_{M,ij} - G_{M,i+1,j}. \tag{13.22b}$$

If the cumulative breakage function $G_{M,ij}$ is a power law then:

$$G_{M,ij} = \left(\frac{x_{i-1}}{x_j}\right)^{\gamma} \tag{13.23}$$

where x_i is the diameter of a child in size class i and x_j is the diameter of a parent in size class j.

By Taylor series:

$$g_{M,ij} = G_{M,ij} - G_{M,i+1,j} = \left(x_{i-1}^{\gamma} - x_i^{\gamma}\right)/x_j^{\gamma} \cong \gamma \Delta x_i x_i^{\gamma-1}/x_j^{\gamma}$$

where $\Delta x_i = x_{i-1} - x_i$. Thus, if $G_{M,ij}$ obtains a power law, $g_{M,ij}$ also approximately obtains a power law.

As noted in Section 5.11, fragmentation often obtains different size distributions in different size ranges. In particular, the smallest fragments in the *dissipative range* obtain one size distribution, the medium fragments in the *semi-inertial range* obtain another size distribution, and the largest fragments in the *energy-containing range* obtain yet another size distribution. In random fragmentation, the smallest dust-like particles are essentially passive. If formed through one or more previous fragmentation events, they may be almost impervious to further change. At most, they may slightly alter the mass, momentum, and energy of the other particles via elastic collisions, similar to Brownian motion. Alternatively, they may act like a sort of sandy grit or a dry graphite lubricant that increases or decreases interparticle friction. Other than this, for the most part, the smallest particles can be safely ignored. Thus, it makes sense to assume that $G_{M,ij}$ is a linear combination of two power laws representing medium and large particles. For example:

$$G_{M,ij} = \phi\left(\frac{x_{i-1}}{x_j}\right)^{\gamma} + (1-\phi)\left(\frac{x_{i-1}}{x_j}\right)^{\beta} \quad (13.24)$$

where $0 \le \phi \le 1$, $\gamma < 1$, and $\beta > 1$; see, e.g., Austin (1999). The first term on the right hand side dominates for $x_{i-1} \ll x_j$ while the two terms are evenly matched for $x_{i-1} \approx x_j$.

As noted in Example 13.4, assuming a power law size distribution is equivalent to assuming that every fragmentation event produces exactly the same number of children. As discussed in the last chapter, there are many real-world situations where every parent produces exactly two children. These include cases where particles are on the verge of either fragmenting or not. In such cases, small random fluctuations occasionally cause a particle to split in two. If small random fluctuations can barely cause particles to fragment into two pieces, they rarely if ever cause particles to fragment into more than two pieces. On the other hand, there are few if any real-world situations where every fragmentation event produces exactly $N > 2$ children. For example, it is hard to image a situation where every fragmentation event produces exactly four children but never two or three or five or six children. At best, such particle systems may produce a fixed number of children per fragmentation event *on average*. Thus, in general, η obtains real instead of rational values. For example, suppose a certain fraction of the fragmentation events in a large system obtain four children with an ideal power law size distribution so that $\eta = 1/3$ while the rest obtain three children with an ideal power law size distribution so that $\eta = 1/2$. Then, on average, the children from all of the individual fragmentation events obtain a *non-ideal* power law size distribution where η is a real number between 1/3 and 1/2. In short, because different fragmentation events tend to produce different outcomes—depending on random variables such as particle mass, momentum, and energy—random fragmentation tends to obtain non-ideal values of η, even in cases with high degrees of spatial homogeneity and isotropy.

13.3 Kernels

Consider the rate kernel $J(x)$ seen in Equation (13.10). By definition, $J(x)$ is *homogeneous* if it is a power law. More specifically:

$$J(x) = J_0 x^{\kappa} \tag{13.25}$$

where J_0 and κ are constants. In this case, κ is called the *degree of homogeneity*. Conversely, the kernel $J(x)$ is *non-homogeneous* if is it not a power law. As seen in Example 13.6, non-homogeneous kernels tend to be special purpose, i.e., each expression for $J(x)$ applies to exactly one application. This book focuses on general purpose results; thus, the remainder of this chapter concerns homogeneous kernels.

Example 13.6 Non-homogeneous random fragmentation kernel for droplets

Consider droplet breakup in mildly agitated liquid-liquid dispersions where ρ is the density of the droplets, σ is the surface tension of the droplets, and ε is the turbulent dissipation rate per unit mass. Coulaloglou & Tavlarides (1977) suggested the following non-homogenous kernel:

$$J(x) = K_0 \varepsilon^{1/3} x^{-2/9} \exp\left[-\frac{K_1 \sigma}{\rho \varepsilon^{2/3}} x^{-5/9}\right]$$

where x is the droplet mass and K_0 and K_1 are constants. This expression only applies under a very specific set of circumstances, namely, cases where: (a.) "the droplets are in a turbulent flow field which is locally isotropic and the droplet size is within the inertial subrange;" see the earlier discussion in Section 8.5; (b.) "the energy spectrum function has a −5/3 dependency on the wave number;" see the earlier discussion in Section 8.5; (c.) "viscous effects are negligible and the drop deforms due to local pressure fluctuations;" and (d.) "an oscillating deformed drop will break if the turbulent kinetic energy transmitted to the droplet by turbulent eddies exceeds the drop surface energy." Under these assumptions, "the fraction of drops breaking is ... proportional to the fraction of turbulent eddies colliding with the droplet that have a turbulent kinetic energy greater than the droplet surface energy." For additional examples of non-homogeneous fragmentation kernels for applications involving droplets and bubbles see, e.g., Liao & Lucas (2009, 2010), Patruno et al. (2009), Becker et al. (2011, 2014), Borka & Jakcobsen (2012) and Maaß & Kraume (2012).

For random fragmentation, if $\kappa > 0$, large particles tend to fragment more frequently than small particles; the fragmentation rate decreases after every fragmentation event; and infinitesimal particles fragment infinitely slowly. On the other hand, if $\kappa < 0$, small particles tend to fragment more frequently than large particles; the fragmentation rate increases after every fragmentation event; and infinitesimal particles fragment infinitely rapidly. This behavior is known as *shattering*. Shattering in fragmentation is related to gelation in aggregation. Like anything that violates conservation of mass, gelation and shattering both allow multiple possible solutions, depending on how hard and strong the spontaneously formed sinks are presumed to be. Shattering will be discussed in more detail in the next section.

Suppose that the volumetric fraction of particles is low. In this case, researchers often assume that random fragmentation mechanisms belong in one of two broad categories. For *one-particle mechanisms* such as thermal degradation, each particle fragments individually without being influenced by any other particle. For one-particle mechanisms, researchers sometimes assume the following:

$$\kappa = j/k \tag{13.26}$$

where k is the number of degrees of freedom and:

$$-k \le j \le k.$$

If $k = d = 3$ then:

$$\kappa \text{ is an integer multiple of } 1/3 \text{ between } -1 \text{ and } 1. \tag{13.27}$$

For one-particle mechanisms, the fragmentation rate $J(x) = J_0 x^\kappa$ is proportional or inversely proportional to a particle size such as diameter, surface area, or mass. More specifically, if $\kappa = 0$, the fragmentation rate does not depend on the particle size. Alternatively:

If $\kappa = 1/3$, the fragmentation rate is proportional to the particle diameter.
If $\kappa = 2/3$, the fragmentation rate is proportional to the particle surface area.
If $\kappa = 1$, the fragmentation rate is proportional to the particle volume.

Similarly:

If $\kappa = -1/3$, the fragmentation rate is inversely proportional to the particle diameter.
If $\kappa = -2/3$, the fragmentation rate is inversely proportional to the particle surface area.
If $\kappa = -1$, the fragmentation rate is inversely proportional to the particle volume.

For more information about one-particle fragmentation mechanisms see, e.g., Valentas et al. (1966), Ouchiyama et al. (2005) and Bridgwater (2007).

As the second broad category, consider *two-particle mechanisms*, where each fragmentation event involves exactly two particles. For example, this occurs when random fragmentation is dominated by random binary collisions. For two-particle mechanisms, researchers sometimes assume the following:

$$\kappa = (j_1 + j_2)/k \tag{13.28}$$

where k is the number of degrees of freedom and:

j_1 ± 1 times the integer dimension of the phase space for particle 1 ($-k \le j_1 \le k$)
j_2 ± 1 times the integer dimension of the phase space for particle 2 ($-k \le j_2 \le k$).

Notice that Equation (13.28) is similar to Equation (9.9). The minor differences are due to the definitions of λ and $\varkappa$. If $k = d = 3$ then:

$$\varkappa \text{ is an integer multiple of } 1/6 \text{ between } -2 \text{ and } 2. \tag{13.29}$$

Notice that Equation (13.29) is a superset of Equation (13.27). In other words, Equation (13.29) allows 25 different values for $\varkappa$, including the seven integer multiples of 1/3 between −1 and 1. Thus, from now on, Equation (13.29) will be used to refer to both one- and two-particle mechanisms.

Table 13.2 gives examples of ideal degrees of homogeneity for random fragmentation taken from the research literature. As these examples show, ideal homogeneous kernels tend to be general purpose, i.e., the same rational value of $\varkappa$ may occur in many different applications. Except for the last two rows, these examples obey Equation (13.29). The last two rows with $\varkappa = 3$ presumably involve some other mechanism, such as a three-particle mechanism, different than the one- and two-particle mechanisms considered here. Notice that most if not all of the examples given in Table 13.2 concern particle systems with relatively low volumetric fractions. Particle systems with low volumetric fractions tend to produce high mixing rates,

Table 13.2 Examples of ideal homogeneous kernels for random fragmentation. In the first column, all values of $\varkappa$ are rational and were obtained analytically. In the second column, x refers to particle mass and J_0 refers to an application-specific constant.

$\varkappa$	J/J_0	Application	References
−2	x^{-2}	Polymer degradation—1st model for shattering	McGrady & Ziff (1987)
−3/2	$x^{-3/2}$	Polymer degradation—2nd model for shattering	McGrady & Ziff (1987)
−1	x^{-1}	Moving-bed attrition, e.g., hoppers, conveyers	Ouchiyama et al. (2005)
−2/3	$x^{-2/3}$	Moving-bed attrition, e.g., hoppers, conveyers	Ouchiyama et al. (2005)
−1/3	$x^{-1/3}$	Moving-bed attrition, e.g., hoppers, conveyers	Ouchiyama et al. (2005)
0	1	Agitated droplet suspensions	Valentas et al. (1966)
0	1	Polymer degradation—transitional	McGrady & Ziff (1987)
0	1	Moving-bed attrition, e.g., hoppers, conveyers	Ouchiyama et al. (2005)
0	1	"Breakup of grains, bubbles, droplets" etc.	Ernst & Pagonabarraga (2007)
1/6	$x^{1/6}$	Batch grinding mills	Kapur (1972); Hoşten (2005)
1/3	$x^{1/3}$	Agitated droplet suspensions	Valentas et al. (1966)
1/3	$x^{1/3}$	Batch grinding mills	Kapur (1972), Das et al. (1995)
1/3	$x^{1/3}$	Agitated floc suspensions	Pandya & Spielman (1982)
1/3	$x^{1/3}$	Fluidized-bed combustion of coal	Peterson (1986)
1/3	$x^{1/3}$	Agitated solid suspensions, e.g., slurries	Barthelmes et al. (2003)
1/3	$x^{1/3}$	Moving-bed attrition, e.g., hoppers, conveyers	Ouchiyama et al. (2005)
2/3	$x^{2/3}$	Agitated droplet suspensions	Valentas et al. (1966)
2/3	$x^{2/3}$	Moving-bed attrition, e.g., hoppers, conveyers	Ouchiyama et al. (2005)
1	x	Agitated droplet suspensions	Valentas et al. (1966)
1	x	Fiber breakage in paper and textile mills	Meyer et al. (1966)
1	x	Polymer degradation	Goren (1968)
1	x	Agitated floc suspensions	Patil & Andrews (1998)
1	x	Polymer degradation	McCoy & Madras (1998, 2003)
1	x	Moving-bed attrition, e.g., hoppers, conveyers	Ouchiyama et al. (2005)
2	x^2	Agitated droplet suspensions	Ramkrishna (1974)
3	x^3	Polymer degradation	Goren (1968)
3	x^3	Fluidized-bed combustion of coal	Peterson (1986)

resulting in high degrees of spatial homogeneity and isotropy. Put another way, because past events affect future events identically regardless of time or location, different particles fragment under similar conditions. Such systems tend to obtain universal rational values for κ. Conversely, particle systems with high volumetric fractions tend to produce low mixing rates, resulting in low degrees of spatial homogeneity and isotropy. Put another way, because the unique effects of previous events tend to linger in the local environment, different particles may fragment under different conditions. As a result, dense particle systems may or may not obtain universal rational values for κ. While ideal rational values of κ are general purpose, non-ideal real values of κ are not, i.e., each real value of κ is associated with one specific application.

To complete this section, Example 13.7 briefly discusses the selection function S in the batch grinding equation, which is similar to the rate kernel J.

Example 13.7 Selection functions for the batch grinding equation

The selection function $S_i = S(x_i)$ in Equation (13.12) is usually assumed to be a power law. In other words,

$$S_i = S_0 x_i^{\alpha} \tag{13.30}$$

where S_0 is a constant and x_i is the diameter of particles associated with size class i. Notice that, while the exponent κ in the rate kernel J often obtains even multiples of 1/6, the exponent α in the selection function S usually does not, except for the special case $\alpha = 0$.

13.4 Shattering

This section concerns *shattering*, i.e., the spontaneous formation of sinks of small particles starting at time $t = 0$. Random fragmentation allows two types of shattering, namely, *strong* and *weak*. For free-field boundary conditions, strong shattering tends to occur when $\kappa < 0$. In this case, the fragmentation rate goes to infinity as the particle size goes to zero. Filippov (1961) says that this leads to "an infinite number of particles of zero mass" which, in turn, causes "the sum of the masses of all particles ... to decrease in the course of time, in spite of the fact that the mass is conserved for each separate splitting." Similarly, McGrady & Ziff (1987) say that "a cascading process occurs ... in which smaller particles break up at increasingly rapid rates, resulting in mass being lost to a phase of 'zero'-size particles. The cascading of the breakup rate, or 'shattering', produces a fractal dust [that] ... is analogous to, but opposite from, the process of gelation found in aggregating systems." Now consider weak shattering. For free-field boundary conditions, weak shattering tends to occur when $\eta \leq 0$. In this case, each fragmentation event creates an infinite number of children, most of which are infinitesimal. Ernst & Szamel (1993) say that weak shattering explains why the "total number of particles could be finite even though the expected number of fragments in every single break-up is infinite."

As seen in Section 13.6, if shattering does not occur, mean-field theory predicts that random fragmentation produces Rosin–Rammler size distributions with $n/m = \eta$ and $l/m = \kappa$:

$$c(M,t) \cong \text{const.}\frac{C_0 M_0}{[M_{\text{avg}}(t)]^2}\left[\frac{M}{M_{\text{avg}}(t)}\right]^{\eta-1} \exp\left\{-\text{const.}\left[\frac{M}{M_{\text{avg}}(t)}\right]^{\kappa}\right\} \tag{13.31}$$

for sufficiently late times and small particles. Notice that if $\kappa > 0$ then:

$$\lim_{M\to 0} c(M,t) \cong \text{const.}\frac{C_0 M_0}{\left[M_{\text{avg}}(t)\right]^2}\left[\frac{M}{M_{\text{avg}}(t)}\right]^{\eta-1}. \tag{13.32}$$

In other words, the size distribution of the infinitesimal particles depends only on the size distribution of the children produced in each individual event, $m(x, y) = (\eta + 1)x^{\eta - 1}/y^{\eta}$, and not on the rate kernel rate $J(x) = J_0 x^{\kappa}$; see also, e.g., Kostoglou et al. (1997). If no shattering occurs, the total mass of the infinitesimal particles is zero. However, if shattering occurs, the total mass of the infinitesimal particles is non-zero. According to mean-field theory, the infinitesimal particles are removed by a spontaneously formed sink and, as a result, the total mass of the particle system decreases with time. In the case of weak shattering, the loss of the infinitesimal particles does not substantially alter the size distribution of the remaining particles, i.e., Equations (13.31) and (13.32) still apply. On the other hand, in the case of strong shattering, the loss of the infinitesimal particles completely changes the size distribution of the remaining particles. More specifically, instead of a power law size distribution with exponent η, the remaining particles obtain a power law size distribution with exponent κ. In essence, when strong shattering occurs, the entire effect of m is concentrated in the smallest particles. When the smallest particles are eliminated, so is the effect of m on the remaining larger particles. Notice that shattering can occur whether or not the smallest particles are actually removed. They can remain physically present in the system provided they do not participate in or substantially affect fragmentation.

13.5 Distributions and Moments

This section provides a brief review of size distributions and moments. For a fuller discussion, see Chapter 3 and Appendix A. The *zero-th moments* of $c(M, t)$ and $f(M, t)$ are as follows:

$$\int_0^\infty c(M,t)\,dM = C(t) \tag{13.33}$$

$$\int_0^\infty f(M,t)\,dM = 1. \tag{13.34}$$

The *first moments* of $c(M, t)$ and $f(M, t)$ are as follows:

$$\int_0^\infty Mc(M,t)\,dM = C(t)M_{avg}(t) \tag{13.35}$$

$$\int_0^\infty Mf(M,t)\,dM = M_{avg}(t). \tag{13.36}$$

For convenience, the following notation will be used throughout the rest of the chapter:

$$C_0 = C(0) \tag{13.37}$$

$$M_0 = M_{avg}(0). \tag{13.38}$$

If conservation of mass applies then:

$$C(t)\,M_{avg}(t) = C_0 M_0 = \text{const.} \tag{13.39}$$

as discussed earlier in Example 12.5. Now consider size distributions. In regular form, Rosin–Rammler size distributions may be written as follows:

$$c(M,t) = C(t)f(M,t) = C(t)A(t)M^{n/m-1}\exp\left[b(t)M^{l/m}\right]. \tag{13.40}$$

As one special case, power law size distributions occur when $l/m = 0$:

$$c(M,t) = C(t)f(M,t) = C(t)A(t)\,M^{n/m-1}. \tag{13.41}$$

As another special case, Weibull size distributions occur when $l/m = n/m$:

$$c(M,t) = C(t)f(M,t) = C(t)A(t)\,M^{n/m-1}\exp\left[-b(t)\,M^{n/m}\right]. \tag{13.42}$$

As a third special case, exponential size distributions occur when $n/m = 1$ and $l/m = 1$

$$c(M,t) = C(t)f(M,t) = C(t)A(t)\exp\left[b(t)M\right]. \tag{13.43}$$

As another possibility, in regular form, log-normal size distributions may be written as follows:

$$f(M,t) = \frac{1}{\sqrt{2\pi\sigma(t)^2}}\frac{1}{M}\exp\left\{-\frac{1}{2\sigma(t)^2}\left[\ln M - a(t)\right]^2\right\}. \tag{13.44}$$

13.6 Full Mean-Field Solutions

13.6.1 Introduction

For aggregation, closed-form time-dependent solutions are only possible for three integer values of λ, namely, 0, 1, or 2. These solutions only apply when the particle volume fraction is low, and spatial homogeneity and isotropy are high. On the other hand, for random fragmentation, closed-form time-dependent solutions are possible for any real values of κ and η. These solutions apply even when the particle volume fraction is high, and the spatial homogeneity and isotropy are low. The only specific value of κ that requires special attention is $\kappa = 0$, which divides shattering from non-shattering cases. The only specific value of η that requires special attention is $\eta = 1$, which was already treated in the last chapter. As the main tradeoff, while analytic solutions for specific integer values of κ and η involve relatively simple transcendental functions such as Bessel functions, analytic solutions for arbitrary real values of κ and η involve more complex transcendental functions such as hypergeometric functions.

13.6.2 Initial Conditions

This section briefly reviews two possible initial conditions. In self-similar form, monodisperse initial conditions may be written as follows:

$$c(M,0) = \frac{C_0}{M_0}\delta\left(\frac{M}{M_0} - 1\right) \tag{13.45}$$

where δ is the Dirac delta function. In addition, in self-similar form, Rosin–Rammler initial conditions may be written as follows:

$$c(M,0) = \text{const.}\,\frac{C_0}{M_0}\left(\frac{M}{M_0}\right)^{\frac{n}{m}-1}\exp\left[-\text{const.}\left(\frac{M}{M_0}\right)^{\frac{1}{m}}\right]. \tag{13.46}$$

Notice that monodisperse initial conditions have the smallest possible size spread, regardless of the metric chosen. Conversely, Rosin–Rammler initial conditions have the largest possible size spread, depending on the metric chosen. In general, mean-field theory predicts that size distributions converge towards those with the largest possible size spreads as time increases, according to some metric implied by the kernel, regardless of the initial conditions.

13.6.3 Kummer's Confluent Hypergeometric Function

Most of the examples in this section involve Kummer's confluent hypergeometric function ${}_1F_1(x, y, z)$. This subsection provides a brief review. By definition, *Kummer's confluent hypergeometric function* is a solution to the following ordinary differential equation:

$$z\frac{d^2}{dz^2}\,{}_1F_1(x,y,z) + (y-z)\frac{d}{dz}\,{}_1F_1(x,y,z) - x\,{}_1F_1(x,y,z) = 0.$$

However, this definition does not really explain Kummer's confluent hypergeometric function. As one way to understand it, it generalizes a wide variety of common algebraic and transcendental (i.e., non-algebraic) functions. For example:

$${}_1F_1(1,2,-z) = \frac{2}{z}\exp\left(-\frac{z}{2}\right)\sinh\left(\frac{z}{2}\right) = \frac{1}{z}\left[1-\exp(-z)\right]$$

where sinh is a hyperbolic sine.

For another example:

$$ {}_1F_1\left(x, x, -z\right) = \exp\left(-z\right). $$

As a third example:

$$ {}_1F_1\left(x, 2x, -z\right) = 4^{x-1/2}\Gamma\left(x + 1/2\right) z^{-(x-1/2)} \exp\left(-\frac{z}{2}\right) I_{x-1/2}\left(\frac{z}{2}\right) $$

where $\Gamma(x + 1/2)$ is gamma function and $I_{x-1/2}\left(z/2\right)$ is a modified Bessel function of the first kind, as discussed earlier in Section 10.4. As a fourth example:

$$ {}_1F_1\left(x, x + 1, -z\right) = |x|\, z^{-x}\left[\Gamma\left(x, z\right) - \Gamma\left(x\right)\right] $$

where $\Gamma(x)$ is a gamma function and $\Gamma(x, z)$ is an upper incomplete gamma function. Table 13.3 shows results for three specific values of x. As a fifth example, if i is a positive integer then:

$$ {}_1F_1\left(-i, y, -z\right) = \frac{i!\Gamma\left(y\right)}{\Gamma\left(y + i\right)} L_i^{(y-1)}(-z) $$

where $\Gamma(x)$ is a gamma function and $L_i^{(y)}(z)$ is a generalized Laguerre function. Table 13.4 shows results for nine specific values of i and y. The classic book by Abramowitz & Stegun (1972) describes Kummer's confluent hypergeometic functions, Bessel functions, upper incomplete gamma functions, Laguerre functions, and many other transcendental functions; alternatively, see the modern online version by Olver et al. (2023).

As another way to understand Kummer's confluent hypergeometric function, consider the following limits:

$$ {}_1F_1\left(x, y, z\right) \cong 1 \text{ as } z \to 0 \tag{13.47} $$

$$ {}_1F_1\left(x, y, z\right) \cong \frac{\Gamma(y)}{\Gamma(x)} z^{x-y} \exp\left(z\right) \text{ as } z \to \infty \tag{13.48} $$

$$ {}_1F_1\left(x, y, z\right) \cong \frac{\Gamma\left(y\right)}{\Gamma\left(y - x\right)} (-z)^{-x} \text{ as } z \to -\infty. \tag{13.49} $$

In addition, if x and y are fixed, ${}_1F_1(x, y, z)$ is often monotone increasing or decreasing for $0 < z < \infty$ or $-\infty < z < 0$. In our application, analytic solutions typically obtain the first limit, Equation (13.47), for sufficiently small times or large particles. This describes unreacted particles, i.e., particles that have not participated in fragmentation and, thus, still retain

Table 13.3 Examples where Kummer's confluent hypergeometric function is equivalent to an expression involving the upper incomplete gamma function.

Hypergeometric Function	Gamma Function	Simplified Function
${}_1F_1(-1/2, 1/2, -z)$	$\frac{1}{2}z^{\frac{1}{2}}\left[\Gamma\left(-\frac{1}{2}, z\right) - \Gamma\left(-\frac{1}{2}\right)\right]$	$\exp\left(-z\right) + \sqrt{\pi z}\,\mathrm{erf}\left(\sqrt{z}\right)$
${}_1F_1(0, 1, -z)$	$\lim_{x\to 0} x\left[\Gamma\left(x\right) - \Gamma\left(x, z\right)\right]$	1
${}_1F_1(1, 2, -z)$	$\frac{1}{z}\left[\Gamma\left(1, z\right) - 1\right]$	$\frac{1}{z}\left[\exp\left(-z\right) - 1\right]$

Table 13.4 Examples where Kummer's confluent hypergeometric function is equivalent to an expression involving the Laguerre function.

Hypergeometric Function	Laguerre Function	Simplified Function
${}_1F_1(-3, 1/2, -z)$	$\frac{16}{5}L_3^{-1/2}(-z)$	$\frac{8}{15}z^3 + 4z^2 + 6z + 1$
${}_1F_1(-2, 1/2, -z)$	$\frac{8}{3}L_2^{-1/2}(-z)$	$\frac{4}{3}z^2 + 4z + 1$
${}_1F_1(-1, 1/2, -z)$	$2L_1^{-1/2}(-z)$	$2z + 1$
${}_1F_1(-3, 1, -z)$	$L_3^0(-z)$	$\frac{1}{6}z^3 + \frac{3}{2}z^2 + 3z + 1$
${}_1F_1(-2, 1, -z)$	$L_2^0(-z)$	$\frac{1}{2}z^2 + 2z + 1$
${}_1F_1(-1, 1, -z)$	$L_1^0(-z)$	$z + 1$
${}_1F_1(-3, 2, -z)$	$\frac{1}{4}L_3^1(-z)$	$\frac{1}{24}z^3 + \frac{1}{2}z^2 + \frac{3}{2}z + 1$
${}_1F_1(-2, 2, -z)$	$\frac{1}{3}L_2^1(-z)$	$\frac{1}{6}z^2 + z + 1$
${}_1F_1(-1, 2, -z)$	$\frac{1}{2}L_1^1(-z)$	$\frac{1}{2}z + 1$

their initial conditions. Conversely, analytic solutions typically obtain either the second or third limit, Equation (13.48) or (13.49), for sufficiently large times and sufficiently small particles. This describes fully-reacted particles, i.e., particles formed through one or more fragmentation events.

As yet another way to understand Kummer's confluent hypergeometric function, consider integral expressions such as the following:

$$\int_0^\infty x^{n-1}{}_1F_1\left(a, b, -kx^l\right) dx = \frac{1}{|l|\, k^{\frac{n}{l}}} \frac{\Gamma(n/l)\Gamma(b)\Gamma(a - n/l)}{\Gamma(a)\Gamma(b - n/l)} \tag{13.50}$$

and:

$$\int_0^y x^{ln-1}{}_1F_1\left[a, 2, k\left(y^l - x^l\right)\right] dx = \frac{1}{kl(a-1)} y^{l(n-1)} \left[{}_1F_1\left(a-1, n, ky^l\right) - 1\right]. \tag{13.51}$$

These integrals mean that closed-form mean-field solutions for random fragmentation, which involve Kummer's confluent hypergeometric function, have reasonably tractable closed-form expressions for their moments. As a final useful observation, consider Kummer's transformation:

$${}_1F_1\left(x, y, -z\right) = \exp\left(-z\right){}_1F_1\left(y - x, y, z\right). \tag{13.52}$$

Kummer's confluent hypergeometric function arises naturally in a number of different applications. For example, McNolty (1973) obtained expressions involving Kummer's confluent hypergeometric function for the density of radar cross sections and infrared sensors; see also McNolty & Hansen (1974, 1979) and Huynen et al. (1975). For another and more relevant example, Khvorostyanov & Curry (1999, 2014) obtained expressions similar to McNolty's for droplet size distributions due to stationary stochastic condensation in clouds. The rest of this section gives a series of closed-form mean-field solutions involving ${}_1F_1$. These solutions are written in terms of $c(M, t)$ or $f(M, t)$, regardless of the notation used in the original treatment.

13.6.4 Non-Shattering Solutions

Example 13.8 considers random fragmentation with $\varkappa = 0$ and $\eta > 0$. Notice that Example 12.6 is a special case of Example 13.8 with $\eta = 1$. In addition, Examples 13.9 and 13.10 consider random fragmentation with $\varkappa > 0$ and $\eta > 0$. Notice that Examples 12.7 and 12.9 are special cases of Example 13.9 with $\varkappa = 1, 2$ and $\eta = 1$.

Example 13.8 Time-dependent closed-form solutions for $\varkappa = 0$ and $\eta > 0$ with monodisperse initial conditions and free-field boundary conditions

For $J(x) = J_0$ and $m(x, y) = (\eta + 1)x^{\eta-1}/y^{\eta}$, McGrady & Ziff (1987) obtained the following time-dependent solution to Equation (13.10):

$$c(M,t) = (C_0/M_0)\exp(-J_0M_0t)\,\delta(M/M_0 - 1) + C_0\exp(-J_0t)\sqrt{(\eta+1)J_0t}\,\frac{I_1\left[2\sqrt{(\eta+1)J_0t\ln(M_0/M)}\right]}{M_0(M_0/M)^{\eta-1}\sqrt{\ln(M_0/M)}} \tag{13.53}$$

where C_0 is the initial total concentration, M_0 is the initial mass of each particle, and $I_1(x)$ is a first-order modified Bessel function of the first kind. If $0 < \eta \leq 1$ then:

$$C(t) = \int_0^{M_0} c(M,t)\,dM = C_0\exp\left(\frac{1}{\eta}J_0t\right) \tag{13.54}$$

is monotone increasing with time. In addition, $\int_0^{M_0} Mc(M,t)dM = C_0M_0$ is constant, i.e., conservation of mass applies. Then according to Equation (13.39):

$$M_{\text{avg}}(t) = \frac{C_0M_0}{C(t)} = M_0\exp\left(-\frac{1}{\eta}J_0t\right). \tag{13.55}$$

For sufficiently small times and large particle sizes, Equation (13.53) becomes:

$$c(M,t) \cong (C_0/M_0)\,\delta(M/M_0 - 1) \tag{13.56}$$

which is a monodisperse size distribution. Conversely, for sufficiently large times and small particle sizes, Equation (13.53) becomes:

$$c(M,t) \cong \frac{C_0\exp(-J_0t)[(\eta+1)J_0t]^{1/4}\exp\left\{2[(\eta+1)J_0t\ln(M_0/M)]^{1/2}\right\}}{(4\pi)^{1/2}M_0(M_0/M)^{\eta-1}[\ln(M_0/M)]^{3/4}}. \tag{13.57}$$

This is true because $I_1(x) \cong \exp(x)/\sqrt{2\pi x}$ as $x \to \infty$. While reminiscent of a log normal size distribution, this limiting case does not correspond to any common form. Thus Ziff & McGrady (1985) call this "curious behavior." However, with different assumptions, random fragmentation with $\varkappa = 0$ may produce a true log normal size distribution; see Example 13.11.

Example 13.9 Time-dependent closed-form solutions for any $\kappa > 0$ and $\eta > 0$ with monodisperse initial conditions and free-field boundary conditions

Filippov (1961) obtained a solution to Equation (13.11). After a substitution of variables, this becomes the following solution to Equation (13.10) for $J(x) = J_0 x^\kappa$ and $m(x,y) = (\eta+1)x^{\eta-1}/y^\eta$:

$$\begin{aligned} c(M,t) &= (C_0/M_0)\exp(-J_0 M_0^\kappa t)\,\delta(M/M_0 - 1) \\ &+ (\eta+1)\,C_0 \exp(-J_0 M_0^\kappa t)\,J_0 M_0^{\kappa-\eta} M^{\eta-1} t\,{}_1F_1\left[1 + (\eta+1)/\kappa, 2, J_0\left(M_0^\kappa - M^\kappa\right)t\right] \end{aligned} \tag{13.58}$$

where C_0 is the initial total concentration and M_0 is the initial mass of each particle. Notice that $M \le M_0$ for all times. If shattering does not occur, then Filippov (1961) showed that:

$$C(t) = C_0 \exp\left(-J_0 M_0^\kappa t\right) {}_1F_1\left(\frac{\eta+1}{\kappa}, \frac{\eta}{\kappa}, J_0 M_0^\kappa t\right) = C_0\, {}_1F_1\left(-\frac{1}{\kappa}, \frac{\eta}{\kappa}, -J_0 M_0^\kappa t\right) \tag{13.59}$$

which is monotone increasing in time. In addition, $\int_0^{M_0} Mc(M,t)\,dM = C_0 M_0$ is constant. For sufficiently small times and large particle sizes, Equation (13.58) becomes:

$$c(M,t) \cong (C_0/M_0)\;\delta(M/M_0 - 1) \tag{13.60}$$

which is a monodisperse size distribution. Conversely, for sufficiently large times and small particle sizes, Equation (13.58) becomes:

$$c(M,t) \cong \frac{C_0 M_0 \kappa}{\Gamma\left[(\eta+1)/\kappa\right]} (J_0 t)^{(\eta+1)/\kappa} M^{\eta-1} \exp\left(-J_0 M^\kappa t\right) \tag{13.61}$$

which is a Rosin–Rammler size distribution with $n/m = \eta$ and $l/m = \kappa$. This is true because ${}_1F_1(x,y,z) \cong \Gamma(y)/\Gamma(x) z^{x-y} \exp(z)$ as $z \to \infty$. Equation (13.61) can be rewritten in a self-similar form as follows:

$$c(M,t) \cong \frac{\kappa b^{\eta/\kappa}}{\Gamma(\eta/\kappa)} \frac{C_0 M_0}{\left[M_{\text{avg}}(t)\right]^2} \left[\frac{M}{M_{\text{avg}}(t)}\right]^{\eta-1} \exp\left\{-b\left[\frac{M}{M_{\text{avg}}(t)}\right]^\kappa\right\} \tag{13.62}$$

or equivalently:

$$f(M,t) = \frac{c(M,t)}{C(t)} \cong \frac{\kappa b^{\eta/\kappa}}{\Gamma(\eta/\kappa)} \frac{1}{M_{\text{avg}}(t)} \left[\frac{M}{M_{\text{avg}}(t)}\right]^{\eta-1} \exp\left\{-b\left[\frac{M}{M_{\text{avg}}(t)}\right]^\kappa\right\} \tag{13.63}$$

where:

$$M_{\text{avg}}(t) = \frac{C_0 M_0}{C(t)} \cong \left(\frac{b}{J_0 t}\right)^{1/\kappa} \tag{13.64}$$

and where $b^{1/\kappa} = \Gamma[(\eta+1)/\kappa]/\Gamma(\eta/\kappa) > 0$. This may be a universal size distribution if κ and η are integer multiples of 1/6. However, this is typically only true for κ and not for η.

Example 13.10 Time-dependent closed-form solutions for any $\kappa > 0$ and $\eta > 0$ with Rosin–Rammler initial conditions and free-field boundary conditions

Saveliev & Gorokhovski (2012) obtained a solution to Equation (13.11). After a substitution of variables, this becomes the following solution to Equation (13.10) for $J(x) = J_0 x^\kappa$ and $m(x,y) = (\eta+1)x^{\eta-1}/y^\eta$:

$$c(M,t) = \frac{C_0 M_0 \kappa}{\Gamma[(\eta+1)/\kappa]}(J_0 t)^{(\eta+1)/\kappa} M^{\eta-1} \exp\left(-J_0 t M^\kappa\right) \tag{13.65}$$

which is a Rosin–Rammler size distribution with $n/m = \eta$ and $l/m = \kappa$. In this expression, $C_0 \to 0$ is the initial total particle concentration and $M_0 \to \infty$ is the initial average particle mass, with specific values chosen so that $0 < C_0 M_0 < \infty$. Notice that if $\kappa > 0$ then:

$$C(t) = \int_0^\infty c(M,t)\, dM = \frac{\Gamma(\eta/\kappa)}{\Gamma[(\eta+1)/\kappa]} C_0 M_0 \left(J_0 t\right)^{1/\kappa} \tag{13.66}$$

which is monotone increasing. In addition, $\int_0^\infty Mc(M,t)\, dM = C_0 M_0$ is constant, i.e., conservation of mass applies. Equation (13.65) can be rewritten in self-similar form as follows:

$$c(M,t) = \frac{\kappa b^{\eta/\kappa}}{\Gamma(\eta/\kappa)} \frac{C_0 M_0}{[M_{\text{avg}}(t)]^2} \left[\frac{M}{M_{\text{avg}}(t)}\right]^{\eta-1} \exp\left\{-b\left[\frac{M}{M_{\text{avg}}(t)}\right]^\kappa\right\} \tag{13.67}$$

$$M_{\text{avg}}(t) = \frac{C_0 M_0}{C(t)} = \left(\frac{b}{J_0 t}\right)^{1/\kappa} \tag{13.68}$$

where $b^{1/\kappa} = \Gamma[(\eta+1)/\kappa]/\Gamma(\eta/\kappa)$. As one way to understand this example, if the initial distribution obtains the broadest possible size spread — in this case, a Rosin–Rammler distribution according to the metric associated with m and J — fragmentation will preserve that initial distribution for all time. Otherwise, fragmentation will progressively drive the distribution towards the broadest possible size spread, as seen in the previous example.

For consistency with the rest of book, this example modified the original notation used by Saveliev & Gorokhovski (2012) in several ways. First, they wrote their governing equation in terms of diameter D rather than mass M. Second, they wrote their solution in terms of f_M instead of c. Third, they wrote their solution in terms of $j_0\tau_0$ instead of $C_0 M_0$ where j_0 is a mass flux and τ_0 is a time. The advantage is that j_0 and τ_0 are both non-zero and finite, unlike C_0 and M_0. The disadvantage is that τ_0 lacks a clear physical interpretation, i.e., it is defined only as "a positive constant." Also note that Saveliev & Gorokhovski (2012) derived a mean-field governing equation using a novel *scaling symmetry* argument in which "at each fragmentation step, a parent particle splits into daughter particles with a partition probability independent of the particle size; i.e., the particle size is modified in a step-by-step manner by multiplication by a random multiplier $0 \le \alpha \le 1$." The result is essentially equivalent to Equation (13.11).

13.6.5 Shattering Solutions

Example 13.11 and 13.12 concern weak shattering due to $\eta \le 0$. Examples 13.13 and 13.14 concern strong shattering due to $\kappa < 0$. Notice that Examples 12.11 and 12.12 are special cases of Example 13.13 with $\kappa = -1, -2$ and $\eta = 1$.

Example 13.11 Time-dependent closed-form solutions for $\kappa = 0$ and $\eta \le 0$ with monodisperse initial conditions and free-field boundary conditions

Kolmogorov (1941) gave the first mean-field solution for random fragmentation; see also Gorokhovski & Herrmann (2008) and Gorokhovski & Saveliev (2003, 2008). While Kolmogorov used the variables $f_M(D, t)$ and $g_M(D)$, this example will use the variables $f(M, t)$ and $g(M)$. Suppose the size of the children is expressed in terms of a probability density function $g(M)$, which does not depend on the size of the parent. As the only restriction, assume that $g(M)$ is well-characterized by its first and second log moments. This includes Rosin–Rammler size distributions as well as many other families of size distributions; see, e.g., Nicolas (2002). Then the following notation for log moments applies:

$$\langle \ln x \rangle = \int_0^1 \ln x \, g(M_0 x)\, dx \quad \text{and} \quad \langle \ln^2 x \rangle = \int_0^1 \ln^2 x \, g(M_0 x)\, dx. \tag{13.69}$$

For example, if $g(M) = (\eta/M_0)(x/M_0)^{\eta-1}$ then:

$$\langle \ln x \rangle = -\frac{1}{M_0 \eta} \quad \text{and} \quad \langle \ln^2 x \rangle = \frac{1}{2 M_0 \eta^2} \quad \text{and} \quad \frac{\langle \ln x \rangle}{\langle \ln^2 x \rangle} = -\frac{\eta}{2}. \tag{13.70}$$

For $J(x) = J_0 = \text{const}$, Kolmogorov (1941) obtained the following mean-field solution:

$$f(M, t) = \frac{1}{\sqrt{2\pi \langle \ln^2 x \rangle J_0 t}} \frac{1}{M} \exp\left\{ -\frac{[\ln (M/M_0) - \langle \ln x \rangle J_0 t]^2}{2 \langle \ln^2 x \rangle J_0 t} \right\} \tag{13.71a}$$

which is a log normal size distribution. Equivalently:

$$f(M, t) = \frac{1}{M_0} \frac{1}{\sqrt{2\pi \langle \ln^2 x \rangle J_0 t}} \exp\left(-\frac{\langle \ln x \rangle^2 J_0 t}{2 \langle \ln^2 x \rangle} \right) \exp\left(-\frac{[\ln (M/M_0)]^2}{2 \langle \ln^2 x \rangle J_0 t} \right) \left(\frac{M}{M_0} \right)^{\frac{\langle \ln x \rangle}{\langle \ln^2 x \rangle} - 1}. \tag{13.71b}$$

For sufficiently small times and large particles, Equation (13.71) becomes:

$$f(M, t) \cong (1/M_0)\, \delta(M/M_0 - 1) \tag{13.72}$$

which is a monodisperse size distribution. Gorokhovski & Saveliev (2003) argue that $C(t) = C_0 \exp[(N - 1) J_0 t)]$ where N is the number of children formed in each fragmentation event.

In addition, they argue that conservation of mass applies only if the third moment of g is proportional to $1/N$. Thus, this example does not obtain conservation of mass, i.e., it involves weak shattering unless g depends on N in just the right way.

Example 13.12 Time-dependent closed-form solutions for any $\kappa > 0$ and $\eta \leq 0$ with monodisperse initial conditions and free-field boundary conditions

Filippov (1961) obtained a solution to Equation (13.11). After a substitution of variables, this becomes the following solution to Equation (13.10) for $J(x) = J_0 x^\kappa$ and $m(x, y) = (\eta + 1)x^{\eta-1}/y^\eta$:

$$\begin{aligned} c(M,t) &= (C_0/M_0)\exp(-J_0 M_0^\kappa t)\,\delta(M/M_0 - 1) \\ &+ (\eta+1)\,C_0 \exp(-J_0 M_0^\kappa t)\,J_0 M_0^{\kappa-\eta} M^{\eta-1}\, t\,{}_1F_1\left[1 + (\eta+1)/\kappa, 2, J_0\left(M_0^\kappa - M^\kappa\right)t\right] \end{aligned} \tag{13.73}$$

where C_0 is the initial total concentration and M_0 is the initial mass of each particle. Notice that Equation (13.73) is exactly the same as Equation (13.58). Ernst & Szamel (1993) showed that:

$$\begin{aligned} C(t) &= \int_0^{M_0} c(M,t)\,dM = C_0 \exp(-J_0 M_0^\kappa t)\,{}_1F_1\left(\frac{\eta+1}{\kappa}, \frac{\eta}{\kappa}, J_0 M_0^\kappa t\right) \\ &- C_0 \exp(-J_0 M_0^\kappa t)\,(J_0 M_0^\kappa t)^{1-\frac{\eta}{\kappa}} \frac{\Gamma\left(\frac{\eta}{\kappa}\right)\Gamma\left(1+\frac{1}{\kappa}\right)}{\Gamma\left(\frac{\eta+1}{\kappa}\right)\Gamma\left(2-\frac{\eta}{\kappa}\right)}\,{}_1F_1\left(1+\frac{1}{\kappa}, 2-\frac{\eta}{\kappa}, J_0 M_0^\kappa t\right). \end{aligned} \tag{13.74}$$

Notice that Equation (13.74) for weak shattering is the same as Equation (13.59) for no shattering, plus an additional term. The additional term ensures that $C(t) \to 0$ as $t \to \infty$. In other words, the total concentration of particles decreases rather than increases over time. This is true because ${}_1F_1(x, y, z) \to \Gamma(y)/\Gamma(x) z^{x-y} \exp(z)$ as $z \to \infty$. Filippov (1961) showed that:

$$\int_0^{M_0} M c(M,t)\,dM = C_0 M_0 \frac{\Gamma\left(1 - \frac{\eta+1}{\kappa}, J_0 M_0^\kappa t\right)}{\Gamma\left(1 - \frac{\eta+1}{\kappa}\right)}. \tag{13.75}$$

In other words, conservation of mass does not apply. While the proof is difficult, Equations (13.74) and (13.75) imply that the total particle concentration and total particle mass are both positive and monotone decreasing if:

$$0 < \kappa < 1 \text{ and } -1 < \eta < \kappa - 1 < 0. \tag{13.76}$$

For sufficiently small times and large particles, Equation (13.73) becomes a monodisperse size distribution. Conversely, for sufficiently large times and small particles, Equation (13.73) becomes:

$$c(M,t) \cong \frac{C_0 M_0 \kappa}{\Gamma[(\eta+1)/\kappa]}(J_0 t)^{(\eta+1)/\kappa} M^{\eta-1} \exp\left(-J_0 M^{\kappa} t\right) \tag{13.77}$$

which is a Rosin–Rammler size distribution with $n/m = \eta$ and $l/m = \kappa$. Notice that Equation (13.77) for random fragmentation with weak shattering is exactly the same as Equation (13.61) for random fragmentation without shattering.

Example 13.13 Time-dependent closed form solutions for $\kappa < 0$ and any η with monodisperse initial conditions and free-field boundary conditions

Filippov (1961) obtained a solution to Equation (13.11). After a substitution of variables, this becomes the following to solution to Equation (13.10) for $J(x) = J_0 x^{\kappa}$ and $m(x,y) = (\eta+1)x^{\eta-1}/y^{\eta}$:

$$\begin{aligned} c(M,t) &= (C_0/M_0)\exp\left(-J_0 M_0^{\kappa} t\right)\delta(M/M_0 - 1) \\ &+ (\eta+1)\,C_0 \exp\left(-J_0 M_0^{\kappa} t\right) J_0 M_0^{\kappa-\eta} M^{\eta-1}\, t\, {}_1F_1\left[1+(\eta+1)/\kappa, 2, J_0\left(M_0^{\kappa} - M^{\kappa}\right)t\right] \end{aligned} \tag{13.78}$$

where C_0 is the initial total concentration and M_0 is the initial mass of each particle. If $\kappa < -1$ then Equation (13.74) applies. However, if $-1 \leq \kappa < 0$ then:

$$C(t) = \int_0^{M_0} c(M,t)\,dM = \infty \tag{13.79}$$

for all $t > 0$. This is because of a non-integrable singularity at $M = 0$. As in previous examples, this can optionally be addressed by introducing a hard sink that removes particles with masses less than $M_{\min}$ where $M_{\min}$ approaches but does not equal 0. Whether or not this is done, it is always true that:

$$\int_0^{M_0} M c(M,t)\,dM = C_0 M_0 \frac{\Gamma\left(1 - \dfrac{\eta+1}{\kappa}, J_0 M_0^{\kappa} t\right)}{\Gamma\left(1 - \dfrac{\eta+1}{\kappa}\right)}. \tag{13.80}$$

Notice that Equation (13.80) is the same as Equation (13.75). For sufficiently small times and large particle sizes, Equation (13.78) becomes a monodisperse size distribution. Conversely, for sufficiently large times and small particle sizes, Equation (13.78) becomes:

$$c(M,t) \cong \frac{|\kappa| C_0 M_0^{\kappa-\eta}}{\Gamma[(\eta+1)/|\kappa|]}\exp\left(-J_0 M_0^{\kappa} t\right)(J_0 t)^{(\eta+1)/|\kappa|} M^{|\kappa|-2} \tag{13.81}$$

which is true because ${}_1F_1(x,y,-z) \cong \Gamma(y)/\Gamma(y-x) z^{-x}$ as $z \to \infty$. Equivalently:

$$c(M,t) \cong \frac{|\kappa|}{\Gamma[(\eta+1)/|\kappa|]}\frac{C_0 M_0}{[M'_{\text{avg}}(t)]^2}\left[\frac{M}{M'_{\text{avg}}(t)}\right]^{|\kappa|-2} \tag{13.82}$$

where:

$$M'_{\text{avg}}(t) \cong M_0 \exp\left(J_0 M_0^{\kappa} t / |\kappa|\right) \left(J_0 M_0^{\kappa} t\right)^{-(\eta+1)/\kappa^2}. \tag{13.83}$$

This is a power law size distribution with $n/m = -\kappa - 1$. Unlike earlier examples, n/m depends only on κ and not on η. This occurs because strong shattering introduces a sink of small particles. Here and in general, sinks due to strong shattering reduce the dependency on η or, in other words, the dependency on child size distributions produced in each event. However, while the exponent n/m does not depend on η, the average $M'_{\text{avg}}(t)$ does depend on η.

Example 13.14 Time-dependent closed-form solutions for $\kappa < 0$ and any η with power law initial conditions and free-field boundary conditions

Saveliev & Gorokhovski (2012) obtained a solution to Equation (13.11). After a substitution of variables, this becomes the following solution to Equation (13.10) for $J(x) = J_0 x^{\kappa}$ and $m(x, y) = (\eta + 1)x^{\eta-1}/y^{\eta}$:

$$c(M,t) = \frac{C_0 M_0 |\kappa|}{M^2}\, {}_1F_1\left[1, 1 - \frac{\eta+1}{\kappa}, -J_0 M^{\kappa} t\right] \tag{13.84}$$

where C_0 is the initial total concentration and M_0 is the initial average particle mass. Notice that if $\kappa < 0$ then:

$$C(t) = \int_0^{\infty} c(M,t)\, dM = \infty$$

$$\int_0^{\infty} Mc(M,t)\, dM = \infty$$

for all $t > 0$. This is due to a non-integrable singularity at $M = 0$. As in previous examples, this can optionally be addressed by introducing a hard sink that removes particles with masses less than $M_{\min}$ where $M_{\min}$ approaches but does not equal 0. Whether or not this is done, for sufficiently small times and large particle sizes, Equation (13.84) becomes:

$$c(M,t) \cong \frac{C_0 M_0 |\kappa|}{M^2} \tag{13.85}$$

which is a power law size distribution with $n/m = -1$. This is true because ${}_1F_1(x, y, z) \cong 1$ as $z \to 0$. For sufficiently large times and small particle sizes, Equation (13.84) becomes:

$$c(M,t) \cong (\eta + 1) C_0 M_0 (J_0 t)^{-1} M^{|\kappa|-2} \tag{13.86}$$

which is true because ${}_1F_1(x, y, -z) \cong \Gamma(y)/\Gamma(y-x)z^{-x}$ as $z \to \infty$. Equivalently:

$$c(M, t) \cong (\eta + 1)|\kappa| \frac{C_0 M_0}{[M'_{\text{avg}}(t)]^2} \left[\frac{M}{M'_{\text{avg}}(t)} \right]^{|\kappa|-2} \tag{13.87}$$

$$M'_{\text{avg}}(t) \cong \left(\frac{1}{|\kappa| J_0 t} \right)^{1/\kappa}. \tag{13.88}$$

Like the previous example, this is a power law size distribution with $n/m = -\kappa - 1$. This example differs from the previous example mainly in terms of $M'_{\text{avg}}(t)$. More specifically, $M'_{\text{avg}}(t)$ now depends only on κ and not on η. This occurs because shattering introduces a stronger sink of small particles than before, which almost completely eliminates the dependence on η. For consistency with other examples, this example modified the original treatment given by Saveliev & Gorokhovski (2012) as described earlier in Example 13.10.

13.7 Late-Time Mean-Field Solutions

For many practical applications, the final state is more important than any intermediate state. As seen in Example 13.15, if conservation of mass applies, mean-field theory for random fragmentation yields general expressions for the average particle mass at sufficiently large times. As seen earlier in Section 11.5, mean-field theory for aggregation yields similar expressions.

Example 13.15 Late-time analytic solutions for the average particle size produced by random fragmentation

In most cases, at sufficiently late times, random fragmentation obtains self-similar size distributions. As one possible self-similar form, suppose that:

$$c(M, t) = \frac{C_0 M_0}{M^2_{\text{avg}}(t)} \tilde{c} \left[\frac{M}{M_{\text{avg}}(t)} \right]. \tag{13.89}$$

To obtain a general expression for $M_{\text{avg}}(t)$, assume that m and J are homogeneous. In other words, assume that:

$$m(x, y) = y^{-1} \tilde{m}(x/y) \tag{13.90}$$

and:

$$J(x) = J_0 x^{\kappa}. \tag{13.91}$$

For free-field boundary conditions, substituting Equations (13.89), (13.90), and (13.91) into Equation (13.10) implies the following:

$$\frac{dM_{\text{avg}}(t)}{dt} = -a_0 J_0 M^{\kappa+1}_{\text{avg}}(t) \tag{13.92}$$

where a_0 is a positive unitless constant; see, e.g., Kapur (1972, eq. 12), Kapur & Fuerstenau (1987, eq. 22), Cheng & Redner (1990, eq. 12) and Gupta (2019, eq. 9). Per Gupta (2019, eq. 10), Equation (13.92) has the following solution:

$$M_{\text{avg}}(t) = M_0\left[\kappa a_0 J_0 M_0^{\kappa} t + 1\right]^{-\frac{1}{\kappa}} \tag{13.93}$$

provided that $\kappa \neq 0$, which is easy to verify through back substitution. If $\kappa > 0$, this implies:

$$M_{\text{avg}}(t) \cong \left[1/(\kappa a_0 J_0 t)\right]^{1/\kappa} \tag{13.94}$$

for sufficiently large times. Similarly, if $\kappa = 0$, Equation (13.92) has the following solution:

$$M_{\text{avg}}(t) = M_0 \exp(-a_0 J_0 t) \tag{13.95}$$

which is easy to verify through back substitution.

The following summarizes some of the key late-time results obtained in the last section. For free-field boundary conditions, if $\kappa > 0$ and $\eta > 0$, mean-field theory obtains a Rosin–Rammler size distribution with $n/m = \eta$ and $l/m = \kappa$. More specifically:

$$c(M,t) \cong \frac{\kappa b^{\eta/\kappa}}{\Gamma(\eta/\kappa)} \frac{C_0 M_0}{[M_{\text{avg}}(t)]^2} \left[\frac{M}{M_{\text{avg}}(t)}\right]^{\eta-1} \exp\left\{-b\left[\frac{M}{M_{\text{avg}}(t)}\right]^{\kappa}\right\} \tag{13.96}$$

where:

$$M_{\text{avg}}(t) \cong (\kappa a_0 J_0 t)^{-1/\kappa} \tag{13.97}$$

for small enough particles and large enough times. In these expressions, a_0 is a positive unitless constant and:

$$b = \left[\frac{\Gamma[(\eta+1)/\kappa]}{\Gamma(\eta/\kappa)}\right]^{\kappa}.$$

If $\kappa = 0$, mean-field theory obtains a log-normal distribution, or at least something reminiscent of a log-normal distribution, depending on the exact assumptions. If $\kappa = 0$ and $\eta > 0$ then:

$$M_{\text{avg}}(t) \cong M_0 \exp(-a_0 J_0 t) \tag{13.98}$$

where a_0 is positive unitless constant. Finally, if $\kappa < 0$ so that strong shattering occurs, mean-field theory obtains a power law size distribution with $n/m = -\kappa - 1$. More specifically:

$$c(M,t) \cong A(\kappa,\eta) \frac{C_0 M_0}{\left[M'_{\text{avg}}(t)\right]^2} \left[\frac{M}{M'_{\text{avg}}(t)}\right]^{-\kappa-2} \tag{13.99}$$

where:

$$M'_{avg}(t) \cong \left(|\kappa|\, a_0 J_0 t\right)^{-1/\kappa} \tag{13.100}$$

for small enough particles and large enough times. Equation (13.99) applies regardless of the strength of the sink formed due to shattering. However, Equation (13.100) only applies for sinks strong enough to eliminate the effects of the children size distributions.

Now, consider periodic boundary conditions. In other words, suppose there is a source of large particles and an equal-but-opposite sink of small particles. At sufficiently late times, periodic boundary conditions cause particle systems to become stationary. If the periodic boundary conditions involve weak soft sources and sinks, mean-field theory may obtain Rosin–Rammler size distributions with $n/m = \eta$ and $l/m = \kappa$ for sufficiently large times and sufficiently small particles. On the other hand, if the periodic boundary conditions involve strong hard sources and sinks, mean-field theory may obtain power law size distributions with $n/m = -\kappa - 1$ for sufficiently large times and sufficiently small particles; see, e.g., Williams (1990, eq. 66) and Elhanbaly (2003, eq. 47). In this case, the mean-field solution depends on $J(x) = J_0 x^{\kappa}$ but not on $m(x, y) = (\eta + 1)x^{\eta-1}/y^{\eta}$. Notice that, as κ increases, the sink in the periodic boundary conditions must work harder to eliminate the dependence on $m(x, y)$. If κ becomes too large, the sink and the equal-but-opposite source may become so strong that they tamp out chaos in the particle system as a whole, resulting in narrow size spreads. As shown in Appendix A, particle size spreads increase as $|n|/m$ decreases. As a rule of thumb, chaos occurs only if $|n|/m \leq 2$. However, according to mean-field solution, $|n|/m = \kappa + 1 > 2$ when $\kappa > 1$. This means that the particle system may no longer be chaotic when $\kappa > 1$ and, thus, mean-field theory may no longer be valid.

13.8 Experimental Validation

As discussed in later chapters, it can be difficult to characterize fragmentation of a single object in experiments. If it is difficult to characterize the fragmentation of a single object, it is much more difficult to characterize the fragmentation of many different objects, especially given that each such event takes place at seemingly random times and locations. If the volumetric fraction of particles is low, it may be possible to observe individual fragmentation events. However, if the volumetric fraction of particles is high, which is more typical of random fragmentation, it may be difficult or impossible observe most of the individual fragmentation events. Instead, the majority of such events occur somewhere in the interior of the particle system, where they are hidden from view by the surrounding material. As a result, in most particle systems experiencing random fragmentation, is difficult or impossible to measure the number and size distribution of the children produced by each fragmentation event. Thus, it must generally be assumed rather than proven that:

$$m(x, y) = (\eta + 1)\, x^{\eta-1}/y^{\eta}$$

in some kind of average sense. In addition, in most particle systems experiencing random fragmentation, is difficult or impossible to measure the rate of fragmentation events. Thus, again, it must be assumed rather than proven that:

$$J(x) = J_0 x^{\kappa}.$$

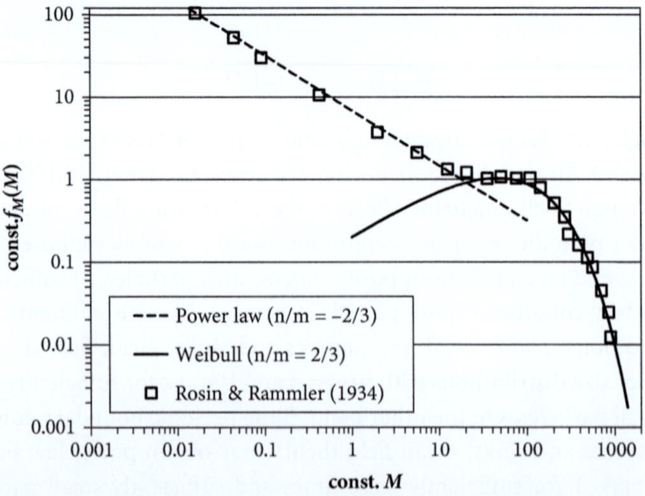

Figure 13.1 Empirical curve fits vs. experimental results for fragment size distributions for stationary ball milling of iron.
Experimental results taken from Rosin & Rammler (1934); see also Brown & Wohletz (1995).

Due to these obstacles, most experimental validation involves curve fitting the as-measured size distribution. Mean-field theory is considered "validated" if it agrees with the experimental curve fits for "reasonable" choices of κ and η. For example, Figure 13.1 shows empirical curve fits to experimental results for stationary ball milling of iron. In this case, the smaller particles obtain a negative Weibull size distribution or, equivalently, a negative power law size distribution with $n/m \cong -2/3$. As one possible explanation, the smaller particles could have been created by random fragmentation with weak shattering, where $\eta \cong \kappa \cong -2/3$. On the other hand, the larger particles obtain a positive Weibull size distribution with $n/m \cong 2/3$. As one possible explanation, the larger particles could have been created by random fragmentation with no shattering, where $n \cong \kappa \cong 2/3$. Notice that $\kappa \cong \pm\, 2/3$ satisfies Equations (13.27) and (13.29). In addition, $\eta \cong \pm\, 2/3$ satisfies Equation (13.20), assuming that each fragmentation event produces either 2.5 or ∞ children on average.

So far, this chapter has focused on mean-field theory in the form of Equation (13.10). However, the vast majority of the relevant research literature focuses on mean-field theory in the form of Equation (13.12), a.k.a., the batch grinding equation. Example 13.16 shows how experimental results are typically used to validate the batch grinding equation.

Example 13.16 Experimental validation of the batch grinding equation

Mazzinghy et al. (2014) used standard sieving techniques to characterize batch-grinding experiments carried out on copper ore using a laboratory-scale 254-mm-diameter ball mill. Based on the experimental size distributions, they back-calculated a selection function $S_i = S_0 x_i^{\alpha}$ where $S_0 = 0.941$tonnes/kW-hr and $\alpha = 0.665$. This selection function is illustrated in Figure 13.2.a. In addition, they back-calculated the following cumulative breakage function:

$$G_{M,ij} = \phi\left(\frac{x_{i-1}}{x_j}\right)^{\gamma} + (1-\phi)\left(\frac{x_{i-1}}{x_j}\right)^{\beta}$$

where $\phi = 0.463$, $\gamma = 0.615$, and $\beta = 3.328$. This cumulative breakage function, which is a linear combination of two power laws, is illustrated in Figure 13.2.b.

Figure 13.3 compares the results of the batch grinding equation, obtained using MTI's MODSIM^{TM} software package, to the experimental results. As expected, the batch grinding equation accurately captures the as-measured product size distributions over a wide range of particle sizes and grind times. As usual, these curve fits only work well for one specific ball mill. Having said this, they can be extrapolated to other ball mills using well-known empirical scaling laws; see, e.g., Mazzinghy et al. (2014, 2015, 2017)

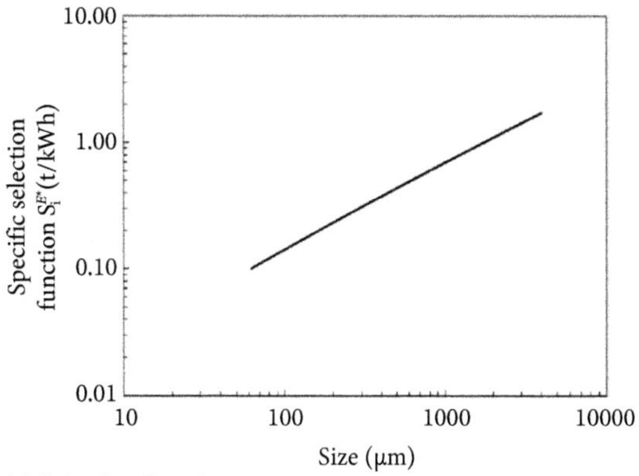

(a) Selection function.

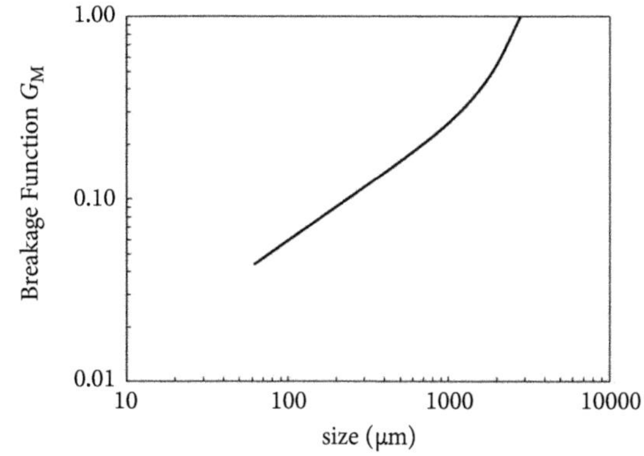

(b) Cumulative breakage function.

Figure 13.2 Selection function and cumulative breakage function back-calculated from batch grinding experiments carried out on copper ore using a laboratory-scale 254-mm-diameter ball mill.

[The notation in Figure 13.2.b has been changed to match the notation used in this book.]
Figures reprinted from Mazzinghy et. al. (2014) with permission from Elsevier.

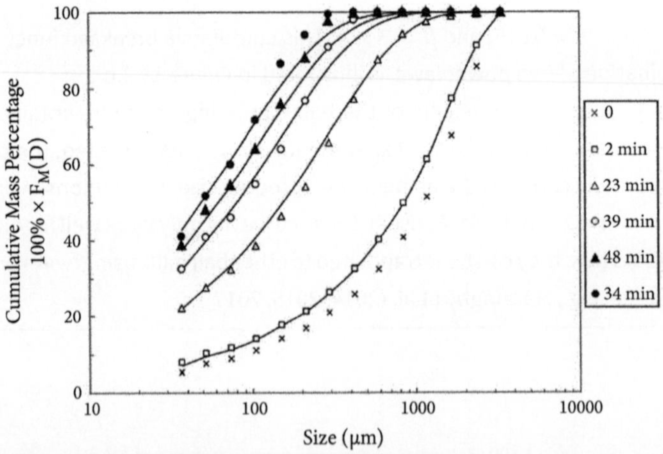

Figure 13.3 Analytical (mean-field theory) vs. experimental results for batch grinding of copper ore using a laboratory-scale 254-mm-diameter ball mill. Experimental results shown as symbols. Mean-field theory results shown as curves.
[The notation in this figure has been changed to match the notation used in this book.]
Figure reprinted from Mazzinghy et. al. (2014) with permission from Elsevier.

This last example is just one among many to appear in the research literature. Based on this body of evidence, Fuerstenau et al. (2004) says the following:

> The very extensive experimental work carried out during the last four decades has proven the validity of the linear grinding model [a.k.a., mean-field theory for random fragmentation] and the size dependence of the breakage rate functions [a.k.a. selection functions] and the normalizability of breakage distribution functions. With confidence, therefore, it is possible to back-calculate these parameters for simulation purposes.

As this quote indicates, there is generally no way to choose the modeling parameters such as α, ϕ, γ, and β in advance. Instead, they are chosen after-the-fact to ensure the best possible agreement between theory and experiment, in a least-squares sense. Rather than true validation, this amounts to calibration. Rather than resulting in analytic models, this results in semi-empirical models; see the earlier discussion in Section 5.5. Such semi-empirical models are often called *population balance models (PBMs)*. According to Tavares (2017), "the operation of a ball mill is a capital- and energy-intensive process, so that even marginal improvements in their efficiency result in immense benefits to the industry." As result, a number of commercial PBM-based software packages have been developed to simulate ball milling operations. Morrell (2019) says that most of these commercial software packages—including MODSIM^{TM}, USIM PAC, METSIM®, and JKSimMet—rely on "the population balance model as their mathematical framework because it elegantly encapsulates the size reduction process." Example 13.16 and 13.17 compare PBMs to empirical and high-fidelity computational models, respectively.

Example 13.17 Empirical vs. semi-empirical methods for modeling batch grinding

Despite their popularity, the commercial PBM software packages naturally have their drawbacks. For example, McIvor et al. (2023) say the following:

> **... A needed input of population balance modeling (PBM), the 'breakage function,' was never intended to be taken literally ... Breakage rates of each size class (the 'selection function') could then be back-calculated, given measured mill feed and product size distributions. This ... ball mill model, comprising the assumed breakage function, ... residence time ..., and back-calculated breakage rates, works quite well ... All three individual elements, however, are false. The model cannot be used to assess changes to the mill itself. While shrouded from proper vetting by complexity, lack of progress ... [has] resulted in discontinuation of almost all PBM research.**

For another example, Yuan et al. (2023) say the following:

> **The breakage rate function and the cumulative breakage distribution function are the core functions of PBM .... However, the calculations of these two key functions are extremely complicated, especially the calculation of the cumulative breakage distribution function, which requires a large number of feature parameters ... These complicated computations make it extremely inconvenient for researchers to use PBM[s] ... Worse still, the excessive number of feature parameters leads to large errors in the calculations ... The Austin empirical formula [Equation 13.24] is a sum of two power functions, which increases the uncertainty of the fits and leads to large errors in its parameters.**

Regarding this last remark, as noted in Section 5.6, choosing three or more free parameters in a size distribution, such as a Rosin–Rammler size distribution, is an ill-posed problem. In practice, PBMs mitigate this issue by optimizing Rosin–Rammler fits to experimental data at many different times at once. However, based on the remarks by Yuan et al. (2023), this approach may not always be completely effective. As another possible approach, PBMs may adjust their optimization routines to encourage exponents in close proximity to known universal values; see, e.g., Figure 13.1.

As an alternative to the semi-empirical PBMs, McIvor et al. (2023) recommend using a type of empirical model known as *apparent cumulative grinding rate models*. More specifically, McIvor et al. (2023) say that these empirical models "should be taught and adopted throughout industry and replace PBM in modeling programs." Indeed, empirical methods are commonly used to model all kinds of fragmentation in all kinds of applications. However, because each such model is limited to a single application, empirical models are outside the scope of this book, which is intended for a broad audience. For more discussion of empirical and semi-empirical models for batch griding, readers may consult textbooks such as King et al. (2012), Gupta & Yan (2016), Litster (2016) and Bilgili et al. (2017).

Example 13.18 High-fidelity computational methods vs. semi-empirical methods for modeling batch grinding

Tavares (2017) compared PBMs to high-fidelity computational approaches, such as the Discrete Element Method (DEM), as discussed earlier in Section 4.7. As the main advantage of DEM, Tavares (2017) notes:

> **DEM makes it possible to predict how collisions vary in magnitude and frequency as the geometry, the ball charge and the mills speed are varied. Therefore, instead of correlating each individual parameter such as mill diameter, filling, speed and ball size, for instance, to the product size, one can directly take into account the record of the collisions and their influence on the particle size distribution in the mill ... [This] is certainly a more logical and intuitive approach than assuming a constant breakage distribution function, as ... in the traditional population balance model formulation.**

As the main disadvantage of DEM, Tavares (2017) notes:

> **... It could take as much as 10GB of computer memory to run ... commercial software such as EDEM (DEM Solutions, UK) ... Moreover, this is only considering a 30cm slice of the mill, which could have more than two billion elements [particles] with sizes above 0.5mm. ....[In addition, DEM must] use some form of the population balance framework to mathematically combine the particles and their associated breakage products into a mass balance per time step.**

In other words, as a practical matter, computational methods for large-scale systems typically rely on mean-field theory, at least to some extent.

13.9 Energy-Size Reduction Relationships

When first introduced to fragmentation, many scientists and engineers express curiosity about energy. This is an urgent practical issue for batch mills. Common questions include the following:

- How much energy is required to produce a given fragment size?
- Where does the energy go? Does it go into surface energy? Does it go into dissipation?

Early researchers suggested different answers to these questions. In particular, Rittinger (1867) suggested that energy goes into surface energy, while Kick (1885) suggested that it goes into dissipation. This led to a long-lasting controversy; see, e.g., Gates (1915, 1922). The mean-field theory results given in this chapter allow us to completely answer the first question and make a start on answering the second question. As the main assumption, suppose that random fragmentation draws power at a steady uniform rate, at least on average. In other words, suppose that:

$$E = Pt \tag{13.101}$$

where t is the time, E is the energy, and P is the power consumed by fragmentation; see, e.g., Kapur & Fuerstenau (1987), Petrakis et al. (2016) and Gupta (2019). For random fragmentation with $\kappa > 0$ and $\eta > 0$, combining Equations (13.97) and (13.101) gives:

$$M_{\text{char}} \cong \left(\frac{bP}{J_0 E}\right)^{1/\kappa} \tag{13.102}$$

where $b = 1/(\kappa a_0)$. This expression applies when $t = E/P$ is large enough. If $\kappa > 0$, this expression says that the larger the net energy consumption, the smaller the characteristic fragment size, which makes intuitive sense. While this book generally expresses results in terms of fragment masses, the relevant research literature concerns fragment diameters. Assume that:

$$M_{\text{char}} = \rho C D_{\text{char}}^{m} \tag{13.103}$$

where ρ is the fragment density, C is a unitless constant that depends on the fragment shape and size distribution, D_{char} is a characteristic fragment diameter, and m is the fragment mass dimension; see, e.g., Laney (2015, 2016). Combining Equations (13.102) and (13.103) gives:

$$D_{\text{char}} \cong \left(\frac{1}{\rho C}\right)^{1/m} \left(\frac{bP}{J_0 E}\right)^{1/m\kappa} \tag{13.104}$$

for sufficiently large times. If the factors that do not involve E are combined as follows:

$$C_0 = \left(\frac{1}{\rho C}\right)^{1/m} \left(\frac{bP}{J_0}\right)^{1/m\kappa} \tag{13.105}$$

then Equation (13.104) becomes:

$$D_{\text{char}} \cong C_0 E^{-1/m\kappa}. \tag{13.106}$$

According to Equation (13.29), in ideal cases, κ is an integer multiple of 1/6 between −2 and 2. Then if $m = 3$ and $\kappa > 0$, Equation (13.106) becomes:

$$D_{\text{char}} \cong C_0 E^{-2/i} \tag{13.107}$$

where $i = 1, 2, \ldots, 12$. Inverting Equation (13.107) leads directly to a set of *energy-size reduction relationships*. These relationships are commonly used in fields such as chemical engineering and mining engineering. Example 13.19 discusses the two most common cases where $i = 1$ and 2. As this example shows, researchers originally discovered energy-size reduction relationships using empirical curve fits to experimental results. These curve fits were done before mean-field theory for random fragmentation had been developed; see, e.g.,

Gilliland (1937), Bond (1952, 1960) and Charles (1957). Population balance models (PBMs) have, to some extent, supplanted energy-size reduction relationships. However, energy-size reduction relationships still have on-going practical value.

Example 13.19 Traditional energy-size reduction relationships

Most fragmentation research attempts to determine fragment sizes resulting from a given set of conditions. However, some fragmentation research concerns the inverse problem. In other words, it attempts to determine the conditions—especially the energy—required to produce a given average fragment size. In this regard, Equation (13.107) implies the following:

$$E \cong C_1 D_{\text{char}}^{-i/2} \tag{13.108}$$

where $C_1 = C_0^{i/2}$ and $i = 1, 2, \ldots, 12$. Equivalently:

$$dE \cong C_2 D_{\text{char}}^{-i/2-1} dD_{\text{char}} \tag{13.109}$$

where $C_2 = -iC_1/2$. Charles (1957) derived this expression empirically assuming that fragments obtain power law size distributions; see also Gilliland (1937). Kapur (1971) derived this expression analytically assuming that fragments obtain power law or Weibull size distributions. In addition, Kapur & Fuerstenau (1987) derived this expression analytically, assuming that fragmentation consumes a constant amount of power. As one example, if $i = 1$, then Equations (13.108) and (13.109) become:

$$E(t) \cong C_1 \left[\frac{1}{\sqrt{D_{\text{char}}(t)}} - \frac{1}{\sqrt{D_{\text{char}}(0)}} \right] \cong \frac{C_1}{\sqrt{D_{\text{char}}(t)}} \tag{13.110}$$

assuming that $D_{\text{char}}(0) \gg D_{\text{char}}(t)$ for sufficiently large times. Bond (1952, 1960) derived this expression empirically. As another example, if $i = 2$, then Equations (13.108) and (13.109) become:

$$E(t) \cong C_1 \left[\frac{1}{D_{\text{char}}(t)} - \frac{1}{D_{\text{char}}(0)} \right] \cong \frac{C_1}{D_{\text{char}}(0)} \tag{13.111}$$

assuming that $D_{\text{char}}(0) \gg D_{\text{char}}(t)$ for sufficiently large times. According to Gilliland (1937), Rittinger (1867) derived this expression empirically, making it arguably the earliest major result not just for random fragmentation, but for fragmentation of any kind. The previous results are typically applied to random fragmentation in mills, grinders, crushers, and so forth; see, e.g., Gupta (2013, 2019), Litster (2016) and Chakraborty (2019).

This section answered one important question: how much energy is required to produce a given average fragment size? Now consider a second important question: where does the energy go? According to Rittinger (1867), the energy expended in fragmentation goes primarily into creating new surface area. If this is true then, using conservation of energy, the constant C_0 in Equations (13.106) and (13.107) can be written in terms of surface tension σ

for liquids or surface energy Γ for solids. On the other hand, according to Kick (1885), the energy expended in fragmentation goes primarily into dissipation. If this is true then, using conservation of energy, the constant C_0 in Equations (13.106) and (13.107) can be written in terms of dynamic viscosity μ for liquids or yield strength Y for solids. In fact, either Rittinger or Kick may be correct depending on the circumstances. This will be discussed later in Chapter 16.

13.10 Conclusions

This chapter generalized the results from the last chapter in two different ways. First, while the last chapter assumed that each fragmentation event creates exactly two children, this chapter allows each fragmentation event to create any number of children. Second, while the last chapter assumed low volumetric fractions, this chapter allows almost any volumetric fraction. As seen in this chapter, mean-field solutions for random fragmentation depend on four things: (a.) the fragmentation rate kernel J; (b.) the size distributions produced by each fragmentation event $m(x, y)$; (c.) the initial conditions; and (d.) the boundary conditions. This chapter found that closed-form mean-field solutions are possible under the following conditions:

(a) the fragmentation rate kernel is a power law;
(b) the child size distribution is a power law or, at least, homogeneous;
(c) the initial conditions are monodisperse, power law, or Rosin–Rammler;
(d) the boundary conditions are free-field or, for late times, periodic.

If all of these conditions are met, mean-field theory produces Rosin–Rammler or power law size distributions, at least for large enough times and small enough particle sizes. Rosin–Rammler size distributions were first discovered by curve fitting results from mineral processing experiments including those involving crushing, grinding, milling, and so forth; see, e.g., Rosin & Rammler (1927a, b; 1933) and Rosin et al. (1933). Similarly, power law size distributions were first discovered by curve fitting results from mineral processing experiments including those involving crushing, grinding, milling, and so forth; see, e.g., Gates (1915), Gaudin (1926) and Schuhmann (1940). This chapter has shown that Rosin–Rammler and power law size distributions—still often thought of as completely empirical—can be rigorously derived using mean-field theory. While it is easy to find experimental evidence to support general forms such as Rosin–Rammler and power law size distributions, it is harder to find experimental evidence supporting specific parameters in those forms, especially as the volumetric fraction of the particle system increases.

According to mean-field theory, shattering occurs when $\kappa < 0$ or $\eta \leq 0$. Shattering in fragmentation is related to gelation in aggregation. As discussed in Section 10.9, gelation removes large particles from an aggregating particle system through a freezing-like or condensation-like process. Similarly, shattering removes small particles from a fragmenting particle system through a boiling-like or evaporation-like process. For example, Ernst & Pagonabarraga (2007) say shattering often "has the character of a continuous phase transition ... [in which] mass instantly 'evaporates' from the system." While there is extensive experimental evidence for gelation, there is little if any direct experimental evidence for shattering. However, without doubt, children from any given fragmentation event may lack enough mass, momentum, and energy to participate in future fragmentation events. While these tiny particles may remain physically present, they become essentially passive, i.e., they are largely irrelevant to the subsequent evolution of the system. Thus, for most practical purposes, the smallest particles may be considered to have been removed from the system by a sink. Whatever their actual cause, sinks of small particles tend to reduce the effects of the child size distributions. In fact, if the sinks are strong enough, they may completely eliminate the effects of the child size distributions on the overall particle size distribution.

References

M. Abramowitz & I. A. Stegun, editors, *Handbook of Mathematical Functions with Formulas, Graphs, and Mathematical Tables, 10th Printing*, National Bureau of Standards, 1972

L. G. Austin, Introduction to the Mathematical Description of Grinding as a Rate Process, *Powder Technology*, 5(1), 1–17, 1971, https://doi.org/10.1016/0032-5910(71)80064-5

L. G. Austin, A Discussion of Equations for the Analysis of Batch Grinding Data, *Powder Technology*, 106(1–2), 71–77, 1999, https://doi.org/10.1016/S0032-5910(99)00047-9

G. Barthelmes, S. E. Pratsinis, & H. Buggisch, Particle Size Distributions and Viscosity of Suspensions Undergoing Shear-Induced Coagulation and Fragmentation, *Chemical Engineering Science*, 58(13), 2893–2902, 2003, https://doi.org/10.1016/S0009-2509(03)00133-7

V. L. Bass, Zur Theorie der Mahlvorgänge [On the Theory of Grinding Processes], *Zeitschrift für angewandte Mathematik und Physik ZAMP*, 5(4), 283–292, 1954, https://doi.org/10.1007/BF01587825

P. J. Becker, F. Puel, R. Henry, & N. Sheibat-Othman, Investigation of Discrete Population Balance Models and Breakage Kernels for Dilute Emulsification Systems, *Industrial & Engineering Chemistry Research*, 50(19), 11358–11374, 2011, https://doi.org/10.1021/ie2006033

P. J. Becker, F. Puel, H. A. Jakobsen, & N. Sheibat-Othman, Development of an Improved Breakage Kernel for High Dispersed Viscosity Phase Emulsification, *Chemical Engineering Science*, 109, 326–338, 2014, https://doi.org/10.1016/j.ces.2014.02.008

E. Bilgili, M. Capece, & A. Afolabi, Modeling of Milling Processes via DEM, PBM, and Microhydrodynamics, in P. Pandey & R. Bharadwaj, editors, *Predictive Modeling of Pharmaceutical Unit Operations*, Elsevier, 2017, https://doi.org/10.1016/B978-0-08-100154-7.00007-7

J. E. L. Bishop & T. M. Searle, Power-Law Asymptotic Mass Distributions for Systems of Accreting or Fragmenting Bodies, *Monthly Notices of the Royal Astronomical Society*, 203(4), 987–1009, 1983, https://doi.org/10.1093/mnras/203.4.987

F. C. Bond, Third Theory of Comminution, *Transactions of the American Institute of Mining, Metallurgical, and Petroleum Engineers*, 193, 484–494, 1952, https://www.onemine.org/document/abstract.cfm?docid=42183&title=The-Third-Theory-Of-Comminution

F. C. Bond, Confirmation of the Third Theory, *Transactions of the American Institute of Mining, Metallurgical, and Petroleum Engineers*, 217, 139–153, 1960

Z. Borka & H. A. Jakobsen, Evaluation of Breakage and Coalescence Kernels for Vertical Bubbly Flows using a Combined Multifluid-Population Balance Model Solved by Least Squares Method, *Procedia Engineering*, 42, 623–633, 2012, https://doi.org/10.1016/j.proeng.2012.07.455

J. Bridgwater, Particle Breakage due to Bulk Shear, in A. D. Salman, M. Ghadiri, & M. J. Hounslow, editors, *Handbook of Powder Technology, Volume 12*, Elsevier, 2007, https://doi.org/10.1016/S0167-3785(07)12006-6

S. R. Broadbent & T. G. Callcott, A Matrix Analysis of Processes Involving Particle Assemblies, *Philosophical Transactions of the Royal Society A: Mathematical, Physical, and Engineering Science*, 249(960), 99–123, 1956, https://doi.org/10.1098/rsta.1956.0016

W. K. Brown & K. H. Wohletz, Derivation of the Weibull Distribution Based on Physical Principles and its Connection to the Rosin–Rammler and Lognormal Distributions, *Journal of Applied Physics*, 78(4), 2758–2763, 1995, https://doi.org/10.1063/1.360073

R. J. Charles, Energy-Size Reduction Relationships in Comminution, *Transactions of the American Institute of Mining, Metallurgical, and Petroleum Engineers*, 208, 80–88, 1957, https://www.onemine.org/document/abstract.cfm?docid=13062

J. Chakraborty, Particle Size Reduction and Enlargement, in T. Chhabra & M. G. Basavaraj, editors, *Coulson and Richardson's Chemical Engineering, Volume 2A: Particulate Systems and Particle Technology*, 6th edition, Butterworth-Heimemann, 2019, https://doi.org/10.1016/B978-0-08-101098-3.00006-8

Z. Cheng and S. Redner, Kinetics of Fragmentation, *Journal of Physics A: Mathematical and General*, 23(7), 1233–1258, 1990, https://doi.org/10.1088/0305-4470/23/7/028

C. A. Coulaloglou & L. L. Tavlarides, Description of Interaction Processes in Agitated Liquid-Liquid Dispersions, *Chemical Engineering Science*, 32(11), 1289–1297, 1977, https://doi.org/10.1016/0009-2509(77)85023-9

P. K. Das, A. A. Khan, & B. Pitchumani, Solution of the Batch Grinding Equation, *Powder Technology*, 85(3), 189–192, 1995, https://doi.org/10.1016/0032-5910(95)03057-9

P. B. Dubovskii, Analytic Solution of the Fragmentation Equation by Methods of the Theory of Adjoint Equations, *Differential Equations*, 36, 1529–1537, 2000, https://doi.org/10.1007/BF02757394

A Elhanbaly, On the Solution of the Integro–Differential Fragmentation Equation with Continuous Mass Loss, *Journal of Physics A: Mathematics and General*, 36(30), 8311–8323, 2003, https://doi.org/10.1088/0305-4470/36/30/308

M. H. Ernst & G. Szamel, Fragmentation Kinetics, *Journal of Physics A: Mathematical and General*, 26(22), 6085–6091, 1993, https://doi.org/10.1088/0305-4470/26/22/011

M. H. Ernst & I. Pagonabarraga, The Nonlinear Fragmentation Equation, *Journal of Physics A: Mathematical and Theoretical*, 40, F331–F337, 2007, https://doi.org/10.1088/1751-8113/40/17/F03

A. F. Filippov, On the Distribution of the Sizes of Particles which Undergo Splitting, *Theory of Probability & Its Applications*, 6(3), 275–294, 1961, https://doi.org/10.1137/1106036

D. W. Fuerstenau, A. De, & P. C. Kapur, Linear and Nonlinear Particle Breakage Processes in Comminution Systems, *International Journal of Mineral Processing*, 74(Supplement), S317–S327, 2004, https://doi.org/10.1016/j.minpro.2004.07.005

A. O. Gates, Kick vs. Rittinger: An Experimental Investigation in Rock Crushing, Performed at Purdue University, *Transactions of the AIME*, 52, 875–909, 1915, http://www.onemine.org/document/abstract.cfm?docid=14056

A. O. Gates, Crushing-Surface Diagrams and Rittinger's Law: A Discussion of Various Experimental Work which Tends to Disprove the Theories of Kick and Stadler, *Engineering and Mining Journal*, 118(8), 324–328, 1922

A. M. Gaudin, An Investigation of Crushing Phenomena, *Transactions of the AIME*, 73, 253–310, 1926, http://www.onemine.org/document/abstract.cfm?docid=164880

E. D. Gilliland, Crushing and Grinding, in W. H. Walker, W. K. Lewis, W. H. McAdams, & E. D. Gilliland, *Principles of Chemical Engineering*, 3rd edition, McGraw-Hill, 1937, https://archive.org/details/in.ernet.dli.2015.205897

S. L. Goren, Distribution of Lengths in the Breakage of Fibres of Linear Polymers, *The Canadian Journal of Chemical Engineering*, 46(3), 185–188, 1968, https://doi.org/10.1002/cjce.5450460309

M. Gorokhovski & M. Herrmann, Modeling Primary Atomization, *Annual Review of Fluid Mechanics*, 40, 343–366, 2008, https://doi.org/10.1146/annurev.fluid.40.111406.102200

M. A. Gorokhovski & V. L. Saveliev, Analyses of Kolmogorov's Model of Breakup and its Application into Lagrangian Computation of Liquid Sprays under Air-Blast Atomization, *Physics of Fluids*, 15(1), 184–192, 2003, https://doi.org/10.1063/1.1527914

M. A. Gorokhovski & V. L. Saveliev, Statistical Universalities in Fragmentation under Scaling Symmetry with a Constant Frequency of Fragmentation, *Journal of Physics D: Applied Physics*, 41, 085405, 2008, https://doi.org/10.1088/0022-3727/41/8/085405

A. Gupta & D. Yan, *Mineral Processing Design and Operations: An Introduction*, 2nd edition, Elsevier, 2016, https://doi.org/10.1016/C2014-0-01236-1

V. K. Gupta, Validation of an Energy–Size Relationship Obtained from a Similarity Solution to the Batch Grinding Equation, *Powder Technology*, 249, 396–402, 2013, https://doi.org/10.1016/j.powtec.2013.09.003

V. K. Gupta, An Appraisal of the Energy-Size Reduction Relationships for Mill Scale-Up Design, *Advanced Powder Technology*, 30(1), 73–84, 2019, https://doi.org/10.1016/j.apt.2018.10.008

B. Hellyer, The Fragmentation of the Asteroids, *Monthly Notices of the Royal Astronomical Society*, 148(4), 383–390, 1970, https://doi.org/10.1093/mnras/148.4.383

Ç. Hoşten, An Alternative Solution to the Grinding Equation in Cumulative Size Distribution Form, *Minerals Engineering*, 18(5), 489–493, 2005, https://doi.org/10.1016/j.mineng.2004.08.016

J. R. Huynen, F. McNolty, & E. Hansen, Component Distributions for Fluctuating Radar Targets, *IEEE Transactions on Aerospace and Electronic Systems*, AES-11(6), 1316–1332, 1975, https://doi.org/10.1109/TAES.1975.308189

P. C. Kapur, The Energy-Size Reduction Relationships in Comminution of Solids, *Chemical Engineering Science*, 26(1), 11–16, 1971, https://doi.org/10.1016/0009-2509(71)86076-1

P. C. Kapur, Self-Preserving Size Spectra of Comminuted Particles, *Chemical Engineering Science*, 27(2), 425–431, 1972, https://doi.org/10.1016/0009-2509(72)85079-6

P. C. Kapur & D. W. Fuerstenau, Energy-Size Reduction Laws Revisited, *International Journal of Mineral Processing*, 20(1–2), 45–57, 1987, https://doi.org/10.1016/0301-7516(87)90016-0

V. I. Khvorostyanov & J. A. Curry, Toward the Theory of Stochastic Condensation in Clouds. Part I: A General Kinetic Equation, *Journal of Atmospheric Sciences*, 56(23), 3985–3996, 1999, https://doi.org/10.1175/1520-0469(1999)056<3985:TTTOSC>2.0.CO;2

V. I. Khvorostyanov & J. A. Curry, *Thermodynamics, Kinetics, and Microphysics of Clouds*, Cambridge University Press, 2014, https://doi.org/10.1017/CBO9781139060004

F. Kick, *Das Gesetz der Proportionalen Widerstände und seine Anwendungen [Principle of Proportional Resistance and Its Application]*, Felix-Verlag, 1885

R. P. King, C. L. Schneider, & E. A. King, *Modeling and Simulation of Mineral Processing Systems*, 2nd edition, Society for Mining, Metallurgy, and Exploration (SME), 2012

A. N. Kolmogorov, O Logarifmicheski Normal'nom Zakone Raspredeleniya Razmerov Chastis pri Droblenii [On the Logarithmic Normal Distribution of Particle Sizes under Grinding], *Doklady Akademii Nauk SSSR*, 31(2), 99–101, 1941; see also *The Logarithmically Normal Law of Distribution of Dimensions of Particles when Broken into Small Parts*, NASA Technical Translation TT F-12, 287, 1969, https://archive.org/details/nasa_techdoc_19690019884

M. Kostoglou, S. Dovas, & A. J. Karabelas, On the Steady-State Size distribution of Dispersions in Breakage Processes, *Chemical Engineering Science*, 52(8), 1285–1299, 1997, https://doi.org/10.1016/S0009-2509(96)00488-5

C. B. Laney, *Transformation and Self-Similarity Properties of Gamma and Weibull Fragment Size Distributions*, Defense Threat Reduction Agency Technical Report DTRA-TR-16-06, December 2015, http://apps.dtic.mil/sti/citations/ADA624878

C. B. Laney, *On the Relationships Between Liquid and Solid Fragmentation*, Defense Threat Reduction Agency Technical Report DTRA-TR-16-19, January 2016, http://apps.dtic.mil/sti/citations/AD1004358

Y. Liao & D. Lucas, A Literature Review of Theoretical Models for Drop and Bubble Breakup in Turbulent Dispersions, *Chemical Engineering Science*, 64(15), 3389–3406, 2009, https://doi.org/10.1016/j.ces.2009.04.026

Y. Liao & D. Lucas, A Literature Review on Mechanisms and Models for the Coalescence Process of Fluid Particles, *Chemical Engineering Science*, 65(10), 2851–2864, 2010, https://doi.org/10.1016/j.ces.2010.02.020

J. Litster, *Design and Processing of Particulate Products*, Cambridge University Press, 2016, https://doi.org/10.1017/CBO9781139017558

A. J. Lynch & C. A. Rowland, *The History of Grinding*, Society for Mining, Metallurgy, and Exploration, 2005

S. Maaß & M. Kraume, Determination of Breakage Rates Using Single Drop Experiments, *Chemical Engineering Science*, 70(5), 146–164, 2012, https://doi.org/10.1016/j.ces.2011.08.027

D. B. Mazzinghy, R. Galéry, C. L. Schneider, & V. K. Alves, Scale Up and Simulation of Vertimill™ Pilot Test Operated with Copper Ore, *Journal of Materials Research and Technology*, 3(1), 86–89, 2014, https://doi.org/10.1016/j.jmrt.2013.11.001

D. B. Mazzinghy, C. L. Schneider, V. K. Alves, & R. Galéry, Vertical Agitated Media Mill Scale-Up and Simulation, *Minerals Engineering*, 73, 69–76, 2015, https://doi.org/10.1016/j.mineng.2014.11.003

D. B. Mazzinghy, J. Lichter, C. L. Schneider, R. Galéry, & J. F. C. Russo, Vertical Stirred Mill Scale-Up and Simulation: Model Validation by Industrial Samplings Results, *Minerals Engineering*, 103–104, 127–133, 2017, https://doi.org/10.1016/j.mineng.2016.11.018

B. J. McCoy & G. Madras, Evolution to Similarity Solutions for Fragmentation and Aggregation, *Journal of Colloid and Interface Science*, 201(2), 200–209, 1998, https://doi.org/10.1006/jcis.1998.5434

B. J. McCoy & G. Madras, Analytical Solution for a Population Balance Equation with Aggregation and Fragmentation, *Chemical Engineering Science*, 58(13), 3049–3051, 2003, https://doi.org/10.1016/S0009-2509(03)00159-3

E. D. McGrady & Robert M. Ziff, "Shattering" Transition in Fragmentation, *Physical Review Letters*, 58(9), 892–895, 1987, https://doi.org/10.1103/PhysRevLett.58.892

R. E. McIvor, J. Finch, & S. Makni, The Truth About Population Balance Modelling, *CIM Journal*, 14(4), 223–238, 2023, https://doi.org/10.1080/19236026.2023.2204708

F. McNolty, Some Probability Density Functions and Their Characteristic Functions, *Mathematics of Computation*, 27(123), 495–504, 1973, https://doi.org/10.2307/2005656

F. McNolty & E. Hansen, Some Aspects of Swerling Models for Fluctuating Radar Cross Section, *IEEE Transactions on Aerospace and Electronic Systems*, AES-10(2), 281–285, 1974, https://doi.org/10.1109/TAES.1974.307928

F. McNolty & E. Hansen, Probability Densities and Characteristic Functions for Fluctuating Targets, *IEEE Transactions on Aerospace and Electronic Systems*, AES-15(4), 474–480, 1979, https://doi.org/10.1109/TAES.1979.308732

Z. A. Melzak, A Scalar Transport Equation, *Transactions of the American Mathematical Society*, 85(2), 547–560, 1957, https://doi.org/10.1090/S0002-9947-1957-0087880-6

R. Meyer, K. E. Almin, & B. Steenberg, Length Reduction of Fibres Subject to Breakage, *British Journal of Applied Physics*, 17(3), 409–416, 1966, https://doi.org/10.1088/0508-3443/17/3/314

S. Morrell, Testing and Calculation for Comminution Machines, in R. C. Dunne, S. Kawatra, & C. A. Young, editors, *SME Mineral Processing & Extractive Metallurgy Handbook*, Society for Mining, Metallurgy & Exploration (SME), 2019

J.-M. Nicolas, Introduction aux Statistiques de Deuxième Espèce: Applications des Logs-Moments et des Logs-Cumulants à l'analyse des lois d'images Radar, *Traitement du Signal*, 19(3), 139–167, 2002; see also J.-M. Nicolas & S. N. Anfinsen, *Introduction to Second Kind Statistics: Application of Log-Moments and Log-Cumulants to the Analysis of Radar Image Distributions*, 2010, http://citeseerx.ist.psu.edu/viewdoc/summary?doi=10.1.1.713.1095

N. Ouchiyama, S. L. Rough, & J. Bridgwater, A Population Balance Approach to Describing Bulk Attrition, *Chemical Engineering Science*, 60(50), 1429–1440, 2005, https://doi.org/10.1016/j.ces.2004.08.037

F. W. J. Olver, A. B. Olde Daalhuis, D. W. Lozier, B. I. Schneider, R. F. Boisvert, C. W. Clark, et al., editors, *NIST Digital Library of Mathematical Functions, Release 1.1.12*, December 15, 2023, http://dlmf.nist.gov

J. D. Pandya & L. A. Spielman, Floc Breakage in Agitated Suspensions: Theory and Data Processing Strategy, *Journal of Colloid and Interface Science*, 90(2), 517–531, 1982, https://doi.org/10.1016/0021-9797(82)90317-4

D. P. Patil & J. R. G. Andrews, An Analytical Solution to Continuous Population Balance Model Describing Floc Coalescence and Breakage—A Special Case, *Chemical Engineering Science*, 53(3), 599–601, 1998, https://doi.org/10.1016/S0009-2509(97)00314-X

L. E. Patruno, C. A. Dorao, H. F. Svendsen, & H. A. Jakobsen, Analysis of Breakage Kernels for Population Balance Modeling, *Chemical Engineering Science*, 64(3), 501–508, 2009, https://doi.org/10.1016/j.ces.2008.09.029

T. W. Peterson, Similarity Solutions for the Population Balance Equation Describing Particle Fragmentation, *Aerosol Science and Technology*, 5(1), 93–101, 1986, https://doi.org/10.1080/02786828608959079

E. Petrakis, E. Stamboliadis, & K. Komnitsas, Evaluation of the Relationship between Energy Input and Particle Size Distribution in Comminution with the Use of Piecewise Regression Analysis, *Particulate Science and Technology*, 35(4), 1168894, 2016, https://doi.org/10.1080/02726351.2016.1168894

D. Ramkrishna, Drop-Breakage in Agitated Liquid-Liquid Suspensions, *Chemical Engineering Science*, 29(2), 987–992, 1974, https://doi.org/10.1016/0009-2509(74)80090-4

J. J. Reid, A Solution to the Batch Grinding Equation, *Chemical Engineering Science*, 20(11), 953–963, 1965, https://doi.org/10.1016/0009-2509(65)80093-8

P. R. V. Rittinger, *Lehrbuch der Aufbereitungskunde [Textbook of Unit Operations]*, von Ernst & Korn, 1867

P. Rosin & E. Rammler, Feinheit und Struktur des Kohlenstaubes [Fineness and Structure of Coal Dust], *Zeitschrift des Vereins Deutscher Ingenieure*, 71, 1–7, 1927a

P. Rosin & E. Rammler, Feinheit und Struktur des Kohlenstaubes unter dem Einfluß von Mühlen und Kohlenart [Fineness and Structure of Coal Dust under the Influence of Mill and Coal Type], *Zement*, 16(35), 820–823, 16(36), 840–844, 16(38), 897–901, 1927b

P. Rosin & E. Rammler, The Laws Governing the Fineness of Powdered Coal, *Journal of the Institute of Fuel*, 7, 29–36, 1933

P. Rosin, E. Rammler, & K. Sperling, Korngrößenprobleme des Kohlenstaubes und ihre Bedeutung für die Vermahlung [Particle Size Problems of Coal Dust and their Importance for Grinding], *Bericht C 52 des Reichskohlenrates* [Report C52 of the National Coal Council], VDI-Verlag, 1933

P. Rosin & E. Rammler, Die Kornzusammensetzung des Mahlgutes im Lichte der Wahrscheinlichkeitslehre [Grain Composition of Ground Material in Light of Probability Theory], *Kolloid-Zeitschrift*, 67(1), 16–26, 1934, https://doi.org/10.1007/BF01439773

V. L. Saveliev & M. A. Gorokhovski, Renormalization of the Fragmentation Equation: Exact Self-Similar Solutions and Turbulent Cascades, *Physical Review E*, 86(6), 061112, 2012, https://doi.org/10.1103/PhysRevE.86.061112

E. V. Schuhmann, Principles of Comminution, Size Distribution and Surface Calculations, *AIME Technical Publication*, 1189, 1–11, 1940, http://www.onemine.org/document/abstract.cfm?docid=9197

K. Sedlatschek & V. L. Bass, Contribution to the Theory of Ball Milling, *Powder Metallurgy Bulletin*, 6, 148–153, 1953

P. T. Spicer & S. E. Pratsinis, Coagulation and Fragmentation: Universal Steady-State Particle-Size Distribution, *AIChE Journal*, 42(6), 1612–1620, 1996, https://doi.org/10.1002/aic.690420612

J. L. Spouge, An Existence Theorem for the Discrete Coagulation-Fragmentation Equations, *Mathematical Proceedings of the Cambridge Philosophical Society*, 96(2), 351–357, 1984, https://doi.org/10.1017/S0305004100062253

J. L. Spouge, An Existence Theorem for the Discrete Coagulation-Fragmentation Equations. II. Inclusion of Source and Efflux Terms, *Mathematical Proceedings of the Cambridge Philosophical Society*, 98(1), 183–185, 1985, https://doi.org/10.1017/S0305004100063362

L. M. Tavares & R. M. De Carvalho, Modeling Breakage Rates of Coarse Particles in Ball Mills, *Minerals Engineering*, 22(7), 650–659, 2009, https://doi.org/10.1016/j.mineng.2009.03.015

L. M. Tavares, A Review of Advanced Ball Mill Modelling, *KONA Powder and Particle Journal*, 34, 106–124, 2017, https://doi.org/10.14356/kona.2017015

K. J. Valentas, O. Bilous, & N. R. Amundson, Analysis of Breakage in Dispersed Phase Systems, *Industrial & Engineering Chemistry Fundamentals*, 5(2), 271–279, 1966, https://doi.org/10.1021/i160018a019

R. J. Weimer & H. C. Rogers, Dynamic Fracture Phenomena in High-Strength Steels, *Journal of Applied Physics*, 50(12), 8025–8030, 1979, https://doi.org/10.1063/1.325989

M. M. R. Williams, An Exact Solution of the Fragmentation Equation, *Aerosol Science and Technology*, 12(3), 538–546, 1990, https://doi.org/10.1080/02786829008959368

C. Yuan, C. Wu, X. Fang, N. Liao, J. Tong, & Y. Li, A New Approach to the Grinding Kinetics of Magnetite Ore Based on the Population Balance Model, *Minerals*, 13(3), 424, 2023, https://doi.org/10.3390/min13030424

R. M. Ziff & E. D. McGrady, The Kinetics of Cluster Fragmentation and Depolymerisation, *Journal of Physics A: Mathematical and General*, 18(15), 3027–3037, 1985, https://doi.org/10.1088/0305-4470/18/15/026

R. M. Ziff, An Explicit Solution to a Discrete Fragmentation Equation, *Journal of Physics A: Mathematical and General*, 25(9), 2569–2576, 1992, https://doi.org/10.1088/0305-4470/25/9/027

14

Fragmentation Cascades Part 3: Mean-Field Theory for Collisional Fragmentation

This is the last of three chapters on mean-field theory for fragmentation cascades. This chapter discusses *collisional fragmentation*, a.k.a., *nonlinear fragmentation*. As seen in Table 14.1, in this problem, two particles collide with enough mass, momentum, and energy that one or both particles fragment to produce somewhere between 2 and ∞ children, over and over again, in a cascade fashion. Example 14.1 describes some common applications in which collisional fragmentation occurs.

Table 14.1 Three common types of fragmentation cascades.

Type	Description	Typical Applications
Binary Fragmentation	Randomly selected particles fragment into exactly two pieces with independently random sizes.	Thermal and oxidative degradation of polymers
Random Fragmentation	Randomly selected particles fragment into two or more pieces with independently random sizes.	Grinding of minerals, ores, and chemicals
Collisional Fragmentation	Random binary collisions cause one or both of the participating particles to fragment into two or more pieces with independently random sizes.	Evolution of objects in the asteroid belt and the Kuiper belt

Example 14.1 Collisional evolution in astrophysical systems

Many astrophysical systems are believed to have first formed through aggregation and then evolved through collisional fragmentation. For example, the asteroid belt is a region of the solar system between the orbits of Mars and Jupiter containing numerous objects, mostly composed of rock or metal, ranging in size from Ceres, with a diameter of about 950 km, down to fine dust. According to O'Brien & Greenberg (2005), "the size distribution of the main belt of asteroids is determined primarily by collisional processes. Large asteroids break up and form smaller asteroids in a collisional cascade." As a more complete explanation, according to Johansen et al. (2015), asteroids originally grew through aggregation of millimeter-sized spherules, known as *chondrules*, followed by fragmentation. More specifically, they say that "the size distribution of smaller asteroids in today's belt was likely filled later with fragments from collisional grinding. The steep size distribution of larger asteroids ends at around 200 km as even larger asteroids would need to accrete chondrules larger than a millimeter in size. At this point, the largest asteroids enter a more democratic growth phase wherein the accretional cross sections are reduced ... The agreement with the observed size

Particle Dynamics with Aggregation and Fragmentation. Culbert B. Laney, Oxford University Press. © Culbert B. Laney (2025).
DOI: 10.1093/9780198920687.003.0014

distribution of asteroids is excellent in the nominal [collisional fragmentation] model, except in the [large size] range around Ceres' mass where it predicts too few objects by a factor of about 2 to 3."

For another example, the Kuiper belt is a region of the solar system beyond the orbit of Neptune containing numerous objects, mostly composed of ices, ranging in size from Pluto, with a diameter of about 2,300 km, down to fine dust. According to Schlichting et al. (2013) "the observed Kuiper Belt size distribution is well matched by coagulation models ... and subsequent collisional evolution. The observed size distribution above ~30 km is primordial, i.e., it has not been modified by collisional evolution ... [However,] the size distribution below ~30 km has been modified by collisions and ... [the observed size distribution] is well matched by collisional evolution models."

Collisional fragmentation involves two main elements: (a) random binary collisions and (b) fragmentation of colliding particles into two or more pieces with independently random sizes. Notice that element (a) also occurs in aggregation while element (b) also occurs in random fragmentation. This combination makes collisional fragmentation one of the hardest problems treated in this book. As a result, this chapter can offer only a few analytic results. As it turns out, those few analytic results are similar to ones already seen in earlier chapters for aggregation, binary fragmentation, and/or random fragmentation. Thus, this chapter essentially ends up reviewing mean-field theory for aggregating and fragmenting particle systems as a whole.

14.1 Governing Equations

14.1.1 Introduction

If a particle system is spatially homogeneous and isotropic, the position field does not affect the statistics of the other random fields. In this case, suppose the velocity, rotation, shape, etc. fields determine an average collision rate, known as the *collision kernel*, which captures the average effects of randomness and chaos. In this approach, mean-field theory represents complex chaotic particle systems by relatively simple deterministic governing equations, where the independent variables are the particle size and time, and the dependent variable is the particle concentration. This chapter assumes that there are an infinite number of particles homogeneously and isotropically distributed in an infinite spatial volume with a finite concentration C. In other words, $0 < C = N/\mathcal{V} < \infty$ in the limit $N \to \infty$ and $\mathcal{V} \to \infty$. This is sometimes called the *thermodynamic limit.* Because there are an infinite number of particles, collisional fragmentation never ends but continues forever. In addition, by assuming spatial homogeneity and isotropy, this chapter implicitly assumes that the children produced in each fragmentation event are instantaneously and evenly distributed over the entire spatial domain. Obviously, such assumptions are not realistic, but they may be approximately true in many cases.

14.1.2 Discrete Governing Equations

For discrete particle systems, let the collision kernel K_{ij} be defined as follows:

> $K_{ij}C_i(t)C_j(t)$ is the average rate per unit volume at which particles of size i and j collide to form particles smaller than size i and j.

In addition, let the child size distribution m_{ijk} be defined as follows:

> m_{ijk} is the average number of children of size i produced by fragmenting collisions between parents of sizes j and k.

In essence, m_{ijk} is a combination of the children produced by parents of sizes j and k. If fragmentation takes place immediately upon collision, the children must be smaller than their parents. This can be written mathematically as follows:

$$m_{ijk} = 0 \text{ unless } i < j \text{ or } i < k.$$

In this case, the mean-field governing equation can be written as follows:

$$\frac{dc_i(t)}{dt} = -c_i(t)\sum_{j=1}^{\infty} K_{ij}c_j(t) + \sum_{k=1}^{\infty}\sum_{j=i+1}^{\infty} m_{ijk}\, K_{jk}c_j(t)\, c_k(t)\,. \tag{14.1}$$

Equation (14.1) may be interpreted as follows:

$$\begin{pmatrix} \textit{rate of change} \\ \textit{in particles of} \\ \textit{size } i \end{pmatrix} = -\begin{pmatrix} \textit{rate at which particles of} \\ \textit{size } i \textit{ break to form particles} \\ \textit{smaller than size } i \end{pmatrix} + \begin{pmatrix} \textit{rate at which particles} \\ \textit{larger than size } i \textit{ break to} \\ \textit{form particles of size } i \end{pmatrix}.$$

Instead of breaking on impact, suppose that the parents aggregate and then, after a brief delay, fragment. Da Costa (2015) says "we can consider this process as made of two consecutive steps: a coagulation step forming an extremely unstable cluster that ... instantaneously breaks into two or more smaller aggregates." Then $m_{ijk} = m_{ikj} = \tilde{m}_{i,j+k}$ and $\tilde{m}_{i,j+k} = 0$ unless $i < j + k$. In this case, Srivastava (1982) obtained the following mean-field governing equation:

$$\frac{dc_i(t)}{dt} = -c_i(t)\sum_{j=1}^{\infty} K_{ij}c_j(t) + \frac{1}{2}\sum_{j=i+1}^{\infty}\sum_{k=1}^{j-1} m_{i,j-k,k}\, K_{j-k,k}c_{j-k}(t)\, c_k(t) \tag{14.2}$$

where the factor of 1/2 prevents double counting the same event; see also Laurençot & Wrzosek (2001), Vigil et al. (2006) and da Costa (2015). To simplify Equations (14.1) and (14.2), researchers have suggested alternatives wherein m depends on only one rather both parents. For example, Krapivsky & Ben-Naim (2003) suggest modified governing equations for cases where only one parent in each binary collision breaks up.

14.1.3 Continuous Governing Equations

For continuous particle systems, let the collision kernel $K(x, y)$ be defined as follows:

> $K(x,y)c(x,y)c(y,t)dy$ is the average rate per unit mass and volume at which parents of masses x and $y + dy$ collide to form particles smaller than mass x and y.

In addition, let $m(x, y, z)$be defined as follows:

> $m(x,y,z)K(y,z)c(z,t)dy\,dz$ is the average number of children of size x produced by fragmenting collisions of parents of sizes $y \pm dy$ and $z \pm dz$.

In essence, $m\,(x, y, z)$ is a combination of the children produced by parents of size x and y. If fragmentation takes place immediately upon collision, the children must be smaller than their parents. This can be written mathematically as follows:

$$m(x,y,z) = 0 \text{ unless } x < y \text{ or } x < z.$$

In this case, Dohnanyi (1967, 1969, 1970) and Dorschner (1970, 1971, 1974) obtained the following mean-field governing equation:

$$\frac{\partial c(x,t)}{\partial t} = -c(x,t)\int_0^\infty K(x,y)c(y,t)dy + \int_0^\infty\int_x^\infty m(x,y,z)K(y,z)c(y,t)c(z,t)dydz. \tag{14.3}$$

Equation (14.3) may be interpreted as follows:

$$\begin{pmatrix}\textit{rate of change}\\ \textit{in particles of}\\ \textit{size } x\end{pmatrix} = -\begin{pmatrix}\textit{rate at which particles of}\\ \textit{size } x \textit{ break to form particles}\\ \textit{smaller than size } x\end{pmatrix} + \begin{pmatrix}\textit{rate at which particles}\\ \textit{larger than size } x \textit{ break to}\\ \textit{form particles of size } x\end{pmatrix}.$$

Instead of breaking on impact, suppose that the parents aggregate and then, after a brief delay, fragment. Wilkins (1982) says the parents "stick together upon colliding; then ... they fragment after their mutual capture, perhaps breaking into many smaller pieces." Then $m(x,y,z) = m(x,z,y) = \tilde{m}(x,y+z)$ and $\tilde{m}(x,y+z) = 0$ unless $x < y + z$. In this case, Safronov (1969) obtained the following mean-field governing equation:

$$\frac{\partial c(x,t)}{\partial t} = -c(x,t)\int_0^\infty K(x,y)c(y,t)dy$$
$$+\frac{1}{2}\int_x^\infty\int_0^z m(x,z-y,y)K(z-y,y)c(z-y,t)c(y,t)dydz \tag{14.4}$$

where the factor of 1/2 prevents double counting; see also, e.g., Barik & Giri (2020).

To simplify Equations (14.3) and (14.4), researchers have suggested alternatives wherein m depends on only one rather than both parents. For example, Cheng & Redner (1988, 1990) and Ernst & Pagonabarraga (2007) suggested modified governing equations for cases where only one parent in each binary collision breaks up; see Example 14.2. As another example, suppose that m is homogeneous with degree -1. In other words:

$$m(ax, ay, az) = a^{-1} m(x, y, z). \tag{14.5a}$$

for any real positive a as seen in, e.g., Cheng & Redner (1988, 1990). Equivalently:

$$m(x, y, z) = y^{-1} \tilde{m}(x/y, z/y) \tag{14.5b}$$

as seen in, e.g., Tanaka et al. (1996). Example 14.3 gives a specific example of such an $\tilde{m}$. In addition to homogeneous m, suppose that collisional fragmentation is *local*, i.e., dominated by collisions between parents of similar sizes. If $y \approx z$, then Equation (14.5b) becomes:

$$m(x, y, z) = y^{-1} \tilde{m}(x/y, z/y) \approx y^{-1} \tilde{m}(x/y, 1). \tag{14.6}$$

Example 14.2 Modified governing equations for cases where only one particle in each binary collision breaks

As one way to simplify the governing equations for collisional fragmentation, suppose that only one particle in each binary collision experiences fragmentation. Krapvisky & Ben-Naim (2003) suggest the following three possibilities: (a) a randomly chosen particle in each binary collision experiences fragmentation; (b) the larger particle in each binary collision experiences fragmentation; (c) the smaller particle in each binary collision experiences fragmentation; see also, e.g., Cheng & Redner (1988, 1990) and Ernst & Pagonabarraga (2007).

For example, suppose that possibility (b) applies, i.e., only the larger particle in each binary collision breaks. According to Ernst & Pagonabarraga (2007), this means that $m(x, y, z)$ is replaced by $\tilde{m}(x, \max(y, z))$ and $K(x, y)$ is replaced by $K(x, y)\theta(x-y)$ where θ is the unit step function. Then Equation (14.3) becomes:

$$\frac{\partial c(x, t)}{\partial t} = -c(x, t) \int_0^x K(x, y)c(y, t)dy + \int_x^\infty \int_0^y \tilde{m}(x, y)\, K(y, z)c(y, t)c(z, t)dz\, dy. \tag{14.7}$$

As the main advantage, Equation (14.7) is substantially simpler than Equation (14.3), because $\tilde{m}$ is a function of only one rather than two parent sizes. As the main disadvantage, it is hard to imagine a real-world scenario in which only the larger particle in each binary collision would break. Thus, it is essentially impossible to validate Equation (14.7) using experimental results. Similar comments apply for possibilities (a) and (c).

Example 14.3 Children with "generalized product type" size distributions

Kostoglou & Karabelas (2006) suggested the following "generalized product type" size distribution:

$$m(x, y, z) = \tilde{m}(x/y, z/y)/y \tag{14.8}$$

where:

$$\tilde{m}(r_1, r_2) = N\frac{(r_1 r_2)^{\eta-1}(1 - r_1 r_2)^{\eta(N-1)-1}}{B[\eta, \eta(N-1)]}. \tag{14.9}$$

In this expression, $\eta > 0$ is a free parameter, $N = N(\eta) \geq 2$ is the "number of fragments per breakage event," and B is an incomplete beta function defined as follows:

$$B(x, y) = \int_0^{\infty} z^{x-1}(1 - z)^{y-1}\,dz.$$

Notice that if $\eta = 1/(N{-}1)$, then $m(x, y, z) = (\eta + 1)x^{\eta-1}/y^{\eta}$. In other words, the "generalized product type" discussed in this example becomes the same as the power law size distribution discussed earlier in Example 13.4.

14.2 Shattering

This section concerns *shattering*, i.e., the spontaneous formation of sinks of small particles. In general, shattering in collisional fragmentation is analogous to gelation in aggregation, as discussed earlier in Section 10.3. As one similarity, shattering in collisional fragmentation and gelation in aggregation both tend to occur after some finite amount of time has passed, depending partly on the collision kernel and the initial conditions. As one difference, shattering in collisional fragmentation creates a sink of small particles, while gelation in aggregation creates a sink of large particles. As another difference, shattering in collisional fragmentation occurs when $\lambda < 1$, while gelation in aggregation occurs when $\lambda > 1$. In any event, like anything that violates conservation of mass, gelation and shattering both allow multiple possible solutions, depending on how hard and strong the spontaneously formed sinks are presumed to be.

14.3 Full Mean-Field Solutions

In general, this chapter assumes that $K(x, y)$ for collisional fragmentation is the same as $K(x, y)$ for aggregation, as discussed in Section 9.3, except for an application-specific constant factor K_0. In particular, this chapter assumes that K is homogeneous with degree

Table 14.2 While mean-field theory produces three closed-form time-dependent solutions for aggregation, mean-field theory produces only two closed-formed time-dependent solutions for collisional fragmentation.

Kernel Name	Equation	Degree of Homogeneity	Aggregation	Collisional Fragmentation
Constant	$K(x, y) = K_0$	$\lambda = 0$	Yes	Yes
Sum	$K(x, y) = K_0(x + y)$	$\lambda = 1$	Yes	No
Product	$K(x, y) = K_0xy$	$\lambda = 2$	Yes	Yes

$-1 \le \lambda \le 3$. As seen in Chapter 10, for aggregation, mean-field theory produces closed-form time-dependent solutions for only three specific collision kernels, namely, the constant, sum, and product kernels. However, collisional fragmentation is a more difficult problem than aggregation. Thus, for collisional fragmentation, mean-field theory produces closed-form time-dependent solutions for only two specific collision kernels, namely, the constant and product kernels. Table 14.2 summarizes the available mean-field theory results for aggregation and collisional fragmentation.

As seen in Chapter 10, obtaining a full closed-form time-dependent solution for aggregation requires making two assumptions: (i.) monodisperse initial conditions i.e., the particles all start off with the same size and (ii.) free-field boundary conditions, i.e., no initial sources or sinks. However, collisional fragmentation is a more difficult problem than aggregation. Thus, obtaining a full closed-form time-dependent solution for collisional fragmentation requires making four assumptions: (i.) monodisperse initial conditions; (ii.) free-field boundary conditions, i.e., no initial sources or sinks; (iii.) conservation of mass, i.e., no new sources or sinks develop over time; (iv.) locality, i.e., results are dominated by collisions between particles of similar sizes. Example 14.4 describes the two known closed-form time-dependent solutions for collisional fragmentation. For aggregation, the solutions develop naturally, and items (iii.) and (iv.) are assessed afterwards, with varying results. By contrast, for collisional fragmentation, items (iii.) and (iv.) are assumed in advance.

Example 14.4 Time-dependent closed-form solutions for sum and product kernels with monodisperse initial conditions and free-field boundary conditions

First, consider the product kernel $K(x,y) = K_0xy$ with $\lambda = 2$. If conservation of mass applies then, per Equation (13.39), $C(t)M_{\text{avg}}(t) = C_0M_0 = \text{const.}$ In addition, if locality applies then, per Equation (14.6), $m(x,y,z) \approx y^{-1}\tilde{m}(x/y, 1)$. Then Equation (14.3) becomes as follows:

$$\frac{\partial c(x,t)}{\partial t} = -K_0C_0M_0xc(x,t) + K_0C_0M_0 \int_x^\infty \tilde{m}\left(\frac{x}{y}, 1\right) c(y,t)dy. \tag{14.10}$$

Notice that Equation (14.10) agrees with Equation (13.10) if $J(x) = K_0C_0M_0x$ and $m(x,y) = \tilde{m}(x/y,1)/y$. Thus, the governing equation for collisional fragmentation has been converted to the governing equation for random fragmentation. Drawing on the results from the last chapter, suppose that $J(x) = K_0C_0M_0x$ and $\tilde{m}(x/y,1)/y = (\eta+1)(x/y)^{\eta-1}$ where $\eta > 0$. Then for monodisperse initial conditions and free-field boundary conditions, Kostoglou & Karabelas (2000) obtained the following time-dependent solution to Equation (14.10):

$$\begin{aligned} c(M,t) &= (C_0/M_0)\exp\left(-K_0C_0M_0^2t\right)\delta\left(M/M_0-1\right) \\ &+(\eta+1)K_0C_0^2M_0t\exp\left(-K_0C_0M_0^2t\right)(M/M_0)^{\eta-1}{}_1F_1\left[\eta+2,2,K_0C_0M_0\left(M_0-M\right)t\right]. \end{aligned} \tag{14.11}$$

Notice that Equation (14.11) agrees with Equation (13.58) after replacing J_0 by $K_0C_0M_0$ and after replacing κ by 1. The other results given in Example 13.9 apply again here including:

$$c(M,t) \cong \frac{C_0M_0}{\Gamma(\eta+1)}(K_0C_0M_0t)^{\eta+1}M^{\eta-1}\exp\left(-K_0C_0M_0Mt\right) \tag{14.12}$$

for sufficiently large times and small particles. This is a gamma size distribution with $n/m = \eta$. Now, consider the constant kernel $K(x,y) = K_0$ with $\lambda = 0$. Under the same assumptions as before, Kostoglou & Karabelas (2000) obtained a time-dependent solution that agrees with Equation (13.53), after replacing J_0 by K_0 and after replacing t by:

$$\tau(t) = \int_0^t C(t')\,dt' = \int_0^t \frac{C_0}{1-(m_0-1)K_0t'}dt' = \frac{C_0}{(1-m_0)K_0}\ln\left[1-(m_0-1)K_0t\right]$$

where $m_0 = \int_0^\infty \tilde{m}(x',1)dx'$ Notice that this solution is only valid when $1-(m_0-1)K_0t > 0$. In other words, this solution is only valid when:

$$t < t_s = \frac{1}{(m_0-1)K_0}. \tag{14.13}$$

According to Kostoglou & Karabelas (2000), "this behavior implies a shattering transition at $t = t_s$. The system no longer conserves the mass because there is a loss of [a] new phase of infinitesimally small particles." For more information about these two closed-form mean-field solutions see, e.g., Lombart et al. (2022).

14.4 Late-Time Mean-Field Solutions

This section concerns late-time mean-field solutions. As seen in Chapter 10, aggregation obtains approximate closed-form late-time solutions in two special cases, namely, *non-local* and *local* cases. Similarly, collisional fragmentation also obtains approximate closed-form late-time solutions in non-local and local cases. These two cases may be described as follows.

1. **In non-local cases, collisional fragmentation is dominated by collisions between particles with substantially different sizes.** This tends to occur when sources add large particles to the system, thus increasing the range of particle sizes. In fact, with or without sources, collisional fragmentation naturally tends to increase the range of particle sizes as time increases. As a result, collisions between unequal-sized particles tend to become increasingly common, just through random chance, unless sinks or other factors are present.
2. **In local cases, collisional fragmentation is dominated by collisions between particles with the same or similar sizes.** This tends to occur when sinks remove small particles from the system, thus reducing the range of particle sizes. Alternatively, this tends to occur when:
   - Collisions between particles of different sizes obtain poor reaction efficiency, e.g., smaller particles tend to glance harmlessly off of larger particles.
   - Small particles entrained in a fluid flow follow streamlines around large particles; only particles large enough to avoid entrainment in the fluid medium can collide.
   - Particles frequently fragment due to impacts with flat solid surfaces. As seen in Figure 14.1, such collisions resemble collisions between two identical particles.

   While mean-field theory can account for sinks, it cannot easily account for these other phenomena. Thus, this section considers only locality due to sinks of small particles, including but not limited to shattering-type sinks that form spontaneously when $\lambda < 1$.

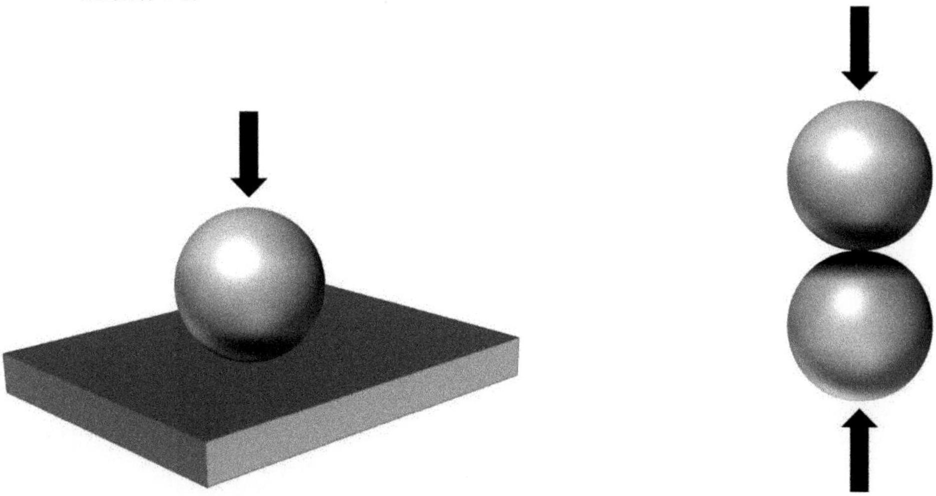

Figure 14.1 Fragmentation due to a collision between a sphere and a flat solid surface (left) resembles a collision between two identical spheres (right).

First, consider non-local cases. As discussed in Section 11.3, for non-local cases, collision kernels tend to obtain approximate forms such as the following:

$$K(x, y) = K_0(x^{\nu}y^{\mu} + x^{\mu}y^{\nu})$$

or

$$K(x,y) = K_0 x^{\nu} y^{\mu}$$

where $\lambda = \mu + \nu$. As seen in Example 14.5, these approximate forms lead to a general result for the late-time average particle size. In addition, as seen in Examples 14.6 and 14.7, these approximate forms bound the possible range of late-time size distributions. More specifically, in non-local cases, collisional fragmentation tends to converge towards Rosin–Rammler size distributions where $n/m \leq \eta$ and $l/m \leq \lambda$. For comparison, the last chapter obtained similar results for random fragmentation, but with exact exponents instead of upper bounds.

Example 14.5 Late-time average size for various collision kernels with free-field boundary conditions

Consider collision kernels such as $K(x,y) = K_0(x^{\nu}y^{\mu} + x^{\mu}y^{\nu})$ and $K(x,y) = K_0 x^{\nu} y^{\mu}$ where $\lambda = \mu + \nu$. Assume that conservation of mass applies, i.e., assume that $C(t)M_{\text{avg}}(t) = C_0 M_0 =$ const. In addition, assume that one particle in each binary collision experiences fragmentation, while the other particle in each binary collision remains intact. In such cases, Ernst & Pagonabarraga (2007, eq. 6) obtained the following equation:

$$\frac{dM_{\text{avg}}(t)}{dt} = -a_0 K_0 C_0 M_0 M_{\text{avg}}^{\lambda}(t) \tag{14.14}$$

where a_0 is a unitless positive constant; see also Cheng & Redner (1990, eq. 41). If $\lambda \neq 1$ Equation (14.14) has the following solution:

$$M_{\text{avg}}(t) = M_0\left[(\lambda - 1)a_0 K_0 C_0 M_0^{\lambda} t + 1\right]^{\frac{1}{1-\lambda}} \tag{14.15}$$

which is easy to verify through back substitution. Then:

$$M_{\text{avg}}(t) \cong \begin{cases} a_1 t^{1/(1-\lambda)} & \lambda > 1,\ t \to \infty \\ M_0 \exp(-a_0 K_0 C_0 M_0 t) & \lambda = 1 \\ a_2 (t_s - t)^{1/(1-\lambda)} & \lambda < 1,\ t \to t_s \end{cases} \tag{14.16}$$

where a_1 and a_2 are constants and where:

$$t_s = \frac{1}{(1-\lambda)\, a_0 K_0 C_0 M_0^{\lambda}} \tag{14.17}$$

is the *shattering time*. Notice that Equation (14.16) is the same as Equation (11.65), except that t_s replaces t_g, and the inequalities on λ are reversed.

Example 14.6 Late-time size distributions for generalized sum kernels with free-field boundary conditions

For non-local cases, collision kernels tend to obtain approximate forms such as $K(x, y) \cong K_0(x^\nu y^\mu + x^\mu y^\nu)$ where $\lambda = \mu + \nu$. At one extreme, suppose that μ or ν is equal to λ. In other words, suppose that $\mu = \lambda$ and $\nu = 0$ or, equivalently, $\mu = 0$ and $\nu = \lambda$. Then $K(x, y)$ becomes the *generalized sum kernel*:

$$K(x,y) = K_0\left(x^\lambda + y^\lambda\right). \tag{14.18}$$

First, consider the case $1 < \lambda \leq 3$. In this case, conservation of mass applies, i.e., shattering does not occur. Suppose that the children obtain a "generalized product type" size distribution with an exponent $\eta > 0$, as discussed earlier in Example 14.3. According to Kostoglou & Karabelas (2006), Equation (14.3) obtains the following particle size distribution:

$$c(M,t) \cong \text{const.} \frac{1}{[M_{\text{avg}}(t)]^2}\left[\frac{M}{M_{\text{avg}}(t)}\right]^{\frac{\eta}{2}-\phi-1} \exp\left\{-\text{const.}\left[\frac{M}{M_{\text{avg}}(t)}\right]^\lambda\right\} \tag{14.19}$$

where:

$$M_{\text{avg}}(t) = \text{const.}\, t^{1/(1-\lambda)}$$

for sufficiently large times and small particles. Notice that Equation (14.19) is a Rosin–Rammler size distribution with $n/m = \eta/2 - \phi$ and $l/m = \lambda$. The expression for ϕ, which depends on η and λ, is complex and is thus omitted. However, as an important special case, if $\eta = 1/(N-1)$ then the "generalized product type" distribution becomes the same as the power law size distribution and $\phi = 0$.

Next, consider the case $0 < \lambda \leq 1$. In this case, conservation of mass only applies prior to shattering. Suppose that only one particle in each binary collision fragments while the other remains intact. In addition, suppose that the children from the single parent obtain power law size distributions. According to Ernst & Pagonabarraga (2007), mean-field theory obtains:

$$c(M,t) \cong \text{const.} \frac{1}{[M_{\text{avg}}(t)]^2}\left[\frac{M}{M_{\text{avg}}(t)}\right]^{\frac{\eta}{2}-1} \exp\left\{-\text{const.}\left[\frac{M}{M_{\text{avg}}(t)}\right]^\lambda\right\} \tag{14.20}$$

where

$$M_{\text{avg}}(t) = \begin{cases} \text{const.}\,(t_s - t)^{1/(1-\lambda)} & 0 < \lambda < 1 \\ M_0 \exp(-a_0 K_0 C_0 M_0 t) & \lambda = 1 \end{cases}$$

for sufficiently large times and small particles. Notice that Equation (14.20) is a Rosin–Rammler size distribution with $n/m = \eta/2$ and $l/m = \lambda$. As this example shows, it is necessary to make increasingly stringent assumptions to obtain closed-form solutions as λ decreases. Regardless of assumptions, there are no known closed-form solutions for $\lambda < 0$.

Example 14.7 Late-time size distributions for generalized product kernels with free-field boundary conditions

For non-local cases, collision kernels tend to obtain approximate forms such as $K(x,y) = K_0 x^{\nu} y^{\mu}$ where $\lambda = \mu + \nu$. The last example considered cases where μ or ν are equal to 0 or λ. At the opposite extreme, suppose that μ and ν share λ evenly, i.e., suppose that $\mu = \lambda/2$ and $\nu = \lambda/2$. Then $K(x,y)$ becomes the *generalized product kernel*:

$$K(x,y) = K_0 x^{\lambda/2} y^{\lambda/2}. \tag{14.21}$$

First, consider the case $1 < \lambda \leq 3$. In this case, conservation of mass applies, i.e., shattering does not occur. Suppose that the children obtain a "generalized product type" size distribution with an exponent $\eta > 0$, as discussed earlier in Example 14.3. According to Kostoglou & Karabelas (2006), Equation (14.3) obtains the following particle size distribution:

$$c(M,t) \cong \text{const.} \frac{1}{[M_{\text{avg}}(t)]^2} \left[\frac{M}{M_{\text{avg}}(t)} \right]^{\eta-1} \exp\left\{ -\text{const.} \left[\frac{M}{M_{\text{avg}}(t)} \right]^{\frac{\lambda}{2}} \right\} \tag{14.22}$$

where $M_{\text{avg}}(t) = \text{const. } t^{1/(1-\lambda)}$ for sufficiently large times and small particles. Notice that Equation (14.22) is a Rosin–Rammler size distribution with $n/m = \eta$ and $l/m = \lambda/2$.

Next, consider the case $0 < \lambda \leq 1$. In this case, conservation of mass only applies prior to shattering. Suppose that only one particle in each binary collision fragments while the other remains intact. In addition, suppose that the children from the single parent obtain power law size distributions. According to Ernst & Pagonabarraga (2007), mean-field theory again obtains Equation (14.22) for sufficiently large times (prior to shattering) and small particles, where $M_{\text{avg}}(t) = \text{const. } (t_s - t)^{1/(1-\lambda)}$ or $M_0 \exp(-a_0 K_0 C_0 M_0 t)$ depending on λ.

Finally, consider the case $\lambda = 0$. Departing temporarily from continuous particle systems, consider discrete particle systems where $K_{ij} = K_0$, $M_i = iM_1$, and M_1 is the monomer mass. Suppose that the larger of the incident particles in each collision splits into exactly two random-sized pieces, while the smaller particle remains intact. According to Krapivsky & Ben-Naim (2003), mean-field theory obtains the following particle size distribution:

$$c_i(t) \cong \frac{\text{const.}}{[M_{\text{avg}}(t)]^2} \left[\frac{M_i}{M_{\text{avg}}(t)} \right]^{-2} \exp\left\{ -\text{const.} \left[\frac{M_i}{M_{\text{avg}}(t)} \right]^{-1} \right\} \tag{14.23}$$

for large enough times (prior to shattering) and small enough particles. Notice that this is a Weibull size distribution with $n/m = -1$. After shattering, "the entire mass [of particles] is transformed into dust at the transition point."

Like the last example, this example is forced to invoke increasingly stringent assumptions as λ decreases: $\lambda = 0$ requires more restrictive assumptions than $0 < \lambda \leq 1$; and $0 < \lambda \leq 1$ requires more restrictive assumptions than $1 < \lambda \leq 3$. Regardless of assumptions, there are no known closed-form solutions for $\lambda < 0$.

Next, consider local cases. Because both colliding particles are the same or similar, both colliding particles produce the same or similar child size distributions, i.e., there is no need to distinguish between the children from the two identical or nearly identical parents. Thus, $m(x, y, z)$ only depends on y, rather than both y and z; c.f. Equation (14.6). As noted earlier, for purposes of mean-field theory, it is generally necessary to assume that local fragmentation is caused by sinks of small particles. The two main options are as follows:

- *Shattering*. If $\lambda < 1$, sinks of small particles tend to form spontaneously after a finite amount of time.[1]
- *Boundary conditions*. Sinks of small particles and, optionally, sources of large particles are present from the very beginning of the problem.

Regarding the latter case, this chapter focuses on periodic boundary conditions, which consist of a sink of small particles and an equal-but-opposite source of large particles. While other boundary conditions generally produce non-stationary solutions, periodic boundary conditions generally produce stationary solutions, at least for large enough times. Example 14.8 describes an application where periodic boundary conditions occur naturally.

Example 14.8 Example of periodic boundary conditions for collisional fragmentation

For the asteroid belt, Dohnanyi (1969) posited a source of large particles as follows:

> **Almost all the mass in the asteroidal belt is ... concentrated in the largest asteroids ... The masses of individual [large] objects in the population decrease with time because of erosion and some objects are violently destroyed from time to time. Redistribution of the comminuted debris produced during erosive and catastrophic collisions constitute a particle creation mechanism.**

In addition, Dohnanyi (1969) posited a sink of small particles as follows:

> **... Particle removal [is] ... due to catastrophic collisions ... Objects [created by such catastrophic collisions] ... will be strongly influenced by radiation forces since they are no longer large enough to experience collisional processes . . . The system itself possesses a 'sink' in the sense that sufficiently small particles are removed by radiation effects [such as] ... radiation damping and radiation pressure.**

[1] Unlike the last chapter, this chapter ignores the possibility that weak shattering can be caused by child size distributions with $-1 < \eta \leq 0$. Instead, this chapter assumes that $0 < \eta \leq 1$. In this range, Ernst & Pagonabarrga (2007) correctly note that "the value of exponent η has no influence on the existence of shattering."

Dohnanyi (1969) says that "masses of the order of 10^{15} kg or less are ... in a steady-state [i.e., stationary] condition because the rate of particle creation equals the rate of particle removal of these masses." Based on these assumptions, Dohnanyi (1969) used mean-field theory to obtain a stationary power law size distribution with $n/m = -5/6$.

Consider local collisional fragmentation with a sink of small particles. In this case, mean-field theory produces approximate late-time solutions for a wide variety of different homogeneous collision kernels. More specifically, according to the best available theory, collisional fragmentation obtains power law or Weibull size distributions with:

$$\frac{n}{m} = -\frac{\lambda + 1}{2} \tag{14.24}$$

for sufficiently large times and small particle sizes. Notice that Equation (14.24) does not depend on the child size distributions, i.e., it does not depend on η. This occurs for reasons similar to those discussed in the last chapter; see Section 13.6.5. Dorschner (1974, eq. 40) proved Equation (14.24) for stationary fragmentation with periodic boundary conditions; see also Tanaka et. (1996). In addition, Ernst & Pagonabarraga (2007, eq. 9) proved Equation (14.24) for non-stationary late-time fragmentation with free-field boundary conditions, where a sink of small particles spontaneously forms due to shattering. In either case, if the particle system is spatially homogeneous and isotropic then, according to Equation (9.15), λ is an integer multiple of 1/3 between −1 and 3. Then Equation (14.24) implies:

$$n/m \text{ is an integer multiple of } 1/6 \text{ between } -2 \text{ and } 0. \tag{14.25}$$

Notice that Equations (14.24) and (14.25) are exactly the same as Equations (11.36b) and (11.37b), respectively. In other words, according to the best available theory, local collisional fragmentation obtains the same late-time universal size distributions as local aggregation, albeit for small rather than large particles. Ernst & Pagonabarraga (2007) call this agreement "remarkable." Example 14.9 discusses two specific instances of Equations (14.24) and (14.25), while Example 14.10 gives a general proof of Equation (14.24).

Example 14.9 Late-time stationary size distributions for collisional fragmentation with $\lambda = 1/3$ and $2/3$

As noted in Example 14.1, the asteroid belt evolved is believed to have evolved into its present state through collisional fragmentation. Assuming collisional fragmentation with $\lambda = 1/3$, Piotrowski (1952, 1953) used a simple form of mean-field theory to suggest that the asteroid belt obtains a power law size distribution with $n/m = -2/3$. Alternatively, assuming collisional fragmentation with $\lambda = 2/3$, Dohnanyi (1969) used a sophisticated form of mean-field theory to suggest that the asteroid belt obtains a power law size distribution with $n/m = -5/6$. Notice that both Piotrowski's and Dohnanyi's results agree with Equations (14.25). As discussed in the next section, modern observations indicate that the asteroid belt actually obtains a power law size distribution with $n/m \cong -1/2$. This does not mean that Piotrowski or

Dohnanyi were wrong. In fact, recent observations indicate that some astrophysical systems, such as the smaller and colder objects in the Kuiper belt, actually do obtain $n/m \cong -2/3$ consistent with Piotrowski's prediction. By the same token, recent observations indicate that some astrophysical systems, such as the largest and most hazardous near-earth asteroids, actually do obtain $n/m \cong -5/6$ consistent with Dohnanyi's prediction. In addition, Dohnanyi's result is a reasonable estimate if all of the objects in the Kuiper belt—hot, cold, small, large—are lumped together. However, it is important to distinguish between different classes of objects, if they evolve independently of one another.

This illustrates a general principle. While aggregation and fragmentation models generally produce legitimate solutions, it can be difficult to predict in advance which solutions apply to which problems. As one possible approach, discussed earlier in Chapter 5, many communities have adopted standard size distributions. Lacking better information, this standard is often considered the best possible estimate. In particular, as noted in Section 5.8, the astrophysics community long ago adopted Dohnanyi's power law size distribution law as a kind of standard. Thus, soon after the Kuiper belt was first proven to exist, Davis & Farinella (1997) suggested that it obtained this standard. The literature survey given in the next section indicates that $n/m \cong -1/2$ might be a better choice of standard.

Example 14.10 Late-time stationary solutions for collisional fragmentation with periodic boundary conditions

Tanaka et. (1996) rewrote Equation (14.3) as follows:

$$\frac{\partial[xc(x,t)]}{\partial t} + \frac{\partial \mathcal{F}(x,t)}{\partial x} = 0. \tag{14.26a}$$

Equivalently, Ernst & Pagonabarraga (2007) rewrote Equation (14.3) as follows:

$$\mathcal{F}(x,t) = -\int_x^{\infty} y\frac{\partial c(y,t)}{\partial t}\,dy. \tag{14.26b}$$

In Equation (14.26), $\mathcal{F}(x,t)$ is the "mass flux across mass x." More specifically:

$$\mathcal{F}(x,t) = -\int_0^{\infty}\int_x^{\infty} yG(x,y,z)K(y,z)\,c(y,t)\,c(z,t)\,dydz \tag{14.27}$$

where $G(x,y,z)$ is the cumulative mass distribution function for children. Notice that:

$$g(x,y,z) = \frac{\partial G(x,y,z)}{\partial x} \tag{14.28}$$

where $g(x,y,z)$ is the "mass fraction of the fragments with masses x to $x+dx$" produced in each collisional fragmentation event.

Assume that G and K are homogeneous. By definition:

$$G(x, y, z) = \tilde{G}(x/y, z/y)$$
$$K(x, y) = x^{\lambda}\tilde{K}(y/x).$$

If $y' = y/x$ and $z' = z/x$, then Equation (14.27) becomes:

$$\mathcal{F}(x,t) = -x^{\lambda+3}\int_0^{\infty}\int_1^{\infty}\tilde{G}(1/y', z'/x')\, y'^{\lambda+1}\tilde{K}(z'/y')\, c(xy',t)\, c(xz',t)\, dy'\, dz'. \quad (14.29)$$

Assuming power law solutions with $c(x,t) = A(t)x^{n/m-1}$, Equation (14.29) becomes:

$$\mathcal{F}(x,t) = -x^{\lambda+1+2n/m}A(t)^2\int_0^{\infty}\int_1^{\infty}\tilde{G}(1/y', z'/x')\, y'^{\lambda+n/m} z'^{n/m}\tilde{K}(z'/y')\, dy'\, dz'. \quad (14.30)$$

In stationary cases, $\partial[xc(x,t)]\partial t = 0$ implies $\partial\mathcal{F}(x,t)/\partial x = 0$. This is true if and only if the exponent of x in Equation (14.30) is zero. In other words, $\lambda + 1 + 2n/m = 0$ or, equivalently, $n/m = -(\lambda + 1)/2$. This agrees with Equation (14.24). Stationary cases require periodic boundary conditions, i.e., a sink of small particles and an equal-but-opposite source of large particles. As noted earlier, periodic boundary conditions limit the possible range of particle sizes, typically leading to locality.

14.5 Experimental Validation

Table 14.3 lists some power law size distributions observed in astrophysical systems. These systems are believed to have evolved primarily through collisional fragmentation. Notice that the power law exponent n/m generally appears to be an even multiple of 1/6 between −2 and 0, to within the accuracy of the measurement techniques employed, consistent with Equation (14.25). Table 14.3 supports the notion than many astrophysical systems evolved through collisional fragmentation or, at least, some combination of aggregation and collisional fragmentation. Furthermore, Table 14.3 supports the notion that aggregation and collisional fragmentation took place under spatially homogeneous and isotropic conditions. As direct evidence for this, Jewitt et al. (2010) say "the expected interval between collisional disruptions of 0.1-km-diameter asteroids in the main belt is about one year, whereas damaging but non-disruptive impacts should be more frequent." Given that the average object in the main asteroid belt travels fast enough to orbit the sun in three to six years, it is reasonable to assume that the products of each fragmentation event have enough time to thoroughly mix with the other objects in the asteroid belt prior to the next fragmentation event. In other words, while each fragmentation event disrupts the spatial homogeneity and isotropy of the asteroid belt, this disruption is only local and temporary. Similar reasoning applies to many other astrophysical systems.

Table 14.3 Selected observations of power law size distributions in astrophysical systems.

Object Class	n/m		References
	Smaller Particles	Larger Particles	
Irregular Satellites of Jupiter, Saturn, etc.	−0.33	–	Bottke et al. (2010)(1)
Kuiper Belt—Hot Objects	−0.33 ± 0.16	−1.43 ± 0.13	Fraser et al. (2014)(2)
Asteroids—Main Belt	−0.447 ± 0.13	−0.537 ± 0.17	Ryan et al. (2015)(3)
	−0.47, −0.52	–	DeMeo & Carry (2013)(4)
	−0.50 ± 0.03	–	Gladman et al. (2009)(5)
	−0.50	–	Masiero et al. (2011)(6)
Asteroids—Near Earth, Hazardous	−0.5 ± 0.067	−0.89 ± 0.23	Mainzer et al. (2012)(7)
Asteroids—Jupiter Trojans	−0.66	–	Grav et al. (2011)(8)
	−0.66 ± 0.1	−1.5 ± 0.3	Emery et al. (2015)(9)
Kuiper Belt—Cold Objects	−0.63 ± 0.1	−2.4 ± 0.5	Fraser et al. (2014)(10)
	−0.63 ± 0.1	–	Moore et al. (2016)(11)
Particles in Saturn's Rings	−0.66	–	Cuzzi et al. (2009, 2010)(12)

(1) Given in the original paper as $q \approx -1$ where $n/m = q/3$. This applies for diameters 8-200 km including all irregular satellites known as of May 1999 (55 for Jupiter, 38 for Saturn, nine for Uranus, and six for Neptune).
(2) Given in the original paper as $q = 5.3 \pm 0.4$ and 2.0 ± 0.5 for large and small objects, respectively, where $n/m = -(q-1)/3$. Based on a survey of earlier papers.
(3) Given in the original paper as $b = 2.34 \pm 0.05$ and 2.61 ± 0.04 for 2–25 km and 5–25 km C-Type asteroids, respectively, where $n/m = -(b-1)/3$. Based on 1,865 asteroids ranging from 0.2 to 169 km.
(4) Given in the original paper as $a = -2.57$ and -2.42 in "middle main belt" and "outer main belt," respectively, where $n/m = (a+1)/3$. Based on "~2,000 bodies."
(5) Given in the original paper as $\alpha = 0.3 \pm 0.02$ where $n/m = -5\alpha/3$. Based on "1087 main-belt [Sub-Kilometer Asteroid Diameter Survey] SKADS asteroids."
(6) Given in the original paper as $a = -2.5$ where $n/m = (a+1)/3$. Based on "over 100,000 Main Belt asteroids."
(7) Given in the original paper as $\alpha = 1.5 \pm 0.20$ and 2.67 ± 0.7 for asteroid diameters less than and greater than 1 km, respectively, where $n/m = -\alpha/3$. Based on ~4,700 ± 1,450 near-Earth asteroids.
(8) Given in the original paper as $\alpha = 2$ where $n/m = -\alpha/3$. Based on 1,660 Jovian Trojans with diameters between 10 and 100 km.
(9) Given in the original paper as $\alpha = 2 \pm 0.3$ and 4.5 ± 0.9 for asteroid diameters of 4–40 km and > 80 km, respectively, where $n/m = -\alpha/3$. Based on a survey of earlier papers.
(10) Given in the original paper as $q = 8.2 \pm 1.5$ and 2.9 ± 0.3 for large and small objects, respectively, where $n/m = -(q-1)/3$. Based on a survey of earlier papers.
(11) Given in the original paper as $q = -2.9 \pm 0.3$ where $n/m = (q+1)/3$. Based on "the overall crater size-frequency distribution for Vulcan Planum [on Pluto's moon Charon] ... in the ~10–100km diameter range."
(12) Given in the original paper as $q \sim 3$ where $n/m = -(q-1)/3$ assuming "the lower radius limit is in the 1–30cm range, and the upper radius limit is in the 2–20m range."

Unfortunately, it is difficult to directly measure or infer λ in astrophysical systems. Thus, it is difficult to validate Equation (14.24). However, it is straightforward to measure n/m. Thus, it is straightforward to validate Equation (14.25). In fact, as expected, Table 14.3 implies that many astrophysical systems obtain universal power law size distributions, where n/m is an even multiple of 1/6 consistent with Equation (14.25). While reported as power law size distributions, some of these could be gamma, Weibull, or Rosin–Rammler size distributions, especially in cases where the estimated power law index n/m varies substantially with particle size. As a common source of confusion, different researchers use different expressions for power law size distributions. The footnotes to Table 14.3 explain how the results given in the original research papers were converted to a consistent form involving n/m; see also Appendix A.2.

14.6 Conclusions

According to mean-field theory, collisional fragmentation cascades typically obtain a hybrid of two different size distributions. The first applies for sufficiently small times and large particle sizes while the second applies for sufficiently large times and small particle sizes. The first is arbitrary and depends on the initial conditions. The second is often universal and depends on the kernel, the boundary conditions, and perhaps the child size distributions. If all of the individual fragmentation events obtain the same universal size distribution, the overall size distribution tends to become a universal. In fact, even when none of the individual fragmentation events obtains a universal size distribution, the overall size distribution may still become a universal, provided that a sufficiently strong sink of small particles is present.

The previous five chapters have made similar observations for aggregation, binary fragmentation, and random fragmentation. In fact, almost every result obtained in this chapter mirrors an earlier result obtained for aggregation, binary fragmentation, or random fragmentation. To emphasize this point, Table 14.4 summarizes late-time mean-field theory results from this and the previous five chapters. Where sources and sinks occur, Table 14.4 assumes that they are hard, i.e., that they impose strict upper and lower limits on the particle sizes. This is because the research literature on fragmentation cascades deals mainly with hard rather than soft sources and sinks. The aggregation results may transition either gradually or abruptly as λ increases; this is because gelation introduces a sink of large particles; see the earlier discussion in Section 11.9. Similarly, the collisional fragmentation results may transition either gradually or abruptly as λ decreases; this is because shattering introduces a sink of small particles. In most cases, if the particle system is spatially homogeneous and isotropic, mean-field theory predicts a universal size distribution, i.e., the exponents are

Table 14.4 Late-time size distributions for aggregation and fragmentation cascades depend on the boundary conditions (i.e., sources, sinks), the kernels (i.e., λ, κ), and sometimes the child size distributions (i.e., η). While the fragmentation results apply for sufficiently small sizes, the aggregation results apply for sufficiently large particle sizes. All sources and sinks are hard.

Phenomenon	Sources/Sinks	Size Distributions (Late Times, Small or Large Sizes)			
		Family	n/m	l/m	Universal
Binary Fragmentation	No	Rosin–Rammler	1	κ	Yes (**)
	Yes (*)	Power Law	$-\kappa-1$	0	Yes (**)
Random Fragmentation	No	Rosin–Rammler	η	κ	No
	Yes (*)	Power Law	$-\kappa-1$	0	Yes (**)
Collisional Fragmentation	No	Rosin–Rammler	$\eta, \eta/2$	$\lambda, \lambda/2$	No
	Yes (*)	Power Law	$-(\lambda+1)/2$	0	Yes (**)
(Collisional) Aggregation	No	Weibull, Gamma	$-\lambda+1$	$-\lambda+1$, 1	Yes (**)
	Yes (*)	Power Law	$-(\lambda+1)/2$	0	Yes (**)

(*) If this column indicates "yes" then a sink spontaneously forms due to shattering or gelation. Alternatively, a sink is present at all times due to the boundary conditions.
(**) If this column indicates "yes" then the exponents n/m and l/m are even multiples of 1/6 between −2 and 2, assuming the particle system is spatially homogeneous and isotropic.

even multiples of 1/6. As discussed in the last chapter, the main exceptions are as follows: dense particle systems which obtain spatial homogeneity and isotropy only in a loose average sense; and mass conserving systems that depend on child size distributions through η. Anything that promotes chaos and randomness also tends to promote spatial homogeneity and isotropy and, thus, universality. This includes external drivers such as vibration, rotation, shaking, stirring, swirling, heating, and so forth. This also includes internal drivers such as elastic or partially elastic collisions between particles.

For completeness, this chapter considered both local and non-local cases. However, as a practical matter, collisional fragmentation almost always involves locality. For example, think of a particle system composed of ball bearings ("BBs") and bowling balls traveling at high speeds through outer space. Putting aside hypervelocity cases, collisions between ball bearings and bowling balls and are unlikely to be of much consequence. At most, the bowling balls might suffer superficial damage such as scoring, pitting, or cratering. Similarly, collisions between two ball bearings are unlikely to be of much consequence. At most, if such collisions cause the ball bearings to stick together, the combination could gradually grow through additional collisions to become as big as the bowling balls. By contrast, collisions between two bowling balls are likely to be highly consequential, leading to large numbers of small fragments with significant amounts of mass, momentum, and energy. The moral of the story is that, for collisional fragmentation, what matters most is collisions between the largest particles with the same or similar sizes. As a result, only the local variants of mean-field theory have been well-validated against experimental results; see Section 14.5. While mathematically interesting, the non-local variants of mean-field theory have generally not been well-validated.

References

P. K. Barik & A. K. Giri, Global Classical Solutions to the Continuous Coagulation Equation with Collisional Breakage, *Zeitschrift für angewandte Mathematik und Physik*, 71, 38, 2020, https://doi.org/10.1007/s00033-020-1261-5

W. F. Bottke, D. Nesvorný, D. Vokrouhlický, & A. Morbidelli, The Irregular Satellites: The Most Collisionally Evolved Populations in the Solar System, *The Astronomical Journal*, 139, 994–1014, 2010, https://doi.org/10.1088/0004-6256/139/3/994

Z. Cheng & S. Redner, Scaling Theory of Fragmentation, *Physical Review Letters*, 60(24), 2450, 1988, https://doi.org/10.1103/PhysRevLett.60.2450

Z. Cheng & S. Redner, Kinetics of Fragmentation, *Journal of Physics A: Mathematical and General*, 23(7), 1233–1258, 1990, https://doi.org/10.1088/0305-4470/23/7/028

F. P. da Costa, Mathematical Aspects of Coagulation-Fragmentation Equations, in J. P. Bourguignon, R. Jeltsch, A. Pinto, M. Viana, editors, *Mathematics of Energy and Climate Change, CIM Series in Mathematical Sciences, Volume 2*, Springer, 2015, https://doi.org/10.1007/978-3-319-16121-1_5

J. N. Cuzzi, R. N. Clark, G. Filacchione, R. G. French, R. Johnson, E. A. Marouf, et al., Chapter 15: Ring Particle Composition and Size Distribution, in M. K. Dougherty, M. L. Esposito, &

S. Krimigis, editors, *Saturn from Cassini-Huygens*, Springer, 2009, https://doi.org/10.1007/978-1-4020-9217-6

J. N. Cuzzi, J. A. Burns, S. Charnoz, R. N. Clark, J. E. Colwell, L. Dones, et al., An Evolving View of Saturn's Dynamic Rings, Supplementary Materials, *Science*, 327(5972), 1470–1475, 2010, https://doi.org/10.1126/science.1179118

D. R. Davis & P. Farinella, Collisional Evolution of Edgeworth–Kuiper Belt Objects, *Icarus*, 125(1), 50–60, 1997, https://doi.org/10.1006/icar.1996.5595

F. E. DeMeo & B. Carry, The Taxonomic Distribution of Asteroids from Multi-Filter All-Sky Photometric Surveys, *Icarus*, 226(1), 723–741, 2013, https://doi.org/10.1016/j.icarus.2013.06.027

J. S. Dohnanyi, Collisional Model of Meteoroids, in J. L. Weinberg, editor, *The Zodiacal Light and the Interplanetary Medium*, NASA SP-150, 1967, https://ntrs.nasa.gov/search.jsp?R=19680018003

J. S. Dohnanyi, Collisional Model of Asteroids and their Debris, *Journal of Geophysical Research*, 74(10), 2531–2554, 1969, https://doi.org/10.1029/JB074i010p02531

J. S. Dohnanyi, On the Origin and Distribution of Meteoroids, *Journal of Geophysical Research*, 75(17), 3468–3493, 1970, https://doi.org/10.1029/JB075i017p03468

J. Dorschner, Theoretische Untersuchungen über den interstellaren Staub. III. Stauberzeugung durch zirkumstellare Zertrümmerungsprozesse, *Astronomische Nachrichten*, 292(2), 79–85, 1970, https://doi.org/10.1002/asna.19702920206

J. Dorschner, Dust Production in Circumstellar Space, *Astronomische Nachrichten*, 293(1), 65–70, 1971, https://doi.org/10.1002/asna.19712930115

J. Dorschner, The Fragmentation of Asteroids—A Synopsis of Analytical Methods, *Astronomische Nachrichten*, 296(3), 141–146, 1974, https://doi.org/10.1002/asna.19742950305

J. P. Emery, F. Marzari, A. Morbidelli, L. M. French, & T. Grav, The Complex History of Trojan Asteroids. in P. Michel, F. E. DeMeo, & W. F. Bottke, editors, *Asteroids IV*, University of Arizona Press, 2015, https://doi.org/10.2458/azu_uapress_9780816532131-ch011

M. H. Ernst & I. Pagonabarraga, The Nonlinear Fragmentation Equation, *Journal of Physics A: Mathematical and Theoretical*, 40, F331–F337, 2007, https://doi.org/10.1088/1751-8113/40/17/F03

W. C. Fraser, M. E. Brown, A. Morbidelli, A. Parker, & K. Batygin, The Absolute Magnitude Distribution of Kuiper Belt Objects, *The Astrophysical Journal*, 782(2), 100, 2014, https://doi.org/10.1088/0004-637X/782/2/100

B. J. Gladman, D. R. Davis, C. Neese, R. Jedicke, G. Williams, J. J. Kavelaars, et al., On the Asteroid Belt's Orbital and Size Distribution, *Icarus*, 202(1), 104–118, 2009, https://doi.org/10.1016/j.icarus.2009.02.012

T. Grav, A. K. Mainzer, J. Bauer, J. Masiero, T. Spahr, R. S. McMillan, et al., WISE/NEOWISE Observations of the Jovian Trojans: Preliminary Results, *The Astrophysical Journal*, 742(1), 40, 2011, https://doi.org/10.1088/0004-637X/742/1/40

D. Jewitt, H. Weaver, J. Agarwal, M. Mutchler & M. Drahus, A Recent Disruption of the Main-Belt Asteroid P/2010 A2, *Nature*, 467, 817–819, 2010, https://doi.org/10.1038/nature09456

A. Johansen, M.-M. Mac Low, P. Lacerda, & M. Bizzarro, Growth of Asteroids, Planetary Embryos, and Kuiper Belt Objects by Chondrule Accretion, *Science Advances*, 1(3), e1500109, 2015, https://doi.org/10.1126/sciadv.1500109

M. Kostoglou & A. J. Karabelas, A Study of the Nonlinear Breakage Equation: Analytical and Asymptotic Solutions, *Journal of Physics A: Mathematical and General*, 33(6), 1221–1232, 2000, https://doi.org/10.1088/0305-4470/33/6/309

M. Kostoglou & A. J. Karabelas, A Study of the Collisional Fragmentation Problem using the Gamma Distribution Approximation, *Journal of Colloid and Interface Science*, 303, 419–429, 2006, https://doi.org/10.1016/j.jcis.2006.08.005

P. L. Krapivsky & E. Ben-Naim, Shattering Transitions in Collision-Induced Fragmentation, *Physical Review E*, 68(2), 021102, 2003, https://doi.org/10.1103/PhysRevE.68.021102

P. Laurençot & D. Wrzosek, The Discrete Coagulation Equations with Collisional Breakage, *Journal of Statistical Physics*, 104(1–2), 193–220, 2001, https://doi.org/10.1023/A:1010309727754

M. Lombart, M. Hutchison, & Y.-N. Lee, Fragmentation with Discontinuous Galerkin Schemes: Non-Linear Fragmentation, *Monthly Notices of the Royal Astronomical Society*, 517, 2012–2027, 2022, https://doi.org/10.1093/mnras/stac2232

A. Mainzer, T. Grav, J. Masiero, J. Bauer, R. S. McMillan, J. Giorgini, et al., Characterizing Subpopulations within the Near-Earth Objects with NEOWISE: Preliminary Results, *The Astrophysical Journal*, 752(2), 110, 2012, https://doi.org/10.1088/0004-637X/752/2/110

J. R. Masiero, A. K. Mainzer, T. Grav, J. M. Bauer, R. M. Cutri, J. Dailey, et al., Main Belt Asteroids with WISE/NEOWISE. I. Preliminary Albedos and Diameters, *The Astrophysical Journal*, 741(2), 68, 2011, https://doi.org/10.1088/0004-637X/741/2/68

J. M. Moore, W. B. McKinnon, J. R. Spencer, A. D. Howard, P. M. Schenk, R. A. Beyer, et al., The Geology of Pluto and Charon through the Eyes of New Horizons: Supplementary Materials, *Science*, 351(6279), 1284–1293, 2016, https://doi.org/10.1126/science.aad7055

D. P. O'Brien & R. Greenberg, The Collisional and Dynamical Evolution of the Main-Belt and NEA Size Distributions, *Icarus*, 178(1), 179–212, 2005, https://doi.org/10.1016/j.icarus.2005.04.001

S. L. Piotrowski, The Collisions of Asteroids, *The Astronomical Journal*, 57(1), 23–24, 1952, https://doi.org/10.1086/106692

S. Piotrowski, The Collisions of Asteroids, *Acta Astronomica, Series A*, 5, 115–138, 1953, https://articles.adsabs.harvard.edu/pdf/1952AJ.....57...23P

E. L. Ryan, D. R. Mizuno, S. S. Shenoy, C. E. Woodward, S. J. Carey, A. Noriega-Crespo, et al., The Kilometer-Sized Main Belt Asteroid Population Revealed by *Spitzer*, *Astronomy & Astrophysics*, 578, A42, 2015, https://doi.org/10.1051/0004-6361/201321375

V. S. Safronov, *Evolyutsiya Doplanetnogo Oblaka i Obrazovanie Zemli i Planet (Evolution of the Protoplanetary Cloud and Formation of the Earth and its Planets)*, Akademii Nauk SSSR, 1969; see also NASA TT F-677, 1972, https://archive.org/details/nasa_techdoc_19720019068

H. E. Schlichting, C. I. Fuentes, & D. E. Trilling, Initial Planetesimal Sizes and the Size Distribution of Small Kuiper Belt Objects, *The Astronomical Journal*, 146(2), 36, 2013, https://doi.org/10.1088/0004-6256/146/2/36

R. C. Srivastava, A Simple Model of Particle Coalescence and Breakup, *Journal of the Atmospheric Sciences*, 39(6), 1317–1322, 1982, https://doi.org/10.1175/1520-0469(1982)039<1317:ASMOPC>2.0.CO;2

H. Tanaka, S. Inaba, & K. Nakazawa, Steady-State Size Distribution for the Self-Similar Collision Cascade, *Icarus*, 123(2), 450–455, 1996, https://doi.org/10.1006/icar.1996.0170

R. R. Vigil, I. Vermeersch, & R. O. Fox, Destructive Aggregation: Aggregation with Collision-Induced Breakage, *Journal of Colloid and Interface Science*, 302(1), 149–158, 2006, https://doi.org/10.1016/j.jcis.2006.05.066

D. Wilkins, A Geometrical Interpretation of the Coagulation Equation, *Journal of Physics A: Mathematical and General*, 15(4), 1175–1178, 1982, https://doi.org/10.1088/0305-4470/15/4/020

15
Unitary Fragmentation Part 1: Basic Principles

This is the first of three chapters on unitary fragmentation. In unitary fragmentation, a single large object decomposes into two or more smaller objects. Following the lead of Grady (2017) and Yarin et al. (2017), this chapter treats unitary fragmentation of liquids and solids in a unified fashion. For solids, unitary fragmentation is traditionally described in terms of stresses and strains. However, for liquids, unitary fragmentation is traditionally described in terms of non-dimensional parameters such as Reynolds and Weber numbers. For simplicity, this chapter will describe both liquids and solids in terms of stresses and strains. For a more complete picture, the next chapter will describe both liquids and solids in terms of Reynolds and Weber numbers.

15.1 Challenges

According to Dias & Parisio (2014), unitary fragmentation "is amongst the toughest problems in the physics of complex systems." There are three main challenges. First, unitary fragmentation can depend on numerous tiny details of the material such as microscopic inclusions, grain boundaries, cracks, voids, bubbles, thermal fluctuations, thin spots, and so forth. Second, unitary fragmentation can depend on periodic instabilities, chaotic instabilities, or both. Third, unitary fragmentation can lead to multiple stages of aggregation and e.g., secondary fragmentation.

Tables 15.1 and 15.2 list some common examples of unitary fragmentation of solids and liquids, respectively. In order to illustrate the challenges with unitary fragmentation, the remainder of this section will discuss two specific examples drawn from these tables. The first example involves the bending of solid rectangular panels; see the case labeled "impact fragmentation" in Table 15.1. In Figures 15.1 and 15.2, a planar airblast impinges on a solid rectangular panel mounted in a frame. In Figure 15.1, a strong, thick concrete panel experiences a low strain rate, i.e., it bends slowly. As it bends, the panel experiences periodic instabilities, which cause it to assume a sinusoidal shape punctuated by a series of nearly evenly spaced, nearly horizontal cracks; see also, e.g., Gladden et al. (2005) and Vandenberghe & Villermaux (2013). Depending on the mounting, loading, and other conditions, the panel eventually separates into two nearly equal sized pieces. Inside the cracks, the strain rate is much higher than elsewhere. Thus, each crack emits a large number of small fragments with seemingly random shapes and sizes. In Figure 15.2, a thin, weak glass panel experiences a high strain rate, i.e., it bows rapidly. As it bows, the panel experiences chaotic instabilities, which cause it to develop an intricate seemingly-random network of cracks. As seen in Figure 15.3, these cracks originate near the outer mounting frame and propagate toward the center, spreading as the panel flexes. By 15 ms, the panel has almost entirely

Particle Dynamics with Aggregation and Fragmentation. Culbert B. Laney, Oxford University Press. © Culbert B. Laney (2025).
DOI: 10.1093/9780198920687.003.0015

Table 15.1 Common types of unitary solid fragmentation.

	Ideal	Non-Ideal
Impact Fragmentation	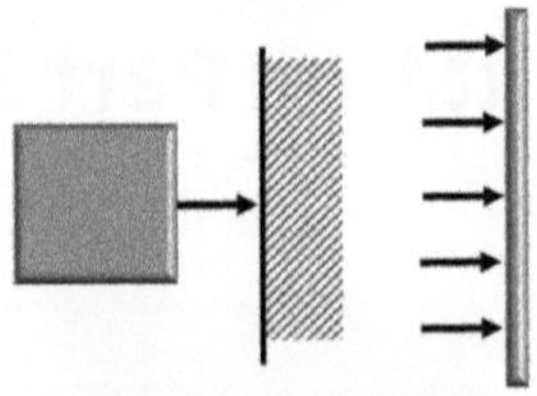	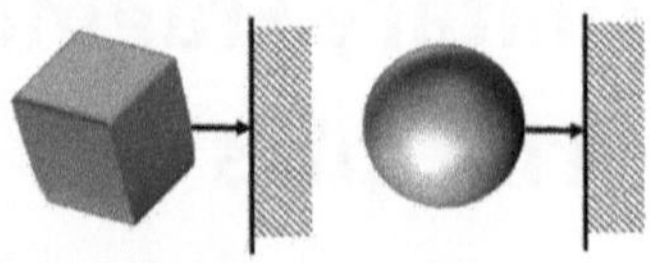
Collisional Fragmentation	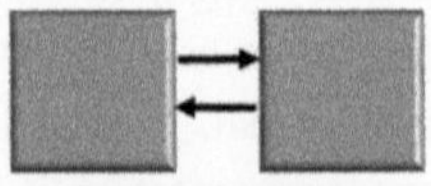	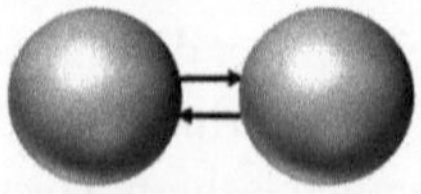
Expanding Disks, Spheres & Spherical Shells	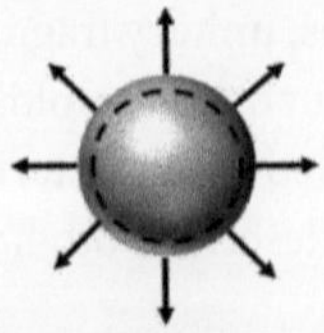	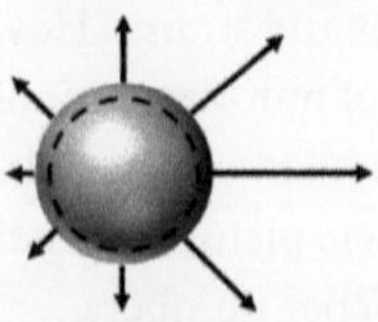
Expanding Rings, Cylinders & Cylindrical Shells	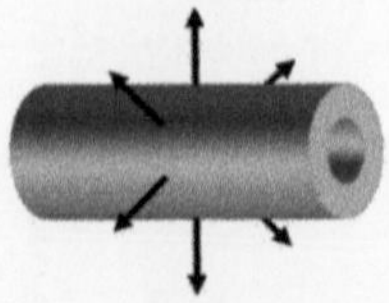	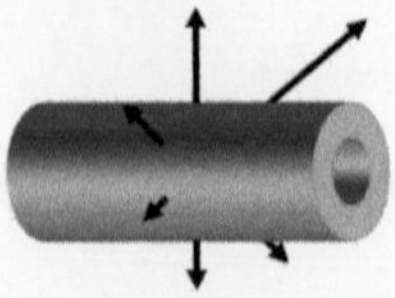
Penetration & Spall	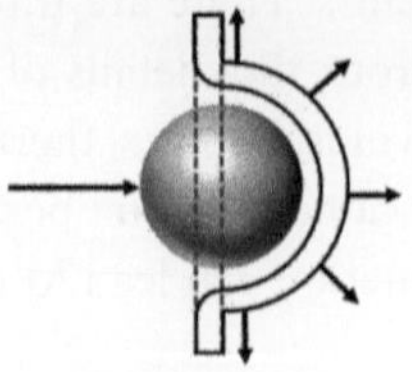	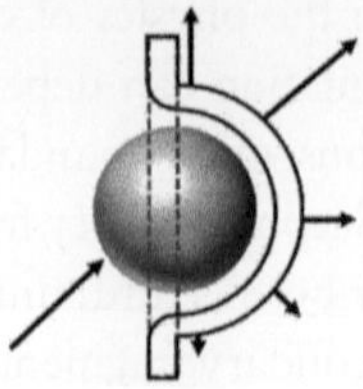
Secondary Fragmentation, e.g., Impact Fragmentation of Primary Fragments	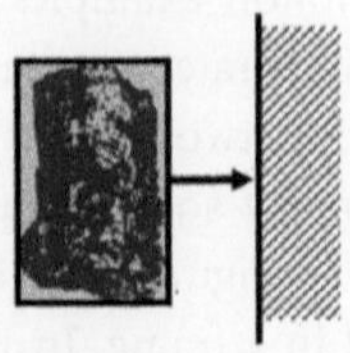	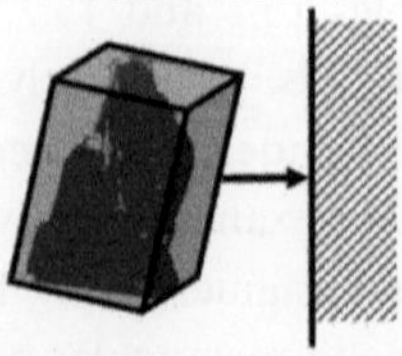

decomposed into fragments with seemingly random sizes and shapes. The fragments initially have large aspect ratios, i.e., they tend to be long and jagged. However, they experience additional rounds of fragmentation, due to a combination of collisions and residual strains, which progressively reduce their sizes and aspect ratios. This may involve a fragmentation cascade, as discussed in the previous three chapters, i.e., fragments break into smaller fragments, which then break into even smaller fragments, and so forth. This book will focus on seemingly random cases with large strain rates as seen in Figure 15.2, rather than partially deterministic cases with mostly small strain rates as seen in Figure 15.1.

Table 15.2 Common types of unitary liquid fragmentation.

	Ideal	Non-Ideal
Round Jet	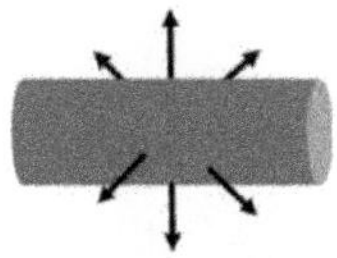	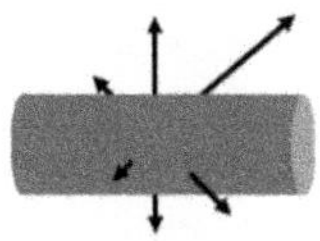
Sheet Jet	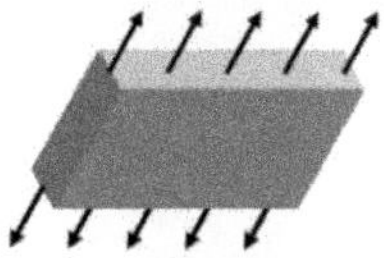	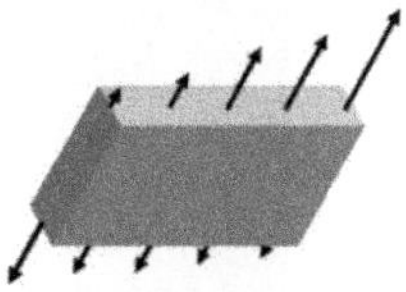
Conical Jet	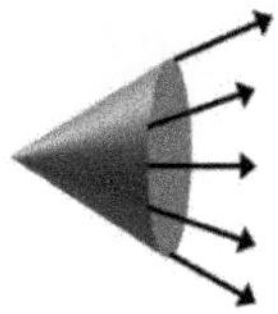	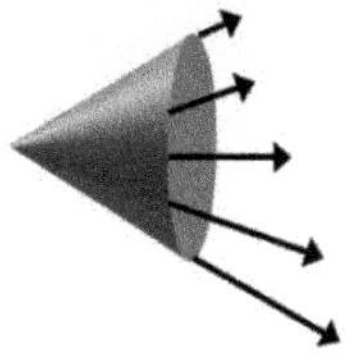
Expanding Gas-Filled Bubble	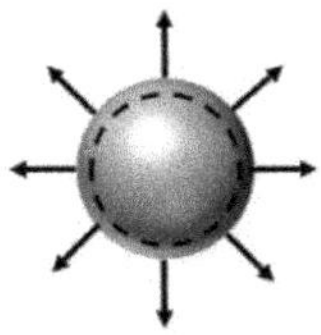	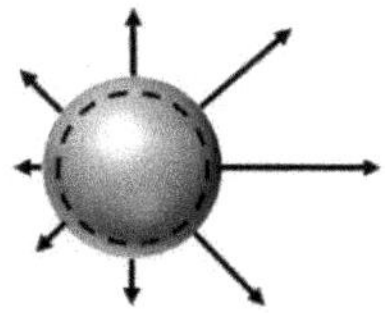
Impact Breakup of Droplets	n/a	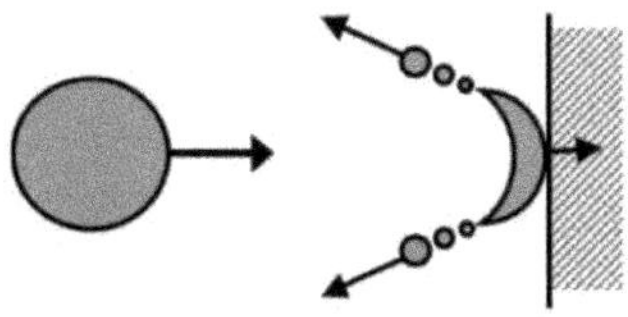
Aerodynamic Breakup of Droplets	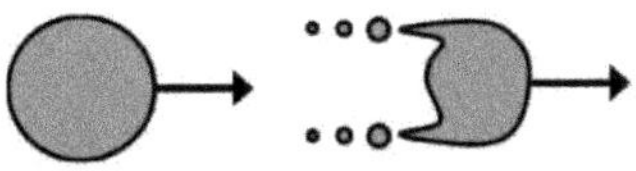	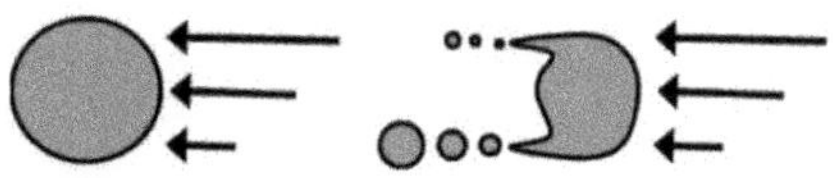

The second example involves the expansion of spherical liquid shells; see the case labeled "expanding gas-filled bubble" in Table 15.2. In Figure 15.4, a spherical liquid shell expands in the radial direction due a hydrogen gas explosion. The shell ultimately fragments due to multiple periodic and/or chaotic instabilities. Periodic instabilities, known *as Rayleigh-Taylor instabilities*, can occur whenever a liquid surface accelerates normally into a gas. In addition, chaotic instabilities can occur due to tensile strains, regardless of the surrounding gas. As seen in Figure 15.4.c, d, these instabilities cause thin spots which, ultimately, develop into circular holes. According to Vledouts et al. (2016), at lower strain rates, periodic instabilities create a network of holes with a "regular character." However, at higher strain rates, chaotic instabilities lead to "nucleating holes at random." As seen in Figure 15.4.e, f, "the holes grow in radius by capillary retraction, accumulating the liquid of the shell into their

(a) 25 ms (b) 42.5 ms (c) 60 ms

Figure 15.1 Side view of an 8′ × 8′ × 5.5″ reinforced concrete panel impacted by a planar shock wave. This case obtains a low strain rate.
Reprinted with permission from Bewick et al. (2017b).

(a) 15 ms (b) 30 ms (c) 50 ms

Figure 15.2 Side view of an 4′ × 6′ × 1/4″ glass panel impacted by a planar shock wave. This case obtains a high strain rate.
Reprinted with permission from Bewick et al. (2018).

rim, which merge with the rim of the neighboring hole, finally leaving the shell as a web of ligaments, which break by capillary instability." At lower strain rates, the liquid ligaments fragment due to periodic instabilities, which pinch off droplets whose diameters are roughly equal to the ligament diameters. These instabilities are generally known as *Rayleigh capillary instabilities*. At higher strain rates, the ligaments fragment due to chaotic instabilities, which

(a) 0 ms

(b) 5 ms

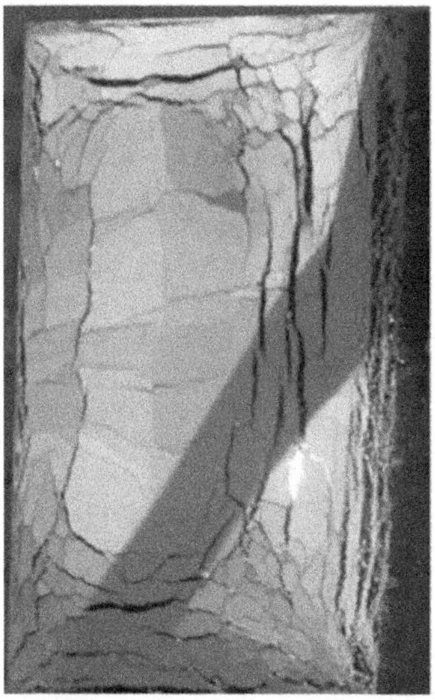
(c) 10 ms

(d) 15 ms

Figure 15.3 Rear view of an $4' \times 6' \times 1/4''$ glass panel impacted by a planar shock wave.
Reprinted with permission from Bewick et al. (2017b).

create droplets whose diameters are far smaller than the ligament diameters. This book will focus on cases with large strain rates and widespread chaos, rather than low strain rates and limited chaos.

In light of these two examples, the three main challenges of unitary fragmentation can now be discussed in more detail. As the first main challenge, unitary fragmentation may depend on the preexisting microstructure of the material. In this regard, Grady (1988) says there are two main cases: *flaw limited* and *energy limited* cases. In flaw-limited cases, material separations always nucleate on pre-existing microscopic flaws. Furthermore, the supply of these pre-existing flaws is limited. If there were more flaws, there would be more fragments.

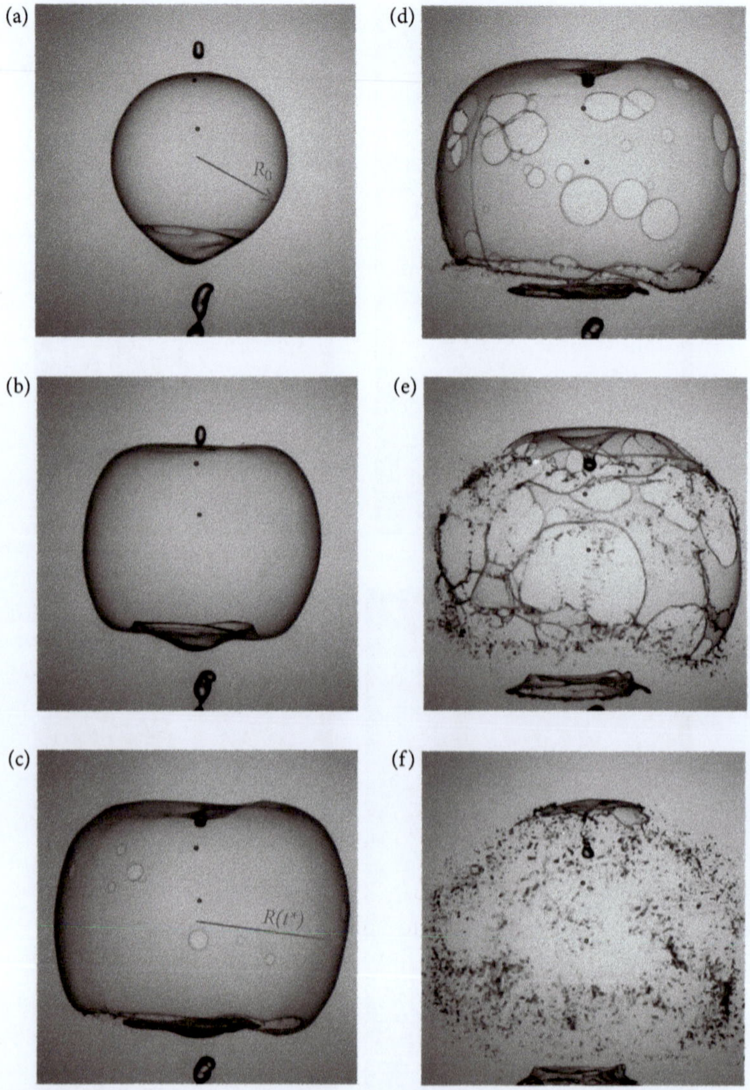

Figure 15.4 A spherical liquid shell expands and fragments due to a gaseous explosion.
Reproduced with permission from Vledouts et al. (2016). Copyright 2016 Cambridge University Press.

In essence, fragmentation plays dot-to-dot, tearing the material apart along perforations. For example, in solid materials, Seaman et al. (1976) say that fragment "nucleation may occur by widening of inherent flaws in the material, cracking of hard inclusions, separation along grain boundaries, or by other mechanisms." In such cases, Antoun et al. (2003) say the main challenge is "to examine the pristine microstructure and ascertain whether it contains pre-existing defects that might serve as damage nucleation sites ... The focus of pre-test characterizations is on determining the nature, characteristic size and geometry of preexisting defects, their distribution density, and whether they are randomly or preferentially oriented." Pioneering early work on flaw-limited fragmentation includes Gilvarry (1961), Seaman et al. (1976) and Curran et al. (1977).

In energy-limited cases, microscopic flaws are present everywhere and in abundance. Even if there were many more flaws, the number of fragments would be exactly the same. As the strain rate increases, the material microstructure changes dynamically, meaning the number of nucleation sites increases. For brittle solids, Ray (2018) says "recent experiments have shown that fracture in brittle materials is preceded by numerous microfractures inside the materials. These tiny ruptures are commonly imperceptible except that the acoustic emissions from these ruptures can be magnified and recorded with the help of ultrasonic detectors." As a result, in energy-limited cases, fragmentation may not depend on the preexisting microstructure much at all, except to the extent that it affects overall material properties such as toughness and strength. Pioneering early work on energy-limited fragmentation includes Mott & Linfoot (1943) and Mott (1947); see also Grady (2006).

Flaw-limited cases typically lead to specific solutions for specific materials under specific conditions. By contrast, energy-limited cases may lead to universal solutions that apply to a wide range of different materials—including brittle solids, ductile solids, inviscid liquids, and viscous liquids—and under a wide range of different conditions. In keeping with its general philosophy, this book will consider only energy-limited fragmentation. In particular, the results shown in Figures 15.2 to 15.4 are energy- rather than flaw-limited. In other words, no advance knowledge of the material microstructure is required to model these cases.

As the second main challenge, unitary fragmentation may depend on periodic instabilities, chaotic instabilities, or both. Periodic instabilities tend to start on the surface of an object and work their way towards the interior. For brittle solids, this includes bending- and buckling-type instabilities; see, e.g., Gladden et al. (2005) and Vandenberghe & Villermaux (2013). For ductile solids and liquids, this includes Rayleigh–Taylor and Kelvin–Helmholtz instabilities; see, e.g., Meng & Colonius (2018). By contrast, chaotic instabilities tend to start in the interior of an object and work their way towards the surface. In some cases, multiple instabilities co-exist and compete with each other, including both periodic and chaotic instabilities. Chaos often leads to universal results that apply equally to any type of material, liquid or solid, under a wide range of different conditions. Thus, this book will focus on chaotic rather than periodic instabilities. However, because they are often so intertwined with chaotic instabilities, this book will also discuss periodic instabilities. As noted previously, periodic instabilities tend to obtain larger average fragment sizes, and smaller fragment size spreads, than chaotic instabilities.

As the third main challenge, unitary fragmentation often leads to multiple stages, each with distinctly different physics. *Primary fragmentation* occurs when particles separate from a large unitary body. *Secondary fragmentation* occurs when a particle produced in an earlier stage experiences fragmentation. Secondary fragmentation occurs because primary fragmentation reduces but does not eliminate the internal and kinetic energy of the surviving material. As illustrated in Figures 15.5 and 15.6, secondary fragmentation in liquids commonly involves two or three stages.[1] At early times following primary fragmentation—when particle concentrations, masses, momenta, and energies are highest—secondary fragmentation may occur due to residual strain rates and collisions between fragments; see Figures 15.5.a and 15.6.a. In some cases, this may take the form of a fragmentation cascade, as discussed in the previous three chapters. Next, at middling times following primary fragmentation, secondary fragmentation may occur due to aerodynamic drag; see Figures 15.5.b and 15.6.b. Finally, at late times following primary fragmentation, secondary

[1] Notice that Figures 15.5 and 15.6 involve simple liquids, whose fragments tend to quickly assume nearly spherical shapes, due to the effects of surface tension. By contrast, non-simple liquids such as viscous melts and non-Newtonian starch solutions may produce highly irregular fragments, more like traditional solids.

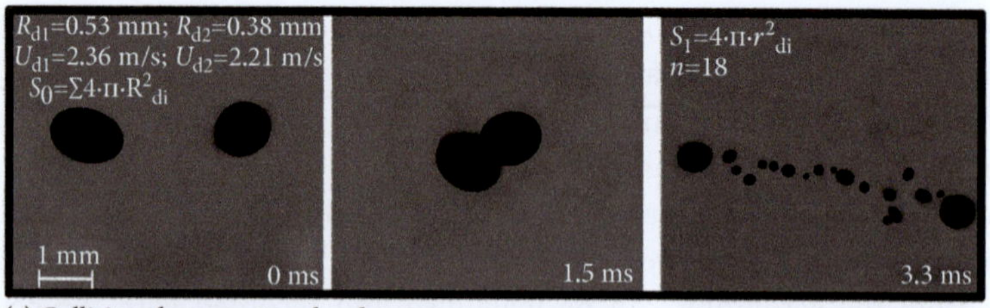

(a) Collisions between two droplets.

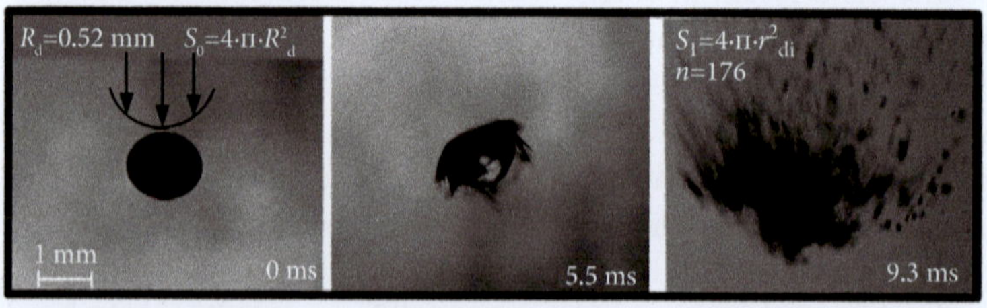

(b) Aerodynamic breakup.

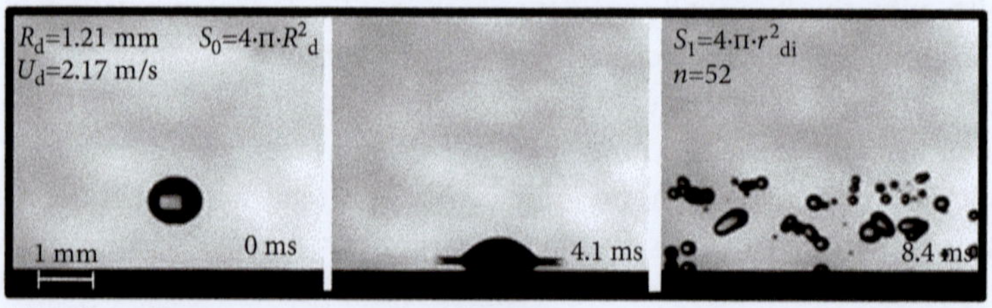

(c) Collision between a droplet and a solid surface.

Figure 15.5 Experimental results showing three types of secondary fragmentation in liquids. Reprinted from Kuznetsov et al. (2019) with permission from Elsevier.

fragmentation may occur due to collisions between fragments and adjacent solid surfaces; see Figures 15.5.c and 15.6.c. Like liquids, solids may also experience secondary fragmentation, e.g., Figure 1.3 shows a solid experiencing aerodynamic breakup. In general, secondary fragmentation of solids is more likely when primary fragmentation causes cracking, localized crumbling, plastic distortion, or other macro- and microscopic damage which reduces the virgin strength of the solid.

As one way to model unitary fragmentation, many computational methods try to capture the true first-principles physics. Thus, fragmentation of brittle solids is often modeled using commercial finite element codes such as LS-DYNA, Abaqus or ANSYS Autodyn that include crack nucleation and propagation models such as X-FEM; see, e.g., Moxnes et al. (2014), Strack et al. (2015), Jaime et al. (2015), Rahbek et al. (2017), Bewick et al. (2017a, b) and An et al. (2017). Similarly, fragmentation of simple liquids is often modeled using commercial Computational Fluid Dynamics (CFD) codes such as ANSYS Fluent or KIVA that include periodic instability models such as Kelvin–Helmholtz and Rayleigh–Taylor (KHRT) models;[2] see, e.g., Jabbari et al. (2013), Sokolov (2018) and Basu et al. (2018 a, b).

[2] Rayleigh–Taylor instabilities occur when a liquid surface accelerates normally into the air (compression), while Kelvin–Helmholz instabilities occur when a liquid surface flows parallel to the air (shear).

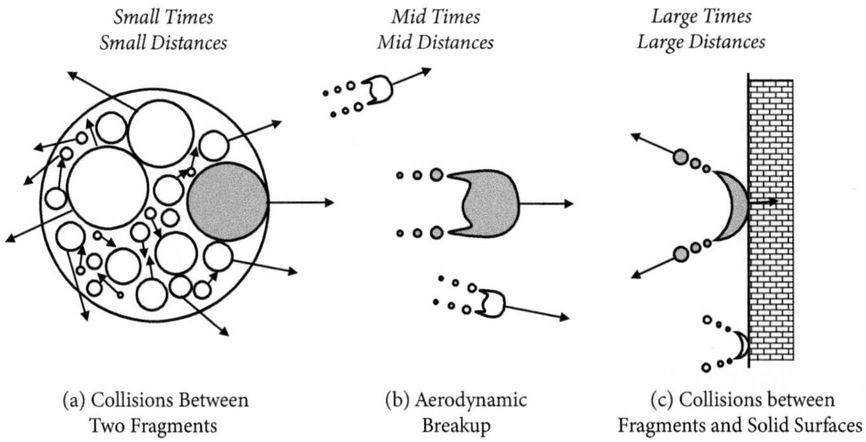

Figure 15.6 Diagram showing three types of secondary fragmentation in liquids.

However, there are significant obstacles to such literal physics-based approaches including the following:

- Before primary fragmentation, this approach may require accurately representing the material microstructure which, at high strain rates, changes continuously.
- During primary fragmentation, it may require accurately representing chaotic instabilities by, say, introducing artificial randomness to overcome artificial dissipation.
- After primary fragmentation, it may require representing vast numbers of particles with seemingly random sizes, shapes, and velocities, spanning many orders-of-magnitude.
- If secondary fragmentation occurs, every particle produced by primary fragmentation becomes a separate unitary fragmentation problem, each just as challenging as the first.

Such factors significantly limit the utility of high-fidelity computational methods in practical applications. For example, Shinjo (2018) says "spray simulation is one of the largest bottlenecks in [the] automotive design processes ... The primary atomization region is the main cause for this situation ... [In the current] conventional approach [to] ... spray simulation ... the near-nozzle primary atomization region is omitted and replaced by some model information which is not necessarily physically accurate." This "model information" includes semi-empirical liquid fragmentation results calibrated using "experiments or direct numerical simulation (DNS)." While the exact language differs, practitioners in many different application areas have generally come to similar conclusions.

Researchers can respond to this situation in two ways. In the first approach, researchers may include more detailed and accurate physics in their computational models. In the second approach, counterintuitively, researchers may include *less* detailed and accurate physics in their computational models. For example, in Discrete Element Methods (DEMs), unitary objects are represented by sets of particles bound together by springs, beams, or other elements; see Section 4.7. In fact, a growing body of evidence indicates that codes with less-realistic physics such as DEMs often outperform codes with more-realistic physics such as FEMs; see the earlier discussion in Chapter 4. However, DEMs still face some of the same fundamental limitations as FEMs. For example, Tavares (2017) notes that "it could take as much as 10GB of computer memory to run ... commercial software such as EDEM (DEM Solutions, UK) ...Moreover, this is only considering a 30 cm slice ... which could have more than two billion elements [particles] with sizes above 0.5mm." Chapter 4 already discussed computational methods for unitary fragmentation. The remainder of this book will focus on analytic methods for unitary fragmentation including mean-field theory, stochastic geometry, maximum entropy theory, and dimensional analysis.

15.2 Strain Rates

From one point of view, unitary fragmentation is caused by mechanical stresses and strains. A *mechanical stress* is an internal force per unit surface or cross-sectional area. A *mechanical strain* is a deformation or change in shape due to a mechanical stress. In general, stresses and strains are tensors. Like all tensors, in a given coordinate system, stresses and strains can be represented by 3×3 matrices of real numbers. In such matrices, the diagonal terms are *tensile* when positive, the diagonal terms are *compressive* when negative, and the off-diagonal terms are *shear*. Tensile strains cause stretching, elongation, or tearing while compressive strains cause contraction, shortening, or crushing. By definition, solids resist shear stresses while fluids do not. Instead, fluids experience continuous shear strains in response to shear stresses. Shear stresses and strains may ultimately lead to tensile stresses and strains. Similarly, compressive stress and strains may ultimately lead to tensile stresses and strains; this important case is commonly known as *spall*; see Example 15.1.

The remainder of this section discusses possible algebraic expressions for simple tensile strains and strain rates. According to Besson et al. (2010) "strain means a change of shape with respect to a reference shape ... When one has to characterize the deformation [strain] ... an infinity of choices appear. The final choice is governed by the strain domain considered

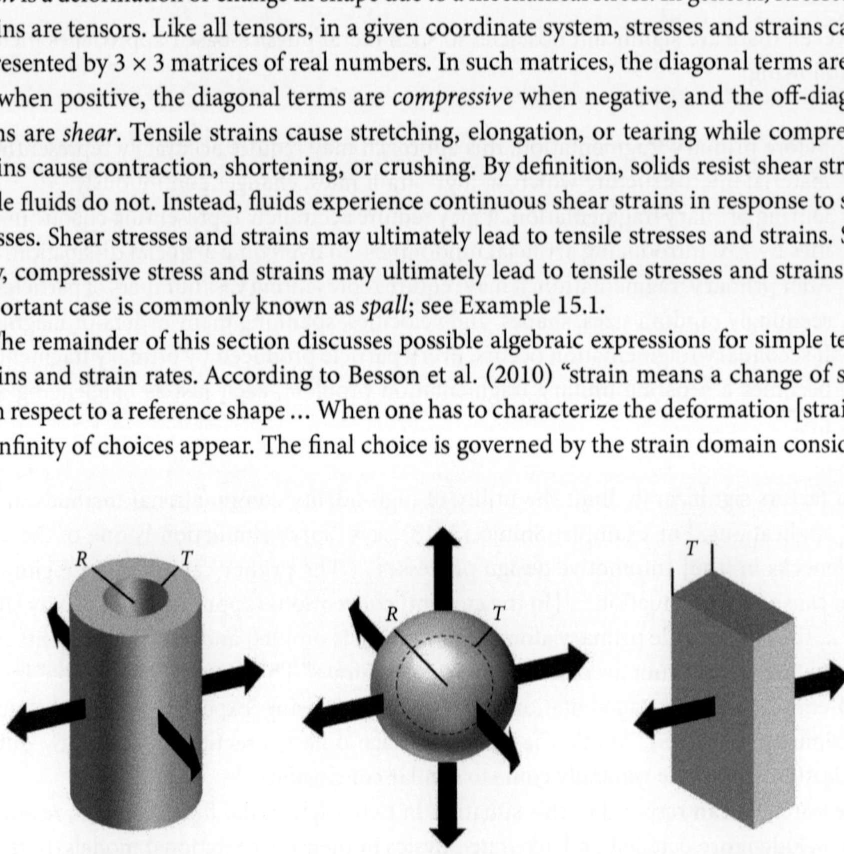

(a) Uniaxial (Cylindrical) (b) Equibiaxial (Spherical) (c) Uniaxial (Cartesian)

Figure 15.7 Common forms of uniform tensile stress in three different coordinate systems.

and ... personal preferences." In general, the expressions given below for strains and strain rates are only *averages*—the actual strains and strain rates vary in a seemingly random fashion in time and space due to the material microstructure, chaotic instabilities, and other factors. This section concerns only uniform, uniaxial forces in simple geometries, which lead to spatial homogeneity and isotropy, at least for large enough strain rates. More specifically, Figure 15.7 illustrates three cases involving pure tensile strain. Examples 15.2, 15.3, and 15.4 give possible expressions for the average strains and strain rates corresponding to Figures 15.7a, 15.7b, and 15.7c, respectively. Notice that, because of the simple geometries, the expressions given in these examples for strains and strain rates are scalars rather than full tensors.

Example 15.1 Introduction to Spall

According to Antoun et al. (2003), "spall is tensile failure that results from the nucleation, growth and coalescence of microfractures and microvoids produced in concentrations of the order of $10^6/\text{cm}^3$ where large stresses are imposed for short times ... The nucleated microcracks or voids grow and coalesce .... [until] the body separates into distinct parts." Grady (1988) notes that "typically, tensile spall within a body will be preceded by a compression process ... Reflection of the [compressive] stress wave at free surfaces and subsequent interaction within the material leads to tensile stress states." As a specific example, Wu et al. (2009) say "attacks using ... explosive devices on reinforced concrete buildings create a rapid release of energy in the form of a shock wave. Most casualties and injuries resulting from such an attack are not caused by the blast itself, but rather by the disintegration and fragmentation of the reinforced concrete member due to concrete spallation on the opposite side of the member and which is propelled at high velocities." Figure 15.8 illustrates the physical processes involved.

For brittle solids, Grady (2017) says spall may involve "cleavage on a plane of fracture of near-atomic dimensions at the crack tip." Alternatively, it may involve "ductile fracture characterized by a process of hole cavitation, plastic ligamentation and surface separation at the crack tip," providing this does not involve substantial amounts of dissipation. For ductile solids, Grady (2017) says spall involves "tensile failure through the plastic growth and coalescence of a collection of cavities." Finally, for inviscid and viscous liquids, Grady (2017) says "the process of spall in simple liquids appears to be that of cavitation and growth of holes or bubbles at sufficient states of transient hydrodynamic tension."

In general, spall is chaotic. More specifically, Antoun et al. (2003) say spall "is an unstable process during which there are no large regions of nearly uniform behavior," e.g., "spherical voids are nucleated at random." Similarly, Grady (2017) says "a state of homogeneous tensile stress is intrinsically unstable and small perturbations in the material ... will lead to the opening of voids and the growth of spall failure." For brittle solids, Grady (2017) says "under the intense dynamic loading of ... tensile spall ... any particular crack is subject to, not only the far field stress, but also the random and chaotic force fields radiating from the many neighboring cracks experiencing similar failure dynamics."

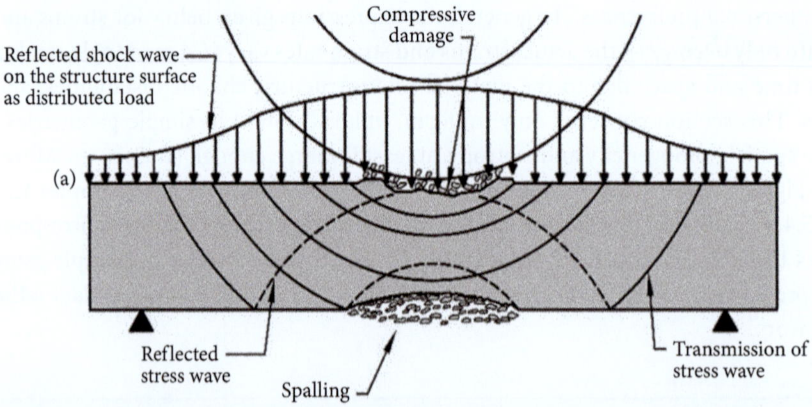

Figure 15.8 Compressive stresses on the one face of a concrete panel lead to spall on the opposite face.
Reprinted with permission from Zhang et al. (2019).

Example 15.2 Uniform uniaxial stress in cylindrical coordinates

Suppose a cylinder, cylindrical shell, circular disk, or circular ring experiences uniform tensile stress in the radial direction; see Figure 15.7a. In this case, the *engineering strain* and *engineering strain rate* may be defined as follows:

$$\varepsilon = \frac{R - R_0}{R_0} \tag{15.1}$$

$$\dot{\varepsilon} = \frac{d\varepsilon}{dt} = \frac{1}{R_0}\frac{dR}{dt} = \frac{V}{R_0} \tag{15.2}$$

where $R_0 = R(0)$ is the initial radius, $R(t)$ is the radius at time t, and $V(t)$ is the radial expansion speed at time t. According to Equation (15.1), ε is the fractional change in the radius. Alternatively, the *true strain* and *true strain rate* may be defined as follows:

$$\varepsilon = \ln \frac{R}{R_0} \tag{15.3}$$

$$\dot{\varepsilon} = \frac{d\varepsilon}{dt} = \frac{R_0}{R}\frac{d}{dt}\left(\frac{R}{R_0}\right) = \frac{V}{R}. \tag{15.4}$$

The previous equations are often used in expanding ring and expanding cylinder tests to determine material strengths at high strain rates; see, e.g., Davis (2004).

Example 15.3 Uniform equi-biaxial stress in spherical coordinates

Suppose a sphere, spherical shell, hemisphere, or hemispherical shell experiences uniform tensile stress in the radial direction; see Figure 15.7.b. In this case, the engineering strain and engineering strain rate are rarely used. Instead, the true strain and strain rate are usually used:

$$\varepsilon = \ln\left(\frac{R}{R_0}\right)^2 = 2\ln\frac{R}{R_0} \tag{15.5}$$

$$\dot{\varepsilon} = \frac{d\varepsilon}{dt} = 2\frac{R_0}{R}\frac{d}{dt}\left(\frac{R}{R_0}\right) = 2\frac{V}{R} \tag{15.6}$$

where $R_0 = R(0)$ is the initial radius, $R(t)$ is the radius at time t, and $V(t)$ is the radial expansion speed at time t. The previous equations are sometimes, albeit infrequently, used in expanding sphere tests to determine material strengths at high strain rates; see, e.g., Buy & Llorca (2002) and Llorca & Buy (2002). As explained by Buy & Llorca (2002) "the interpretation of the results may be difficult because of some parasitical phenomena." For additional perspective, section 6.3.1 of Grady (2006) treats general biaxial strains, allowing for different rates in different directions.

Example 15.4 Uniform uniaxial stress in Cartesian coordinates

Suppose a rectangular object experiences uniform tensile stress along its smallest dimension; see Figure 15.7c. In this case, the *engineering strain* and *engineering strain rate* may be defined as follows:

$$\varepsilon = \frac{T - T_0}{T_0} \tag{15.7}$$

$$\dot{\varepsilon} = \frac{d\varepsilon}{dt} = \frac{1}{T_0}\frac{dT}{dt} = \frac{V}{T_0} \tag{15.8}$$

where $T_0 = T(0)$ is the initial thickness, $T(t)$ is the thickness at time t, and $V(t)$ is the axial expansion speed at time t. According to Equation (15.7), ε is the fractional change in the thickness. Alternatively, the *true strain* and *true strain rate* may be defined as follows:

$$\varepsilon = \ln\frac{T}{T_0} \tag{15.9}$$

$$\dot{\varepsilon} = \frac{d\varepsilon}{dt} = \frac{T_0}{T}\frac{d}{dt}\left(\frac{T}{T_0}\right) = \frac{V}{T}. \tag{15.10}$$

As another alternative, suppose a rectangular object experiences uniform tensile stress along its largest dimension L instead of its smallest dimension T; see Figure 15.9a. In this case, the same expressions apply, after replacing T by L and T_0 and L_0, respectively. The resulting expressions are often used in tensile tests to determine material strengths at low to moderate strain rates; see, e.g., Davis (2004). In such tests, Davis (2004) notes that "as long as the deformation is uniform, ... the true ... strain can be calculated from the engineering quantities." As a final alternative, suppose a rectangular object experiences uniform compressive stress along its largest dimension L; see Figure 15.9b. In most cases, as seen in Example 15.5, the response involves a non-uniform strain tensor with tensile, compressive, and shear components.

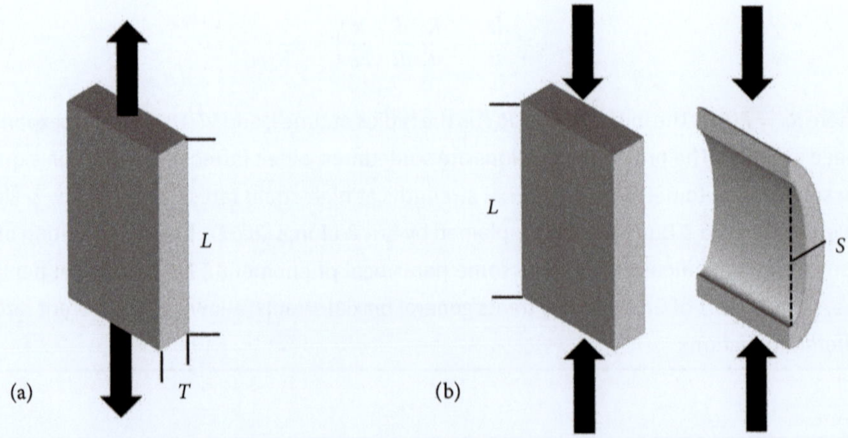

Figure 15.9 Two common forms of stress and strain in Cartesian coordinates.

Example 15.5 Bending strains on panels and bars

Suppose that a flat panel or bar bends due to lateral loads as seen in Figure 15.1 or compressive loads as seen in Figure 15.9b. In such cases, the true strain and true strain rate are rarely used. Instead, the engineering strain and strain rate are usually used. In general, the panel or bar experiences a range of different tensile, compressive, and shear stresses and strains. However, in the simplest case, the panel or bar bends sinusoidally, such that the maximum tensile strain and strain rates lie along the center line of the tensile face. Then the maximum engineering tensile strain and strain rate are as follows:

$$\varepsilon = \frac{6TS}{L^2} \tag{15.11}$$

$$\dot{\varepsilon} = \frac{d\varepsilon}{dt} = \frac{6T}{L^2}\frac{dS}{dt} = \frac{6TV}{L^2} \tag{15.12}$$

where L is the initial length of the panel, T is the initial thickness of the panel, $S(t)$ is the lateral displacement of the center of the panel at time t, and $V(t)$ is the panel center speed at time t. The previous equations were developed for use in flexure tests to determine the strength of certain types of materials, such as ceramics, which cannot be well-characterized by standard tensile tests; see, e.g., Baratta (1984), eq. 3–11 in Harris (1999), and table 7–2 in Wiederhorn et al. (2011). As one issue with such flexure tests, Davis (2004) notes that, because the maximum stress occurs on the surface, flexure tests may experience "preferential failure from surface flaws" unrelated to the bulk material properties. However, the larger the strain rates, the less such pre-existing flaws matter. As another issue, bending causes not only tensile strains but also shear and compressive strains. This creates measurement problems for "materials relatively weak in shear" or "that fail in compression first."

15.3 Conclusions

This chapter has discussed some of the basic principles of unitary fragmentation. Building on this foundation, the next two chapters describe possible analytic models for the average fragment size and the fragment size distribution. Throughout all three chapters, liquids and solids are treated in a semi-unified fashion. According to Buchanan (2017), "the real fundamental difference between liquids and solids isn't structural, but dynamic: liquids can flow and solids cannot. More technically, the difference lies in the response to a shear stress—solids may deform, elastically, but ultimately stand up to the stress. In contrast, liquids cannot, and instead deform continuously." Solid fragmentation is traditionally described in terms of stresses and strains as described in books such as Davison et al. (1996), Antoun et al. (2003), Grady (2006) and Lambert et al. (2019). By contrast, liquid fragmentation is traditionally described in terms of non-dimensional parameters such as Reynolds numbers and Weber numbers as described in books such as Bayvel & Orzechowski (1993), Lin (2003), Ashgriz (2011), Lefebvre & McDonell (2017) and Basu et al. (2018a, b). However, as seen in this chapter, both liquids and solids can be described in terms of stresses and strains. Similarly, as seen in the next chapter, both liquids and solids can be described in terms of non-dimensional parameters such as Reynolds numbers and Weber numbers.

This book generally assumes that aggregation and fragmentation cascades are never-ending random processes. By contrast, this book generally assumes that unitary fragmentation ends after a fixed, finite amount of time, possibly to be followed by fragmentation or aggregation cascades. If unitary fragmentation is spatially homogeneous and isotropic—at least at the instant the unitary body is converted into particles—it may be treated as a random *event* rather than as a random *process*. In other words, unitary fragmentation can be modeled as having only two states, e.g., "before" vs. "after" or "input" vs. "output," without considering intermediate states. In that sense, unitary fragmentation can be treated like other classic random events such as coin flips and dice rolls.

Even when unitary fragmentation is spatially homogeneous and isotropic, different size ranges tend to obtain different size distributions. One possible scenario is as follows:

- *"Large,"* a.k.a., *energy-containing range*. Fragments are formed primarily by tensile strain. A significant percentage (> 10%) of these fragments form in less than three dimensions, i.e., they have smooth surfaces inherited from the original object.
- *"Medium,"* a.k.a., *semi-inertial range*. Fragments are formed primarily by tensile strain. Most of these fragments (> 90%) form in three-dimensions, i.e., they do not have any smooth surfaces inherited from the original object.

- *"Small,"* a.k.a., *dissipative range*. Fragments are formed primarily by shear and compressive strain. Unlike the other two ranges, dissipation is always an important factor in this range.

Of course, not every unitary fragmentation event obtains all three ranges—some only obtain one or two. This should become clearer over the course of the next two chapters.

References

H. M. An, H. Y. Liu, H. Han, X. Zheng, & X. G. Wang, Hybrid Finite-Discrete Element Modelling of Dynamic Fracture and Resultant Fragment Casting and Muck-Piling by Rock Blast, *Computers and Geotechnics*, 81, 322–345, 2017, https://doi.org/10.1016/j.compgeo.2016.09.007

T. Antoun, L. Seaman, D. R. Curran, G. I. Kanel, S. V. Razorenov, & A. V. Utkin, *Spall Fracture*, Springer, 2003, https://doi.org/10.1007/b97226

N. Ashgriz, editor, *Handbook of Atomization and Sprays: Theory and Applications*, Springer, 2011, https://doi.org/10.1007/978-1-4419-7264-4

F. Baratta, Requirements for Flexure Testing of Brittle Materials, in S. Freiman & C. H. Michael, editors, *Methods for Assessing the Structural Reliability of Brittle Materials*, ASTM International, 1984, https://doi.org/10.1520/STP35245S

S. Basu, A. K. Agarwal, & C. Patel, editors, *Droplets and Sprays: Applications for Combustion Propulsion*, Springer, 2018a, https://doi.org/10.1007/978-981-10-7449-3

S. Basu, A. K. Agarwal, & C. Patel, editors, *Droplets and Spray Transport: Paradigms and Applications*, Springer, 2018b, https://doi.org/10.1007/978-981-10-7233-8

L. Bayvel & Z. Orzechowski, *Liquid Atomization*, Taylor & Francis, 1993, https://doi.org/10.1201/9780203748787

J. Besson, G. Cailletaud, J.-L. Chaboche, & S. Forest, *Non-Linear Mechanics of Materials*, Springer, 2010, https://doi.org/10.1007/978-90-481-3356-7

B. Bewick, G. Rolater, J. Bui, M. Barsotti, A. Ziemba, M. Sanai, et al., *Fragmentation of Solid Materials Using Shock Tubes. Part 1: First Test Series in a Small Diameter Shock Tube*, Defense Threat Reduction Agency Technical Report DTRA-TR-17-21, January 2017a, http://apps.dtic.mil/sti/citations/AD1025099

B. Bewick, G. Rolater, K. Marchand, A. Ziemba, M. Sanai, & C. Laney, *Fragmentation of Solid Materials Using Shock Tubes. Part 2: First Test Series in a Large Diameter Shock Tube*, Defense Threat Reduction Agency Technical Report DTRA-TR-17-50, December 2017b, http://apps.dtic.mil/sti/citations/AD1043784

B. Bewick, G. Rolater, K. Marchand, A. Ziemba, M. Sanai, & C. Laney, *Fragmentation of Solid Materials Using Shock Tubes. Part 4: Second Test Series in a Large Diameter Shock Tube*, Defense Threat Reduction Agency Technical Report DTRA-TR-19-5, November 2018, https://apps.dtic.mil/sti/citations/AD1065932

M. Buchanan, A Matter of Responding to Stress, *Nature Physics*, 13, 620, 2017, https://doi.org/10.1038/nphys4203

F. Buy & F. Llorca, The Expanding Shell Test: Numerical Simulation of the Experiment, in M. D. Furnish, N. N. Thadhani, & Y. Horie, editors, *Shock Compression of Condensed Matter, AIP Conference Proceedings*, 620, 2002, https://doi.org/10.1063/1.1483546

D. R. Curran, L. Seaman, & D. A. Shockey, Dynamic Failure in Solids, *Physics Today*, 30(1), 46–55, 1977, https://doi.org/10.1063/1.3037367

J. R. Davis, editor, *Tensile Testing*, 2nd edition, ASM International, 2004, https://doi.org/10.31399/asm.tb.tt2.9781627083553

L. Davison, D. E. Grady, & M. Shahinpoor, editors, *High-Pressure Shock Compression of Solids II: Dynamic Fracture and Fragmentation*, Springer-Verlag, 1996, https://doi.org/10.1007/978-1-4612-2320-7

L. Dias & F. Parisio, Minimal Fragmentation of Regular Polygonal Plates, *Physical Review E*, 90(3), 032405, 2014, https://doi.org/10.1103/PhysRevE.90.032405

J. J. Gilvarry, Fracture of Brittle Solids. I. Distribution Function for Fragment Size in Single Fracture (Theoretical), *Journal of Applied Physics*, 32(3), 391–399, 1961, https://doi.org/10.1063/1.1736016

J. R. Gladden, N. Z. Handzy, A. Belmonte, & E. Villermaux, Dynamic Buckling and Fragmentation in Brittle Rods, *Physical Review Letters*, 94(3), 035503, 2005, https://doi.org/10.1103/PhysRevLett.94.035503

D. E. Grady, The Spall Strength of Condensed Matter, *Journal of Mechanics and Physics of Solids*, 36(3), 353–384, 1988, https://doi.org/10.1016/0022-5096(88)90015-4

D. Grady, *Fragmentation of Rings and Shells: The Legacy of N. F. Mott*, Springer, 2006, https://doi.org/10.1007/b138675

D. E. Grady, *Physics of Shock and Impact, Volume 1: Fundamentals and Dynamic Failure*, IOP Publishing, 2017, https://doi.org/10.1088/978-0-7503-1254-7

D. C. Harris, *Materials for Infrared Windows and Domes: Properties and Performance*, SPIE, 1999, https://doi.org/10.1117/3.349896

F. Jabbari, M. Jadidi, R. Wuthrich, & A. Dolatabadi, A Numerical Study of Suspension Injection in Plasma-Spraying Process, *Journal of Thermal Spray Technology*,23, 3–13, 2013, https://doi.org/10.1007/s11666-013-0030-9

M. C. Jaime, Y. Zhou, J.-S. Lin, & I. K. Gamwo, Finite Element Modeling of Rock Cutting and its Fragmentation Process, *International Journal of Rock Mechanics and Mining Sciences*, 80, 137–146, 2015, https://doi.org/10.1016/j.ijrmms.2015.09.004

G. V. Kuznetsov, N. E. Shlegel, Y. Solomatin, & P. A. Strizhak, Combined Techniques of Secondary Atomization of Multi-Component Droplets, *Chemical Engineering Science*, 209, 115199, 2019, https://doi.org/10.1016/j.ces.2019.115199

D. E. Lambert, C. L. Pasiliao, B. Erzar, B. Revil-Baudard, & O. Cazacu, editors, *Dynamic Damage and Fragmentation*, Wiley, 2019, https://doi.org/10.1002/9781119579311

A. H. Lefebvre & V. G. McDonell, *Atomization and Sprays*, 2nd edition, CRC Press, 2017, https://doi.org/10.1201/9781315120911

S. P. Lin, *Breakup of Liquid Sheets and Jets*, Cambridge University Press, 2003, https://doi.org/10.1017/CBO9780511547096

F. Llorca & F. Buy, The Contribution of the Expanding Shell Test to the Modeling of Elastoplasticity at High Strain Rates, in M. D. Furnish, N. N. Thadhani, & Y. Horie, editors, *Shock*

Compression of Condensed Matter, AIP Conference Proceedings, 620, 2002, https://doi.org/10.1063/1.1483546

J. C. Meng & T. Colonius, Numerical Simulation of the Aerobreakup of a Water Droplet, *Journal of Fluid Mechanics*, 835, 1108–1135, 2018, https://doi.org/10.1017/jfm.2017.804

N. F. Mott & E. H. Linfoot, *A Theory of Fragmentation*, U. K. Ministry of Supply, AC3348, January 1943 [reproduced in section II.1 of Grady (2006)], https://doi.org/10.1007/978-3-540-27145-1_9

N. F. Mott, Fragmentation of Shell Cases, *Proceedings of the Royal Society A: Mathematical, Physical and Engineering Sciences*, 189(1018), 300–308, 1947, https://doi.org/10.1098/rspa.1947.0042

J. F. Moxnes, A. K. Prytz, Ø. Frøyland, S. Klokkehaug, S. Skriudalen, E. Friis, et al., Experimental and Numerical Study of the Fragmentation of Expanding Warhead Casings by Using Different Numerical Codes and Solution Techniques, *Defence Technology*, 10(12), 161–176, 2014, https://doi.org/10.1016/j.dt.2014.05.009

D. B. Rahbek, J. W. Simons, B. B. Johnsen, T. Kobayashi, & D. A. Shockey, Effect of Composite Covering on Ballistic Fracture Damage Development in Ceramic Plates, *International Journal of Impact Engineering*, 99, 58–68, 2017, https://doi.org/10.1016/j.ijimpeng.2016.09.010

P. Ray, Statistical Physics Perspective of Fracture in Brittle and Quasi-Brittle Materials, *Philosophical Transactions of the Royal Society A: Mathematical, Physical and Engineering Sciences*, 377(2136), 20170396, 2018, https://doi.org/10.1098/rsta.2017.0396

L. Seaman, D. R. Curran, & D. A. Shockey, Computational Models for Ductile and Brittle Fracture, *Journal of Applied Physics*, 47(11), 4814–4826, 1976, https://doi.org/10.1063/1.322523

J. Shinjo, Recent Advances in Computational Modeling of Primary Atomization of Liquid Fuel Sprays, *Energies*, 11(11), 2971, 2018, https://doi.org/10.3390/en11112971

A. Sokolov, L. D. Rubio, J. F. Brady, & I. S. Aranson, Instability of Expanding Bacterial Droplets, *Nature Communications*, 9, 1322, 2018, https://doi.org/10.1038/s41467-018-03758-z

O. E. Strack, R. B. Leavy, & R. M. Brannon, Aleatory Uncertainty and Scale Effects in Computational Damage Models for Failure and Fragmentation, *International Journal for Numerical Methods in Engineering*, 102(3–4), 468–495, 2015, https://doi.org/10.1002/nme.4699

L. M. Tavares, A Review of Advanced Ball Mill Modelling, *KONA Power and Particle Journal*, 34, 106–124, 2017, https://doi.org/10.14356/kona.2017015

N. Vandenberghe & E. Villermaux, Geometry and Fragmentation of Soft Brittle Impacted Bodies, *Soft Matter*, 9(34), 8162–8176, 2013, https://doi.org/10.1039/c3sm50789k

A. Vledouts, J. Quinard, N. Vandenberghe, & E. Villermaux, Explosive Fragmentation of Liquid Shells, *Journal of Fluid Mechanics*, 788, 246–273, 2016, https://doi.org/10.1017/jfm.2015.716

S. Wiederhorn, R. Fields, S. Low, G.-W. Bahng, A. Wehrstedt, J. Hahn, et al., Mechanical Properties, in H. Czichos, T. Saito, & L. Smith, editors, *Springer Handbook of Metrology and Testing*, Springer, 2011, https://doi.org/10.1007/978-3-642-16641-9_7

C. Wu, R. Nurwidayati, & D. J. Oehlers, Fragmentation from Spallation of RC Slabs due to Airblast Loads, *International Journal of Impact Engineering*, 36(12), 1371–1376, 2009, https://doi.org/10.1016/j.ijimpeng.2009.03.014

A. L. Yarin, I. V. Roisman, & C. Tropea, *Collision Phenomena in Liquids and Solids*, Cambridge University Press, 2017, https://doi.org/10.1017/9781316556580

C. Zhang, G. Gholipour, & A. A. Mousavi, Nonlinear Dynamic Behavior of Simply-Supported RC Beams Subjected to Combined Impact-Blast Loading, *Engineering Structures*, 181, 124–142, 2019, https://doi.org/10.1016/j.engstruct.2018.12.014

16

Unitary Fragmentation Part 2: Average Sizes

This chapter describes analytic models for the characteristic particle sizes produced in unitary fragmentation. The next chapter describes analytical models for particle size distributions. Both chapters use mean-field theory to set the stage. However, mean-field theory is ultimately too limiting. Thus, this chapter supplements mean-field theory with alternatives such as energy analysis, stability analysis, and dimensional analysis. Similarly, the next chapter supplements mean-field theory with alternatives such as stochastic geometry and maximum entropy theory. While most of this book assumes chaos or randomness, this chapter is an exception. Of the four modeling approaches discussed here, only mean-field theory assumes chaos or randomness.

16.1 Mean-Field Theory Models

As one possible model of unitary fragmentation, suppose that a given body is divided into N random pieces, and each of those pieces is divided into N random pieces, and so forth. In this case, unitary fragmentation is the same as random fragmentation as discussed in Chapter 13. According to mean-field theory, random fragmentation depends on four things:

- *Rate kernels.* As usual, this section assumes that the rate kernels are power laws, i.e., $J(x) = J_0 x^{\kappa}$ where $J_0 > 0$ is constant. Furthermore, this section assumes $\kappa > 0$.
- *Child size distributions.* As usual, this section assumes that the child size distributions are power laws, i.e., $m(x, y) = (\eta+1)x^{\eta-1}/y^{n}$ where $0 < \eta = 1/(N-1) \leq 1$.
- *Initial conditions.* As usual, this section assumes that the effects of the initial conditions drop out at sufficiently late times, due to the effects of chaos or randomness.
- *Boundary conditions.* This section assumes free-field boundary conditions. Because of this assumption, and because $\kappa > 0$, conservation of mass applies.

As discussed in the last chapter, unitary fragmentation often obtains a uniform steady strain rate, at least on average. In such cases, it is reasonable to assume that the fragmentation rate J is proportional to the strain rate $\dot{\varepsilon}$. In other words:

$$J_0 = C_0 \dot{\varepsilon} \tag{16.1}$$

where C_0 is constant. Then, according to Equation (13.97), mean-field theory obtains the following result:

$$M_{\text{char}} \cong \left(\frac{b}{t_0 C_0 \dot{\varepsilon}} \right)^{1/\kappa} \tag{16.2}$$

where M_{char} is a characteristic fragment mass, $b = 1/(\kappa a_0)$ is a constant, and t_0 is the time at which fragmentation ends.

Particle Dynamics with Aggregation and Fragmentation. Culbert B. Laney, Oxford University Press. © Culbert B. Laney (2025).
DOI: 10.1093/9780198920687.003.0016

While this book generally expresses results in terms of fragment masses, the relevant research literature concerns fragment diameters. Assume that:

$$M_{\text{char}} = \rho C D_{\text{char}}^{m}. \tag{16.3}$$

In this expression, ρ is the fragment density, C is a unitless factor that depends on the fragment shape and size distribution, and m is the fragment mass dimension; see, e.g., Laney (2016). Combining Equation (16.2) and (16.3) gives:

$$D_{\text{char}} \cong \left(\frac{1}{\rho C}\right)^{1/m} \left(\frac{b}{t_0 C_0 \dot{\varepsilon}}\right)^{1/m\kappa}. \tag{16.4}$$

If the factors that do not involve $\dot{\varepsilon}$ are combined as follows:

$$C_1 = \left(\frac{1}{\rho C}\right)^{1/m} \left(\frac{b}{t_0 C_0}\right)^{1/m\kappa} \tag{16.5}$$

then Equation (16.4) becomes:

$$D_{\text{char}} \cong C_1 \dot{\varepsilon}^{-1/m\kappa}. \tag{16.6}$$

According to Equation (13.29), in ideal cases, κ is an integer multiple of 1/6 between −2 and 2. If $m = 3$ and $\kappa > 0$ then Equation (16.6) becomes:

$$D_{\text{char}} \cong C_1 \dot{\varepsilon}^{-2/i} \tag{16.7a}$$

where $i = 1, 2, \ldots, 12$. An alternative and arguably better notation is as follows:

$$D_{\text{char}} \sim \dot{\varepsilon}^{-2/i} \tag{16.7b}$$

where "~" means "is proportional to." This notation is used to draw attention to some factors in a product while ignoring others. Recall that Section 13.9 obtained a similar result by assuming that fragmentation consumes power at a constant rate, at least on average. In fact, unitary fragmentation sometimes does consume power P at a roughly constant rate; see, e.g., Bergstrom (1963) and Tavares (1999). However, the vast majority of the relevant research assumes that unitary fragmentation involves constant strain rates rather than constant power consumption.

As the main drawback, mean-field results are expressed in terms of mean-field parameters. In other words, to determine D_{char} using Equation (16.4), it is necessary to determine mean-field parameters such as J_0, κ, and η. While these mean-field parameters have reasonably clear physical meanings for random fragmentation, they do not have clear physical meanings for unitary fragmentation. Rather, they are artifacts of assuming that unitary fragmentation occurs due to iteratively subdividing particles into a fixed number of pieces ($N = 1 + 1/\eta$) at a finite rate ($J_0 x^{\kappa}$) over a finite period of time (t_0). The remainder of this chapter concerns alternative modeling approaches such as conservation of energy, stability analysis, and dimensional analysis. Just like mean-field theory, these alternatives obtain $D_{char} \cong C_1 \dot{\varepsilon}^{-2/i}$ where i is a positive integer. However, instead of being expressed in terms of artificial mean-field parameters, C_1 is expressed in terms of measurable physical properties. For liquids, these physical properties typically include surface tension or viscosity. For solids, these physical properties typically include fracture toughness or yield strength.

16.2 Non-Mean-Field-Theory Models

This section considers the following three types of analytic models:

First, *energy analysis* obtains average diameters by applying conservation of energy before and after unitary fragmentation. In general, energy analysis must account for three or four different forms of energy. However, at high enough strain rates, energy analysis generally only needs to account for two dominant forms of energy, which substantially simplifies the results.

Second, *stability analysis* assumes that the average fragment diameter is proportional to the wavelength of the most unstable wave. This approach assumes that unitary fragmentation occurs due to periodic surface instabilities such as Rayleigh–Taylor and Kelvin–Helmholtz instabilities. As discussed in the last chapter, these types of periodic instabilities tend to occur at relatively low strain rates.

Third, *dimensional analysis* obtains average diameters by combining strain rates and material properties to form units of length. Such combinations are generally validated or calibrated using experimental results. Unlike energy analysis and stability analysis, dimensional analysis does not require making any assumptions about the size of the strain rates or the types of instabilities.

In all of these approaches, the model results depend on whether or not unitary fragmentation involves substantial amounts of dissipation. When a substantial amount of dissipation does not occur, liquids may be called *inviscid* while solids may be called *brittle*. For inviscid liquids, the model results depend on the surface tension σ with units of dyne/cm, erg/cm^2, g/s^2, etc. For brittle solids, the model results depend on the surface energy per unit of exposed surface area Γ with units such as dyne/cm, erg/cm^2, or g/s^2. While Γ is difficult to measure directly, it can be related to easily measurable quantities as follows:

$$\Gamma = \frac{K_f^2}{2\rho c^2} = \frac{G_f}{2} \tag{16.8}$$

where ρ is the density, c the elastic wave speed, K_f is the dynamic fracture toughness with units such as g/(cm$^{1/2}$ s^2), and $G_f = K_f^2/\rho c^2$ is the dynamic fracture energy with units such as erg/cm^2.

When a substantial amount of dissipation does occur, liquids may be called *viscous* while solids may be called *ductile*. For viscous liquids, the model results depend on the dynamic viscosity μ with units such as g/(cm-s). For ductile solids, the model results depend on the tensile yield strength Y with units such as g/(cm-s^2). The models given here assume that μ, Y, σ, and Γ are static material properties. However, in reality, they are dynamic properties. For example, Grady (1988) notes that K_f typically decreases and Y typically increases as $\dot{\varepsilon}$ increases. To take this a step further, the models given here implicitly assume that viscous, inviscid, ductile, and brittle are inherent material properties. However, looking at it from another point of view, they are dynamic properties. More specifically, in this chapter inviscid and brittle responses are typical for fragmentation driven solely by tension, while viscous and ductile response are typical for fragmentation driven solely by shear.

Table 16.1 gives examples of simple analytic models for the average fragment diameters of inviscid liquids and brittle solids obtained using energy analysis, stability analysis, and dimensional analysis. Notice that these models all agree with Equation (16.7). Also notice that the most common value is $i = 3$, which is sometimes called *Grady's law*. While Table 16.1 sometimes gives specific values for the leading coefficient such as $(24)^{1/3}$, the leading coefficient is generally a complex function of the fragment shapes and sizes. Thus, to avoid leaving the wrong impression, Table 16.1 sometimes omits the leading coefficient and other factors. In these cases, instead of an equal sign, Table 16.1 uses the notation "~."

Table 16.1 Analytic expressions for average fragment diameters for inviscid liquids and brittle solids written in terms of strain rate $\dot{\varepsilon}$. The variable L refers to a length scale. All other variables are defined in the main text.

i	$-2/i$	Expression	Material	Shape	Reference
1^1	-2	$D'_{M,\text{avg}} \sim \frac{\sigma}{\rho L^2 \dot{\varepsilon}^2}$	Inviscid liquid	Planar sheet	Lefebvre (1992), eq. (10)
2^2	−1	$D'_{M,\text{avg}} \sim \left(\frac{\sigma L}{\rho}\right)^{1/2} \frac{1}{L\dot{\varepsilon}}$	Inviscid liquid	Planar sheet	Chaussonnet et al. (2016), eq. (20)
3^1	−2/3	$D_{\text{avg}} = \left(\frac{24\Gamma}{\rho\dot{\varepsilon}^2}\right)^{1/3}$	Brittle solid	Cylindrical shell	Mott & Linfoot (1943), eq. 2
3^1	−2/3	$D_{\text{avg}} = \left(\frac{40\Gamma}{\rho\dot{\varepsilon}^2}\right)^{1/3}$	Brittle solid	Spherical shell	Grady (1982), eq. 12
3^1	−2/3	$D_{\text{avg}} = \left(\frac{48\Gamma}{\rho\dot{\varepsilon}^2}\right)^{1/3}$	Brittle solid	Any 3D volume	Grady (1988), eq. 14
3^1	−2/3	$D_{\text{avg}} = \left(\frac{48\sigma}{\rho\dot{\varepsilon}^2}\right)^{1/3}$	Inviscid liquid	Any 3D volume	Grady (1988), eq. 36
3^3	−2/3	$D'_{M,\text{avg}} \sim \left(\frac{\sigma}{\rho\dot{\varepsilon}^2}\right)^{1/3}$	Inviscid liquid	Cylindrical jet	Sallam, Dai, & Faeth (2002), eq. 2
4^2	−1/2	$D_{\text{avg}} \sim \left(\frac{\sigma L}{\rho\dot{\varepsilon}^2}\right)^{1/4}$	Inviscid liquid	Circular disk	Villermaux & Bossa (2011), eq. 4.17

Continued

Table 16.1 *Continued*

i	$-2/i$	Expression	Material	Shape	Reference
4^2	$-1/2$	$D_{avg} \sim \left(\frac{\sigma^3}{\rho^3 L^3 \dot{\varepsilon}^2}\right)^{1/4}$	Inviscid liquid	Spherical shell	Vledouts et al. (2016), eq. 5.13
8^3	$-1/4$ [-0.2463]	$M_{avg} \sim \dot{\varepsilon}^{-0.739}$ $D_{avg} \sim M_{avg}^{1/3} \sim \dot{\varepsilon}^{-0.2463}$	Brittle solid	Cylindrical shell	Mott, Wilkinson, & Wise (1944), eq. 5

(1) Energy analysis.
(2) Stability analysis.
(3) Dimensional analysis with empirical calibration.

Energy analysis can be used to obtain many of the results given in Table 16.1, even those originally obtained with stability analysis or dimensional analysis; see, e.g., Grady & Kipp (1980), Grady & Olsen (2003) and Laney (2016). Examples 16.1 to 16.5 describe energy analysis in more detail.

Example 16.1 Conservation of energy for unitary fragmentation

In general, during unitary fragmentation, kinetic and elastic energy decrease, while surface energy and dissipation increase. Then conservation of energy says:

$$E_{\text{kinetic}} + E_{\text{elastic}} + E_{\text{surface}} + Q = \text{const.} \tag{16.9a}$$

where E_{kinetic} is the kinetic energy, E_{elastic} is the elastic energy, E_{surface} is the surface energy, and Q is the dissipation. Alternatively:

$$-(\Delta E_{\text{kinetic}} + \Delta E_{\text{elastic}}) = \Delta E_{\text{surface}} + \Delta Q \tag{16.9b}$$

where Δ refers to the change due to unitary fragmentation. Just prior to the onset of fragmentation, all of the elastic energy that can contribute to fragmentation has often been converted to kinetic energy. Thus, as noted by Grady (1988), elastic energy can often be neglected in favor of kinetic energy. Then:

$$-\Delta E_{\text{kinetic}} \cong \Delta E_{\text{surface}} + \Delta Q. \tag{16.10}$$

In the earliest widely cited work on fragmentation, Rittinger (1867) suggested that the energy expended in fragmentation is proportional to the fragment surface area. In other words, $\Delta E_{\text{surface}} \gg \Delta Q$. Then Equation (16.10) becomes:

$$-\Delta E_{\text{kinetic}} \cong \Delta E_{\text{surface}}. \tag{16.11}$$

This occurs in cases where there is little or no dissipation, which is true for inviscid fluids and brittle solids. For inviscid liquids, $\Delta E_{\text{surface}}$ depends on the surface tension σ with units of dyne/cm, erg/cm^2, g/s^2, etc. For brittle solids, $\Delta E_{\text{surface}}$ depends on the surface energy per unit of exposed surface area Γ with units of dyne/cm, erg/cm^2, g/s^2 etc.

While Rittinger (1867) suggested that the energy expended on fragmentation is proportional to the fragment *surface area*, Kick (1885) suggested that the energy expended on fragmentation is proportional to fragment *volume*. Assuming that dissipation occurs over a volume, this is equivalent to saying that expended energy is proportional to dissipation. For example, Grady (1988) says "the mechanism of energy dissipation ... [is] the activation, growth and linkup of nominally spherical voids ... [with] local plastic dissipation during void growth. Some fracture energy may reside in the surface energy of the newly created voids, but this will be negligible in most circumstances." If $\Delta Q \gg \Delta E_{\text{surface}}$, then Equation (16.10) becomes:

$$-\Delta E_{\text{kinetic}} \cong \Delta Q. \tag{16.12}$$

This occurs in cases with substantial amounts of dissipation such as viscous fluids and ductile solids. For viscous liquids, ΔQ is proportional to a length scale times the dynamic viscosity μ, with units such as g/(cm-s). For ductile solids, ΔQ is proportional to a length scale times the yield strength Y, with units such as g/(cm-s^2).

Example 16.2 Conservation of energy for inviscid liquids and brittle solids

For inviscid liquids and brittle solids, Equation (16.11) can be written as follows:

$$\iiint \frac{1}{2}V^2 dM - \sum_{i=1}^{N} \frac{1}{2}U_i^2 M_i = \sum_{i=1}^{N} \Gamma_i \mathscr{A}_i - \Gamma \mathscr{A} \tag{16.13}$$

where the left hand side is the change in kinetic energy and the right hand side is the change in surface energy. The integral on the left hand side is taken over the fragmenting body. The sums are taken over the number of particles N created by fragmentation. In addition:

- V is the velocity of the unitary body prior to fragmentation
- U_i is the velocity of the *i*-th particle created by fragmentation
- M is the mass of the unitary body prior to fragmentation
- M_i is the mass of the *i*-th particle created by fragmentation
- $\mathscr{A}$ is the surface area of the unitary body prior to fragmentation
- $\mathscr{A}_i$ is the surface area of the *i*-th particle created by fragmentation
- Γ is the surface energy per unit area of the unitary body prior fragmentation
- Γ_i is the surface energy per unit area of the *i*-th particle created by fragmentation

Imagine that a unitary body is subdivided into material parcels that will, after fragmentation, become individual particles. By definition, these parcels have the same mass—but not necessarily the same shape or surface area—as the particles they will become. Then:

$$\iiint \frac{1}{2} V^2 dM = \sum_{i=1}^{N} \frac{1}{2} V_i^2 M_i \tag{16.14}$$

where V_i is the velocity of the i-th material parcel just before it breaks away to become the i-th particle. Assume the resulting detached fragments have much less kinetic energy than the attached parcels, i.e., $\sum_{i=1}^{N} V_i^2 M_i/2 \gg \sum_{i=1}^{N} U_i^2 M_i/2$. In addition, assume the resulting detached fragments have much more surface area than the original unitary body, i.e., $\sum_{i=1}^{N} \Gamma_i \mathcal{A}_i \gg \Gamma \mathcal{A}$. Then Equation (16.13) becomes:

$$\sum_{i=1}^{N} \frac{1}{2} V_i^2 M_i \cong \sum_{i=1}^{N} \Gamma_i \mathcal{A}_i. \tag{16.15}$$

This book generally measures particle sizes by mass M_i. However, Equation (16.15) involves both the particle mass M_i and the particle surface area $\mathcal{A}_i$. To reduce these two size metrics to one, it is helpful to introduce the particle diameter D_i. According to Equations (6.2) and (6.3), regular (non-fractal) particles obtain $\mathcal{A}_i = c_{\mathcal{A}} D_i^{m-1}$ and $M_i = c_M \rho D_i^m$ where $c_{\mathcal{A}}$ and c_M are constants and m is an integer dimension (1, 2, or 3). To complete the model, it is necessary to introduce expressions for V_0 and Γ_i. In general, V_i is proportional to powers of D_i and $\dot{\varepsilon}$. Similarly, Γ_i is proportional to powers of D_i; see Examples 16.3 and 16.4.

Example 16.3 Conservation of energy for inviscid liquids where V_i is proportional to $\dot{\varepsilon}$ and Γ_i is constant

Consider Example 16.2. As the simplest possible case, suppose that:

$$V_i = \dot{\varepsilon} L/2 \tag{16.16}$$

$$\Gamma_i = \sigma \tag{16.17}$$

where $\dot{\varepsilon}$ is a constant strain rate, σ is a constant surface tension, and L is a constant length scale. For example, Grady (1981) suggests letting L be the initial length of a linear body. Other possibilities include: the thickness of a cylindrical shell, e.g., Mott et al. (1944); and the thickness of a flat sheet, e.g., Lefebvre (1992). Then Equation (16.15) becomes:

$$\frac{\sum_{i=1}^{N} D_i^m}{\sum_{i=1}^{N} D_i^{m-1}} \cong \frac{8 c_{\mathcal{A}} \sigma}{c_M \rho \dot{\varepsilon}^2 L^2}. \tag{16.18}$$

According to Equation (3.37), the left hand side is the Sauter mean diameter $D'_{M,\mathrm{avg}}$. Then if $c_1 = 8c_{\mathcal{A}}/c_M$, Equation (16.18) can be written as follows:

$$D'_{M,\,\mathrm{avg}} \cong \frac{c_1 \sigma}{\rho \dot{\varepsilon}^2 L^2}. \tag{16.19}$$

Example 16.4 Conservation of energy for brittle solids where V_i is proportional to $\dot{\varepsilon}D_i$ and Γ_i is a constant

Consider Example 16.2. As one possible case, suppose that:

$$V_i = \dot{\varepsilon}D_i/2 \tag{16.20}$$

$$\Gamma_i = \Gamma \tag{16.21}$$

where $\dot{\varepsilon}$ is a constant strain rate and Γ is a constant surface energy per unit surface area. Then Equation (16.15) becomes:

$$\frac{\sum_{i=1}^{N} D_i^{m+2}}{\sum_{i=1}^{N} D_i^{m-1}} \cong \frac{8c_{\mathscr{A}}\Gamma}{c_M \rho \dot{\varepsilon}^2}. \tag{16.22}$$

As shown by Laney (2016), the left hand side can be written as $c_2 D_{\text{avg}}^3$ where c_2 is a constant that depends on the fragment shapes and the fragment size distribution. If $C = 8c_{\mathscr{A}}/c_2 c_M$, then Equation (16.22) becomes:

$$D_{\text{avg}} = \left(\frac{C\Gamma}{\rho\dot{\varepsilon}^2}\right)^{1/3} \tag{16.23}$$

which is known as Grady's law. While usually treated like a constant, C is actually a function of the fragment shapes and sizes; see Example 16.5 for details.

Example 16.5 The average fragment size depends on the shape and size distributions

Consider Example 16.4. For inviscid liquids, Equation (16.23) becomes:

$$D_{\text{char}} = D_{\text{avg}} = \left(C\sigma/\rho\dot{\varepsilon}^2\right)^{1/3}. \tag{16.24}$$

As noted earlier, C generally depends on the size and shape of the fragments. First, consider the effects of fragment sizes. In particular, suppose that unitary fragmentation produces a Weibull size distribution where:

$$f(D) = \frac{nb}{D_{\text{char}}}\left(\frac{D}{D_{\text{char}}}\right)^{n-1} \exp\left[-b\left(\frac{D}{D_{\text{char}}}\right)^{n}\right] \tag{16.25}$$

and where $n > 0$. In addition, suppose all of the fragments have the same shape. Then Equation (1.1) says $M = \text{const.}\,\rho D^m$. In this case, Laney (2016) showed that:

$$C = 48\frac{\Gamma(1+1/n)^6\Gamma(1+2/n)^3}{\Gamma(1+3/n)^4} \tag{16.26}$$

where Γ is the gamma function.

As seen in Figure 16.1, C increases monotonically as n increases, reaching an asymptote of 48. Chaos typically induces a wide range of fragment sizes, which corresponds to small values of n. Conversely, periodicity and other deterministic mechanisms typically induce a narrow range of fragment sizes, which corresponds to large values of n. While $C = 48$ only applies to periodic fragmentation, the research literature commonly assumes $C = 48$ in all cases; see, e.g., Grady (1988). This is partly because most experiments only measure fragments above some minimum size, meaning they overestimate D_{avg} and thus overestimate C. Now, consider the effects of fragment shapes. Equation (16.26) only applies if the fragments have identical shapes. In other cases, C tends to obtain larger values, e.g., Grady & Kipp (1993) report that C can easily be as large as 720.

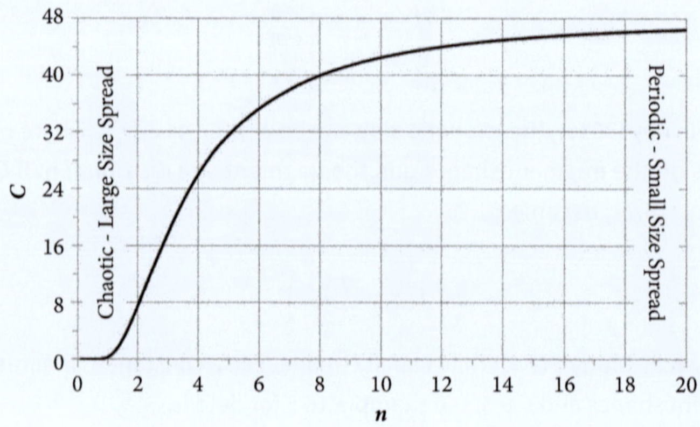

Figure 16.1 For Weibull size distributions, the coefficient C in Equation (16.24) for the average fragment diameter is a function of the exponent n in Equation (16.25).

So far, the results in this section have been expressed in terms of strain rates. However, as it turns out, they can equally well be expressed in terms of non-dimensional *Weber numbers*. For liquids, one possible definition is as follows:

$$We = \frac{\rho V^2 L}{\sigma} \tag{16.27}$$

where σ is the liquid surface tension, ρ is the liquid density, V is a liquid velocity scale, and L is a liquid length scale. Depending on the exact circumstances, the Weber number may be defined in a variety of different ways; see Table 16.2. For example, suppose a laminar air stream drives a turbulent liquid layer across a solid flat plate. Then possible velocity scales V include: the average streamwise air velocity; the average cross-stream air velocity; the average streamwise liquid velocity, the average cross-stream liquid velocity, or some metric of the turbulent liquid velocity fluctuations. In addition, possible length scales L include the thickness of the liquid layer at the end of the plate, the thickness of the liquid layer at some point downstream of the plate, the turbulent integral length scale of the liquid at the end of the plate, the turbulent integral length scale at some point downstream of the plate, the length of the flat plate, the thickness of the flat plate, or the surface roughness of the flat plate.

Table 16.2 Common variants for the definition of the Weber number, $We = \rho V^2 L/\sigma$.

	Liquid	Air	Solid
Surface Tension σ	✓	–	–
Density ρ	✓	✓	–
Velocity Scale V	$\checkmark^{1,2}$	$\checkmark^{1,2}$	–
Length Scale L	$\checkmark^{1,2}$	$\checkmark^{1,2}$	$\checkmark^{2}$

(1) May be constant (e.g., initial length, initial velocity) or variable (e.g., length at a given point in time and/or space, velocity at a given point in time and/or space).
(2) May be a macroscale (e.g., overall dimensions of liquid, solid, or gas) or a microscale (e.g., liquid turbulence integral length scale, solid surface roughness).

This is not as complicated as it may sound. A number of specific examples will be given later in the chapter, which will illustrate how length and velocity scales may be selected from among the myriad possibilities.

Table 16.3 gives some examples of average fragment sizes for inviscid liquids expressed in terms of Weber numbers. In all cases, these expressions are equivalent to the ones that appeared earlier in Table 16.1, assuming $\dot{\varepsilon}$ is a constant times V/L as in Examples 15.2, 15.3, and 15.4.

Table 16.3 Analytic expressions for average fragment diameters for inviscid liquids expressed in terms of Weber numbers. In this table, L is a length scale, which may or may not be the same as the length scale used to form the Weber number.

i	Expression	Shape	Reference
1^1	$D'_{M,\text{avg}} \cong 1,714L/We$	Planar sheet	Lefebvre (1992)
2^2	$D'_{M,\text{avg}} \cong 1.4L/We^{1/2}$	Planar sheet	Chaussonnet et al. (2016)
3^3	$D'_{M,\text{avg}} \sim L/We^{1/3}$	Cylindrical jet	Sallam, Dai & Faeth (2002)
4^2	$D_{\text{avg}} \cong L/We^{1/4}$	Circular disk	Villermaux & Bossa (2011)

(1) Energy analysis.
(2) Stability analysis.
(3) Dimensional analysis with empirical calibration.

Unitary fragmentation can only occur if the characteristic fragment diameter is substantially smaller than the largest dimension of the original body. Thus, the expressions given in Tables 16.1 and 16.3 are valid provided that:

$$D'_{M,\text{avg}} \ll L_{\max} \quad (16.28.a)$$

or similarly:

$$D_{\text{avg}} \ll L_{\max} \quad (16.28.a)$$

where $L_{\max}$ is the largest dimension of the original body. This means that the expressions given in Table 16.1 are valid *provided that the strain rate is sufficiently large.* Equivalently, the expression given in Table 16.3 are valid *provided that the Weber number is sufficiently large*. At lower strain rates, alternative expressions may apply. For example, after accounting for both surface energy and elastic potential energy, Marquez et al. (2016, 2018) obtained the following expression for the count mean diameter of fragments produced by brittle solids:

$$D_{\text{avg}}^3 + \frac{12s^2}{\rho^2 c^2 \dot{\varepsilon}^2} D_{\text{avg}} - \frac{24K_f^2}{\rho^2 c^2 \dot{\varepsilon}^2} = 0 \qquad (16.29)$$

where s is similar to the spring constant in Hooke's law. Notice that this expression approaches Grady's law as the strain rate increases. For other examples see, e.g., Zhou et al. (2005, 2006a, b).

Turning from inviscid liquids and brittle solids, Table 16.4 gives examples of simple analytic models for average fragment diameters of viscous liquids, ductile solids, and ductile-brittle solids obtained using energy analysis and dimensional analysis. While the results given in Table 16.1 typically concern fragmentation driven solely by tension, the results given in Table 16.4 concern fragmentation driven at least partly by shear.

Table 16.4 Analytic expressions for average fragment diameters for viscous liquids, brittle solids, and ductile-brittle solids written in terms of train rate $\dot{\varepsilon}$.

i	$-2/i$	Expression	Material	Shape	References
2^1	-1	$D_{\text{avg}} \sim \left(\frac{Y}{\rho}\right)^{1/2} \frac{1}{\dot{\varepsilon}}$	Ductile solid	Cylindrical shell	Mott (1947), eq. 5
2^1	-1	$D_{\text{avg}} \sim \left(\frac{Y}{\rho}\right)^{1/2} \frac{1}{\dot{\varepsilon}}$	Ductile solid	Any 3D volume	Grady (1988), eq. 24
4^1	$-1/2$	$D_{\text{avg}} = \left(\frac{8\mu}{\rho\dot{\varepsilon}}\right)^{1/2}$	Viscous liquid	Any 3D volume	Grady (1988), eq. 39
4^1	$-1/2$	$D_{\text{avg}} \sim \left(\frac{\Gamma}{Y\dot{\varepsilon}}\right)^{1/2}$	Ductile-brittle solid	Ring	(*) Stránský & Jirásek (2011), eq. 4
5^2	$-2/5$ [−0.4264]	$D_{\text{avg}} \sim \left(\frac{\Gamma^{3/2}}{\rho^{1/2} Y\dot{\varepsilon}}\right)^{2/5}$	Ductile-brittle solid	Ring	(*) Zhou, Molinari & Ramesh (2005), eq. 12

(1) Energy analysis.
(2) Dimensional analysis with empirical calibration.
(*) Rarely observed.

As it turns out, the first three expressions given in Table 16.4 can equally well be expressed in terms of non-dimensional *Reynolds numbers*. For liquids, one possible definition is as follows:

$$Re = \frac{\rho V L}{\mu} \qquad (16.30)$$

where μ is the liquid dynamic viscosity, ρ is the liquid density, V is a liquid velocity scale, and L is a liquid length scale. Just like the Weber number, the Reynolds number may be defined in many different ways; the earlier discussion applies again here. As a specific example, suppose that $\dot{\varepsilon}$ is a constant times V/L where V is a velocity scale and L is a length scale such as radius or thickness; see Examples 15.2, 15.3, and 15.4. Then the third row of Table 16.4 can be rewritten as follows:

$$D_{\text{avg}} = \text{const.}\, L/Re^{1/2}.$$

As noted earlier, unitary fragmentation can only occur if the characteristic fragment diameter is substantially smaller than the largest dimension of the original body. Thus, the expressions given in the first three row of Table 16.4 are valid provided that $D_{\text{avg}} \ll L_{\text{max}}$ where L_{max} is the largest dimension of the original body. For example, $D_{\text{avg}} = (8\mu/\rho\dot{\varepsilon})^{1/2}$ is valid *provided that the strain rate is sufficiently large.* Equivalently, $D_{\text{avg}} = \text{const.}\ L/Re^{1/2}$ is valid *provided that the Reynolds numbers is sufficiently large.*

To summarize, inviscid and brittle cases naturally involve lower limits on Weber numbers, while viscous and ductile cases naturally involve lower limits on Reynolds numbers. However, for consistency and simplicity, viscous and brittle cases are often expressed in terms of lower limits on the Weber number plus upper limits on the *Ohnesorge number* where:

$$Oh = \frac{\mu}{\sqrt{\rho\sigma L}} = \frac{\sqrt{We}}{Re}. \qquad (16.31)$$

Because $Re = \sqrt{We}/Oh$, a lower limit on We plus an upper limit on Oh implies a lower limit on Re. This book generally prefers to work in terms of strain rates rather than non-dimensional parameters such as Weber numbers, Reynolds numbers, and Ohnesorge numbers. Strain rates offer two main advantages: (1.) they apply equally to liquids and solids of all kinds; (2) they often involve unique objective length and velocity scales based on the physics of each problem.

16.3 Experimental Validation Part I: Primary Fragmentation in Cylindrical Geometries

This is the first of four sections on experimental validation. This section concerns primary fragmentation due to uniaxial strain in cylindrical coordinates; see Figure 15.7a. More specifically, Example 16.6 concerns fragmentation of solid cylindrical shells, Example 16.7 concerns fragmentation of solid cylindrical rings, and Example 16.8 concerns fragmentation of round liquid jets.

Example 16.6 Analytic vs. experimental results for primary fragmentation of cylindrical metal shells due to high-explosive detonation

Pearson (1990) says explosively loaded cylindrical steel shells evolve in three phases:

- *Phase 1. Elastic-Plastic Expansion.* The shell diameter increases by, e.g., 30%. "This phase comes to an end when the first sign of fracture appears on the outer surface ... These fractures first appear as hairline markings oriented in the axial direction."

- *Phase 2. Plastic Expansion/Fracture*. "The cylinder or case continues expanding ... Fractures lengthen and open up ... [and] extend inward toward the inner surface ... The overall appearance may be somewhat like that of an expanding latticework."
- *Phase 3. Fragmentation*. At first, "the fragments ... are quite narrow compared to their lengths ... After the formation of these long, narrow fragments ... the relative motions of the fragments tend to reduce ... [them] into ... shorter pieces."

If the cylindrical shell is open at both ends, the parts closest to the explosion fragment first, i.e., different parts of the shell may experience different phases at any given time; see, e.g., Stronge et al. (1989). However, if the cylindrical shell is capped at both ends, different parts of the shell tend to experience the same phase at any given time; see Figure 16.2.

At first glance, the situation appears to be hopelessly complex. However, despite this, a number of researchers have found reliable algebraic equations for estimating the mean fragment sizes. In the earliest such work, for a cylindrical metal shell subject to steady uniform radial expansion, Mott & Linfoot (1943) used energy analysis to find the following expression for the count mean diameter:

$$D_{\text{avg}} = \left(\frac{24R^2\Gamma}{\rho V^2}\right)^{1/3} = \left(\frac{24\Gamma}{\rho\dot{\varepsilon}^2}\right)^{1/3} = \left(\frac{24K_f^2}{\rho^2c^2\dot{\varepsilon}^2}\right)^{1/3} \tag{16.32}$$

where $\dot{\varepsilon} = V/R$ as in Example 15.2, V is the shell velocity, R is the initial shell radius, ρ is the shell density, Γ is the "energy per unit area required to form a crack" in the shell, c is the elastic wave speed of the shell, and K_f is the dynamic fracture toughness of the shell. To complete the picture, Gurney (1943) found a simple expression for V; see Example 7.4.

Brittle solids may produce an enormous number of fragments, most of which are extremely small. For example, Weimer & Rogers (1979) report that explosively-driven steel cylinders produced up to 18,910 fragments. With a great deal of time, effort, and patience, they were able to verify that Equation (16.32) agreed with their experimental results, at least in terms of the dependence on fracture toughness K_f; see Figure 16.3.

Given the difficulty of capturing and measuring extremely large numbers of extremely small fragments, most experiments impose some minimum fragment size. However, such artificial lower bounds may cause experiments to substantially overestimate the true count mean diameter. As one common approach for overcoming this problem, experiments sometimes report characteristic diameters that are relatively insensitive to small fragments, such as the mass mean diameter, the Sauter mean diameter, or the size modulus. The results are similar to Equation (16.32) except for the multiplicative constant. As another common approach, the length of the cylinder may be reduced. The shorter the cylinder, the fewer the fragments. In the extreme, this leads to cylindrical rings, as discussed in the next example.

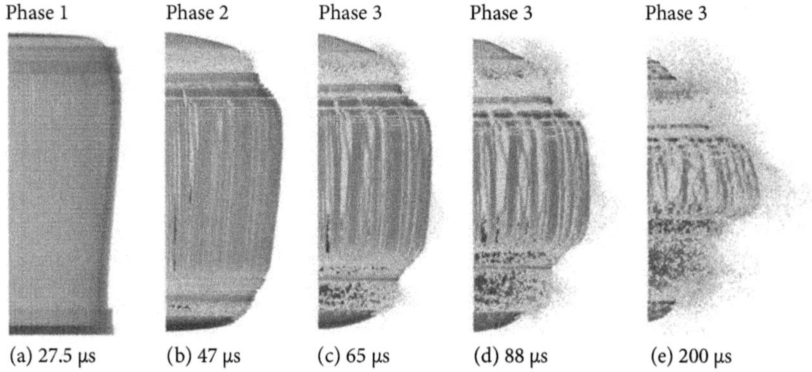

Figure 16.2 Computational results for fragmentation of an explosively-driven closed cylindrical steel shell. TNT was detonated at the top of the shell, just below the end cap. The initial outer diameter was 11.6cm and the initial wall thickness was 6mm. These results were produced using Smoothed Particle Hydrodynamics (SPH) as discussed earlier in Section 5.5.
Figure reprinted from Kong et al. (2013) with permission from Elsevier.

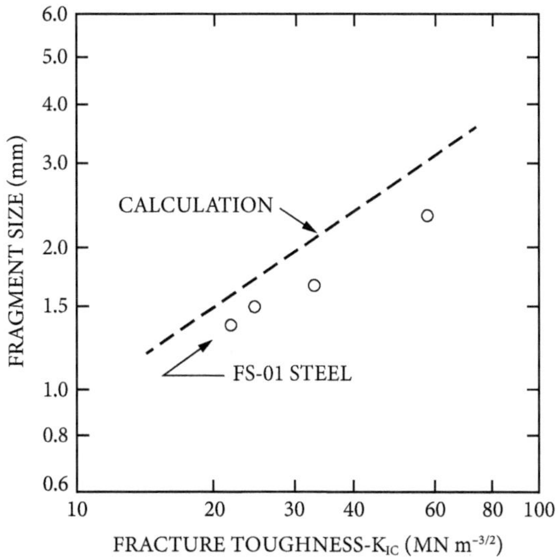

Figure 16.3 Analytical (dashed line) vs. experimental (circles) results for the count mean diameters of fragments from an explosively driven steel cylindrical shell. The analytical results come from Grady (1982). The experimental results come from Weimer & Rogers (1979). Improved agreement can be obtained by slightly adjusting the modeling parameters such as the fracture toughness.
Figure reprinted from Grady (1982) with the permission of AIP Publishing.

Example 16.7 Analytic vs. experimental results for primary fragmentation of cylindrical metal rings

Consider rapidly expanding cylindrical metal rings. Grady & Olsen (2003) used energy analysis to obtain the following expression for the count mean diameter:

$$D_{\text{avg}} = \left(\frac{24R^2\Gamma}{\rho V^2}\right)^{1/3} = \left(\frac{24\Gamma}{\rho\dot{\varepsilon}^2}\right)^{1/3} = \left(\frac{24K_f^2}{\rho^2c^2\dot{\varepsilon}^2}\right)^{1/3} \tag{16.33}$$

where $\dot{\varepsilon} = V/R$ as in Example 15.2, V is the ring radial velocity, R is the initial ring radius, ρ is the ring density, Γ is the "fracture resistance ... characterizing the dissipation in the fracture growth process" in the ring, c is the elastic wave speed of the ring, and K_f is the dynamic fracture toughness of the ring. To validate this expression, Grady & Olsen (2003) conducted a series of experiments in which "U6N metal rings approximately 30mm in diameter and with a 0.75mm square cross section were accelerated by a pulsed magnetic field to radial velocities of 50-300m/s." Each of these experiments produced only a small number of fragments, varying between 8 and 45 per ring. To obtain statistically significant results, a number of such experiments were conducted. As seen in Figure 16.4, the experimental results agree well with Equation (16.33).

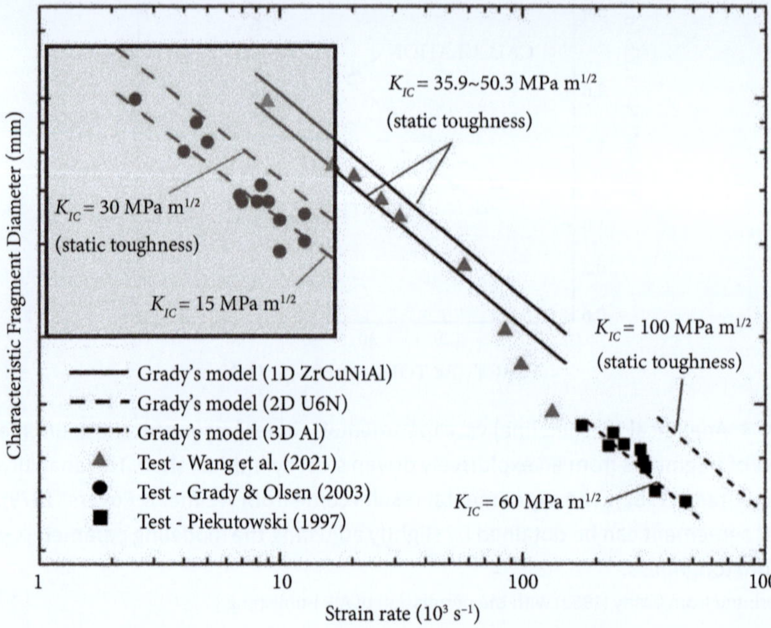

Figure 16.4 Analytical (dashed lines) vs. experimental (circles) results for the characteristic diameters of fragments produced by expanding metal rings. In addition to the results of interest, highlighted in the shaded box, this figure shows two other sets of results.
[The notation in this figure has been changed to match the notation used in this book.].
Figure reproduced from Wang et al. (2021) with permission from Springer Nature.

Example 16.8 Analytic vs. experimental results for primary fragmentation of free round liquid jets with fully developed turbulence

Suppose a fully turbulent liquid jet exits a round nozzle, where it meets a stagnant gas. After the liquid jet exits the nozzle, its surface roughens, and its average radius increases due to the turbulence-induced cross-stream velocity; see Figure 16.5. In this case, Faeth et al. (1995) say that "primary breakup occurs due to the formation of ligaments and other irregular liquid elements along the surface of the liquid core ... The outcome of primary breakup frequently is irregular drops or ligaments." Faeth et al. (1995) used dimensional analysis, validated and calibrated using experimental results, to obtain the following expression for the Sauter mean diameter of the primary liquid fragments:

$$\frac{D'_{M,\,avg}(x)}{\Lambda} = 0.65\left(\frac{x}{\Lambda We_{\Lambda}^{1/2}}\right)^{2/3} \tag{16.34}$$

where $0 < x \le x^*$ is the downstream distance from the jet exit, $x^* = \text{const.}\sqrt{\rho U^2 R_0^3/\sigma^2}$ is the jet length, Λ is the "radial integral length scale of the turbulence," $We_{\Lambda} = \rho U^2 \Lambda/\sigma$ is a Weber number, ρ is the liquid density, σ is the liquid surface tension, and U is the streamwise liquid velocity. Equation (16.34) agrees well with a wide range of experimental results; see Figure 16.6. Notice that Equation (16.34) can be rewritten as follows:

$$D'_{M,\,avg}(x) = 0.65\left[\frac{\sigma}{\rho\dot{\varepsilon}(x)^2}\right]^{1/3} \tag{16.35}$$

where $\dot{\varepsilon}(x) = U/x$. Suppose that the median jet surface is a perfect cone with radius $R(x)$. As seen in Figure 16.5, $R(x) = R_0 + (V/U)x$ where V is the radial root-mean-square (r.m.s.) liquid velocity at the surface of the cone. Faeth et al. (1996) argue that $V/U \approx 0.06$ "relatively independent of position ... for fully developed turbulent pipe flow." In this case, $\dot{\varepsilon}(x) = U/x \cong V/R(x)$, which agrees with Example 15.2. Then Equation (16.35) is just Grady's law. From one point of view, the jet is composed of thin circular rings, which expand as they flow downstream. When a ring on the surface fragments, the ring underneath takes its place. For additional information on this example see, e.g., Ruff et al. (1991), Tseng et al. (1992), Wu & Faeth (1993), Wu et al. (1992, 1995), Faeth (1996) and Sallam et al. (2002).

As the last example shows, stress and strain in the radial direction may obtain the same result as stress and strain in the longitudinal direction. When in doubt, it is usually safest to assume that the stresses and strains occur across the shortest rather than the longest dimension of a given body.

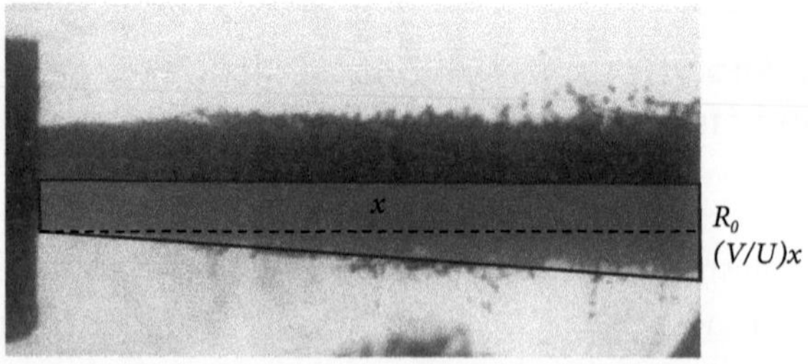

Figure 16.5 Primary atomization of a fully turbulent round liquid jet.
The photograph seen in this figure is reprinted from Faeth et al. (1995) with permission from Elsevier.

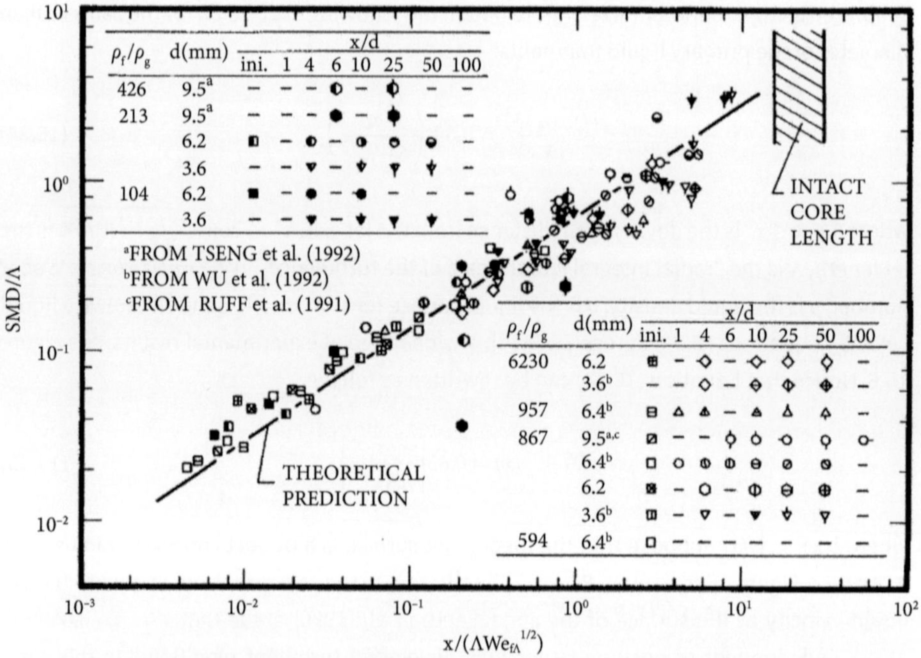

Figure 16.6 Analytical (solid line) vs. experimental (symbols) results for Sauter mean diameters of fragments from a fully turbulent round liquid jet. In this figure, Λ refers to the turbulent integral length scale, ρ_f refers to the liquid density, ρ_g refers to the gas density, and *d* refers to the jet exit diameter. Tests include water, n-heptane, and various glycerol-water mixtures injected into helium, air, and Freon 12 at pressures of 1 and 2 atm.
Figure reprinted from Faeth (1996) with permission from Elsevier.

16.4 Experimental Validation Part II: Primary Fragmentation in Spherical Geometries

This is the second of four sections on experimental validation. This section concerns primary fragmentation due to equibiaxial strain in spherical coordinates; see Figure 15.7b. More specifically, Examples 16.9 and 16.10 concern fragmentation of spherical and hemispherical solid shells, respectively. In addition, Examples 16.11, 16.12, and 16.13 concern fragmentation of spherical liquid shells. As these examples show, it can be difficult to obtain perfectly symmetric spherical expansion. In addition, it can be difficult to measure the resulting fragments. However, these issues tend resolve themselves as the shell thickness decreases. Thus, the best-available experimental validations concern thin spherical shells; see Examples 16.10 and 16.13. In fact, the thin liquid shells discussed in Example 16.13 resemble those produced by soap bubble wands. While often dismissed as merely children's toys, they have become the focus of serious study; see, e.g., Panizza & Courbin (2016) and Salkin et al. (2016).

Example 16.9 Analytic vs. experimental results for primary fragmentation of spherical solid shells due to high-explosive detonations

If the loading is asymmetric, spherical metal shells will fragment much like cylindrical metal shells, as discussed in Example 16.6. In this case, Gray et al. (2005) says linear cracks initially occur in the direction of the detonation wave due to "seeding of strain localizations that occur when the detonation wave sweeps from one end [of the shell] to the other." In addition, for spherical glass shells subjected to asymmetric gaseous explosions, Wittel et al. (2006) say that "the break-up is observed to start at ... 'hot spots' with random positions on the surface, from which long straight cracks radiate without any apparent branching. This mechanism results in a large number of long thin fragments, making them unstable against out-of-plane bending and sequential breaking."

On the other hand, if the loading is perfectly symmetric, spherical shells tend to fragment in a completely different way. For example, in one set of experiments, Gray et al. (2005) say "no effort was spared to make the fielded design as close to the idealized [spherical] geometric configuration as possible ... [The] design ... caused the motion of the shell to approximate ... closely a spherical expansion." According to Gray et al. (2005), such cases evolve in three phases:

- *Phase 1. Elastic-Plastic Expansion*. The shell diameter increases by, e.g., 200%. This phase comes to an end once the shell develops significant "strain localizations" or "localized thinning" at uniformly random spots.
- *Phase 2. Plastic Expansion/Fracture*. The shell continues expanding. "Some strain localizations dominate and coalesce to form the fragments, while the growth of others is arrested as a natural consequence of the competition among localizations."
- *Phase 3. Fragmentation*. Uniformly loaded spherical shells produce compact fragments with aspect ratios near unity. This contrasts with non-uniformly loaded spherical shells, which initially produce elongated fragments with large aspect ratios.

Assuming a spherical brittle shell subjected to steady uniform radial expansion, Grady (1982) used energy analysis to find the following expression:

$$D_{avg} = \left(\frac{C\Gamma}{\rho\dot{\varepsilon}^2}\right)^{1/3} \tag{16.36}$$

where $\dot{\varepsilon} = 2V/R$ as in Example 15.3.

To complete the picture, Gurney (1943) found the following result:

$$V = \frac{G}{\sqrt{\frac{M_{\mathrm{sph}}}{M_{H.E.}} + \frac{3}{5}}} \tag{16.37}$$

where G is a constant that depends on the type of explosive, M_{sph} is the spherical shell mass, and $M_{H.E.}$ is the high-explosive mass; see also, e.g., Kennedy (1998) and Grady (2017). While there is no reason to doubt Equation (16.36), it is difficult to validate. For relevant experiments see, e.g., Garg & Siekmann (1966), Olive et al. (1979), Ivanov et al. (2001), Balagansky et al. (2001), Gray et al. (2005), Potocki & Hull (2010), Bolis et al. (2013) and Callahan et al. (2023). Unfortunately, none of these experiments obtained results suitable for validating Equation (16.36). The next example describes an easier-to-execute approach.

Example 16.10 Analytic vs. experimental results for primary fragmentation of solid hemispherical shells due to high-speed impacts

As seen in the previous example, while high-explosives can be used to fragment spherical metal shells, extreme care is required to obtain perfectly symmetrical conditions. As a simpler alternative, suppose a metal sphere impacts and penetrates a metal sheet. In this scenario, no special care is required to create symmetrical hemispherical conditions in the sheet. In particular, Piekutowski (1995, 1997) conducted tests in which aluminum spheres were fired at aluminum sheets using a two-stage light gas gun. The sphere diameters were between 6.35 and 15.88mm, the sheet thicknesses were between 0.25 and 4.80mm, and the impact speeds were between 1.98 and 7.38km/s. In many cases, during or after penetration, Piekutowski (1995, 1997) found that "a hemispherical shell of [sheet] fragments spalled from the rear surface of the sphere." To characterize these fragments, he used fine-source, soft flash x-rays to observe between 10 and 3,600 fragments with diameters greater than 0.25mm. According to Grady's law:

$$D_{\mathrm{avg}} = \left(\frac{24K_f^2}{\rho^2 c^2 \dot{\varepsilon}^2}\right)^{1/3} \tag{16.38}$$

where $\dot{\varepsilon} = 2\,V/R$ as in Example 15.3, V is "the radial expansion velocity of the spall fragments measured from the radiographs," and R is the hemisphere radius. If $\rho = 2{,}791$ kg/m^3, $c = 6{,}402$ m/s, and $K_f = 30$ MPa-m$^{1/2}$, Piekutowski (1995) found that the experimental median diameter was "one and one-half to two times larger" than the one obtained using Equation (16.38). By contrast, if $\rho = 2{,}700$kg/m^3, $c = 5{,}180$ m/s, and $K_f = 60$ MPa-m$^{1/2}$, Wang et al. (2021) obtained good agreement between Piekutowski's experimental data and Equation (16.38); see Figure 16.7.

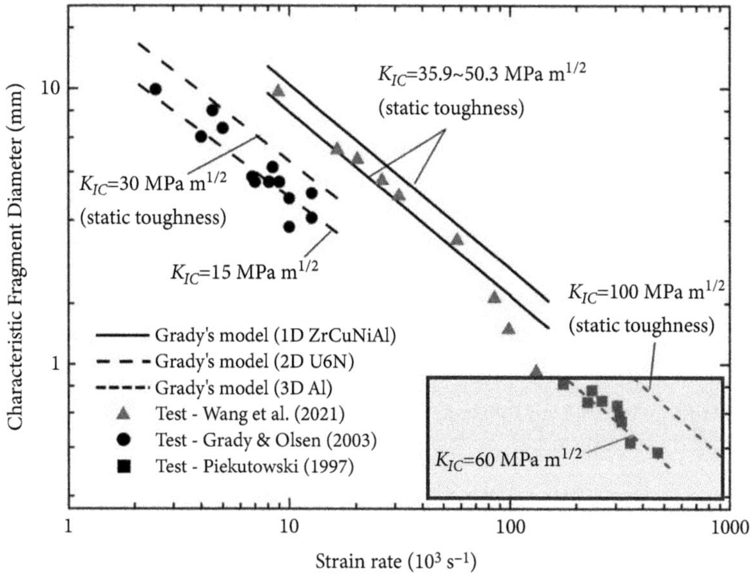

Figure 16.7 Analytical (dashed lines) vs. experimental (squares) results for the characteristic diameters of fragments produced by expanding metal hemispheres. In addition to the results of interest, highlighted in the shaded box, this figure shows two other sets of results.
[The notation in this figure has been changed to match the notation used in this book.]
Figure reproduced from Wang et al. (2021) with permission from Springer Nature.

Example 16.11 Analytic vs. experimental results for primary fragmentation of spherical liquid shells due to high-explosive detonations

Milne et al. (2017) studied a spherically symmetric high-explosive detonated in the center of a spherical glass tank filled with water. Using video analysis, Milne et al. (2017) observed large primary fragments with a relatively narrow size spread. As seen in Figure 16.8, each of the large primary fragments leaves a dense trail of small children droplets in its wake, leading to "characteristic jets or fingers." Based on Grady (1988), Milne et al. (2017) suggested that the primary fragments obtained the following count mean diameter:

$$D_{\text{avg}} \approx \left(C\sigma/\rho\dot{\varepsilon}^2\right)^{1/3} \tag{16.39}$$

where $\dot{\varepsilon} = 2V/R$. As discussed in Example 16.5, the factor C depends on the size and shape of the fragments. Regarding fragment sizes, Milne et al. (2017) observed "a tight distribution of [jet] lengths suggesting a similar distribution of primary fragment sizes ... The ... measured [size distribution] functions ... are more akin to a monomodal than a Mott distribution." In this case, if the primary fragments have identical shapes, Example 16.5 says that $C = 48$. However, Milne et al. (2017) found that with this value of C, the experimental results obtained "much larger primary fragments ... [They were] roughly an order of magnitude larger" than

Equation (16.39) would predict. This implies that that primary fragments did not have identical shapes. In fact, interactions between the water and the hot byproducts of high-explosive detonation presumably resulted in highly irregular bubble-filled primary fragments, which only gradually assumed roughly spherical shapes due to aerodynamic forces and surface tension. As noted in Example 16.5, in such cases, Grady & Kipp (1993) report that C may easily be as large as 720.

In the next example, the high-explosive is replaced by a passive penetrator. This results in smaller fragments with wider size ranges. In this example, the primary fragments are large and tend to be on the surface of the aerosol cloud, making them relatively easy to observe. In the next example, the exact opposite is true—the primary fragments are small and tend to be in the interior of an initially dense cloud, making them relatively difficult to observe.

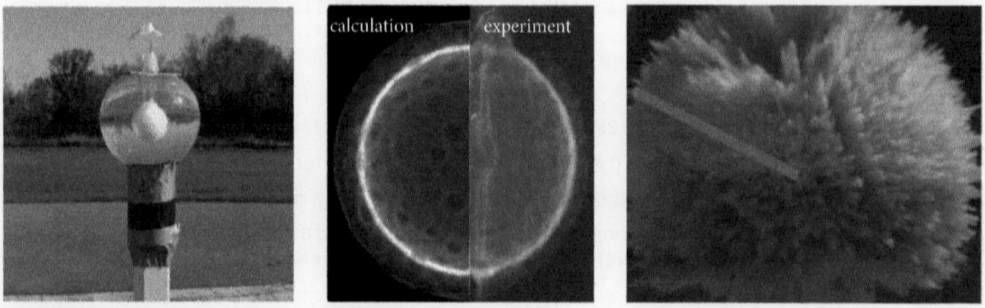

(a) Initial conditions. (b) Early time results. (c) Late time results.

Figure 16.8 Experimental results for explosively-driven water initially contained in a spherical glass bowl.
Reprinted with permission from Milne et al. (2017).

Example 16.12 Analytic vs. experimental results for primary fragmentation of liquid shells due to high-speed impacts

Suppose that a high-speed projectile penetrates a thin-walled liquid-filled container. According to Fourest et al. (2014), a rapidly tumbling projectile can transfer a large fraction of its kinetic energy to the fluid in the container, causing "a perfect spherical cavity [like those] ... observed in underwater explosions ... On the other hand, a projectile, that does not tumble, transfers its kinetic energy more slowly and creates a more elongated cavity shape in its wake." In either case, the energy transfer causes some or all of the liquid to exit the container. If liquid exits at high enough speeds, it will tend to aerosolize. If the liquid is a fuel, the aerosol may deflagrate or explode, dramatically increasing the resulting damage.

This problem is commonly known as *hydrodynamic ram*. Hydrodynamic ram occurs in many different problems such as the following:

- Meteorite and debris impacts on satellite thruster fuel tanks. According to Zhao et al. (2020), "the satellite tank filled with liquid hydrazine, which accounts for a large part of the satellite mass, is one of the most vulnerable components."
- Projectile impacts on aircraft fuel tank. According to Wang et al. (2020), the "fuel tank is ... one of the important vulnerable components for aircraft ... Current air defense weapons usually attack the fuel tanks by ... inert metal projectiles."

In *low damage cases*, the container remains intact except for projectile penetration holes and, perhaps, some bulging. In this scenario, the liquid exits the container through the penetration holes, forming roughly round jets. Depending on the exact circumstance, models such as the one discussed earlier in Example 16.8 may apply; see, e.g., Lecysyn et al. (2009), Tatman et al. (2018). In *medium damage cases*, the container wall partially splits open. For thin-walled cylindrical containers, the tear typically starts at the penetration hole and travels towards the end caps; see the left hand side of Figure 16.9. In some cases, a pre-existing seam in the sidewall may split open. Either way, the liquid exits the container through narrow slits, forming roughly planar sheets. Depending on the exact circumstance, models such as the one discussed later in Example 16.14 may apply. Finally, in *high damage cases*, the container walls fail catastrophically; see the right hand side of Figure 16.9. In this case, the liquid tends to form a roughly spherical expanding shell, regardless of the original shape of the container. Depending on the exact circumstances, models such as the one discussed in Example 16.11 may apply; see, e.g., Borg et al. (2001, 2006) and Westlake et al. (2004). In fact, the best available evidence indicates that:

$$D_{\text{avg}} \approx \left(24\sigma/\rho\dot{\varepsilon}^2\right)^{1/3} \tag{16.40}$$

where $\dot{\varepsilon} = 2\,V/R$, which is just Grady's model. Unfortunately, the best available evidence is indirect. The usual measurement techniques such as particle image velocimetry (PIV), particle tracking velocimetry (PTV), and digital in-line holography (DIH) only apply once droplet concentrations decrease to sufficiently low levels, either through natural spatial divergence or by artificial diversion into a sampling device. By this time, the liquid fragments have generally experienced a substantial amount of gravitational settling, as well as secondary breakup due to aerodynamic forces and collisions. Thus, the average size of the original early-time primary fragments must generally be inferred from limited samples of evolved late-time fragments. Despite such challenges, for "drop distributions collected using two ... witness cards," Borg et al. (2006) reports that Grady's law "predicted the correct median drop size."

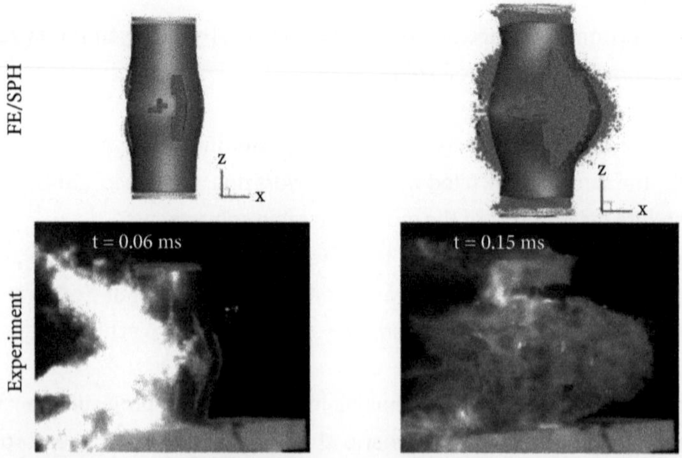

Figure 16.9 Computational results (top) vs. experimental results (bottom) for medium-energy (left) and high-energy (right) hydrodynamic ram. The hydrodynamic ram occurs due to a high-speed projectile penetrating a thin-walled liquid-filled cylindrical container. The computational results were obtained with Smoothed Particle Hydrodynamics (SPH) as discussed in Section 4.4. In this figure, medium damage, as shown on the left, progresses to high damage, as shown on the right. However, in other cases, the container only ever sustains medium damage.

Figure reprinted with permission from Matura et al. (2015).

Example 16.13 Analytic vs. experimental results for primary fragmentation of thin spherical liquid shells due to gaseous chemical combustion

As seen earlier in Figure 15.4, Vledouts et al. (2016) studied "the forced radial expansion of a [thin] spherical liquid shell by an exothermic chemical reaction." In these experiments, "a violently accelerated motion ... triggers the instability of the shell, developing thickness modulations, ultimately piercing it in a number of places." In this case, Vledouts et al. (2016) used stability analysis to obtain the following expression for the count mean diameter of the resulting liquid fragments:

$$D_{avg}/D_0 \approx \pi\, We^{-3/4} \qquad (16.41)$$

where $We = \rho V_f^2 L/\sigma$, V_f is the flame speed, ρ is the shell density, σ is the surface tension, L is the initial shell thickness, and D_0 is the initial shell diameter. As seen in Figure 16.10, Equation (16.41) obtains good agreement with selected experimental results for Weber numbers between 10 and 1,000. According to equation 4.25 in Vledouts et al. (2016):

$$R = R_0 + \frac{V_f^3 t^3}{D_0^2}, \quad V = \frac{dR}{dt} = \frac{3V_f^3 t^2}{D_0^2}, \quad V_f = \left(\frac{D_0^2 V}{3t^2}\right)^{1/3}$$

where t is the time at which the liquid shell becomes unstable, V is the radial velocity of the liquid shell at time t, R is the radius of the liquid shell at time t, and $R_0 = D_0/2$. If $\dot{\varepsilon} = 2\,V/R$ as in Example 15.3, then Equation (16.41) can be rewritten as follows:

$$D_{\text{avg}} \approx \pi \left(\frac{9\sigma^3 t^4}{4\rho^3 L^3 R^2 \dot{\varepsilon}^2}\right)^{1/4} \tag{16.42}$$

which agrees with Equation (16.7) with $i = 4$, assuming that $tR^{-1/2}$ is roughly constant. While this is not exactly true, Vledouts et al. (2016), eq. 3.3, says $t(R - R_0)^{-1/3}$ is constant.

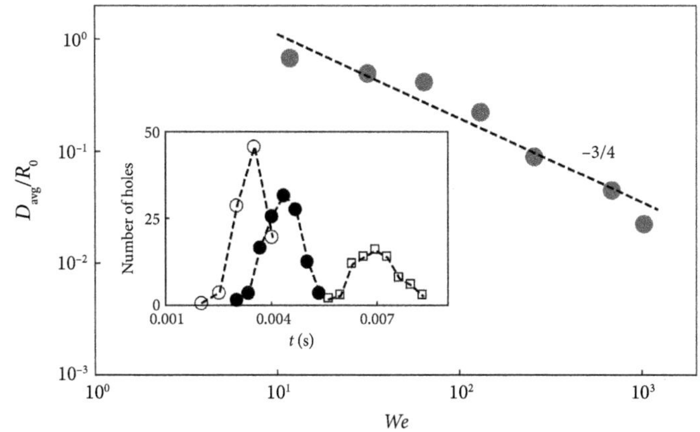

Figure 16.10 Analytical (dashed line) vs. experimental (circles) results for the count mean diameter of fragments from an expanding thin spherical liquid shell.
[The notation in this figure has been changed to match the notation used in this book.].
Reproduced with permission from Vledouts et al. (2016). Copyright 2016 Cambridge University Press.

16.5 Experimental Validation Part III: Primary Fragmentation in Cartesian Geometries

This is the third of four sections on experimental validation. This section concerns primary fragmentation due to uniaxial strain in Cartesian coordinates. This section is limited to planar liquid sheets. As one possibility, suppose the liquid sheet initially experiences only streamwise strains, i.e., all flow is parallel to the sheet. As a specific example, suppose a laminar liquid sheet is driven along the surface of a flat plate by an airstream parallel to the plate. If the liquid film separates cleanly from the trailing edge of the plate, without clinging or accumulating, Lefebvre (1992) says "the initial interaction between the liquid and the coflowing air is lethargic enough to allow ample time ... for the sheet to become wavy and for these waves to increase in amplitude until disintegration occurs according to the classical wavy-sheet mechanism." In this case, stability analysis leads to a full time-dependent solution. The solution is complex and is omitted. By contrast, Examples 16.14, 16.15, and 16.16 concern liquid sheets that experience significant cross-stream strains, i.e., some flow is perpendicular to the sheet. For more such examples see, e.g., table 6.11 in Lefebvre & McDonell (2017) and table 7 in Chaussonnet et al. (2020).

Example 16.14 Analytic vs. experimental results for primary fragmentation of free planar liquid jets with fully developed turbulence

Sallam et al. (1999) studied "annular liquid jets, having aspect ratios greater than 23 to approximate plane free jets ... [with the] annular passages having length/hydraulic diameter ratios greater than 40:1 both to avoid developing cavitating flows and to ensure fully developed turbulent flow at the jet exit." After meeting stagnant air, Sallam et al. (1999) found that such jets develop in two phases:

- Phase 1. "First of all, the liquid surface becomes rough, with protruding ligaments and other irregularities, relatively close to the jet exit ... Even though the liquid surface becomes roughened close to the jet exit, the first appearance of drops ... is deferred for a time. This occurs because the surface energy required to form very small drops is not available at the smallest scales of turbulence."
- Phase 2. "Subsequently, the diameter and length of the liquid surface roughness elements progressively increase with increasing distance from the jet exit ... After the onset of drop formation, ... drop diameters [produced by] turbulent primary breakup progressively increase with increasing distance from the jet exit, generally paralleling the corresponding increase of the scale of ligaments;" see Figure 16.11.

Sallam et al. (1999) used dimensional analysis, validated and calibrated using experimental results, to obtain the following expression for the Sauter mean diameter of the liquid fragments:

$$\frac{D'_{M,\,avg}(x)}{\Lambda} = 0.45\left(\frac{x}{\Lambda We_\Lambda^{1/2}}\right)^{2/3} \tag{16.43}$$

where x is the downstream distance from the jet exit, Λ is the integral length scale of the turbulence, $We_\Lambda = \rho U^2\Lambda/\sigma$ is a Weber number, ρ is the liquid density, σ is the liquid surface tension, and U is the streamwise mean velocity. Notice that Equation (16.43) can be rewritten as follows:

$$D'_{M,\,\mathrm{avg}}(x) = 0.45\left[\frac{\sigma}{\rho\dot{\varepsilon}(x)^2}\right]^{1/3} \tag{16.44}$$

where $\dot{\varepsilon}(x) = U/x$. Suppose that the median jet surface is a perfect triangular wedge with thickness $T(x)$. Then $T(x) = T_0 + (V/U)x$ where V is the cross-stream mean velocity. Per Sallam et al. (1999), V/U is a small constant. In this case, $\dot{\varepsilon}(x) = U/x \cong V/T(x)$ for large enough x, which agrees with Example 15.4. Then Equation (16.44) is just Grady's law.

Sallam et al. (1999) conducted a series of tests where the "conditions consisted of fully developed turbulent plane water jets injected into still air at standard temperature and pressure with Reynolds numbers of 91,000–424,000, Weber numbers of 13,000–151,000 and Ohnesorge numbers of 0.0009–0.0012. Pulsed shadowgraphy was used to measure ... drop sizes after primary breakup and the location of the onset of primary breakup." As seen in Figure 16.12, Equation (16.43) agrees well these experimental results. As this figure shows, the results are similar to those for free turbulent round jets as discussed in Example 16.8; also, the results are similar to those for plane turbulent wall jets; see Dai et al. (1997, 1998).

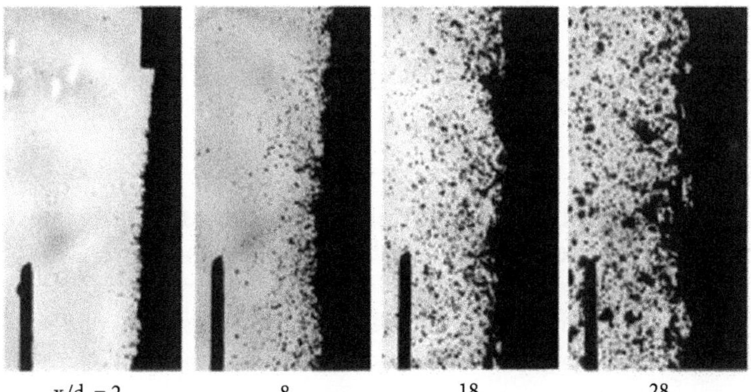

Figure 16.11 Primary atomization of a fully turbulent free planar liquid jet. In this figure, x is the distance from the jet exit and d_h is the "hydraulic diameter."
Figure reprinted from Sallam et al. (1999) with permission from Elsevier.

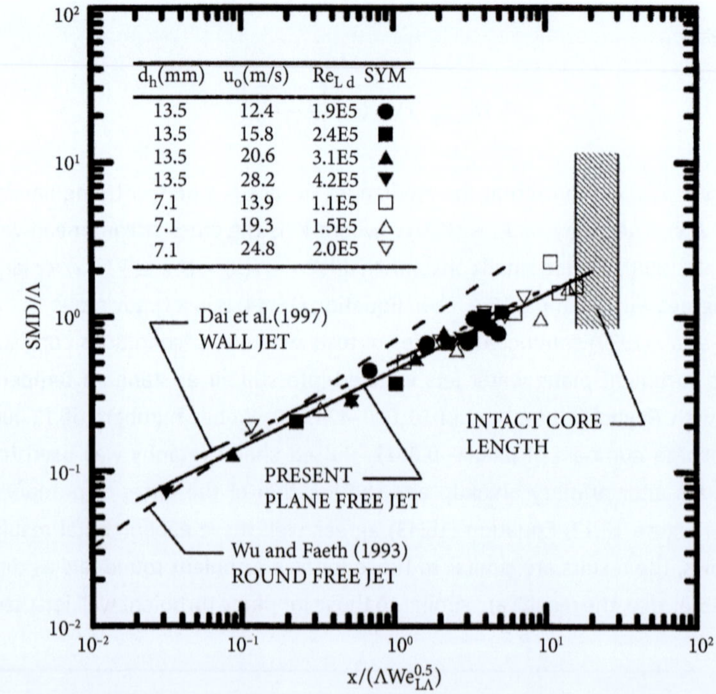

Figure 16.12 Analytical (solid line) vs. experimental (symbols) results for Sauter mean diameters of fragments from a fully turbulent planar liquid jet. Also shown are results for round free jets and wall jets. In this figure, d_h is the "hydraulic diameter" and u_0 is the streamwise jet speed.

Figure reprinted from Sallam et al. (1999) with permission from Elsevier.

Example 16.15 Analytic vs. experimental results for fragmentation of planar liquid jets driven by an oblique airstream

In prefilming airblast atomizers, a high-speed airstream drives a thin turbulent liquid film across a flat plate. As one possibility, Lefebvre (1992) says "the atomizing air impacts the liquid at an appreciable angle ... [so that] the liquid is shattered almost instantly" once it passes the sharp trailing edge of the plate. In this case, Lefebvre (1992) used energy analysis to find the following expression for the Sauter mean diameter:

$$D'_{M,\text{avg}} \cong \frac{3}{\dfrac{1}{T} + \dfrac{0.007\,\rho V_{\text{air}}^2}{4\sigma\left(1 + \dot{M}/\dot{M}_{\text{air}}\right)}} = \frac{3T}{1 + 0.00175\,\dfrac{\rho V^2 T}{\sigma}} = \frac{3T}{1 + 0.00175\,We}. \tag{16.45a}$$

In this expression, the liquid sheet has density ρ, surface tension σ, mass flux $\dot{M}$, thickness T, streamwise velocity V, and Weber number $We = \rho V^2 T/\sigma$. Similarly, the air has mass flux $\dot{M}_{\text{air}}$ and velocity V_{air}.

This expression assumes the air and liquid flows are related as follows:

$$V_{\text{air}} = V\sqrt{1 + \dot{M}/\dot{M}_{\text{air}}}.$$

Notice that, if the Weber number is sufficiently large, Equation (16.45a) becomes:

$$D'_{M,\text{avg}} \cong 1,714\,T/We = 1,714\,\sigma/(\rho V^2) = 1,714\,\sigma/(\rho T^2 \dot{\varepsilon}^2) \tag{16.45b}$$

where $\dot{\varepsilon} = V/T$ as in Example 15.4. Notice that Equation (16.45b) agrees with Equation (16.7) if $i = 1$. As seen in Figure 16.13, Lefebvre (1992) found that Equation (16.45) agrees well with a several different experimental results. More recently, Urbán et al. (2017) found similar agreement, while better distinguishing between primary and secondary atomization.

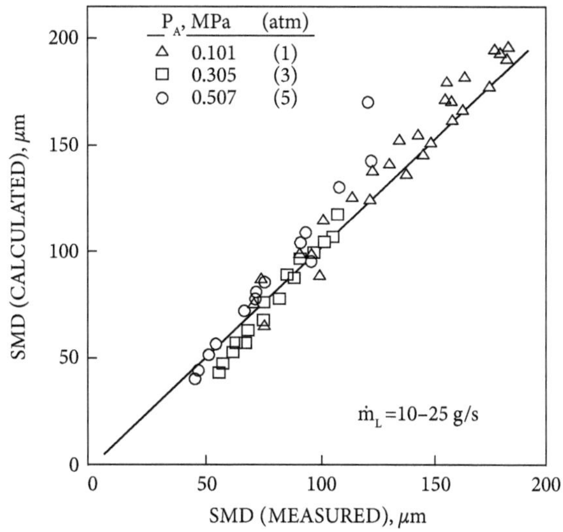

Figure 16.13 Analytical (y-axis) vs. experimental (x-axis) results for the Sauter mean diameter (SMD) of fragments from a 102-μm-thick layer of water passing through a prefilming airblast atomizer.

Reproduced with permission from Lefebvre (1992).

Example 16.16 Analytic vs. experimental results for fragmentation of planar liquid jets driven by a parallel airstream

Like the previous example, this example concerns prefilming airblast atomizers. In the previous example, the airstream was at an oblique angle to the flat plate. By contrast, in this example, the airstream is parallel to the flat plate. However, the liquid now clings to the sharp trailing edge of the flat plate prior to separation, which induces a cross-stream component of velocity. According to Chaussonnet et al. (2016), the liquid film "accumulates at the tip and in the wake of the prefilmer [and] ... is then fragmented by the gas flow into bags and ligaments;" see also, e.g., Holz et al. (2019). In this case, Chaussonnet et al. (2016) used stability analysis, validated and calibrated using experimental results, to obtain the following expression for the Sauter mean diameter:

$$D'_{M,\,avg} \cong 1.4\frac{T}{We^{1/2}} = 1.4\left(\frac{\sigma T}{\rho}\right)^{1/2}\frac{1}{V} = 1.4\left(\frac{\sigma T}{\rho}\right)^{1/2}\frac{1}{T\dot{\varepsilon}} \tag{16.46}$$

where $\dot{\varepsilon} = V/T$, ρ is the liquid density, σ is the liquid surface tension, and V is the streamwise velocity. In the previous example, T was the thickness of the liquid sheet. By contrast, in this example, T is the thickness of the plate's trailing edge. This implies that the strain rate V/T is induced during the brief time that the liquid dwells on the trailing edge. For future reference, the Weber number can be written in several different ways including the following:

$$We = \frac{\rho_{air}\,T}{\sigma}\left(\frac{\sqrt{\rho}}{\sqrt{\rho}+\sqrt{\rho_{air}}}V_{air}\right)^2 = \frac{\rho T}{\sigma}\left(\frac{\sqrt{\rho_{air}}}{\sqrt{\rho}+\sqrt{\rho_{air}}}V_{air}\right)^2 = \frac{\rho T V^2}{\sigma} \tag{16.47}$$

where the airstream has density ρ_{air} and velocity V_{air}. Notice that Equation (16.46) agrees with Equation (16.7) if $i = 2$. As seen in Figure 16.14, Chaussonnet et al. (2016, 2020) found that Equation (16.46) agrees well with experimental results for We between 10 and 400.

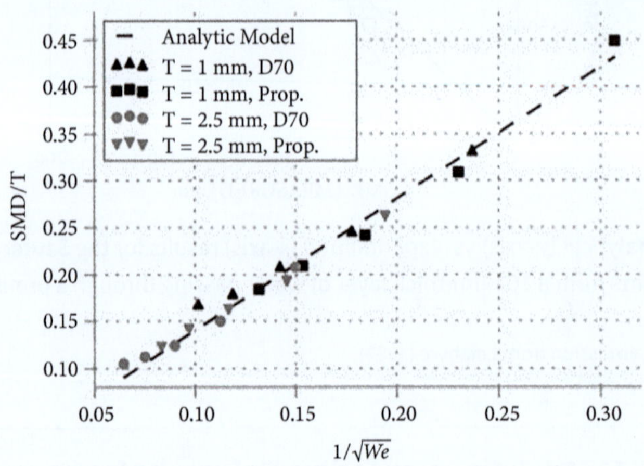

Figure 16.14 Analytical (dashed line) vs. experimental (symbols) results for the Sauter mean diameter (SMD) of fragments from a prefilming airblast atomizer.
[The notation in this figure has been changed to match the notation used in this book.].
Reprinted from Chaussonnet et al. (2006) with permission from Elsevier.

16.6 Experimental Validation Part IV: Secondary Fragmentation

The previous three sections concerned primary breakup. This section concerns secondary breakup or, at least, cases that resemble those observed in secondary breakup. First, consider secondary breakup of solids. In some cases, solid fragments experience secondary breakup due to aerodynamic forces; see, e.g., Figure 16.15. However, this is relatively rare—it generally only occurs when the primary fragments are heavily damaged. More often, solids experience secondary breakup due to collisions between two primary fragments and/or collisions between primary fragments and adjacent solid surfaces. While it is hard to study collisions between two primary fragments, it is relatively easy to study collisions between primary fragments and solid surfaces. Fortunately, these two problems are closely related; see, e.g., Figure 14.1.

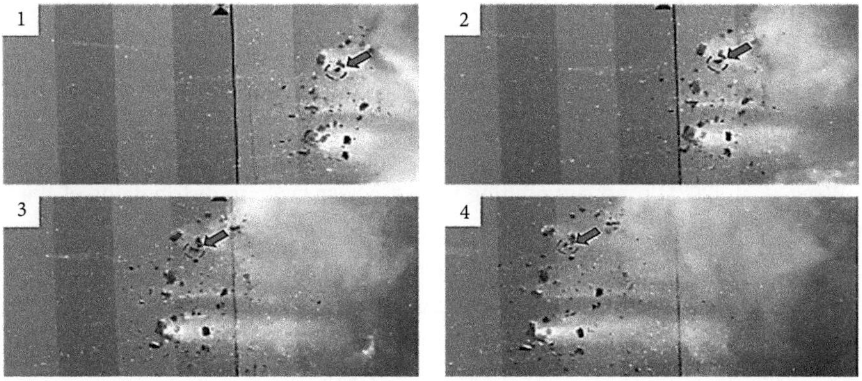

Figure 16.15 Fragments from a blast-shattered concrete disk experience secondary breakup due to aerodynamic forces. In this figure, the primary fragments travel from right to left, leaving a cloud of small dust-like secondary fragments their wake. In addition, in a least one case, a large fragment decomposes into smaller but still easily visible fragments, as indicated by the arrow.
Figure reproduced with permission from Sanai et al. (2023).

As one possible way to study collisions, Example 16.17 describes gun-launched projectiles impacting rigid steel targets. Notice that these projectiles have regular geometric shapes, instead of the irregular jagged shapes typical of primary fragments. This may or may not be an important distinction, depending on the exact circumstances. Unlike the other examples given in this chapter, this example describes both the characteristic fragment sizes *and* the fragment size distributions. Comparing these leads to some important conclusions. First, the size distribution may transition as the strain rate increases. In this example, the size distribution changes continuously as a function of strain rate, until the strain rate exceeds some threshold. Below this threshold, the results are non-ideal. Above this threshold, the results are ideal. Second, the characteristic sizes may transition as the strain rate increases. In this example, below some threshold, the characteristic sizes obey Grady's model. Above this threshold, the characteristic sizes obey Lefebvre's model. Third, although they both transition, the size distribution and characteristic size transition at different strain rates. Fourth and finally, analytic models for characteristic fragment sizes may apply both in ideal and non-ideal cases.

Example 16.17 Analytic vs. experimental results for fragmentation of gun-launched zirconium glass cylinders due to impacts with rigid steel targets

Wang et al. (2021) studied fragmentation of small compact projectiles composed of $Zr_{55}Cu_{30}Ni_5Al_{10}$ bulk metallic glass (BMG). The projectiles were cylinders with a length of 10mm and a diameter of 10mm. The projectiles were fragmented by shooting them at rigid steel targets. More specifically, Wang et al. (2021) conducted four low-velocity (< 300m/s) impact tests using a one-stage light gas gun and six high-velocity (> 300m/s) impact tests using a black powder ballistic gun. In the low-velocity tests, fragments were captured in an unlined, clear acrylic chamber to allow for high-speed photography. In the high-velocity tests, the fragments were captured in a hemispherical metal chamber, where the inner walls were lined to reduce secondary fragmentation. Both the acrylic and metal chambers were filled with argon gas to prevent the zirconium glass fragments from combusting in air. To determine the fragment sizes, the fragments "were separated through a series of sized sieving screens and weighed within each size range by a high-precision electronic scale with an accuracy of 0.001 g."

Wang et al. (2021) used curve fitting to find power law size distributions of the form $\mathcal{M}(D) = AD^k$ where $\mathcal{M}(D)$ is the total mass of the fragments with diameters less than D. Following standard sieving practices, as described earlier in Example 3.4, Wang et al. (2021) chose $D_{\text{char}} = D_{100}$. As one possible model, a variation of Grady's law obtains the following:

$$D_{\text{char}} = \left(\frac{48\Gamma}{\rho\dot{\varepsilon}^2}\right)^{1/3} = \left(\frac{24K_f^2}{\rho^2c^2\dot{\varepsilon}^2}\right)^{1/3} \tag{16.48}$$

where $\dot{\varepsilon} = V/L$ as in Example 15.4. In this expression, the projectile has velocity V, length $L = 10$mm, density $\rho = 6.714$g/cm^3, wave speed c where $\rho c^2 = 81.6$GPa, and dynamic fracture toughness $K_f = 38$MPa-m$^{1/2}$. The expression $\dot{\varepsilon} = V/L$ assumes that the initial compressive strain rate in the projectile leads to an equal-but-opposite tensile strain rate just prior to fragmentation. As seen in Figures 16.4 and 16.7, the experimental results agree well with Equation (16.48), but only for small enough strain rates. For larger strain rates, a variation of Lefebvre's model discussed earlier in Example 16.15 obtains the following:

$$D_{\text{char}} = \frac{10\Gamma}{\rho L^2\dot{\varepsilon}^2} = \frac{5K_f^2}{\rho^2c^2L^2\dot{\varepsilon}^2} \tag{16.49}$$

where $\dot{\varepsilon} = V/L$ just like before. As seen in Figure 16.16, all four low-velocity impact tests and two of the six high-velocity impact tests agree well with Grady's model. On the other hand, four of the six high-velocity impact tests agree well with LeFebvre's model.

In addition to characteristic fragment sizes, Wang et al. (2021) also reported fragment size distributions. As seen in Figure 16.17, the four low-velocity impact tests obtained non-ideal power law size distributions where n/m decreased from −0.38 to −0.61 as the strain rate increased. In other words, the size uniformity of the fragments as measured by n/m increased as the strain rate increased. On the other hand, the six high-velocity impact tests obtained nearly ideal power law size distributions where $n/m \cong -5/6$ regardless of the strain rate. This indicates that a high degree of spatial homogeneity and isotropy occurred once the average strain rate exceeded some threshold.

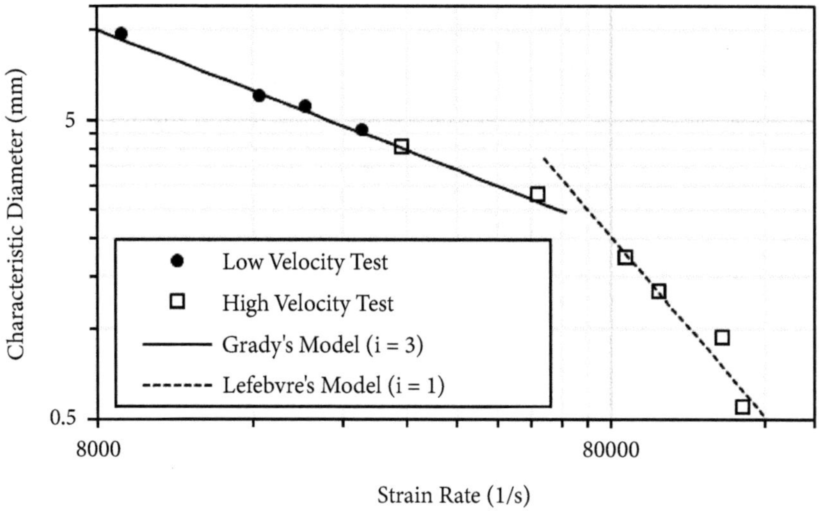

Figure 16.16 Analytical (lines) vs. experimental (symbols) results for the characteristic fragment diameter of a zirconium glass projectile. Notice that the characteristic fragment diameter decreases as the strain rate increases. The test results are based on the size modulus obtained from sieve data as reported by Wang et al. (2021).

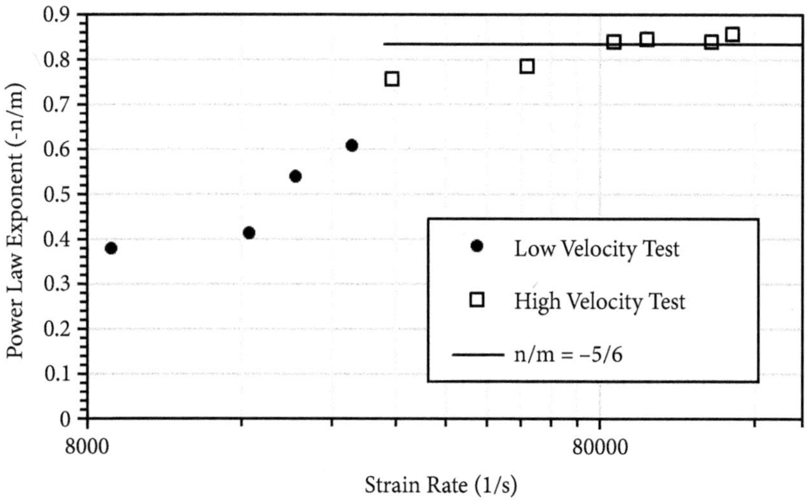

Figure 16.17 Analytical (line) vs. experimental (symbols) results for fragment size distributions from a zirconium glass projectile as a function of strain rate. The fragment size distributions transition from non-ideal to ideal as the strain rate increase, i.e., the power law exponent obtains a horizontal asymptote. The test results are based on curve fits to sieve data as reported by Wang et al. (2021).

Now, consider secondary breakup of liquids. Like solid fragments, liquid fragments may experience secondary breakup due to collisions with solid surfaces; see Example 16.18. In addition, because liquid primary fragments tend to be substantially weaker than solid primary fragments, they often experience secondary breakup due to aerodynamic forces. In this case, as the first step, the primary fragments tend to assume a spherical shape due surface tension, i.e., they become droplets. As the second step, the droplets flatten into discs. As the third step, the flattened discs fragment in a way that depends on dimensionless parameters such as the Weber number and Ohnesorge number. In this context, the Weber number is often defined as follows:

$$We = \rho_{\text{air}} V_0^2 D_0 / \sigma \tag{16.50}$$

where ρ_{air} is the air density, D_0 is the initial liquid droplet diameter, and V_0 is the initial relative speed between the droplet and the air. Notice that Equation (16.50) is the same as Equation (16.27) except that the air density ρ_{air} replaces the liquid density ρ. Figure 16.18 shows one possible schematic for the droplet breakup mechanism vs. Weber number; see also, e.g., Chou et al. (1997) and Sharma et al. (2022). Figure 16.18 assumes that the droplet has zero initial speed while the air has a constant uniform speed V_0. This is equivalent to assuming that the droplet has initial speed V_0 while the air is stagnant. In some cases, aerodynamic breakup occurs through one or more sudden events. For example, for bag breakup, the bag bursts and then the rim breaks up. By contrast, for shear-driven breakup such as sheet stripping, the droplet breaks up gradually over an extended period-of-time. In practice, shear-driven breakup such as sheet stripping is the most common form of aerodynamic breakup. Examples 16.19 and 16.20 discuss shear-driven breakup in more detail. Notice that Example 16.20 is the only example in this chapter where shear stress leads to viscous results—in all other examples, tensile stress leads to inviscid results.

Example 16.18 Analytic vs. experimental results for fragmentation of liquid droplets colliding with solid surfaces

Researchers have studied droplets landing on surfaces for almost 150 years beginning with Worthington (1876, 1877); see also Josserand & Thoroddsen (2016) and Yarin et al. (2017). There are many possible scenarios. For the scenario seen in Figure 16.19. Villermaux & Bossa (2011) found that the breakup occurs in three main phases:

- *Phase 1. Parent Droplet Goes from Round to Flat.* "The drop crushes ... and feeds a radially developing sheet ...The details of drop deformation when it hits the solid [surface] are essentially decoupled from the subsequent dynamics of the sheet itself."
- *Phase 2. Ligament Formation.* "The average sheet radius increases up to a maximal value ... while the sheet rim corrugates. These corrugations are the base for radial ligaments," a.k.a., "fingers."
- *Phase 3. Fragmentation.* The "radial ligaments ... ultimately fragment into smaller drops ... The final drop size distribution in the spray is ... a linear superposition of ... distributions characteristic of ligament breakup."

Using stability analysis, Villermaux & Bossa (2011) found the following:

$$D_{avg} \cong D_0 We^{-1/4} \tag{16.51}$$

where $We = \rho V_0^2 D_0/\sigma$ is a Weber number greater than 100, D_0 is the initial droplet diameter, and V_0 is the initial droplet velocity. As seen in Figure 16.20, Equation (16.51) agrees well with experimental data for We between 100 and 30,000. As an alternative approach, suppose the droplets are formed through tensile stretching of radial ligaments. As one possible outcome of energy analysis:

$$D_{avg} = \text{const.}\left(\frac{\sigma D_0}{\rho \dot{\varepsilon}^2}\right)^{1/4} \tag{16.52}$$

where $\dot{\varepsilon} = \text{const.}\, V_0/D_0$ as seen in Example 15.2. Notice that Equation (16.52) agrees with Equation (16.51). In addition, Equation (16.52) is the same as Equation (16.7) with $i = 4$.

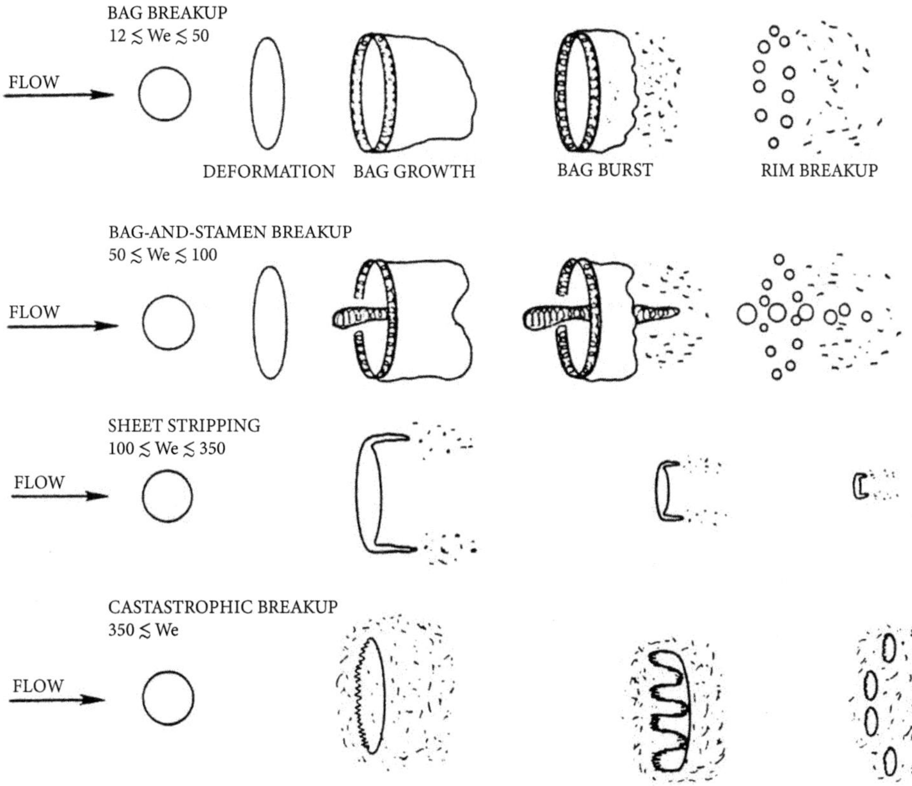

Figure 16.18 Some common types of aerodynamic breakup of liquid droplets. Different sources cite different ranges of Weber numbers depending, among other factors, on Ohnesorge number.

Figure reprinted from Pilch & Erdman (1987) with permission from Elsevier.

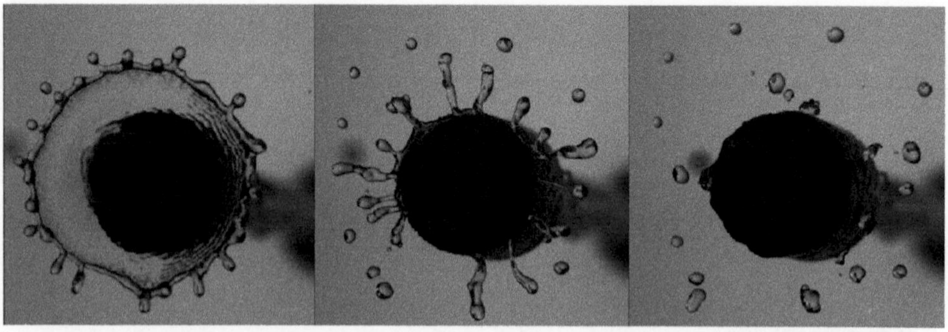

Figure 16.19 Experimental result for an initially spherical liquid droplet colliding with a flat solid surface. The solid surface has a small diameter. This reduces the effects of any surface roughness, while making it easy to see what happens at the droplet's perimeter.
Reproduced with permission from Villermaux & Bossa (2011). Copyright 2011 Cambridge University Press.

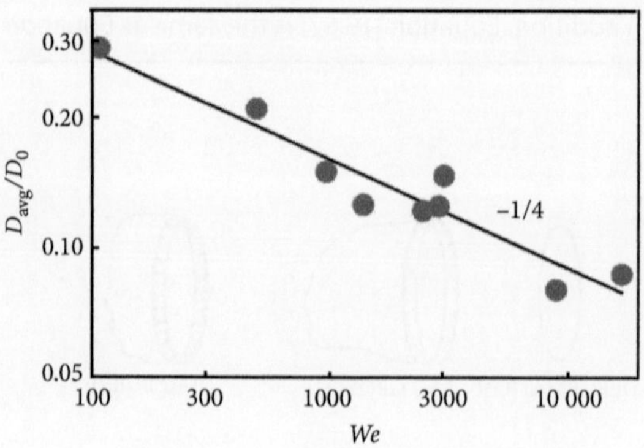

Figure 16.20 Analytical (line) vs. experimental (circles) results for the count mean diameter of fragments from a liquid droplet impacting a solid surface. The analytic results were calculated using Equation (16.51).
Reproduced with permission from Villermaux & Bossa (2011). Copyright 2011 Cambridge University Press. [The notation in this figure has been changed to match the notation used in this book.]

Example 16.19 Shear-driven aerodynamic breakup of liquid droplets

Consider a spherical liquid droplet traveling through still air. Aerodynamic forces will generally cause such droplets to fragment, either gradually or all at once. In particular, in some cases, shearing causes the surface of the parent droplet to progressively shed a trail of smaller children. As seen in Figures 16.21 and 16.22, this generally occurs in three phases:

- *Phase 1. Parent Droplet Goes from Round to Flat.* Chou et al. (1997) say "the drop deforms into a flattened shape because the liquid is drawn toward the drop periphery where the pressure is low due to acceleration of the gas flow over the drop surface."

- *Phase 2. Sheet and Ligament Formation.* According to Guildenbecher et al. (2017), "the high gas-velocity at the periphery of the deformed drop results in the formation of thin sheet-like membranes." In addition, according to Chou et al. (1997), "there is an extensive system of ligaments protruding from the periphery of the parent drop."
- *Phase 3. Fragmentation.* According to Guildenbecher et al. (2017), "as time progresses ... sheets are stretched downstream where they fragment." In addition, according to Chou et al. (1997), "numerous individual drops [are] present near the downstream end of the ligaments as a result of ligament breakup."

Figures 16.21 and 16.22 illustrate cases without and with a coherent boundary layer on the surface of the parent droplet, respectively. The latter is sometimes called *sheet stripping breakup*. At low Ohnesorge numbers, Pilch & Erdman (1987) say that sheet stripping breakup only occurs for Weber numbers greater than 100. However, more recently, researchers have found that something resembling sheet stripping breakup can occur at much lower Weber numbers; see, e.g., Guildenbecher et al. (2017).

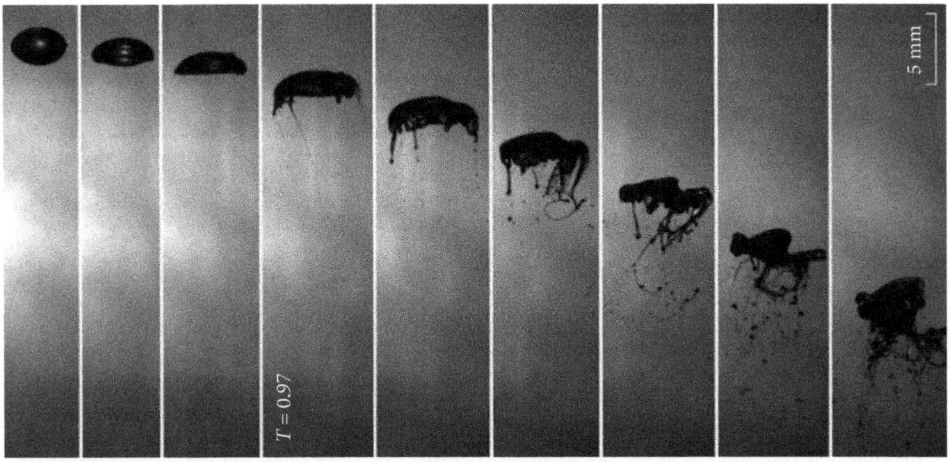

Figure 16.21 Experimental results for a liquid droplet without a coherent boundary layer experiencing gradual shear-induced breakup.
Reprinted from Theofanous & Li (2008) with the permission of AIP Publishing.

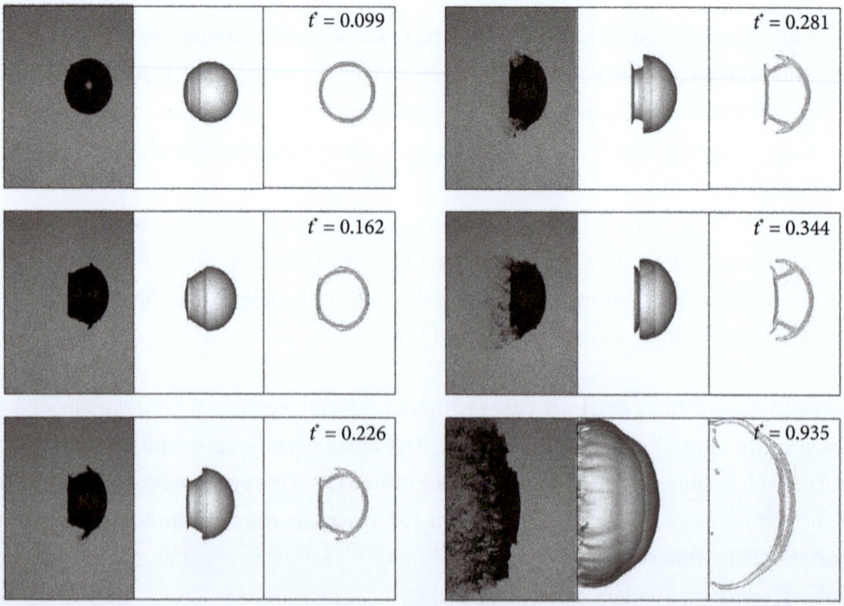

Figure 16.22 Experimental (left column) vs. computational (center and right columns) results for a liquid droplet with a boundary layer experiencing gradual shear-induced breakup. This is sometimes called sheet stripping breakup.

Experimental results reprinted from Theofanous et al. (2012) with the permission of AIP Publishing. Computational results reproduced from Meng & Colonius (2018). Copyright 2017 Cambridge University Press.

Example 16.20 Analytic vs. experimental results for fragmentation of liquid droplets due to sheet stripping

As seen in the last example, sheet stripping appears to be hopelessly complex. However, based on curve fits to experimental results, Chou et al. (1997) found that parent droplets breakup during the time interval $0.273\,t^* \leq t \leq t^*$ where:

$$t^* = 5.5\,(D_0/V_0)\sqrt{\rho/\rho_{\mathrm{air}}} \tag{16.53}$$

and where D_0 is the initial diameter of the liquid droplet, V_0 is the initial velocity of the liquid droplet, ρ is the density of the liquid droplet, and ρ_{air} is the density of the surrounding air. For $0.273\,t^* \leq t \leq t^*$, Chou et al. (1997) found that the droplet fragments obtain the following Sauter mean diameter:

$$D'_{M,\,\mathrm{avg}}(t) = \min[2.0(\mu t/\rho)^{1/2}, 0.09\,D_0] \tag{16.54}$$

where μ is the dynamic viscosity of the liquid droplet.

For small enough times, Chou et al. (1997) argue that the children droplets have diameters equal, on average, to the thickness $\delta(t)$ of a laminar viscous boundary layer trailing from the surface of a parent droplet. In other words, $D'_{M,\,avg}(t) = \text{const.}\delta(t)$. According to Schlichting & Gersten (2017):

$$\delta(t) = \text{const.}\sqrt{\mu L(t)/[\rho U(t)]} \tag{16.55}$$

where L is the length of the trailing boundary layer and $U(t)$ is the streamwise velocity at the edge of the trailing boundary layer, i.e., the velocity of the liquid droplet. According to equation 13 in Chou et al. (1997), $U(t) = \text{const.}\ t$. Assuming that $L(t) = \text{const.}\ t^2$:

$$D'_{M,\,avg}(t) = \text{const.}\ \delta(t) = \text{const.}\ (\mu t/\rho)^{1/2} \tag{16.56}$$

which agrees with Equation (16.54) at early times. As an alternative proof, a variant of the third row of Table 16.4 says:

$$D'_{M,\,avg}(t) = \text{const.}\ [\mu/\rho\dot{\varepsilon}(t)]^{1/2} \tag{16.57}$$

where $\dot{\varepsilon}(t) = V(t)/\delta(t)$ and $V(t) = d\delta(t)/dt$ is the mean cross-stream velocity. Because $\dot{\varepsilon}(t) = V(t)/\delta(t) = 1/2t = \text{const.}\ U(t)/L(t)$, Equations (16.54) and (16.57) agree at early times. At later times, Chou et al. (1997) found that $D'_{M,\,avg}(t)$ equals $0.09\,D_0$ due to a "quasi-steady shear breakup mechanism at small Ohnesorge numbers" involving "stabilization of the flow within the drop" and/or "complete development of the boundary layer near the surface."

As seen in Figure 16.23, Equation (16.54) agrees well with experimental results under a wide variety of conditions. More specifically, according to Chou et al. (1997) the "test conditions included Weber numbers of 125–375, Ohnesorge numbers of 0.003–0.040, liquid/gas density ratios of 670–990 and Reynolds numbers of 3,000–12,000." More recently, Guildenbecher et al. (2017) studied droplet breakup for Weber numbers of 14 and 55, well below the traditional threshold for sheet stripping. They still obtained good agreement between Equation (16.54) and their experimental results; see their figure 13.

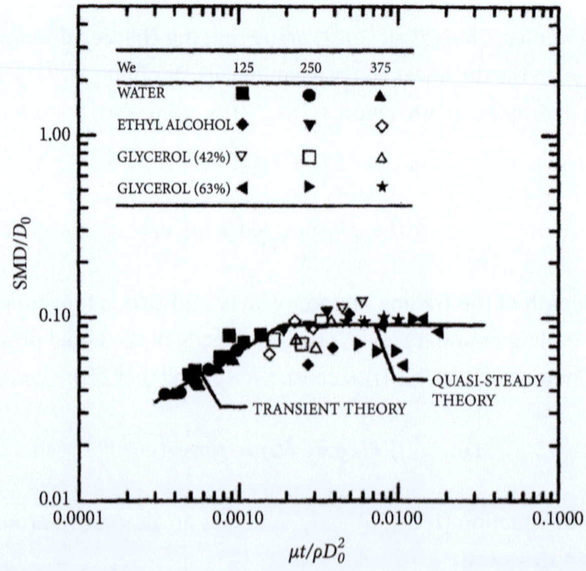

Figure 16.23 Analytical (lines) vs. experimental (symbols) results for the Sauter mean diameter (SMD) of fragments from an airborne liquid droplet exposed to a high-speed airstream. The analytic results were calculated using Equation (16.54).
[The notation in this figure has been changed to match the notation used in this book.]
Reprinted from Chou et al. (1997) with permission from Elsevier.

As seen in Example 16.21, the models given in this section for secondary breakup may be combined with the models given in previous sections for primary breakup.

Example 16.21 Analytical results for primary breakup due to uniform expansion of spherical liquid shell and secondary breakup due impact with a solid surface

Examples 16.11 and 16.18 give the following expressions:

$$D_{\text{avg},1} = \text{const.} \left(\frac{\sigma R^2}{\rho V^2}\right)^{1/3}$$

$$D_{\text{avg},2} = \text{const.} \left(\frac{\sigma D_{\text{avg},1}^3}{\rho V^2}\right)^{1/4}.$$

Combining the previous two expression gives the following:

$$D_{\text{avg},2} = \text{const.} \left(\frac{\sigma R}{\rho V^2}\right)^{1/2}.$$

So far, this section has focused on the secondary breakup of individual primary fragments. However, in many cases, the same models describe the secondary breakup of entire sets of primary fragments, provided they all meet the same or similar fates. If rapid mixing occurs following secondary fragment formation, it may make sense to integrate in time and/or space to obtain a single characteristic size. For example, consider primary fragmentation of free round liquid jets with fully developed turbulence. As seen in Example 16.8, the characteristic size of the resulting fragments depends on the distance downstream from the jet exit. In addition, consider secondary fragmentation of liquid droplets due aerodynamic sheet stripping. As seen in Example 16.20, the characteristic size of the resulting fragments depends on the time of formation. Integrating the results for primary breakup over space, and integrating the results for secondary breakup over time, leads to a single overall characteristic size for the entire set of secondary fragments. In this scenario, because primary breakup depends on surface tension, while secondary breakup depends on viscosity, the overall characteristic size depends on both surface tension and viscosity. Thus, when vigorous mixing occurs, the secondary fragments may appear to be the result of a single homogeneous, isotropic brittle-ductile event.

16.7 Conclusions

This chapter discussed analytic models for average fragment diameters, assuming a single large object decomposes into a large number of small pieces, either in a deterministic or random fashion. Instead of "average diameter," this chapter often uses the term "characteristic diameter" to include things like the size modulus. Table 16.5 lists four common choices for characteristic diameters. For other possible choices, see, e.g., Kapur & Fuerstenau (1987). As noted in the last chapter, unitary fragmentation often produces three different size ranges. When this occurs, a given characteristic diameter can only represent one size range. Except for count mean diameter, the characteristic diameters listed in Table 16.5 tend to favor the largest fragments and thus tend to represent the largest energy-containing range.

Table 16.5 Common choices for characteristic diameters.

D_{char}	Name	Notes
D_{avg}	count mean diameter	Defined by Equation (3.34) and Table 3.1.
$D_{M,\text{ avg}}$	mass mean diameter	Defined by Equation (3.36) and Table 3.1.
$D'_{M,\text{ avg}}$	Sauter mean diameter	Defined by Equation (3.37) and Table 3.1.
D_{100}	size modulus	Defined by Equation (3.17), i.e., $F_M(D_{100}) = 1$[1]

(1) Generally found by extrapolating curve fits to sieve results as described in Example 3.4.

As shown by Laney (2016), it is generally straightforward to convert from one characteristic diameter to another. Similarly, it is generally straightforward to convert from any characteristic diameter to any characteristic mass or vice versa. The three main conditions are as follows: the two characteristic sizes in question must be finite; the shapes of the particles must be known; and the full size distribution of the fragments must be known. When these three conditions are met, any two characteristic diameters tend to differ by a unitless constant factor. Similarly, as seen in Equation (16.3), any characteristic mass M_{char} tends to equal a unitless constant times ρD_{char}^{m} where D_{char} is a characteristic diameter. This is important because, except for this chapter, most of this book expresses results in terms of masses rather than diameters.

This chapter has shown that simple analytic models often capture characteristic fragment sizes surprisingly well, even in cases that initially appear to be highly complex. This is especially true when large, complicated problems can be reduced to a series of smaller, simpler problems that each obtain constant strain rates in Cartesian, cylindrical, or spherical coordinates. In such cases, this chapter has shown that characteristic fragment diameters are products of four things:

1. *Powers of strain rates.* More specifically, characteristic diameters are proportional to $\dot{\varepsilon}^{-2/i}$ where $\dot{\varepsilon}$ is the constant strain rate and i is a positive integer between 1 and 12.
2. *Powers of material properties.* More specifically, characteristic diameters are proportional to powers of density, viscosity, surface tension, fracture toughness, etc.
3. *Powers of characteristic lengths.* More specifically, characteristic diameters are proportional to a power of a length, measured in the direction that strain is applied.
4. *Function of the fragment sizes and shapes.* This function is usually set equal to a constant; however, this may lead to gross inaccuracies; see Examples 16.5 and 16.11.

Except for the last factor, characteristic fragment diameters can be modeled separately from fragment size distributions. Essentially, characteristic sizes reflect the specific details of each problem while size distributions reflect semi-generic geometric aspects.

References

A. Balagansky, Yu., A. Karanik, V. A. Agureikin, A. V. Vinogradov, & A. I. Balagansky, Fracture Behavior of Explosively Loaded Spherical Molded Steel Shells, *Theoretical and Applied Fracture Mechanics*, 36(2), 165–173, 2001, https://doi.org/10.1016/S0167-8442(01)00066-0

B. H. Bergstrom, Energy and Size Distribution Aspects of Single Particle Crushing, in C. Fairhurst, editor, *Rock Mechanics: Proceedings of the Fifth Symposium on Rock Mechanics held at the University of Minnesota*, May 1962, Pergamon Press, 1963

C. Bolis, D. Counilh, J. M. Lagrange, & P. A. Frugier, Fragmentation of a Titanium Alloy Shell in Expansion: From Experiment to Simulation, *Procedia Engineering*, 58, 672–677, 2013, https://doi.org/10.1016/j.proeng.2013.05.077

J. P. Borg, D. Grady, & J. R. Cogar, Instability and Fragmentation of Expanding Liquid Systems, *International Journal of Impact Engineering*, 26(1-10), 65–76, 2001, https://doi.org/10.1016/S0734-743X(01)00066-5

J. P. Borg, T. J. Downs, & A. Lloyd, High Strain Rate Fragmentation of Liquid Systems at Atmospheric Pressure, *International Journal of Impact Engineering*, 33(1–12), 119–125, 2006, https://doi.org/10.1016/j.ijimpeng.2006.09.043

M. Callahan, D. Sun, M. A. Linne, A. S. Wu, G. H. Campbell, B. Friedman, et al., Explosive Fragmentation of Additively Manufactured Stainless Steel, *Journal of Applied Physics*, 134(15), 155105, 2023, https://doi.org/10.1063/5.0170223

G. Chaussonnet, O. Vermorel, E. Riber, & B. Cuenot, A New Phenomenological Model to Predict Drop Size Distribution in Large-Eddy Simulations of Airblast Atomizers, *International Journal of Multiphase Flow*, 80, 29–42, 2016, https://doi.org/10.1016/j.ijmultiphaseflow.2015.10.014

G. Chaussonnet, S. Gepperth, S. Holz, R. Koch, & H.-J. Bauer, Influence of the Ambient Pressure on the Liquid Accumulation and on the Primary Spray in Prefilming Airblast Atomization, *International Journal of Multiphase Flows*, 125, 103229, 2020, https://doi.org/10.1016/j.ijmultiphaseflow.2020.103229

W.-H. Chou, L.-P. Hsiang, & G. M. Faeth, Temporal Properties of Drop Breakup in the Shear Breakup Regime, *International Journal of Multiphase Flow*, 23(4), 651–669, 1997, https://doi.org/10.1016/S0301-9322(97)00006-2

Z. Dai, L.-P. Hsiang, & G. M. Faeth, Spray Formation at the Free Surface of Turbulent Bow Sheets. *Proceedings of the Twenty-First Symposium on Naval Hydrodynamics*, National Academy Press, 1997, https://nap.nationalacademies.org/catalog/5870/twenty-first-symposium-on-naval-hydrodynamics

Z. Dai, W.-H. Chou, & G. M. Faeth, Drop Formation due to Turbulent Primary Breakup at the Free Surface of Plane Liquid Wall Jets, *Physics of Fluids*, 10(5), 1147–1157, 1998, https://doi.org/10.1063/1.869639

G. M. Faeth, L.-P. Hsiang & P.-K. Wu, Structure and Breakup Properties of Sprays, *International Journal of Multiphase Flow*, 21 (Supplement), 99–127, 1995, https://doi.org/10.1016/0301-9322(95)00059-7

G. M. Faeth, Spray Combustion Phenomena, *International Symposium on Combustion*, 26(1), 1593–1612, 1996, https://doi.org/10.1016/S0082-0784(96)80383-3

T. Fourest, J.-M. Laurens, E. Deletombe, J. Dupas, & M. Arrigoni, Analysis of Bubbles Dynamics Created by Hydrodynamic Ram in Confined Geometries using the Rayleigh–Plesset Equation, *International Journal of Impact Engineering*, 73, 66–74, 2014, https://doi.org/10.1016/j.ijimpeng.2014.05.008

S. K. Garg & J. Siekmann, On Fracture of a Thin Spherical Shell under Blast Loading, *Experimental Mechanics*, 6(1), 39–44, 1966, https://doi.org/10.1007/BF02327112

D. E. Grady, Application of Survival Statistics to the Impulsive Fragmentation of Ductile Rings, in M. A. Meyers & L. E. Murr, editors, *Shock Waves and High-Strain-Rate Phenomena in Metals*, Plenum Press, 1981, https://doi.org/10.1007/978-1-4613-3219-0_11

D. E. Grady, Local Inertial Effects in Dynamic Fragmentation, *Journal of Applied Physics*, 53(1), 322–325, 1982, https://doi.org/10.1063/1.329934

D. E. Grady, The Spall Strength of Condensed Matter, *Journal of Mechanics and Physics of Solids*, 36(3), 353–384, 1988, https://doi.org/10.1016/0022-5096(88)90015-4

D. Grady, *Fragmentation of Rings and Shells: The Legacy of N. F. Mott*, Springer, 2006, https://doi.org/10.1007/b138675

D. E. Grady, *Physics of Shock and Impact, Volume 1: Fundamentals and Dynamic Failure*, IOP Publishing, 2017, https://doi.org/10.1088/978-0-7503-1254-7

D. E. Grady & M. E. Kipp, Continuum Modeling of Explosive Fracture in Oil Shale, *International Journal of Rock Mechanics and Mining Sciences and Geomechanics Abstracts*, 17(3), 147–157, 1980, https://doi.org/10.1016/0148-9062(80)91361-3

D. E. Grady & M. E. Kipp, Dynamic Fracture and Fragmentation, in J. R. Asay & M. Shahinpoor, editors, *High-Pressure Shock Compression of Solids*, Springer-Verlag, 1993, https://doi.org/10.

1007/978-1-4612-0911-9_8 [see also D. E. Grady & M. E. Kipp, Fragmentation of Solids Under Dynamic Loading, in T. Wierzbicki & N. Jones, editors, *Structural Failure*, Wiley, 1989]

D. E. Grady & M. L. Olsen, A Statistics and Energy Based Theory of Dynamic Fragmentation, *International Journal of Impact Engineering*, 29, 293–306, 2003, https://doi.org/10.1016/j.ijimpeng.2003.09.026

G. T. Gray, P. J. Maudlin, L. M. Hull, Q. K. Zuo, & S.-R. Chen, Predicting Material Strength, Damage, and Fracture: The Synergy between Experiment and Modeling, *Journal of Failure Analysis and Prevention*, 5(3), 7–17, 2005, https://doi.org/10.1361/15477020523725

D. R. Guildenbecher, J. Gao, J. Chen, & P. E. Sojka, Characterization of Drop Aerodynamic Fragmentation in the Bag and Sheet-Thinning Regimes by Crossed-Beam, Two-View, Digital In-Line Holography, *International Journal of Multiphase Flow*, 94, 107–122, 2017, https://doi.org/10.1016/j.ijmultiphaseflow.2017.04.011

R. W. Gurney, *The Initial Velocities of Fragments from Bombs, Shells, and Grenades*, Ballistic Research Laboratories Report 405, 1943, http://apps.dtic.mil/sti/citations/ADA289704

S. Holz, S. Braun, G. Chaussonnet, R. Koch, & H.-J. Bauer, Close Nozzle Spray Characteristics of a Prefilming Airblast Atomizer, *Energies*, 12(14), 2835, 2019, https://doi.org/10.3390/en12142835

A. I. Ivanov, M. A. Syrunin, A. G. Fedorenko, & A. P. Tsoi, Fragmentation of Spherical Shells under Internal Blast Loading, *Strength of Materials*, 33, 150–156, 2001, https://doi.org/10.1023/A:1010432432000

C. Josserand & S. T. Thoroddsen, Drop Impact on a Solid Surface, *Annual Review of Fluid Mechanics*, 48, 365–391, 2016, https://doi.org/10.1146/annurev-fluid-122414-034401

P. C. Kapur & D. W. Fuerstenau, Energy-size Reduction Laws Revisited, *International Journal of Mineral Processing*, 20(1–2), 45–57, 1987, https://doi.org/10.1016/0301-7516(87)90016-0

J. E. Kennedy, The Gurney Model for Explosive Output for Driving Metal, in J. A. Zukas & W. P. Walters, editors, *Explosive Effects and Applications*, Springer, 1998, https://doi.org/10.1007/978-1-4612-0589-0_7

F. Kick, *Das Gesetz der Proportionalen Widerstände und seine Anwendungen [Principle of Proportional Resistance and its Application]*, Felix-Verlag, 1885

X. Kong, W. Wu, J. Li, P. Chen, & F. Liu, A Numerical Investigation on Explosive Fragmentation of Metal Casing Using Smoothed Particle Hydrodynamic Method, *Materials & Design*, 51, 729–741, 2013, https://doi.org/10.1016/j.matdes.2013.04.041

C. B. Laney, *Effects of Mixtures on Liquid and Solid Fragment Size Distributions*, Defense Threat Reduction Agency Technical Report DTRA-TR-16-52, May 2016, http://apps.dtic.mil/sti/citations/AD1008708

N. Lecysyn, A. Dandrieux, F. Heymes, L. Aprin, P. Slangen, L. Munier et al., Ballistic Impact on an Industrial Tank: Study and Modeling of Consequences, *Journal of Hazardous Materials*, 172(2-3), 587–594, 2009, https://doi.org/10.1016/j.jhazmat.2009.07.086

A. H. Lefebvre, Energy Considerations in Twin-Fluid Atomization, *Journal of Engineering for Gas Turbines and Power*, 114(1), 89–96, 1992, https://doi.org/10.1115/1.2906311

A. H. Lefebvre & V. G. McDonell, *Atomization and Sprays*, 2nd edition, CRC Press, 2017, https://doi.org/10.1201/9781315120911

A. M. Marquez, C. H. Braithwaite, T. P. Weihs, N. M. Krywopusk, D. J. Gibbins, K. S. Vecchio, et al., Fragmentation and Constitutive Response of Tailored Mesostructured Aluminum Compacts, *Journal of Applied Physics*, 119, 145903, 2016, https://doi.org/10.1063/1.4945813

A. M. Marquez, Z. Li., C. H. Braithwaite, T. P. Weihs, M. Krywopusk, D. J. Gibbins, & M. A. Meyers, Fragmentation and Mechanical Performance of Tailored Nickel-Aluminum Laminate Compacts, *Materials Science & Engineering A*, 727, 123–132, 2018, https://doi.org/10.1016/j.msea.2018.04.027

P. Matura, G. Heilig, M. Lueck, & M. Sauer, Simulation and Experiments of Hypervelocity Impact in Containers with Fluid and Granular Fillings, *Procedia Engineering*, 103, 365–372, 2015, https://doi.org/10.1016/j.proeng.2015.04.034

J. C. Meng & T. Colonius, Numerical Simulation of the Aerobreakup of a Water Droplet, *Journal of Fluid Mechanics*, 835, 1108–1135, 2018, https://doi.org/10.1017/jfm.2017.804

A. Milne, A. Longbottom, D. L. Frost, J. Loiseau, S. Goroshin, & O. Petel, Explosive Fragmentation of Liquids in Spherical Geometry, *Shock Waves*, 27, 383–393, 2017, https://doi.org/10.1007/s00193-016-0671-y

N. F. Mott & E. H. Linfoot, *A Theory of Fragmentation*, U. K. Ministry of Supply, AC3348, January 1943 [reproduced in section II.1 of Grady (2006)], https://doi.org/10.1007/978-3-540-27145-1_9

N. F. Mott, J. H. Wilkinson, & T. H. Wise, *Fragmentation of Service Projectiles*, U.K. Ministry of Supply, A.C. 6338, May 1944 [reproduced in section II.6 of Grady (2006)], https://doi.org/10.1007/978-3-540-27145-1_14

N. F. Mott, Fragmentation of Shell Cases, *Proceedings of the Royal Society A: Mathematical, Physical and Engineering Sciences*, 189(1018), 300–308, 1947, https://doi.org/10.1098/rspa.1947.0042

F. Olive, J. Marilleau, R. Loichot, C. Peyre, & A. Nicaud, High Speed Fracture of Metals Studied By Cinematographic, Magnetic and Radiographic Methods, *Proceedings of the SPIE*, 189, 805–809, 1979, https://doi.org/10.1117/12.957702

J. Pearson, *A Fragmentation Model for Cylindrical Warheads*, Naval Weapons Center Technical Publication NWC TP 7124, December 1990, http://apps.dtic.mil/sti/citations/ADA238153

A. J. Piekutowski, Fragmentation of a Sphere Initiated by Hypervelocity Impact with a Thin Sheet, *International Journal of Impact Engineering*, 17(4–6), 627–638, 1995, https://doi.org/10.1016/0734-743X(95)99886-V

A. J. Piekutowski, Effects of Scale on Debris Cloud Properties, *International Journal of Impact Engineering*, 20(6–10),639–650, 1997, https://doi.org/10.1016/S0734-743X(97)87451-9

M. Pilch & C. A. Erdman, Use of Breakup Time Data and Velocity History Data to Predict the Maximum Size of Stable Fragments for Acceleration-Induced Breakup of a Liquid Drop, *International Journal of Multiphase Flow*, 13(6), 741–757, 1987, https://doi.org/10.1016/0301-9322(87)90063-2

P. Panizza & L. Courbin, Blowing Bubbles by the Numbers, *Physics Today*, 69(7), 78–79 2016, https://doi.org/10.1063/PT.3.3246

M. L. Potocki & L. Hull, Predictive Simulation and Experimental Confirmation of the Onset of Instability of Explosively Driven Shells, in R. Chmielewski & L. Kruszka, editors, *Proceedings of the 7th International Symposium on Impact Engineering (ISIE 2010)*, Warsaw, Poland: Military University of Technology, 2010, https://www.osti.gov/biblio/1011478 or https://www.researchgate.net/publication/282123596_Tensile_behavior_of_concrete_under_high_strain_rate ("ISIE2010 e-Book of Proceedings.pdf")

P. R. V. Rittinger, *Lehrbuch der Aufbereitungskunde [Textbook of Unit Operations]*, von Ernst & Korn, 1867

G. A. Ruff, L. P. Bernal, & G. M. Faeth, Structure of the Near-Injector Region of Non-Evaporating Pressure-Atomized Sprays, *Journal of Propulsion and Power*, 7(2), 221–230, 1991, https://doi.org/10.2514/3.23315

L. Salkin, A. Schmit, P. Panizza, & L. Courbin, Generating Soap Bubbles by Blowing on Soap Films, *Physical Review Letters*, 116(7), 077801, 2016, https://doi.org/10.1103/PhysRevLett.116.077801

K. A. Sallam, Z. Dai, & G. M. Faeth, Drop Formation at the Surface of Plane Turbulent Liquid Jets in Still Gases, *International Journal of Multiphase Flow*, 25(6–7), 1161–1180, 1999, https://doi.org/10.1016/S0301-9322(99)00042-7

K. A. Sallam, Z. Dai, & G. M. Faeth, Liquid Breakup at the Surface of Turbulent Round Liquid Jets in Still Gases, *International Journal of Multiphase Flow*, 28(3), 427–449, 2002, https://doi.org/10.1016/S0301-9322(01)00067-2

M. Sanai, W. Webber, & G. Aiken, Fragmentation of Building Façade Materials Exposed to High Pressure, in *Proceedings of the 6th International Conference on Protective Structures (ICPS6)*, International Association of Protective Structures, 2023

H. Schlichting & K. Gersten, *Boundary-Layer Theory*, Springer, 2017, https://doi.org/10.1007/978-3-662-52919-5

S. Sharma, N. K. Chandra, S. Basu, & A. Kumar, Advances in Droplet Aerobreakup, *The European Physical Journal Special Topics*, 2022, https://doi.org/10.1140/epjs/s11734-022-00653-z

J. Stránský & M. Jirásek, Modeling of Dynamic Fragmentation: One-Dimensional Case, in V. Fuis, editor, *Proceedings of the 17th International Conference, Engineering Mechanics*, Academy of Sciences of the Czech Republic, 2011

W. J. Stronge, M. Xiaoqing, & Z. Lanting, Fragmentation of Explosively Expanded Steel Cylinders, *International Journal of Mechanical Sciences*, 31(11–12), 811–823, 1989, https://doi.org/10.1016/0020-7403(89)90026-X

L. Tatman, A. J. Lingenfelter, & D. Liu, *Flow Properties through a Ballistically Generated Orifice During a Hydrodynamic Ram Event*, 2018 AIAA/ASCE/AHS/ASC Structures, Structural Dynamics, and Materials Conference, 8–12 January 2018, Kissimmee, Florida, AIAA 2018-0232, https://doi.org/10.2514/6.2018-0232

L. M. Tavares, Energy Absorbed in Breakage of Single Particles in Drop Weight Testing, *Minerals Engineering*, 12(1), 43–50, 1999, https://doi.org/10.1016/S0892-6875(98)00118-6

T. G. Theofanous & G. J. Li, On the Physics of Aerobreakup, *Physics of Fluids*, 20, 052103, 2008, https://doi.org/10.1063/1.2907989

T. G. Theofanous, V. V. Mitkin, C. L. Ng, C-H. Chang, X. Deng, & S. Sushchikh, The Physics of Aerobreakup. II. Viscous Liquids, *Physics of Fluids*, 24, 022104, 2012, https://doi.org/10.1063/1.3680867

L. K. Tseng, P. K. Wu, & G. M. Faeth, Dispersed-Phase Structure of Pressure-Atomized Sprays at Various Gas Densities, *Journal of Propulsion and Power*, 8(6), 157–1166, 1992, https://doi.org/10.2514/3.11457

A. Urbán, M. Zaremba, M. Malý, V. Józsa, & J. Jedelský, Droplet Dynamics and Size Characterization of High-Velocity Airblast Atomization, *International Journal of Multiphase Flow*, 95, 1–11, 2017, https://doi.org/10.1016/j.ijmultiphaseflow.2017.02.001

E. Villermaux & B. Bossa, Drop Fragmentation on Impact, *Journal of Fluid Mechanics*, 668, 412–435, 2011, https://doi.org/10.1017/S002211201000474X

A. Vledouts, J. Quinard, N. Vandenberghe, & E. Villermaux, Explosive Fragmentation of Liquid Shells, *Journal of Fluid Mechanics*, 788, 246–273, 2016, https://doi.org/10.1017/jfm.2015.716

H.-F. Wang, J.-W. Xie, C. Ge, H.-G. Guo, & Y.-F. Zheng, Experimental Investigation on Enhanced Damage to Fuel Tanks by Reactive Projectiles Impact, *Defence Technology*, 2020, https://doi.org/10.1016/j.dt.2020.03.017

C. T. Wang, Y. He, C. Ji, Y. He, L. Guo, & Y. Meng, Dynamic Fragmentation of a Zr-based Metallic Glass Under Various Impact Velocities, *Journal of Materials Science*, 56, 2900–2911, 2021, https://doi.org/10.1007/s10853-020-05495-5

R. J. Weimer & H. C. Rogers, Dynamic Fracture Phenomena in High-Strength Steels, *Journal of Applied Physics*, 50(12), 8025–8030, 1979, https://doi.org/10.1063/1.325989

W. Westlake, B. Keith, V. Smith, & S. Diehl, SERPENT: A Counterproliferation-Counterforce Analysis Tool, *NBC Report*, Fall/Winter 2004, https://www.hsdl.org/?view&did=481844

F. K. Wittel, F. Kun, B. H. Kröplin, & H. J. Herrmann, Study on the Fragmentation of Shells, *International Journal of Fracture*, 140(1–4), 243–254, 2006, https://doi.org/10.1007/s10704-005-3995-7

A. M. Worthington, On the Forms Assumed by Drops of Liquids Falling Vertically on a Horizontal Plate, *Proceedings of the Royal Society of London*, 25, 261–272, 1876, https://www.jstor.org/stable/113486

A. M. Worthington, A Second Paper on the Forms Assumed by Drops of Liquids Falling Vertically on a Horizontal Plate, *Proceedings of the Royal Society of London*, 25, 498–503, 1877, https://www.jstor.org/stable/113511

P.-K. Wu & G. M. Faeth, Aerodynamic Effects on Primary Breakup of Turbulent Liquids, *Atomization and Sprays*, 3(3), 265–289, 1993, https://doi.org/10.1615/AtomizSpr.v3.i3.20

P.-K. Wu, L.-K. Tseng, & G. M. Faeth, Primary Breakup in Gas/Liquid Mixing Layers for Turbulent Liquids, *Atomization and Sprays*, 2(3), 295–317, 1992, https://doi.org/10.1615/AtomizSpr.v2.i3.60

P.-K. Wu, R. F. Miranda, & G. M. Faeth, Effects of Initial Flow Conditions on Primary Breakup of Nonturbulent and Turbulent Round Liquid Jets, *Atomization and Sprays*, 5(2), 175–196, 1995, https://doi.org/10.1615/AtomizSpr.v5.i2.40

A. L. Yarin, I. V. Roisman, & C. Tropea, *Collision Phenomena in Liquids and Solids*, Cambridge University Press, 2017, https://doi.org/10.1017/9781316556580

B.-L. Zhao, J.-G. Zhao, C.-Y. Cui, Y.-S. Duan, & Y. Wang, Growth Model of Cavity Generated by the Projectile Impacting Liquid-Filled Tank, *Defence Technology*, 16(3), 609–616, 2020, https://doi.org/10.1016/j.dt.2019.09.013

F. Zhou, J.-F. Molinari, & K. T. Ramesh, A Cohesive Model Based Fragmentation Analysis: Effects of Strain Rate and Initial Defects Distribution, *International Journal of Solids and Structures*, 42(18), 5181–5207, 2005, https://doi.org/10.1016/j.ijsolstr.2005.02.009

F. Zhou, J.-F. Molinari, & K. T. Ramesh, Effects of Material Properties on the Fragmentation of Brittle Materials, *International Journal of Fracture Mechanics*, 139, 169–196, 2006a, https://10.1007/s10704-006-7135-9

F. Zhou, J.-F. Molinari, & K. T. Ramesh, Characteristic Fragment Size Distributions in Dynamic Fragmentation, *Applied Physics Letters*, 88, 261918, 2006b, https://doi.org/10.1063/1.2216892

17

Unitary Fragmentation Part 3: Size Distributions

The last chapter discussed average sizes for unitary fragmentation. This chapter discusses size distributions about those averages. More specifically, this chapter discusses models derived using mean-field theory, stochastic geometry, and maximum entropy theory; see the earlier discussion in Chapter 5. Mean-field theory treats unitary fragmentation like a random process; stochastic geometry treats unitary fragmentation like a random event; and maximum entropy theory makes no assumption either way; see the earlier discussion at the end of Chapter 15.

17.1 Common Size Distributions

As it turns out, unitary fragmentation often produces Weibull size distributions. If $n > 0$, as one possible self-similar form, Weibull size distributions can be written as follows:

$$f(D) = \frac{n}{D_{\text{avg}}}\Gamma\left(1+\frac{1}{n}\right)^{n}\left(\frac{D}{D_{\text{avg}}}\right)^{n-1}\exp\left[-\Gamma\left(1+\frac{1}{n}\right)^{n}\left(\frac{D}{D_{\text{avg}}}\right)^{n}\right]. \tag{17.1a}$$

According Equation (1.1), $M = c_1\rho D^m$. Then Equation (17.1a) is equivalent to the following:

$$f(M) = \frac{n}{mM_{\text{avg}}}\Gamma\left(1+\frac{m}{n}\right)^{n/m}\left(\frac{M}{M_{\text{avg}}}\right)^{n/m-1}\exp\left[-\Gamma\left(1+\frac{m}{n}\right)^{n/m}\left(\frac{M}{M_{\text{avg}}}\right)^{n/m}\right]. \tag{17.1b}$$

According to Equations (3.14) and (3.15), $F(y) = \int_0^y f(x)dx$ and $f(x) = dF(x)/dx$. Then:

$$F(D) = 1 - \exp\left[-\Gamma(1+1/n)^{n}(D/D_{\text{avg}})^{n}\right] \tag{17.1c}$$

$$F(M) = 1 - \exp\left[-\Gamma(1+m/n)^{n/m}(M/M_{\text{avg}})^{n/m}\right]. \tag{17.1d}$$

If $n < 0$, as one possible self-similar form, Weibull size distributions can be written as follows:

$$f(D) = \frac{|n|}{D'_{\text{avg}}}\Gamma\left(1+\frac{1}{|n|}\right)^{|n|}\left(\frac{D}{D'_{\text{avg}}}\right)^{n-1}\exp\left[-\Gamma\left(1+\frac{1}{|n|}\right)^{|n|}\left(\frac{D}{D'_{\text{avg}}}\right)^{n}\right]. \tag{17.2a}$$

Equivalently:

$$f(M) = \frac{|n|}{mM'_{\text{avg}}}\Gamma\left(1+\frac{m}{|n|}\right)^{|n|/m}\left(\frac{M}{M'_{\text{avg}}}\right)^{n/m-1}\exp\left[-\Gamma\left(1+\frac{m}{|n|}\right)^{|n|/m}\left(\frac{M}{M'_{\text{avg}}}\right)^{n/m}\right]. \tag{17.2b}$$

Particle Dynamics with Aggregation and Fragmentation. Culbert B. Laney, Oxford University Press. © Culbert B. Laney (2025).
DOI: 10.1093/9780198920687.003.0017

In practice, negative Weibull size distributions are usually approximated by power law size distributions. In fact, negative Weibull size distributions equal power law size distributions in the limit $D \to \infty$ and $M \to \infty$, i.e., for large enough fragments.

17.2 Mean-Field Theory

As one possible model of unitary fragmentation, suppose that part of the original body is divided into N randomly sized pieces, and each of those pieces is divided into N randomly sized pieces, and so forth. In this case, unitary fragmentation is the same as random fragmentation as discussed in Chapter 13. According to mean-field theory, random fragmentation depends on four things:

- *Rate kernels.* As usual, this section assumes that the rate kernels are power laws, i.e., $J(x) = J_0 x^\kappa$ where $J_0 > 0$ is constant. Furthermore, this section assumes $\kappa < 0$.
- *Child size distributions.* As usual, this section assumes that the child size distributions are power laws, i.e., $m(x, y) = (\eta+1)x^{\eta-1}/y^\eta$ where $0 < \eta = 1/(N-1) \leq 1$.
- *Initial conditions.* As usual, this section assumes that the effects of the initial conditions drop out at sufficiently late times, due to the effects of chaos.
- *Boundary conditions.* This section assumes free-field boundary conditions. However, because $\kappa < 0$, shattering occurs. Thus, conservation of mass does not apply.

These assumptions are the same as those made in the last chapter except that κ is negative instead of positive. This sign reversal introduces a hard sink of small "dusty" particles that reduces or eliminates the effects of the child size distributions. This assumes the smallest particles are removed from the system or, at least, do not participate in or substantially affect fragmentation. In this case, according to Equations (13.29) and (13.99), random fragmentation obtains a power law size distribution where:

$$n/m \text{ is an integer multiple of } 1/6 \text{ between } -1 \text{ and } 1. \tag{17.3}$$

In many cases, unitary fragmentation is followed by secondary aggregation or fragmentation; see Chapter 15. In fact, these secondary cascades may begin even before primary fragmentation ends. For example, a large fragment may start to break up into smaller fragments, even before it has completely separated from the original body. For another example, if tension turns into compression, a partially separated fragment may rejoin the original unitary body. Thus, for plastic materials, Timár et al. (2010) say that "broken particle contacts are able to reconnect when compressed against each other, leading eventually to the healing of cracks." Assuming that secondary aggregation occurs due to something like binary collisions, Chapter 11 predicts power law, gamma, or Weibull size distributions where:

$$n/m \text{ is an integer multiple of } 1/6 \text{ between } -2 \text{ and } 2 \tag{17.4}$$

for non-local cases and:

$$n/m \text{ is an integer multiple of } 1/6 \text{ between } -2 \text{ and } 0 \tag{17.5}$$

for local cases.

Similarly, assuming that secondary fragmentation occurs due to something like binary collisions, Chapter 14 predicts power law or Weibull size distributions where:

$$n/m \text{ is an integer multiple of } 1/6 \text{ between } -2 \text{ and } 0 \tag{17.6}$$

for local cases; c.f. Equation (14.25). There is no known similar result for non-local cases.

17.3 Stochastic Geometry

Fragmentation researchers have proposed two main geometric models: random packing and random tessellation. First, consider random packing. As described by Åström (2006), this approach assumes that all fragments have the exact same shape—such as spheres or cubes—while their sizes vary randomly to fill the available space. In all known cases, this approach obtains non-ideal size distributions. For example, Aste (1996) discusses random two-dimensional packings that involve simple shapes such as circles, squares, diamonds, equilateral triangles, etc. This leads to power law size distributions where n/m is any real number between $-1/2$ and -1. For another example, Lind et al. (2008) discuss random Apollonian sphere packings. This leads to power law size distributions where n/m is any real number between $-2/3$ and -1.

Next, consider random tessellation. This approach allows both the fragment sizes and shapes to vary randomly, provided the results fit together like puzzle pieces. Figure 15.3 shows a typical real-world fragmentation pattern. For comparison, Figure 17.1 shows four examples of random tessellation. In some treatments, random tessellation leads to a full-range size distribution. However, to make the modeling easier, random tessellation usually focuses on the limit of small and/or large fragments. For example, suppose that the true size distribution is as follows:

$$f(M) = \frac{1}{M_{\text{avg}}}\left(\frac{2M}{M_{\text{avg}}}\right)^{-1/2} \exp\left[-\left(\frac{2M}{M_{\text{avg}}}\right)^{1/2}\right]. \tag{17.7}$$

Then, random tessellation might find the following approximate size distribution:

$$f(M) \cong \frac{1}{M_{\text{avg}}}\left(\frac{2M}{M_{\text{avg}}}\right)^{-1/2} \tag{17.8}$$

which applies for small enough M. In some cases, random tessellations subdivide the original unitary body into many different pieces. In other cases, to make the modeling easier, random tessellation subdivides the original unitary body into only two random pieces. Parisio & Dias (2011) dubbed this *minimal fragmentation*; see also Brodskii et al. (2012), Dias & Parisio (2014), Brodskii & Kulik (2016). In some but not all cases, random tessellation is equivalent to mean-field random fragmentation; see the caption to Figure 17.1. In particular, minimal fragmentation is generally equivalent to mean-field binary fragmentation with $N = 2$.

With this background in mind, Tables 17.1 and 17.2 show some typical results of random tessellation. In all cases, n/m is ideal, i.e., it is an integer multiple of 1/6. However, *this only occurs if the tessellation is sufficiently random.* For example, Ben-Naim & Krapivsky (2019)

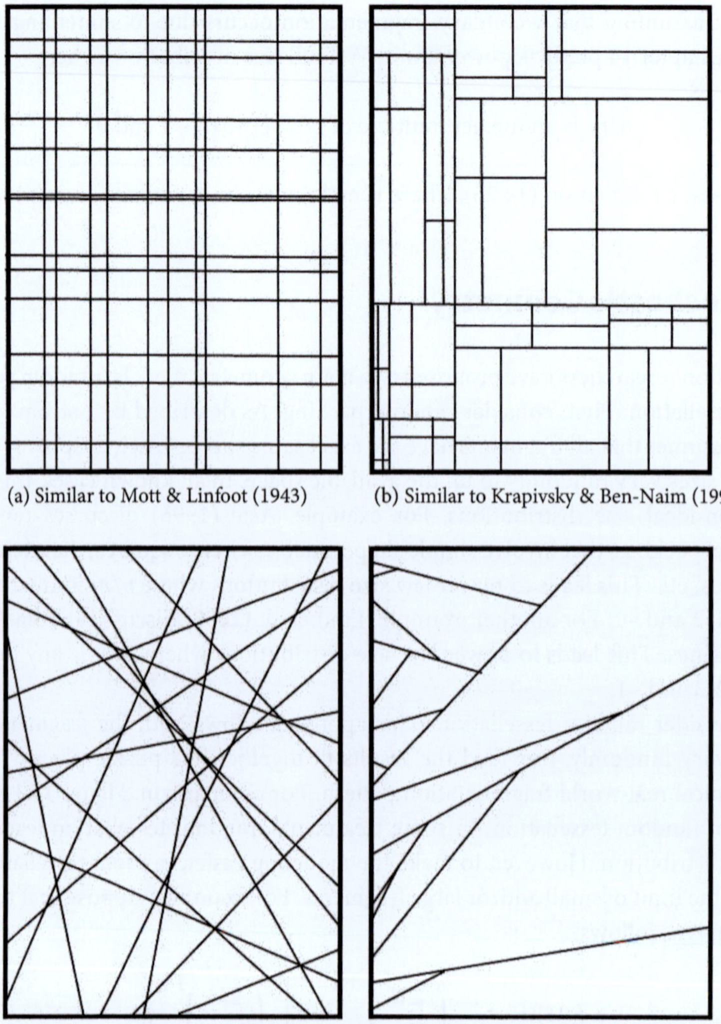

(a) Similar to Mott & Linfoot (1943) (b) Similar to Krapivsky & Ben-Naim (1994)

(c) Similar to Grady & Kipp (1985) (d) Similar to Dos Santos et al. (2007)

Figure 17.1 Examples of random tessellations meant to represent fragmentation. Notice that pattern (b.) can be obtained through random fragmentation, i.e., by dividing part of the original object into four pieces, by dividing each of those pieces into four pieces, and so forth. By contrast patterns (a.), (c.), and (d.) cannot be obtained through random fragmentation.

studied tessellations in which "both the vertical size and the horizontal size of a rectangle are discrete variables." In other words, there were only a finite number of possible cell dimensions. While a random process was used to pick among the finite possibilities, there was still not as much randomness as when the cell dimensions are chosen from among an infinite continuum of values. As a result, Ben-Naim & Krapivsky (2019) report that "the fragment size [sometimes] becomes deterministic in the final state as all fragments have the same size. In this sense, the final state can be trivial."

Table 17.1 Examples of random tessellations that lead to power law size distributions.

Small Frag n/m	Large Frag n/m	m	Object Shape	Subdivision Procedure	References
−2/3	–	3	Any	Single random surface assuming the two halves created by the surface obtain "geometrical similarity."	Gilvarry (1961, 1964)
−2/3	–	3	Cube	Multiple random walks with random starting points, moving vertically, horizontally, or laterally at each step.	Inaoka & Takayasu (1996, 1999)
−2/3	–	3	Cube	Linear cracks "propagate with a constant speed." Allows for random crack splitting and crack merging.	Kekäläinen et al. (2007)
−1/2	–	2	Square	Linear cracks "propagate with a constant speed." Allows for random crack splitting and crack merging.	Kekäläinen et al. (2007)
−1/2	–	2	Square	Multiple random walks on a square lattice, with random starting points, moving vertically or horizontally at each step.	Spahn et al. (2014)
−1/3	–	2	Square	Linear cracks "undergo random transverse displacements as they move." Allows for random crack splitting and crack merging.	Dhar (2015)
1/6	2	3	Flat Disk, Concave Lens	Single line where "the probability of fracture depends on the area of the [newly] formed surface."	Brodskii & Kulik (2016)
1/3	–	2	Circle or Polygon	Single line with a uniformly random position and angle.	Dias & Parisio (2014)
1/2	–	2	Square or Triangle	Multiple lines with uniformly random positions and angles. Lines terminate at intersection points.	dos Santos, Donangelo & Souza (2007)
1/2	1	2	Square	Single line with a uniformly random position and angle.	Parisio & Dias (2011)
1/2	1	2	Triangle	Single line with a uniformly random position and angle.	Dias & Parisio (2014)
2/3	0, 1	2	Circle	Two straight lines originating at the center of a circle with a uniformly random angle between them.	Brodskii, Tur, & Yanovsky (2012)
2/3	–	2	Circle or Polygon	Single line with a uniformly random position and non-uniformly random angle.	Dias & Parisio (2014)
1	–	2	Circle or Polygon	Single line with a uniformly random position and non-uniformly random angle.	Dias & Parisio (2014)

Table 17.2 Examples of random tessellations that lead to self-similar gamma size distributions.

n/m	m	Object Shape	Subdivision	References
1	1	Line	Multiple points with uniformly random positions.	Lienau (1936a, b, c)
1(1)	2	Rectangle	Multiple vertical and horizontal lines with uniformly random positions.	Mott & Linfoot (1943)
1	2	Rectangle	Multiple lines with uniformly random positions and angles.	Grady & Kipp (1985)
1(2)	2	Rectangle	One vertical and one horizontal line with uniformly random positions. Iterate.	Krapivsky & Ben Naim (1994)
1(2)	2	Rectangle	"Daughter volumes are a uniformly distributed proportion of mother volume."	Cowan (2010)
5(3)	2	Any	"Random Voronoi polygons."	Kiang (1966)
7(3)	3		"Random Voronoi polyhedral."	Kiang (1966)
7/2(3)	2	Any	"Poisson Voronoi cells" in 2D.	Járai-Szabó & Néda (2007)
5(3)	3		"Poisson Voronoi cells" in 3D.	Járai-Szabó & Néda (2007)

(1) Rough approximation.
(2) For sufficiently large cells.
(3) Large values of n/m indicate that the cell sizes are relatively uniform and, thus, may not be realistic for chaotic fragmentation.

While there is no known general proof, the results given in Table 17.1 suggest that unitary fragmentation may obtain power law or, perhaps, Weibull size distributions where:

$$n/m \text{ is an integer multiple of } 1/6 \text{ between } -1 \text{ and } 1. \tag{17.9}$$

The results given in Table 17.2 suggests that unitary fragmentation may obtain gamma size distributions with values of n/m as large as 7. However, such large values of n/m are indicative of highly uniform size distributions which are, in turn, indicative of deterministic rather than random tessellations. As noted in Chapters 1 and 2, this book focuses on random or seemingly random processes and events, where values of n/m larger than two are unlikely to occur; see also, e.g., Grady & Kipp (1985) and section 2.2.3.2 of Grady (2017).

17.4 Maximum Entropy Theory

As a third possible analytic approach, this section concerns *maximum entropy theory*. As discussed in Section 5.3, there are many possible definitions of entropy. For example:

$$S = -\int_0^\infty \ln[f(M)]\,f(M)dM. \tag{17.10}$$

However entropy is defined, the goal is to choose a probability density function f that maximizes the entropy S subject to constraints such as the following:

$$\int_0^\infty f(M)dM = 1 \tag{17.11}$$

$$\int_0^\infty Mf(M)dM = M_{\text{avg}}. \tag{17.12}$$

In most cases, maximum entropy theory is used to derive general families of size distributions. For example, Lienhard & Meyer (1967) and Dumouchel (2006, 2009) use maximum entropy theory to derive Rosin-Rammler size distributions. Less commonly, maximum entropy theory is used to derive specific size distributions. For example, Grady & Kipp (1985) and Li & Tankin (1987) maximized Equation (17.10) subject to Equations (7.11) and (7.12) to obtain exponential size distributions where:

$$f(M) = \frac{1}{M_{\text{avg}}}\exp\left(-\frac{M}{M_{\text{avg}}}\right). \tag{17.13}$$

Recall that exponential size distributions are the same as Weibull or gamma size distributions with $n/m = 1$. More generally, as shown in Chapter 19, maximum entropy theory can be used to derive Weibull and gamma size distributions such that:

$$n/m \text{ is an integer multiple of } 1/3 \text{ between } -1 \text{ and } 1. \tag{17.14}$$

Comparing this result with Equation (17.3) derived using mean-field theory and Equation (17.9) derived using stochastic geometry, it appears that maximum entropy theory predicts only about half of the possible universal values for n/m. Section 19.5 will explain why maximum entropy theory generally only produces a subset of the possible universals.

A number of researchers have described the close theoretical relationships between random tessellations and maximum entropy theory; see, e.g., Kumar & Kumaran (2005), Berestycki & Pitman (2007) and Berdichevsky (2012). To the extent that a given tessellation disagrees with maximum entropy theory, the tessellation may simply not be sufficiently random. For example, Grady & Kipp (1985) compared seven different tessellations. The more random tessellations, such as those shown in Figure 17.1.b and 17.1.c, obtained good agreement with maximum entropy theory. By contrast, the less random tessellations, such as Possion-Voroni tessellations in two- and three-dimensions, did not obtain good agreement with maximum entropy theory.

17.5 Experimental Validation of Primary Fragmentation Models

This is the first of two sections on experimental validation. This section concerns primary fragmentation. More specifically, Example 17.1 concerns primary fragmentation of solids due to uniaxial strain in cylindrical coordinates, Example 17.2 concern primary fragmentation of consolidated granular materials due to equibiaxial strain in spherical coordinates, and Example 17.3 concerns primary fragmentation of liquids due to uniaxial strain in Cartesian coordinates.

Example 17.1 Analytic vs. experimental results for primary fragmentation of cylindrical metal shells due to high-explosive detonation

This is a continuation of Example 16.6. Based on empirical curve fits to experimental results, many researchers have obtained Weibull size distributions; see Table 17.3. As this table shows, in ideal or nearly ideal cases, the Weibull exponent n/m tends to be approximately an integer multiple of 1/6 between −1 and 1.

Table 17.3 Fragment size distributions obtained in experiments involving explosively-driven cylindrical metal shells. Based on physical collection with a soft-catch collection device. (a) Positive Weibull size distributions.

n/m	m	Test Object	Collector	Reference
1/3 [small frags]	3	"3″ U.P. and 3.7″ A.A. shell"	sawdust (U.P.) strawboard (A.A.)	Mott & Linfoot (1943)
1/3	3	1.9 cm-thick Armco iron	"large volume of sawdust"	Mock & Holt (1981, 1983); Laney (2016b)
1/3 (0.29, 0.31, 0.35)	3	6 mm-thick mild, hard, and heat-treated steel	0.5 mm-thick steel, 1cm Styrofoam	Arnold & Rottenkolber (2008)
1/2 (0.48, 0.55)	2	2 mm-thick mild steel and aluminum alloy	0.5 mm-thick steel, 1cm Styrofoam	Arnold & Rottenkolber (2008)
1/2 [large frags]	2	"3″ U.P. and 3.7″ A.A. shell"	sawdust (U.P.) strawboard (A.A.)	Mott & Linfoot (1943)
1/2	-	18 tests of shells and bombs circa 1943–44	sand, sawdust, or strawboard	Mott et al. (1944)
1/2 [small frags]	3	2.23 cm-thick HF-1 steel	"large volume of sawdust"	Mock & Holt (1981, 1983); Laney (2016b)
2/3 [large frags]	2	2.23 cm-thick HF-1 steel	"large volume of sawdust"	Mock & Holt (1981, 1983); Laney (2016b)
2/3 (0.65–0.73)	3	105 mm and 150 mm-diameter steel shells	immersion in water	Held (1990a, b)
2/3 (0.67)	3	8 mm-thick heat-treated AerMet 100 steel alloy	"soft recovery" with clay, Celotex	Grady et al. (2001); Wilson et al. (2001)
5/6 (0.85)	3	8 mm-thick as-received AerMet 100 steel alloy	"soft recovery" with clay, Celotex	Grady et al. (2001); Wilson et al. (2001)

continued

Table 17.3 *continued*

(b) Negative Weibull size distributions, a.k.a., power law size distributions.

n/m	m	Test Object	Collector	Reference
−2/3 (−0.66)[1] [small frags]	2	0.2 mm-thick alumina tube	(not specified)	Katsuragi, Ihara, & Honjo (2005)
−2/3	3	33 mm-thick 7075-T6 aluminum alloy	immersion in sand	Edwards & Deal (2011); Laney (2016b)
−1/2	3	20 mm-thick 7075-T6 aluminum alloy	immersion in sand	Edwards & Deal (2011); Laney (2016b)
−1/2 (−0.47,−0.51)	2	2 mm-thick alumina	immersion in water	Bannikova et al. (2014, 2016)
−1/3	3	25 mm-thick 7075-T6 aluminum alloy	immersion in sand	Edwards & Deal (2011); Laney (2016b)

(1) Excludes low strain rates that led to "one-dimensional-like fragments that retained the long tube geometry."

Example 17.2 Analytic vs. experimental results for primary fragmentation of spherical granular shells due to explosive detonation

Examples 16.9, 16.11, and 16.13 discussed expanding spherical shells produced by explosives in solids or liquids. This example discusses expanding spherical shells produced by explosives in granular materials, such as sand or metallic powder. According to Loiseau et al. (2018), such cases evolve in three phases:

- *Phase 1: Compression.* "Sufficiently high shock pressures can induce varying degrees of particle consolidation within the shell. Thin shells comprised of soft/weak materials may be compacted throughout most of their thickness, whereas thick shells of tough/hard materials may only consolidate close to the charge."
- *Phase 2: Expansion and primary fragmentation.* "While this partially consolidated shell is driven outward by the expanding detonation products, rarefaction waves release the shock pressure. Consolidated material fragments due to tensile loading from the rarefactions and hoop stress ... Fragment sizes [are] on the order of the compacted shell thickness," i.e., they are much larger than the original grains of powder. Fragment speeds are well-predicted by Gurney's law, after modifications to account for the initial voice space inherent to granular materials.
- *Phase 3: Jetting and secondary fragmentation.* "Jet-like structures then form as the shell radially expands ... The jet-like structures emerge from the detonation product fireball and shed unconsolidated and weakly consolidated particles in their wakes due to aerodynamic drag;" see, e.g., Figure 17.2.

In Phase 1, there are three main scenarios. In the lowest energy cases, *compaction* involves a decrease in void space due to rearrangement, plastic distortion, and fragmentation of particles. In such cases, particles still retain their individual identities; see Figure 17.3a. Where two particles are in contact over an extended surface, they may obtain weak bonds due to van der Waals or Coulomb forces. In addition, if moisture is present, they may obtain weak bonds due to surface tension. On the other hand, if extensive plastic distortion locks two particles together, they may obtain strong bonds. In medium energy cases, *welding* occurs when shear stress overwhelms material strength at the interface between two particles. In such cases, the near-surface region of both particles flows hydrodynamically. This flow may obtain periodic instabilities, which lock the particles together through a wavy interface; see Figure 17.3b. Alternatively, this flow may obtain chaotic instabilities, which completely obliterate the original boundaries between the particles. In the highest energy cases, *fusion* occurs when the interface between two particles melts due to heat from friction, hot byproducts of detonation, void collapse, distortion, etc. For more information on consolidation see, e.g., Hegazy & Blazynski (1986), Raybould & Blazynski (1987), Meyers et al. (1999) and Nesterenko (2001).

Xue et al. (2013) conducted six experiments involving spherical explosive charges surrounded by quartz sand. The sand was placed into concentric plastic spheres with diameters of 40 mm and 120 mm. The plastic containers were chosen to be thin (2 mm) and brittle "to ensure that the interference with the blast profile was as low as possible." Based on video analysis of the resulting jets, Xue et al. (2013) obtained Weibull size distributions where $n/m \approx 1$ or 11/6; see Figure 17.4. Notice that n/m is relatively large, which indicates that the fragment size spread is relatively small. In addition, n/m is an even multiple of $1/6$, which implies that fragmentation is ideal. Xue et al. (2013) report that adding "even a small amount of interstitial oil [or water] ...substantially reduce[d] the size of the fragments," and increased n/m.

Figure 17.2 Experimental results for explosively-driven sand initially contained in a glass sphere. Notice that time increases from left to right.
Reproduced from Frost et al. (2012) with the permission of AIP Publishing.

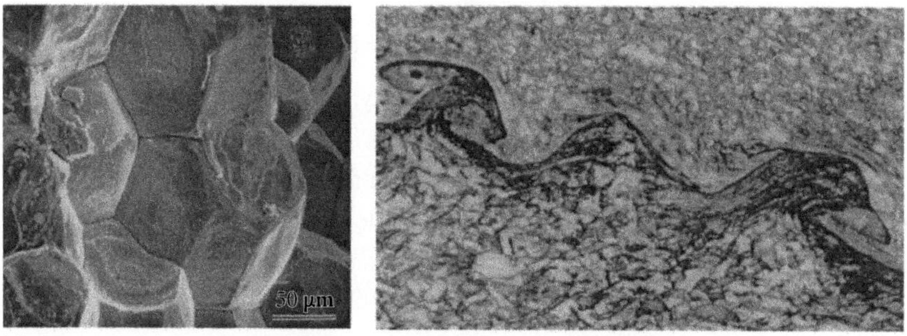

(a) Compaction (a.k.a. densification) (b) Welding due to shear strain.

Figure 17.3 Two possible types of consolidation. In part (b), the welding occurs due to a Rayleigh-Taylor instability at the interface between two flat plates moving roughly parallel to one another.

Reprinted from Meyers et al. (1999) with permission from Elsevier.

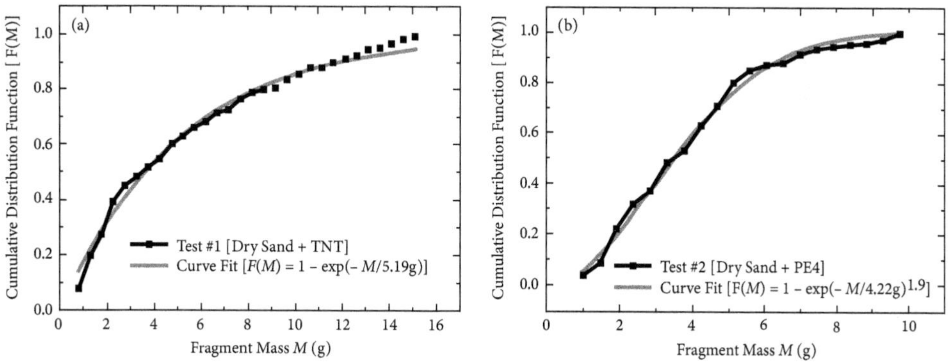

Figure 17.4 Analytical (grey lines) vs. experimental (black lines with squares) results for the fragment size distributions obtained from explosively-driven spherical powder shells. Experimental results are based on the "jet diameter measured from high-speed snapshots and the average jet mass derived from the number of jets."

[The notation in these figures has been changed to match the notation used in this book.]
Reproduced from Xue et al. (2013) with permission from Springer Nature.

Example 17.3 Analytic vs. experimental results for fragmentation of planar liquid jets driven by a parallel airstream

Suppose a high-speed airstream drives a thin turbulent liquid film across a flat plate, such that the liquid clings to the sharp trailing edge. As discussed earlier in Example 16.16, Chaussonnet et al. (2016) obtained the following expression for the Sauter mean diameter:

$$D'_{M,\,\mathrm{avg}} \cong 1.4TWe^{-1/2} \tag{17.15}$$

where $We = \rho V^2 T/\sigma$, ρ is the liquid density, $V = V_{\text{air}}/\left(1+\sqrt{\rho/\rho_{\text{air}}}\right)$ is a characteristic liquid velocity, V_{air} is the air velocity, ρ_{air} is the air density, and T is the thickness of trailing edge of the flat plate. For Weibull size distributions with $n > 0$, the Sauter mean diameter seen in Equation (17.15) may be converted to the mass mean diameter seen in Equation (17.1) using the following expression:

$$D_{\text{avg}} = \frac{\Gamma(1+1/n)\Gamma(1+2/n)}{\Gamma(1+3/n)} D'_{M,\text{avg}}. \tag{17.16}$$

Based on empirical curve fits to experimental results, Chaussonnet et al. (2016) obtained Weibull size distributions with $n > 0$. According to Figure 17.5, n is a linear function of $We^{-1/2}$ just like D_{avg}. Because n varies continuously with We, these results are *non-ideal*. In other words, in these experiments, the strain rates that control primary fragmentation were not spatially homogeneous and isotropic.

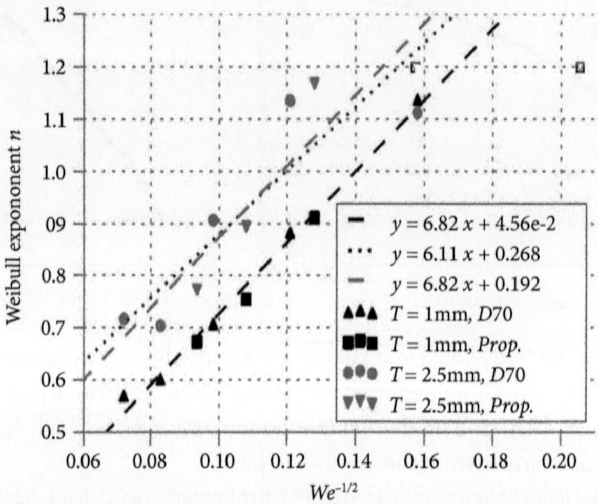

Figure 17.5 Analytical (dashed lines) vs. experimental (symbols) results for the fragment size distributions from a prefilming airblast atomizer.
[The notation in this figure has been changed to match the notation used in this book.]
Reprinted from Chaussonnet et al. (2016) with permission from Elsevier.

17.6 Experimental Validation of Secondary Fragmentation Models

This is the second of two sections on experimental validation. This section concerns secondary breakup or, at least, cases that resemble those observed in secondary breakup. More specifically, Example 17.4 concerns secondary fragmentation of solids due to collisions with rigid surfaces, Example 17.5 concerns secondary fragmentation of liquids due to collisions with rigid surfaces, and Example 17.6 concerns secondary fragmentation of liquids due to aerodynamic drag.

Example 17.4 Analytic vs. experimental results for solid objects colliding with solid surfaces

In *Taylor impact tests*, a small gun-launched projectile impacts a rigid flat surface, sometimes called an *anvil*. Taylor impact tests typically involve small projectiles, between 1 and 20 mm in diameter. In the original technique, the projectile speeds were kept relatively low, so that the projectile only deformed. However, it has become increasingly common to use higher speeds, so that the projectile fragments. Many Taylor impact tests obtain non-ideal Weibull size distributions. For example, Pan & Render (1996) found that 12.7 mm ice spheres with impact speeds between 100 and 175 m/s obtained Weibull size distributions. Their conclusions were based on two-dimensional photographic measurements of splatter on a glass witness plate; see also Vargas et al. (2018). For another example, Cheong et al. (2003, 2004) and Salman et al. (2002) found that 12.7 mm-diameter glass spheres and 3.15–7.15 mm-diameter alumina spheres with impact speeds between 5 and 35 m/s obtained Weibull size distributions with continuously variable exponents. Their results were based on physical collection with a “fabric cage enclosing the target,” which helped reduce secondary fragmentation.

Taylor impact tests impose strict limits on the projectile’s size, shape, and strength. In particular, the projectile must be small enough to fit through the barrel of a gun, streamlined enough to fly straight, and strong enough to survive rapid acceleration. As an alternative, consider *drop shatter tests*. In drop shatter tests, an object is dropped or thrown onto a rigid flat surface such as a concrete floor; see, e.g., Paluszny et al. (2016). Alternatively, to control the impact orientation, the object may be attached to a parachute or induced to slide down a steeply inclined plane; see, e.g., Bowman & Take (2015), Haug et al. (2016) and Zhao et al. (2018). Drop shatter tests allow for unlimited sizes, shapes, and strengths, provided only that the test object is weak enough to break on impact at or below its terminal velocity.

Table 17.4 shows some results of drop shatter tests where the objects landed on the floor with a semi-random orientation. As a result of the semi-random orientation, and the relatively low impact velocity, these tests obtained non-ideal size distributions. For comparison, Table 17.5 shows some results of drop shatter tests where the objects landed in a carefully controlled fashion, with a flat face exactly parallel to the floor. As a result of the controlled orientation, these tests obtained ideal size distributions. In addition, Table 17.5 shows an example in which an ovoid shell impacts the floor at a random orientation angle. In this case, the shell tends to buckle progressively, starting from the impact point. Assuming the shell maintains a constant speed, the vast majority of the surface tends to experience the same lateral rates-of-strain, excluding the leading and trailing edges. This is confirmed by experimental photographs; see Figure 4.4. As a result, most fragments form under nearly identical conditions, resulting in ideal size distributions.

Table 17.4 Non-ideal size distributions obtained in drop-shatter tests.

Configuration	Examples			
	n/m	m	Notes	References
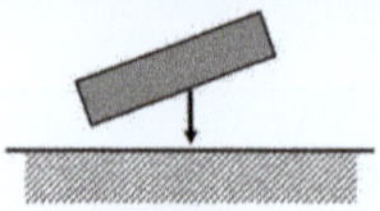	−0.05 ± 0.05 −0.08 ± 0.05 (Gamma)	3 3	Gypsum bars, 180 × 12 × 5 mm Gypsum disks, 320 × 5 mm *"Fragmented by throwing them onto a hard floor."*	Oddershede et al. (1993)
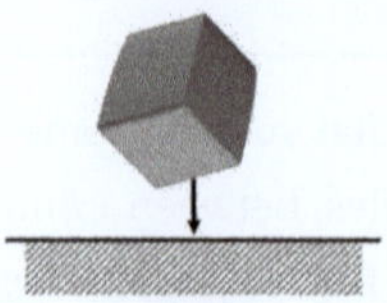	−0.56 ± 0.05 (Power law)	3	Gypsum cubes, 74 × 74 mm *"Fragmented by throwing them onto a hard floor."*	Oddershede et al. (1993)
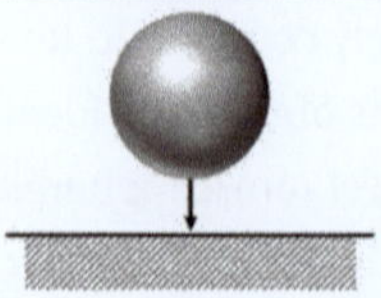	−0.60 ± 0.05 (Power Law)	3	Gypsum balls, 120 mm diam. *"Fragmented by throwing them onto a hard floor."*	Oddershede et al. (1993)

Table 17.5 Ideal size distributions obtained in drop-shatter tests.

Configuration	Examples			
	n/m	m	Notes	References
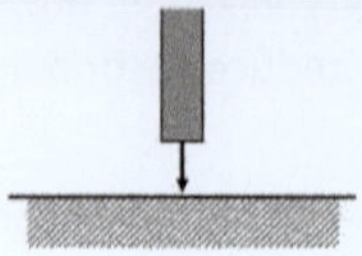	−1/6 (−0.17 ± 0.05) (Power law)	3	Gypsum plate, 195 × 151 × 4 mm *"Thrown vertically down ... onto the floor."*	Oddershede et al. (1993)
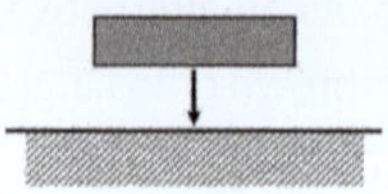	−1/3 [small] −1/6 [large] (Power Law)	3 3	Gypsum disks, 10.65 × 2.28 cm *"Parachute was [used] to make the disks fall with the flat side down and to evenly impact the floor."*	Åström et al. (2004), Laney (2016b)
	−1/2 (−0.50) −2/3 (−0.67) (Power Law)	3 3	Clay plates, 0.4–2 cm thick. *Dropped onto a hard floor from a height of 2 m.*	Meibom & Balslev (1996)
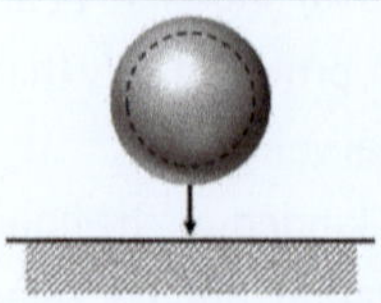	−1/3 (−0.35 ± 0.02) (−0.32 ± 0.03) (Power Law)	2	Eggshells, ~0.5 mm thick. *"Catapulted against the ground ... using rubber bands ... Carried out inside a soft plastic bag so that secondary fragmentation ... did not occur."*	Wittel et al. (2004, 2005, 2006)

Example 17.5 Analytic vs. experimental results for liquid droplets colliding with solid surfaces

Consider a spherical liquid droplet impacting a non-responding flat surface. Depending on the exact circumstances, different researchers have suggested log-normal, gamma, Weibull, and other types of size distributions. This example will focus on Weibull size distributions. As discussed earlier, Weibull size distributions may be written as follows:

$$f(D) = \frac{nb}{D_{\text{avg}}}\left(\frac{D}{D_{\text{avg}}}\right)^{n-1} \exp\left[-b\left(\frac{D}{D_{\text{avg}}}\right)^{n}\right]$$

where $b = \Gamma(1 + 1/n)^n$ assuming that $n > 0$. As one possibility, by curve fitting experimental results, Mundo et al. (1995) and Stanton & Rutland (1996) obtained *non-ideal* Weibull size distribution where:

$$n = 2.71 - 9.25 \times 10^{-4}\, We \tag{17.17}$$

and where $We = \rho V_0^2 D_0/\sigma$ is a Weber number, D_0 is the initial droplet diameter, V_0 is the component of the initial droplet velocity normal to the solid surface, ρ is the liquid density, and σ is the liquid surface tension. For more information on this and other non-ideal models see, e.g., Cossali et al. (2005), Moreira et al. (2010), and Moreira & Panão (2011).

As another possibility, as seen in Table 17.6, some researchers have obtained *ideal* Weibull size distributions where n is an integer multiple of $1/2$ or, equivalently, n/m is an integer multiple of $1/6$ (assuming $m = 3$). For example, as seen in Figure 17.6, Xu et al. (2007) found the following:

$$f(D) = \frac{1}{D_{\text{avg}}} \exp\left[-\frac{D}{D_{\text{avg}}}\right] \tag{17.18}$$

which is a Weibull size distribution with $n = 1$. In this case, Xu et al. (2007) found that D_{avg} depends on surface roughness but not impact velocity. By contrast, Example 16.18 found that D_{avg} depends on impact velocity but not surface roughness. This occurred because, in the earlier example, the drop splash extended beyond the edges of the solid surface, which reduced the opportunity for surface roughness to influence the outcome.

Table 17.6 Based on experimental results, the fragments from a liquid droplet impacting a solid surface may obtain an ideal Weibull size distribution. These results assume that $m = 3$.

n/m	n	Droplet Properties	Solid Surface Properties	Reference
1/3	1	60–150 μm water or ethanol droplets, 12–18 m/s	Rotating 150 mm steel disc	Mundo et al. (1995); Bai et al. (2002)
1/3	1	3.1–3.4 mm ethanol droplets, 2–8 m/s	Glass slide with a textured epoxy layer	Xu (2007); Xu et al. (2007)
2/3	2	16–135 μm water or diesel droplets, 4–26 m/s	Metal sphere, 93 mm in diameter	Roisman, Horvat, & Tropea (2006)

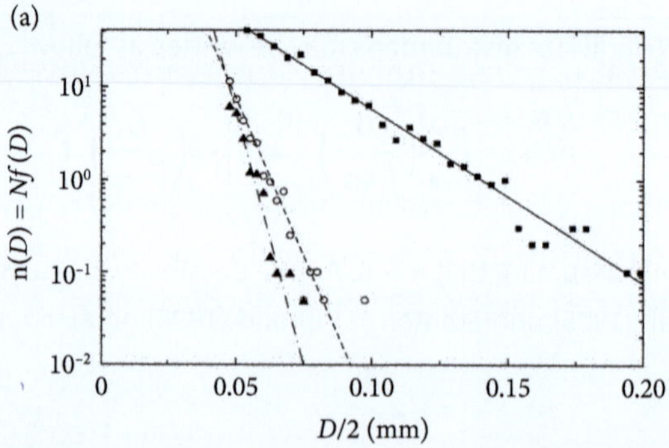

(a) Effects of surface roughness. In this figure, D_{avg} was approximately equal to 0.008 mm, 0.012 mm, and 0.046 mm when the surface roughness was 3 μm (triangles), 5 μm (circles), and 16 μm (squares), respectively. Because the size distribution was not normalized using D_{avg}, which is the usual practice in this book, the slope in the linear-log plane changes with D_{avg} rather than remaining the same.

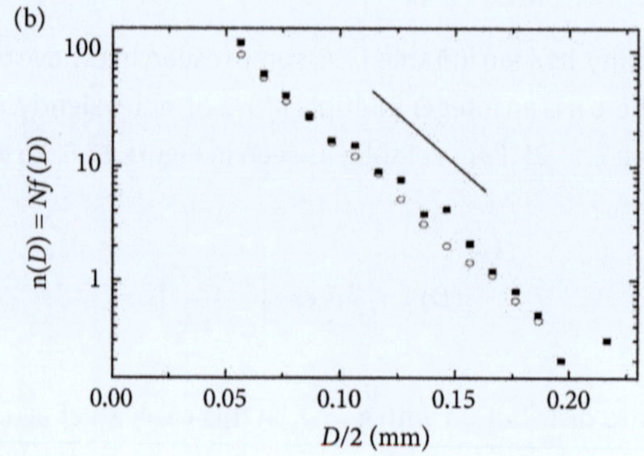

(b) Effects of impact velocity. In this figure, Davg is approximately equal to 0.05 mm for impact velocities of 4.3±0.1 m/s (circles) and 5.2±0.1 m/s (squares).

Figure 17.6 Analytical (lines) vs. experimental (symbols) results for the fragment size distributions from a liquid droplet impacting a solid surface. The lines are Weibull size distributions with $n = 1$ or, equivalently, $n/m = 1/3$.
[The notation in these figures has been changed to match the notation used in this book.] Figures reprinted with permission from Xu et al. (2007). Copyright 2007 American Physical Society.

Example 17.6 Analytic vs. experimental results for gradual aerodynamic breakup of liquid droplets

Consider a spherical liquid droplet traveling through still air, as discussed earlier in Examples 16.19 and 16.20. As one possibility, based on empirical curve fits to experimental data, Chou et al. (1997) suggested the following root normal size distribution:

$$f_M(D,t) = \frac{A_0}{D'_{M,\text{avg}}(t)} \frac{1}{\sqrt{2\pi\sigma_0^2}} \sqrt{\frac{D'_{M,\text{avg}}(t)}{D}} \exp\left[-\frac{1}{2\sigma_0^2}\left(\sqrt{\frac{D}{D'_{M,\text{avg}}(t)}} - 1\right)^2\right] \tag{17.19}$$

where $\sigma_0 = 0.24 \pm 20\%$ and $D'_{M,\text{avg}}(t)$ is given by Equation (16.54) or (16.57). When σ_0 equals 0.24, this is sometimes called *Simmons universal root normal size distribution*; see Simmons (1977). As another possibility, based on empirical curve fits to experimental data, Laney (2015a) suggested the following root normal size distribution:

$$f(D,t) = \frac{A_1}{D_{\text{avg}}(t)} \frac{1}{\sqrt{2\pi\sigma_1^2}} \sqrt{\frac{D_{\text{avg}}(t)}{D}} \exp\left[-\frac{1}{2\sigma_1^2}\left(\sqrt{\frac{D}{D_{\text{avg}}(t)}} - \sqrt{1-\sigma_1^2}\right)^2\right] \tag{17.20}$$

where $\sigma_1 = 0.0086, 0.125, 0.24, 0.34, 0.47, 0.72$, or some other discrete value. For large enough fragments, Equations (17.19) and (17.20) are essentially identical. However, for small fragments, Equation (17.20) is substantially better behaved than Equation (17.19). More specifically, Equation (17.20) reduces the severity of the singularity at $D = 0$ relative to Equation (17.19). As a third possibility, based on empirical curve fits to experimental data, Laney (2015b, 2016a) suggested the following Weibull size distribution:

$$f(D,t) = \frac{|n|\, b}{D_{\text{avg}}(t)} \left[\frac{D}{D_{\text{avg}}(t)}\right]^{n-1} \exp\left[-b\left[\frac{D}{D_{\text{avg}}(t)}\right]^{n}\right] \tag{17.21}$$

where n is an integer multiple of 1/2 between −6 and 6. If $n > 0$, then Weibull size distributions are essentially identical to root normal size distributions, at least for large enough fragments. For example, Equation (17.19) with $\sigma_0 = 0.24$ is almost the same as Equation (17.21) with $n = 1.25$. For another example, Equation (17.20) with $\sigma_1 = 0.24$ is almost the same as Equation (17.21) with $n = 2$. On the other hand, if $n < 0$, then Weibull size distributions are entirely different than root normal size distributions. Instead, Weibull size distributions converge toward power law size distributions for large enough fragments.

Figure 17.7 shows experimental results for a single droplet falling through the air under the influence of gravity. In this case, a Weibull size distribution with $n = 1$ or, equivalently, $n/m = 1/3$, obtains a good fit to the experimental results. In the cloud physics community, this Weibull size distribution is known as the Marshall–Palmer universal size distribution; see Section 5.7.

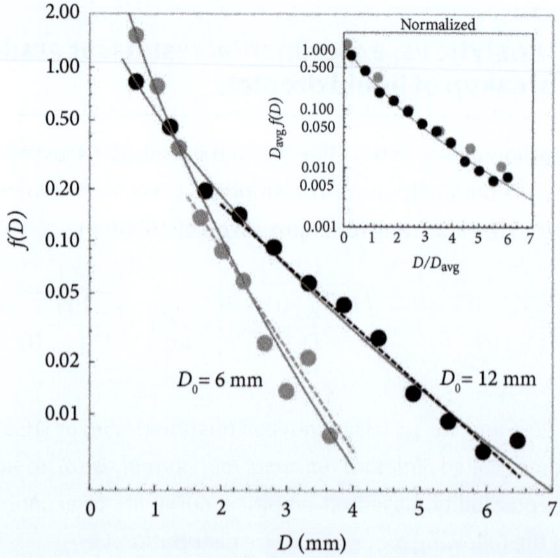

Figure 17.7 Analytical (dashed lines and solid curves) vs. experimental (circles) results for the fragment size distributions from a liquid droplet falling through the air under influence of gravity. The straight dashed lines are exponential size distributions. To obtain a better fit to the experimental results, the solid curves are Weibull size distributions with $n/m = 1/3$.
[The notation in this figure has been changed to match the notation used in this book.]
Reproduced from Villermaux & Bossa (2009) with permission from Springer Nature.

So far, this section has focused on the breakup of individual primary fragments. However, in many cases, the same models describe the breakup of entire sets of primary fragments, provided they all meet the same or similar fates. In fact, as discussed earlier in Section 1.5, secondary fragmentation tends to overwrite the size distributions created by primary fragmentation. As a result, secondary fragmentation may obtain ideal size distributions—e.g., Weibull size distributions where n/m is an integer multiple of 1/6—regardless of whether or not primary fragmentation obtains ideal size distributions. Put another way, the average fragment sizes following secondary breakup generally "remember" primary breakup, as discussed in the last chapter. By contrast, the fragment size distributions about those averages often completely "forget" primary breakup. Such "forgetting" is a common occurrence in chaotic systems. However, such "forgetting" only occurs when all of the primary fragments meet the same or similar fates. In some cases, the smallest primary fragments will not experience secondary breakup, which obviously prevents them from "forgetting" the primary fragment size distribution. In other cases, the smallest primary fragments will experience a different form of secondary breakup than the largest primary fragments.

To demonstrate these principles, Example 17.7 surveys some experimental results for secondary fragmentation in liquid spray systems. Alternatively, these principles may be demonstrated computationally or analytically. More specifically, suppose that $f_2(x) = \int f_1(y) f_{12}(x|y)dy$ where $f_1(y)$ is the probability density function for the primary fragments, $f_2(x)$ is the probability density function for the secondary fragments, and $f_{12}(x|y)$ is the conditional probability that a primary fragment of size y will produce a secondary fragment of size x, as seen in Examples 17.4, 17.5, and 17.6. While there is no known general proof, trial-and-error shows that f_2 usually obtains a familiar self-similar form, such as a Weibull size distribution, just like f_1 and f_{12}.

Example 17.7 Analytic vs. experimental results for secondary fragmentation in liquid spray systems

Table 17.7 shows curve fits to experimental results reported in the research literature for various liquids passing through commercially available spray nozzles. While other communities commonly use Weibull size distributions, the liquid spray community commonly uses root normal size distributions. Thus, Table 17.7 reports curve fits obtained using both Weibull and root normal size distributions. As seen in Figure 17.8, a Weibull size distribution with $n/m = 2/3$ obtains a good fit to the experimental results obtained by Tishkoff (1979) and Li & Tankin (1987) over the entire as-measured size range, which spans nearly two orders-of-magnitude. On the other hand, in this same figure, a root normal size distribution with $\sigma_0 = 0.17$ obtains a good fit to the experimental results for the largest droplets, but dramatically underestimates the results for the smallest droplets. As seen in Figure 17.9, a Weibull size distribution with $n/m = -3/2$ obtains a good fit to the experimental result obtained by Pimentel et al. (2010) for the largest fragments, whose sizes span about one order-of-magnitude. On the other hand, root normal size distributions do not obtain a particularly good fit over any size range; thus, the figure shows only a Weibull size distribution. In this case, it seems likely that the smallest liquid fragments did not experience secondary breakup.

Simmons (1977) used root normal size distributions to curve fit over 2,000 experimental results from about 100 different spray nozzles. He obtained $\sigma_0 = 0.24 \pm 20\%$, i.e., σ_0 mostly fell between 0.20 and 0.29. The overall average, $\sigma_0 = 0.24$, has become known as *Simmons' universal root normal size distribution*. However, according to the definition used in this

Table 17.7 Experimental results for the fragment size distributions produced by various liquid spray systems. Based on surveys by Paloposki (1994), Pimentel (2010) and Laney (2015a, b).

$n/m^{(*)}$	$n^{(*)}$	$\sigma_0{}^{(**)}$	No. Tests	Liquid	Nozzle Type	Reference
1/3	1	0.29	2	water	hollow cone	Tate & Olson (1962)
			1	water	swirl cone	Tate & Olson (1962)
1/2	3/2	0.20	4	β-naphthol	swirl jet (1)	Turner & Moulton (1953)
2/3	2	0.17	4	water	hollow cone (2)	Tishkoff (1979)
			4	water	hollow cone (3)	Li & Tankin (1987)
−3/2	−9/2	n/a	4	jet fuel	pressure injector (4)	Pimentel et al. (2010)
			9	Jet fuel	air-assist atomizer (5)	Pimentel et al. (2010)

(*) Curve fits obtained using Weibull size distributions as seen in Equations (17.1) and (17.2).
(**) Curve fits obtained using root normal size distributions as seen in Equation (17.19).
(1) Delavan Company. Model WDA 20–80°.
(2) Spraying Systems Company. Models 1/4 LN and 1/8 ASS.
(3) Spraying Systems Company. Model 1/4 TTG.
(4) Bosch GmbH. Common-rail pressure injector system.
(5) BETE Company. Model XA-PR200.

book, Simmons' distribution is actually an average of two different universals, closely corresponding to $n/m \cong 0.5 \times (1/3 + 1/2)$. Similarly, Empie et al. (1993, 1995, 1997) curve fitted 140 experimental results. They obtained $\sigma_0 \cong 0.20 \pm 40\%$. Clearly, the results depend on the exact selection of experiments.

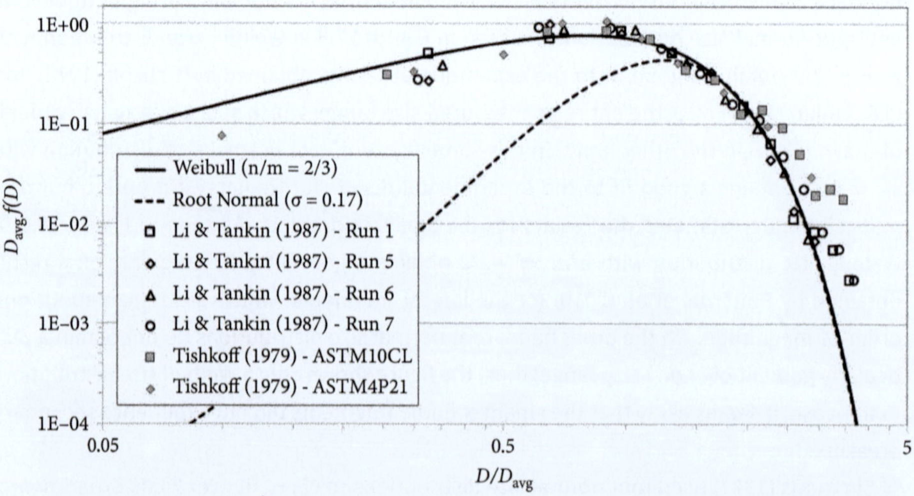

Figure 17.8 Analytical (solid and dashed curves) vs. experimental (symbols) results for the fragment size distributions from a selection of liquid spray systems. In this figure, the root normal size distribution has been renormalized to obtain the best possible agreement with the larger droplets.

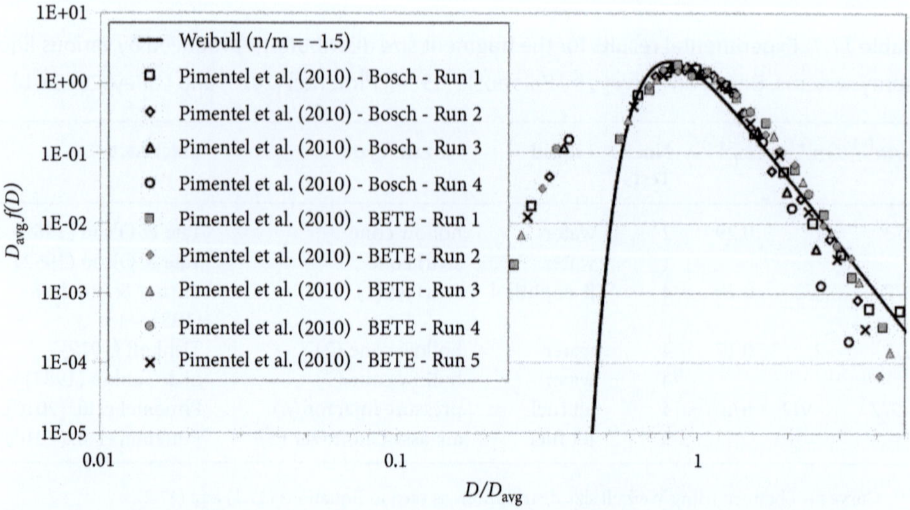

Figure 17.9 Analytical (solid curve) vs. experimental (symbols) results for the fragment size distributions from a selection of liquid spray systems. In this figure, the Weibull size distribution has been renormalized to obtain the best possible agreement with the larger droplets.

17.7 Conclusions

This chapter considered three different analytic theories for unitary fragmentation—mean-field theory, stochastic geometry, and maximum entropy theory—which use roughly consistent assumptions and obtain roughly consistent results. More specifically, they all tend to obtain power law, gamma, or Weibull size distributions. Under ideal conditions, where the strain rates are spatially homogeneous and isotropic, n/m is an integer multiple of 1/3 or 1/6 between −2 and 2, regardless of the exact conditions. On the other hand, under non-ideal conditions, n/m depends continuously on the exact conditions, e.g., it may be a linear function of the inverse Weber number.

As seen in Sections 17.2 and 17.6, secondary aggregation and secondary fragmentation tend to obtain the same size distributions as primary fragmentation. More specifically, in ideal cases, primary fragmentation, secondary fragmentation, and secondary aggregation all tend to obtain power law, gamma, or Weibull size distributions where n/m is an integer multiple of 1/3 or 1/6 between −2 and 2. Recall that secondary fragmentation may occur in several stages, some of which involve cascades and some of which do not; see Figures 15.5 and 15.6. As it turns out, every stage tends to affect the average fragment size but only the last stage tends to affect the size distribution about that average. Because secondary fragmentation is difficult to study, there is only limited direct evidence for this statement; however, there is substantial indirect evidence. As one challenge, it is not always possible to fully separate primary and secondary fragmentation, e.g., the latter may start before the former ends. This means that experimental results for primary fragmentation often include at least some secondary fragmentation. Some tests take elaborate precautions to reduce or prevent secondary fragmentation such as soft-catch devices; others take little or no precautions. When secondary fragmentation or secondary aggregation occur, a new size distribution may completely overwrite the original size distribution; see, e.g., Section 1.5. Possible combinations include: ideal primary fragmentation followed by non-ideal secondary fragmentation or aggregation; non-ideal primary fragmentation followed by ideal secondary fragmentation or aggregation; and so forth. Regardless, it is the final stage—not any of the earlier stages—that ultimately determines whether the size distribution is ideal or non-ideal.

References

W. Arnold & E. Rottenkolber, Fragment Mass Distribution of Metal Cased Explosive Charges, *International Journal of Impact Engineering*, 35(12), 1393–1398, 2008, https://doi.org/10.1016/j.ijimpeng.2008.07.049

T. Aste, Circle, Sphere, and Drop Packings, *Physical Review E*, 53(3), 2571–2579, 1996, https://doi.org/10.1103/PhysRevE.53.2571

J. A. Åström, R. P. Linna, J. Timonen, P. F. Møller, & L. Oddershede, Exponential and Power-Law Mass Distributions in Brittle Fragmentation, *Physical Review E*, 70, 026104, 2004, https://doi.org/10.1103/PhysRevE.70.026104

J. A. Åström, Statistical Models of Brittle Fragmentation, *Advances in Physics*, 55(3–4), 247–278, 2006, https://doi.org/10.1080/00018730600731907

C. X. Bai, H. Rusche, & A. D. Gosman, Modeling of Gasoline Spray Impingement, *Atomization and Sprays*, 12(1–3), 1–27, 2002, https://doi.org/10.1615/AtomizSpr.v12.i123.10

I. Bannikova, S. Uvarov, M. Davydova, & O. Naimark, Study of Ceramic Tube Fragmentation under Shock Wave Loading, *Procedia Materials Science*, 3, 592–597, 2014, https://doi.org/10.1016/j.mspro.2014.06.098

I. Bannikova, S. Uvarov, & O. Naimark, The Effect of Initial Porosity of a Sample Under Electric Explosion Loading, in E, S. Gorkunov, V. E. Panin & S. Ramasubbu, editors, *Mechanics, Resource and Diagnostics of Materials and Structures: Proceedings of the 10th International Conference on Mechanics, Resource and Diagnostics of Materials and Structures (MRDMS-2016)*, AIP Conference Proceedings 1785, American Institute of Physics, 2016, https://doi.org/10.1063/1.4967060

E. Ben-Naim & P. L. Krapivsky, Jamming and Tiling in Fragmentation of Rectangles, *Physical Review E*, 100, 032122, 2019, https://doi.org/10.1103/PhysRevE.100.032122

V. L. Berdichevsky, Universal Grain Size Distribution, Most Chaotic Microstructures and Tessellation Condition, *International Journal of Engineering Science*, 57, 24–35, 2012, https://doi.org/10.1016/j.ijengsci.2012.03.037

N. Berestycki & J. Pitman, Gibbs Distributions for Random Partitions Generated by a Fragmentation Process, *Journal of Statistical Physics*, 127(2), 381–418, 2007, https://doi.org/10.1007/s10955-006-9261-1

E. T. Bowman & W. A. Take, The Runout of Chalk Cliff Collapses in England and France—Case Studies and Physical Model Experiments, *Landslides*, 12(2), 225–239, 2015, https://doi.org/10.1007/s10346-014-0472-2

R. Ye. Brodskii, A. V. Tur, & V. V. Yanovsky, Weak Fragmentation of Round Plates, *Functional Materials*, 19(4), 525–532, 2012, http://functmaterials.org.ua/contents/19-4/

R. Y. Brodskii & T. V. Kulik, Fragmentation of Thin Lenses, *Functional Materials*, 23(2), 279–297, 2016, https://doi.org/10.15407/fm23.02.279

G. Chaussonnet, O. Vermorel, E. Riber, & B. Cuenot, A New Phenomenological Model to Predict Drop Size Distribution in Large-Eddy Simulations of Airblast Atomizers, *International Journal of Multiphase Flow*, 80, 29–42, 2016, https://doi.org/10.1016/j.ijmultiphaseflow.2015.10.014

Y. S. Cheong, A. D. Salman, & M. J. Hounslow, Effect of Impact Angle and Velocity on the Fragment Size Distribution of Glass Spheres, *Powder Technology*, 138(2–3), 189–200, 2003, https://doi.org/10.1016/j.powtec.2003.09.010

Y. S. Cheong, G. K. Reynolds, A. D. Salman, & M. J. Hounslow, Modelling Fragment Size Distribution using Two-Parameter Weibull Equation, *International Journal of Mineral Processing*, 74S, S227–S237, 2004, https://doi.org/10.1016/j.minpro.2004.07.012

W.-H. Chou, L.-P. Hsiang, G. M. Faeth, Temporal Properties of Drop Breakup in the Shear Breakup Regime, *International Journal of Multiphase Flow*, 23(4), 651–669, 1997, https://doi.org/10.1016/S0301-9322(97)00006-2

G. E. Cossali, M. Marengo, & M. Santini, Single-Drop Empirical Models for Spray Impact on Solid Walls: A Review, *Atomization and Sprays*, 15(6), 699–736, 2005, https://doi.org/10.1615/AtomizSpr.v15.i6.50

R. Cowan, New Classes of Random Tessellations Arising from Iterative Division of Cells, *Advances in Applied Probability*, 42(1), 26–47, 2010, https://doi.org/10.1239/aap/1269611142

D. Dhar, Fragmentation of a Sheet by Propagating, Branching and Merging Cracks, *Journal of Physics A: Mathematical and Theoretical*, 48(17), 175001, 2015, https://doi.org/10.1088/1751-8113/48/17/175001

L. Dias & F. Parisio, Minimal Fragmentation of Regular Polygonal Plates, *Physical Review E*, 90(3), 032405, 2014, https://doi.org/10.1103/PhysRevE.90.032405

F. P. M. Dos Santos, R. Donangelo, & S. R. Souza, Schematic Models for Fragmentation of Brittle Solids in One and Two Dimensions, *Physica A: Statistical Mechanics and its Applications*, 374(2), 680–690, 2007, https://doi.org/10.1016/j.physa.2006.08.058

C. Dumouchel, A New Formulation of the Maximum Entropy Formalism to Model Liquid Spray Drop-Size Distribution, *Particle & Particle Systems Characterization*, 23(6), 468–479, 2006, https://doi.org/10.1002/ppsc.200500989

C. Dumouchel, The Maximum Entropy Formalism and the Prediction of Liquid Spray Drop-Size Distribution, *Entropy*, 11(4), 713–747, 2009, https://doi.org/10.3390/e11040713

M. R. Edwards & C. Deal, Effect of Heat Treatment on Small Scale Fragmentation of Aluminium Alloy, *Materials Science and Technology*, 27(1), 332–338, 2011, https://doi.org/10.1179/026708309X12506933873341

H. J. Empie, S. J. Lien, W. Yang, & D. B. Samuels, *Kraft Black Liquor Delivery Systems Report Number 4*, Institute of Paper Science and Technology Progress Report, January 1993, http://hdl.handle.net/1853/535

H. J. Empie, S. J. Lien, W. Yang, & T. N. Adams, Spraying Characteristics of Commercial Black Liquor Nozzles, *TAPPI Journal*, 78(1), 121–128, 1995, https://imisrise.tappi.org/TAPPI/Products/95/JAN/95JAN121.aspx

H. J. Empie, S. J. Lien, & W. Yang, Drop Size Modification in Black Liquor Sprays From Commercial Nozzles Using Pulsed Flow, *Atomization and Sprays*, 7(5), 457–466, 1997, https://doi.org/10.1615/AtomizSpr.v7.i5.10

D. L. Frost, Y. Grégoire, O. Petel, S. Goroshin, & F. Zhang, Particle Jet Formation During Explosive Dispersal of Solid Particles, *Physics of Fluids*, 24, 091109, 2012, https://doi.org/10.1063/1.4751876

J. J. Gilvarry, Fracture of Brittle Solids. I. Distribution Function for Fragment Size in Single Fracture (Theoretical), *Journal of Applied Physics*, 32(3), 391–399, 1961, https://doi.org/10.1063/1.1736016

J. J. Gilvarry, Fragment Size in Single Fracture—A Review of Theory and Experiment, *Wear*, 227–243, 7(3), 1964, https://doi.org/10.1016/0043-1648(64)90086-9

D. E. Grady & M. E. Kipp, Geometric Statistics and Dynamic Fragmentation, *Journal of Applied Physics*, 58(3), 1210–1222, 1985, https://doi.org/10.1063/1.336139

D. E. Grady, L. T. Wilson, D. R. Reedal, L. D. Kuhns, M. E. Kipp, & J. W. Black, Comparing Alternate Approaches in the Scaling of Naturally Fragmenting Munitions, in I. R. Crewther, editor, *Proceedings of the 19th International Symposium on Ballistics*, IBS, 2001, http://ciar.org/ttk/mbt/papers/symp_19/WM06_591.pdf

D. E. Grady, *Physics of Shock and Impact, Volume 1: Fundamentals and Dynamic Failure*, IOP Publishing, 2017, https://doi.org/10.1088/978-0-7503-1254-7

Ø. T. Haug, M. Rosenau, K. Leever, & O. Oncken, On the Energy Budgets of Fragmenting Rockfalls and Rockslides: Insights from Experiments, *Journal of Geophysical Research: Earth Surface*, 121(7), 1310–1327, 2016, https://doi.org/10.1002/2014JF003406

A. A. Hegazy & T. Z. Blazynski, Some Aspects of Shock Consolidation of Polymeric, PVC-Metallic and PVC-Silica Powder Mixtures, *Journal of Materials Science*, 21(12), 4262–4268, 1986, https://doi.org/10.1007/BF01106540

M. Held, Fragment Mass Distribution of HE Projectiles, *Propellants, Explosives, Pyrotechnics*, 15(6), 254–260, 1990a, https://doi.org/10.1002/prep.19900150606

M. Held, Fragment Mass Distribution of Debris, In *Minutes of the Twenty-Fourth Explosives Safety Seminar, Volume II*, Department of Defense Explosive Safety Board, 1990b, http://apps.dtic.mil/sti/citations/ADA532197

H. G. Houghton, Spray Nozzles, in J. H. Perry, editor, *Perry's Chemical Engineers' Handbook*, 2nd edition, McGraw-Hill, 1941

H. Inaoka & H. Takayasu, Universal Fragment Size Distribution in a Numerical Model of Impact Fracture, *Physica A*, 229(1), 5–25, 1996, https://doi.org/10.1016/0378-4371(96)00048-9

H. Inaoka & H. Takayasu, Application of Statistical Physics to Impact Fragmentation, *Physica A*, 274(1–2), 300–309, 1999, https://doi.org/10.1016/S0378-4371(99)00426-4

F. Járai-Szabó & Z. Néda, On the Size Distribution of Poisson Voronoi Cells, *Physica A: Statistical Mechanics and its Applications*, 385(2), 518–526, 2007, https://doi.org/10.1016/j.physa.2007.07.063

H. Katsuragi, S. Ihara, & H. Honjo, Explosive Fragmentation of a Thin Ceramic Tube Using Pulsed Power, *Physical Review Letters*, 095503, 2005, https://doi.org/10.1103/PhysRevLett.95.095503

P. Kekäläinen J. A. Åström, & J. Timonen, Solution for the Fragment-Size Distribution in a Crack-Branching Model of Fragmentation, *Physical Review E*, 76, 026112, 2007, https://doi.org/10.1103/PhysRevE.76.026112

T. Kiang, Random Fragmentation in Two and Three Dimensions, *Zeitschrift für Astrophysik*, 64, 433–439, 1966, http://adsabs.harvard.edu/full/1966ZA 64 433K

P. L. Krapivsky & E. Ben-Naim, Scaling and Multiscaling in Models of Fragmentation, *Physical Review E*, 50(5), 3502–3507, 1994, https://doi.org/10.1103/PhysRevE.50.3502

V. S. Kumar & V. Kumaran, Voronoi Cell Volume Distribution and Configurational Entropy of Hard-Spheres, *The Journal of Chemical Physics*, 123(11), 114501, 2005, https://doi.org/10.1063/1.2011390

C. B. Laney, *Improved Root Normal Size Distributions for Liquid Atomization*, Defense Threat Reduction Agency Technical Report DTRA-TR-16-03, November 2015a, http://apps.dtic.mil/sti/citations/ADA624919

C. B. Laney, *Transformation and Self-Similarity Properties of Gamma and Weibull Fragment Size Distributions*, Defense Threat Reduction Agency Technical Report DTRA-TR-16-06, December 2015b, http://apps.dtic.mil/sti/citations/ADA624878

C. B. Laney, *On the Relationships Between Liquid and Solid Fragmentation*, Defense Threat Reduction Agency Technical Report DTRA-TR-16-19, January 2016a, http://apps.dtic.mil/sti/citations/AD1004358

C. B. Laney, *Effects of Mixtures on Liquid and Solid Fragment Size Distributions*, Defense Threat Reduction Agency Technical Report DTRA-TR-16-52, May 2016b, http://apps.dtic.mil/sti/citations/AD1008708

X. Li & R. S. Tankin, Droplet Size Distribution: A Derivation of a Nukiyama–Tanasawa Type Distribution Function, *Combustion Science and Technology*, 56(1–3), 65–76, 1987, https://doi.org/10.1080/00102208708947081

C. C. Lienau, Random Fracture of Brittle Solids, *Journal of the Franklin Institute*, 221(4), 485–494, 1936a, https://doi.org/10.1016/S0016-0032(36)90309-5

C. C. Lienau, Random Fracture of Brittle Solids (Continued), *Journal of the Franklin Institute*, 221(5), 673-686, 769–780, 1936b, https://doi.org/10.1016/S0016-0032(36)91108-0

C. C. Lienau, Random Fracture of Brittle Solids (Concluded), *Journal of the Franklin Institute*, 221(6), 769–780, 1936c, https://doi.org/10.1016/S0016-0032(36)90526-4

J. H. Lienhard & P. L. Meyer, A Physical Basis for the Generalized Gamma Distribution, *Quarterly of Applied Mathematics*, 25(3), 330–334, 1967, https://doi.org/10.1090/qam/99884

P. G. Lind, R. M. Baram, & H. J. Herrmann, Obtaining the Size Distribution of Fault Gouges with Polydisperse Bearings, *Physical Review E*, 77(2), 021304, 2008, https://doi.org/10.1103/PhysRevE.77.021304

D. W. Loebker & H. J. Empie, Independently Controlled Drop Size in Black Liquor Sprays into the Kraft Recovery Boiler Using Effervescent Atomization, *Journal of Pulp and Paper Science*, 27(1), 18–25, 2001, http://hdl.handle.net/1853/3145

J. Loiseau, Q. Pontalier, A. M. Milne, S. Goroshin, & D. L. Frost, Terminal Velocity of Liquids and Granular Materials Dispersed by a High Explosive, *Shock Waves*, 28, 473–487, 2018, https://doi.org/10.1007/s00193-018-0822-4

A. Meibom & I. Balslev, Composite Power Laws in Shock Fragmentation, *Physical Review Letters*, 76(14), 2492–2494, 1996, https://doi.org/10.1103/PhysRevLett.76.2492

M. A. Meyers, D. J. Benson, & E. A. Olevsky, Shock Consolidation: Microstructurally-Based Analysis and Computational Modeling, *Acta Materialia*, 47(7), 2089–2108, 1999, https://doi.org/10.1016/S1359-6454(99)00083-X

W. Mock & W. H. Holt, *Computational and Experimental Determination of Fragmentation for Naturally Fragmenting Warheads*, U.S. Naval Surface Warfare Center Technical Report, NSWC TR 80-238, May 1981, http://apps.dtic.mil/sti/citations/ADA110444

W. Mock & W. H. Holt, Fragmentation Behavior of Armco Iron and HF-1 Steel Explosive-Filled Cylinders, *Journal of Applied Physics*, 54(5), 2344–2351, 1983, https://doi.org/10.1063/1.332346

A. L. Moreira, A. S. Moita, & M. R. O. Panão, Advances and Challenges in Explaining Fuel Spray Impingement: How Much of Single Droplet Impact Research is Useful?, *Progress in Energy and Combustion Science*, 36(5), 554–580, 2010, https://doi.org/10.1016/j.pecs.2010.01.002

A. L. Moreira & M. R. O. Panão, Spray-Wall Impact, in N. Ashgriz, editor, *Handbook of Atomization and Sprays*, Springer, 2011, https://doi.org/10.1007/978-1-4419-7264-4_21

N. F. Mott & E. H. Linfoot, *A Theory of Fragmentation*, U. K. Ministry of Supply, AC3348, January 1943 [reproduced in section II.1 of Grady (2006), https://doi.org/10.1007/978-3-540-27145-1_9]

N. F. Mott, J. H. Wilkinson, & T. H. Wise, *Fragmentation of Service Projectiles*, U.K. Ministry of Supply, A.C. 6338, May 1944 [reproduced in section II.6 of Grady (2006), https://doi.org/10.1007/978-3-540-27145-1_14]

C. Mundo, M. Sommerfeld, & C. Tropea, Droplet-Wall Collisions: Experimental Studies of the Deformation and Breakup Process, *International Journal of Multiphase Flow*, 21(2), 151–173, 1995, https://doi.org/10.1016/0301-9322(94)00069-V

V. F. Nesterenko, Advanced Materials Treatment by Shock Waves. in *Dynamics of Heterogeneous Materials: High-Pressure Shock Compression of Condensed Matter*, Springer, 2001, https://doi.org/10.1007/978-1-4757-3524-6_6

L. Oddershede, P. Dimon, & J. Bohr, Self-Organized Criticality in Fragmenting, *Physical Review Letters*, 71(19), 3107–3110, 1993, https://doi.org/10.1103/PhysRevLett.71.3107

T. Paloposki, Drop Size Distributions in Liquid Sprays, *Acta Polytechnica Scandinavica, Mechanical Engineering Series 114*, Helsinki, Finland, 1994; see also *Proceedings of the Sixth International Conference on Liquid Atomization and Spray Systems (ICLASS 94)*, Begell House, 1994, https://doi.org/10.1615/ICLASS-94.1160

A. Paluszny, X. H. Tang, M. Nejati, & R. W. Zimmerman, A Direct Fragmentation Method with Weibull Function Distribution of Sizes Based on Finite- and Discrete Element Simulations, *International Journal of Solids and Structures*, 80, 38–51, 2016, https://doi.org/10.1016/j.ijsolstr.2015.10.019

H. Pan & P. M. Render, Impact Characteristics of Hailstones Simulating Ingestion by Turbofan Aeroengines, *Journal of Propulsion and Power*, 12(3), 457–462, 1996, https://doi.org/10.2514/3.24057

F. Parisio & L. Dias, Minimal Fragmentation Problem, *Physical Review E*, 84(3), 035101(R), 2011, https://doi.org/10.1103/PhysRevE.84.035101

R. Pimentel, R. Stowe, P. Harris, A. DeChamplain, & D. Kretschmer, Spray Characterization Based on the Pearson System of Frequency Curves, *Atomization and Sprays*, 20(5), 365–386, 2010, https://doi.org/10.1615/AtomizSpr.v20.i5.10

D. Raybould & T. Z. Blazynski, Non-Metallic Materials Under Shock Loading, in T. Z. Blazynski, editor, *Materials at High Strain Rates*, Springer, 1987, https://link.springer.com/book/9781851660674

I. V. Roisman, K. Horvat, & C. Tropea, Spray Impact: Rim Transverse Instability Initiating Fingering and Splash, and Description of a Secondary Spray, *Physics of Fluids* 18, 102104, 2006, https://doi.org/10.1063/1.2364187

A. D. Salman, C. A. Biggs, J. Fu, I. Angyal, M. Szabó, & M. J. Hounslow, An Experimental Investigation of Particle Fragmentation Using Single Particle Impact Studies, *Powder Technology*, 128(1), 36–46, 2002, https://doi.org/10.1016/S0032-5910(02)00151-1

H. C. Simmons, The Correlation of Drop-Size Distributions in Fuel Nozzle Sprays—Part I: The Drop-Size/Volume-Fraction Distribution, *Journal of Engineering for Power*, 99(3), 309–314, 1977, https://doi.org/10.1115/1.3446488

F. Spahn, E. Vieira Neto, A. H. F. Guimarães, A. N. Gorban, & N. V. Brilliantov, A Statistical Model of Aggregate Fragmentation, *New Journal of Physics*, 16(1), 013031, 2014, https://doi.org/10.1088/1367-2630/16/1/013031

D. W. Stanton & C. J. Rutland, Modeling Fuel Film Formation and Wall Interaction in Diesel Engines, *SAE Transactions*, 103, 808–824, 1996, www.jstor.org/stable/44736319 [See also SAE Technical Paper 960628, 1996, https://doi.org/10.4271/960628]

R. W. Tate & E. O. Olson, Spray Droplet Size of Pressure-Atomizing Burner Nozzles, *ASHRAE Journal*, 4(3), 39–43, 1962

G. Timár, J. Blömer, F. Kun, & H. J. Herrmann, New Universality Class for the Fragmentation of Plastic Materials, *Physical Review Letters*, 104(9), 095502, 2010, https://doi.org/10.1103/PhysRevLett.104.095502

J. M. Tishkoff, *GMR Laser-video Imaging System Results for the ASTM 'Round-Robin' Nozzle Test*, General Motors Research Publication GMR-3098, 1979

G. M. Turner & R. W. Moulton, Drop Size Distributions from Spray Nozzles, *Chemical Engineering Progress*, 49(4), 185–190, 1953, https://archive.org/details/sim_chemical-engineering-progress_1953-04_49_4/page/184

M. M. Vargas, C. R. Ruggeri, M. Pereira, & D. Revilock, *Fragment Size Distribution for Ice Particle Impacts on a Glass Plate*, 2018 Atmospheric and Space Environments Conference, AIAA 2018-4225, https://doi.org/10.2514/6.2018-4225

E. Villermaux & B. Bossa, Single-Drop Fragmentation Determines Size Distribution of Raindrops, *Nature Physics*, 5, 697–702, 2009, https://doi.org/10.1038/nphys1340

L. T. Wilson, D. R. Reedal, L. D. Kuhns, D. E. Grady, & M. E. Kipp, Using a Numerical Fragmentation Model to Understand the Fracture and Fragmentation of Naturally Fragmenting Munitions of Differing Materials and Geometries, in I. R. Crewther, editor, *Proceedings of the 19th International Symposium on Ballistics*, IBS, 2001, http://ciar.org/ttk/mbt/papers/symp_19/WM16_671.pdf

F. K. Wittel, F. Kun, H. J. Herrmann, & B. H. Kröplin, Fragmentation of Shells, *Physical Review Letters*, 93(3), 035504, 2004, https://doi.org/10.1103/PhysRevLett.93.035504

F. K. Wittel, F. Kun, H. J. Herrmann, & B. H. Kröplin, Breakup of Shells Under Explosion and Impact, *Physical Review E*, 71(1), 016108, 2005, https://doi.org/10.1103/PhysRevE.71.016108

F. K. Wittel, F. Kun, B. H. Kröplin, & H. J. Herrmann, Study on the Fragmentation of Shells, *International Journal of Fracture*, 140(1–4), 243–254, 2006, https://doi.org/10.1007/s10704-005-3995-7

L. Xu, Liquid Drop Splashing on Smooth, Rough, and Textured Surfaces, *Physical Review E*, 75, 056316, 2007, https://doi.org/10.1103/PhysRevE.75.056316

L. Xu, L. Barcos, & S. R. Nagel, Splashing of Liquids: Interplay of Surface Roughness with Surrounding Gas, *Physical Review E*, 76, 066311, 2007, https://doi.org/10.1103/PhysRevE.76.066311

K. Xue, F. Li, & C. Bai, Explosively Driven Fragmentation of Granular Materials, *The European Physical Journal E*, 36, 95, 2013, https://doi.org/10.1140/epje/i2013-13095-x

T. Zhao, G. B. Crosta, G. Dattola, & S. Utili, Dynamic Fragmentation of Jointed Rock Blocks During Rockslide-Avalanches: Insights From Discrete Element Analyses, *Journal of Geophysical Research: Solid Earth*, 123(4), 3250–3269, 2018, https://doi.org/10.1002/2017JB015210

18

Maximum Entropy Theory for Discrete Particle Systems

This chapter concerns maximum entropy theory for discrete particle systems experiencing aggregation and fragmentation. To introduce this topic, Example 18.1 gives the traditional definition of entropy—which involves particle energies—as it appears in statistical thermodynamics. Example 18.2 describes an alternative definition of entropy—which involves particle sizes—as it appears in aggregation and fragmentation. This alternative definition was first given by Flory (1942) and Stockmayer (1943); see Example 5.5. This chapter is based on the Flory–Stockmayer definition of entropy.

Example 18.1 The traditional definition of entropy in discrete particle systems concerns particle energies

Consider a large system of identical particles such as a molecular gas; see Example 2.1. Assume that this particle system can be treated using statistical thermodynamics as described by, e.g., Carey (1999). Among other things, this means that particles can exchange momentum and energy, but cannot experience aggregation or fragmentation, e.g., the particles cannot stick to each other. In this case, the microstate is the mass, momentum, energy, and position of each particle in a system. In addition, the macrostate is the pressure, total mass, total energy, and total volume of the system.

The entropy is defined to be a constant times the logarithm of the number of microstates leading to a given macrostate. The most probable macrostate is the one with the largest entropy, i.e., the macrostate that can be achieved by the greatest number of microstates, subject to constraints. By this definition, entropy is related to the particle energies. For example, Lambert (2007) says that entropy is a measure of particle "energy spreading out." Particle masses, momenta and positions matter only to the extent that they affect energy spread.

The second law of thermodynamics says that, for closed systems, entropy always increases. By contrast, Hübler & Crutchfield (2010) note that "in open systems the entropy can increase or decrease over time" because the particle system can exchange mass, momentum, and energy with its environment. Thus, depending on the circumstances, entropy in a particle system may be either maximized or minimized.

From one point of view, entropy is related to particle positions rather than particle energies. This point of view leads to concepts such as *positional entropy* and *disorder*. Indeed, entropy tends to increase when the number of positional microstates increases. However, according to Lambert (2007), this is only because there are "a larger number of arrangements

Particle Dynamics with Aggregation and Fragmentation. Culbert B. Laney, Oxford University Press. © Culbert B. Laney (2025).
DOI: 10.1093/9780198920687.003.0018

for the system's energy." Thus, Lambert (2007) advocates avoiding concepts such as positional entropy and disorder. Similarly, Martyushev (2013) states "the connection between order and ... entropy is an evident misunderstanding ... For non-experts, the 'order' of the structure is the systematicness [or the] correctness in the arrangement ... For physicists and chemists, the 'order' of the structure is evaluated based on the complexity of prediction, which is related specifically to the number of microstates."

Example 18.2 An alternative definition of entropy in discrete particle systems concerns particle sizes

Consider a system of identical particles which can exchange mass, momentum, and energy with each other and with the environment. Unlike the previous example, assume the particles can experience aggregation and fragmentation, e.g., the particles can stick to each other. In this case, the microstate is the mass, momentum, energy, and position of each particle, and the macrostate is the size distribution of the particle clusters.

The size entropy is defined to be a constant times the logarithm of the number of microstates leading to a given macrostate. The most probable macrostate is the one with the largest entropy, i.e., the macrostate that can be achieved by the greatest number of microstates, subject to constraints. While energy played the leading role in the last example, size plays the leading role in this example. Particle position, momentum, and energy matter only affect entropy to the extent that they affect the size spread. For example, in aggregation, the microstate can only change if two or more particles are in the same position, so that they can experience a perfectly inelastic collision. For another example, for some types of binary fragmentation, such as polymers experiencing thermal degradation, the microstate can only change if the internal energy of a cluster is sufficiently large. By this definition, the minimum size entropy occurs for monodisperse size distributions, e.g., molecular gases. Conversely, the maximum size entropy occurs for uniform size distributions, i.e., when there is a broad range of cluster sizes, each equally likely to occur, between some maximum and minimum. In general, the size entropy lies somewhere between these two extremes.

Examples 18.1 and 18.2 imply that there is an analogy between particle energies and particle sizes. Similar analogies have appeared earlier in the book. For example, Sections 8.4 and 8.5 argued that universal *energy spectra* in self-organized particle systems are analogous to universal *size distributions* in regular aggregating and fragmenting particle systems. For another example, Example 9.1 argued that the degrees of freedom for particle energies in ideal molecular gases are analogous to the degrees of freedom for particle sizes in ideal aggregating particle systems.

Maximum entropy theory for discrete particle systems was introduced earlier in Section 5.3. For another perspective, Beck & Schögl (1993) say the following about chaotic systems: "In broad terms ... if in a probability distribution we suppress the correlations between subsystems [e.g., particles] without changing the probabilities of the single microstates of the subsystems, we lose information, and the entropy is expected to increase. This means the entropy takes on its maximum for the uncorrelated distribution."

As a result, maximum entropy theory commonly appears in treatments of chaotic systems, including but certainly not limited to the aggregating and fragmenting particle systems studied in this book. For a discussion of maximum entropy theory in other chaotic applications see, e.g., McCauley (1993), Williams (1997) and Cencini et al. (2010).

18.1 Combinatorics and Constrained Optimization

According to Stockmayer (1943), the total number of ways to obtain a given size distribution by combining a finite number of monomers is as follows:

$$\Omega = N_{\text{total}}! \prod_{i=i_{\min}}^{i_{\max}} \frac{1}{n_i!}\left(\frac{W_i}{i!}\right)^{n_i} \tag{18.1}$$

where:

N_{total}	total number of monomers
n_i	size distribution, i.e., number of particles composed of i monomers
W_i	number of different ways to form a particle composed of i monomers
$i_{\min}$	minimum number of monomers in a cluster (≥ 1)
$i_{\max}$	maximum number of monomers in a cluster ($\leq N_{\text{total}}$)

In other words, Ω is the number of microstates leading to a given macrostate. Notice that Ω is sometimes called the *partition function*. As a useful definition, let:

$$Y_i = \frac{W_i}{i!}. \tag{18.2}$$

Notice that $i!$ is the number of different ways to arrange i different objects in a linear sequence. Thus Y_i is a ratio of two counts involving i objects.

If all cases are equally probable, the goal is to maximize Ω or, equivalently, $\ln \Omega$. Choosing the latter for convenience, the *size entropy* may be defined as follows:

$$S = \ln \Omega = \sum_{i=i_{\min}}^{i_{\max}} [n_i \ln Y_i - \ln n_i!] + \sum_{i=1}^{N_{\text{total}}} \ln i. \tag{18.3}$$

According to Sterling's approximation, $\ln n_i! \approx n_i \ln n_i - n_i$ for large enough n_i. Then:

$$S \cong -\sum_{i=i_{\min}}^{i_{\max}} n_i \ln \frac{n_i}{Y_i} + \text{const.} \tag{18.4}$$

where the additive constant depends on the total number of particles N and the total number of monomers N_{total}. Notice that, if all cases are not equally probable, then maximizing the partition function Ω is not the same as maximizing the entropy S. More specifically, if there is a selection bias, then Matsoukas (2014) says the partition function and "entropy in the cluster ensemble are distinct properties, and between the two, it is the partition function that is of fundamental importance." Examples 18.3 and 18.4 show how to derive size distributions using maximum entropy theory for finite and infinite particles systems, respectively, assuming no selection bias.

Example 18.3 Derivation of size distributions in finite systems of particles using maximum entropy theory

Consider a system with N_{total} monomers, each with mass M_1, grouped into N particles of average mass M_{avg}. The size entropy S may be defined as follows:

$$S = -\sum_{i=i_{\min}}^{i_{\max}} n_i \ln \frac{n_i}{Y_i} + \text{const.} \tag{18.5}$$

Some possible constraints on the entropy are as follows:

$$\sum_{i=i_{\min}}^{i_{\max}} n_i = N \tag{18.6}$$

$$\frac{1}{N}\sum_{i=i_{\min}}^{i_{\max}} M_i n_i = M_{\text{avg}} \tag{18.7}$$

$$\frac{1}{N}\sum_{i=i_{\min}}^{i_{\max}} \frac{n_i}{M_i} = \frac{1}{M'_{\text{avg}}}. \tag{18.8}$$

As it turns out, Equation (18.7) is just conservation of mass. To prove this, recall $M_i = iM_1$ and $NM_{\text{avg}} = N_{\text{total}} M_1$. Then Equation (18.7) can be rewritten as follows:

$$\sum_{i=i_{\min}}^{i_{\max}} i n_i = N_{\text{total}}. \tag{18.9}$$

If Equation (18.5) is maximized subject to Equations (18.6) and (18.7), then:

$$\begin{aligned} n_i &= Y_i \exp(-\alpha_1 - \alpha_2 i) \\ &= AY_i \exp(-ai) \end{aligned} \tag{18.10}$$

where α_1 and α_2 are Lagrange multipliers and a and A are constants. According to Flory (1941a, b, c, 1942), $Y_i = 1$ for particle aggregation. Then Equation (18.10) is an exponential size distribution or, equivalently, a gamma size distribution with $n/m = 1$. For comparison, if Equation (18.5) is maximized subject to Equations (18.6) and (18.8) then:

$$\begin{aligned} n_i &= Y_i \exp(-\beta_1 - \beta_2/i) \\ &= BY_i \exp(-b/i) \end{aligned} \tag{18.11}$$

where β_1 and β_2 are Lagrange multipliers and b and B are constants. Unlike before, there is no known way to derive an expression for Y_i. Thus, maximum entropy theory does not produce a useful result in this case.

Example 18.4 Derivation of size distributions in infinite particle systems using maximum entropy theory

The previous example concerned finite particle systems. This example concerns infinite particle systems. In this case, suppose that size entropy is defined as follows:

$$S = -\text{const.} \sum_{i=1}^{\infty} f_i \ln \frac{f_i}{Y_i} + \text{const.} \tag{18.12}$$

where f_i is the fraction of particles composed of i monomers. Possible constraints on the entropy are as follows:

$$\sum_{i=1}^{\infty} f_i = 1 \tag{18.13}$$

$$\sum_{i=1}^{\infty} M_i f_i = M_{\text{avg}} \tag{18.14}$$

$$\sum_{i=1}^{\infty} M_i^2 f_i = \left(M^2\right)_{\text{avg}} \tag{18.15}$$

$$\sum_{i=1}^{\infty} M_i^3 f_i = \left(M^3\right)_{\text{avg}}. \tag{18.16}$$

If Equation (18.12) is maximized subject to Equations (18.13), (18.14), (18.15), and (18.16) then:

$$f_i(t) = B_i(t) \exp\left[-b_1(t)i - b_2(t)i^2 - b_3(t)i^3\right] \tag{18.17}$$

where $B_i(t) = \text{const.}\, Y_i(t)$. In this expression, $B_i(t)$, $b_1(t)$, $b_2(t)$, $b_3(t)$ should be chosen so that Equations (18.13), (18.14), (18.15), and (18.16) apply. For comparison, using mean-field theory for unsteady collisional aggregation with $\lambda = 2$, Lushnikov (2004) derived the following result:

$$f_i(t) = B_i(t) \exp\left(-2(t_c - t)^2 i - \text{const.}\,(t_c - t)\, i^2 - \text{const.}\, i^3\right) \tag{18.18}$$

$$B_i(t) = \text{const.} \left[i^{-5/2} + \text{const.} \sqrt{t_c - t}\,\theta\,(t_c - t)\right] \tag{18.19}$$

where t_c is a critical time and θ is the Heaviside step function. Notice that Equation (18.18) is the same as Equation (18.17) but now with specific expressions for $B_i(t)$, $b_1(t)$, $b_2(t)$, $b_3(t)$. Despite the dual evidence of mean-field theory and maximum entropy theory, there is no known computational or experimental evidence indicating that Equations (18.17) or (18.18) actually occur in practice. This illustrates the importance of carefully validating analytic solutions against experimental results to guard against non-physical outcomes.

Table 18.1 summarizes some of the solutions that can be obtained with maximum entropy theory for aggregation in discrete particle systems. Table 18.1.a assumes a finite particle system where each monomer supports a small number of bonds with other monomers; thus, the size distributions are written in terms of n_i. On the other hand, Table 18.1.b assumes an infinite particle system where each monomer supports a large number of bonds with other monomers; thus, the size distributions are written in terms of f_i. With one exception, the results given in Table 18.1 become gamma size distributions with $n/m = 1, -1/2$, and $-3/2$ for sufficiently late times and large particles. As seen earlier in the book, these gamma size distributions often appear in computational, experimental, and analytic results for particle aggregation. By contrast, Poisson distributions such as the one given in Table 18.1.b rarely appear in computational, experimental, or analytic results for particle aggregation.

Comparing the results of this chapter with those of Chapter 10, it appears that both maximum entropy theory and mean-field theory for non-stationary aggregation lead to the same gamma size distributions with $n/m = 1, -1/2$, or $-3/2$. Thus, it is not surprising to discover that there is a connection between partition functions in maximum entropy theory and collision kernels in mean-field theory. More specifically, Spouge (1983) derived the following expression:

$$W_j = j!\,Y_j = \frac{1}{2(j-1)}\sum_{i=1}^{j-1}\binom{j}{i}W_i W_{j-i} K_{i,j-i} \tag{18.20}$$

where K_{ij} is "the number of different bonds that may be formed between" a pair of particles of size M_i and M_j. This definition of K_{ij} is essentially equivalent to the definition of discrete collision kernels given in Chapter 9. As seen in Chapter 9, K_{ij} is homogeneous if:

$$K_{ai,\,aj} = a^{\lambda} K_{ij} \tag{18.21}$$

where a is an integer and λ is the *degree of homogeneity*. It generally requires elaborate combinatorial reasoning to derive Y_j directly; see, e.g., Stockmayer (1943). While still far from trivial, it may be easier to derive K_{ij} and then convert it to Y_j using Equation (18.20). For example, for so-called ARB$_{f\text{-}1}$ polymers, Spouge (1983) suggested the following:

$$K_{ij} = 2 + (k-2)(i+j) \tag{18.22}$$

where $\lambda = 1$ and k is the number of bonds of a given type that each monomer can support. Notice that Equation (18.22) becomes a sum kernel for large enough k. Table 18.1 gives the corresponding expression for Y_j, which leads to a gamma size distribution with $n/m = -1/2$. For another example, for so-called RA$_f$ polymers, Spouge (1983) suggested the following:

$$K_{ij} = [2 + (k-2)i][2 + (k-2)j] \tag{18.23}$$

where $\lambda = 2$ and k is the number of bonds of a given type that each monomer can support. Notice that Equation (18.23) becomes a product kernel for large enough k. Table 18.1 gives the corresponding expression for Y_j, which leads to a gamma size distribution with $n/m = -3/2$.

Table 18.1.a Examples of maximum entropy theory for aggregation in discrete systems with small k. In this table, k is the number of bonds that each monomer can support.

Type (n/m)	Variables	Size Distribution	References
Gamma (1)	$Y_i = 1$	$n_i = B\exp(-bi)$	Flory (1942)
Gamma (−1/2)	$Y_i = \dfrac{((k-1)i)!}{i!((k-2)i+1)!}$ $k = 3, 4, 5...$	$n_i = \dfrac{B(k)((k-1)i)!\exp[-b(k)i]}{i!((k-2)i+1)!}$ $\approx B'(k)i^{-\frac{3}{2}}\exp\left[-b'(k)i\right]^{(*)}$	Spouge (1983)
Gamma (−3/2)	$Y_i = \dfrac{k^i((k-1)i)!}{i!((k-2)i+2)!}$ $k = 3, 4, 5...$	$n_i = \dfrac{B(k)k^i((k-1)i)!\exp[-b(k)i]}{i!((k-2)i+2)!}$ $\approx B'(k)i^{-\frac{5}{2}}\exp\left[-b'(k)i\right]^{(*)}$	Stockmayer (1943); Spouge (1983)

Table 18.1.b Examples of maximum entropy theory for aggregation in discrete systems with large or infinite k.

Type (n/m)	Variables	Size Distribution	References
Gamma (1)	$Y_i = 1$	$f_i = B\exp(-bi)$	Flory (1941a, b, c)
Gamma (−1/2)	$Y_i = \dfrac{i^{i-1}}{i!}$	$f_i = B\dfrac{i^{i-1}}{i!}\exp(-bi)$ $\approx \dfrac{B}{\sqrt{2\pi}}i^{-\frac{3}{2}}\exp(-(b-1)i)^{(*)}$	Spouge (1983)
Gamma (−3/2)	$Y_i = \dfrac{i^{i-2}}{i!}$	$f_i = B\dfrac{i^{i-2}}{i!}\exp(-bi)$ $\approx \dfrac{B}{\sqrt{2\pi}}i^{-\frac{5}{2}}\exp(-(b-1)i)^{(*)}$	Spouge (1983)
Poisson	$Y_i = \exp(-\nu)\dfrac{\nu^{i-1}}{(i-1)!}$ $\nu = M_{avg}/M_1 > 1$	$f_i = \exp(-\nu)\dfrac{\nu^{i-1}}{(i-1)!}$ $\approx \dfrac{1}{\sqrt{2\pi\nu}}i^{-i+\frac{1}{2}}\exp[(\ln\nu-1)i-\nu]^{(*)}$	Flory (1942)

(*) For large enough i, as obtained by Stirling's approximation.

Table 18.1 concerns non-stationary aggregation with a focus on late times prior to gelation, if it occurs. In fact, the same results may apply at sufficiently large times following gelation, if it occurs; see, e.g., van Dongen & Ernst (1984) and Hendriks (1984). In addition, similar results may apply for reverse aggregation, a.k.a., binary fragmentation, either alone or in combination with forward aggregation. In particular, in such cases, Griffith (1943), Cohen & Benedek (1982), Spouge (1983), Van Dongen & Ernst (1984) and Hendriks (1984) used maximum entropy theory to derive gamma size distributions with $n/m = 1, -1/2$ and $-3/2$ at late times.

18.2 Conclusions

This chapter discussed maximum entropy theory for discrete particle systems. Maximum entropy theory provides the essential information—the observable macrostate—without the need to fully characterize the exact microstate. Many researchers who have studied maximum entropy theory for discrete particle systems have tried to account for the dynamic time-dependent aspects of aggregation—and, less frequently, various combinations of aggregation and binary fragmentation—including the net effect of rate kernels. In other words, they have tried to inject the key elements of mean-field theory into maximum entropy theory. As a result, maximum entropy theory selects the same three late-time non-stationary solutions as mean-field theory, namely, gamma size distributions with $n/m = 1, -1/2$, and $-3/2$; see Chapter 10. By making some additional assumptions about locality vs. non-locality, mean-field theory can produce a wide variety of other late-time analytic solutions; see Chapter 11. However, maximum entropy theory has no obvious way of understanding locality vs. non-locality. Thus, maximum entropy theory cannot produce a realistic range of late-time analytic results. In addition, while mean-field theory produces full time-dependent analytic solutions in selected cases, maximum entropy theory does not. In short, while maximum entropy theory for discrete particle systems provides an interesting theoretical perspective, it is substantially less powerful than mean-field theory.

This chapter has highlighted the distinct character of discrete particle systems. This distinct character stems partly from the fact that each monomer may only support a finite number of linkages, a.k.a., bonds. This means that discrete particles are closely related to graphs and trees, where monomers play the role of *vertices* (a.k.a. *nodes* or *points*) and linkages play the role of *edges* (a.k.a. *lines* or *paths*). By contrast, continuous particles do not involve linkages and are not closely related to graphs and trees. Mean-field theory generally makes it easy to avoid the differences between discrete and continuous particle systems. In particular, the mean-field governing equations for aggregation and fragmentation are almost the same, after making the obvious transliteration between summation and integration.[1] By contrast, maximum entropy theory makes it almost impossible to avoid the differences between discrete and continuous systems. In fact, the treatment of discrete particle systems given in this chapter bears little resemblance to the treatment of continuous particle systems given in the next chapter.

This book focuses on results that apply broadly, regardless of the specific application. Thus, this chapter focused on results that apply to both discrete and continuous particle systems. In particular, the universal gamma size distributions with $n/m = 1, -1/2$, and $-3/2$ described in this chapter occur in both discrete and continuous particle systems. However, while the discrete variant of maximum entropy theory can be used to derive these universals, the continuous variant generally cannot (the universal gamma size distribution with $n/m = 1$ is a rare case of overlap). Similarly, the next chapter will focus on results that, as it

[1] The differences are, of course, still present in mean-field theory. However, they tend to appear in the solution rather than in the governing equations; see, e.g., Example 12.6.

turns out, apply to both discrete and continuous particle systems. However, while the continuous variant of maximum entropy theory can be used to derive these results, the discrete variant generally cannot. The combination of the discrete and continuous variants provides a clearer, more complete picture than either one alone.

References

C. Beck & F. Schögl, *Thermodynamics of Chaotic Systems: An Introduction*, Cambridge University Press, 1993, https://doi.org/10.1017/CBO9780511524585

V. P. Carey, *Statistical Thermodynamics and Microscale Thermodynamics*, Cambridge University Press, 1999, https://doi.org/10.1017/CBO9780511626395

M. Cencini, F. Cecconi, & A. Vulpiani, *Chaos: From Simple Models to Complex Systems*, World Scientific, 2010, https://doi.org/10.1142/7351

R. J. Cohen & G. B. Benedek, Equilibrium and Kinetic Theory of Polymerization and the Sol-Gel Transition, *The Journal of Physical Chemistry*, 86(19), 3696–3714, 1982, https://doi.org/10.1021/j100216a005

P. J. Flory, Molecular Size Distribution in Three Dimensional Polymers. I. Gelation, *Journal of the American Chemical Society*, 63(11), 3083–3090, 1941a, https://doi.org/10.1021/ja01856a061

P. J. Flory, Molecular Size Distribution in Three Dimensional Polymers. II. Trifunctional Branching Units, *Journal of the American Chemical Society*, 63(11), 3091–3096, 1941b, https://doi.org/10.1021/ja01856a062

P. J. Flory, Molecular Size Distribution in Three Dimensional Polymers. III. Tetrafunctional Branching Units, *Journal of the American Chemical Society*, 63(11), 3096–3100, 1941c, https://doi.org/10.1021/ja01856a063

P. J. Flory, Random Reorganization of Molecular Weight Distribution in Linear Condensation Polymers, *Journal of the American Chemical Society*, 64(9), 2205–2212, 1942, https://doi.org/10.1021/ja01261a049

L. Griffith, A Theory of the Size Distribution of Particles in a Comminuted System, *Canadian Journal of Research*, 21A(6), 57–64, 1943, https://doi.org/10.1139/cjr43a-005

E. M. Hendriks, Cluster Size Distributions in Equilibrium, *Zeitschrift für Physik B Condensed Matter*, 57(4), 307–314, 1984, https://doi.org/10.1007/BF01470421

A. Hübler & J. P. Crutchfield, Order and Disorder in Open Systems, *Complexity*, 16(1), 6–9, 2010, https://doi.org/10.1002/cplx.20344

F. L. Lambert, Configurational Entropy Revisited, *Journal of Chemical Education*, 84(9), 1548–1550, 2007, https://doi.org/10.1021/ed084p1548

A. A. Lushnikov, From Sol to Gel Exactly, *Physical Review Letters*, 93(19), 2004, https://doi.org/10.1103/PhysRevLett.93.198302

L. M. Martyushev, Entropy and Entropy Production: Old Misconceptions and New Breakthroughs, *Entropy*, 15(4), 1152–1170, 2013, https://doi.org/10.3390/e15041152

T. Matsoukas, Statistical Thermodynamics of Clustered Populations, *Physical Review E*, 90, 022113, 2014, https://doi.org/10.1103/PhysRevE.90.022113

J. L. McCauley, *Chaos, Dynamics, and Fractals: An Algorithmic Approach to Deterministic Chaos*, Cambridge University Press, 1993, https://doi.org/10.1017/CBO9780511564154

J. L. Spouge, Equilibrium Polymer Size Distributions, *Macromolecules*, 16(1), 121–127, 1983, https://doi.org/10.1021/ma00235a024

W. H. Stockmayer, Theory of Molecular Size Distribution and Gel Formation in Branched-Chain Polymers, *The Journal of Chemical Physics*, 11(2), 45–55, 1943, https://doi.org/10.1063/1.1723803

P. G. J. van Dongen & M. H. Ernst, Size Distribution in the Polymerisation Model A_fRB_g, *Journal of Physics A: Mathematical and General*, 17(11), 2281–2297, 1984, https://doi.org/10.1088/0305-4470/17/11/023

G. P. Williams, *Chaos Theory Tamed*, CRC Press, 1997, https://doi.org/10.1201/9781482295412

19

Maximum Entropy Theory for Continuous Particle Systems

This chapter concerns maximum entropy theory for continuous particle systems experiencing aggregation and fragmentation. While maximum entropy theory for discrete particle systems draws on statistical thermodynamics, as discussed in the last chapter, maximum entropy theory for continuous particle systems draws on Shannon's information theory; see, e.g., Cover & Thomas (2006) and Golan (2018). The first relevant research was done by Grady & Kipp (1985) and Sellens & Brzustowski (1985, 1986). Grady & Kipp (1985) focused on fragment size distributions produced by small-scale explosions, e.g., "explosive-filled cylinders." Within a few years, Englman et al. (1987, 1988) and Englman (1991) extended this approach to applications as diverse as "shattered crystal pieces, droplets in sprays, atomic or molecular clusters, space debris and fragmented nuclei." By contrast, Sellens & Brzustowski (1985, 1986) focused on joint size-velocity distributions in liquid sprays. Shortly thereafter, Li & Tankin (1987) adapted this approach to pure size distributions. In this chapter, size entropy will be defined as follows:

$$S = -\text{const.} \int_0^\infty \ln\left[f_Y(X)\right] f_Y(X)\, dX + \text{const.} \tag{19.1}$$

where X and Y are particle size metrics such as mass, $f_Y(X)$ is a Y-weighted probability density function as discussed in Section 3.3, and the notation "const." refers to unnamed constants as discussed in Section 1.3. Then maximizing S subject to constraints leads to an expression for $f_Y(X)$. To understand such results, it is first necessary to convert them to a common form. For example, Example 19.1 shows how to convert expressions for $f_Y(X)$ into expressions for $f(M)$.

In principle, maximum entropy theory can be tailored to specific applications through the choice of X and Y. However, because X and Y tend to be common particle size metrics such as mass, there is nothing very application-specific about them. Similarly, in principle, maximum entropy theory can be tailored to specific applications through the choice of constraints. However, because the constraints tend to involve common average sizes—or, less commonly, general physical laws such as conservation of mass, momentum and energy—there is nothing very application-specific about them. Thus, for continuous systems, maximum entropy theory tends to produce generic results. In particular, the results apply equally to aggregation and fragmentation. This makes maximum entropy theory for continuous systems, as discussed in this chapter, different than maximum entropy theory for discrete systems, as discussed in the last chapter. In particular, while maximum entropy theory for discrete systems can capture dynamic aspects specific to aggregation, such as collision kernels, maximum entropy theory for continuous systems cannot.

Particle Dynamics with Aggregation and Fragmentation. Culbert B. Laney, Oxford University Press. © Culbert B. Laney (2025).
DOI: 10.1093/9780198920687.003.0019

Example 19.1 Changes of variables in size distributions

Consider $F_Y(M)$ and $F_Y(X)$ where X and Y are particle size metrics such as mass M or diameter D. Notice that:

$$F_Y(M) = F_Y(X). \tag{19.2}$$

which is true by the definition of F_Y. Differentiating both sides of Equation (19.2) yields:

$$f_Y(M) = \left|\frac{dX}{dM}\right| f_Y(X). \tag{19.3}$$

If X and Y are geometric size metrics then by definition:

$$X = \text{const.}\, M^{l/m} \text{ and } Y = \text{const.}\, M^{k/m} \tag{19.4}$$

for some l and k. According to Equation (3.8), $f_Y(M) = Yf(M) / \int_0^\infty Y(x)f(x)\,dx$ so that:

$$f(M) = \text{const.}\, M^{\frac{l-k}{m}-1} f_Y(X) = \text{const.}\, M^{\frac{l-k}{m}-1} f_Y\left(M^{l/m}\right) \tag{19.5}$$

where the multiplicative constant is chosen to ensure $\int_0^\infty f(M)\,dM = 1$.

19.1 Families of Size Distributions

Instead of specific size metrics such as mass or diameter, suppose that X or Y or both are chosen to be *families* of size metrics. For example, suppose that:

$$X = \text{const.}\, M^{n/m}$$

where M is the particle mass and n/m is any number. If X or Y are families of size metrics, then maximum entropy theory produces families of size distributions. For example, Example 19.2 and 19.3 use maximum entropy theory to derive general expressions for Weibull and gamma size distributions, respectively. Building on Examples 19.2 and 19.3, Example 19.4 uses maximum entropy theory to derive general expressions for Rosin–Rammler size distributions, which include but are not limited to Weibull and gamma size distributions.

According to Domouchel (2006), Lienhard & Meyer (1967) were the first to use maximum entropy theory to derive Rosin–Rammler distributions, a.k.a. generalized gamma distributions. Instead of aggregation or fragmentation, Lienhard & Meyer (1967) used maximum entropy theory to model the "occurrence of a specified event, such as the failure of a component or system." Thus, instead of particle sizes, X is the event time and Y is proportional to "the number of ways in which N distinguishable occurrences of the event can take place," similar to Y_i in the last chapter. In later work, Dumouchel (2006) suggested an alternative derivation of Rosin–Rammler size distributions, a.k.a. "a three-parameter Generalized Gamma Distribution." In particular, Domouchel (2006) used maximum entropy theory to model "the drop-size distribution of liquid sprays [as] ... a consequence of the action of both destabilizing and stabilizing forces such as aerodynamic and surface tension forces, respectively."

As it turns out, maximum entropy theory can be used to derive an almost unlimited variety of continuous distributions. This includes most if not all of the standard distributions seen in probability and statistics; see, e.g., table 17.1 in Cover & Thomas (2006) and table 19.1 in Park & Bera (2009). In addition, this includes many non-standard distributions; see, e.g., Example 19.5. However, computational, experimental, and mean-field theory evidence indicate that most of the resulting non-standard size distributions rarely if ever occur in aggregation, fragmentation, or any other application for that matter, except perhaps in some approximate sense.

Example 19.2 Derivation of Weibull size distributions using maximum entropy theory

Suppose $X = \text{const.}\, M^{n/m}$ and $Y = \text{const.}$ Then Equation (19.1) may be written as follows:

$$S = -\text{const.} \int_0^\infty \ln\,[f(X)]\, f(X)\, dX + \text{const.} \tag{19.6}$$

Possible constraints on $f(X)$ are as follows:

$$\int_0^\infty f(X)\, dX = 1 \tag{19.7}$$

$$\int_0^\infty X f(X)\, dX = X_{\text{avg}} \tag{19.8}$$

$$\int_0^\infty \frac{f(X)}{X}\, dX = \frac{1}{X'_{\text{avg}}}. \tag{19.9}$$

If $n > 0$ and Equation (19.6) is maximized subject to Equations (19.7) and (19.8) then:

$$f(X) = \frac{1}{X_{\text{avg}}} \exp\left(-\frac{X}{X_{\text{avg}}}\right) \tag{19.10}$$

which can be proven using Lagrange multipliers. According to Equation (19.5) with $l = n$ and $k = 0$, Equation (19.10) can be rewritten as follows:

$$f(M) = \frac{\text{const.}}{M_{\text{avg}}} \left(\frac{M}{M_{\text{avg}}}\right)^{\frac{n}{m}-1} \exp\left[-\text{const.}\left(\frac{M}{M_{\text{avg}}}\right)^{\frac{n}{m}}\right] \tag{19.11}$$

where the constants are chosen so that $\int_0^\infty f(M)\, dM = 1$ and $\int_0^\infty M f(M)\, dM = M_{\text{avg}}$. Notice that this is a Weibull size distribution with $n > 0$.

If $n < 0$ and Equation (19.6) is maximized subject to Equations (19.7) and (19.9) then:

$$f(X) = \frac{1}{X'_{avg}} \exp\left(-\frac{X}{X'_{avg}}\right). \tag{19.12}$$

Using Equation (19.5) as before, Equation (19.12) can be rewritten as follows:

$$f(M) = \frac{\text{const.}}{M'_{avg}} \left(\frac{M}{M'_{avg}}\right)^{\frac{n}{m}-1} \exp\left[-\text{const.}\left(\frac{M}{M'_{avg}}\right)^{\frac{n}{m}}\right] \tag{19.13}$$

where the constants are chosen so that $\int_0^\infty f(M)\,dM = 1$ and $\int_0^\infty (1/M) f(M)\,dM = 1/M'_{avg}$. Notice that this is a Weibull size distribution with $n < 0$. Table 19.1 gives some specific examples of positive and negative Weibull size distributions, as given by Equations (19.11) and (19.13), respectively, including conversions between $f(D)$ and $f(M)$.

Table 19.1 Specific examples of Weibull size distributions which can be derived using maximum entropy theory for continuous particle systems. See also Appendix A.6.

X	f(M)	f(D)
D	$\frac{b}{mM_{avg}}\left(\frac{M}{M_{avg}}\right)^{\frac{1}{m}-1} \exp\left[-b\left(\frac{M}{M_{avg}}\right)^{\frac{1}{m}}\right]$, $b = \Gamma(1+m)^{\frac{1}{m}}$	$\frac{1}{D_{avg}} \exp\left(-\frac{D}{D_{avg}}\right)$
M	$\frac{1}{M_{avg}} \exp\left(-\frac{M}{M_{avg}}\right)$	$\frac{bm}{D_{avg}}\left(\frac{D}{D_{avg}}\right)^{m-1} \exp\left[-b\left(\frac{D}{D_{avg}}\right)^{m}\right]$, $b = \Gamma\left(1+\frac{1}{m}\right)^{m}$
D^{-1}(*)	$\frac{b}{mM'_{avg}}\left(\frac{M}{M'_{avg}}\right)^{-\frac{1}{m}-1} \exp\left[-b\left(\frac{M}{M'_{avg}}\right)^{-\frac{1}{m}}\right]$, $b = \Gamma(1+m)\frac{1}{m}$	$\frac{1}{D'_{avg}}\left(\frac{D}{D'_{avg}}\right)^{-2} \exp\left[-\left(\frac{D}{D'_{avg}}\right)^{-1}\right]$
M^{-1}(**)	$\frac{1}{M'_{avg}}\left(\frac{M}{M'_{avg}}\right)^{-2} \exp\left[-\left(\frac{M}{M'_{avg}}\right)^{-1}\right]$	$\frac{bm}{D'_{avg}}\left(\frac{D}{D'_{avg}}\right)^{-m-1} \exp\left[-b\left(\frac{D}{D'_{avg}}\right)^{-m}\right]$, $b = \Gamma\left(1+\frac{1}{m}\right)^{m}$

(*) In the third row, the original exponential form is as follows:

$$f(D^{-1}) = \frac{1}{\{D^{-1}\}'_{avg}} \exp\left(-\frac{D^{-1}}{\{D^{-1}\}'_{avg}}\right).$$

(**) In the fourth row, the original exponential form is as follows:

$$f(M^{-1}) = \frac{1}{\{M^{-1}\}'_{avg}} \exp\left(-\frac{M^{-1}}{\{M^{-1}\}'_{avg}}\right).$$

Example 19.3 Derivation of gamma size distributions using maximum entropy theory

Suppose that the size entropy is defined as follow:

$$S = -\text{const.} \int_0^\infty \ln [f_Y(M)] f_Y(M)\, dM + \text{const.} \tag{19.14}$$

where $Y = \text{const.}\ M^{-n/m+1}$ with $n > 0$. Possible constraints on $f_Y(M)$ are as follows:

$$\int_0^\infty f_Y(M)\, dM = 1 \tag{19.15}$$

$$\int_0^\infty M f_Y(M)\, dM = M_{Y\,\text{avg}}. \tag{19.16}$$

If Equation (19.14) is maximized subject to Equations (19.15) and (19.16), then:

$$f_Y(M) = \frac{1}{M_{Y\text{avg}}} \exp\left(-\frac{M}{M_{Y\text{avg}}}\right) \tag{19.17}$$

which can be proven using Lagrange multipliers. According to Equation (19.5) with $l/m = 1$ and $k/m = -n/m + 1$, Equation (19.17) can be rewritten as follows:

$$f(M) = \frac{\text{const.}}{M_{\text{avg}}} \left(\frac{M}{M_{\text{avg}}}\right)^{n/m-1} \exp\left(-\text{const.}\ \frac{M}{M_{\text{avg}}}\right) \tag{19.18}$$

where the constants are chosen so that $\int_0^\infty f(M)\, dM = 1$ and $\int_0^\infty M f(M)\, dM = M_{\text{avg}}$. Notice that this is a gamma size distribution. Alternatively, suppose that:

$$S = -\text{const.} \int_0^\infty \ln \left[f_Y\left(M^{-1}\right)\right] f_Y\left(M^{-1}\right) dM^{-1} + \text{const.} \tag{19.19}$$

where $Y = \text{const.}\ M^{-n/m-1}$ with $n < 0$. Possible constraints on $f_Y(M^{-1})$ are as follows:

$$\int_0^\infty f_Y\left(M^{-1}\right) dM^{-1} = 1 \tag{19.20}$$

$$\int_0^\infty M f_Y\left(M^{-1}\right) dM^{-1} = \left\{M^{-1}\right\}'_{Y\text{avg}}. \tag{19.21}$$

While the details are omitted, this yields *inverse gamma size distributions*, which are also known as *Pearson type V size distributions*. Inverse gamma size distributions are discussed in Appendix A.7.

Example 19.4 Derivation of Rosin–Rammler size distributions using maximum entropy theory

Combining the last two examples, suppose the size entropy is defined as follows:

$$S = -\,\text{const.}\int_0^\infty \ln\left[f_Y(X)\right]f_Y(X)\,dX + \text{const.} \tag{19.22}$$

where $X = M^{l/m}$ and $Y = \text{const.}\,M^{(l-n)/m}$. Possible constraints on $f_Y(X)$ are as follows:

$$\int_0^\infty f_Y(X)\,dX = 1 \tag{19.23}$$

$$\int_0^\infty Xf_Y(X)\,dX = X_{Y\,\text{avg}} \tag{19.24}$$

$$\int_0^\infty f_Y(X)/X\,dX = 1/X'_{Y\,\text{avg}}. \tag{19.25}$$

If l and n are positive and Equation (19.22) is maximized subject to Equations (19.23) and (19.24) then:

$$f_Y(X) = \frac{1}{X_{Y\,\text{avg}}}\exp\left(-\frac{X}{X_{Y\,\text{avg}}}\right) \tag{19.26}$$

which can be proven using Lagrange multipliers. According to Equation (19.5) with $k/m = (l-n)/m$, Equation (19.26) can be rewritten as follows:

$$f(M) = \frac{\text{const.}}{M_{\text{avg}}}\left(\frac{M}{M_{\text{avg}}}\right)^{\frac{n}{m}-1}\exp\left[-\,\text{const.}\left(\frac{M}{M_{\text{avg}}}\right)^{\frac{l}{m}}\right]. \tag{19.27}$$

where the constants are chosen so that $\int_0^\infty f(M)\,dM = 1$ and $\int_0^\infty Mf(M)\,dM = M_{\text{avg}}$. Notice that this is a Rosin-Rammler size distribution. Similarly, if l and n are negative and Equation (19.22) is maximized subject to Equations (19.23) and (19.25) then:

$$f_Y(X) = \frac{1}{X'_{Y\,\text{avg}}}\exp\left(-\frac{X}{X'_{Y\,\text{avg}}}\right). \tag{19.28}$$

As before, this exponential size distribution can be rewritten as follows:

$$f(M) = \frac{\text{const.}}{M'_{\text{avg}}}\left(\frac{M}{M'_{\text{avg}}}\right)^{\frac{n}{m}-1}\exp\left[-\,\text{const.}\left(\frac{M}{M'_{\text{avg}}}\right)^{\frac{l}{m}}\right] \tag{19.29}$$

where the constants are chosen so that $\int_0^\infty f(M)\,dM = 1$ and $\int_0^\infty (1/M)f(M)\,dM = 1/M'_{\text{avg}}$.

Example 19.5 Derivation of other size distributions using maximum entropy theory

As one possible definition of entropy, suppose that:

$$S = -\text{const.} \int_0^\infty \ln[f(D)] f(D)\, dD + \text{const.} \tag{19.30}$$

which is the same as Equation (19.6) with $X = D$. Consider the following constraints:

$$\int_0^\infty f(D)\, dD = 1 \tag{19.31}$$

$$\int_0^\infty \frac{f(D)}{D} = \frac{1}{D'_{\text{avg}}} \tag{19.32}$$

$$\int_0^\infty D^2 f(D)\, dD = \left(D^2\right)_{\text{avg}} \tag{19.33}$$

$$\int_0^\infty D^3 f(D)\, dD = \left(D^3\right)_{\text{avg}}. \tag{19.34}$$

These constraints assume that $m = 3$. According Ahmadi & Sellens (1993), if Equation (19.30) is maximized subject to Equations (19.31), (19.32), (19.33), and (19.34) then:

$$f(D) = \text{const.} \exp\left(-\text{const.}\, D^{-1} - \text{const.}\, D^2 - \text{const.}\, D^3\right). \tag{19.35}$$

As usual, the four free constants should be chosen so that Equations (19.31), (19.32), (19.33), and (19.34) apply. For small fragments, Equation (19.35) becomes:

$$f(D) \approx \text{const.} \exp\left(-\text{const.}\, D^{-1}\right). \tag{19.36}$$

For middle-sized fragments, Equation (19.35) becomes:

$$f(D) \approx \text{const.} \exp\left(-\text{const.}\, D^2\right). \tag{19.37}$$

where D^2 is proportional to surface area $\mathscr{A}$. For large fragments, Equation (19.35) becomes:

$$f(D) \approx \text{const.} \exp\left(-\text{const.}\, D^3\right). \tag{19.38}$$

where D^3 is proportional to mass M. This example is related to Example 18.4 from the last chapter. As one difference, this example does not involve anything similar to Y_i.

19.2 Specific Size Distributions

For discrete particle systems, developing specific size distributions f_i requires first developing specific expressions for Y_i, which may be extremely difficult. For example, after a long and complex proof based on combinatorics, Stockmayer (1943) obtained the following result:

$$Y_i = \frac{k^i((k-1)i)!}{i!((k-2)i+2)!}$$

where i is proportional to the particle mass. As described by Kovac & Fixman (2017), to develop this expression, Stockmayer (1943) relied on assistance from some of the leading scientists of his day, especially Maria Goeppert Mayer and Paul Flory, both of whom later received Nobel Prizes.

By comparison, for continuous particle systems, developing specific weighting factors Y is straightforward. In most cases, both X and Y are simple size metrics such as particle mass or diameter. As one possibility, suppose that:

$$X = \text{const.}\ M^{l/m} \text{ and } Y = \text{const.}$$

where M is the particle mass and l/m is some rational number. For regular three-dimensional particles, i.e., for $m = 3$, the only obvious physical choices for l/m are as follows:

$l/m = 0$	X is a constant
$l/m = \pm 1/3$	X is proportional to the diameter D or inverse diameter D^{-1}
$l/m = \pm 2/3$	X is proportional to the surface area $\mathscr{A}$ or inverse surface area $\mathscr{A}^{-1}$
$l/m = \pm 1$	X is proportional to the mass M or inverse mass M^{-1}

Then, using the approach given in Example 19.2, maximum entropy theory obtains seven Weibull size distributions where $n/m = l/m$ is 0, ±1/3, ±2/3, or ±1. In other words:

$$n/m \text{ is an integer multiple of } 1/3 \text{ between } -1 \text{ and } 1. \tag{19.39}$$

As another possibility, suppose that $X = M$ and $Y = \text{const.}\ M^{k/m}$. Like before, the only obvious physical choices for k/m are 0, ±1/3, ±2/3, or ±1. Then, using the approach given in Example 19.3, maximum entropy theory obtains seven gamma size distributions where $n/m = -k/m+1$ is 0, 1/3, 2/3, 1, 4/3, 5/3, or 2. In other words:

$$n/m \text{ is an integer multiple of } 1/3 \text{ between } 0 \text{ and } 2. \tag{19.40}$$

As a third possibility, suppose that $X = M^{-1}$ and $Y = \text{const.}\ M^{k/m}$. Like before, the only obvious physical choices for k/m are 0, ±1/3, ±2/3, or ±1. Then, using the approach given in Example 19.3, maximum entropy theory obtains seven negative gamma size distributions where $n/m = -k/m - 1$ is 0, −1/3, −2/3, −1, −4/3, −5/3, or −2. In other words:

$$n/m \text{ is an integer multiple of } 1/3 \text{ between } -2 \text{ and } 0. \tag{19.41}$$

For whatever reason, maximum entropy theory has apparently not been previously used to obtain the specific gamma or inverse gamma size distributions given by Equations (19.40) and (19.41). However, maximum entropy theory has been used to obtain some of the specific Weibull size distributions given by Equation (19.39). For example, Grady & Kipp (1985) and Li & Tankin (1987, 1988) used maximum entropy theory to derive Weibull size distributions with $n/m = 1$. For another example, Liu et al. (1995) and Liu (1995) used maximum entropy theory to derive Weibull size distributions with $n/m = 1/3$ and $2/3$; see also Liu & Hallett (1997, 1998) and Liu et al. (2002). After comparing the latter predictions to observations of droplet sizes in clouds, Khvorostyanov & Curry (2008) found that the "exponential tail is larger than the often observed value" and thus "these solutions have only been rarely applied." Equation (19.39) partially addresses such concerns by showing that maximum entropy theory can obtain Weibull size distributions with seven rather than only two values for n/m. With so many different distributions to choose from, the chances of obtaining a good fit to a given set of observations improve significantly. For more information on maximum entropy theory in cloud physics applications see, e.g., Yano et al. (2016, 2018), Yano (2019) and Wu & McFarquhar (2018, 2019).

19.3 Velocity Distributions

While this chapter has so far focused on particle size distributions, maximum entropy theory can also be applied to particle velocity distributions and joint particle size-velocity distributions. In the best known approach, Sellens & Brzustowski (1985, 1986) used maximum entropy theory to obtain joint particle size-velocity distributions $f(D, V)$. In this approach, conservation of mass, momentum, and energy were used as constraints. Many other researchers have since developed variants of this approach; see, e.g., Li & Tankin (1988), Sellens (1989), Li et al. (1991), Chin et al. (1991, 1995), Kim et al. (2003), Dumouchel (2009), Nath et al. (2011) and Dasgupta et al. (2018, 2019). Most of these variants assume that each constraint leads to a free parameter, which can be chosen arbitrarily to match experimental or computational data. For example, maximizing the entropy for $f(D, V)$ subject to six constraints—namely, conservation of mass, conservation of x-, y-, z-momentum, conservation of kinetic energy, and conservation of surface energy—leads to an expression with six free parameters. With so many free parameters, the resulting form can be well-fitted to almost any set of computational or experimental data; see Section 5.6. However, this book takes the stance that the free parameters should be used to ensure that the resulting size distribution obtains the assumed constraints. According to this stance, each constraint makes it harder rather than easier to match computational or experimental data.

Paltridge (2009) suggested that maximum entropy theory will "remain entirely in the realm of academic discussion until someone ... identifies a complex system whose behavior cannot be predicted by any [other] means." As discussed earlier in Chapter 7, Weibull distributions often offer good fits to experimental data for fragment speed. Maximum entropy theory offers a possible analytic proof of this empirical observation. As it turns out, Weibull speed distributions can be derived using maximum entropy theory, much like Weibull size distributions can be derived using maximum entropy theory, as shown in Example

19.2. In fact, there is currently no other known way of deriving Weibull speed distributions for aggregating and fragmenting particle systems. Thus, velocity distributions and joint size-velocity distributions in aggregating and fragmenting particle systems seems to fit Paltridge's criteria of "a complex system whose behavior cannot be predicted by any [other] means."

19.4 Validation

As a possible experimental validation, consider the liquid spray results discussed in Section 5.10. According to Figure 5.7, liquid sprays obtain power law size distributions where n/m is an integer multiple of 1/6 between −2 and 0. More specifically, n/m is −1/6 (4.5%), −1/3 (8.5%), −1/2 (13.5%), −2/3 (16%), −5/6 (18.5%), −1 (24.5%), −7/6 (11%), −4/3 (2%), −3/2 (1%), or −11/6 (0.5%). As one possibility, maximum entropy theory predicts Weibull size distributions where n/m is an even multiple of 1/3 between −1 and 1. As seen in Figure 3.3, negative Weibull size distributions are approximately equal to power law size distributions. Thus, maximum entropy theory approximately agrees with the experimental results in three cases, namely, $n/m = -1/3$ (8.5%), −2/3 (16%), and −1 (24.5%). As this comparison shows, an empirical model derived from a well-populated database of experimental results generally offers more useful results than an analytic model based on maximum entropy theory. An appropriately selected experimental database provides most if not all of the likely size distributions, customized for that specific application, along with an estimate for how often each of those size distributions occurs. By contrast, maximum entropy theory provides a generic set of size distributions, with no sense how often they could occur. In some cases, maximum entropy theory predicts sizes distributions that never occur in a given application. In other cases, maximum entropy theory omits size distributions that do occur in a given application.

As an analytic validation, consider the mean-field theory results for late-time non-local aggregation given in Section 11.3. With free-field boundary conditions, mean-field theory predicts gamma size distributions where n/m is an even multiple of 1/6 between −2 and 0; see Equation (11.7). In this case, as one possibility, maximum entropy theory predicts gamma size distributions where n/m is an even multiple of 1/3 between 0 and 2; see Equation (19.40). Then maximum entropy theory consistently disagrees with mean-field theory. On the other hand, with periodic boundary conditions, mean-field theory predicts Weibull size distributions where n/m is an even multiple of 1/6 between −2 and 2; see Equation (11.9). In this case, as one possibility, maximum entropy theory predicts Weibull size distributions where n/m is an even multiple of 1/3 between −1 and 1. Then maximum entropy theory consistently agrees with mean-field theory or, more precisely, maximum entropy theory is a subset of mean-field theory. Where maximum entropy theory disagrees with mean-field theory, it does not automatically mean that maximum entropy is theory is wrong and mean-field theory is right. However, mean-field theory is more likely to be right because it can account for important physical factors such as binary collisions (with and without locality), sources, and sinks that maximum entropy theory cannot account for. In particular, mean-field theory can naturally account for both one- and two-particle mechanisms; however, maximum entropy theory can only naturally account for one-particle mechanisms; see the earlier discussion in Section 13.3.

19.5 Conclusions

This chapter has discussed maximum entropy theory for continuous particle systems. Maximum entropy theory can be used to derive an almost unlimited variety of different distributions depending on the exact definitions of X and Y; see, e.g., table 17.1 in Cover & Thomas (2006) and Table 19.1 in Park & Bera (2009). However, if X and Y are *geometric* quantities, i.e., power laws in M, then maximum entropy theory obtains *only* Rosin–Rammler distributions including, as special cases, Weibull and gamma distributions. If the power law exponents for M are integer multiples of 1/3—which is typical of geometric quantities in three-dimensions—then so are the exponents in the resulting Rosin–Rammler distributions.

For discrete particle systems, maximum entropy theory captures non-stationary aspects of aggregation and fragmentation, such as rate kernels. As a practical matter, this severely limits the range of possible solutions. In fact, for aggregation in discrete particle systems, maximum entropy theory produces only three gamma size distributions where n/m is 1, −1/2, or −3/2. Notice that n/m is always an integer multiple of 1/2. While a few other size distributions have also been obtained, such as the Poisson size distributions, these are generally not observed in aggregating or fragmenting particle systems. By contrast, for continuous particle systems, maximum entropy theory captures only stationary aspects of aggregation and fragmentation, e.g., it does not account for rate kernels. In fact, for continuous particle systems, maximum entropy theory cannot even distinguish between aggregation and fragmentation. As a practical matter, this allows for a wide range of possible solutions, including both standard and non-standard size distributions, the latter of which are unlikely to ever occur. However, simple geometric considerations narrow the outcomes to Weibull and gamma size distributions where n/m equals 0, ±1/3, ±2/3, ±1, and so forth. Notice that n/m is always an integer multiple of 1/3.

As one way to look at things, maximum entropy theory simply suggests some typical universal size distributions, common to aggregation and fragmentation of all types. If the underlying physics is well-understood, other analytic approaches can offer better suggestions. However, if the basic physics is not well-understood, maximum entropy theory is a reasonable place to start. As it turns out, in a number of important applications, it can be difficult to tell if a given particle system evolves primarily through aggregation, primarily through fragmentation, or through some combination of the two. For example, in the cloud physics community, some researchers assume that aggregation determines the size distributions of raindrops, e.g., Pumir & Wilkinson (2016). However, other researchers assume that fragmentation determines the size distributions of raindrops, e.g., Villermaux & Bossa (2009). Maximum entropy theory does not distinguish between aggregation and fragmentation. Thus, a number of researchers have used maximum entropy theory to model the size distributions of raindrops, as discussed earlier in Section 19.2, albeit with mixed results.

While maximum entropy theory does not require any information about a given system, it also does not produce any information about that system. In other words, it cannot clarify the nature of the underlying physics. In this regard, Paltridge (2009) says that maximum entropy theory may provide "a good description of an actual result—for instance that a complex system seems automatically to select a particular ... state from a spectrum of potential states—but it does not of itself provide an obvious reason why such selection should take place." Because it does not require or provide any physical insights, maximum entropy theory by itself cannot provide any indication of whether it is or is not a valid modeling approach. For example, according to Example 11.15, raindrop formation may be driven by gelation. While maximum entropy theory for discrete systems as discussed in the last chapter can understand gelation, at least to some extent, maximum entropy theory for continuous systems as discussed in this chapter cannot. Thus, if rain formation is driven by gelation, maximum entropy theory produces results as usual, with no warning about the potential lack of physicality.

To overcome the limitations of maximum entropy theory, many researchers have suggested combining it with experimental results. While maximum entropy theory by itself is generic, the combination with experimental results may be application-specific. For example, Ahmadi & Sellens (1993) used the four free constants in Equation (19.35) to optimize agreement between maximum entropy theory and selected experimental results for liquid spray systems. However, because Equation (19.35) does not have a universal form, it may be difficult to choose constants that apply for a wide range of different conditions. In addition, when the free constants are used for experimental calibration, they can no longer be used to ensure that the assumed constraints apply, which undermines the whole basis of maximum entropy theory. In short, while maximum entropy theory for continuous particle systems provides an interesting theoretical perspective, it is generally of limited practical value, except possibly for velocity and joint velocity-size distributions where there is no known analytic alternative.

References

M. Ahmadi & R. W. Sellens, A Simplified Maximum-Entropy-Based Drop-Size Distribution, *Atomization and Sprays*, 3(3), 291–310, 1993, https://doi.org/10.1615/AtomizSpr.v3.i3.30

L. P. Chin, P. G. LaRose, R. S. Tankin, T. Jackson, J. Stutrud, & G. Switzer, Droplet Distributions from the Breakup of a Cylindrical Liquid Jet, *Physics of Fluids A: Fluid Dynamics*, 3(8), 1897–1906, (1991), https://doi.org/10.1063/1.857919

L. P. Chin, P. C. Hsing, R. S. Tankin, & T. Jackson, Comparisons between Experiments and Predictions Based on Maximum Entropy for the Breakup of a Cylindrical Liquid Jet, *Atomization and Sprays*, 5(6), 603–620, 1995, https://doi.org/10.1615/AtomizSpr.v5.i6.50

T. M. Cover & J. A. Thomas, *Elements of Information Theory*, 2nd edition, Wiley, 2006, https://doi.org/10.1002/047174882X

D. Dasgupta, S. Nath, & D. Bhanja, Dual-Mode Nonlinear Instability Analysis of a Confined Planar liquid Sheet Sandwiched between Two Gas Streams of Unequal Velocities and Prediction of Droplet Size and Velocity Distribution using Maximum Entropy Formulation, *Physics of Fluids*, 30, 044104, 2018, https://doi.org/10.1063/1.5022346

D. Dasgupta, S. Nath, & D. Bhanja, A Study on Dual Role of Viscosity on the Stability of a Viscous Planar Liquid Sheet Surrounded by Inviscid Gas Streams of Equal Velocities, and Prediction of Resulting Droplet Distribution using Maximum Entropy Formulation, *Physics of Fluids*, 31(7), 074103, 2019, https://doi.org/10.1063/1.5097692

C. Dumouchel, A New Formulation of the Maximum Entropy Formalism to Model Liquid Spray Drop-Size Distribution, *Particle & Particle Systems Characterization*, 23(6), 468–479, 2006, https://doi.org/10.1002/ppsc.200500989

C. Dumouchel, The Maximum Entropy Formalism and the Prediction of Liquid Spray Drop-Size Distribution, *Entropy*, 11(4), 713–747, 2009, https://doi.org/10.3390/e11040713

R. Englman, N. Rivier, & Z. Jaeger, Fragment-Size Distribution in Disintegration by Maximum-Entropy Formalism, *Philosophical Magazine Part B*, 56(6), 751–769, 1987, https://doi.org/10.1080/13642818708215309

R. Englman, N. Rivier, & Z. Jaeger, Size-Distribution in Sudden Breakage by the Use of Entropy Maximization, *Journal of Applied Physics*, 63(9), 4766–4768, 1988, https://doi.org/10.1063/1.340114

R. Englman, Fragments of Matter from a Maximum-Entropy Viewpoint, *Journal of Physics: Condensed Matter*, 3(9), 1019–1053, 1991, https://doi.org/10.1088/0953-8984/3/9/001

A. Golan, *Foundations of Info-Metrics: Modeling, Inference, and Imperfect Information*, Oxford University Press, 2018, https://10.1093/oso/9780199349524.001.0001

D. E. Grady & M. E. Kipp, Geometric Statistics and Dynamic Fragmentation, *Journal of Applied Physics*, 58(3), 1210–1222, 1985, https://doi.org/10.1063/1.336139

V. I. Khvorostyanov & J. A. Curry, Analytical Solutions to the Stochastic Kinetic Equation for Liquid and Ice Particle Size Spectra. Part I: Small-Size Fraction, *Journal of the Atmospheric Sciences*, 65(7), 2025–2043, 2008, https://doi.org/10.1175/2007JAS2484.1

W. T. Kim, S. K. Mitra, X. Li, L. A. Prociw, & T. C. J. Hu, A Predictive Model for the Initial Droplet Size and Velocity Distributions in Sprays and Comparison with Experiments, *Particle & Particle Systems Characterization*, 20, 135–149, 2003, https://doi.org/10.1002/ppsc.200390011

J. Kovac and M. Fixman, Walter Hugo Stockmayer, *Biographical Memoirs of the National Academy of Sciences,* The National Academies Press, 2017

X. Li & R. S. Tankin, Droplet Size Distribution: A Derivation of a Nukiyama–Tanasawa Type Distribution Function, *Combustion Science and Technology*, 56 (1–3), 65–76, 1987, https://doi.org/10.1080/00102208708947081

X. Li & R. S. Tankin, Derivation of Droplet Size Distribution in Sprays Using Information Theory, *Combustion Science and Technology*, 60(4–6), 345–357. 1988, https://doi.org/10.1080/00102208808923992

X. Li, L. P. Chin, R. S. Tankin, T. Jackson, J. Stutrud, & G. Switzer, Comparison Between Experiments and Predictions Based on Maximum Entropy Theory for Sprays from a Pressure Atomizer, *Combustion and Flame*, 86(1–2), 73–89, 1991, https://doi.org/10.1016/0010-2180(91)90057-I

J. H. Lienhard & P. L. Meyer, A Physical Basis for the Generalized Gamma Distribution, *Quarterly of Applied Mathematics*, 25(3), 330–334, 1967, https://doi.org/10.1090/qam/99884

Y. Liu, L. You, W. Yang, & F. Liu, On the Size Distribution of Cloud Droplets, *Atmospheric Research*, 35(2–4), 201–216, 1995, https://doi.org/10.1016/0169-8095(94)00019-A

Y. Liu, On the Generalized Theory of Atmospheric Particle Systems, *Advances in Atmospheric Sciences*, 12(4), 419–438, 1995, https://doi.org/10.1007/BF02657003

Y. Liu & J. Hallett, The 1/3 Power Law Between Effective Radius and Liquid-Water Content, *Quarterly Journal of the Royal Meteorological Society*, 123(542), 1789–1795, 1997, https://doi.org/10.1002/qj.49712354220

Y. Liu & J. Hallett, On Size Distributions of Cloud Droplets Growing by Condensation: A New Conceptual Model, *Journal of the Atmospheric Sciences*, 55(4), 527–536, 1998, https://doi.org/10.1175/1520-0469(1998)055<0527:OSDOCD>2.0.CO;2

Y. Liu, P. H. Daum, & J. Hallett, A Generalized Systems Theory for the Effect of Varying Fluctuations on Cloud Droplet Size Distributions, *Journal of the Atmospheric Sciences*, 59(14), 2279–2290, 2002, https://doi.org/10.1175/1520-0469(2002)059<2279:AGSTFT>2.0.CO;2

S. Nath, A. Datta, A. Mukhopadhyay, S. Sen, & T. J. Tharakan, Prediction of Size and Velocity Distributions in Sprays formed by the Breakup of Planar Liquid Sheets using Maximum Entropy Formalism, *Atomization and Sprays*, 21(6), 483–501 (2011), https://doi.org/10.1615/AtomizSpr.2011003709

G. W. Paltridge, A Story and a Recommendation about the Principle of Maximum Entropy Production, *Entropy*, 11(4), 945–948, 2009, https://doi.org/10.3390/e11040945

S. Y. Park & A. K. Bera, Maximum Entropy Autoregressive Conditional Heteroskedasticity Model, *Journal of Econometrics*, 150, 219–230, 2009, https://doi.org/10.1016/j.jeconom.2008.12.014

A. Pumir & M. Wilkinson, Collisional Aggregation Due to Turbulence, *Annual Review of Condensed Matter Physics*, 7, 141–170, 2016, https://doi.org/10.1146/annurev-conmatphys-031115-011538

R. W. Sellens & T. A. Brzustowski, A Prediction of the Drop-Size Distribution in a Spray from First Principles, *Atomisation and Spray Technology*, 1(2), 89–102, 1985

R. W. Sellens & T. A. Brzustowski, A Simplified Prediction of Droplet Velocity Distribution in a Spray, *Combustion and Flame*, 65(3), 273–279, 1986, https://doi.org/10.1016/0010-2180(86)90041-6

R. W. Sellens, Prediction of the Drop Size and Velocity Distribution in a Spray Based on the Maximum Entropy Formalism, *Particle & Particle Systems Characterization*, 6(1–4), 17–27, 1989, https://doi.org/10.1002/ppsc.19890060103

W. H. Stockmayer, Theory of Molecular Size Distribution and Gel Formation in Branched-Chain Polymers, *The Journal of Chemical Physics*, 11(2), 45–55, 1943, https://doi.org/10.1063/1.1723803

E. Villermaux & B. Bossa, Single-Drop Fragmentation Determines Size Distribution of Raindrops, *Nature Physics*, 5, 697–702, 2009, https://doi.org/10.1038/nphys1340

W. Wu & G. M. McFarquhar, Statistical Theory on the Analytical Form of Cloud Particle Size Distributions, *Journal of Atmospheric Sciences*, 75(8), 2801–2814, 2018, https://doi.org/10.1175/JAS-D-17-0164.1

W. Wu & G. M. McFarquhar, Reply to "What is the Maximum Entropy Principle? Comments on 'Statistical Theory on the Function Form of Cloud Particle Size Distributions'", *Journal of the Atmospheric Sciences*, 76(12), 3961–3963, 2019, https://doi.org/10.1175/JAS-D-18-0374.1

J.-I. Yano, A. J. Heymsfield & V. T. J. Phillips, Size Distributions of Hydrometeors: Analysis with the Maximum Entropy Principle, *Journal of the Atmospheric Sciences*, 73(1), 95–108, 2016, https://doi.org/10.1175/JAS-D-15-0097.1

J.-I. Yano, A. J. Heymsfield, & A. Bansemer, Determination of the Ice Particle Size Distributions Using Observations as the Integrated Constraints, *Journal of the Atmospheric Sciences*, 75(3), 787–804, 2018, https://doi.org/10.1175/JAS-D-17-0145.1

J.-I. Yano, What Is the Maximum Entropy Principle? Comments on 'Statistical Theory on the Functional Form of Cloud Particle Size Distributions', *Journal of the Atmospheric Sciences*, 76(12), 3955–3960, 2019, https://doi.org/10.1175/JAS-D-18-0223.1

20
Conclusions

This book concerns dynamic particle systems that experience aggregation and fragmentation. The first part of the book, Chapters 1 to 8, surveyed a wide variety of different particle systems with a wide range of different particle sizes, particle shapes, particle velocities, particle energies, particle compositions, medium compositions, particle-particle forces, and particle-medium forces. As a result of these wide ranges, the particle systems evolved through a wide variety of different physical mechanisms. These included instantaneous events such as binary collisions as well as gradual processes such as evaporation and condensation. These also included dissipation which causes many systems—such as granular gases, turbulent fluids, and groups of living organisms—to self-organize. Despite their many differences, the particle systems described in the first part of the book had at least a few things in common. For example, they all involved randomness or seeming-randomness, of the sort that can be described by traditional probability theory. For another example, the particles were all composed of the same substance, i.e., mixed systems were not considered. With only a few things in common, the first part of the book showed that different particle systems may obtain surprisingly similar results.

The second part of the book, Chapters 9 to 19, concentrated on size distributions in five key problems:

1. *Aggregation cascades* in which colliding or proximate pairs of particles merge to form a single particle, over and over again, in a cascade fashion.
2. *Binary fragmentation cascades* in which randomly selected particles divide into exactly two randomly sized pieces, over and over again, in a cascade fashion.
3. *Random fragmentation cascades* in which randomly selected particles divide into two or more randomly sized pieces, over and over again, in a cascade fashion.
4. *Collisional fragmentation cascades* in which colliding pairs of particles divide into two or more randomly sized pieces, over and over again, in a cascade fashion.
5. *Unitary fragmentation* in which a single object divides into randomly sized pieces, either all at once or gradually over time (i.e., an instantaneous event or an extended process).

As this list shows, while there is only one major type of aggregation cascade, there are multiple types of fragmentation cascades. In theory, aggregation cascades could involve collisions between exactly three, exactly four, or some random number of particles. However, in practice, when multiple particles converge, two will meet slightly before the rest. Thus, what appears to be simultaneous collisions between three or more particles is actually a series of binary collisions occurring in rapid succession. While this book considered three types of fragmentation cascades, there are actually many other

Particle Dynamics with Aggregation and Fragmentation. Culbert B. Laney, Oxford University Press. © Culbert B. Laney (2025).
DOI: 10.1093/9780198920687.003.0020

possibilities. As one such possibility, as a variant of random fragmentation, randomly selected particles may break up into children with fixed deterministic sizes. For example, this can occur in polymer degradation, where end-chain scission produces children composed of exactly one, two, three, etc. monomers. As another such possibility, particle systems may experience various combinations of aggregation, binary fragmentation, random fragmentation, and collisional fragmentation. This book focused on the five problems listed previously, on the belief that these adequately illustrate the core principles of aggregation and fragmentation, and that other possibilities are just variations on a theme.

The three types of fragmentation cascades listed previously consist of a series of unitary fragmentation events. Thus, at first glance, it would make sense to discuss unitary fragmentation before binary, random and collisional fragmentation. However, this book instead discussed binary, random and collisional fragmentation before unitary fragmentation. As it turns out, a great deal can be learned about fragmentation cascades by using simple placeholder models for unitary fragmentation, e.g., power law size distributions. In fact, based on such simple placeholders, the book discovered that fragmentation cascades do not necessarily depend on the underlying unitary fragmentation events. This is true whenever sinks remove particles with small masses—and, by implication, with small momenta and energies—in a sufficiently vigorous way. While many mechanisms may create such sinks, the most common explanation is that particles can no longer participate in fragmentation once they become too small. Even when they remain physically present in the particle system, such small particles do not engage in fragmentation cascades or substantially influence their outcome. At most, in a process similar to Brownian motion, the smallest particles participate in elastic binary collisions that slightly alter the momentum and energy of the largest particles, which makes them a beneficial source of randomness. While unitary fragmentation is important for understanding fragmentation cascades, the opposite is also true—unitary fragmentation can be modeled as a random fragmentation cascade with a sink of small particles. Like random fragmentation in dense systems, it is difficult to directly observe unitary fragmentation. In both cases, the outermost material blocks the view of the interior. However, it seems likely that, while they may appear to be instantaneous events, many unitary fragmentation events are actually extended processes that involve binary fragmentation, random fragmentation, collisional fragmentation, and perhaps other types of cascades starting at the earliest times.

For all five key problems listed previously, the book generally assumed that particle systems are spatially homogeneous and isotropic. If the particle system has a low volumetric fraction, this is true in a rigorous sense, as described in Chapter 2. If the particle system has a high volumetric fraction—in other words, if it is dense enough to randomly trap and release mass, momentum, and energy via "avalanches" or "catastrophes"—this is true only in a loose average sense. Most analytic methods, and many computational methods, implicitly assume spatial homogeneity and isotropy, at least for limited times and spatial regions. For example, in mean-field theory, these assumptions lead directly to integro-differential governing equations for the particle size distribution as a function of time. In many cases, these integro-differential equations admit stationary or, less commonly, non-stationary closed-form solutions. This is especially true at late times, where normalizations with characteristic particle sizes simplify the results and expose their universal aspects.

While this book focused on five key problems, this material has broad implications for dynamic particle systems of all kinds. For example, consider self-organized groups of living organisms and life-like self-propelled colloids. While they do not evolve through binary collisions, these systems can still often be well-described by mean-field theory for combined aggregation and fragmentation. As a result, they may obtain exactly the same late-time mean-field solutions as normal particle systems; see, e.g., Table 5.6. For another example, once the main aggregation and fragmentation processes and events have ended, many particle systems experience one or more rounds of *secondary aggregation* or *secondary fragmentation* due, e.g., to residual stresses in the particles, interactions between the particles in the system, interactions between the particles and their medium, or interactions between the particles and adjacent solid boundaries. While they involve different mechanisms, secondary aggregation and secondary fragmentation may obtain results similar to primary aggregation and primary fragmentation. In fact, secondary aggregation and fragmentation may overwrite the results of the original primary aggregation and fragmentation processes and events, possibly converting ideal cases to non-ideal cases, or vice versa.

To validate analytic and computational models for aggregation and fragmentation, this book reviewed a wide variety of experimental results. For unitary fragmentation, such experiments can be as simple as dropping a brittle object onto a hard floor. Thus, many such results are readily available in the research literature. In fact, some of the most famous expressions surveyed in this book were originally discovered by curve fitting experimental results for fragmentation. This includes, for example, power law and Weibull size distributions. While such results are still sometimes considered to be empirical, there is now an overwhelming body of analytic theory to support them. In fact, on the whole, it is probably fair to say that analytic models, computational models, and experimental results for particle aggregation and fragmentation generally agree with one another. This is true, despite the fact that many experimental results for aggregation involve some degree of fragmentation, which is not always fully accounted for by analytic and computational models. This is because, as it turns out, combinations of aggregation and fragmentation tend to obtain the same types of size distributions as aggregation or fragmentation alone.

This book mentioned fluid dynamics a number of times. Fluid dynamics is important to aggregation and fragmentation for several reasons. First, the particles themselves may be liquid. If so, this affects how the particles experience aggregation and fragmentation. For example, liquid particles often obtain compact spherical shapes after merging, while solid particles often obtain non-compact fractal shapes. For another example, liquid particles may fragment due to boundary layers, instabilities, and turbulence induced by drag forces, while solid particles do not. Second, even if the particles themselves are solid, they may be suspended in a fluid medium such as air or water. In such cases, for small enough particles, boundary layers, instabilities, and turbulence in the fluid medium may affect particle aggregation and fragmentation. In some cases, the fluid medium fundamentally determines the outcomes of aggregation and fragmentation, e.g., the particle size distributions may depend entirely on the turbulent kinetic energy spectrum; see, e.g., Yeoh et al. (2014) and Podgórska (2020).

Third and most importantly, dense particle systems often behave something like non-Newtonian liquids while non-dense particle systems often behave something like ideal or non-ideal gases. As a result, there are a number of analogies between aggregating particle systems, fragmenting particle systems, and fluid dynamics. For example, this book discussed analogies such as the following:

- There is an analogy between particle aggregation and reverse fluid turbulence cascades. In addition, there is an analogy between particle fragmentation and forward fluid turbulence cascades. For more information on these analogies, see Section 8.5.
- There is an analogy between universal energy distributions in fluid turbulence such as Kolmogorov's 5/3 law and universal size distributions in aggregating and fragmentating particle systems. For more information this analogy, see Sections 8.5 and 8.6.
- There is an analogy between ratios of specific heats in ideal gases and degrees of homogeneity in aggregating and fragmenting particle systems. In fact, all three cases involve Maxwell's concept of degrees of freedom; see Sections 9.3 and 13.3.
- There is an analogy between gelation in aggregation, shattering in fragmentation, and shocks in fluids. In all three cases, discontinuities in the solution form spontaneously after a finite time. For more information on this analogy, see Section 10.9.
- There is an analogy between gelation in aggregation and liquid phase changes such as freezing; see Section 10.9. In addition, there is an analogy between shattering in fragmentation and liquid phase changes such as evaporation; see Section 13.10.

These analogies potentially allow the particle aggregation and fragmentation community to take advantage of results from the fluid dynamics community. In fact, because of these analogies, simple cases of particle aggregation, particle fragmentation, and fluid flows may obtain the same scalar governing equations; see Example 11.4. While literally thousands of books have been written about fluid dynamics, only a few books have been written about particle aggregation and fragmentation. Instead, aggregation and fragmentation are usually studied as appendages to other topics. This book has tried to depict particle dynamics with aggregation and fragmentation as a worthy independent field of study, arguably the equal of fluid dynamics.

As a fundamental tenet, the models discussed in this book assume that aggregating and fragmenting particle systems are random or chaotic. For aggregating systems, the most common source of chaos is binary collisions, especially partially or fully elastic collisions, which cause the participating particles to rebound in seemingly random directions and at seemingly random speeds. In other words, it is not the successful "sticky" collisions that create chaos in aggregating particle systems; rather it is the failed "non-sticky" collisions. For fragmenting systems, common sources of chaos include elastic binary collisions, solid instabilities such as those inherent to crack propagation, liquid instabilities such as Rayleigh–Taylor and Kelvin–Helmholtz instabilities, and pre-existing material microstructure such as impurities, inclusions, grain boundaries, cracks, voids, bubbles, thermal fluctuations, thin spots, and so forth.

Table 20.1 compares the ideal size distributions obtained from the various analytic models discussed this book. As this table shows, while some analytic models predict that n/m is an integer multiple of 1/3, most predict that n/m is an integer multiple of 1/6, consistent with a vast body of experimental and computational results. In addition, as this table shows, different analytic models predict different ranges for n/m. Both the range and the discrete intervals within that range depend on the modeling assumptions, such as one- vs. two-particle mechanisms. In one-particle mechanisms, each particle evolves in isolation,

Table 20.1 Summary of ideal analytic results for the exponent n/m appearing in Weibull, gamma, and power law size distributions for aggregating and fragmenting particle systems. All results are for chaotic/random, spatially homogeneous, and isotropic particle systems. This table assumes that all size distributions are written as $f(M)$.

(a) Some analytic approaches produce a relatively small number of possible universal solutions.

Theory	Phenomenon	Type	Delta	Range	No.	Equation
Mean-Field Theory	Binary & Random Fragmentation *(one particle)*	Power Law	1/3	[−1, 0]	4	(13.27) (13.99)
Maximum Entropy	Aggregation	Gamma	1/3	[0, 2]	7	(19.40)
	Fragmentation *(any type)*	Weibull	1/3	[−1, 1]	7	(19.39)

(b) Some analytic approaches produce a relatively large number of possible universal solutions.

Theory	Phenomenon	Type	Delta	Range	No.	Equation
Mean-Field Theory	Aggregation *(non-local)*[1]	Gamma	1/6	[−2, 0]	13	(11.7)
		Weibull	1/6	[−2, 2]	25	(11.9)
Mean-Field Theory	Aggregation *(local)*[1]	Gamma	1/6	[−2, 0]	13	(11.37b)
		Power Law	1/6	[−2, 0]	13	(11.37b)
Mean-Field Theory	Binary & Random Fragmentation *(two particle)*	Power Law	1/6	[−1, 1]	13	(13.29) (13.89)
Mean-Field Theory	Collisional Fragmentation *(local)*[1]	Weibull	1/6	[−2, 0]	13	(14.25)
		Power Law	1/6	[−2, 0]	13	(14.25)
Stochastic Geometry	Unitary Fragmentation	Weibull	1/6	[−1, 1]	13	(17.9)
		Power Law	1/6	[−1, 1]	13	(17.9)

(1) Two-particle mechanisms in aggregation and fragmentation are considered local or non-local, depending on whether interactions between similar or dissimilar particles predominate.

without reference to other particles. In two-particle mechanisms, particles evolve in pairs, e.g., through binary collisions. As discussed in Section 13.3, one-particle mechanisms tend to make n/m an integer multiple of 1/3, while two-particle mechanisms tend to make n/m an integer multiple of 1/6.

In Table 20.1, the exponent n/m always falls within a limited range, where the lower limit is never less than −2 and the upper limit is never greater 2. This is not a coincidence. Instead, this book made specific assumptions to ensure that $|n|/m \leq 2$; see Equation (3.60). The rationale is that larger values of $|n|/m$ are associated with smaller fragment size spreads, which are unlikely to occur in chaotic or random systems. As described in Appendix A, chaos, randomness, and particle size spreads tend to decrease as $|n|/m$ increases. The models studied in this book fundamentally assume chaos or randomness, which implies that there must be some upper limit on $|n|/m$. However, it is hard to say that chaos ends and determinism begins once $|n|/m$ exceeds any specific value such as two. The upper limit could just as well have been chosen to three, four, or more. In fact, values as large as seven or eight occasionally appear in the relevant research literature. As a practical matter, changing the upper limit on $|n|/m$ from one to two makes a dramatic difference, while the changing the upper limit to three, four, or more makes only a minor difference. In other words, the effect of changing $|n|/m$ by a fixed increment decreases as $|n|/m$ increases. As $|n|/m$ increases, chaos and randomness gradually decrease. At the same time, size distributions gradually lose their distinction and vary almost continuously with increments in $|n|/m$. While other thresholds could have been adopted, $|n|/m \leq 2$ captures most of the relevant results.

Notice that all of the size distributions listed in Table 20.1—including power law, Weibull, and gamma size distributions—are examples of Rosin–Rammler size distributions. With the exceptions of power law and exponential distributions, Rosin–Rammler distributions are not a part of classic probability and statistics. However, as it turns out, any Rosin–Rammler size distribution can be rewritten as either a power law or exponential size distribution. This requires only a simple change of independent and dependent variables; see Examples 19.1 to 19.4. Thus, Rosin–Rammler size distributions are hardly exotic. It is convenient to allow the general form of the distribution to change while the independent and dependent variables stay the same. However, it is equally valid to keep the general form of the distribution the same—forcing it to always be, say, a power law or exponential size distribution—while the independent and dependent variables change. This book also discussed other possible families of size distributions, such as log-normal and root-normal size distributions. However, in most cases, analytic theory indicates that these are only approximations to the true Rosin–Rammler size distributions. Having said this, log normal size distributions or their close cousins may occur on the borderline between shattering and non-shattering fragmentation; see, e.g., Examples 12.6, 13.8, and 13.11. In addition, root normal size distributions may be useful in hybrid cases, e.g., where the particle size distribution transitions between a gamma size distribution typical of aggregation for small particles and a Weibull size distribution typical of fragmentation for large particles.

The remainder of this chapter will compare the strengths and weaknesses of the various computational and analytic models discussed in the book. Along the way, the remainder of this chapter will review many of the main themes of the book. This book discussed three analytic modeling approaches that were originally developed for aggregation, namely, mean-field theory, maximum entropy theory, and dimensional analysis. While originally developed for aggregation, these same approaches also apply to fragmentation of all types. In addition, this book discussed several analytic modeling approaches that were originally developed for unitary fragmentation including stochastic geometry, energy analysis, and instability analysis. While originally developed for unitary fragmentation, these techniques can sometimes also be applied to aggregation and fragmentation of other types.

In ideal cases, this book has shown that the various analytic modeling approaches roughly agree with each other, after accounting for the differing assumptions. Not surprisingly, given that they more or less agree, the available theoretical evidence indicates that these analytic approaches are closely related to one another, except possibly near discontinuities due to gelation or shattering. Of course, there are some important differences. For example, mean-field theory can provide full time-dependent results, while the other analytic modeling approaches discussed in this book can only provide late-time results. For another example, by varying the modeling parameters, mean-field theory, dimensional analysis, and stochastic geometry can obtain complete or nearly complete sets of universal exponents such as n or n/m. By contrast, maximum entropy theory can only obtain subsets of universal exponents, e.g., those produced by both aggregation and fragmentation. As partial compensation, maximum entropy theory offers a nearly unlimited variety of general parametric forms, at least for continuous particle systems. At the other extreme, dimensional analysis is strictly limited to power law size distributions. Because of this limitation, and because it has no unique advantages over other possible approaches, dimensional analysis is relatively rare.

In addition to analytic models, the aggregation community has modified and adapted a number of general-purpose computational modeling approaches for discrete systems such as Molecular Dynamics (MD), N-Body (NB), Brownian Dynamics (BD), Monte-Carlo (MC), and Event-Driven Molecular Dynamics (EDMD) methods. Similarly, the fragmentation community has modified and adapted a number of general-purpose computational modeling approaches for continuous systems such as Finite Volume Methods (FVMs), Finite Element Methods (FEMs), Discrete Element Methods (DEMs), and Smoothed Particle Hydrodynamics (SPH) methods. In ideal cases, this book has found that the various computational modeling approaches generally agree with one another, assuming they include a sufficient amount of chaos or randomness.

In ideal cases, most of the models discussed in this book obtain universal results for selected aspects of the solutions. For example, regular bonded particle systems often obtain universal size, shape, and speed distributions. In addition, self-organized unbonded particle systems often obtain universal kinetic energy spectra. Each such universal occurs when a set of underlying assumptions is met. In analytic models, the underlying assumptions are often stated in abstract mathematical terms, e.g., ones involving a rate kernel in mean-field theory or a constraint in maximum entropy theory, rather than in terms of easily observed physical traits. Thus, it can be difficult to say whether the underlying assumptions apply to a given real-world problem. A common practice is to extrapolate known analytic, empirical, and semi-empirical solutions to related problems, especially to those that share the same basic geometry, e.g., cylindrical metal shells, round liquid jets, spherical liquid droplets, etc. By contrast, in high-fidelity computational models, the underlying assumptions are usually stated in clear physical terms, such as the yield strength of a solid material or the attractive force between particles. This makes it easier to associate a specific problem with a specific solution. As another significant advantage, high-fidelity computational models tend to obtain complete solutions including position, size, shape, velocity, and rotation distributions. By contrast, analytic, empirical, and semi-empirical models tend to obtain only partial solutions, e.g., just a size distribution.

Because they offer such overwhelming advantages, high-fidelity computational methods have long since displaced analytic models in many fields. However, this is not the case for problems such as aggregation and fragmentation, with large numbers of elements evolving in a chaotic fashion. As one of the key effects of chaos, it causes the range of particle sizes to increase progressively over time. More specifically, for aggregation, the average particle size tends to increase, and the spread about the average tends to increase, unless there is a sink of large particles. Similarly, for fragmentation, the average particle size tends to decrease, and the spread about the average tends to increase, unless there is a sink of small particles. While analytic, empirical, and semi-empirical models can easily support large or even infinite size spreads, computational models cannot. If the size spread is narrow, computational models can accurately represent physical systems at a relatively low cost. More specifically, meshless methods require only a relatively small number of nodes, and mesh-based methods require only a relatively small number of cells. However, as the size spread increases, computational methods become increasingly costly, i.e., meshless methods may require an extremely large number of nodes, and mesh-based methods may require an extremely large number of cells.

As one possible approach, computational methods can use enough nodes or cells to resolve the entire range of size scales present in the actual physical system, from the smallest to the largest. In the fluid turbulence community, this brute-force approach is known as Direct Numerical Simulation (DNS). As noted by Elger et al. (2019), DNS is "unrealistic for 99.9% of ... problems because the required computer time is too large for today's computers." As another possible approach, to narrow the range of particle sizes, computational methods may impose an artificial sink of large particles for aggregation or an artificial sink of small particles for fragmentation. This approach makes two assumptions: (a.) the narrowed range does not alter the fundamental behavior of the system and (b.) particles outside of the computational range do not substantially influence the particles inside the computational range, except in easy-to-account-for ways. Regarding assumption (a.), artificially limiting the range of particles may reduce the amount of chaos present. This is especially true when the smallest, most chaotic scales are missing. Thus, to obtain a realistic level of chaos, computational methods generally must artificially introduce chaos or randomness and/or artificially decrease numerical and physical dissipation that would otherwise damp out chaos. In addition, artificially limiting the range of particles may convert a collisional system from non-local to local, i.e., it may convert a system that is dominated by collisions between differently sized particles to a system that is dominated by collisions between similarly sized particles. To prevent this, computational methods must generally allow particle size ranges of at least 1.5 or 2 orders-of-magnitude. Regarding assumption (b.), for aggregation, this approach assumes a pure reverse cascade from small to large across the cutoff point. Similarly, for fragmentation, this approach assumes a pure forward cascade from large to small across the cutoff point. In the fluid turbulence community, this approach is known as *Implicit Large Eddy Simulation (ILES)* or *Monotonically Integrated Large Eddy Simulation (MILES)*; see, e.g., Grinstein (2016). In classic ILES and MILES, the larger scales neither know nor care what happens at the missing smaller scales, provided the larger scales lose energy to the missing smaller scales at the physically correct rate.

For the meshless computational methods typically used for aggregation, there is no practical alternative to making assumptions (a.) and (b.). However, for the mesh-based computational methods that are often used for fragmentation, there is a common alternative. More specifically, the larger particles may be represented by a computational model while the smaller particles may be represented by analytic, empirical, or semi-empirical *subgrid-scale (SGS) models*. This approach takes advantage of the fact that, in many fragmentation problems, the largest scales obtain little to no chaos. To the extent that the largest scales are deterministic, they are unique to a given problem and thus are best treated with a computational model. At the other extreme, the smallest scales tend to be highly chaotic, and possibly ideal, and thus can be well-captured by analytic, empirical, or semi-empirical models.

Most real-world problems involve multiple challenges, including but not limited to a large number of elements with a wide range of scales. Practical engineering applications tend to be large, complex, and multifaceted. As a result, commercial codes are generally forced to use analytic, empirical or semi-empirical models for at least some parts of a given problem. Of course, to the extent that a computational model relies on analytic, empirical, or semi-empirical models, it is not general purpose. Instead, it is restricted to cases where the specific assumptions underlying the chosen analytical, empirical, or semi-empirical models apply. A single book such as this one could not possibly address the many challenges faced in all of the problems where aggregation and fragmentation occur—rather, it has focused on the challenges that many or most such problems have in common.

More specifically, for simplicity and generality, this book has focused on ideal cases which are chaotic regardless of time, space, and particle size. Throughout the book, the term "ideal" refers to spatial homogeneity and isotropy. These are statistical terms that roughly mean "well-mixed and spatially uniform on average." In ideal cases, analytic methods can obtain general results that apply at all scales, quickly and efficiently, with exact rational exponents, without becoming bogged down in the exact details of any specific problem. Analytic methods generally assume that particle properties such as size and shape are statistically independent. This means that distributions for particle properties such as size and shape can be obtained separately from one another. However, this statistical independence is generally only true for exactly one definition of particle size. Unfortunately, lacking a way of knowing in advance what the right definition of particle size is for a given problem, most analytic models ultimately predict a finite set of size distributions. Down selecting from this finite set to a unique solution requires knowing or assuming something about the problem. Thus, analytic methods are generally supplemented by empirical methods, which draw on databases of results for problems with similar features. The simplest approach involves using a community standard such as those listed in Tables 5.2, 5.4, and 8.2.

Table 20.2 summarizes some of the main challenges faced when modeling particle systems with aggregation and fragmentation. Despite these challenges, modern researchers routinely obtain realistic solutions for aggregation and fragmentation. However, while these solutions may accurately describe *some* problem, they do not always accurately describe the *specific* problem of interest. In essence, while practitioners can generate any number of valid physically-realizable solutions, they often have no fully reliably means of predicting in advance which of these solutions applies to a given problem. Of course, once the true solution is known through experiments or observations, it is relatively easy to infer one or more sets of modeling assumptions that accurately yield the true solution. However, for most practitioners, the goal is prediction and not post-diction.

Not surprisingly, the simpler the problem, the more likely that a given modeling effort will succeed. This book has focused on the simplest types of aggregation and fragmentation—ideal fully-chaotic problems. Modeling efforts for this class of problems not only have a relatively high probability of success, many of the results apply to a broad range of problems in a wide range of fields including aerospace engineering, astrophysics, civil engineering, chemical engineering, meteorology, mining engineering, and mechanical engineering.

Table 20.2 Common challenges with modeling aggregation and fragmentation.

1. *Mapping Specific Problems to Specific Solutions.*	No model can be expected to provide an accurate solution unless the problem is correctly specified. There are on-going challenges in describing real-world aggregation and fragmentation problems in ways that models can understand. This is especially true for the chaotic aspects. No model can be expected to provide an accurate solution unless it fully solves the problem. However, many models assume rather than solve for certain aspects of the solution, e.g., particle shapes and velocities. The results are valid only for problems where the underlying assumptions are valid.
2. *Wide Range of Scales*	Due to chaos, particle size spreads tend to increase with time. Even when the particles initially all have exactly the same size, their sizes may eventually span many orders-of-magnitude. When this occurs, computational methods can generally only be used for a limited subset of scales, depending on the computational resources available.
3. *Blending Ideal Chaotic, Non-Ideal Chaotic, and Deterministic Elements*	Unless the problem is unusually simple, it will generally involve a blend of ideal chaotic, non-ideal chaotic, and deterministic aspects. The exact blend may be a function of time, space, and scale. The ideal chaotic aspects may have universal solutions that can be found with analytic models. However, the non-ideal chaotic and deterministic aspects have specific solutions that only apply to a single problem or a relatively narrow range of problems.

References

D. F. Elger, B. A. LeBret, & C. T. Crowe, *Engineering Fluid Mechanics*, 12th edition, Wiley, 2019

F. F. Grinstein, editor, *Coarse Grained Simulation and Turbulent Mixing*, Cambridge University Press, 2016, https://doi.org/10.1017/CBO9781316480243

W. Podgórska, *Multiphase Particulate Systems in Turbulent Flows: Fluid-Liquid and Solid-Liquid Dispersions*, CRC Press, 2020, https://doi.org/10.1201/9781315118383

G. H. Yeoh, C. P. Cheung, & J. Tu, *Multiphase Flow Analysis Using Population Balance Modeling: Bubbles, Drops and Particles*, Butterworth-Heienemann, 2014, https://doi.org/10.1016/C2011-0-05568-0

APPENDIX A

Different Forms of Common Size Distributions

The research literature on aggregation and fragmentation uses a wide variety of different forms to express particle size distributions. Because of these different forms, size distributions that look the same may actually be different. Conversely, size distributions that look different may actually be the same. This is a common source of confusion. To clarify the situation, this appendix offers a quick reference guide to the forms commonly used to express particle size distributions. Before continuing with this discussion, a few key definitions from the main text will be repeated here for convenience. First, recall that D refers to the particle diameter and M refer to the particle mass. From Section 1.3, recall that:

$$M = \text{const.}\, \rho D^m$$

where ρ is the density, $1 \leq m \leq d$ is the particle *mass dimension*, and d is the integer *spatial dimension*. From Section 3.3, recall that f and f_M are probability density functions where:

$$f_M(x) = \text{const.}\, Mf(x)$$

and where $x = D$ or M. Instead of cumulative distribution functions, as seen in the main text, this appendix uses *complementary cumulative distribution functions* defined as follows:

$$\bar{F}(x) = 1 - F(x) = \int_x^\infty f(x')dx'$$

$$\bar{F}_M(x) = 1 - F_M(x) = \int_x^\infty f_M(x')dx'$$

where $x = D$ or M.

A.1 Sizes Distributions that Look the Same but are Actually Different

There are four types of gamma size distributions commonly seen in the research literature. In regular form, these can be written as follows:

Type I: $f_M(D) = A D^q \exp(-bD)$
Type II: $f(D) = A D^{n-1} \exp(-bD)$
Type III: $f_M(M) = A M^\alpha \exp(-bM)$
Type IV: $f(M) = A M^{r-1} \exp(-bM)$

While there is no standard notation, these expressions are typical. Although these expressions may look similar, they are actually different from each other; see, e.g., Laney (2016a, b). In other words, while these four types of gamma size distributions may be approximately the same over a limited range of particle sizes, there is no choice of q, n, α, r that makes them exactly the same.

A.2 Size Distributions that Look the Same and Actually are the Same

There are four types of power law size distributions commonly seen in the research literature:

Type I: $f_M(D) = A D^q$
Type II: $f(D) = A D^{n-1}$
Type III: $f_M(M) = A M^\alpha$
Type IV: $f(M) = A M^{r-1}$

As before, while there is no standard notation, these expressions are typical. These expressions look similar and, in fact, they are the same when the following relationships apply:

$$q = n + m - 1 \tag{A.1}$$

$$r = \alpha \tag{A.2}$$

$$\alpha + 1 = (q + 1)/m \tag{A.3}$$

$$r = n/m \tag{A.4}$$

For example, in the astrophysics community, one well-known size distribution is a Type IV power law size distribution with $r = -5/6$; see Dohnanyi (1967, 1969, 1970). According to Equation (A.2), this is the same as a Type III power law size distribution with $\alpha = -5/6$. According to Equation (A.4), if $m = 3$, this is the same as a Type II power law size distribution with $n = -5/2$. According to Equation (A.1), this is the same as a Type I power law size distribution with $q = -1/2$.

There are four types of Weibull size distributions commonly seen in the research literature:

Type I: $f_M(D) = A D^q \exp\left(-bD^{q+1}\right)$
Type II: $f(D) = A D^{n-1} \exp\left(-bD^n\right)$
Type III: $f_M(M) = A M^\alpha \exp\left(-bM^{\alpha+1}\right)$
Type IV: $f(M) = A M^{r-1} \exp\left(-bM^r\right)$

As before, while there is no standard notation, these expressions are typical. As shown in Laney (2015, 2016a, b), Type I and Type III Weibull size distributions are the same if:

$$\alpha + 1 = (q + 1)/m. \tag{A.5}$$

In addition, Type II and Type IV Weibull size distributions are the same if:

$$r = n/m. \tag{A.6}$$

While Type I and II Weibull size distributions may be approximately the same over a limited range of particle sizes, there is no choice of q or n that makes them exactly the same. A similar observation applies to Type III and IV Weibull size distributions.

A.3 Size Distributions that Look Different but are Actually the Same

The remainder of this appendix concerns size distributions that look different but are actually the same. In particular, this appendix shows how to express the most commonly seen size distributions in the following four self-similar forms:

$$f(D) = \tilde{f}(D/D_{avg})/D_{avg} \qquad f(M) = \approx{f}(M/M_{avg})/M_{avg}$$
$$\bar{F}(D) = \tilde{F}(D/D_{avg}) \qquad \bar{F}(M) = \approx{F}(M/M_{avg})$$

in cases where D_{avg} and M_{avg} exist and are finite.

In addition, this appendix shows how to express the most commonly seen size size distributions in the following four different self-similar forms:

$$f(D) = \tilde{f}'(D/D'_{\text{avg}})/D'_{\text{avg}} \qquad f(M) = \overset{\approx}{f}'(M/M'_{\text{avg}})/M'_{\text{avg}}$$

$$\bar{F}(D) = \tilde{f}'(D/D'_{\text{avg}}) \qquad \bar{F}(M) = \overset{\approx}{F}'(M/M'_{\text{avg}})$$

in cases where D'_{avg} and M'_{avg} exist and are finite. *Unless stated otherwise, all of the expressions given below are for Type IV size distributions.* This distinction may or may not make any difference, depending on the size distribution. More specifically, as noted previously, it always makes a difference for gamma size distributions, never makes a difference for power law size distributions, and sometimes make a difference for Weibull size distributions.

A.4 Simple Ways of Deciding if Size Distributions are the Same or Different

Consider the following four metrics of particle size spread:

$$R_D = \frac{D_{\text{avg}}}{D'_{\text{avg}}} \qquad S_D = -\int_0^\infty \ln[f(D)]f(D)\,dD$$

$$R_M = \frac{M_{\text{avg}}}{M'_{\text{avg}}} \qquad S_M = -\int_0^\infty \ln[f(M)]f(M)\,dM$$

As a rough rule of thumb, if two size distributions have similar values of R_D, then the two size distributions are similar. Conversely, if two size distributions have different values of R_D, then the two size distributions are different. Similar remarks apply to R_M, S_D, and S_M. Notice that R_D, R_M, S_D, and S_M are measures of particle size spread. In other words, they are small for monodisperse systems with narrow particle size spreads. Conversely, they are large for uniformly distributed systems with broad particle size spreads. As discussed in Chapter 19, S_D and S_M are forms of *size entropy*. While not as well-known as size entropy, expressions for R_D and R_M tend to be simpler and more intuitive than expressions for S_D and S_M. In general, instead of specifying a size distribution by the parameters that appear in a given form, it can be clearer to specify a size distributions using R_D or R_M; see, e.g., Laney (2015, 2016a, b). While the values of parameters such as n/m depend on the chosen form, the values of R_D and R_M are the exactly the same for a given size distribution, no matter what form it takes.

A.5 Rosin–Rammler Size Distributions

Rosin–Rammler size distributions were first described by Rosin & Rammler (1927a, b); see, e.g., Schubert & Wächtler (1987) and Stoyan (2013). Rosin–Rammler size distributions are also known as *Nukiyama–Tanasawa size distributions* or *generalized gamma size distributions*; see, e.g., Nukiyama & Tanasawa (1938a, b) and Stacy (1962), respectively. Tables A.1 and A.2 show eight self-similar forms for Rosin–Rammler size distributions.

Special cases of Rosin–Rammler size distributions include the following:

- Weibull size distributions with $l = n$ (Type II and IV)
- Gamma size distributions with $l = 1$ (Type II) or $l = m$ (Type IV)
- Power law size distributions with $l = 0$ (Types I, II, III, IV)

These three special cases will be discussed in the following three sections.

Table A.1a Four self-similar forms for Rosin–Rammler size distributions where the characteristic size is D_{avg} or M_{avg}. Each form has three free parameters: n, l, and the average particle size. All four forms shown are equivalent and are valid for $n > 0, l > 0$.

	$x = D$	$x = M$
f	$\frac{\lvert l\rvert\, b_D^{n/l}}{\Gamma\left(\frac{n}{l}\right) D_{avg}} \left(\frac{D}{D_{avg}}\right)^{n-1} \exp\left[-b_D\left(\frac{D}{D_{avg}}\right)^{l}\right]$	$\frac{\lvert l\rvert\, b_M^{n/l}}{m\Gamma\left(\frac{n}{l}\right) M_{avg}} \left(\frac{M}{M_{avg}}\right)^{\frac{n}{m}-1} \exp\left[-b_M\left(\frac{M}{M_{avg}}\right)^{\frac{l}{m}}\right]$
$\bar{F}$	$\frac{\Gamma\left[\frac{n}{l}, b_D\left(\frac{D}{D_{avg}}\right)^{l}\right]}{\Gamma\left(\frac{n}{l}\right)}$	$\frac{\Gamma\left[\frac{n}{l}, b_M\left(\frac{M}{M_{avg}}\right)^{\frac{l}{m}}\right]}{\Gamma\left(\frac{n}{l}\right)}$

Note: $\Gamma(x)$ is the gamma function and $\Gamma(x, y)$ is the incomplete upper gamma function.

Table A.1b Expressions associated with Table A.1a.

	$x = D$	$x = M$
b_x	$\left[\frac{\Gamma\left(\frac{n+1}{l}\right)}{\Gamma\left(\frac{n}{l}\right)}\right]^{l}$	$\left[\frac{\Gamma\left(\frac{n+m}{l}\right)}{\Gamma\left(\frac{n}{l}\right)}\right]^{\frac{l}{m}}$
R_x	$\frac{\Gamma\left(\frac{n+1}{l}\right)\Gamma\left(\frac{n-1}{l}\right)}{\Gamma\left(\frac{n}{l}\right)^2}$	$\frac{\Gamma\left(\frac{n+m}{l}\right)\Gamma\left(\frac{n-m}{l}\right)}{\Gamma\left(\frac{n}{l}\right)^2}$
S_x	$\frac{n}{l} - \frac{n-1}{l}\psi\left(\frac{n}{l}\right) + \ln\left[\frac{D_{avg}}{\lvert l\rvert b_D^{1/l}}\Gamma\left(\frac{n}{l}\right)\right]$	$\frac{n}{l} - \frac{n-m}{l}\psi\left(\frac{n}{l}\right) + \ln\left[\frac{mM_{avg}}{\lvert l\rvert b_M^{m/l}}\Gamma\left(\frac{n}{l}\right)\right]$

Note: $\Gamma(x)$ is the gamma function and $\psi(z) = d[\text{In}\,\Gamma(z)]/dz$ is the digamma function.
Source: The expressions for b_x and R_x are taken from Laney (2016a, b). The expressions for S_x are taken from Khodabin & Ahmadabadi (2010)

Table A.2a Four self-similar forms for Rosin–Rammler size distributions where the characteristic size is D'_{avg} or M'_{avg}. Each form has three free parameters: n, l, and the average particle size. All four forms shown are equivalent and are valid for $n < 0, l < 0$.

	$x = D$	$x = M$
f	$\frac{\lvert l\rvert\, b_D^{n/l}}{\Gamma\left(\frac{n}{l}\right) D'_{avg}} \left(\frac{D}{D'_{avg}}\right)^{n-1} \exp\left[-b_D\left(\frac{D}{D'_{avg}}\right)^{l}\right]$	$\frac{\lvert l\rvert\, b_M^{n/l}}{m\Gamma\left(\frac{n}{l}\right) M'_{avg}} \left(\frac{M}{M'_{avg}}\right)^{\frac{n}{m}-1} \exp\left[-b_M\left(\frac{M}{M'_{avg}}\right)^{\frac{l}{m}}\right]$
$\bar{F}$	$\frac{\Gamma\left[\frac{n}{l}, b_D\left(\frac{D}{D'_{avg}}\right)^{l}\right]}{\Gamma\left(\frac{n}{l}\right)}$	$\frac{\Gamma\left[\frac{n}{l}, b_M\left(\frac{M}{M'_{avg}}\right)^{\frac{l}{m}}\right]}{\Gamma\left(\frac{n}{l}\right)}$

Note: $\Gamma(x)$ is the gamma function and $\Gamma(x, y)$ is the incomplete upper gamma function.

Table A.2b Expressions associated with Table A.2a.

	$x = D$	$x = M$
b_x	$\left[\dfrac{\Gamma\left(\frac{n}{l}\right)}{\Gamma\left(\frac{n-1}{l}\right)}\right]^l$	$\left[\dfrac{\Gamma\left(\frac{n}{l}\right)}{\Gamma\left(\frac{n-m}{l}\right)}\right]^{\frac{l}{m}}$
R_x	$\dfrac{\Gamma\left(\frac{n+1}{l}\right)\Gamma\left(\frac{n-1}{l}\right)}{\Gamma\left(\frac{n}{l}\right)^2}$	$\dfrac{\Gamma\left(\frac{n+m}{l}\right)\Gamma\left(\frac{n-m}{l}\right)}{\Gamma\left(\frac{n}{l}\right)^2}$
S_x	$\frac{n}{l} - \frac{n-1}{l}\psi\left(\frac{n}{l}\right) + \ln\left[\dfrac{D'_{avg}}{\lvert l\rvert b_D^{1/l}}\Gamma\left(\frac{n}{l}\right)\right]$	$\frac{n}{l} - \frac{n-m}{l}\psi\left(\frac{n}{l}\right) + \ln\left[\dfrac{m M'_{avg}}{\lvert l\rvert b_M^{m/l}}\Gamma\left(\frac{n}{l}\right)\right]$

Note: $\Gamma(x)$ is the gamma function and $\psi(z) = d[\ln \Gamma(z)]/dz$ is the digamma function.
Source: The expressions for b_x and R_x are taken from Laney (2016a, b). The expressions for S_x are taken from Khodabin & Ahmadabadi (2010).

A.6 Weibull Size Distributions ($l = n$)

Weibull distributions were described by Weibull (1939a, b, 1951); see also Rinne (2008). Weibull distributions come in two main variants: *positive* Weibull distributions with $n > 0$ and *negative* or *inverse* Weibull distributions with $n < 0$; see, e.g., Keller et al. (1985), Calabria & Pulcini (1989), Musleh & Helu (2014), Oluyede & Yang (2014) and Lai (2014, section 2.6). Positive Type I and III Weibull size distributions are also known as Rosin–Rammler–Sperling–Bennett (RRSB) size distributions; see, e.g., Rosin & Rammler (1933), Rosin, Rammler, & Sperling (1933), Bennett (1936), Schubert & Wächtler (1987) and Stoyan (2013). This section focuses on Type II and IV Weibull size distributions which, as noted earlier, are the same. Tables A.3 and A.4 show four self-similar forms for positive and negative Weibull size distributions, respectively. Notice that Tables A.3 and A.4 are the same as Tables A.1 and A.2 with $l = n$.

Table A.3a Four self-similar forms for positive Type II and IV Weibull size distributions where the characteristic size is D_{avg} or M_{avg}. Each form has two free parameters, namely, n and the average particle size. All four forms shown are equivalent and are valid for $n > 0$.

	$x = D$	$x = M$
f	$\dfrac{nb_D}{D_{avg}}\left(\dfrac{D}{D_{avg}}\right)^{n-1}\exp\left[-b_D\left(\dfrac{D}{D_{avg}}\right)^n\right]$	$\dfrac{nb_M}{mM_{avg}}\left(\dfrac{M}{M_{avg}}\right)^{\frac{n}{m}-1}\exp\left[-b_M\left(\dfrac{M}{M_{avg}}\right)^{\frac{n}{m}}\right]$
$\bar{F}$	$\exp\left[-b_D\left(\dfrac{D}{D_{avg}}\right)^n\right]$	$\exp\left[-b_M\left(\dfrac{M}{M_{avg}}\right)^{\frac{n}{m}}\right]$

Table A.3b Expressions associated with Table A.3a.

	$x = D$	$x = M$
b_x	$\Gamma\left(1+\frac{1}{n}\right)^n$	$\Gamma\left(1+\frac{m}{n}\right)^{\frac{n}{m}}$
R_x	$\Gamma\left(1+\frac{1}{n}\right)\Gamma\left(1-\frac{1}{n}\right)$	$\Gamma\left(1+\frac{m}{n}\right)\Gamma\left(1-\frac{m}{n}\right)$
S_x	$1+\left(1-\frac{1}{n}\right)\gamma+\ln\left[\frac{D_{\text{avg}}}{nb_D^{1/n}}\right]$	$1+\left(1-\frac{m}{n}\right)\gamma+\ln\left[\frac{mM_{\text{avg}}}{nb_M^{m/n}}\right]$

Note: $\gamma = -\psi(1) = 0.557721566...$ is Euler's constant.
Source: The expressions for b_x and R_x are taken from Laney (2015; 2016a, b). The expressions for S_x are taken from Cover & Thomas (2006).

Table A.4a Four self-similar forms for negative Type II and IV Weibull size distributions where the characteristic size is D'_{avg} or M'_{avg}. Each form has two free parameters, namely, n and the average particle size. All four forms shown are equivalent and are valid for $n < 0$.

	$x = D$	$x = M$
f	$\frac{\lvert n\rvert\, b_D}{D'_{\text{avg}}}\left(\frac{D}{D'_{\text{avg}}}\right)^{n-1}\exp\left[-b_D\left(\frac{D}{D'_{\text{avg}}}\right)^n\right]$	$\frac{\lvert n\rvert\, b_M}{m\,M'_{\text{avg}}}\left(\frac{M}{M'_{\text{avg}}}\right)^{\frac{n}{m}-1}\exp\left[-b_M\left(\frac{M}{M'_{\text{avg}}}\right)^{\frac{n}{m}}\right]$
$\bar{F}$	$\exp\left[-b_D\left(\frac{D}{D'_{\text{avg}}}\right)^n\right]$	$\exp\left[-b_M\left(\frac{M}{M'_{\text{avg}}}\right)^{\frac{n}{m}}\right]$

Table A.4b Expressions associated with Table A.4a.

	$x = D$	$x = M$
b_x	$\Gamma\left(1+\frac{1}{\lvert n\rvert}\right)^{\lvert n\rvert}$	$\Gamma\left(1+\frac{m}{\lvert n\rvert}\right)^{\frac{\lvert n\rvert}{m}}$
R_x	$\Gamma\left(1+\frac{1}{n}\right)\Gamma\left(1-\frac{1}{n}\right)$	$\Gamma\left(1+\frac{m}{n}\right)\Gamma\left(1-\frac{m}{n}\right)$
S_x	$1+\left(1-\frac{1}{n}\right)\gamma+\ln\left[\frac{D'_{\text{avg}}}{\lvert n\rvert b_D^{1/n}}\right]$	$1+\left(1-\frac{m}{n}\right)\gamma+\ln\left[\frac{m\,M'_{\text{avg}}}{\lvert n\rvert b_M^{m/n}}\right]$

Note: $\gamma = -\psi(1) = 0.557721566$ is Euler's constant.
Source: The expressions for b and R are taken from Laney (2015, 2016a, b). The expressions for S are taken from Cover & Thomas (2006).

A.7 Gamma Size Distributions (l = 1 or m)

This section concerns Type II and IV gamma size distributions with $l = 1$ and m, respectively. Type IV gamma size distributions are also known as *Schulz* or *Schulz–Zimm size distributions*; see, e.g., Schulz (1939), Zimm (1948), Hiemenz & Lodge (2007) and Rodriguez et al. (2014).

Tables A.5 and A.6 show four self-similar forms for Type II and IV gamma size distributions, respectively. Tables A.5 and A.6 assume that $n > 0$. If $n < 0$, gamma size distributions have a severe singularity for small particle sizes. To avoid this singularity, a minimum particle size such as $D_{\min}$ or $M_{\min}$ should be imposed. Unfortunately, if there is minimum particle size greater than zero, it is impossible to give analytic expressions for A_x b_x, R_x, S_x.

If $n < 0$, possible alternatives are Type II and IV *inverse gamma size distributions* with $l = -1$ and $-m$, respectively. These are also known as *Pearson Type V distributions*; see, e.g., Tanizaki (2004), Lynch (2007), Florescu (2015) and Glen (2017). Because l and n have the same sign, these distributions avoid the severe singularity at $D = 0$ or $M = 0$. Despite this, these size distributions are rarely if ever used to describe aggregating and fragmentating particle systems.

Table A.5a Four self-similar forms for Type II gamma size distributions where the characteristic size is D_{avg} or M_{avg}. Each form has two free parameters, namely, n and the average particle size. All four forms shown are equivalent and are valid for $n > 0$.

	$x = D$	$x = M$
f	$\dfrac{b_D^n}{\Gamma(n)D_{\text{avg}}}\left(\dfrac{D}{D_{\text{avg}}}\right)^{n-1}\exp\left[-b_D\dfrac{D}{D_{\text{avg}}}\right]$	$\dfrac{b_M^n}{m\Gamma(n)M_{\text{avg}}}\left(\dfrac{M}{M_{\text{avg}}}\right)^{\frac{n}{m}-1}\exp\left[-b_M\left(\dfrac{M}{M_{\text{avg}}}\right)^{\frac{1}{m}}\right]$
$\bar{F}$	$\dfrac{\Gamma\left[n, b_D\dfrac{D}{D_{\text{avg}}}\right]}{\Gamma(n)}$	$\dfrac{\Gamma\left[n, b_M\left(\dfrac{M}{M_{\text{avg}}}\right)^{\frac{1}{m}}\right]}{\Gamma(n)}$

Note: $\Gamma(x)$ is the gamma function and $\Gamma(x, y)$ is the incomplete upper gamma function.

Table A.5b Expressions associated with Table A.5a.

	$x = D$	$x = M$
b_x	n	$\left[\dfrac{\Gamma(n+m)}{\Gamma(n)}\right]^{\frac{1}{m}}$
R_x	$\dfrac{n}{n-1}$	$\dfrac{\Gamma(n+m)\Gamma(n-m)}{\Gamma(n)^2}$
S_x	$n-(n-1)\psi(n)+\ln\left[\dfrac{D_{\text{avg}}\Gamma(n)}{b}\right]$	$n-(n-m)\psi(n)+\ln\left[\dfrac{mM_{\text{avg}}\Gamma(n)}{b^m}\right]$

Note: $\Gamma(x)$ is the gamma function and $\psi(z) = d[\text{In}\,\Gamma(z)]/dz$ is the digamma function.
Source: The expressions for b_x and R_x are taken from Laney (2015, 2016a, b). The expressions for R_x are taken from Cover & Thomas (2006).

Table A.6a Four self-similar forms for Type IV gamma size distributions where the characteristic size is D_{avg} or M_{avg}. Each form has two free parameters, namely, n and the average particle size. All four forms shown are equivalent and are valid for $n > 0$.

	$x = D$	$x = M$
f	$\dfrac{m b_D^{n/m}}{\Gamma\left(\frac{n}{m}\right) D_{avg}} \left(\dfrac{D}{D_{avg}}\right)^{n-1} \exp\left[-b_D \left(\dfrac{D}{D_{avg}}\right)^m\right]$	$\dfrac{b_M^{n/m}}{\Gamma\left(\frac{n}{m}\right) M_{avg}} \left(\dfrac{M}{M_{avg}}\right)^{\frac{n}{m}-1} \exp\left[-b_M \dfrac{M}{M_{avg}}\right]$
$\bar{F}$	$\dfrac{\Gamma\left[\frac{n}{m}, b_D \left(\frac{D}{D_{avg}}\right)^m\right]}{\Gamma\left(\frac{n}{m}\right)}$	$\dfrac{\Gamma\left[\frac{n}{m}, b_M \frac{M}{M_{avg}}\right]}{\Gamma\left(\frac{n}{m}\right)}$

Note: $\Gamma(x)$ is the gamma function and $\Gamma(x, y)$ is the incomplete upper gamma function.

Table A.6b Expressions associated with Table A.6a.

	$x = D$	$x = M$
b_x	$\left[\dfrac{\Gamma\left(\frac{n+1}{m}\right)}{\Gamma\left(\frac{n}{m}\right)}\right]^m$	$\dfrac{n}{m}$
R_x	$\dfrac{\Gamma\left(\frac{n+1}{m}\right)\Gamma\left(\frac{n-1}{m}\right)}{\Gamma\left(\frac{n}{m}\right)^2}$	$\dfrac{n}{n-m}$
S_x	$\dfrac{n}{m} - \dfrac{n-1}{m}\psi\left(\dfrac{n}{m}\right) + \ln\left[\dfrac{D_{avg}\Gamma(n/m)}{m b_D^{1/m}}\right]$	$\dfrac{n}{m} - \left(\dfrac{n}{m} - 1\right)\psi\left(\dfrac{n}{m}\right) + \ln\left[\dfrac{M_{avg}\Gamma(n/m)}{b_M}\right]$

Note: $\Gamma(x)$ is the gamma function and $\psi(z) = d[\ln \Gamma(z)]/dz$ is the digamma function.
Source: The expressions for b_x and R_x are taken from Laney (2015). The expressions for S_x are taken from Cover & Thomas (2006).

A.8 Power Law Size Distributions ($l = 0$)

This section concerns power law size distributions with $l = 0$. As noted earlier, power laws commonly appear in Type I, II, III, and IV forms. However, this difference is only superficial—the size distribution itself is the same. For simplicitly, this section focuses on Type II and IV power law size distributions. Notice that power law size distributions are commonly known as *Gaudin–Schuhmann* or *Gates–Gaudin–Schuhmann* distributions; see Gates (1915), Gaudin (1926) and Schuhmann (1940). This is especially true of Type I power law size distributions.

If $n > 0$, power law size distributions have a severe singularity at large particle sizes, i.e., as $D \to \infty$ or $M \to \infty$. Thus, positive power law size distributions should only be used below a maximum particle size such as D_{max} or M_{max}. On the other hand, if $n < 0$, power law size distributions have a severe singularity at small particle sizes, i.e., there is a vertical asymptote as $D \to 0$ or $M \to 0$. Thus, negative power law size distributions should only be used above a minimum particle size such as D_{min} or M_{min}. Tables A.7 and A.8 show some examples of self-similar forms for power law size distributions. Table A.7 assumes that both a minimum and maximum particle size exist. These are sometimes called *Funk–Dinger* or *Dinger–Funk size distributions*; see Example A.1. By contrast, Table A.8 assumes that either a minimum or a maximum particle size exists, which allows for simpler but less general expressions. These are sometimes called *Fuller size distributions*; see Example A.1.

Table A.7a Self-similar forms for power laws where the characteristic size is an average.

	$x = D$		$x = M$	
f	$\frac{n b_D}{D_{avg}}\left(\frac{D}{D_{avg}}\right)^{n-1}$	$n > 0$	$\frac{n b_M}{m M_{avg}}\left(\frac{M}{M_{avg}}\right)^{\frac{n}{m}-1}$	$n > 0$
	$\frac{\lvert n\rvert b_D}{D'_{avg}}\left(\frac{D}{D'_{avg}}\right)^{n-1}$	$n < 0$	$\frac{\lvert n\rvert b_M}{m M'_{avg}}\left(\frac{M}{M'_{avg}}\right)^{\frac{n}{m}-1}$	$n < 0$
$\bar{F}$	$-b_D\left(\frac{D}{D_{avg}}\right)^{n} + \frac{D^n_{max}}{D^n_{max} - D^n_{min}}$	$n > 0$	$-b_M\left(\frac{M}{M_{avg}}\right)^{\frac{n}{m}} + \frac{M^{n/m}_{max}}{M^{n/m}_{max} - M^{n/m}_{min}}$	$n > 0$
	$b_D\left(\frac{D}{D'_{avg}}\right)^{n} + \frac{D^n_{max}}{D^n_{max} - D^n_{min}}$	$n < 0$	$b_M\left(\frac{M}{M'_{avg}}\right)^{\frac{n}{m}} + \frac{M^{n/m}_{max}}{M^{n/m}_{max} - M^{n/m}_{min}}$	$n < 0$

Table A.7b Expressions associated with Table A.7a.

	$x = D$		$x = M$	
b_x	$\frac{D^n_{avg}}{D^n_{max} - D^n_{min}}$	$n > 0$	$\frac{M^{n/m}_{avg}}{M^{n/m}_{max} - M^{n/m}_{min}}$	$n > 0$
	$\frac{D'^n_{avg}}{\lvert D^n_{max} - D^n_{min}\rvert}$	$n < 0$	$\frac{M'^{n/m}_{avg}}{\lvert M^{n/m}_{max} - M^{n/m}_{min}\rvert}$	$n < 0$
x_{avg}	$\frac{n}{n+1}\frac{D^{n+1}_{max} - D^{n+1}_{min}}{D^n_{max} - D^n_{min}}$		$\frac{n/m}{n/m+1}\frac{M^{n/m+1}_{max} - M^{n/m+1}_{min}}{M^{n/m}_{max} - M^{n/m}_{min}}$	
x'_{avg}	$\frac{n-1}{n}\frac{D^{n}_{max} - D^{n}_{min}}{D^{n-1}_{max} - D^{n-1}_{min}}$		$\frac{n/m-1}{n/m}\frac{M^{n/m}_{max} - M^{n/m}_{min}}{M^{n/m-1}_{max} - M^{n/m-1}_{min}}$	
R_x	$\frac{n-1}{n+1}\frac{D^{n+1}_{max} - D^{n+1}_{min}}{D^{n-1}_{max} - D^{n-1}_{min}}$		$\frac{n/m-1}{n/m+1}\frac{M^{n/m+1}_{max} - M^{n/m+1}_{min}}{M^{n/m-1}_{max} - M^{n/m-1}_{min}}$	
S_x	$\text{sign}(n)\left(1 - \frac{1}{n}\right) - \ln\frac{n}{D^n_{max} - D^n_{min}}$ $-(n-1)\frac{\ln(D_{max})D^n_{max} - \ln(D_{min})D^n_{min}}{D^{n-1}_{max} - D^{n-1}_{min}}$		$\text{sign}(n)\left(1 - \frac{m}{n}\right) - \ln\frac{n/m}{M^{n/m}_{max} - M^{n/m}_{min}}$ $-\left(\frac{n}{m} - 1\right)\frac{\ln(M_{max})M^{n/m}_{max} - \ln(M_{min})M^{n/m}_{min}}{M^{n/m-1}_{max} - M^{n/m-1}_{min}}$	

Table A.8a Self-similar forms for power laws where the characteristic size is a maximum or minimum.

	$x = D$		$x = M$	
f	$\frac{n}{D_{max}}\left(\frac{D}{D_{max}}\right)^{n-1}$	$n > 0$	$\frac{n}{mM_{max}}\left(\frac{M}{M_{max}}\right)^{\frac{n}{m}-1}$	$n > 0$
	$\frac{\lvert n\rvert}{D_{min}}\left(\frac{D}{D_{min}}\right)^{n-1}$	$n < 0$	$\frac{\lvert n\rvert}{mM_{min}}\left(\frac{M}{M_{min}}\right)^{\frac{n}{m}-1}$	$n < 0$
$\bar{F}$	$1-\left(\frac{D}{D_{max}}\right)^{n}$	$n > 0$	$1-\left(\frac{M}{M_{max}}\right)^{\frac{n}{m}}$	$n > 0$
	$\left(\frac{D}{D_{min}}\right)^{n}$	$n < 0$	$\left(\frac{M}{M_{min}}\right)^{\frac{n}{m}}$	$n < 0$

Table A.8b Expressions associated with Table A.8a.

	$x = D$		$x = M$	
x_{avg}	$\frac{n}{n+1}D_{max}$	$n > 0$	$\frac{n}{n+m}M_{max}$	$n > 0$
	$\frac{n}{n+1}D_{min}$	$n < -1$	$\frac{n}{n+m}M_{min}$	$n < -m$
x'_{avg}	$\frac{n-1}{n}D_{max}$	$n > 1$	$\frac{n-m}{n}M_{max}$	$n > m$
	$\frac{n-1}{n}D_{min}$	$n < 0$	$\frac{n-m}{n}M_{min}$	$n < 0$
R_x	$\frac{n^2}{n^2-1}$	$\lvert n\rvert > 1$	$\frac{n^2}{n^2-m^2}$	$\lvert n\rvert > m$
	∞	$\lvert n\rvert \leqslant 1$	∞	$\lvert n\rvert \leqslant m$
S_x	$1-\frac{1}{n}-\ln\frac{n}{D_{max}}$	$n > 0$	$1-\frac{m}{n}-\ln\frac{n}{mM_{max}}$	$n > 0$
	$1-\frac{1}{n}-\ln\frac{\lvert n\rvert}{D_{min}}$	$n < 0$	$1-\frac{m}{n}-\ln\frac{\lvert n\rvert}{mM_{min}}$	$n < 0$

Example A.1 Size distributions for particle packing

As discussed in Section 17.3, particle packing is a possible way of modeling unitary fragmentation. As a specific application of particle packing, consider the aggregates used to make concrete, such as gravel and crushed stone. If there is a maximum but not a minimum size, Fuller & Thompson (1907) found that an aggregate approaches the greatest possible density when it obtains a power law size distribution of the following form:

$$\bar{F}_M(D) = 1 - \left(\frac{D}{D_{\max}}\right)^k$$

where $k = n + m > 0$. This is a power law size distribution like those listed in Table A.8. Alternatively, if there is both a maximum and a minimum size, Funk et al. (1980) and Funk & Dinger (1994) found that granular materials approach the greatest possible density when:

$$\bar{F}_M(D) = \frac{D_{\max}^k - D^k}{D_{\max}^k - D_{\min}^k}$$

which is a power law size distribution like those listed in Table A.7. Over the years, researchers have suggested a variety of different values for k. For example, Fuller & Thompson (1907) suggested $k = 1/2$. Alternatively, Andreasen & Andersen (1930) suggested that k is between 1/3 to 1/2 with an average of 0.37. As the latter example indicates, the power law exponents for particle packing generally do *not* obtain universal values; see also Section 17.3. For more information on power law size distributions for concrete aggregate packing see, e.g., Fennis & Walraven (2012) and Sohail et al. (2018).

A.9 Conversion Techniques

This section explains how to derive the results given earlier in Tables A.1 to A.8. The following four equations are the workhorses that are used convert from one form to another:

$$f(D) = -\frac{d\bar{F}(D)}{dD} \tag{A.7}$$

$$f(M) = -\frac{d\bar{F}(M)}{dM} \tag{A.8}$$

$$\bar{F}(D) = \bar{F}(M) \tag{A.9}$$

$$f(M) = \frac{dD}{dM}f(D). \tag{A.10}$$

In Equation (A.10), notice that Equation (1.1) implies the following:

$$dD/dM = (1/m)(c_1\rho)^{-1/m}M^{1/m-1}.$$

Example 19.1 explains how to convert from one regular form to another. Example A.2 explains how to convert from one self-similar form to another.

Example A.2 Conversion techniques for self-similar size distributions

Consider a conversion from $f(D) = \tilde{f}(D/D_{\text{avg}})/D_{\text{avg}}$ to $\tilde{\tilde{f}}(M/M_{\text{avg}})/M_{\text{avg}}$. As one possible approach, consider the ratio $D_{\text{avg}}/M_{\text{avg}}$. It is generally possible to derive an analytic expression for this ratio, similar to those given previously for R_D and R_M. Then D_{avg} can be replaced by $M_{\text{avg}}(D_{\text{avg}}/M_{\text{avg}})$ wherever it appears in the original expression. While this seems like it would be simple and straightforward, it is actually quite tedious. As a better alternative, D/D_{avg} can be replaced by $(M/M_{avg})^{1/m}$ wherever it appears in the original expression. According to Equation (A.10), the original expression for $f(D)$ should be multiplied by dD/dM. However, in order to keep everything in self-similar form, it is preferable to multiply $f(D)$ by:

$$\left(\frac{1}{M_{avg}}\right)^{\frac{1}{m}-1}\frac{dD}{dM}=\frac{1}{m}\left(\frac{M}{M_{avg}}\right)^{\frac{1}{m}-1}.$$

In the original expression, there must be two free parameters, e.g., A and b. In the original expression, they should be chosen to ensure that $\int f(D)\,dD = 1$ and $\int Df(D)\,dD = D_{avg}$. In the converted expression, there must also be two free parameters. In the converted expression, they should be chosen to ensure that $\int f(M)\,dM = 1$ and $\int Mf(M)\,dM = M_{avg}$

References

A. H. M. Andreasen & J. Andersen, Über die Beziehung zwischen Kornabstufung und Zwischenraum in Produkten aus losen Körnern (mit einigen Experimenten), *Kolloid-Zeitschrift*, 50(3), 217–228, 1930, https://doi.org/10.1007/BF01422986

J. G. Bennett, Broken Coal, *Journal of the Institute of Fuel*, 10, 22–39, 1936

R. Calabria & G. Pulcini, Confidence Limits for Reliability and Tolerance Limits in the Inverse Weibull Distribution, *Reliability Engineering & System Safety*, 24(1), 77–85, 1989, https://doi.org/10.1016/0951-8320(89)90056-2

J. S. Dohnanyi, Collisional Model of Meteoroids, in J. L. Weinberg, editor, *The Zodiacal Light and the Interplanetary Medium*, NASA SP-150, 1967 https://ntrs.nasa.gov/search.jsp?R=19680018003

J. S. Dohnanyi, Collisional Model of Asteroids and their Debris, *Journal of Geophysical Research*, 74(10), 2531–2554, 1969, https://doi.org/10.1029/JB074i010p02531

J. S. Dohnanyi, On the Origin and Distribution of Meteoroids, *Journal of Geophysical Research*, 75(17), 3468–3493, 1970, https://doi.org/10.1029/JB075i017p03468

S. A. A. M. Fennis & J. C. Walraven, Using Particle Packing Technology for Sustainable Concrete Mixture Design, *Heron*, 57(2), 73–102, 2012, http://heronjournal.nl/57-2/1.html

I. Florescu, *Probability and Stochastic Processes*, Wiley, 2015

W. Fuller & S. E. Thompson, The Laws of Proportioning Concrete, *Transactions of the American Society of Civil Engineers*, 59(2), 67–143, 1907, https://doi.org/10.1061/TACEAT.0001979

J. E. Funk, D. R. Dinger, & J. E. J. Funk, *Coal Grinding and Particle Size Distribution Studies for Coal-Water Slurries at High Solids Content*, Alfred University Research Foundation Report for Empire State Electric Energy Research Corporation (ESEERCO), 1980

J. E. Funk & D. R. Dinger, Introduction to Predictive Process Control, in *Predictive Process Control of Crowded Particulate Suspensions*, Springer, 1994, https://doi.org/10.1007/978-1-4615-3118-0_1

A. O. Gates, Kick vs. Rittinger: An Experimental Investigation in Rock Crushing, Performed at Purdue University, *Transactions of the AIME*, 52, 875–909, 1915, http://www.onemine.org/document/abstract.cfm?docid=14056

A. M. Gaudin, An Investigation of Crushing Phenomena, *Transactions of the AIME*, 73, 253–310, 1926, http://www.onemine.org/document/abstract.cfm?docid=164880

A. G. Glen, On the Inverse Gamma as a Survival Distribution, in A. G. Glen & A. G. Leemis, editors, *Computational Probability Applications*, Springer, 2017, https://doi.org/10.1007/978-3-319-43317-2

A. Z. Keller, M. T. Goblin, & N. R. Farnworth, Reliability Analysis of Commercial Vehicle Engines, *Reliability Engineering*, 10(1), 15–25, 1985, https://doi.org/10.1016/0143-8174(85)90039-3

M. Khodabin & A. Ahmadabadi, Some Properties of Generalized Gamma Distribution, *Mathematical Sciences*, 4(1), 9–28, 2010, https://www.researchgate.net/publication/50198287

C.-D. Lai, *Generalized Weibull Distributions*, Springer, 2014, https://doi.org/10.1007/978-3-642-39106-4

C. B. Laney, *Transformation and Self-Similarity Properties of Gamma and Weibull Fragment Size Distributions*, Defense Threat Reduction Agency Technical Report DTRA-TR-16-06, December 2015, http://apps.dtic.mil/sti/citations/ADA624878

C. B. Laney, *On the Relationships Between Liquid and Solid Fragmentation*, Defense Threat Reduction Agency Technical Report DTRA-TR-16-19, January 2016a, http://apps.dtic.mil/sti/citations/AD1004358

C. B. Laney, *Effects of Mixtures on Liquid and Solid Fragment Size Distributions*, Defense Threat Reduction Agency Technical Report DTRA-TR-16-52, May 2016b, http://apps.dtic.mil/sti/citations/AD1008708

S. M. Lynch, *Introduction to Applied Bayesian Statistics and Estimation for Social Scientists*, Springer, 2007, https://doi.org/10.1007/978-0-387-71265-9 [section 3.5.2]

R. M. Musleh & A. Helu, Estimation of the Inverse Weibull Distribution Based on Progressively Censored Data: Comparative Study, *Reliability Engineering and System Safety*, 131, 216–227, 2014, https://doi.org/10.1016/j.ress.2014.07.006

S. Nukiyama & Y. Tanasawa, An Experiment on the Atomization of Liquid by Means of an Air Stream, *Transactions of the Society of Mechanical Engineers, Japan*, 4(14), 128-135, 1938a, https://doi.org/10.1299/kikai1938.4.128

S. Nukiyama & Y. Tanasawa, An Experiment on the Atomization of Liquid by Means of an Air Stream (the 2nd Report), *Transactions of the Society of Mechanical Engineers, Japan*, 4(15), 138–143, 1938b, https://doi.org/10.1299/kikai1938.4.15_138

B. O. Oluyede & T. Yang, Generalizations of the Inverse Weibull and Related Distributions with Application, *Electronic Journal of Applied Statistical Analysis*, 7(1), 94–116, 2014, https://doi.org/10.1285/i20705948v7n1p94

H. Rinne, *The Weibull Distribution: A Handbook*, CRC Press, 2008, https://doi.org/10.1201/9781420087444

F. Rodriguez, C. Cohen, C. K. Ober, & L. A. Archer, *Principles of Polymer Systems*, 6th edition, CRC Press, 2014

P. Rosin & E. Rammler, Feinheit und Struktur des Kohlenstaubs (Fineness and Structure of Coal Dust), *Zeitschrift des Vereins Deutscher Ingenieure*, 71, 1–7, 1927a

P. Rosin & E. Rammler, Feinheit und Struktur des Kohlenstaubes unter dem Einfluß von Mühlen- und Kohlenart (Fineness and Structure of Coal Dust under the Influence of Mill and Coal Type), *Zement*, 16(35), 820–823, 16(36), 840–844, 16(38), 897–901, 1927b

P. Rosin & E. Rammler, The Laws Governing the Fineness of Powdered Coal, *Journal of the Institute of Fuel*, 7, 29–36, 1933

P. Rosin, E. Rammler, & K. Sperling, Korngrößenprobleme des Kohlenstaubes und ihre Bedeutung für die Vermahlung [Particle Size Problems of Coal Dust and their Importance for Grinding], *Bericht C 52 des Reichskohlenrates* [Report C52 of the National Coal Council], VDI-Verlag, 1933

H. Schubert & E. Wächtler, Erich Rammler—A Pioneer of Particle Technology, *Particle & Particle Systems Characterization*, 4(1–4), 45–48, 1987, https://doi.org/10.1002/ppsc.19870040110

E. V. Schuhmann, Principles of Communition, Size Distribution and Surface Calculations, *AIME Technical Publication*, 1189, 1–11, 1940, http://www.onemine.org/document/abstract.cfm?docid=9197

G. V. Schulz, Über die Kinetik der Kettenpolymerisationen. V. Der Einfluß verschiedener Reaktionsarten auf die Polymolekularität, *Zeitschrift für Physikalische Chemie, Abteilung B, Chemie der Elementarprozesse*, B43, 25–46, 1939, https://doi.org/10.1515/zpch-1939-4304

M. G. Sohail, B. Wang, A. Jain, R. Kahraman, N. G. Ozerkan, B. Gencturk, M. Dawood, & A. Belarbi, Advancements in Concrete Mix Designs: High-Performance and Ultrahigh-Performance Concretes from 1970 to 2016, *Journal of Materials in Civil Engineering*, 30(3), 04017310, 2018, https://doi.org/10.1061/(ASCE)MT.1943-5533.0002144

E. W. Stacy, A Generalization of the Gamma Distribution, *The Annals of Mathematical Statistics*, 33, 1187–1192, 1962, https://doi.org/10.1214/aoms/1177704481

D. Stoyan, Weibull, RRSB or Extreme-Value Theorists?, *Metrika*, 76(2), 153–159, 2013, https://doi.org/10.1007/s00184-011-0380-6

H. Tanizaki, *Computational Methods in Statistics and Econometrics*, CRC Press, 2004, https://doi.org/10.1201/9780203022023

W. Weibull, A Statistical Theory of the Strength of Materials, *Ingeniörs Vetenskaps Akademiens Handligar*, 151, 1–45, 1939a, https://www.scribd.com/document/353394483

W. Weibull, The Phenomenon of Rupture in Solids, *Ingeniörs Vetenskaps Akademiens Handligar*, 153, 1–55, 1939b, https://www.scribd.com/document/348631487

W. Weibull, A Statistical Distribution Function of Wide Applicability, *Journal of Applied Mechanics*, 18(3), 293–297, 1951, https://doi.org/10.1115/1.4010337

B. H. Zimm, Apparatus and Methods for Measurement and Interpretation of the Angular Variation of Light Scattering: Preliminary Results on Polystyrene Solutions, *The Journal of Chemical Physics*, 16(12), 1099–1115, 1948, https://doi.org/10.1063/1.1746740

Index

Note: The following abbreviations have been used: *f* = figures/photographs, n = notes, *t* = tables

α-crystallin (eye lens protein) 278*t*
Aarseth, J. 68
Abaqus (commercial finite element code) 391
Abrahamson, J. 213*t*
Abramowitz, M. 336
Abulwafa, E.M. et al 209, 239, 250*t* n4a
Adam, M. et al 280*t*
aerodynamics *vi*, 2, 5*f*, 125, 126, 278, 282, 487
  secondary fragmentation 431*f*, 434, 435*f*, 436, 437*f*, 440, 456, 460, 463–4, 465*f*
  unitary fragmentation 386*t*, 390, 391*f*, 392*f*, 422, 423
aeronautics 73, 198
aerosol dynamics *xv*, 124*t*, 125*t*, 130*t*, 214, 218, 250*t*, 281, 422–3
aerospace engineering *vi*, *xv*, 81*t*, 423, 507
aggregation *see* primary aggregation; reverse aggregation; secondary aggregation; secondary aggregation, incomplete unitary aggregation; unitary aggregation
aggregation cascades *see* full mean-field solutions; late-time mean-field solutions; mean-field governing equations
aggregation rate *see* collision/collection kernel
Ahmadabadi, A. 512*t*, 513*t*
Ahmadi, M. 491, 495
Ahrens, T.J. 149*f*
AIP Publishing 151*f*, 415*f*, 437*f*, 457*f*
Airy functions 232
Aizenman, M. 295
Alder, J. 71
Aldous, D.J. 104, 226
ALE *see* Arbitrary Lagrangian–Eulerian
Alger, M. 292–3
aluminum/alumina 456*t*
American Chemical Society 314*f*
American Geophysical Union 154*f*
American Institute of Physics *xv*
American Meteorological Society 146*f*
American Physical Society 7*f*, 83*f*, 162*f*, 244*f*, 246*f*, 248*f*, 463*f*
ammonium nitrate–fuel oil (ANFO) 123
analytic methods *xviii*, *xix–xx*, 1, 16, 94, 121, 364, 500–2, 503*t*, 504–6, 507
  binary fragmentation 297*t*, 311*f*, 313*f*, 314*f*
  dimensional analysis 113–14
  energy and energy distributions 194, 197*t*, 201
  examples of results for aggregation 123, 124–5*t*, 126
  examples of results for fragmentation 126, 127–8*t*, 129
  examples of simultaneous results for aggregation/fragmentation 123, 129, 130–1*t*
  late-time mean-field solutions 257, 259, 275, 276, 282, 284
  maximum entropy theory 110–13, 480, 481, 483, 493, 494, 495, 496
  mean-field theory 103–6, 220, 232–3, 240, 243*f*, 244*f*, 246*f*, 248*f*, 250*t*, 251
  particle velocity distributions 170, 175, 176*f*
  random fragmentation 331*t*, 335, 336, 345–6, 349, 350*f*, 354
  shapes and shapes distributions 156, 164
  stochastic geometry 106–7, 108*f*, 109–10
    see also empirical methods; semi-empirical methods; unitary fragmentation
AnaTec Fine Particle Analyzer 157, 158*f*
Andersen, J. 516
Andreasen, A.H.M. 516
Andreotti, B. et al 185
Andrews, J.R.G. 130*t*, 331*t*
ANFO *see* ammonium nitrate–fuel oil
anhydride cured epoxies 280*t*
anisotropic fragments 160
ANSYS Autodyn (commercial finite element code) 391
ANSYS Fluent (Computational Fluid Dynamics code) 391
Antoun, T. et al 389, 393, 394
Aplite dust 157*f*
Apollonian sphere packings (random) 450
apparent cumulative grinding rate models 351
Applied Research Associates (ARA) 5*f*
ARA *see* Applied Research Associates
Arbitrary Lagrangian–Eulerian (ALE) 92, 93*t*
Argon gas 279
Arnold, W. 76*t*, 105, 455*t*
artificial dissipation *see* dissipation and dissipative range
artificial viscosity 94

Aschwanden, M.J. et al 30, 32, 33, 127*t*
Ashby, Ross 30
aspect ratios 385, 419, 426
  shapes and shape distributions 147–8, 149*f*, 150, 151–2*f*, 154–5*f*, 159–60, 161, 163, 164
asphaltene 278*t*
asteroids 6, 126, 292*t*, 320*t*, 323
  collisional fragmentation 363*t*, 364, 375, 376–7, 378, 379*t*
Åström, A. et al 81*t*, 84–5, 86*f*, 117, 127*t*, 450, 461*t*
  astrophysics *vi*, *xv*, *xvi*, 6, 33, 128*t*, 507, 510
  aggregation cascades 215*t*, 217–18, 228, 250*t*
  analytic/empirical model results for aggregation/fragmentation 124*t*, 126, 127*t*, 130*t*
  collisional fragmentation 363*t*, 364, 376, 377, 378, 380–1
  computational methods 68–9, 70*t*, 71, 72*t*, 74, 81*t*
  fragmentation cascades 292*t*, 320*t*, 323
atmospheric convection 28
atmospheric sciences *xv*, 5
atmospheric turbulence 258
atomic bonding time 10
atomic charge sets 133
atomization (primary/secondary) 114–15, 418*f*, 427*f*, 428, 429–30*f*, 459*f*, 466*t*
atoms *xv*, *xvii*, 9, 69, 125
attractors 28
Atwood, I. et al 87*f*
Austin empirical formula 351
autocorrelated 23
autocovariance 180, 181
avalanches 32*t*, 33, 319, 500
averages 53–4

β-lactoglobulin (globular protein) 278*t*
Babić, K. 192
Babick, F. 14
Babinsky, E. 120
bag breakup 434, 435*f*
Bai, C.X. et al 462*t*
Bak, P. et al 32–3, 195
Bak, T.A. 295
Baldassarri, A. et al 190, 191*t*
ball bearings (stainless steel) 3*f*, 28, 381
ballistic aggregation 71, 170, 186
Ballistic-Limited (BL) 155, 156
ballistic trajectories 71, 155
ball milling (iron) 116*f*, 320*t*, 323, 347, 348*f*, 349*f*, 350*f*, 351–2, 355
Ball, R.C. et al 155*f*, 266, 267
Balslev, I. 461*t*
Banasiak, J. et al 1, 226
Bannikova, I. et al 456*t*
Barré, L. 278*t*
Barrow, D. 130*t*
Barthelmes, G. et al 331*t*
Barth, H.G. 247
basalt 88*f*
Bassi, F.A. et al 278*t*
Bass, V.L. 323–4, 325
Bastea, S. 68, 69, 70*t*
Bastian, N. et al 129
Batchelor, G.K. 197*t* n1, 6, 200
batch grinding 320, 324–5, 331*t*, 332, 348, 349*f*, 350*f*, 351–2
Baum, H.R. 250*t* n3
Baxter, W. 24
Beck, C. 477–8
Becker, R. 130*t*, 294
Behera, B. 81*t*
bench blasting 121, 122*f*, 123
bending/buckling-type instabilities/strains 16, 90, 92*t*, 389, 397*f*, 419
Bendon, D.A. 76*t*
Benedek, G.B. 124*t*, 482
Ben-Naim, E. 127*t*, 365–6, 367, 374, 451*f*, 453*t*
Bennett, G. 54, 57
Berera, A. 192
Bergstrom, B.H. 127*t*
Bergstrom, H. 13–14
Berry, E.X. 213*t*
Bertalanff, Ludwig von 30
Bertoin, J. 1, 226
Bess, D. 127*t*, 129
Bessel functions of the first kind 232–3, 234, 236, 238, 239, 300, 335, 336, 338
Besson, J. et al 393
Bewick, B. et al 4*f*, 5*f*, 48*f*, 79*f*, 387*f*
  velocity and velocity distributions, 174, 175*f*, 176*f*
biaxial strains *see* uniform uniaxial/equi-biaxial stress/strains
Billingham, N.C. 292
binary accretion 218
binary aggregation 302, 315
binary collisions *xviii*, *xxi*, 2, 15 n2, 27, 33, 71, 93*t*, 185, 499, 501, 502, 503
  collisional fragmentation 366, 367, 372, 373
  mean-field theory 207, 210, 449–50
  random fragmentation 321, 330–1
  velocity and velocity distributions 170, 171, 494
binary fragmentation 87, 292*t*, 315, 363*t*, 450, 499, 500, 503*t*
  experimental validation 308, 309*t*, 310, 311*f*, 312, 313–14*f*
  full mean-field solutions 299–307
  governing equations 293–6
  kernels 296, 297*t*
  late-time mean-field solutions 307–8
  maximum entropy theory 477, 482, 483
  random fragmentation and 319, 320*t*, 322, 326
  shattering 297–8
  size distributions/moments 298–9, 380*t*
    see also collisional fragmentation; random fragmentation
bin limits 49–50
biological cell growth processes 117
biosciences *xv*, 30, 131*t*

Biswas, A. et al 171–2
blasting *see* drilling and blasting
Blatz, P.J. 130*t*, 295
Blum, J. 257
BMG *see* bulk metallic glass
Bodrova, S. 130*t*
body forces *see* long-range body forces
Bolton, J. 170
Boltzmann constant 170, 171, 175, 214
Boltzmann–Fokker–Planck equation 175
Boltzmann, L. 22, 110
Bonabeau, E. et al 131*t*
bonded particle systems 6, 7*f*, 16, 130, 186, 195–6, 199, 200, 310, 505
Bond, F.C. 354
Borg, J.P. et al 423
Bossa, B. 406*t*, 411*t*, 434–5, 436*f*, 465*f*
Bottke, W.F. et al 379*t* n1
bottom charges 123
Bouchaud, E. et al 280*t*
boundary conditions 68, 163, 164, 354, 501
  collisional fragmentation 375, 379, 380*t*
  energy and energy distributions 188, 189, 192
  late-time mean-field solutions 257, 259, 261, 264–5, 268–70, 274–5, 282–3, 284*f*, 285, 346
  unitary fragmentation 402, 437*f*, 438*f*, 439, 449
    see also free-field boundary conditions; periodic boundary conditions
bound particles 80
Boussinesq's eddy-viscosity closure 196
branched structures 156, 163, 213*t*, 214*t*, 247, 280*t*, 419
Braun, S. et al 75
breakage functions 321–3, 325–9, 331*t*, 348, 349, 350–1, 352, 368, 419
breakup 13, 44, 73, 81, 164, 331*t*, 463–4, 465*f*
  analytical/empirical methods 115, 125, 127, 128*t*, 129
  primary fragmentation 423, 426, 427
  secondary fragmentation 431*f*, 432, 433*f*, 434, 435–8*f*, 439, 440*f*
Brilliantov, V. (et al) 31, 71, 130*t*, 282
brittle solids 27, 150
  computational methods 74, 80, 89, 92*t*
  unitary fragmentation 389, 391, 404, 405–6*t*, 407, 409, 412*t*, 413–14, 419, 440
Brock, J.R. 213*t*, 218, 261, 265, 266, 285
Brodskii, R.Ye. 452*t*
Broide, M.L. 242, 243*f*, 244*f*, 278*t*
Brownian Dynamics (BD) 9, 10, 25, 105, 328, 500, 505
  aggregation cascades 213*t*, 214, 242, 285
  computational methods 71, 72*t*, 73, 93*t*
Brzustowski, Thomas A. 113, 485, 493
bubbles (aggregation/fragmentation) 215, 386*t*
Buchanan, M. 398
buckling *see* bending/buckling-type instabilities/strains
bulk metallic glass (BMG) 432, 433*f*
Bunde, A. 186*f*
buoyancy 193
Burton, C. et al 185, 189
Bushell, C. et al 10
Bushuev, G. et al 69, 70*t*, 133
butterfly effect 28, 249
Buy, F. 395

Cady, F. 117
Camacho, J. 268
Cambon, C. 193, 198
Cambridge University Press 390*f*, 424*f*, 436*f*, 438*f*
Campbell, C.S. 92*t*
Camp, T.R. 213*t*
capillary instabilities 387, 388
Carey, V.P. 476
Carmona, A. et al 80, 81, 87, 89*f*, 90, 91, 92*t*
Carnevale, F. et al 170
Carpinteri, A. 117, 127*t*
Carry, B. 379*t* n4
Cartesian geometries 425–6, 427–30*f*, 441
Cartesian (uniaxial) coordinate systems 394*f*, 396, 397*f*, 454
cascades (aggregation/fragmentation) 4, 5–6, 15, 16, 126, 127, 332, 499, 502, 506
  analytical/empirical methods 106, 107, 108
  energy and energy distributions 193, 194, 195–6, 198
  unitary fragmentation 385, 390, 398, 449, 468
    see also full mean-field solutions; late-time mean-field solutions; mean-field governing equations
catastrophes 32*t*, 33, 126, 319, 435*f*, 500
cavitation 75, 115, 394
Cecconi, F. et al 25
cells 67, 106, 506
cellulose degradation 107, 109
Celotex 455*t*
centrifugal forces 91, 320
Cepuritis, R. et al 157*f*
CFD *see* Computational Fluid Dynamics
chain-end degradation 293, 294, 309, 310, 311, 312, 315, 500
chains 106–7, 301
Chakrabarti, A. et al 69, 70*t*
Chakraborty, B. 211
Chandraeskhar, S. 105
chaos theory and systems *vii*, *xvii*, *xviii*, *xxi*, 64, 249, 477–8, 502, 503*t*, 504–7, 508*t*
  analytical/empirical methods 103, 105, 133
  computational methods 68, 69, 71, 74, 75, 77, 78, 80, 92, 93*t*, 94
  energy and energy distributions 185, 191, 192, 193, 194, 199, 220
  fragmentation cascades 293, 299, 319, 347, 364, 380
  mean-field governing equations 207, 211, 212, 216, 219, 220
  properties of systems 22*t*, 23*f*, 24*f*, 25, 26*f*, 27–8, 29*t*, 31, 32*t*, 34, 37, 38
  unitary fragmentation 384, 386*t*, 389, 392, 393, 394, 402, 449, 457, 464

characteristic (fragment) diameters 123, 353, 411, 413, 414, 416*f*, 421*f*, 433*f*, 441*t*, 442
charged-particle aggregation experiments 279
Charles, R.J. 354
Chaussonnet, G. et al 405*t*, 411*t*, 430*f*, 458, 459*f*
Chayut, M. 112
chemistry and chemical engineering *vi*, *xv*, 30, 198, 292*t*, 293, 320*t*, 353, 425, 477, 507
Cheng, Z. 367
Chen, J. 80
Cheong, Y.S. et al 460
Chernin, D. 26
Chervanyov, I. 130*t*
children 2, 8, 27, 129, 130, 500
  binary fragmentation 293, 294, 295, 296, 315, 319
  collisional fragmentation 363, 364, 365, 366, 368, 373, 374, 376, 377, 379, 380*t*
  random fragmentation 320–3, 325, 326–8, 332, 341, 344, 346–8, 354, 355
  unitary fragmentation 402, 421, 436, 449
chondrules 363
Chou, W.-H. et al 436, 437, 438, 439, 440*f*, 463–4
Chu, C.-R. 197*t* n1
Chung, G.Y. 309*t*
civil engineering *xv*, 126, 198, 507
Clauset, A. 117
clay 4*f*, 461*t*
closed-form solutions
  aggregation cascades 229–40, 250, 257
  binary fragmentation 299–307, 315
  collisional fragmentation 369*t*, 370, 373, 374
  random fragmentation 324, 326, 334–5, 337, 338–45, 355
closed particle systems 3, 476
cloud physics *xv*, 5, 124*t*, 125*t*, 153, 154*f*, 464, 492, 495
  aggregation cascades 228, 250*t*, 276, 281–2, 285
Clough, W. 73
clusters and clustering *xv*, 5, 6, 9, 14, 125, 130, 171, 297*t*
  aggregation cascades 238, 239, 241*f*, 245, 248, 269, 270, 280*t*
  basic properties of particles 28, 31, 33
  computational methods 68, 69–70, 75
  energy and energy distributions 179, 185, 186*f*, 187*f*, 188*f*, 189–91, 194–5
  maximum entropy theory 477, 479, 485
CMD *see* count median diameter
coagulation equation (Smoluchowski) 207–8, 214*t*, 266, 272, 273
coagulation models 364, 365
coalescence time 215, 217, 268, 393, 394, 419
coefficient of spatial correlation 181
coefficient of temporal correlation *see* autocorrelation
Cohen, R.J. 124*t*, 242, 243*f*, 244*f*, 278*t*, 482
Colby, R.H. et al 280*t*
Cole, G.H.A. 209
collapse-driven fragmentation 126, 278
collisional fragmentation 292*t*, 499, 500, 503*t*
  experimental validation 378, 379*t*
  fragmentation cascades 320*t*, 321, 363*t*, 364, 379, 380*t*, 381, 385*t*
  full mean-field solutions 368, 369*t*, 370
  governing equations 364–8
  late-time mean-field solutions 370, 371*f*, 372–8
  shattering 368
    see also binary fragmentation; random fragmentation
collision/collection kernels (aggregation rate) 481, 485
  aggregation cascades 242, 252, 284, 380*t*
  late-time mean-field solutions 257, 258, 259, 261*t*, 262–6, 273, 281
  mean-field governing equations 207, 208, 210–12, 213*t*, 214–16, 220
collision frequency function *see* collision/collection kernel (aggregation rate)
collisions xix, 3, 5*f*, 6, 9–10, 31, 37, 63, 170, 364, 499, 502, 506
  analytical/semi-empirical/empirical methods 104, 114, 124, 126, 129
  chaos theory 22*t*, 23*f*, 24, 25, 26*f*
  computational methods 68–9, 71, 80, 92*t*, 93*t*
  energy and energy distributions 185, 188–9, 190
  late-time mean-field solutions 260, 265, 267, 268, 271–2, 281, 282, 283
  maximum entropy theory 477, 480
  mean-field governing equations 207, 210, 211, 213–15*t*, 217
  random fragmentation 320, 321, 330, 351
  unitary fragmentation 385, 390, 392*f*, 431, 459, 460, 461*t*, 462*t*, 463*t*
colloids *xviii*, 1, 9*f*, 14, 155, 156, 297*t*
  analytical/empirical methods 105, 124*t*, 125*t*, 130*t*, 131*t*, 133
  computational methods 69, 70*t*, 71, 72*t*
  full mean-field solutions 231, 241, 242, 243*f*, 244*f*, 245*t*, 246*f*, 250*t*, 251
  late-time mean-field solutions 276, 277–8*t*, 279, 280
Colonius, T. 438*f*
column charges 123
combinatorics 478–81, 482*t*, 491
combined aggregation-fragmentation 282
combustion *v*, 114, 331*t*, 425
compaction *see* secondary aggregation
complementary cumulative distribution functions 509
complementary error function 62, 305
Complete Neglect of Differential Overlap (CNDO) charge set 69
complex systems 29
compressible turbulence (variable density) 193, 195
compressive strains/stresses 393, 394, 396*f*, 397*f*, 398, 432, 449, 456
Computational Fluid Dynamics (CFD) 391
Computational Gas dynamics (Laney) *vi*
computational models/methods *xviii*, *xix–xx*, 1, 16, 32, 49, 67, 93*t*, 94, 500–2, 504–7, 508*t*
  aggregation cascades 220, 240, 251, 269, 276, 282, 284–5
  analytical/empirical methods and 103, 116, 132, 133

energy and energy distributions 189, 191*t*, 201
high-fidelity methods for particle aggregation 68–9, 70*t*
high-fidelity methods for particle fragmentation 73–5, 76*t*, 77, 78–9*f*, 351–2
maximum entropy theory 480, 481, 487, 493
particle shape distributions 154, 155*f*, 156, 164
simplified methods for aggregation 70–1, 72*t*, 73
simplified methods for fragmentation 79–80, 81*t*, 82, 83–4*f*, 85, 86–90*f*, 91, 92*t*
unitary fragmentation 391, 392, 393, 415*f*, 422–3, 424*f*, 438*f*, 465
velocity distributions 173, 175
Computerized Particle Analyzer (CPA-4) (W.S. Tyler) 47*f*
computer science 186
concentration 114, 225, 226, 390
binary fragmentation 299, 300, 301, 302, 304, 307, 312, 314
distributions/averages 44–5, 46*f*, 47*f*, 52, 63*t*
mean-field governing equations 207, 208, 210
random fragmentation 244, 321, 338, 339, 340, 342, 343, 364
concrete *v*, 2, 5*f*, 176f
computational methods 75, 76–7, 78–9*f*, 92*t*
unitary fragmentation 384, 387*f*, 394, 396*f*, 431*f*
condensation 258–9, 281, 285, 355, 499
conditional probability density function 169, 465
confluent hypergeometric function (Kummer's) 232, 234, 335, 336*t*, 337t
Connaughton, C. et al 268, 276, 282
conservation of (kinetic/surface) energy 172, 354, 404, 406–8, 409, 493
conservation of mass 23, 24, 31, 219–220, 274, 275, 402, 449
binary fragmentation 298, 299, 300, 301, 302–3, 304, 305, 306, 307
collisional fragmentation 368, 369–70, 372, 373, 374, 380
computational methods 68, 73, 74, 79, 80
full mean-field solutions 227, 228, 229, 230, 231, 234, 235, 237, 238, 239, 240, 249
maximum entropy theory 479, 485, 493
random fragmentation 322, 323, 326, 329, 334, 338, 340, 342, 345
constant (rate) kernels
fragmentation cascades 294, 302, 331*t*, 370
full mean-field solutions 227, 229–34, 250
late-time mean-field solutions 257, 275–6
constrained optimization 110, 477, 478–81, 482*t*, 505
continuous governing equations 208–10, 295–6, 322–5, 366–8
continuous particle systems 6–8, 11, 16, 63*t*, 324, 374, 483, 485–6, 495–6, 505
aggregation cascades 210, 216, 225, 235
analytical methods 107–8, 111, 112, 113–14, 129
binary fragmentation 296 300, 301, 302, 306
families of size distributions 486–7, 488*t*, 489–91, 504
specific size distributions 491–3
validation 493–4
velocity distributions 493
see also discrete particle systems; maximum entropy theory
Continuous Surface Cap Model *see* Material Model 159
continuum regime 213*t*, 295
convection 266
CONVERGE 115
conversion techniques 517, 518*t*, 519*t*
Cooney, R.P. 292–3
correlation/anticorrelation 23, 180, 181, 182, 187*f*, 188*f*, 190
Coulaloglou, C.A. 329–30
Coulombic interactions/forces 69, 80, 457
counting-based measurement techniques 44, 321, 366
count mean diameter (CMD) 55, 421, 424*f*, 425, 436*f*, 441*t*
Coupier, D. 106
covalently bonded atoms/molecules 125
covariances *see* variances/covariances (energy)
Cover, T.M. 514*t*, 515*t*, 516*t*, 517*t*
Cowan, R. 453*t*
crack propagation 15, 22, 27, 74, 80, 174, 502
unitary fragmentation 391, 394, 419, 452*t*
cratering 126
cross-stream strains 410, 417, 426, 427, 429–30, 439
Crutchfield, J.P. 3*f*, 28, 476
crystallography 133, 183, 189, 213*t*, 250*t*, 485
computational methods 69, 70*t*, 92*t*
shapes and shape distributions 153, 154*f*, 162
cumulative breakage function 327–8, 348, 349*f*, 351
cumulative distribution functions *xvii*, 49–52, 63*t*, 157, 199, 377, 509
Cundall, A. 80
Cunningham, B. 123
Curran, D.R. (et al) 150, 161, 389
Curry, J.A. 258–9, 337, 492
curve fits *v*, 51, 85, 93*t*, 132, 283, 501
fragmentation cascades 311*f*, 347, 348*f*, 350*f*, 353, 355
particle distributions 160–1, 164, 175, 176, 189, 190
secondary fragmentation 432, 433*f*, 438
semi-empirical/empirical methods 103, 116*f*, 117, 118, 119, 120–2*f*, 123, 233
unitary fragmentation 441*f*, 455*t*, 458*f*, 459, 463, 464, 465, 466
Cuzzi, J.N. et al 379*t* n12
cybernetics 30
cylinders and cylindrical shells *v*, *xv*, 12, 150, 151*f*, 172, 173*f*, 185, 485
analytical/empirical methods 119, 120–2*f*
size distributions 454, 455–6*t*
unitary fragmentation 394*f*, 395, 405–6*t*, 408, 411*t*, 412*t*, 413–14, 415–16*f*, 419, 422–3, 424*f*, 432, 433*f*, 441, 461*t*
cylindrical geometries 413–14, 415–16*f*, 417, 418*f*

Da Costa, F.P. 213*t*, 365
Dagorn, L. 131*t*
Dai, Z. 405*t*, 411*t*, 428*f*
Daphalapurkar, N.P. et al 74
Daraboina, N. 309*t*
Das, S.K. et al 170–1, 188, 189, 190, 191*t*, 331*t*
daughters 2, 292, 340, 453
Davidson, Peter A. *vi*, *xxii* 163, 192, 193, 194, 199, 200
Davis, A. 109
Davis, D.R. 173, 377
Davis, J.R. 396, 397
Deal, C. 456*t*
De Carvalho, R.M. 320
Degond, P. et al 131*t*
  degradation 107, 109, 323
  binary fragmentation 293, 309*t*, 310, 311*f*, 312, 313*f*, 314*f*, 315, 363*t*
  chain-end degradation 293, 294, 309, 310, 311, 312, 315, 500
  random fragmentation 320*t*, 323, 331*t*
  thermal degradation *xviii*, 477
degrees of freedom (Maxwell's concept) *xviii*, *xxi*, 28–9, 211, 212, 330–1, 477, 502
deliberately self-organized systems 29, 30–1
DeMeo, F.E. 379 n4
DEMs *see* Discrete Element Methods
DEM Solutions (UK) 352, 393
dendrimers 163
dense particle systems 80, 501–2
density *v*, 63, 153, 299, 380, 389, 458, 462, 500, 509
  energy and energy distributions 179–81, 187*f*, 188–9, 195, 198
  experimental validation 403, 405, 410, 411*t*, 412, 414, 417, 418*f*, 425, 427–8, 430, 434, 442
  random fragmentation 320, 321, 324, 332, 353
  secondary fragmentation 438, 439
deoxyribonucleic acid (DNA) 31
De Regt, H.W. 212
Derjaguin–Landau–Verwey–Overbeek (DLVO) potential 69
Deshpande, S. et al 27
determinism/non-determinism vi, *vii*, 15–16, 64, 179, 198, 319, 364, 500, 504, 506–7, 508*t*
  aggregation cascades 216, 219, 258
  binary fragmentation 294, 299, 315
  computational methods 67–8, 70, 75, 77, 80, 92, 93*t*
  unitary fragmentation 385, 410, 441, 453
deterministic cluster-finding algorithms 185, 198
detonation *see* explosions
Dhar, D. 452*t*
diameters 10, 11, 174, 387, 441*t*, 459, 462, 486, 492, 509
  aggregation cascades 212, 214, 215, 243, 246*f*
  Cartesian geometries 425, 426, 427*f*, 428*f*
  distributions/averages 43, 55*t*, 56, 57, 60, 62*f*, 63
  energy and energy distributions 198, 199, 200
  particle shape distributions 147, 157, 158*f*, 159, 160, 165
  random fragmentation 324, 328, 330, 332, 352, 353
  secondary fragmentation 432, 433*f*, 434, 435*f*, 436*f*, 438
  unitary fragmentation 403–4, 405–6*t*, 408, 410*f*, 411–12*t*, 413–14, 415–16*f*, 418*f*, 421*t*, 424*f*
Dias, L. 384, 450, 452*t*
diatomic gases 212
dielectric degree of homogeneity 279
dielectric liquid (castor oil) 28
diffusion 105, 213*t*, 220, 266, 281, 309
Diffusion-Limited Cluster-Cluster/Cluster-Monomer Aggregation (DLCA/DLA) 10, 14, 15, 25, 133, 155, 241, 242, 243*f*, 244*f*, 245*t*, 246f
Diffusion-Limited (DL) (colloidal particles theory) 69, 105, 155, 156
Diggle, J. 103
digital in-line holography (DIH) 423
dimensional analysis *xx*, 170, 196, 272, 353, 393, 504, 505, 509
  analytical/empirical methods 113–14, 124*t*, 127*t*
  particle shape distributions 146*f*, 147, 159–60, 163
  unitary fragmentation 403, 404, 405, 412*t*, 417, 426, 434
  see also size and size distributions
Dinger, D.R. 516
Dirac delta function 183, 209, 216, 229
Direct Numerical Simulation (DNS) 392, 506
Discrete Element Methods (DEMs) *vii*, 173, 191*t*, 269, 351–2, 392, 505
  computational methods 80, 81*t*, 82, 83–4*f*, 85, 86–90*f*, 91, 92–3*t*, 94
discrete governing equations 207–8, 293–5, 321–2, 365–6
discrete particle systems 6, 7*f*, 8, 10–11, 16, 52, 324, 374, 476–8, 483, 485, 491, 494–5, 505
  aggregation cascades 225, 235, 282
  analytical methods 106–7, 110, 112–13, 130
  binary fragmentation 296, 298, 300, 301, 302, 306
  combinatorics and constrained optimization 478–81, 482*t*
  computational methods 63*t*, 69
  see also maximum entropy theory: continuous particle systems
disorder (entropy theory) 476, 477
dissipation and dissipative range 94, 133, 215, 218, 266, 499, 506
  basic properties of particles 29*t*, 31, 32*t*
  energy and energy distributions 185, 192, 193, 194, 196, 197*t* n7
  random fragmentation 328, 329, 352, 354
  unitary fragmentation 392, 394, 398, 404, 405, 406, 407
distributions *see* concentration distributions; probability density functions; cumulative distribution functions; number distributions; size distributions
DLCA/DLA *see* Diffusion-Limited Cluster-Cluster/Cluster-Monomer Aggregation
DNA *see* deoxyribonucleic acid
DNS *see* direct numerical simulation

Dohnanyi, J.S. 127*t*, 128*t*, 366, 375, 376, 377
Domaradzki, J.A. 197*t*
Domokos, G. et al 146*f*, 149*f*, 150
Donangelo, R. 452*t*
Döring, W. 130, 294
Dorschner, J. 366, 376
Dos Santos, F.P.M. et al 451*f*, 452*t*
Dostal, H. 213*t*
drag (shape dependent) 155
Dresden, M. 105
drilling and blasting *xv*, 117, 126, 174, 431*f*
driven granular gases 185, 189, 191*t*, 193, 195
droplets *see* liquids
drop shatter tests 460, 461*t*
drop-size distribution 487
dry ice *see* ice and ice crystals
ductile materials 150, 389, 394, 405, 407, 412*t*, 413, 440
Dumouchel, C. 454, 486, 487
Dunnavan, L. et al 146*f*, 154, 155*f*, 162–3
Durda, D. et al 12, 149*f*
dynamic fracture toughness 404, 405, 414, 415*f*, 416, 432, 441
dynamic light scattering 245
dynamic particle systems 29, 105, 501
dynamic viscosity 194, 354, 405, 407, 412, 438
Dziekan, P. 281

early time clustering regime 190, 191*t*
earth sciences *vi*, *xv*
Ebeling, W. 30–1
eddies *see* fluid turbulence
EDEM software 352, 393
EDMD *see* Event-Driven Molecular Dynamics
Edwards, M.R. 456*t*
eggshells
Eigen, M. 30
Einstein, Alfred 105
ejecta (geologic materials) 126
Elaissari, A. 277*t*
elastic collisions xxi, 71, 170, 171, 185, 190, 328, 380, 502
  basic properties of particles 23, 24, 25, 27, 31, 38
elastic potential energy 13, 406, 412
elastic wave speed 405, 414
electrical engineering 30
electromagnetic/electrostatic forces 3*f*, 6, 9, 28, 69, 285
electron microscopy 74
electrostatic repulsion 242
element erosion models 77
Elger, D.F. et al 506
Elsinga, E. 196
Emery, J.P. et al 379*t* n9
Empie, H.J. et al 127*t*, 466
empirical methods *xx*, 1, 103, 164, 175, 197*t*, 201, 233, 493, 501, 505–7
  examples of results for aggregation 123, 124–5*t*, 126
  examples of results for fragmentation 126, 127–8*t*, 129
  modelling 115, 116*f*, 117–19, 120–2*f*
  random fragmentation 347, 348*f*, 350–1, 353, 354, 355
  see also analytic methods; semi-empirical methods; size distributions
energetic materials *xv*, 1
energy analysis 404, 405, 412*t*, 414, 419, 428, 435, 504
energy and energy distributions *v*, *xvii*, *xviii*, *xxi* 3, 5, 201, 249, 329, 363, 381, 485, 493, 499, 500, 502
  basic principles of particles 23, 27, 29*t*, 32*t*, 33, 37
  binary fragmentation 293, 299
  computational methods 68, 69, 73, 74, 75, 79, 80
  discrete particle systems 476–8
  fluid turbulence 179, 191–6, 197*t*, 198
  granular gases 179, 183, 185, 186–8*f*, 189–90, 191*t*
  kinetic energy and size distributions 179, 198–200
  mean-field governing equations 212, 217, 219
  random fragmentation 319, 320, 329, 355
  saturation 87, 88
  structure factors and functions 179, 183–4
  unitary fragmentation 402, 404, 407, 422, 441*t*
  variances/covariances 179–83
energy-containing range 133, 193, 328, 398
energy limited fragmentation 388, 389, 390*f*
energy-size reduction theory (Von Rittinger) 196, 352–4
engineering strain/strain rates 395, 396, 397*f*
Englman, R. et al 485
ensemble and ensemble averages 44, 49, 54, 179, 181–2, 192
Enskog–Boltzmann equation 175
entropy theory *xix*, *xx*, 16, 476–8
Enukashvili, I.M. 214*t*
equibiaxial (spherical) coordinate systems 394*f*, 395
equilibrium *see* thermal equilibrium
Erdman, C.A. 435*f*, 437
Erns*t*, M.H. (et al) 482
  aggregation cascades 239, 262–3, 271, 274
  fragmentation cascades 332–3, 342, 355, 367, 372, 373–4, 375 n1, 376, 377–8
error functions 62, 303, 305
Esipov, E. 171
Eulerian computational methods 67, 92, 249
evaporation 355, 499, 502
Event-Driven Molecular Dynamics (EDMD) 170, 171, 505
  computational methods 71, 72*t*, 93*t*, 94
  energy and energy distributions 185–6, 188–9, 189–190, 191*t*
expected values 179
experimental methods and validation *xviii*, 1, 16, 49, 189, 196, 367, 501, 502, 507
  aggregation cascades 276–8*t*, 279, 280*t*, 281–2
  analytical/empirical methods 116*f*, 120*f*, 121*f*, 122*f*, 132
  binary fragmentation and validation 308, 309*t*, 310, 311*f*, 312, 313, 314*f*
  Cartesian geometries 425–6, 427–30*f*
  computational methods 82, 83*f*, 84*f*, 86

experimental methods and validation (*Continued*)
continuous particle systems 493–4
cylindrical geometries 413–14, 415–16*f*, 417, 418*f*
fragmentation cascades 315, 347, 348–50*f*, 351–2, 378, 379*t*, 381
full mean-field solutions 240, 241*f*, 242, 243–4*f*, 245*t*, 246*f*, 247, 248*f*, 251
late-time mean-field solutions 269, 275, 284–5
maximum entropy theory 480, 481, 487, 494, 495
particle shape distributions 154, 155, 156, 158, 160–1, 162*f*, 164
particle velocity distributions 173, 173, 175*f*, 176*f*
random fragmentation 321, 324, 350*f*, 353, 355
secondary fragmentation 431*f*, 432, 433*f*, 434, 435–8*f*, 439, 440*f*
size distributions of primary fragmentation 454, 455–6*t*, 457–9*f*
size distributions of secondary fragmentation 459, 460, 461–2*t*, 463*f*, 464, 465*f*, 466*t*, 467*f*, 468
spherical geometries 418–20, 421–2*f*, 423, 424*f*, 425
unitary fragmentation 391*f*, 404, 415–16*f*, 417, 418*f*, 419–20, 421–2*f*, 423, 424*f*, 425, 432, 433*f*
explicit/implicit sources/sinks 259
explosions 394, 485
analytical/empirical methods 109–10, 118, 119, 122*f*, 126, 127*t*, 128*t*
computational methods 76*t*, 84*f*, 85
detonation waves (high-explosive) 172, 413–14, 415*f*
particle shape distribution 149*f*, 150, 151*f*, 152*f*, 153, 160–1, 162*f*
particle velocity distributions 172, 173*f*, 174
unitary fragmentation 413–14, 415*f*, 419–20, 421, 422*f*, 455–6*t*, 457*f*, 458*f*, 464
exponential initial conditions 230–1, 234, 236, 242, 302–5, 310, 312
exponential size distributions 72, 116, 118, 334, 454, 465*f*, 490, 504
aggregation cascades 216, 230, 232, 236, 263, 284
analytical methods 104, 107, 108, 111, 113
binary fragmentation 295, 302, 303, 305, 310
distributions/averages 56, 60–1, 64*t*
Extended Finite Element Method (X-FEM) 74, 391
extended vortex tubes/filaments (rotating clusters) *xviii*, 190
extensive quantity 45
extent of reaction 247

Faeth, Gerard M. *vii*, 405*t*, 411*t*, 417, 418*f*, 428*f*
Falkovich, G. 201
Family, F. et al 297*t*
far from equilibrium 29*t*
Farinella, P. 377
FDEM *see* Finite and Discrete Element Method
Feistel, R. 31
FEMs *see* Finite Element Methods
Fenistein, D. 278*t*
Feynman, Richard 201
Field, G.B. 124*t*, 217–18
Filippov, Aleksei Fedorovich 106, 339, 342–4
Findeisen, W. 214*t*
Finite and Discrete Element Method (FDEM) 89, 90*f*, 92*t*
Finite Element Methods (FEMs) 67, 73–5, 76*t*, 78*f*, 79–80, 89*f*, 92*t*, 93*t*, 391–2, 505
finite particle systems *see* infinite/finite particle systems
Finite Volume Methods (FVMs) 67, 81, 93*t*, 505
fire science 250*t*
first-principles methods *xx*, 68, 92, 94, 103, 391
fission *see* binary fragmentation
5/3 law of turbulence (Kolmogorov) *v*, *xxi*, 197, 198, 200, 502
Fixman, M. 112, 491
flaw limited fragmentation 388–9
flexure tests 397
floc suspensions 331*t*
Flory, Paul J. 201
analytical/empirical methods 107, 111, 112, 113, 124*t*, 125*t*
maximum entropy theory 476, 479, 482*t*, 492
fluctuation spectrum 179, 181–2, 183, 190, 191*t*, 192, 195, 197–200, 328
fluid dynamics *xvii*, *xx–xxi*, 25, 125, 163, 183–4, 249, 257, 266, 371, 501, 502
fluidized-bed combustion (coal) 331*t*
fluid turbulence xxi, 133, 179, 191–6, 197*t*, 198, 199–201, 213*t*, 214*t*, 506
Foerster, Heinz von 30
forced locality
late-time mean-field solutions 257, 258, 268, 269, 270, 276, 277, 279, 280, 282
mean-field governing equations 210, 219, 220
'forgetting' 3*f*, 22*t*, 23, 24, 26, 27, 28, 219, 232
formulated products/processes 1, 81*t*
Fourest, T. et al 422
Fourier transforms 183, 184
fractals 72*t*, 225, 231, 245, 279, 332, 501
basic properties of particles 28, 32*t*, 33
particle types 3*f*, 7*f*, 8, 9*f*, 10
shapes and shape distributions 156, 158, 164
fracture energy 89, 320, 405, 407
fracture toughness *see* dynamic fracture toughness
fragmentation *see* primary fragmentation; secondary fragmentation; unitary fragmentation
fragmentation cascades *see* binary fragmentation; collisional fragmentation; random fragmentation
fragment stripper 174
fragment-tracking algorithm 77
Fraser, W.C. et al 379*t* n2&10
free-field boundary conditions 402, 449, 494
binary fragmentation 297, 300–8
collisional fragmentation 369–70, 372–4, 376
full mean-field solutions 225, 228, 229–32, 234–40
late-time mean-field solutions 257, 259, 260, 262, 263, 268, 269, 270, 274–5, 277, 283, 284*f*, 285
mean-field governing equations 216–17, 218*f*, 220

random fragmentation 332, 338, 339, 340, 341–5, 346, 355
see also boundary conditions; periodic boundary conditions
freely cooling/freely decaying 171–2, 185–6, 188–9, 189–90, 191*t*, 193, 195, 197*t*
free-merging aggregation
late-time mean-field solutions 257, 258, 261*t*, 262, 266, 277
mean-field governing equations 210, 218, 219, 220
free molecular regime 213*t*
Friedlander, S.K. 25, 114, 124*t*, 214, 283
Frost, D.L. et al 457*f*
Fry, D. et al 155*f*
Fuchs, N.A. 213*t*
Fuerstenau, D.W. et al 349, 354
Fujiwara, A. et al 149*f*, 150
Fuller size distributions *see* gamma size distributions
Fuller, W. 515–16
full mean-field solutions 225–6, 250*t*, 251
closed-form solutions for constant kernels 227, 229–34
binary fragmentation 299–307, 315
closed-form solutions for product kernels 227, 228, 237–40
closed-form solutions for sum kernels 227, 235–6
collisional fragmentation 368, 369*t*, 370
computational/experimental validation 240, 241*f*, 242, 243–4*f*, 245*t*, 246*f*, 247, 248*f*
gelation 228, 248–9
moments 226–7
random fragmentation 334–5, 336–7*t*, 338–45
size distributions 227
see also cascades; late-time mean-field solutions; mean-field governing equations
fully reacted/unreacted particles 232, 235, 236, 250, 300, 336, 337
Funk, J.E. et al 516
Funk–Dinger/ Dinger–Funk size distributions *see* power law size distributions
Furuhashi, S. 131*t*
fusion 457
FVMs *see* Finite Volume Methods

GA-8 Gilsonic UltraSiever® Sonic Sifter (Gilson) 46*f*
gamma functions 336*t*, 410
gamma size distributions 210–11, 315, 503*t*, 504, 509, 511, 513, 514–17*t*
analytical methods 111, 112, 113, 124*t*, 127*t*, 130, 131*t*, 132, 133
collisional fragmentation 370, 378, 380*t*
computational methods 70*t*, 72*t*
distributions/averages 56, 60, 62, 64*t*
energy and energy distributions 197*t*, 200
full mean-field solutions 225, 227, 228, 231–2, 234, 235, 238–40, 242, 246, 247–8, 250*t*
late-time mean-field solutions 259–60, 262–4, 268–70, 276–7, 278*t* n1, 279, 280*t*, 282, 283, 284*f*, 285
maximum entropy theory 476, 477, 479, 481, 482*t*, 483, 486, 488–9, 492, 494, 495
semi-empirical/empirical methods 116, 117, 118, 120, 124*t*, 127*t*, 132
unitary fragmentation 449–50, 453, 454, 461*t*, 468
Gates, O. 117
Gaudin, M. 117
Gaudin–Schuhmann/Gates–Gaudin–Schuhmann distributions *see* power law size distributions
Gaussian cutoff function (shifted) 283, 284
Gaussian with zero mean 69
gel (larger particles) 2, 3
full mean-field solutions 228, 237, 238, 239, 247, 248–9
late-time mean-field solutions 258, 269, 270, 271, 278, 279, 280*t*, 285
gelation (time/point) 186, 482, 495, 502, 505, 506
fragmentation cascades 297, 298, 329, 332, 355, 368, 380
full mean-field solutions 227, 228, 237, 238, 239, 240, 241*f*, 248–9
late-time mean-field solutions 259–60, 262, 265–6, 269–70, 274–6, 277–8*t*, 279, 280*t*, 281–3, 285
generalized gamma size distributions *see* Rosin–Rammler size distributions
generalized product kernel 368, 373, 374
general systems theory 30
generating functions *xx*, 269, 270
Genzale, C.L. 114–15
geophysics 153, 154*f*
Giblin, I. 173
Gibson, B. et al 68
Gilliland, E.D. 354
Gilson Company Inc 46*f*
Gilvarry, J.J. 127*t*, 389, 452*t*
Gimel, J.C. et al 278*t*
Gingold, A. 74
Ginot, F. et al 30, 130, 131*t*
Gladman, B.J. et al 379*t* n5
Glansdorff, P. 31
glass 13, 16, 176*f*, 191*t*, 200, 292
unitary fragmentation 384, 387*f*, 419, 431, 432, 433*f*, 462*t*
glass shells
Godunov, S. 79
Golan, A.M. 110, 213*t*
gold 14, 245*t*, 246*f*, 278*t*
Goldhirsch, I. 31, 185, 188
Golovin, A.M. 124*t*, 235, 250*t* n1
Goren, S.L. 323, 331*t*
Gorokhovski, M.A. 340, 341–2, 344–5
Goto, M. et al 150
governing equations *see* mean-field governing equations
Grabowski, W.W. 281
Grady, Dennis E. 14, 76*t*, 485, 492
analytical/empirical methods 109, 110, 111, 113, 127*t*

Grady, Dennis E (*Continued*)
unitary fragmentation 384, 394–5, 405*t*, 406–8, 410, 412*t*, 415–16*f*, 419, 421*f*, 422, 431, 451*f*, 453*t*, 454, 455*t*
Grady's Law/Model 405, 409, 417, 420, 421*f*, 423, 427, 432, 433*f*
Graham, R. 31
granular flow 33, 191, 195
granular gases 7*f*, 170–2, 175, 499
basic properties of particles 24*f*, 30, 31, 33
computational methods 70*t*, 71, 72*t*
energy and energy distributions 179, 183, 185, 186–8*f*, 189–90, 191*t*, 194, 200
granular solids 33, 454, 456, 457*f*
gravitational collision-coalescence 281
gravitational settling rates 157, 213*t*, 214*t*, 215*t*, 217, 245, 272, 277, 423
gravity and microgravities *xvii*, 5, 6, 9, 26, 68, 133, 149, 155, 185, 241, 320, 464, 465*f*
late-time mean-field solutions 257, 278, 279, 285
Grav, T. et al 379*t* n8
Gray, G.T. et al 419
Greenberg, R. 363
Griffith, Laurence 112–13, 482
grinding *see* milling and grinding (comminution)
Grinstein, F. 80
Gudowska-Nowak, E. et al 105
Gueron, S. 131*t*
Guildenbecher, D.R. et al 437, 439
gun-launched projectiles 431–2, 433*f*, 460, 461*t*
Gurney, R.W. 172, 414, 420, 456
gypsum 461*t*
gyration (radius of) 156, 158, 159, 160, 165

Haken, H. 31
Hansen, J.-P. 71, 72*t*
hardening 181
harmonic mean 159
Harris, C.C. et al 127*t*
Hartmann, K. 131
Havlin, S. 186*f*
Hayakawa, Y. 131*t*
heat and heat capacities 212, 502
Heaviside step function 74, 480
Heinrich, G. 130*t*
Heisenberg, Werner 201
Held, M. 455*t*
Hellyer, B. 323
hemispherical shells 420, 421*f*
Hendriks, E.M. et al 250*t* n2a, 270
Herrmann, H.J. 92*t*
Hidy, G.M. 213*t*
high-density particle systems 33, 189
high-fidelity computational methods 67, 72, 92, 93*t*, 94, 131, 133, 392, 505
particle aggregation 68–9, 70*t*
particle fragmentation 73–5, 76*t*, 77, 78–9*f*, 81
random fragmentation 350, 351–3
high/low formal accuracy 68 n2
high-speed photography/video 14, 157, 174, 422*f*, 432, 458*f*, 460
high speed projectiles 12*f*, 13*f*, 126, 149*f*, 173–4
unitary fragmentation 422–3, 424*f*, 431–2, 433*f*, 455*t*, 460, 461*t*
Hilborn, C. 28
Hiroe, T. et al 150
Holt, W.H. 76*t*, 455*t*
analytical/empirical methods 119, 120–2*f*
particle shape distributions 149*f*, 150, 151*f*, 152*f*
homogeneity *xviii*, 22, 168, 481, 500, 502, 503*t*, 507
analytical/empirical methods 103, 104, 113–14, 116, 121, 123, 126, 131, 132–3
basic properties of particles 34–7, 38
binary fragmentation 293, 314, 315
collisional fragmentation 364, 367, 368, 369*t*, 375, 376, 377–8, 380
computational methods 69, 72, 75, 81, 88
distributions/averages 43, 44, 47, 49, 50, 52, 54, 56
energy and energy distributions 181, 182, 184, 191*t*, 193, 196, 197*t*, 198, 200
full mean-field solutions 225, 228, 250, 251
kernels 296, 297*t*
late-time mean-field solutions 257, 260, 277, 279, 280, 282–3, 284*f*
mean-field governing equations 207, 210–11, 213–15*t*, 220
random fragmentation 319, 326, 329–30, 331*t*, 332, 335, 345, 355
unitary fragmentation 393, 394, 398, 433, 440, 459, 468
Honjo, H. 456*t*
Hooke's Law 82, 412
Ho, R.D.J.G. 192
Horvai, P. et al 124*t*, 196, 210, 266
Horvat, K. 462*t*
Hoshen–Kopelman cluster-finding algorithm 185, 186*f*
Hoşten, C. 331*t*
Hounslow, M.J. 209–10, 234, 250*t* n5
Hübler, A.H. 3*f*, 28, 476
Hummel, M. et al 24*f*
Hünenberger, H. 68
Hunt, R. 124*t*
hydrodynamic instabilities 115
hydrodynamic ram 423, 424*f*
hydrosol dynamics *xv*, 124*t*
hyperbolic partial differential equations 249
hypergeometric functions 335, 336*t*, 337*t*

ice and ice crystals 153, 154*f*, 162, 185, 189
ideal/non-ideal fragmentation 385*t*, 431, 451, 455*t*, 501
ideal/non-ideal gases *xxi*, 29, 198, 477, 501, 502
ideal size distribution 12–14, 38, 126, 133, 282, 460, 461*t*, 462, 464, 468, 505
Ihara, S. 456*t*
ILES *see* Implicit Large Eddy Simulations

impact fragmentation 117, 149, 173, 218
computational methods 87, 88, 89 90*f*, 91, 92*t*
unitary fragmentation 384, 385*t*, 386*t*, 420, 421*f*, 422–3, 424*f*, 432, 433*f*, 460, 461*t*, 462, 463*f*
Implicit Large Eddy Simulations (ILES) 506
Inaba, S. et al 215*t*
Inaoka, H. 127*t*, 452*t*
incomplete unitary aggregation 2
incompressible turbulence (constant density) 193, 195
indistinguishable/distinguishable particles 210
inelastic collisions *xxi*, 27, 32*t*, 188, 217, 477
inertial range 193–4, 196, 197*t*, 198, 199, 214*t*
infinite/finite particle systems 228 n1, 398, 404, 451, 502, 506, 507
fragmentation cascades 299, 306, 314–15, 329, 332, 333, 364, 375
late-time mean-field solutions 266, 279, 285
maximum entropy theory 478, 479, 480, 481, 482*t*
infrared sensors 337
initial conditions 37, 38, 192
binary fragmentation 300–7, 310, 312, 314, 315
collisional fragmentation 368, 369–70, 379
full mean-field solutions 225, 228, 229–32, 234–40, 245, 250
late-time mean-field solutions 257, 258, 259, 263, 264, 273
mean-field governing equations 215–16, 219 n2, 220
random fragmentation 335, 336, 338, 340, 354, 355
unitary fragmentation 402, 414, 449
instabilities 75, 114–15, 215, 501, 502, 504
unitary fragmentation 384, 386*t*, 387, 389, 391, 404, 457, 458*f*
integral scales 193
intensive quantity 45
International Space Station 279
interstellar dust aggregation 10
inverse gamma size distributions 489, 492
inviscid flow *vii*, 249
unitary fragmentation 389, 394, 404, 405–6*t*, 407–8, 409, 411*t*, 413, 434
iron shells 149*f*, 150, 151*f*, 172, 173*f*
Iroshnikov, P.S. 197*t*
Isobe, M. 189, 190, 191*t*, 200
isotropy *xviii*, *xxi*, 22, 103, 104, 500, 503*t*, 507
aggregation cascades 225, 250, 257, 282, 284
analytical/empirical methods 116, 121, 123, 126, 131, 132, 133
basic properties of particles 34–7, 38
binary fragmentation 293, 314, 315
collisional fragmentation 376, 378, 380
computational methods 69, 72, 75, 81, 88
distributions/averages 43, 44, 47, 49, 50, 52, 54
energy and energy distributions 181, 182, 184, 191*t*, 192, 196, 197*t*, 200
mean-field governing equations 207, 213*t*, 214*t*
random fragmentation 319, 329, 332, 335, 364
unitary fragmentation 393, 398, 433, 440, 459, 468
Ivlev, A.V. 279

'jamming' 33, 189
Járai-Szabó, F. 453*t*
Jaupart, C. 131
Jeans, H. 22
Jeffrey, D.J. 114, 271–2
Jellinek, H.H.G. 293
jets (liquid) 386*t*, 413, 417, 418*f*, 426, 427–30*f*, 440, 456, 458*f*, 459*f*
Jewitt, D. 378
Jiménez, J. 196
Jirásek, M. 412*t*
JKSimMet software 350
Johansen, A. et al 363
Johnson, L. et al 127*t*, 129
joint particle size-velocity distributions 493, 496
joint probability density functions 34–5, 36–7, 168–9, 180, 182
Jungblut, S. et al 156
Jun, K. 28

Kadono, T. et al 149*f*
Kaminski, E. 131
Kaneko, K. 29
Kapur, P.C. 331*t*, 354
Karabelas, A.J. 370, 373
Karagozian & Case model *see* Material Model 72R3
Katsuragi, H. 456*t*
Katul, G. 197*t* n1
Kautz, R. 28
Kekäläinen, P. et al 452*t*
Kellay, H. et al 191*t*
Kelvin–Helmholtz (KH) instability model 75, 115, 389, 391 n2, 502
Kempf, S. et al 10, 72*t*
kernels 402, 449, 505
binary fragmentation 293, 294, 295, 296, 297*t*, 302, 315
collisional fragmentation 364, 365, 366, 368, 371*f*, 372–3, 375, 379
full mean-field solutions 22, 227, 228, 229–40, 241*f*, 247–8, 250, 251
late-time mean-field solutions 257, 259, 260, 261*t*, 262, 263, 265–8, 275–6, 279, 281–5
maximum entropy theory 481, 483, 485, 495
mean-field governing equations 207, 208, 210–12, 213–15*t*, 216, 220
random fragmentation 319, 321, 322, 329–30, 331*t*, 332, 335, 354, 355, 364, 365
Khodabin, M. 512*t*, 513*t*
Khoei, R. 74
KHRT computer models *see* Kelvin–Helmholtz instability model; Rayleigh–Taylor instability model
Khvorostyanov, V.I. 258–9, 337, 492
Kiang, T. 453*t*
Kick, F. 352, 354, 407
Kida, S. et al 197*t*
Killingbeck, J.P. 209
Kim, W. et al 14–15

Kim, Y.C. 309*t*
kinematic viscosity 193
kinetic energy *v*, 28, 38, 103, 113, 175, 185, 493, 501, 505
  aggregation cascades 220, 259, 282
  computational methods 68, 86, 87–9*f*
  energy and energy distributions 179, 185, 189–90, 191*t*, 192–6, 197*f*, 198–200
  fragmentation cascades 298, 320, 329–30
  unitary fragmentation 389–90, 406, 407, 422
  universal size distributions 12*f*, 13*f*, 14, 16
King, R.P. et al 1, 105–6
Kipp, Martin E. 485, 492
  analytical/empirical methods 110, 111, 113, 127*t*
  unitary fragmentation 410, 422, 451*f*, 453*t*, 454
KIVA (Computational Fluid Dynamics code) 115, 391
Klett, J.D. 124*t*, 285
Koch, R. et al 75
Kolesnichenko, A.V. 192
Kolmogorov, Andrei Nokolaevich *v*, *xxi*, 105, 341, 502
  energy and energy distributions 193, 194, 196, 197*t*, 198, 200, 201
Koltzenburg, S. et al 163
Kong, X. et al 76*t*, 77, 415*f*
Konopka, U. et al 279
Kostoglou, M. 370, 373
Kovac, J. 112, 491
Kraichnan, R.H. 197 n4&7, 200
Krapivsky, P.L. et al 127*t*, 226, 365–6, 367, 374, 451*f*, 453*t*
Kronecker delta function 208
Kroupa, P. 127*t*, 128
Kruse, T.M. et al 308
Kuhn, Werner 107, 109
Kuiper belt 292*t*, 320*t*, 363*t*, 364, 376, 377, 379*t*
Kulik, T.V. 452*t*
Kumar, G.S. 309*t*
Kumar, V.R. 309*t*
Kummer, Ernst 335, 336*t*, 337*t*
Kunamaneni, S. et al 280*t*
Kun, F. 81*t*, 92*t*, 173, 174
  particle shape distributions 150, 152–3, 160–1, 162*f*
Kuznetsov, M. 123, 391*f*
Kuz–Ram model 123

ladder polymers 163
Lagrange multipliers (calculus) 111, 479, 480, 487, 489, 490
Lagrangian computational methods 67, 73, 92, 93*t*
Laguerre function 336, 337*t*
Lambert, F.L. 476–7
Lamb, Horace 201
laminar air streams 410, 426
LAMMPS *see* large-scale atomic/molecular massively parallel simulator
Laney, C.B. 62, 76*t*, 131, 132*f*
  unitary fragmentation 409–10, 441, 455–6*t*, 461*t*, 464, 466*t*, 513–17*t*
Lange, A. 149*f*
Langevin equation 71
Laplace transform 269
Lapp, A. et al 280*t*
large-scale atomic/molecular massively parallel simulator (LAMMPS) 69
laser diffraction 157
La Spina, A. 149, 150, 164
late-time mean-field solutions 216, 257–9, 283, 284*f*, 285, 501, 505
  aggregation combined with fragmentation 282
  average particle sizes 273–6, 372
  binary fragmentation 307–8
  boundary conditions 259
  collisional fragmentation 370, 371*f*, 372–8, 380*t*
  common size distributions 259–60
  computational/experimental validation 276, 277–8*t*, 279, 280*t*, 281–2
  random fragmentation 345–7
  size distributions for local aggregation 267–73
  size distributions for non-local aggregation 260, 261*t*, 262–7
  see also full mean-field solutions; mean-field governing equations, cascades
late-time size distributions 236, 250*t*, 251, 257, 376–7, 380*t*
late-time stationary/non-stationary solutions 377–8, 483
late-time turbulent kinetic energy spectra 197*f* n1–7
latex particles 245*t*, 246, 277*t*, 278*t*
Lattuada, M. 248–9
Laurençot, P. 130*t*
least-squares fitting 117
Lee, C.-H. et al 215
Lee, M.H. 276
Lefebvre, A.H. 405*t*, 411*t*, 426, 428, 429*f*, 431, 432, 433*f*
Leibler, L. 280t
Leliwa-Kopystynski, J. et al 173
length scales 410, 411*t*, 412, 413, 417, 418*f*, 427, 441
Lennard–Jones potential 69
Levich, B.G. 214*t*
Levin, S.A. 131*t*
Levy, S. 27
Leyvraz, F. 226, 228, 237–8, 250*t* n3a, 270–1, 276
Libersky, L.D. 74, 76*t*
Lienau, C. 108, 127*t*, 453*t*
Lienhard, J.H. 454
lifelike self-propelled colloids 30, 130, 131*t*, 501
ligament formation 75, 387–8, 394, 417, 426, 430, 434–5, 437
Li, G.J. 437*f*
Lind, P.G. et al 450
Lindsay, P.J. 157
linear chains 3*f*, 163, 213*t*, 247, 280*t*, 295, 301
linear fragmentation *see* random fragmentation
linearized stability analysis 115
linear polymers 163

Linfoot, E.H. 57, 85
  analytical/empirical methods 108, 109, 118, 119, 127*t*, 128*t*, 129
  unitary fragmentation 389, 405*t*, 414, 451*f*, 453*t*, 455*t*
linkages (chains) 295, 300–1, 310, 483
Lin, M.Y. 28, et al 14, 124*t*, 125*t*, 245*t*, 246*f*, 266, 278*t*
Linna, P. et al 84–5, 86*f*
liquid atomization *vii*, *xv*, 74–5, 126, 127*t*, 128*t*, 495
  liquids *v*, *xvii*, 2, 8, 28, 73, 79, 485, 499, 501, 502
  aggregations cascades 211, 215, 258, 264, 281
  average sizes 434, 435*f*, 436*f*, 437*f*, 438*f*, 439, 440*f*
  basic principles of unitary fragmentation 384, 386*t*, 389, 390*f*, 391*f* n2, 392*f*, 394, 398
  Cartesian geometries 425, 426, 427*f*, 428*f*, 429–30*f*
  fragmentation cascades 323, 329–30, 331*t*, 337, 354, 392
  particle distributions 114–15, 157, 164, 188, 193
  size distributions 454, 456*t*, 459–60, 461, 462*t*, 463*f*, 464, 465*f*, 466*t*, 467*f*
  unitary fragmentation 404, 405–6*t*, 407–8, 410, 411*t*, 412*t*, 413, 417, 418*f*, 421, 422*f*, 423, 424*f*, 434, 435*f*, 436*f*, 437*f*, 438*f*, 439
Litster, J. 1
Liu, Y. et al 124*t*, 492
living organisms xvi, 130, 131*t*
Li, X. 454, 466*t*, 467*f*, 485, 492
Li, X.Y. 28, 127*t*, 132, 281
Llorca, F. 395
local aggregation *see* non-local/local aggregation
log moments 341
log normal size distributions 105, 106, 284, 461, 504
  distributions/averages 56, 61, 64*t*
  fragmentation cascades 299, 308, 314, 334, 338, 341, 346
Loiseau, J. et al 456
Long, A.B. 214*t*
long-range body forces 22, 26, 27, 68, 69, 93*t*
Long, R.R. 197*t* n5
Lorenz, E. 249
Lorenz, N. 28, 94
low-density particle systems *xx–xxi*, 33, 189
Loyalka, S.K. 213*t*
LS-DYNA (commercial finite element code) 76, 77, 78*f*, 391
lucky particles 228
Lucy, B. 74
Lumley, John *vi*
Lushnikov, A.A. 228, 237–8, 250*t* n3a, 260, 268, 285, 480
Lusignan, C.P. et al 247, 248*f*, 280*t*, 283
Lynch, A.J. 196, 320

McComb, W. David *vi*, 197, 198
McCoy, B.J. 130*t*, 293, 309*t*, 312, 313*f*, 331*t*
McGrady, E.D. 293–4, 332, 338
  binary fragmentation 293–4, 297*t*, 300, 301, 302–7
  random fragmentation 331*t*, 332, 338
machine learning techniques 132
Mach number 249
McIvor, R.E. et al 350, 351
McKay, A.T. 233
Mackintosh, F.C. 188, 189
McLeod, J.B. 124*t*, 237–9, 250*t* n1a
McNolty, F. 337
Madras, G. (et al) 130*t*, 293, 308, 309*t*, 312, 313*f*, 314*f*, 331*t*
Ma, G. et al 87, 89*f*, 90*f*, 92*t*
magnetohydrodynamics 197*t*
Magnotti, M. 114–15
Mahnke, R. 30–1
Mainzer, A. et al 379*t* n7
Makino, J. et al 215*t*
Malvern Mastersizer 2000S system 157, 158*f*
Mandelbrot, B. 8
Manley, S. et al 278–9
Ma, Q. et al 131*t*
marginal probability density functions 168–9
Marimuthu, A 309*t*
Markutsya, S. et al 69, 70*t*, 72*t*, 73
Marmottant, P. 127*t*
Marov, Y. 192
Marquez, A.M. et al 412
Marshall, S. 124*t*, 125*t*
Marshall, W.R. 62
Marshall–Palmer universal size distribution 464
Martin, E. (et al) 277*t*
Martyushev, L.M. 477
Marusic, I. 196
Masiero, J.R. et al 379*t* n6
mass *v*, *xvii*, *xviii*, 3, 5, 8, 10, 11, 194, 200, 500, 509
  analytical/empirical methods 11, 116, 129
  basic properties of particles 23, 27, 29*t*, 31, 32*t*, 33, 37
  binary fragmentation 293, 295, 297*t*, 298, 300–5, 307, 310
  collisional fragmentation 363, 377, 381
  computational methods 68, 72, 73, 74, 79, 80
  distributions/averages 43, 48*f*, 50, 52, 54, 55*t*, 56, 58, 60, 61, 63*t*
  full mean-field solutions 229, 229, 243*f*, 245, 247
  late-time mean-field solutions 259, 264, 268, 269, 270–1, 274
  maximum entropy theory 476, 477, 486, 491, 492
  mean-field governing equations 210, 211, 217, 219–20
  particle shape distributions 147, 150, 152*f*, 153*f*, 158, 161, 163, 164–5
  particle velocity distributions 169, 170, 173, 174, 175*f*
  random fragmentation 319, 322–5, 328–30, 331*t*, 244, 338–40, 342–3, 345, 352–3, 355
  unitary fragmentation 390, 403, 407, 408, 414, 420, 428, 441*t*, 459
mass histogram 45
mass mean diameter 55, 129
mass-weighted size distributions 50–1, 63, 64, 321, 323

Material Model 72R3 (Karagozian & Case model) 76, 78
Material Model 159 (Continuous Surface Cap Model) 76, 77, 78
mathematics and mathematicians 1, 30, 49, 130*t*, 198, 201, 215*t*, 226, 250*t*
Mathieu, J. 191, 192
Matsoukas, T. 111, 241*f*, 248, 266, 298, 479
Matsui, T. et al 12*f*
Matthews, D. 46, 63
Matura, P. et al 424*f*
Matuttis, G. 80
Ma, W. et al 74
maximum entropy theory *xx*, 16, 175, 393, 402, 448, 453–4, 467, 503*t*, 504, 505
  analytical/empirical methods 110–13, 121, 124*t*, 127*t*, 129, 133
    see also continuous particle systems; discrete particle systems
maximum likelihood estimation (MLE) 117
Maxwellian condensation (regular) 258
Maxwell, J.C. 28, 212
Maxwell's concept of degrees of freedom *xviii*, *xxi*, 28–9, 211, 212, 330–1, 477, 502
Maxwell–Boltzmann (MB) speed distribution 170, 171, 172
Maxwell–Boltzmann velocity 189
Mayer, Joseph E. 112
Mayer, Maria Goeppert 112, 492
Mazza, M.G. 69, 70*t*
Mazzinghy, D.B. et al 348, 349*f*, 350*f*
MD *see* Molecular Dynamics
mean-field governing equations 226, 250*t*, 500, 504, 505
  binary fragmentation 293–6, 299, 300, 312
  boundary conditions 216–17, 218*f*, 219
  collisional fragmentation 292*t*, 364–8, 369*t*
  conservation of mass 219–20
  continuous governing equation 208–10
  discrete governing equation 207–8
  initial conditions 215–16
  kernels (collison/collection) 207, 208, 210–12, 213–15*t*
  random fragmentation 319, 320*t*, 321–5, 340
  uniqueness 220
    see also cascades; full mean-field solutions; late-time mean-field solutions
mean-field theory *xx*, 16, 171, 175, 393, 500, 501, 503*t*, 505
  analytical/empirical methods 103–6, 109, 110, 124*t*, 127*t*, 129, 130*t*, 131*t*, 133
  maximum entropy theory 480, 481, 483, 487, 494
  unitary fragmentation 402–4, 448, 449–50, 454, 467
    see also binary fragmentation; collisional fragmentation; full mean-field solutions; late-time mean-field solutions; mean-field governing equations; random fragmentation
mechanical degradation 293, 323
mechanical engineering *vi*, *xv*, 110, 198, 507
mechanical stress/strain 393
Meibom, A. 461*t*
melamine-formaldehyde particles 279
Melzak, Z.A. 322–3
Meng, J.C. 438*f*
Menon, G. 220
mesh-based/mesh-free computational methods 67, 73–4, 76, 79–80, 92, 93*t*, 506–7
meteorologists and meteorology 198, 507
METSIM software 350
Meyer, P.L. 454
Meyer, R. et al 323, 331*t*
Meyers, M.A. et al 458*f*
Michaelides, E. et al 211
Michikami, T. et al 12, 13, 149*f*
microcomputed tomography (x-ray) 157
microgravity tests 172
micro-/macro-states 476, 477, 478, 482–3
Micromeritics SediGraph 5100 system 157, 158*f*
microscales (Kolmogorov) 193, 194
midpoint-chain scission 293
Miklashevich, I. 27
MILES *see* Monotonically Integrated Large Eddy Simulation
milling and grinding (comminution) 105–6, 320*t*, 347, 355, 363*t*
Milne, A, et al 421, 422*f*
mineral processing industry 1, 292*t*, 320*t*, 355
minimal fragmentation 450
mining engineering *vi*, *xv*, 81*t*, 85, 117, 123, 126, 127*t*, 149, 353, 507
mitosis *see* binary fragmentation
MLE *see* maximum likelihood estimation
MMD *see* mass median diameter
Mock, W. 76*t*, 455*t*
  analytical/empirical methods 119, 120–2*f*
  particle shape distributions 149*f*, 150, 151*f*, 152*f*
'modified B3LYP charge set' 69
MODSIM^{TM} software (MTI) 348, 350
Mo, J. 25
molar mass distribution 248, 310
molecular chains *xviii*, 293
molecular chaos (*Stoßzahlansatz*) 22*t*, 23, 24, 25, 26, 170, 172, 192
molecular clouds 1, 124*t*, 126
Molecular Dynamics (MD) 133, 156, 248, 505
  computational methods 67, 68, 69, 70*t*, 73, 80, 92, 93*t*
molecular gases 23, 28, 29, 170, 185, 212, 476, 477
molecular weight/molecular weight distribution 247, 293, 311*f*
molecules 69, 125
moles 310
Molinari, J.F. 27, 412*t*
moments 226–7, 228, 233
  binary fragmentation 298–9, 300
  random fragmentation 333–4, 337, 341–2

momentum *v*, *xvii*, *xviii*, 3, 5, 194, 195, 390, 500
aggregation cascades 217, 219, 249
basic properties of particles 23, 24, 27, 29*t*, 31, 32*t*, 37
computational methods 68, 71, 73, 74, 79, 80
fragmentation cascades 293, 299, 319, 328, 329, 355, 363, 381
maximum entropy theory 476, 477, 485, 493
Monaghan, J.J. 74
monoatomic gases 212
monodisperse initial conditions 369–70
aggregation cascades 225, 229–30, 235, 237–8, 242, 264
binary fragmentation 295, 300–4, 305–7, 310, 314
random fragmentation 335, 338–9, 341–4, 355
monodisperse size distributions 216, 263, 304, 477
monomers 6, 8, 24*f*, 52, 60, 154, 162, 170, 175, 374, 500
aggregation cascades 207, 208, 247, 257, 264, 265, 267, 282
analytical methods 106–7, 110, 111, 130, 240
binary fragmentation 293, 294, 295, 301, 310, 315
computational methods 69–70, 71, 73
energy and energy distributions 179, 181, 183, 185, 186*f*, 187*f*, 188*f*, 189–91, 194
maximum entropy theory 478, 479, 481, 482*t*, 483
monotonically decaying functions 193
Monotonically Integrated Large Eddy Simulation (MILES) *vii*, 80, 506
Monte-Carlo (MC) methods 71, 72*t*, 93*t*, 104, 172, 240, 241*f*, 281, 505
Moore, J.M. et al 379*t* n11
Moran, L. 131*t*
Morfill, G.E. 279
Mori, S. 247
Morrell, S. 350
Morris, T. 109
most probable size distribution 110
mothers 2, 292, 453
Mott, Nevill F. 57, 85, 201
analytical/empirical methods 108, 109–10, 118, 119, 127*t*, 128*t*
unitary fragmentation 389, 405–6*t*, 412*t*, 413, 451*f*, 453*t*, 455*t*
Mott plots 119
Mott–Linfoot universal size distribution 128*t*, 129, 421, 464
Moulton, R.W. 466*t*
moving-bed attrition 331*t*
MTI 348
Mulholland, G.W. 250*t* n3
Mullis, Kary *viii*
multiphase flow *xvii*
Mundo, C. et al 462*t*
Murray, D. 76, 77
Myagkov, N. 76*t*

Nakazawa, K. 215*t*
NASA *see* National Aeronautics and Space Administration
Nazarenko, S.V. 124*t*
N-Body (NB) systems 26, 68–9, 70*t*, 93*t*, 505
Néda, Z. 453*t*
neural networks 132
Newton, Isaac 68
Nexopart GmbH & Co. 47*f*
Niwa, S. 131*t*
nodes and node splitting 67, 73, 82, 506
non-covalent forces 125
non-deterministic *see* deterministic/non-deterministic computational methods
non-equilibrium thermodynamics 31
non-ideal size distribution 12, 13, 14, 15, 38, 86, 459, 460, 461*t*, 462
nonlinear fragmentation *see* collisional fragmentation
non-local/local aggregation/fragmentation 319, 419, 449–50, 483, 494
collisional fragmentation 367, 369, 370, 371*f*, 372–5, 376, 378, 380–1
late-time mean-field solutions 257, 260, 261*t*, 262–73, 282, 284*f*, 285
mean-field governing equations 210, 211, 212, 219
non-Newtonian liquids 501
non-stationary aggregation/fragmentation 4 n1, 259, 260, 262–5, 268, 308, 320, 376, 481, 482, 494
non-stationary closed-form solutions 235, 237–40, 375, 500
non-stationary mean-field solutions *see* full mean-field solutions
normal size distributions 304, 305
nuclear engineering/physics *xv*, *xvii*
Nukiyama–Tanasawa size distributions *see* Rosin–Rammler size distributions
number density *see* concentration distributions
number distributions 52, 63*t*
number-weighted size distributions 45–6, 51, 63*t*, 64, 321
numerical dissipation 69, 71, 74

O'Brien, D.P. 363
oceanographers *xv*, 198
Oddershede, L. et al 33, 461*t*
O'Hare, E. 76*t*, 77, 78–9*f*
Ohnesorge numbers 413, 427, 434, 435*f*, 437, 439
Ohta, T. 189
Olafsen, J.S. 24
Olivier, B.J. 14, 278*t*
Olsen, M.L. 416*f*, 421*f*, 466*t*
one-particle mechanisms (fragmentation) 330, 331, 332, 499
one-point/one-time probability density function 34, 35–6, 37, 181
one-point/two-time probability density function 34–5, 36, 37, 180, 182
OpenFOAM 115
open particle systems 3, 476
Opper, M. 104
oscillatory systems 28, 267
Ouchiyama, N. et al 331*t*

oxidative degradation 311–12, 313*f*, 320*t*
Oyama, N. et al 33, 191*t*

packing fraction 188, 189
Pagonabarraga, I. 331*t*, 355, 367, 372, 373–4, 375 n1, 376, 377–8
pair annihilation 191
Pál, G. et al 173, 174
Palmer, W.M. 124*t*, 125*t*
Paloposki, T. 131, 466*t*
Paltridge, G.W. 493, 495
Pandya, J.D. 323, 331*t*
Pan, H. 460
Paolicchi, P. 149, 150, 164
paper mills 331*t*
parallel trajectories 174
parameters 120, 132, 146*f*, 147
Pardo, A. 115–16
parents 2, 8, 27, 130
  binary fragmentation 292, 293, 294, 295, 296
  collisional fragmentation 365–6, 367, 373, 374
  random fragmentation 321, 322, 323, 325, 326–8, 340, 341
  secondary fragmentation 434, 436, 437, 438
Parisio, F. 384, 450, 452*t*
partial differential equations 265–6
partially elastic collisions 24, 31, 38, 71, 171, 185, 380, 502
particle concentration 33, 216
particle dynamics 2, 104, 219–20
  different fields/applications *xvi*, *xviii*
  modelling challenges with aggregation/fragmentation 507, 508*t*
  sizes of 10–11, 51
  types of 6, 7*f*
particle image velocimetry (PIV) 423
particle physics *xv*, *xvii*
particle sizes *see* sizes and size distributions
particle tracking velocimetry (PTV) 423
partition functions 110, 111, 478, 479, 481
Pask, Gordon 30
Pathak, N. et al 7*f*, 72*t*, 185–6
Patil, D.P. 130*t*, 331*t*
Patterson, D. et al 112
Patton, E.V. et al 280*t*
Paul, S. 170
Pawlowska, H. 281
PBMs *see* population balance models
Pearson, B.R. 199
Pearson, J. 413–14
Pearson type V size distributions 489, 514
PEC *see* Protection Engineering Consultants
Pefferkorn, E. 277*t*
Pego, R.L. 131*t*, 220
Peng, G. 189
Percival, B. 157
perfectly elastic/inelastic collisions 27, 71, 170, 188–9, 190
periodic boundary conditions 28, 217–18, 346–7, 355, 410, 494
  aggregation cascades 259, 261, 264, 265, 268, 269, 282, 285
  collisional fragmentation 375, 376, 377, 378
    see also boundary conditions; free-field boundary conditions
periodic instabilities 384, 386*t*, 387, 389, 391, 404, 457
Peterson, T.W. 331*t*
pharmaceuticals *xv*, 1
phase transitions 248, 298, 355, 502
physical collection (fragments) 14, 44
physical space 211
physics and physicists 30, 198, 201, 250*t*, 391, 392–3, 477
Piekutowski, A.J. 416*f*, 420, 421*f*
Pielichowski, K. et al 293
Pierce, F. et al 71, 72*t*
Pilch, M. 435*f*, 437
Pimentel, R. et al. 131, 132, 466*t*, 467*f*
Piotrowski, L. 127*t*, 376
Piskunov, V.N. 268, 297*t*
PIV *see* particle image velocimetry
planet formation *see* astrophysics
plasmas 128*t*
plastic distortion 457
Poincaré, H. 23, 28
Poisson size distribution 107, 481, 482*t*, 495
Poisson Voronoi tessellations 453*t*, 454
polycaprolactone (PCL) 313, 314*f*
Poly Classic 69
polydimethylsiloxane (PDMS) 280*t*
polyesters 248*f*, 280*t*
polyethylene 293, 309*t*
polymers and polymer science *xvi*, *xviii* 1, 2, 163, 477, 481, 500
  analytical/empirical methods 109, 112, 124*t*, 125*t*, 130*t*
  binary fragmentation 292*t*, 293, 294, 297*t*, 309*t*, 312, 315, 363*t*
  computational methods 72*t*, 90–1, 92*t*
  full mean-field solutions 228, 247–8, 250*t*, 251
  late-time mean-field solutions 276, 279, 280*t*
  mean-field governing equations 213*t*, 214*t*
  random degradation 320*t*, 323, 331*t*
polystyrene 14
  binary fragmentation 309*t*, 310, 311*f*, 312, 313*f*
  full mean-field solutions 242, 243*f*, 244*f*, 245*f*, 246*f*
  late-time mean-field solutions 277*t*, 278*t*, 280*t*
polyurethane 280*t*
Ponson, L. 27
Pope, Stephen B. *vi*, 184–200
population balance models (PBMs) (equations) 104
  fragmentation cascades 296, 321, 322, 349, 350–2, 353
  mean-field governing equations 207–8, 209
Porod's law 189, 190–1
Pöschel, T. et al 31, 71, 171, 172, 185
positional entropy 476–7
Posner, J.D. 131*t*
Potapov, A. 92*t*

power law size distributions 173–4, 493–4, 500–1, 503*t*, 504–5, 510–11, 514, 515*t*, 517*t*, 518*t*
analytical/empirical methods 113–14, 116*f*, 117–18, 124*t*, 127*t*, 131–2, 133
binary fragmentation 297, 298, 306, 307, 308
collisional fragmentation 368, 373, 375, 376, 377, 378, 379*t*, 380*t*
computational methods 72*t*, 76*t*, 81*t*, 82, 84*f*, 85, 86, 87, 90, 91, 92*t*
distributions/averages 11, 12*f*, 13*f*, 14, 32*t*, 33, 56, 57, 58, 64*t*
energy and energy distributions 186, 188, 189, 190, 191, 194, 197*t*, 200
full mean-field solutions 228, 237, 239, 240
late-time mean-field solutions 260, 264–5, 268–72, 276, 277–8*t* n1, 279, 280*t* n2, 282–3, 284, 285
mean-field governing equations 210, 218, 220
particle shape distributions 160, 161, 162–3, 164
random fragmentation 325, 326–9, 332, 333, 334, 344–8, 351, 354, 355
unitary fragmentation 402, 432, 433, 449–50, 452–3*t*, 455–6*t*, 461*t*, 464, 468
Prasad, V. et al 185
Pratsinis, S.E. 322
prefectly elastic 23
Prigogine, Ilya 31
primary aggregation 6, 14–15, 501
primary breakup 13, 129, 164, 439, 440*f*, 464
primary fragmentation 2, 6, 13*f*, 14, 389, 390, 391, 392, 468, 501
Cartesian geometries 425–6, 427–30*f*
cylindrical geometries 413–14, 415–16*f*, 417, 418*f*
spherical geometries 418–20, 421–2*f*, 423, 424*f*, 425
probability density functions 34–7, 79, 106–7, 116, 453–4, 465, 485
distributions/averages 47, 48*f*, 49–50, 63*t*
energy and energy distributions 179, 180, 181, 182, 198, 199, 200
particle shape distributions 162*f*, 168–9
random fragmentation 323–4, 341
probability distribution functions 16, 110, 116, 199
probability theory *xvi*, 61, 105–6, 169, 179, 186, 192, 269, 499, 504
density functions *xvii*, 10, 14
process and product engineering *xv*
production range *see* energy-containing range
product kernels 481
collisional fragmentation 369–70, 373–4
full mean-field solutions 227, 228, 237–40, 241*f*, 247–8, 250, 251
late-time mean-field solutions 257, 268, 275–6, 282, 285
progeny fragments 2, 292, 320
Protection Engineering Consultants (PEC) 47*f*
Proudman, I. 197*t* n1
Pruessner, G. 32, 33
PTV *see* particle tracking velocimetry
Pugno, N. 117, 127*t*
pulsed shadowgraphy 427
Pulvermacher, B. 213*t*
Pumir, A. 220, 281
Puri, S. 191*t*

quasi-ideal impact fragmentation 87, 88, 90–1, 92*t*
quasi-ideal size distributions 87, 88, 92*t*
Quinn, T.R. 70*t*

Radjai, F. 195
Raff, R. 213*t*
Rahman, A. 68
rain initiation *see* cloud physics
Raizen, M.G. 25
Ramabhadran, T.E. et al 209, 231–2
Ramesh, K.T. 412*t*
Ramkrishna, D. 331*t*
Rammler, E. 57, 123
Randles, W. 74, 76*t*
random-chain degradation 293, 294, 309*t*, 310, 311, 312, 313, 315
random fragmentation 292*t*, 294, 354–5, 394, 499–500, 502, 503*t*, 504
breakage functions 325–9
collisional fragmentation and 363*t*, 364, 369, 380*t*
distributions/moments 333–4
energy-size reduction relationships 352–4
experimental validation 347, 348–50*f*, 351–2
full mean-field solutions 334–5, 336–7*t*, 338–45
governing equations 319, 320*t*, 321–5
late-time mean-field solutions 345–7
rate kernels 329–30, 331*t*, 332
shattering 320, 332–3
unitary fragmentation 402, 449, 450, 451*f*
see also binary fragmentation; collisional fragmentation
randomness and seemingly random *v*, *vi*, *xvii*, 1, 2, 3*f*, 10, 15–16, 37, 64, 499, 500, 503, 505–6
aggregation cascades 233, 247, 258
analytical methods 103–4, 106, 107, 108, 110–13, 133
binary fragmentation 293, 299, 309
chaos theory 22*t*, 23, 24, 25, 26, 27–8
computational methods 67–9, 71–2, 74–5, 77–8, 82, 85, 91–2, 93*t*, 94
energy and energy distributions 179, 185
fragmentation cascades 364, 380
mean-field governing equations 207, 219, 220
self-organization 29*t*, 31, 32*t*
unitary fragmentation 402, 448
random packing 450
random scalar processes 34, 179–81, 183
random tessellations 450, 451*f*, 452–3*t*, 454
random vector processes 35, 181–3
random walks 452*t*
Rangarajan, P. et al 308, 309*t*
rapid/slow convergence 68 n3
rarefaction waves 456
Rarity, J.G. et al 277*t*
rate kernels *see* kernels

rate/s-of-strain 73, 74, 82
Rayleigh capillary instabilities 387, 388
Rayleigh–Taylor instability model 75, 115, 386*t*, 389, 391 n2, 458*f*, 502
Ray, P. 27, 74, 389
Ray, S. 292–3
reaction efficiency 257, 371
Reaction-Limited Cluster-Cluster/Cluster-Monomer Aggregation (RLCA/RLA) 10, 14, 133, 155, 277
  full mean-field solutions 241, 242, 243*f*, 244*f*, 245*t*, 246*f*
Reddy, N. 73
Redner, S. 367
regular particles 7*f*, 8
reinforced concrete 76–7, 78–9*f*, 394
relaxation-time approximations 209, 234
Render, P.M. 460
restructuring *see* secondary aggregation
reverse aggregation *xix*, 130*t*, 294, 482
Reynolds numbers
  energy and energy distributions 192, 193, 194, 197, 198
  unitary fragmentation 384, 398, 412, 413, 427, 439
Rittinger, P.R.V. 352, 354, 406–7
RLCA/RLA *see* Reaction-Limited Cluster-Cluster/Cluster-Monomer Aggregation
rock crushing/blasting 117, 121, 123, 157
Rodgers, G.J. 209, 239, 250*t*
Rogers, H.C. 321, 414, 415*f*
Roisman, I.V. 462*t*
root-mean-square-average 153, 156, 159, 170
root normal size distributions 56, 61, 62*f*, 64*t*, 127*t*, 132, 463–4, 465–6, 467*f*, 504
Rosin, P. 57, 123
Rosin–Rammler size distributions 194, 200, 454, 504, 511, 512–13*t*
  aggregation cascades 247, 259, 263
  analytical/empirical methods 106, 116*f*, 120, 121
  binary fragmentation 298, 304, 307–8
  collisional fragmentation 371–2, 373, 378, 380*t*
  distributions/averages 56–61, 64*t*
  maximum entropy theory 486–7, 489–90, 494
  random fragmentation 333–5, 339, 340, 341, 343, 346, 347, 348*f*, 351, 355
  velocity distributions 170, 171, 175
Rosin–Rammler–Sperling–Bennett (RRSB) size distributions 123, 513
Rotach, M.W. 192
rotation and rotation rates *xviii*, *xix*, 6, 37, 44, 104, 133, 165, 168, 505
  energy and energy distributions 185, 187, 188*f*, 189, 190, 191, 200
  fragmentation cascades 293, 319, 364, 380
  mean-field governing equations 207, 211, 212, 220
Rottenkolber, E. 76*t*, 455*t*
Rottereau, M. et al 72*t*
roughness (surfaces) 410, 411*t*, 426, 436*f*, 462, 463*f*
Roux, S. 195
Rowland, C.A. 196, 320
RRSB *see* Rosin–Rammler–Sperling–Bennett size distributions
Ruckenstein, E. 213*t*
Ruff, G.A. et al 418*f*
runaway growth 228
Runge–Kutta numerical method 68
Rutland, C.J. 462
Ryan, E.L. et al 379*t* n3
Ryan, E.V. 173

Saad, D. 104
Sachse, M. et al 173
Saffman, P.G. 197*t* n1, 213*t*, 214*t*
Safronov, V.S. 124*t*, 215*t*, 236, 250*t* n3, 366
Sagaut, P. 193, 198
Saito, O. 296
Sallam, K.A. (et al) 405*t*, 411*t*, 426, 427*f*, 428*f*
Salman, A.D. et al 460
Salpeter, E. 127*t*, 128*t*
Salpeter–Kroupa size distribution 127–8
Sanai, M. et al 431*f*
sandpiles 32–3, 456*t*, 457*f*, 458*f*
Saslaw, W.C. 124, 217–18
satellite vulnerability *xv*, 423
Satoh, A. 71
Sauter mean diameter (SMD) 55, 129, 458–9
  experimental validation 408, 414, 417, 418*f*, 426, 428*f*, 429*f*, 441*t*
  secondary fragmentation 430*f*, 438, 440*f*
Saveliev, V.L. 340, 341–2, 344–5
scaled size distributions *see* self-similarity
Schaefer, W. 9*f*, 156
Schlichting, H.E. et al 364
Schmitt, F.G. 196
Schögl, F. 477–8
Schosseler, F. 280*t*
Schoutens, E. 12, 127*t*, 128*t*
Schowalter, W.R. 214*t*
Schuhmann, V. 117, 355, 514
Schulz/Schulz–Zimm size distributions *see* gamma size distributions
Schulz, V. 125t
Schumann, T.E.W. et al 124*t*, 208, 213*t*, 216, 230–1, 250*t* n3
scission *see* binary fragmentation
Scott, J. 191, 192
Scott, W.T. 231–2, 250*t* n3, 275–6
Seaman, L. et al 389
SEC *see* size exclusion chromatography
secondary aggregation (compaction) 5, 6, 14–15, 185–6, 245, 278, 457, 458*f*, 468, 501
secondary breakup 129, 164, 423, 434, 435*f*, 440, 464
secondary fragmentation 2, 3, 5*f*, 6, 13–14, 82, 89, 278, 385*t*, 501
  experimental validation 431*f*, 432, 433*f*, 434, 435–8*f*, 439, 440*f*
  unitary fragmentation and 389, 390 n1, 391*f*, 392*f*, 398, 465, 466*t*, 467*f*, 468
second-order partial differential equations 265–6

second-order structure function 184, 199
sedimentation 133, 157, 218, 241, 272, 277, 278, 281, 285
Sedlatschek, K. 325
Seebaugh, R. 12, 127*t*, 128*t*
Seebaugh–Schoutens size distribution 12, 129
seemingly random *see* chaos
selection bias 111, 479
selection functions 323, 325, 332, 348, 349*f*, 350
self-affine character 153
self-organization 16, 130, 131*t*, 477, 499, 501, 505
  basic properties of particles 22, 29*t*, 30–1, 37
  energy and energy distributions 179, 185–6, 188–2, 194, 198, 200
self-organized criticality 31, 32*t*, 33–4, 37
self-similarity 7–8, 73, 171
  binary fragmentation 303, 304, 305
  common size distributions 510–11, 512*t*, 513*t*, 515*t*, 517*t*, 518*t*, 519
  distributions/averages 54–6, 56–7, 60, 61–2
  full mean-field solutions 231, 236, 238, 239–40
  late-time mean-field solutions 262, 264, 273–4, 275
  random fragmentation 335, 339, 340, 345
  unitary fragmentation 448, 453*t*, 465
Sellens, Richard W. 113, 485, 491, 493, 495
semi-empirical methods *xx*, 1, 103, 114–15, 201, 349, 350–2, 392, 505, 506, 507
  see also analytic methods; empirical methods; size distributions
semi-inertial range 133, 328, 398
Seneta, E. 105
Shadloo, S. et al 74
Shannon's information theory 111, 485
shapes and shape distributions *v*, *xviii*, *xix*, *xx*, *xxi*, 2, 3*f*, 168, 186, 364, 499, 501, 505, 508*t*
  analytical/empirical methods 103, 104, 120, 133
  basic properties of particles 22, 28, 38, 43
  computational methods 67, 69, 75
  different particles and average shape 149*f*, 150, 151–5*f*
  different particles and different shapes 156, 157–8*f*, 159–61, 162*f*, 163–5
  distributions/averages 43, 45–6, 47*f*, 56, 146*f*, 147
  fragmentation cascades 293, 319, 353
  mean-field governing equations 207, 211, 220
  types of 7*f*, 8*f*, 9, 10, 16, 198
  unitary fragmentation 392, 403, 409, 410*f*, 411*t*, 431, 442, 450, 452–3*t*
shattering/non-shattering 485, 502, 504, 505, 506
  binary fragmentation 295, 297*t*, 298, 299, 300, 306, 307, 308
  collisional fragmentation 368, 371, 372, 373, 374, 375 n1, 376, 380
  random fragmentation 320, 329, 331*t*, 332–3, 335, 338–45, 347, 355
  unitary fragmentation 428, 431*f*, 449, 460
shear and shear stresses 191*t*, 193, 285, 297*t*, 457, 458*f*
  basic principles of unitary fragmentation 393, 396, 397*f*, 398
  experimental validation 405, 412*t*, 434, 436, 437*f*, 438*f*, 439
sheet stripping (breakup) 434, 435*f*, 436, 437*f*, 438*f*, 439, 440
shells *see* cylinders and cylindrical shells
Shinjo, J. 392
shock and shock waves *v*, 48*f*, 174, 175*f*, 176*f*, 249, 394, 396*f*, 456, 502
Shukla, N.B. 309*t*
sieves and sieving 4*f*, 157, 432, 433*f*
  distributions/averages 44, 45–6, 47*f*, 48, 50–1, 63
  random fragmentation 321, 324–5, 348
sifters 46*f*
Simha, R. 295, 300, 301
Simmons, C. 127*t*, 128*t* 129, 132, 466
Simmons universal root normal size distribution 463, 466
Simon, V. et al 113
Singh, C. 69, 70*t*
Singh, P. 209, 239–40, 250*t*
sinks *see* sources and sinks
Sivalingam, G. 309*t*, 313, 314*f*
size and size distributions *xviii*, *xix*, *xxi*, 168, 499, 502, 504, 505, 507, 508*t*, 511
  average particle sizes 273–6
  binary fragmentation 295, 298–9, 300, 301–4, 305, 306, 307, 311*f*, 313*f*, 314, 315
  collisional fragmentation 363, 364, 368, 370–1, 372, 373, 374, 376–79, 380*t*
  common distributions 56–7, 58*f*, 59*f*, 60–1, 62*f*, 63*t*, 64*t*, 258, 259
  computational methods 69, 72, 75, 80, 81*t*, 86–8*f*
  concentration distributions 44, 45–6, 47*f*
  continuous particle systems 485, 486–7, 488*t*, 489–93
  different forms of 509–11, 512–19*t*
  distributions/averages 43, 50–1, 52, 63*t*
  energy and energy distributions 185, 198–200
  full mean-field solutions 225–7, 230–1, 237, 239, 242, 244*f*, 246–7, 248*f*, 249–51, 300
  late-time mean-field solutions 257, 258, 279, 280*t*, 282–3, 284, 285
  local/non-local aggregation 260, 261*t*, 262–73
  maximum entropy theory 477, 478, 479, 480, 481, 482*t*, 486, 490–1, 494–5
  mean-field governing equations 207, 210, 212, 216, 219, 220
  particles 3*f*, 12, 14, 15, 16, 22, 30, 38
  particle shape distributions and 146*f*, 147, 160, 161
  random fragmentation 319, 321, 325, 328, 332–4, 338–47, 348*f*, 349–55
  research literature 125*t*, 126, 127*t*, 128*t*, 129
  self-similarity 7–8, 54–6, 73
  see also analytical methods; dimensional analysis; semi-empirical methods, empirical methods; unitary fragmentation; Weibull size distributions
size exclusion chromatography (SEC) 247, 279, 310, 312, 313

SMD *see* Sauter mean diameter
smeared crack approach 76
Smirnov, V.I. 268, 285
Smith, J.M. 309*t*
Smoluchowski, Marian von *xxi*, 105, 124*t*, 125*t*, 201, 207, 213*t*, 229–30, 250*t* n1, 281, 284
Smol'yakov, A.V. 192
Smoothed Particle Hydrodynamics (SPH) 505
  computational methods 73, 74–5, 76*t*, 93*t*
  unitary fragmentation 415*f*, 422–3, 424*f*
soap bubble wands 419
Society for Industrial and Applied Mathematics *xv*
soft-catch devices/techniques 13, 455*t*, 468
Sojka, P.E. 120
solid state physics 183, 189
solids *v*, *xvii*, *xx*, 2, 8, 27, 157, 164, 211, 354, 371*f*, 501, 502, 505
  analytical/empirical methods 128*t*, 131, 132
  basic principles of unitary fragmentation 384, 385*t*, 387*f*, 388*f*, 389, 391*f*, 392*f*, 393, 398
  computational methods 73, 74, 80, 83, 86*f*, 87*f*, 90*f*
  secondary fragmentation 431, 434, 435*f*, 436*f*, 439, 440*f*, 454, 456, 459, 460, 461*t*, 462*t*
  sizes and size distributions 44–5, 404, 405–6*t*, 407, 409, 410, 411–12*t*, 413, 414, 419–20, 421*f*
sol (smaller particles) 2, 228, 237, 270, 271, 280
Sommer, F. et al 46
soot aggregation 71
Sorensen, C.M. 14, 278*t*
Sornette, D. 7
Soto, R. 23
sources and sinks (hard/soft) 6, 449, 494, 500, 506
  binary fragmentation 297, 299, 306, 308
  collisional fragmentation 368, 369, 370–1, 374, 375, 376, 378, 380*t*
  full mean-field solutions 228, 238, 239–40
  late-time mean-field solutions 257–8, 259–62, 264–7, 269–73, 275, 276, 277, 282, 285
  mean-field governing equations 208–9, 217, 219, 220
  random fragmentation 320, 329, 332, 333, 343, 344, 345, 346–7, 355
Souza, S.R. 452*t*
Spahn, F. et al 452*t*
spall 385*t*, 393–4, 396*f*, 420
spatial covariances/correlations 180, 181, 182, 187*f*, 188–9, 190–1, 192
spatial dimension 8, 9, 509
spatial homogeneity *xxi*, 103–4, 168, 393, 398, 433, 459, 468, 500, 503*t*, 507
  aggregation cascades 207, 250, 257, 282–3, 284
  analytical/empirical methods 116, 121, 123, 126, 131*t*, 132–3
  basic properties of particles 35, 36, 37
  binary fragmentation 293, 314, 315
  collisional fragmentation 376, 378, 380
  computational methods 69, 72, 75, 81, 88
  distributions/averages 43, 44, 47, 49, 50, 52, 54, 56
  energy and energy distributions 181, 182, 184, 185, 191*t*, 196, 197*t*, 200
  random fragmentation 319, 329, 332, 335, 364
speed *xvi*, 3, 12*f*, 13*f*, 15, 22, 43, 56, 434, 505
  computational methods 63*t*, 70–1, 82, 92*t*
  particle velocity distributions 168, 169, 170–2, 173, 174, 176*f*
Sperling, K. 57, 123
SPH *see* Smoothed Particle Hydrodynamics
spherical geometries 418–20, 421–2*f*, 423, 424*f*, 425, 454
spherical volume equivalent 159, 165
Spicer, P.T. 322
Spielbauer, M. et al 127*t*
Spielman, L.A. 323, 331*t*
splitting *see* binary fragmentation
spontaneous fragmentation *see* random fragmentation
spontaneously self-organized systems 29*t*, 30–1
Spouge, J.L. 124*t*, 321, 481, 482*t*
Springer, K. et al 87*f*
Springer Nature 9*f*, 83*f*, 416*f*, 421*f*, 458*f*, 465*f*
square-root-log plane 119
Sreenivasan, K. 201
Srivastava, R.C. 365
stability analysis 404, 405, 425, 426, 430, 435, 458–9
stability and instability 67 n1, 68, 75, 402, 404
Standard Breakup Model (National Aeronautics and Space Administration) 128*t*, 129
Stanton, D.W. 462
star polymers 163
stationary/nearly stationary 4 n1, 12, 217, 218, 308, 495, 500
  collisional fragmentation 375, 376, 377, 378
  energy and energy distributions 180, 182, 185, 189, 191*t*
  late-time mean-field solutions 259, 261, 264–7, 271–3
  random fragmentation 337, 346, 347, 348*f*
statistical independence 22*t*, 27, 150, 164, 507
  energy and energy distributions 180, 181, 200
  velocity distributions 168, 169, 170, 174
statistical mechanics 110, 111, 210, 298
steady *see* stationary
Stegun, I.A. 336
Stein, P.C. 213*t*
Stein, T.H.M. 124*t*
Sterling, W.J. (et al) 309*t*, 310, 311*f*
Stirling's approximation 235, 478–9, 482*t*
stochastic condensation *xvi*, 258–9, 264, 337
stochastic geometry *xx*, 503*t*, 504, 505
  analytical/empirical methods 106–7, 108*f*, 109–11, 118, 124*t*, 127*t*
  unitary fragmentation 393, 402, 448, 450, 451*f*, 452–3*t*, 454, 467
Stockmayer, Hugo 111, 124*t*, 214*t*, 248, 476, 478, 482*t*, 491–2
Strack, O.D.L. 80
strain rates v, 89, 150, 214*t*
  basic principles of unitary fragmentation 384–6, 389–90, 393, 394*f*, 395, 396*f*, 397*f*, 398

sizes and size distributions 402–4, 405–6*t*, 408–11, 412*t*, 413, 416*f*, 417, 419, 421*f*, 426, 430–2, 433*f*, 441, 459–60, 468
Stránský, J. 412*t*
strength v, 37, 87, 93*t*, 391
stresses 37, 89, 91, 501
unitary fragmentation 393, 394*f*, 395, 396*f*, 397*f*, 398, 417
strong shattering 332, 333, 341, 344
structure factors/functions 113, 179, 183–4, 188, 189, 190, 191, 194, 198, 199
subgrid-scale (SGS) 73, 93*t*, 94, 506
sum kernels 227, 235–6, 250, 257, 275–6, 282, 283, 369*t*, 370, 372–3
Suneel et al 280*t*
surface areas and energy 83*f*, 183, 211
maximum entropy theory 487, 491, 493
random fragmentation 320, 329, 330, 352, 354
roughness 410, 411*t*, 426, 436*f*, 462, 463*f*
secondary fragmentation 434, 436*f*, 439, 440*f*
shapes and shape distributions 147, 152, 153*f*, 155*f*, 158, 452*t*
unitary fragmentation 394, 397, 404, 406–7, 408–10, 411*t*, 412, 417, 422, 425–6, 428, 430, 441, 457
Swee*t*, B. 113
sympathetic detonations 126
Szamel, G. 332, 342

Taguchi, H. 191*t*, 200
Takagi, Y. et al 12
Takayasu, H. 127*t*, 452*t*
Tame, J. 22, 23
Tanaka, H. 215*t*, 377
Tankin, R.S. 127*t*, 132, 454, 466*t*, 467*f*, 485, 492
Tanner, X. 73, 79
Tate, R.W. 62, 466*t*
Tatsumi, T. 197*t* n3
Täuber, U.C. 104
Tavares, L.M. 320, 321, 349–50, 351–2, 392–3
Tavlarides, L.L. 329–30
Taylor impact tests 460, 461*t*
Taylor series 265, 267
Tchen, C.-M. 197*t*
temporal correlation/covariance *see* autocorrelation
temporal homogeneity 88, 113–14, 168, 185, 198
basic properties of particles 34–5, 36, 37
distributions/averages 43, 44, 47, 49, 50, 52, 54, 56
tensile strains/stresses
basic principles of unitary fragmentation 393, 394*f*, 395, 396, 397*f*, 398
sizes and unitary fragmentation 412*t*, 432, 434, 435, 449, 456
tensile strength 89, 90, 91, 92*t*, 93*t*, 405
tensors 182, 183, 184, 393, 394
Terranova, B. et al 73
tessellations (random) 106, 108–9
textile mills 331*t*
Thadhani, N. 113
Theofanous, T.G. (et al) 437*f*, 438f
thermal degradation/decomposition *xviii*, 477
binary fragmentation 292*t*, 293, 309–10, 311*f*, 313, 314*f*, 363*t*
random fragmentation 320*t*, 323, 330
thermal equilibrium 189, 190, 384
thermodynamics and thermodynamic limit 225, 228 n1, 248, 314–15, 364, 476
thermostat algorithms 68, 190, 191*t*
Thomas, J.A. 514*t*, 515*t*, 516*t*, 517*t*
Thompson, S.E. 515–16
three-parameter Generalized Gamma Distribution *see* Rosin–Rammler size distributions
three-particle mechanisms 332
Tikhonov-type regularization methods 121
Timár, G. et al 87, 89*f*, 90, 91, 92*t*, 449
time dependent solutions 198, 426, 483, 505
aggregation cascades 225, 229–40, 250, 257
binary fragmentation 299–307, 312
collisional fragmentation 364, 369*t*, 370
random fragmentation 322, 323, 324, 334–5, 338–45
Tishkoff, J.M. 466*t*, 467*f*
Tkachenko, V.M. 192
Tobolsky, A.V. 130*t*, 295
Townsend, A.A. 199
Trappe, V. et al 280*t*
tricholorobenzene solution 312, 313*f*
Trizac, E. 71, 72*t*
Tropea, C. 462*t*
Trubnikov, B.A. 215*t*
true strain/strain rates 395, 396, 397*f*
Tschudi, H.R. 270–1
Tseng, L.K. et al 418*f*
Tsuda, I. 29
Tur, A.V. 452*t*
turbulence *v*, *vi*, *vii*, *xv*, *xviii*, *xx*, 9, 163, 215, 258, 329, 499, 501
analytical/empirical methods 105, 115, 133
computational methods 75, 79, 94
fluids *xxi*, 133, 179, 191–6, 197*t*, 198, 199–201, 215, 502, 506
unitary fragmentation 410, 417, 418*f*, 426, 427*f*, 428*f*, 440, 458, 459*f*
Turbulence: An Introduction for Scientists and Engineers (Davidson) *xxii*
turbulent kinetic energy spectra 194, 197*t*, 198, 200
Turcotte, L. 131
Turner, G.M. 466*t*
Turner, J.S. et al 73, 213*t*, 214*t*
2/3 law of turbulence (Kolmogorov) 105, 196–7
two-particle mechanisms (fragmentation) 330–1, 332
two-point/one-time joint probability density functions 35, 36, 180, 182

ultrasound-assisted impact fragmentation 88
unbonded particles 6, 7*f*, 16, 24, 130, 186, 198, 200, 505
uncorrelated 23, 478
uniaxial (cylindrical/Cartesian) coordinate systems 394*f*, 395, 396, 413, 418, 425, 454
uniform compressive stresses 396

uniform uniaxial/equi-biaxial stress/strains 395, 396
uniqueness 220
unitary aggregation 2, 3*f*, 6
unitary fragmentation 149*f*, 172, 398, 467–8, 499, 500, 501, 503*t*, 504
  analytical/empirical methods 107, 108, 110, 113, 127, 129, 133
  average sizes of secondary fragmentation 431*f*, 432, 433*f*, 434, 435–8*f*, 439, 440*f*
  basic properties 33, 37
  Cartesian geometries 425–6, 427–30*f*, 441
  challenges of 384, 385–6*t*, 387–8*f*, 389, 390–2*f*, 393
  common size distributions 46, 441*t*, 442, 448–9
  computational methods 73, 74–5, 76*t*, 79, 80, 81*t*
  cylindrical geometries 413–14, 415–16*f*, 417, 418*f*, 441
  maximum entropy theory 448, 453–4
  mean-field theory models 402–4, 448, 449–50
  non-mean-field-theory models 404, 405–6*t*, 407–9, 410*f*, 411–12*t*, 413
  particle definitions 2, 3, 4*f*, 6, 10, 15, 16
  size distributions of primary fragmentation 454, 455–6*t*, 457–9*f*
  size distributions of secondary fragmentation 459, 460, 461–2*t*, 463*f*, 464, 465*f*, 466*t*, 467*f*
  spherical geometries 418–20, 421–2*f*, 423, 424*f*, 425, 441
  stochastic geometry 448, 450, 451*f*, 452*t*, 453*t*
  strain rates 384, 385, 393, 394*f*, 395, 396*f*, 397*f*
  see also analytical methods; size and size distributions
universal aggregate shapes 153
universal energy spectra 477
universal exponents 160, 161, 188, 190, 191, 505
universal gamma size distribution 111, 483
universal power law size distribution 12*f*, 13*f*, 14–15, 30, 33, 117, 378
universal shapes 149, 153, 155, 164
universal speed/velocity distributions 171, 174, 175
universals (size distributions) *xix*, 15, 33, 38, 54, 502
  aggregation/fragmentation cascades 250, 257, 282–3, 339, 351, 379, 380
  analytical/empirical methods 116, 124, 126
  binary fragmentation 302, 304, 306, 307, 308
  maximum entropy theory 477, 495
  unitary fragmentation 389, 454
universal turbulent kinetic energy spectra *xxi*
unsteadiness *see* non-stationary
Urbán, A. et al 429
USIM PAC software 350

vacuum *xvii*
Valentas, K.J. et al 323, 331*t*
Valtonen, M.J. 26
van der Waals forces xvii, 6, 9, 68, 69, 125, 457
Van Dongen, P.G.J. 262–4, 271, 274, 482
Van Noije, C. et al 191*t*
Van Zon, J.S. 188, 189
vaporization 258, 264, 309
variances/covariances (energy) 179–83, 184
Varoqui, R. 277*t*
velocity and velocity distributions *v*, *xviii*, *xix*, *xxi*, 16, 44, 175, 176*f*, 364, 381, 499, 505, 508*t*
  analytical/empirical methods 103, 104, 133
  basic properties of particles 32*t*, 35, 38
  chaos theory 22*t*, 23, 24, 26, 27
  computational methods 68, 69, 72, 75, 80, 89, 90, 91
  energy and energy distributions 181–4, 185, 187*f*, 188, 189–91, 192, 193, 194, 198–9, 200
  joint probability density functions 168–9
  maximum entropy theory 485, 493, 496
  mean-field governing equations 207, 211, 212, 214, 220
  secondary fragmentation 430, 432, 433*f*, 435, 437, 438, 439
  size distributions 459, 460, 462, 463*f*
  unitary fragmentation 392, 394, 407–8, 410, 411*t*, 412–14, 416–17, 420, 425, 428
  velocity distributions for aggregation 170–2
  velocity distributions for fragmentation 172, 173*f*, 174, 175*f*
velocity autocorrelation function 23, 25
vibrational energy *xviii*, 27, 37, 45, 380
video cameras *see* high-speed photography/video
Villermaux, E. 127*t*, 406*t*, 411*t*, 434–5, 436*f*, 465*f*
viscoelastic liquids/solids 280
viscosity 32*t*, 214, 354
  energy and energy distributions 192, 193–4, 196
  unitary fragmentation 389, 394, 404–5, 407, 412*t*, 413, 434, 438, 440–1
Vledouts, A. (et al) 386, 390*f*, 406*t*, 424*f*, 425
volumetric fraction of particles 37, 248, 319, 320, 323, 330, 332, 334–5, 354–5, 500
Von Hoerner, S. 68
Von Rittinger, Peter 196
vortex shapes 163, 185, 190–1, 192, 193, 195, 200
Voutilainen, A. et al 120–1

Wainwright, T.E. 71
Wallace, C. 70*t*
Walline, K.S. 309*t*
Wang, C.S. 283
Wang, C.T. et al 416*f*, 420, 421*f*, 423, 432, 433*f*
Wang, L.-P. 281
Wang, M. 309*t*
Wang, Y.-F. et al 71
Warren, B. 155*f*
water *see* liquids
Watson, H.W. 212
wavelengths 198, 404
wave numbers 194, 198
wave vectors 183
weak shattering 332–3, 341, 342, 343, 375 n1
Weber numbers 384, 398, 462, 468
  Cartesian geometries 425, 427, 428, 429, 430
  cylindrical geometries 410, 411*t*, 412, 413, 417
  secondary fragmentation 434, 435*f*, 437, 439

Weibull size distributions 218, 501, 503*t*, 504, 510, 511, 513–14*t*
analytical/semi-empirical/empirical methods 116–19, 120–2*f*, 124*t*, 127*t*, 132, 133
collisional fragmentation 374, 375, 378, 380*t*
computational methods 70*t*, 74, 76*t*, 85, 92*t*
distributions/averages 56–8, 59*f*, 60, 62*f*, 64*t*
late-time mean-field solutions 259, 260–1, 265–8, 282–5
maximum entropy theory 486, 487, 488*t*, 492, 493–4, 495
particle velocity distributions 174, 175, 176*f*
random fragmentation 334, 347, 354
secondary fragmentation 459, 460, 461, 462*t*, 464, 465*f*, 466, 467*f*, 468
unitary fragmentation 409, 410*f*, 448–50, 452–3, 454, 455*t*, 456*t*, 457
see also size and size distributions
Weimer, R.J. 321, 414, 415*f*
welding 457, 458*f*
well-mixed systems 38
Westbrook, D. et al 146*f*, 153, 154*f*
White, G. 293
White, W.H. 114, 208, 273
Wickham-Eade, E. et al 88*f*
Wiener, Norbert 30
Wilcoxon, J.P. 278*t*
Wilkins, D. 366
Wilkinson, J.H. 406*t*
Wilkinson, M. 220, 281
Williams, A. 79
Williams, M.M.R. 213*t*
Wilson, L.T. et al 455*t*
Wise, T.H. 406*t*
witness plates 44, 426, 464
Wittel, F.K. et al 152–3, 160–1, 419, 461*t*
computational methods 81*t*, 82, 83*f*, 84*f*, 88, 89*f*
Worthington, A.M. 434
Wu, C. et al 394
Wu, P.-K. et al 418*f*, 428*f*
Wurm, G. 257

X-FEM *see* Extended Finite Element Method
Xue, K. et al 457, 458*f*
Xu, L. et al 462*t*, 463*f*

Yano, I. et al 118, 121
Yanovsky, V.V. et al 131, 452*t*
Yarin, A.L. et al 384
yield strength 354, 404, 405, 407, 505
Yorke, J.A. 28
Young's modulus 82, 85, 89
Yuan, C. et al 350–1
Yu, P. et al 172

Zeichner, G.R. 214*t*
Zeiss Xradia 500 Versa system 157, 158*f*
Zhang, C. et al 396*f*
Zhang, J. 29, 30
Zhao, B.-L. et al 423
Zhou, F. 412*t*
Zhou, J. et al 27, 127*t*
Zhou, W. 76*t*
Ziff, R.M.
binary fragmentation 293–4, 295, 297*t*, 300, 301, 302–7
mean-field governing equations 213*t*, 214*t*
random fragmentation 331*t*, 332, 338
Zvyagina, E.V. 215*t*